Handbook of
Applied Solid State Spectroscopy

D. R. Vij

Handbook of Applied Solid State Spectroscopy

With 347 Figures

Springer

D. R. Vij
Kurukshetra University
India

With 347 figures

Library of Congress Control Number: 2006923765

ISBN-10: 0-387-32497-6 Printed on acid-free paper.
ISBN-13: 978-0387-32497-5

Printed in United States of America. (TB/EB)

9 8 7 6 5 4 3 2 1

springer.com

PREFACE

Solid state spectroscopy is currently a burgeoning field finding applications in many branches of science, including physics, chemistry, biosciences, surface science, and materials science. After surveying the current developments taking place in the field, the contributing authors considered it appropriate to produce a volume containing various spectroscopic techniques currently in practice around the world—all under one cover.

The purpose of the proposed edited volume is to bring together the different techniques applied to various aspects of solid state spectroscopy, which continue to receive attention worldwide. The authors expect this volume to become a valuable source in the hands of not only the mature and experienced reader but also the new entrant in different areas described in this compendium. The proposed works are of particular interest to graduate students, teachers, and researchers in the field of physics, chemistry, and materials science, including solid state biosciences. This volume should serve both as text and reference to readers.

The handbook comprises 16 chapters, each in the form of overviews of latest works by authors who are leading world experts in the field of solid state spectroscopy. The contributions are from leading international physical and chemical spectroscopists on the latest advances in the solid state spectroscopic techniques as applied in various branches of science and industry. Various techniques presented in the book lay stress on fundamentals, experimental techniques and procedures, applications, and new developments.

Efforts have been made to develop this volume as an international standard, one which will provide authentic and quantitative information to readers. The handbook represents a valuable contribution by outstanding researchers in their individual fields. It has an international flavor, with contributions drawn from the United States, Germany, Canada, Switzerland, Poland, Australia, India, Brazil, Iran, Denmark, and Sweden.

It is my pleasant duty to express my gratitude to various authors who made this handbook possible by contributing respective chapters in the fields of their expertise. I am indebted to all those authors and publishers who freely granted permissions to reproduce their copyright works. The patience of staff at Kluwer Academic Publishers and Springer-Verlag is also thankfully acknowledged. Last but not the least, I wish to express my deep appreciation to my wife Dr. Meenakshi Vij and my daughters Surabhi and Monika for their indispensable support, encouragement, and patience during the writing and editing of this handbook.

D.R. Vij, Editor
Kurukshetra, India

CONTENTS

CONTRIBUTORS

Chris Arumainayagam, Wellesley College, MA 02481, USA, Tel: +1-781-283-3326
Email: carumain@wellesley.edu

Baldassare Di Bartolo, Department of Physics, Boston College, Chestnut Hill, MA 02467, USA, Tel: +1-617-552-3601
E-mail: dibartob@bc.edu

Tito José Bonagamba, Institute de Física de São Carlos, Universidade de São Paulo – Brazil, Tel: +55 16 3373 9871
E-mail: tito@if.sc.usp.br

S.M. Butorin, Department of Physics, Uppsala University, Uppsala, Sweden

J.M Cadogan, School of Physics, The University of New South Wales, Sydney, NSW2052, Australia, Tel: +61-2-9385-5203
E-mail: J.Cadogan@unsw.edu.au

John Collins, Department of Physics, Wheaton College, Norton, MA 02766, USA, Tel: +1-508-286-3976
Email: jcollins@wheatoncollege.edu

Antony R. Crofts, Department of Biochemistry, University of Illinois at Urbana-Champaign, Urbana, IL 61801 USA, Tel: +1-217 333-2043
E-mail: a-crofts@life.uiuc.edu

Eduardo Ribeiro deAzevedo, Institute de Física de São Carlos, Universidade de São Paulo - Brazil, Tel: +55 16 3373 8086
E-mail: azevedo@if.sc.usp.br

Sergei Dikanov, Department of Veterinary Clinical Medicine, University of Illinois at Urbana-Champaign, Urbana, IL 61801, USA, Tel: +1-217-333-3776
E-mail: dikanov@uiuc.edu

L.C. Duda, Department of Physics, Uppsala University, Uppsala, Sweden

Esmaeil Enjilela, Faculty of Mechanical Engineering, K.N. Toosi University of Technology, Tehran, Iran

Nolan T. Flynn, Wellesley College, Wellesley, MA 02481, USA, Tel: +1-781-283-3097; FAX: +1-781-283-3642
Email: nflynn@wellesley.edu

Albert Furrer, Laboratory for Neutron Scattering, ETH Zurich &PSI Villigen, CH-5232 Villigen PSI, Switzerland, Tel:+41-56-310-2088

G. Geipel, Institute of Radiochemistry, Forschungszentrum Rossendorf, Dresden, Germany, Tel: +49-351 260 2306
E-mail: G.Geipel@fz-ossendorf.de

Richard P. Gunawardane, Wellesley College, Wellesley, MA 02481, USA, Tel: +1-781 283 3530
E-mail:rgunawar@Wellesley.edu

J.-H. Guo, Lawrence Berkeley National Laboratory, Berkeley, CA 94720, USA

Yu Guo, Department of Physics and College of Optics and Photonics–CREOL, University of Central Florida, Orlando, FL 32816-2385, USA

K.W. Hipps, Department of Chemistry and Materials Science Program, Washington State University, Pullman, WA 99164-4630, USA, Tel: +1-509-335-3033
E-mail: hipps@wsu.edu

Farhang Honarvar, Faculty of Mechanical Engineering, K.N. Toosi University of Technology, Tehran, Iran, Tel: +98-21-734-3300
Email: honarvar@mie.utoronto.ca

Neena Jaggi, Kurukshetra University; Kurukshetra, India Tel:+91-1744-291-379
E-mail: neena_jaggi@rediffmail.com

Lowell D. Kispert, Chemistry Department, The University of Alabama, Tuscaloosa, Alabama, USA, Tel: +1-205-348-7134
E-mail: lkispert@bama.ua.edu

Hsiao-Lu Denise Lee, Department of Chemistry, Stanford University, Stanford, CA 94305-5080, USA
Tel: +1-650-723-2501
Email: hsiaolu@stanford.edu

E.J. Nordgren, Department of Physics, Uppsala University, Uppsala, Sweden, Tel: +46-18471-3554
E-mail: joseph@fysik.uu.se

Lidia Piekara-Sady, Institute of Molecular Physics, Polish Academy of Sciences, 60-179 Poznan, Poland
E-mail: lpsady@tlen.pl

Andrew Podlesnyak, Laboratory for Neutron Scattering ETHZ & PSI, CH-5232 Villigen PSI, Switzerland, Tel: +41 56 310 4613
E-mail: andrew.podlesnyak@psi.ch

D.H Ryan, Department of Physics, McGill University, Montreal (Quebec)H3A 2T8, Canada, Tel: +1-514-398-6534
E-mail: dominic@physics.mcgill.ca

Alfons Schulte, Department of Physics and College of Optics and Photonics – CREOL, University of Central Florida, Orlando, FL 32816-2385, USA, Tel: +1-407-823-5196
E-mail: afs@physics.ucf.edu

Bryan H. Suits, Physics Department, Michigan Technological University, Houghton, MI 49931, USA Tel: +1-906-487-2093
E-mail: suits@mtu.edu

Erik W. Thulsrup, Dept. of Life Sciences and Chemistry, Roskilde University POB 260, DK-4000 Roskilde, Denmark, Tel: +45-4674-2709
E-mail: ewt@ruc.dk

Peter W. Thulstrup, Department of Natural Sciences, The Royal Veterinary and Agricultural University, Th. 40. DK-1871 Frederiksberg, Denmark, Tel: +45-3528-2464
E-mail: pwt@kvl.dk

D.R. Vij, Kurukshetra University; Kurukshetra, India Tel: +91-1744-238-816
E-mail: drvkuk@rediffmail.com

CHAPTER 1

NUCLEAR MAGNETIC RESONANCE SPECTROSCOPY

Eduardo Ribeiro deAzevedo and Tito José Bonagamba
Instituto de Física de São Carlos – Universidade de São Paulo – Brazil

1.1 INTRODUCTION

After the development during the last 30 years of methods such as magic-angle spinning, multiple-resonance experiments (cross-polarization, high power decoupling, etc.), multiple pulses techniques, and multidimensional spectroscopy, pulsed solid state nuclear magnetic resonance (NMR) is among the most important experimental tools for studying physical and chemical properties of new materials in the solid-state.

In this chapter, we discuss some basic aspects of the NMR methodology and show some applications of the methods for studying molecular dynamics and local structure in solid materials. Despite the fact that the methods and applications presented in this chapter are predominantly related to organic materials (polymers and derivatives, and small organic molecules) with some extensions to inorganic materials (glasses, ionic conductors, etc.), several basic aspects of advanced solid state NMR discussed here can be extended for understanding other NMR techniques necessary for studying other kinds of solid materials [1–9]. Due to the limitations of space, our discussions will be focused to NMR methods for studying an important class of solid materials, namely, organic materials, particularly polymers and derivatives, which allows the application of a large number of different techniques for understanding their molecular dynamics and local conformation. In writing the introduction of this chapter we have used as references several classic books of modern solid state NMR [1–9] and quantum mechanics [10].

1.2 PROPERTIES OF NUCLEAR SPINS

Nuclear magnetic resonance is a phenomenon that occurs for nuclear spins that possess magnetic moment and angular momentum $\vec{\mu}$ and $\vec{J}$, respectively.

The relationship between both moments is $\vec{\mu} = \gamma \vec{J}$, where γ is the gyromagnetic ratio. In terms of quantum mechanics the angular momentum $\vec{J}$ is defined by $\vec{J} = \hbar \vec{I}$, where $\vec{I}$ is a dimensionless operator denominated nuclear spin. The magnetic moment is quantized according to the usual quantization of the angular momentum, i.e., the magnitude of the magnetic moment is given by $|\vec{\mu}| = \gamma\hbar\sqrt{I(I+1)}$ [11]. The allowed components of $\vec{\mu}$ in the direction of a magnetic field along the *z* direction (Zeeman interaction) are given by $\mu_z = \gamma\hbar I_z$. I_z is the spin quantum number taking $(2I+1)$ values from –*I* to +*I,* and *I* can take only integer or semi-integer values 1/2, 1, 3/2, 2,.... Additionally, nuclei presenting $I > 1/2$ have also an electric quadrupole moment.

1.3 NUCLEAR SPIN INTERACTIONS IN SOLIDS

The nuclear spin Hamiltonian involves interactions that are related to the magnetic and electric properties of the nuclei and the experimental conditions of the experiment. Some terms determine the intensity of the signal, while others define the position and the shape of the spectral lines [1, 2, 7].

The nuclear spin Hamiltonian can be separated into two important groups: *internal* and *external interactions*. The external interactions are associated to the coupling of the nuclear spin magnetic moment $\vec{\mu} = \gamma\hbar\vec{I}$ with the static magnetic field $\vec{B}_0 = B_0\hat{z}$ (*Zeeman effect*) and with the oscillating magnetic radio frequency (rf) field applied perpendicularly to the static field

$$\vec{B}_{\text{rf}} = B_1(t)\left[\cos\left(\omega_{\text{rf}}t + \phi(t)\right)\hat{x} + \sin\left(\omega_{\text{rf}}t + \phi(t)\right)\hat{y}\right], \tag{1.1}$$

which induce the transitions between adjacent levels. The Hamiltonians for both interactions are expressed by:

$$H_z = -\vec{\mu}\cdot\vec{B}_0 = \hbar\left(-\gamma B_0\right)I_z = \hbar\omega_0 I_z \tag{1.2}$$

and

$$H_{\text{rf}} = -\vec{\mu}\cdot\vec{B}_{\text{rf}} \ . \tag{1.3}$$

For special excitation procedures (transition-selective, multitransition, or multiple quantum excitations) the rf field may be amplitude-, frequency-, and/or phase-modulated through the manipulation of the parameters $B_1(t)$, ω_{rf}, and $\phi(t)$, respectively. Both external interactions dominate the behavior of the spin system, but in general they do not contain the structural and dynamic information, which are directly related to the internal nuclear spin interactions.

The internal interactions are associated to the couplings of the nuclear spins with, for example, magnetic fields originated from the orbital motion of the electrons induced by the static magnetic field (*chemical shift interaction*), other nuclear spins through their magnetic dipole moments (*dipolar interaction*), and the electric field gradients through their electric quadrupole moments (*quadrupolar interaction*). Therefore, the internal Hamiltonian can be written as a sum of several independent terms, which will be described in the following items, and where we will assume that all the spin interactions can be represented by second rank Cartesian tensors in the following way [2]:

$$H = \vec{I} \cdot \tilde{A} \cdot \vec{S} = \begin{bmatrix} I_x & I_y & I_z \end{bmatrix} \cdot \begin{bmatrix} A_{xx} & A_{xy} & A_{xz} \\ A_{yx} & A_{yy} & A_{yz} \\ A_{zx} & A_{zy} & A_{zz} \end{bmatrix} \cdot \begin{bmatrix} S_x \\ S_y \\ S_z \end{bmatrix}. \tag{1.4}$$

Chemical Shift. The chemical shift Hamiltonian is given by:

$$H_{\mathrm{CS}} = \vec{\mu} \cdot \left(\tilde{\sigma} \cdot \vec{B}_0\right) = \gamma\hbar\left(\vec{I} \cdot \tilde{\sigma} \cdot \vec{B}_0\right), \tag{1.5}$$

where $\tilde{\sigma}$ is a second order tensor that represents the proportionality factor between $\vec{B}_0$ and the magnetic fields originated along the directions x, y, and z from the orbital motion of the electrons induced by the static magnetic field. Assuming that $\vec{B}_0 = B_0\hat{z}$, the chemical shift Hamiltonian results in

$$H_{\mathrm{CS}} = \gamma\hbar\left(I_x\sigma_{xz} + I_y\sigma_{yz} + I_z\sigma_{zz}\right)B_0 . \tag{1.6}$$

The induced magnetic fields are very small compared to $\vec{B}_0$, consequently the chemical shift parameters σ_{ij} are of the order of 10^{-6}, resulting in shifts of the Zeeman energy levels on the order of parts per million (ppm). Therefore, the chemical shift can be treated as a first order perturbation with respect to the Zeeman interaction, and the effective Hamiltonian can be truncated and written as follows [7]:

$$H_{\mathrm{CS}} = \gamma\hbar\left(I_z\sigma_{zz}\right)B_0 . \tag{1.7}$$

Taking into consideration both Zeeman and chemical shift interactions gives the full Hamiltonian:

$$H = H_z + H_{\mathrm{CS}} = -\hbar\left(\gamma B_0\right)\left(1 - \sigma_{zz}\right)I_z . \tag{1.8}$$

Dipolar Interaction. The classical interaction energy between two magnetic moments $\vec{\mu}_1$ and $\vec{\mu}_2$ is [6]

$$E = \frac{\vec{\mu}_1 \cdot \vec{\mu}_2}{r^3} - \frac{3\left(\vec{\mu}_1 \cdot \vec{r}\right)\left(\vec{\mu}_2 \cdot \vec{r}\right)}{r^5}, \tag{1.9}$$

where $\vec{r}$ is the distance vector between $\vec{\mu}_1$ and $\vec{\mu}_2$. To write the dipolar Hamiltonian we take $\vec{\mu}_1 = \gamma\hbar\vec{I}_1$ and $\vec{\mu}_2 = \gamma\hbar\vec{I}_2$, where we assumed that the gyromagnetic factors and spins may be different. The general dipolar contribution to the Hamiltonian for N spins then becomes

$$H_{\mathrm{D}} = \gamma_1\gamma_2\hbar^2 \sum_{j=1}^{N}\sum_{\substack{k=1\\k\neq j}}^{N}\left[\frac{\vec{I}_j\cdot\vec{I}_k}{r_{jk}^3} - \frac{3\left(\vec{I}_j\cdot\vec{r}_{jk}\right)\left(\vec{I}_k\cdot\vec{r}_{jk}\right)}{r_{jk}^5}\right]. \tag{1.10}$$

The dipolar Hamiltonian can also be represented by

$$H_{\mathrm{D}} = \sum_{j=1}^{N}\sum_{k=1}^{N}\vec{I}_j\cdot\tilde{D}\cdot\vec{I}_k, \tag{1.11}$$

where $\tilde{D}$ is defined by

$$D_{mn} = \gamma_1\gamma_2\hbar^2\,\frac{\vec{e}_m\cdot\vec{e}_n - 3\left(\vec{e}_m\cdot\vec{r}_{jk}/\left|\vec{r}_{jk}\right|\right)\cdot\left(\vec{e}_n\cdot\vec{r}_{jk}/\left|\vec{r}_{jk}\right|\right)}{r_{jk}^3} \tag{1.12}$$

and $\vec{e}_1$, $\vec{e}_2$, and $\vec{e}_3$ are the basis vectors of the coordinate system in which the tensor $\tilde{D}$ is represented.

Taking only two spins $\vec{I}_1$ and $\vec{I}_2$ and expressing I_{1x}, I_{2x}, I_{1y}, and I_{2y} in terms of the raising and lowering operators I_1^+, I_1^-, I_2^+, and I_2^-, and expressing the Cartesian coordinates in terms of spherical coordinates, we can write the dipolar Hamiltonian in a form that is more convenient:

$$H_{\mathrm{D}} = \frac{\gamma_1\gamma_2\hbar^2}{r^3}\left(A+B+C+D+E+F\right), \tag{1.13}$$

where

$$\begin{aligned}
A &= I_{1z}I_{2z}\left(1-3\cos^2\theta\right) & D &= -\frac{3}{2}\left(I_1^-I_{2z}+I_{1z}I_2^-\right)\left(\sin\theta\cos\theta\right)e^{-i\phi}\\
B &= -\frac{1}{4}\left(I_1^+I_2^-+I_1^-I_2^+\right)\left(1-3\cos^2\theta\right) & E &= -\frac{3}{4}I_1^+I_2^+\sin^2\theta e^{-i2\phi}\\
C &= -\frac{3}{2}\left(I_1^+I_{2z}+I_{1z}I_2^+\right)\left(\sin\theta\cos\theta\right)e^{-i\phi} & F &= -\frac{3}{4}I_1^-I_2^-\sin^2\theta e^{i2\phi}
\end{aligned} \tag{1.14}$$

and θ is the angle between the distance vector $\vec{r}$ and $\vec{B}_0$.

While the Zeeman interaction corresponds to the interaction of a nuclear moment with magnetic fields of the order 1–10 teslas, the dipolar interaction corresponds to an interaction with a field of about 1 gauss. Therefore, the dipolar interaction can be also treated as a first order perturbation with respect to the Zeeman interaction and the effective dipolar Hamiltonian can be truncated to the first two elements *A* and *B*:

$$H_{\mathrm{D}} = \frac{\gamma_1\gamma_2\hbar^2}{r^3}\left(A+B\right). \tag{1.15}$$

If the dipolar interaction occurs between unlike spins (heteronuclear dipolar interaction), only the term *A* is necessary for the first order perturbation calculations:

$$H_{\mathrm{D}} = -\frac{\gamma_1\gamma_2\hbar^2}{r^3}\left(3\cos^2\theta-1\right)I_{1z}I_{2z}. \tag{1.16}$$

Quadrupolar Interaction. When the nuclei have spin > ½, they possess also the electric quadrupole moment that interact with the electric field gradients present in the spin molecular surroundings [1, 7]. This Hamiltonian can be written in the following form:

$$H_Q = \frac{eQ}{6\hbar I(2I-1)} \sum_{\alpha,\beta=1}^{3} V_{\alpha\beta} \left[\frac{3}{2}\left(I_\alpha I_\beta + I_\beta I_\alpha\right) - \delta_{\alpha\beta} I^2 \right] = \frac{eQ}{6\hbar I(2I-1)} \vec{I} \cdot \tilde{V} \cdot \vec{I}\,, \quad (1.17)$$

where eQ, I, and $V_{\alpha\beta}$ are, respectively, the electric quadrupule moment, the nuclear spin quantum number, and the second spatial (α,β) derivative of the electrical potential at the site of the nucleus.

1.3.1 General Structure of the Internal Hamiltonians

The internal Hamiltonians have a common structure [1]. They can be expressed as:

$$H = C \sum_{\alpha,\beta=1}^{3} I_\alpha \cdot R_{\alpha\beta} \cdot A_\beta = C \sum_{\alpha,\beta=1}^{3} R_{\alpha\beta} \cdot T_{\beta\alpha}\,, \quad (1.18)$$

where C is a constant equal to $\gamma\hbar$, $\gamma_1\gamma_2\hbar^2$, and $eQ/6\hbar I(2I-1)$, respectively, for chemical shift, dipolar, and quadrupolar interactions. As already observed, $R_{\alpha\beta}$ is a second rank Cartesian tensor whose components depend on the specific characteristics of each interaction. Finally, $T_{\beta\alpha}$ are *dyadic products* constructed from two vectors, one of which is always a nuclear spin vector and another that can be the same nuclear spin vector (homonuclear dipolar or quadrupolar interactions), another nuclear spin vector (heteronuclear dipolar interaction), or the external magnetic field (chemical shift).

The second rank Cartesian tensor $\tilde{R}$ can be decomposed into three components [1]:

$$\tilde{R} = \tilde{R}^{(0)} + \tilde{R}^{(1)} + \tilde{R}^{(2)}\,. \quad (1.19)$$

The first term is the trace of the tensor $\tilde{R}$:

$$\tilde{R}^{(0)} = \frac{1}{3}\mathrm{Tr}\left(\tilde{R}\right)\tilde{1} \quad (1.20)$$

and represents the isotropic contribution of each interaction, where $\tilde{1}$ is the *identity tensor*. In the case of internal Hamiltonians, it is equal to zero for dipolar and quadrupolar interactions and equal to σ_{iso} for the isotropic chemical shift.

The components of the traceless antisymmetric $\tilde{R}^{(1)}$ and symmetric $\tilde{R}^{(2)}$ constituents are given by

$$R^{(1)}_{\alpha\beta} = \frac{1}{2}\left(R_{\alpha\beta} - R_{\beta\alpha}\right) \quad \text{and} \quad R^{(2)}_{\alpha\beta} = \frac{1}{2}\left(R_{\alpha\beta} + R_{\beta\alpha}\right) - \frac{1}{3}\sum R_{\alpha\alpha}\,. \quad (1.21)$$

Some interactions contain, in principle, all the three components. However, only the isotropic and traceless symmetric parts are measurable by NMR. Therefore, we will disregard any eventual antisymmetric constituents and treat the second rank Cartesian tensor $\tilde{R}$ as symmetric.

For each interaction there exists a *principal axes system* (PAS), where their respective tensors $\tilde{R}$ become diagonal. The diagonal elements in the PAS are called *principal components*: R_{xx}, R_{yy}, and R_{zz}. It is often convenient to introduce instead of the three parameters $R_{\alpha\alpha}$, three new parameters:

$$\tau = \frac{1}{3}\left(R_{xx} + R_{yy} + R_{zz}\right) \text{ (isotropic term – tensor trace)}, \tag{1.22}$$

$$\delta = R_{zz} - \tau \text{ (anisotropy factor)}, \tag{1.23}$$

$$\eta = \frac{R_{yy} - R_{xx}}{\delta} \text{ (asymmetry parameter)}. \tag{1.24}$$

The tensor trace τ is different from zero only for the chemical shift interaction and in this case is named *isotropic chemical shift*. For the dipolar interaction the anisotropy factor is $\delta = 1/r^3$ and the asymmetry parameter is $\eta = 0$.

In terms of $R_{\alpha\alpha}$ and the new parameters τ, δ, and η, we have in the PAS:

$$\tilde{R}(PAS) = \begin{bmatrix} R_{xx} & 0 & 0 \\ 0 & R_{yy} & 0 \\ 0 & 0 & R_{zz} \end{bmatrix} = \tau \begin{bmatrix} 1 & 0 & 0 \\ 0 & 1 & 0 \\ 0 & 0 & 1 \end{bmatrix} + \delta \begin{bmatrix} -(1+\eta)/2 & 0 & 0 \\ 0 & -(1-\eta)/2 & 0 \\ 0 & 0 & 1 \end{bmatrix}. \tag{1.25}$$

1.3.2 Behavior of Internal Hamiltonians under Rotations

All the techniques used for high resolution solid state NMR to average out nuclear spin interactions rely on rotations of the internal Hamiltonians [1]. Due to this, it is worthwhile to represent the Hamiltonians using irreducible spherical tensor operators, denoted by $T_{l,m}$ and $R_{l,m}$, which can be obtained from the respective Cartesian tensors $R_{\alpha\beta}$ and $T_{\alpha\beta}$. Using these operators we can rewrite the general Hamiltonian in the following way:

$$H = C\sum_{l}\sum_{m=-l}^{l}(-1)^m R_{l,-m} \cdot T_{l,m}. \tag{1.26}$$

Because we are dealing with second rank Cartesian tensors $\tilde{R}$, only the $R_{l,m}$ with $l = 0,2$ are different from zero. If we consider that the $\tilde{R}$ tensors are represented in their principal axes systems, only components with $m = 0,\pm 2$ are non-zero. Denoting the components of the irreducible spherical tensor operators $R_{l,m}$ in their respective PAS's by $\rho_{l,m}$, we obtain the following relations: $\rho_{0,0} = \tau$, $\rho_{2,0} = \sqrt{3/2}\delta$, and $\rho_{2,\pm 2} = \eta\delta/2$. Normally, the $T_{l,m}$ elements are expressed in the laboratory frame. To express the $R_{l,m}$ elements also in the laboratory frame we have to make use of the *Wigner rotation matrices* $\wp^{l}_{m',m}(0,\theta,\phi)$:

$$R_{l,m} = \sum_{m'=-l}^{l} \wp^{l}_{m',m}(0,\theta,\phi)\, \rho_{l,m'}, \qquad (1.27)$$

where θ and ϕ are the Euler angles by which the laboratory frame can be brought into coincidence with the principal axes system. The first Euler angle was chosen equal to zero because the Hamiltonian is invariant under rotations around $\vec{B}_0$.

Using this procedure and getting only the first order perturbation terms ($m = 0$) of the general secular Hamiltonian, we obtain [12]:

$$H^{\mathrm{sec}} = \sqrt{3}C\tau T_{0,0} + \sqrt{\frac{3}{2}}C\delta T_{2,0}\left[\frac{3\cos^2\theta - 1}{2} + \frac{1}{2}\eta\sin^2\theta\cos 2\phi\right]. \qquad (1.28)$$

The $T_{2,0}$ spherical tensor components are summarized in Table 1.1 for each spin interaction. Because dipolar and quadrupolar tensor traces are equal to zero, the only important $T_{0,0}$ component is that of the chemical shift interaction. For this interaction, $T_{0,0} = -I_z B_0 / \sqrt{3}$.

Table 1.1 Irreducible tensor components $T_{2,0}$.

Interaction	Chemical Shift	Dipolar	Quadrupolar
$T_{2,0}$	$\sqrt{2/3}I_z B_0$	$\sqrt{1/6}\left(3I_{1z}I_{2z} - \vec{I}_1 \cdot \vec{I}_2\right)$	$\sqrt{1/6}\left(3I_z^2 - \vec{I}\cdot\vec{I}\right)$

From Table 1.1 and equation (1.28) we obtain the secular chemical shift Hamiltonian:

$$H_{\mathrm{cs}}^{\mathrm{sec}} = \hbar\omega_0 I_z\left\{\sigma + \delta\left[\frac{3\cos^2\theta - 1}{2} + \frac{1}{2}\eta\ \sin^2\theta\ \cos 2\phi\right]\right\}. \qquad (1.29)$$

Similar Hamiltonians can be obtained for the dipolar and quadrupolar interactions.

One common property of the Hamiltonians shown in equations (1.28) and (1.29) is that all of them are diagonal in the I_Z basis and have the same angular dependence. Thus, the transition energies and, consequently, the anisotropic part of the NMR frequencies, can be written in a general form for all spin interactions as

$$\omega(\theta,\phi) = \omega_{\mathrm{iso}} + \omega_{\mathrm{aniso}}(\theta,\phi) = \omega_{\mathrm{iso}} + \delta\left(\frac{3\cos^2\theta - 1}{2} + \frac{\eta\sin^2\theta\cos 2\phi}{2}\right). \qquad (1.30)$$

The anisotropy of the internal interaction in the solid state makes NMR a unique tool for probing molecular orientation, local structure, relative conformation between sites, or molecular motion. Many of the solid state NMR techniques used for studying these properties are based on the anisotropy of the internal interactions, making this issue crucial for understanding solid state NMR.

1.4 QUANTUM MECHANICAL CALCULATIONS

In elementary quantum mechanics, it is customary to study systems whose state is perfectly known (*pure state*) in terms of wave functions. For nuclear spins in magnetic fields, where the state of the system is not perfectly known (*statistical mixture of states*), it is more convenient to describe it in terms of the density operator $\rho(t)$. Due to this, the density operator description of spin systems, which combines quantum mechanics and statistical mechanics, will be employed, and it will be shown how the quantum mechanical time evolutions can be calculated for both simple and advanced NMR experiments [7, 10].

The density operator describes the state of the system, while the Hamiltonian represents the interactions, which act to change the state of the system [7]. They are related through the von Neumann equation:

$$\frac{d}{dt}\rho(t) = \frac{1}{i\hbar}\left[H(t),\rho(t)\right]. \tag{1.31}$$

If A is an observable component, the mean value of A at the instant t is obtained from the expression

$$\langle A\rangle(t) = \mathrm{Tr}\{\rho(t)A\}. \tag{1.32}$$

If $H(t)$ and $\rho(t)$ commute, the system characterized by ρ does not change with time. If they do not commute and H is time-independent ($\partial H/\partial t = 0$), a formal solution for the von Neumann equation is given by:

$$\rho(t) = e^{-i(H/\hbar)t}\rho(0)e^{i(H/\hbar)t}. \tag{1.33}$$

The operator

$$U(t) = e^{-i(H/\hbar)t}, \tag{1.34}$$

which compels the density operator to change in time, is often termed *propagator*.

In the case of NMR experiments, the density operator evolves under different time-independent Hamiltonians during distinct intervals of time, and we can easily calculate the time evolution of ρ as follows:

$$\rho(t) = \underbrace{e^{-i(H_n/\hbar)t_n}\ldots\left\{\underbrace{e^{-i(H_3/\hbar)t_3}\left[\underbrace{e^{-i(H_2/\hbar)t_2}\left(\underbrace{e^{-i(H_1/\hbar)t_1}\rho(0)e^{i(H_1/\hbar)t_1}}_{event\ 1}\right)e^{i(H_2/\hbar)t_2}}_{event\ 2}\right]e^{i(H_3/\hbar)t_3}}_{event\ 3}\right\}\ldots e^{i(H_n/\hbar)t_n}}_{event\ n}. \tag{1.35}$$

This equation will be used below to introduce the NMR concepts in a fully quantum mechanical way, allowing the connection to several valid classical pictures of the precessing magnets, as used in several introductory NMR texts.

1.4.1 Quantum Mechanical Description of NMR

1.4.1.1 System in Thermodynamic Equilibrium

The first example to be analyzed is that of a system in thermodynamic equilibrium with a reservoir at absolute temperature T. It can be shown that its density operator, under these conditions, is given by [10]:

$$\rho = \frac{e^{-H/kT}}{Z}, \tag{1.36}$$

where H is the Hamiltonian operator of the system, k is the *Boltzmann constant*, and Z is the *partition function*, which makes the trace of ρ equal to 1:

$$Z = \mathrm{Tr}\left\{e^{-\mathrm{H}/kT}\right\}. \tag{1.37}$$

In NMR, the dominant interaction for the Hamiltonian is the Zeeman interaction [7]

$$H_z = -\gamma\hbar \mathrm{I}_z B_0 = \hbar\omega_0 \mathrm{I}_z. \tag{1.38}$$

For temperatures above 1 K and under magnetic fields currently available, we always have

$$\left|\hbar\omega_0/kT\right| << 1 \tag{1.39}$$

so that we obtain the high temperature approximation:

$$\rho = e^{-H_z/kT}/Z \approx \left(\tilde{1} + \frac{\hbar\omega_0 \mathrm{I}_z}{kT}\right). \tag{1.40}$$

The identity operator $\tilde{1}$ commutes with any Hamiltonian H and is time-independent. Due to this, only the partial density operator

$$\Delta\rho = \frac{\hbar\omega_0 \mathrm{I}_z}{kT} = \alpha \mathrm{I}_z \tag{1.41}$$

evolves under equation (1.31). Therefore, for our subsequent discussions we will consider it as the density operator ($\rho \equiv \Delta\rho$).

In order to further simplify our next calculations, we are also going to consider $\alpha = 1$, so that the thermodynamic equilibrium density operator will be represented only by I_z. For all the NMR applications, we will have as the equilibrium state

$$\rho_{\mathrm{eq}} \approx I_z. \tag{1.42}$$

We are now equipped with the necessary physical and mathematical background to start discussing NMR in quantum mechanical terms [4].

1.4.1.2 Zeeman Interaction

As already discussed, the Hamiltonian for the Zeeman interaction is expressed by

$$H_z = -\vec{\mu}\cdot\vec{B}_0 = \hbar(-\gamma B_0) I_z = \hbar\omega_0 I_z. \tag{1.43}$$

When the magnetic field $\vec{B}_0$ is static, the Hamiltonian is time-independent, and we can analyze the spin motion according to equations (1.33) and (1.34). Considering that in equilibrium $\rho(0) = I_z$ and that the propagator is given by:

$$U(t) = e^{-i(H_z/\hbar)t} = e^{i\gamma B_0 I_z t} = e^{-i\omega_0 I_z t}, \tag{1.44}$$

the matrix density at the instant t is

$$\rho(t) = e^{-i\omega_0 I_z t}\rho(0)e^{i\omega_0 I_z t} = e^{-i\omega_0 I_z t}\mathrm{I}_z e^{i\omega_0 I_z t}. \tag{1.45}$$

Since we are interested in the evolution of the magnetization, which is proportional to the angular momentum $\vec{I}$, we are going to calculate the expectation values for I_z, I_x, and I_y as a function of time using the expression $\langle I_\alpha\rangle(t) = \mathrm{Tr}\{I_\alpha(t)\rho(0)\} = \mathrm{Tr}\{e^{i\omega_0 I_z t}I_\alpha e^{-i\omega_0 I_z t}\rho(0)\}$, where $\alpha = x, y, z$.

$\langle I_z\rangle(t)$ is given by

$$\begin{aligned}\langle I_z\rangle(t) &= \mathrm{Tr}\{I_z(t)\rho(0)\} = \mathrm{Tr}\{e^{i\omega_0 I_z t}\mathrm{I}_z e^{-i\omega_0 I_z t}\rho(0)\}\\ &= \mathrm{Tr}\{e^{i\omega_0 I_z t}e^{-i\omega_0 I_z t}I_z\rho(0)\} = \mathrm{Tr}\{I_z\rho(0)\} = \langle I_z\rangle(0),\end{aligned} \tag{1.46}$$

i.e., the z-component of the angular momentum is time-independent.

Instead of calculating separately the expectation values for I_x and I_y, we will compute the expectation values for $I_+ = I_x + iI_y$ and $I_- = I_x - iI_y$:

$$\langle I_\pm\rangle(t) = \mathrm{Tr}\{I_\pm(t)\rho(0)\} = \mathrm{Tr}\{e^{i\omega_0 I_z t}\mathrm{I}_\pm e^{-i\omega_0 I_z t}\rho(0)\}. \tag{1.47}$$

From 1.47 it is straightforward to obtain [4]:

$$\begin{aligned}\langle I_+\rangle(t) &= e^{i\omega_0 t}\langle I_+\rangle(0)\\ \langle I_-\rangle(t) &= e^{-i\omega_0 t}\langle I_-\rangle(0).\end{aligned} \tag{1.48}$$

Separating in the components $I_x(t)$ and $I_y(t)$:

$$\begin{aligned}\langle I_x\rangle(t) &= \langle I_x\rangle(0)\cos(\omega_0 t) - \langle I_y\rangle(0)\sin(\omega_0 t)\\ \langle I_y\rangle(t) &= \langle I_x\rangle(0)\sin(\omega_0 t) + \langle I_y\rangle(0)\cos(\omega_0 t).\end{aligned} \tag{1.49}$$

The oscillations of I_x and I_y correspond to rotations around the magnetic field with *Larmor frequency* $\omega_0 = -\gamma B_0$. In conclusion, the time-dependence of the expectation value of the total angular momentum $\langle I\rangle$ corresponds to a precession around the static magnetic field. Another way of expressing this important result is through the precession equation

$$\frac{d}{dt}\langle\vec{I}\rangle = (-\gamma B_0\hat{z})\wedge\langle\vec{I}\rangle = (\vec{\omega}_0)\wedge\langle\vec{I}\rangle. \tag{1.50}$$

Considering that $\vec{\mu} = \gamma\hbar\vec{I}$ and that the bulk magnetization is given by $\vec{M} = \sum_i\langle\vec{\mu}_i\rangle$, we can also write the following precession equations:

$$\frac{d}{dt}\langle\vec{\mu}\rangle = \vec{\omega}_0\wedge\langle\vec{\mu}\rangle \quad\text{and}\quad \frac{d}{dt}\vec{M} = \vec{\omega}_0\wedge\vec{M}. \tag{1.51}$$

These precession equations obtained with quantum mechanical arguments correspond to the normal classical view presented in basic books, which are very useful for understanding the fundamental aspects of NMR experiments.

1.4.1.3 Rotating Frame

Since the Larmor precession is a constant frequency factor for all the interactions that affect the NMR spectra, we usually eliminate it by moving to a frame that rotates around the *z*-axis with a frequency $\omega_R \approx \omega_0$ [7]. This new reference system is called the *rotating frame* (x', y', $z' = z$).

Formally, any density operator $\rho(t) = e^{-i\omega_0 t I_z}\rho(0)e^{i\omega_0 t I_z}$ observed from the frame rotating with frequency ω_R can be written as

$$\begin{aligned}\rho_R(t) &= e^{i\omega_R t I_z}\rho(t)e^{-i\omega_R t I_z} = e^{i\omega_R t I_z}\left(e^{-i\omega_0 t I_z}\rho(0)e^{i\omega_0 t I_z}\right)e^{-i\omega_R t I_z} \\ &= e^{i(\omega_R-\omega_0)tI_z}\rho(0)e^{-i(\omega_R-\omega_0)tI_z}\end{aligned} \tag{1.52}$$

and, considering the "ideal" or "resonant" case where $\omega_R = \omega_0$:

$$\rho_R(t) = \rho(0), \tag{1.53}$$

where the spins no longer precess around the magnetic field. Henceforth, we will use the rotating frame to analyze the effects of the internal interactions and rf fields on the spin system.

1.4.1.4 The Effects of Radio Frequency Pulses in the Rotating Frame

Radio frequency (rf) fields $\vec{B}_{rf}$ are of central importance in pulsed NMR and fundamental for the implementation of advanced experiments [7]. They are generated by the probe coil, oscillate perpendicularly to the static magnetic field with frequencies close to the Larmor frequency ($\omega_{rf} \approx \gamma B_0$), and have amplitude B_1 of the order of 10 G. For the typical magnetic fields used in modern NMR spectrometers (~ 10 T), the Larmor frequencies for ^{13}C and ^1H are approximately 100 and 400 MHz, respectively, which are included in the radio frequency range. Due to this, we call these oscillating fields *rf fields*. Being a time-dependent perturbation with energy close to the separation between adjacent Zeeman energy levels, an rf field is responsible for the excitation of the nuclei. Normally, we express the rf fields as

$$\vec{B}_{rf} = B_1(t)\left[\cos\left(\omega_{rf}t + \phi(t)\right)\hat{x} + \sin\left(\omega_{rf}t + \phi(t)\right)\hat{y}\right] \tag{1.54}$$

and for special excitation procedures (transition-selective, multitransition, or multiple quantum excitations) the rf field may be amplitude-, frequency-, and/or phase-modulated, through the manipulation of the parameters $B_1(t)$, ω_{rf}, and/or $\phi(t)$, respectively. In the rotating frame, considering that $\omega_{rf} = \gamma B_0$, they can be represented by the simpler equation:

$$\vec{B}_{\text{rf}} = B_1(t)\left[\cos(\phi(t))\hat{x}' + \sin(\phi(t))\hat{y}'\right]. \tag{1.55}$$

The phase ϕ will define the direction of the rf field in the rotating frame along the $x'y'$ plane. For example, for $\phi = 0°$, 90°, 180°, and 270°, we will have the following respective arbitrary orientations for B_1: x', y', $-x'$, and $-y'$. As it is going to be shown in the applications of NMR, the possibility of applying the rf fields along different directions in the rotating frame is of fundamental importance for manipulating the spins during complex pulse sequences (*phase cycling*). Taking into consideration the precession equations (equations (1.47) and (1.49)), and that in the rotating frame the only external magnetic field observed by the nuclei is the rf field, the magnetization will precess around $\vec{B}_{\text{rf}}$. The precession or flip angle can be set directly by adjusting the duration and/or the amplitude of the rf irradiation. Usually, we adjust the duration in order to apply rf pulses to flip the magnetization with angles defined by

$$\varphi = \gamma B_1 t_{\text{p}}, \tag{1.56}$$

which are normally chosen to be $\pi/2$ or π.

In terms of quantum mechanics the rf Hamiltonian in the rotating frame is given by

$$H_{\text{rf}} = -\gamma\hbar I_\alpha B_{1\alpha}, \tag{1.57}$$

where α can be, for example, x', y', $-x'$, and $-y'$, identifying the B_1 field orientation and the component of the angular momentum in the rotating frame.

Now, we can calculate the evolution of the density matrix under the rf pulse for any initial state $\rho(0) = I_{a'}$ [4]:

$$\rho(t) = e^{i(\gamma B_{1\alpha})I_\alpha t_{\text{p}}} \mathrm{I}_{\alpha'} e^{-i(\gamma B_{1\alpha})I_\alpha t_{\text{p}}}. \tag{1.58}$$

The effect of the rf pulse is to introduce a rotation of the operator $I_{a'}$ around the rf field $\vec{B}_{1\alpha}$ by a flip angle $\varphi_\alpha = \gamma B_{1\alpha} t_{\text{p}}$. For example, if we apply the rf pulse on I_z with the rf field along the x' direction of the rotating frame $B_{1x'}$ during t_{p} so that $\varphi_{x'} = \pi/2$, we will have as a result a $\pi/2-$ rotation of I_z around x' bringing it to the y' direction, according to the left-hand rule. Using the density matrix evolution we have:

$$\begin{aligned}\rho(t_{\text{p}}) &= e^{i\hbar(\gamma B_{1x'})I_{x'}t_{\text{p}}} \mathrm{I}_z e^{-i\hbar(\gamma B_{1x'})I_{x'}t_{\text{p}}} \\ &= I_z \cos(\pi/2) + I_{y'} \sin(\pi/2) = I_{y'}.\end{aligned} \tag{1.59}$$

If we duplicate t_{p} or $B_{1x'}$ we will have $\varphi_{x'} = \pi$, or the spin inversion: $\rho(t_{\text{p}}) = -I_z$. We can generalize these results stating that in the rotating frame the effect of the rf pulses can be represented by rotation matrices around x', y', $-x'$, and $-y'$, or any oblique axis, given by the left-hand rule.

1.4.2 The NMR Signal – Zeeman Interaction

After applying, for example, a $\pi/2$ pulse on I_z, the operator I_y will evolve only under the static magnetic field $B_0\hat{z}$. This evolution, from the point of view of an observer located in the laboratory frame, will be given by [4]

$$\begin{aligned}\rho(t) &= e^{i(\gamma B_0)I_z t} I_y e^{-i(\gamma B_0)I_z t} \\ &= I_y \cos(\omega_0 t) + I_x \sin(\omega_0 t).\end{aligned} \tag{1.60}$$

This means that, as discussed before for the precession of the equilibrium magnetization, the angular momentum operator will perform a precession motion around $B_0\hat{z}$ in the xy-plane. Considering that the magnetic moment of the nuclear spin is given by $\vec{\mu} = \gamma\hbar\vec{I}$ and that for a large number of identical spins the total magnetization is $\vec{M} = \sum_i \langle\vec{\mu}_i\rangle$, we can extend the precession motion for the expected transverse magnetization:

$$\langle M_{xy}(t)\rangle = \langle M_y\rangle\cos(\omega_0 t) + \langle M_x\rangle\sin(\omega_0 t). \tag{1.61}$$

Normally, the same coil that generates the rf field is used for detecting the NMR signal. In this case the precession of the bulk magnetization generates, through Faraday's law, an electromotive force. This alternating voltage oscillates with the Larmor frequency and is denoted *free induction decay* (FID). When the spins are evolving under the internal Hamiltonians, they will produce signals with several different frequencies, which are detected simultaneously by the coil as a superposition of all the individual frequency FIDs. To separate all the frequencies, the FID is Fourier-transformed, resulting, for example, in the spectra shown in Figure 1.4.

1.5 HIGH RESOLUTION SOLID STATE NMR METHODS

To introduce the simplest high resolution solid state NMR methods we are going to restrict our discussion to organic samples containing only ^{13}C and ^{1}H nuclei.

In liquid samples the molecules typically execute fast and random motions so that the anisotropic components of the above discussed spin interactions are averaged out and the only remaining contribution for the NMR spectra comes from the isotropic terms, resulting in spectra composed of very well defined resonance lines. Given that these motions are restricted in solid samples, the anisotropic components are not averaged or are only partially averaged. Since the resonance of each nucleus depends on the local field at its site, and the intensity of the these local fields depends on the orientation of the neighboring nuclei, of these electron clouds, and of the electric gradient fields, there will be a considerable spread in the resonance frequencies resulting in broad spectra, where it is difficult to distinguish the different isotropic chemical shifts. Another problem for obtaining spectra from solid samples for rare nuclei (low natural abundance) with small gyromagnetic ratio, such as ^{13}C, is associated with the low sensitivity related to two problems: small NMR signal and long spin-lattice relaxation times [8]. In order to circumvent these problems three methods were proposed: dipolar (homo and heteronuclear) decoupling and magic-angle spinning (MAS) for improving the resolution of the spectra, and cross-polarization for increasing the sensitivity.

1.5.1 Dipolar Decoupling

The dipolar decoupling technique is used to suppress the magnetic dipolar interaction between nuclear spins. When observing rare ^{13}C nuclei, the homonuclear dipolar interaction (^{13}C-^{13}C) is negligible due to their small natural abundance and the only important dipolar interaction comes from the abundant ^{1}H nuclei. To suppress the ^{13}C-^{1}H dipolar interaction, the method called *heteronuclear decoupling* is used and was proposed by Sarles and Cotts in 1958 [13]. It consists in the continuous wave (CW) irradiation with rf at the ^{1}H resonance frequency while observing the ^{13}C signal. In order to effectively suppress the ^{13}C-^{1}H dipolar interaction, it is necessary to apply rf B_1 fields satisfying the condition $\gamma B_1 > \Delta\nu_{\text{dip}}$, where $\Delta\nu_{\text{dip}}$ is the line width due to the heteronuclear dipolar interaction. In the cases of residual line widths arising from insufficient proton decoupling power, a simple two-pulse phase modulation (TPPM) scheme greatly improves the quality of the spectra [14]. In the case of homonuclear dipolar interaction, normally ^{1}H-^{1}H dipolar interaction, the method employed is called *homonuclear decoupling*. There are several ways to suppress this interaction and the most common are those based on multiple pulses, such as WaHuHa [15], MREV-8 [16, 17], and BR-24 [18], or on the Lee-Goldburg method [19].

1.5.2 Magic-Angle Spinning (MAS)

Because this method is of fundamental importance for the advanced techniques to be presented later in this chapter, the magic-angle spinning technique will now be discussed in more detail. In 1959 Andrew et al. [20] and Lowe [21] proposed, independently, this method to suppress the magnetic dipolar interaction. In 1962, Andrew and Eades [22] showed that MAS could also be applied to eliminate other anisotropic interactions. To introduce this method, let us take as an example the magnetic dipolar interaction between the nuclei ^{13}C and ^{1}H. The dipolar magnetic field produced by ^{1}H along the static magnetic field on the ^{13}C site is given by $B_{\text{dip}} \approx (\mu_H / r^3)(3\cos^2\theta - 1)$, where r is the modulus of the distance vector connecting both nuclei, and θ is the angle between $\vec{r}$ and $\vec{B}_0$. The term $(3\cos^2\theta - 1)$ describes the anisotropy of the dipolar interaction. If this term were equal to zero, the dipolar interaction would be eliminated. One way to do this is to choose $\theta = \theta_m \approx 54.74^\circ$, where θ_m is called the *magic-angle*. Of course, it is impossible to put all the spin pairs aligned along this specific orientation, except in very special cases. The method proposed to reach this condition consists in spinning the sample around the magic-angle orientation with a rotation frequency $\nu_r > \Delta\nu_{dip}$. In this situation, all the internuclear vectors $\vec{r}$ will be, on the average, along the magic direction resulting in an average dipolar field $\langle B_{dip}\rangle = \langle(\mu_H/r^3)(3\cos^2\theta_m - 1)\rangle \approx 0$. The same procedure can be applied for suppressing other nuclear spin interactions but, in the general case, it is necessary to use the orientation of the principal axis of the interaction tensor, and the rotation frequency should be $\nu_r > \Delta\nu$, where $\Delta\nu$ is the line-broadening introduced by the specific interaction. In the case of ^{13}C-^1H dipolar interaction, $\Delta\nu_{dip}$ is in the range of 1 kHz to 100 kHz. For ^{13}C chemical shift anisotropy, since this interaction depends on the intensity of the magnetic field, choosing, for example, $B_0 \approx 10$ T, $\Delta\nu_{csa}$ is in the range of 1 kHz to 10 kHz. For usual magic-angle spinning systems, the maximum ν_r is around 25 kHz. In this situation, it would be relatively easy to suppress the ^{13}C chemical shift anisotropy from the ^{13}C spectra, but it would be difficult to suppress the ^{13}C-^1H dipolar interaction. For suppressing the ^{13}C-^1H dipolar interaction, it is necessary to apply both heteronuclear decoupling and MAS. In the cases where it is impossible to spin the sample with $\nu_r > \Delta\nu$, spinning sidebands will be observed in the spectrum. This topic will be discussed in detail after equation (1.69).

Let us now analyze MAS using the general secular Hamiltonian [12]:

$$\begin{aligned} H^{sec} &= C\left(T_{0,0}R_{0,0} + T_{2,0}R_{2,0}\right) \\ &= \sqrt{3}C\tau T_{0,0} + \sqrt{\frac{3}{2}}C\delta T_{2,0}\left[\frac{3\cos^2\theta - 1}{2} + \frac{1}{2}\eta\,\sin^2\theta\cos 2\phi\right]. \end{aligned} \tag{1.62}$$

In order to introduce the effects of spinning the sample around the magic-angle [12], we have to express the $R_{l,m}$ tensor in a coordinate system associated to the rotor, Figure 1.1, where its components will be denoted by $\nu_{l,m}$.

In Figure 1.1 three reference systems can be observed: the laboratory system (x, y, z), the rotor system (RS) (x_{RS}, y_{RS}, z_{RS}), and the temporary system (1, 2, 3), the latter representing the transition between laboratory and rotor systems. To connect the laboratory and rotor systems we have to employ the Wigner rotation equation:

$$R_{l,m} = \sum_{m'=-l}^{l} \wp^{l}_{m',m}(\Psi,\chi,\Phi)\, \nu_{l,m'} . \tag{1.63}$$

Because $H^{\sec}$ is invariant under rotations around $\vec{B}_0$, the Euler angle Ψ, can be chosen equal to zero. The angle between the rotor spinning axis and $\vec{B}_0$, χ will be chosen later as the magic-angle. The angle Φ represents the instantaneous phase $\omega_r t$, during the rotation of the sample with frequency ω_r around z_{RS}.

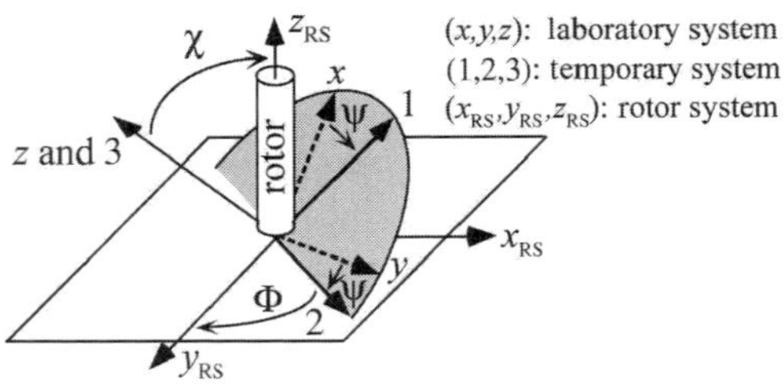

Figure 1.1 Laboratory, temporary, and rotor coordinate systems.

Taking all these assumptions, the secular Hamiltonian can be written as [12]

$$\begin{aligned} H^{\sec} &= CT_{0,0}\nu_{0,0} + CT_{2,0} \sum_{m'=-2}^{2} \wp^{2}_{m',0}(0,\chi,\omega_r t)\nu_{2,m'} \\ &= CT_{0,0}\nu_{0,0} + CT_{2,0}\left[\frac{1}{2}\left(3\cos^2\chi - 1\right)\nu_{2,0}\right] \\ &+ CT_{2,0}\left[\sqrt{\frac{3}{8}}\sin 2\chi\left(\nu_{2,1}e^{i\omega_r t} - \nu_{2,-1}e^{-i\omega_r t}\right) + \sqrt{\frac{3}{8}}\sin^2\chi\left(\nu_{2,2}e^{i2\omega_r t} + \nu_{2,-2}e^{-i2\omega_r t}\right)\right]. \end{aligned} \tag{1.64}$$

Now, $\nu_{l,m}$ can be diagonalized by a new Wigner rotation to its principal axes system (PAS), where its components will be denoted by $\rho_{l,m}$ and written in the following way:

$$\nu_{l,m} = \sum_{m'=-l}^{l} \wp^{l}_{m',m}(\alpha,\beta,\gamma)\,\rho_{l,m} . \tag{1.65}$$

The angles α, β, and γ are the Euler angles connecting PAS and RS. Using the equation above we have [12]

$$\begin{aligned} H^{\sec} &= C\rho_{0,0}T_{0,0} + \sqrt{\frac{3}{8}}\,\delta C T_{2,0}\left(3\cos^2\chi - 1\right) \\ &\times\left[\frac{1}{2}\left(3\cos^2\beta - 1\right) + \frac{\eta}{2}\left(\sin^2\beta\cos 2\gamma\right)\right] + \sqrt{\frac{3}{2}}\,\delta C T_{2,0}\,\xi(t), \end{aligned} \tag{1.66}$$

where

$$\xi(t) = C_1\cos\omega_r t + S_1\sin\omega_r t + C_2\cos 2\omega_r t + S_2\sin 2\omega_r t \tag{1.67}$$

and

$$C_1 = \frac{1}{2}\sin 2\chi \sin\beta\left[\cos\beta\left(\eta\cos 2\gamma - 3\right)\cos\alpha - \eta\ \sin 2\gamma\sin\alpha\right] \tag{1.68}$$

$$S_1 = \frac{1}{2}\sin 2\chi \sin\beta\left[\cos\beta\left(3 - \eta\cos 2\gamma\right)\sin\alpha - \eta\sin 2\gamma\cos\alpha\right]$$

$$C_2 = \frac{1}{2}\sin^2\chi\left\{\left[\frac{3}{2}\sin^2\beta + \frac{\eta}{2}\cos 2\gamma\left(1+\cos^2\beta\right)\right]\cos 2\alpha - \eta\cos\beta\sin 2\gamma\sin 2\alpha\right\}$$

$$S_2 = \frac{1}{2}\sin^2\chi\left\{-\left[\frac{3}{2}\sin^2\beta + \frac{\eta}{2}\cos 2\gamma\left(1+\cos^2\beta\right)\right]\sin 2\alpha - \eta\cos\beta\sin 2\gamma\cos 2\alpha\right\}.$$

From equation (1.66) it can be observed that when the sample is spun around the magic-angle ($\chi = \theta_{\mathrm{m}}$) with frequency ω_r, then

$$H^{\mathrm{sec}} = C\rho_{0,0}T_{0,0} + \sqrt{\frac{3}{2}}\,\delta C T_{2,0}\,\xi(t)\,. \tag{1.69}$$

The second term of equation (1.69) will be suppressed only if $\nu_{\mathrm{r}} > \Delta\nu$. If this condition is not satisfied, the spectrum will present spinning sidebands. Before discussing in detail the spinning sidebands, let us illustrate how they are formed using the ^{13}C-^{1}H dipolar interaction to exemplify this phenomenon. Normally, each individual internuclear vector $\vec{r}$ is tilted by an angle ε relatively to the z_{RS}-axis, Figure 1.2a. When spinning the sample around the magic-angle, the orientation $\theta(\omega_r t)$ of the internuclear vector relative to $\vec{B}_0$ will oscillate within the interval $(\theta_{\mathrm{m}} - \varepsilon, \theta_{\mathrm{m}} + \varepsilon)$. Since the ^{1}H dipolar magnetic field on the ^{13}C site depends on $\theta(\omega_{\mathrm{r}}t)$ as $\left[3\cos^2\theta(\omega_{\mathrm{r}}t) - 1\right]$, it will also oscillate. Therefore, the resonance frequency of each ^{13}C will periodically oscillate with a period $t_{\mathrm{r}} = 2\pi/\omega_{\mathrm{r}}$, returning to the original frequency every complete turn.

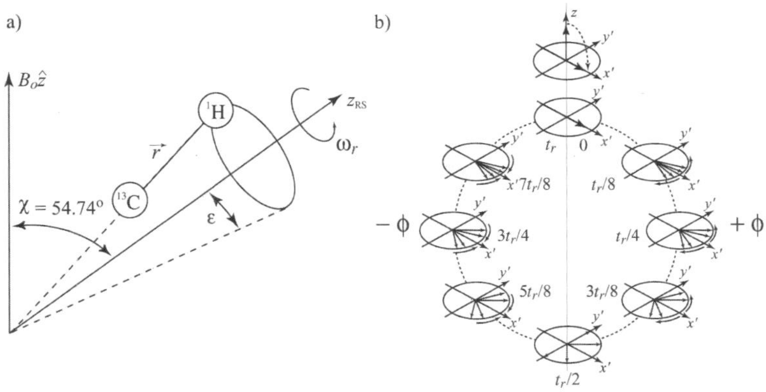

Figure 1.2 a) Sample rotation with the internuclear vector aligned along a direction different from the rotation axis (z_{RS}). b) Schematic representation of magnetization dephasing due to MAS and rotational echo.

In the real case of an amorphous or polycrystalline sample, there will be ^{13}C-^{1}H pairs whose internuclear vectors will make all the possible ε angles with respect to z_{RS}-axis. Hence, during the rotation period t_r, there will be an interval of resonance frequencies for each individual ^{13}C-^{1}H pair $\omega:[\omega_{mín}, \omega_{máx}]_{\varepsilon}$. After the ^{13}C nuclei excitation (Figure 1.2b) there will be a coherent transverse magnetization consisting of the vector sum of all individual ^{13}C magnetic moments, which under MAS will evolve with frequencies that will oscillate within the specific interval $\omega:[\omega_{mín}, \omega_{máx}]_{\varepsilon}$. At $t_r/2$, the ^{13}C magnetic moments will spread out in the transverse plane resulting in a strong decay in the NMR signal. However, at t_r all the ^{13}C magnetic moments will have the same resonance frequencies they had after the excitation, resulting in the refocalization of the spins, as in a spin-echo experiment. This echo effect will occur every complete rotation period, resulting in a train of echoes called *rotational echoes*.

When t_r is longer than the transverse relaxation time $T_2 = 2\pi/\Delta\omega$, several rotational echoes will show up in the detected NMR signal separated by t_r. The respective NMR spectrum, obtained after the Fourier transformation of this train of rotational echoes, will result not only in the spectral lines expected for each isotropic chemical shift of the ^{13}C belonging to the molecule, but also in several replicas of them. These replicas are named *spinning sidebands* and are equally separated from the real isotropic chemical shift frequencies by multiples of the spinning frequency ν_r. When $t_r < T_2$, the rotational echoes disappear and the corresponding spectrum will contain only the isotropic chemical shift frequencies. This, together with the magic-angle condition, ensures the suppression of the anisotropic components of the secular Hamiltonian.

Knowing the nature of the rotational echoes, we now return to the analysis of MAS using the secular Hamiltonian under the magic-angle condition [12], now only for the case of chemical shift anisotropy:

$$H_{cs}^{sec} = \hbar(-\gamma B_0)\sigma I_z + \hbar(-\gamma B_0)\delta\xi(t) I_z . \qquad (1.70)$$

In this case, each individual ^{13}C nucleus i will present a chemical shift resonance frequency given by

$$\omega_i = \omega_0\sigma_i + \omega_0\delta_i\xi_i(t) \qquad (1.71)$$

and will evolve in the transverse plane with the following phase:

$$\phi_i(t) = \int_0^t \omega_i(t')\,dt' = \int_0^t \omega_0\sigma_i\,dt' + \int_0^t \omega_0\delta_i\xi_i(t')\,dt' . \qquad (1.72)$$

One important property of equation (1.72) is that the second integral vanishes at each rotation period, which means the total accumulated phase due to the anisotropic internal interactions is zero for one rotation period (Figure 1.2b).

$$\phi_{\text{aniso}}(t) = \int_0^{t_r} \omega_{\text{aniso}}(t')\,dt' = 0 . \tag{1.73}$$

This condition can be perturbed by external interactions, like rf pulses or molecular rotations, making the MAS spectrum susceptible to spin manipulation or motion effects.

1.5.3 Cross-Polarization (CP)

As already discussed, it is experimentally difficult to obtain spectra from solid samples for rare nuclei (low natural abundance) or with small gyromagnetic ratios, such as ^{13}C. The low sensitivity is associated to two problems: very small NMR signal and long spin-lattice relaxation times T_1 [5, 8]. In order to circumvent these problems the cross-polarization method was proposed [23]. This method is based on the transference of polarization from the abundant ^{1}H spins, with short spin-lattice relaxation times T_1, to the rare ^{13}C nuclei. After this polarization transfer, the rare nucleus signal intensity is increased by a factor $\gamma_{\text{abundant}}/\gamma_{\text{rare}}$ when compared with the excitation with a $\pi/2$ pulse. In the case of ^{13}C and ^{1}H this factor is about 4. Despite the fact that the NMR experiment is performed for the ^{13}C nuclei, the repetition rate for signal averaging is determined by the short ^{1}H T_1, because the ^{13}C magnetization is now defined by the ^{1}H nuclei. In order to establish an efficient contact between ^{1}H and ^{13}C nuclei for the polarization transfer, it is necessary to satisfy the Hartmann-Hahn condition $\gamma_H B_{1H} = \gamma_C B_{1C}$. This means that the rf fields have to be applied simultaneously in such a way that the precession frequencies of both nuclei around the respective resonant rf fields are the same. As a result, a resonance exchange of energy between ^{13}C and ^{1}H can readily occur through a mutual spin flip mechanism [8].

1.5.4 The CP-MAS Experiment

The combination of heteronuclear decoupling, MAS, and CP techniques in only one experiment was proposed in 1976 by Schaefer and Stejskal [24] and marked the birth of high resolution solid state NMR spectroscopy for rare nuclei. Figure 1.3 shows the procedure to perform this clever experiment.

To summarize the experiment, one applies $\pi/2$ pulse to ^{1}H, which is followed by a change of 90° in the phase of the ^{1}H rf field relative to the first pulse leading to the spin-locking of the abundant spin system. Still under the spin-locking condition, which was used to lower the ^{1}H spin temperature much below the lattice temperature, the Hartmann-Hahn condition is established, and this situation is kept until the polarization transfer from ^{1}H to ^{13}C spins is complete. After that, the observation of the ^{13}C FID is performed under heteronuclear decoupling. In order to increase the signal-to-noise ratio of ^{13}C FID, the CP-MAS experiment is repeated as many times as it is necessary for the total ^{1}H spin-lattice relaxation between acquisitions to occur. MAS is continuously applied during the experiment.

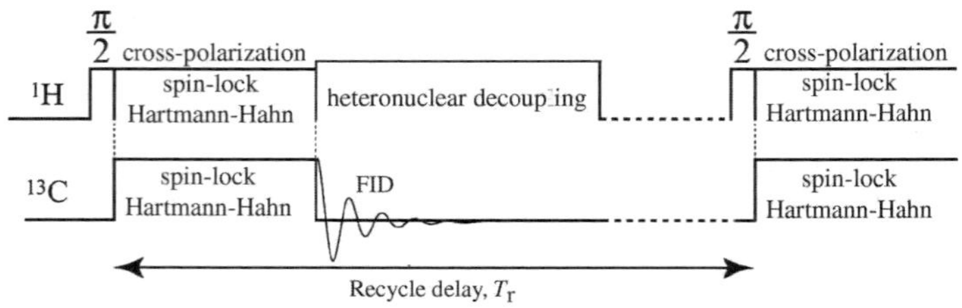

Figure 1.3 CP-MAS pulse sequence. The sample is under MAS during the entire experiment.

1.5.5 NMR Spectra

It can be observed that the secular Hamiltonians obtained above (equation (1.28)) are independent of rotations around the static magnetic field $\vec{B}_0 = B_0\hat{z}$, since the Hamiltonians are invariant under rotations around the principal axis of the laboratory frame. However, the same Hamiltonians depend, in general, on both θ and φ. θ is the angle between the principal axis of the PAS and the principal axis of the laboratory frame (which corresponds to the direction of the static magnetic field). φ is the rotation angle around the principal axis of the PAS and is important only when $\eta \neq 0$. Therefore, it is fundamental that one be acquainted with the orientation of the interaction tensor relative to a molecular frame so one may analyze the Hamiltonians and the corresponding spectra.

The first term of the secular Hamiltonian is independent of the orientation of the PAS relative to the laboratory frame and, due to this, is called the *isotropic term*. This term is different from zero for the chemical shift interaction and equal to zero for dipolar and quadrupolar interactions. The second term, which depends both on θ and φ, is known as the *anisotropic term* of the interaction, and contains all the necessary information for studying dynamics and conformation in solid materials.

The measurement of the tensor parameters τ, δ, and η, as well as the orientation angles θ and φ is one of the main purposes of NMR spectroscopy. There are several important factors governing the NMR experiment to access these parameters:

- the relative intensity of the internal Hamiltonians compared to the Zeeman Hamiltonian;
- the relative intensities of the internal Hamiltonians;
- the manner in which the internal Hamiltonians manifest themselves in NMR spectra;

- the natural (molecular motions) or intentional (advanced NMR methods) averaging of the internal interactions in both ordinary $\vec{r}$ and/or spin $\vec{I}$ spaces.

Keeping in mind all this information, we have several ways of obtaining the NMR spectra:

- High resolution NMR in liquids: Due to the fast Brownian molecular motion, the anisotropic terms of the internal Hamiltonians are averaged out and the spectra are composed of narrow lines and defined only by the isotropic terms.
- Wide-line NMR in amorphous or polycrystalline solids: Due to the anisotropic components of the internal Hamiltonians, which contribute differently for each tensor orientation, the spectra are very broad due to the superposition of the NMR lines (powder pattern spectra, Figure 1.4).
- High resolution NMR in single crystals: Despite the contribution of the anisotropic components, the spectra are again composed of narrow lines because there is only one orientation relative to the external static field for all the tensors.
- High resolution NMR in amorphous or polycrystalline solids: Due to the utilization of advanced solid state NMR methods, such as hetero- or homonuclear decoupling or magic-angle spinning, the anisotropic terms of the internal Hamiltonians are averaged out and the spectra are composed of narrow lines and defined only by the isotropic terms.

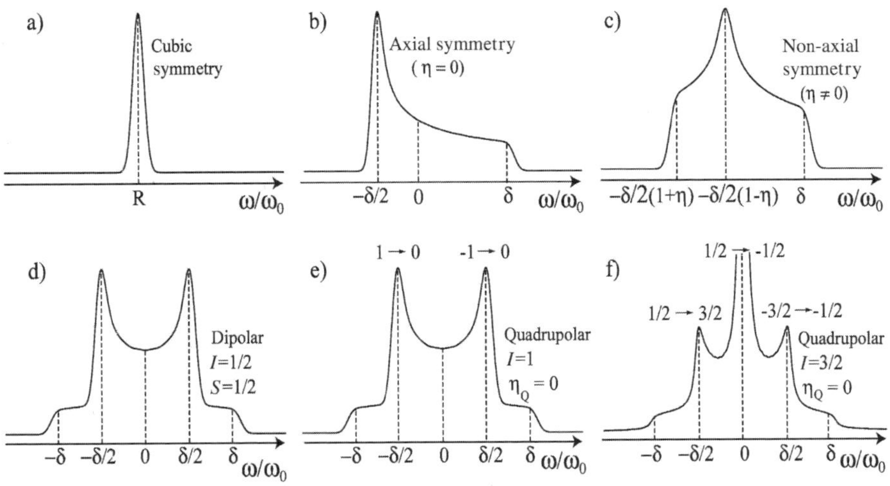

Figure 1.4 Powder-pattern spectra: Chemical shift under a) cubic, b) axial $(\eta = 0)$, and c) non-axial $(\eta \neq 0)$ symmetries; d) dipolar interaction between two spins 1/2 distant from each other of fixed distance r; and e), f) quadrupolar interaction for spins 1 and 3/2.

1.6 PRINCIPLES OF TWO-DIMENSIONAL SPECTROSCOPY

There is a wide variety of NMR methods based on the time acquisition (t_2) of a series of FIDs, where each one is acquired with a different value of a time-dependent variable evolving during a second period of time (t_1) before detection [9]. This has been termed *two-dimensional (2D) spectroscopy* and was first suggested in 1971 by Jenner [25] and implemented in 1976 by Aue, Bartholdi, and Ernst [26]. Most of these experiments include an additional time interval before FID detection, the *mixing time* (t_m) [27]. The general pulse sequences for obtaining 2D spectra are illustrated in Figure 1.5. In both cases, a and b, the *preparation period* involves, for example, the generation of transverse magnetization through a $\pi/2$ pulse or cross-polarization procedure. During the *evolution period* t_1, the magnetization evolves under a specific spin interaction, and during the *FID detection period* t_2, the magnetization is observed evolving under the action of the same or another spin interaction. To detect the 2D time signal $S(t_1,t_2)$, taking into account the magnetization evolution during t_1 and t_2, a series of experiments must be performed with a systematic variation of the evolution time t_1, by adding increments Δt_1 and acquiring the corresponding FIDs during t_2. These FIDs acquired as a function of t_1 will have amplitudes or phases modulated by the interaction acting during the evolution period. As a consequence, $S(t_1, t_2)$ will contain the information of the frequencies defined by the interactions independently acting during t_1 and t_2. The 2D Fourier transform of $S(t_1, t_2)$, namely, $S(\omega_1, \omega_2)$, will be the probability map of finding a frequency ω_1 in the first period and a frequency ω_2 in the detection period [7]. In order to perform the double Fourier transformation and get 2D spectra without phase distortions, some special procedures have to be performed, for example: the off-resonance technique [7], the time-proportional-phase incrementation method (TPPI) [28], and hypercomplex data acquisition [29]. More specific interpretations of the spectral intensity will be given for each 2D experiment described in this chapter [7].

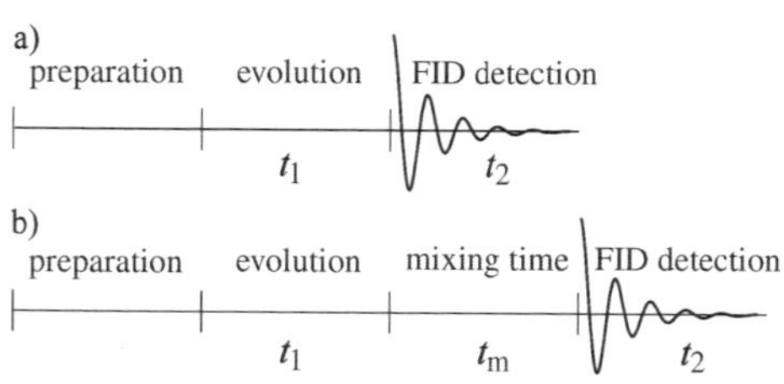

Figure 1.5 General timing of two-dimensional NMR experiments.

In the mixing time period, which is not mandatory for all 2D experiments, one can employ suitable pulses for effecting coherence transfer, monitor cross-relaxation, or chemical exchange, etc. [30].

1.7 MOLECULAR DYNAMICS AND LOCAL MOLECULAR CONFORMATION IN SOLID MATERIALS

Molecular dynamics and local conformation have important effects on mechanical, transport, and optical properties of polymers [31], activity of proteins [32], stability of pharmaceuticals [33], transport properties in zeolites [34], behavior of amorphous materials near the glass transition [35,36], ion transport in organic and inorganic ionic conductors [37–39], and other structural properties of organic and inorganic systems [40, 41]. Solid state NMR provides powerful techniques for elucidating details of segmental dynamics and local conformation in solid materials [7]. NMR methods allow the study of dynamics occurring in a wide range of frequencies with a high degree of detail. Fast dynamics (~100 MHz) can be characterized to some extent by NMR relaxation time measurements [6]. Dynamics occurring within the kHz frequency scale can be studied by line shape analysis [42] or dipolar-chemical shift correlation methods [43]. Slow dynamics (1 Hz – 1 kHz) can be studied by the so-called *exchange NMR experiments*, where relatively slow segmental reorientations are observed in terms of changes of orientation-dependent NMR frequencies [7]. Concerning structural studies, NMR also provides a series of methods capable of producing reliable measurements of torsion angles between localized sites, short and intermediate range structure, as well as interatomic distances [44–48]. In the following sections some of these methods and their applications for studying molecular dynamics and local molecular conformation will be discussed. Because NMR relaxation methods have been extensively reviewed and there are many specialized books that address this topic [6, 49], they will not be discussed in this chapter. We will concentrate our attention on the NMR methods capable of yielding information about molecular dynamics and local structure of amorphous solids (mostly organic solids).

1.7.1 Line Shape Analysis

The analysis of the behavior of the NMR line shapes as a function of temperature is one of the most traditional procedures for probing molecular dynamics using solid state NMR. To provide a better understanding of this statement, let us remember that only under static conditions the NMR Hamiltonians can be written as in equation (1.28), which results in the NMR powder patterns shown in Figure 1.4. In other words, if there is molecular motion, regular powder patterns will only be observed if its correlation time τ_c is much longer than the inverse of the anisotropy parameter of the spin interaction, i.e., $\tau_c >> 2\pi/\delta$ (slow exchange regime). At larger motional rates, i.e., smaller τ_c, ($\tau_c \sim 2\pi/\delta$, the intermediate exchange regime) the NMR spectrum is not simply defined by the static anisotropic NMR frequency, equation (1.30), but it also depends on the time scale and geometry of the molecular motion [50]. In the limit where many frequency changes occur during the signal acquisition, $\tau_c << 2\pi/\omega$, (the fast exchange limit) [49] a fully averaged powder pattern with the shape defined by the characteristics of the molecular motion is observed. The characteristic shape of the spectrum obtained in the fast exchange regime is dictated by the amplitude and geometry of the motion. In general, the NMR line shape in the fast exchange limit can be obtained simply by calculating the effective precession frequency (average frequency) under the anisotropic spin interaction [7, 49, 51]. This effective precession frequency is calculated according to an average tensor that depends on the motion geometry. Because the average tensor is also a second rank tensor, its anisotropy can also be characterized by two parameters, namely $\overline{\eta}$ and $\overline{\delta}$.

The dependence of $\overline{\eta}$ and $\overline{\delta}$ on the geometry of the motion is demonstrated in Figures 1.6a and b, for the cases of chemical shift anisotropy ($I = 1/2$ nuclei) and first order quadrupolar ($I = 1$ nuclei) interactions, respectively.

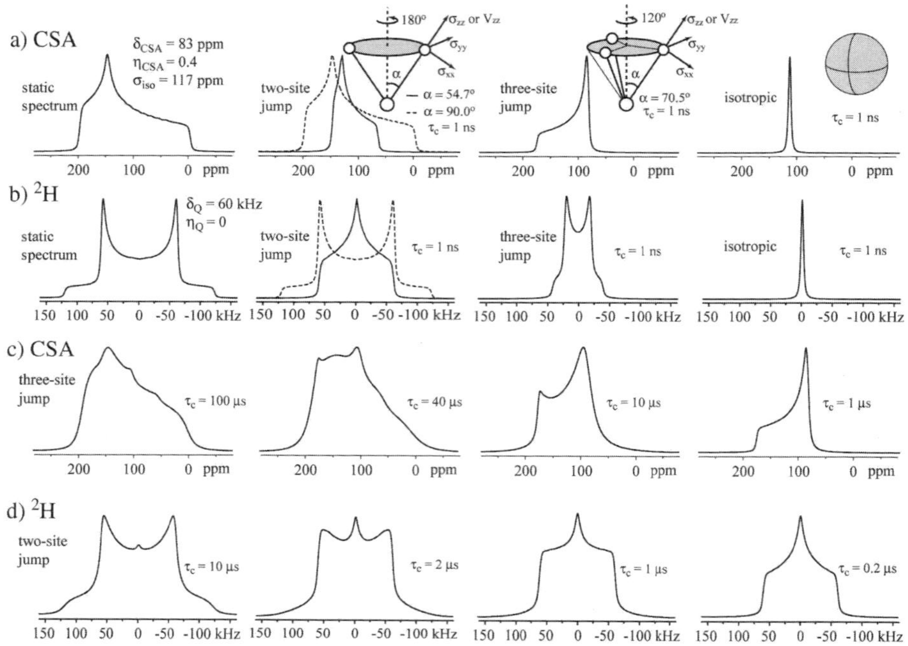

Figure 1.6 Series of powder pattern line shapes simulated for different kinds of molecular motions: a) $I = \frac{1}{2}$ (CSA) and b) $I = 1$ (^2H quadrupolar). c) Chemical shift anisotropy powder pattern for three-site jump motion with $\alpha = 70.5°$ (methyl rotations) and different correlation times. d) ^2H powder pattern for two-site jump motion with $\alpha = 54.7°$ and different correlation times.

Different spectra are observed according to the motional geometry and amplitude. Two general features are: i) for any uniaxial motion, such as the three-site jump shown in Figure 1.6, the resulting average tensor will be axially symmetric because the rotationally averaged x and y principal values are equal; ii) for isotropic motions all the three principal values are averaged out and a single isotropic line is observed in the fast limit [7].

For spin 1/2 systems under the influence of the Zeeman and chemical shift interactions, the density matrix of the spin system behaves exactly like macroscopic magnetizations. Hence, if many molecular orientations can exist simultaneously, the concept of magnetization isochromats can be used for describing the set of different nuclear spins with the same precession frequency, i.e., under identical chemical shielding. If we consider the same kind of chemical species, the spin system in an amorphous or polycrystalline material can be described by a set of magnetization isochromats, M_i, each one corresponding to a given molecular orientation. Under the influence of molecular motion during the precession of the transverse magnetization, the molecular orientations, and consequently the precession frequencies, change stochastically, producing significant modifications in the resulting NMR spectrum.

Thus, the time evolution of the transverse magnetization isochromats depends not only on their individual precession frequencies but also on the probability of exchange between different isochromats. Assuming a stationary Markov process, the time evolution of each transverse magnetization isochromat can be calculated with the following equation [7]:

$$\frac{d\vec{m}(t)}{dt} = \left(i\tilde{\omega} + \tilde{\Pi}\right)\vec{m}(t), \tag{1.74}$$

where $\vec{m}(t)$ is a vector where each component is a complex number representing the isochromat magnetization vector in the xy-plane, $M_i(t)$. $\tilde{\omega}$ is a diagonal matrix, which is composed of the elements $\omega_{jl} = \delta_{jl}\omega_j$, where ω_j is the precession frequency of the jth magnetization isochromat and depends on the molecular orientation according to equation (1.30). The matrix $\tilde{\Pi}$ is the so-called *exchange matrix*, composed of off-diagonal elements Π_{jl} that represent the transition rate k from the state l (one given orientation) to the state j (another orientation), and diagonal elements Π_{jj} that represent the total transition rate from a state j to any other state with a negative sign. Note that the transition rates are directly associated with the correlation time of the motion (for example, for a two-site jump $k = 1/(2\tau_c)$). To obtain the total magnetization it is necessary to sum over all isochromats. This is done by multiplying the complex magnetization vector $\vec{m}(t)$ by the row vector $\vec{1} = (1,1,1,1,...,1)$. Using such procedure, from the formal solution of equation (1.74), the total magnetization vector can be written as [7]

$$\vec{M}(t) = \vec{1}\cdot\left\{\exp\left[\left(i\tilde{\omega} + \tilde{\Pi}\right)t\right]\vec{m}(0)\right\}. \tag{1.75}$$

Thus, the resulting magnetization depends on the number of accessible frequencies and on the exchange rates (correlation times) through the exchange matrix. To include the initial distribution of magnetization isochromats, the NMR signal under the influence of the motion is calculated as the powder averaging of equation (1.75), i.e. [7, 50],

$$S(t) = \left\langle \vec{1} \cdot \left\{ \exp\left[\left(i\tilde{\omega} + \tilde{\Pi} \right) t \right] \vec{m}(0) \right\} \right\rangle . \tag{1.76}$$

Figure 1.6c shows a set of simulated CSA powder patterns as a function of the correlation time, considering jumps between three molecular sites. For motional rates, $k = 1/3\tau_c$, of the order of the chemical shift anisotropy parameter, δ_{CSA}, the powder spectrum changes considerably with τ_c, making it possible to estimate this parameter from the line shapes. In the fast exchange regime, $\tau_c << 2\pi/\delta_{CSA}$, the spectrum also has a powder pattern, but with distinct CSA parameters ($\bar{\eta}_{CSA}$ and $\bar{\delta}_{CSA}$). Nevertheless, there are some intrinsic difficulties in obtaining such information from CSA line shapes: i) a small deviation in the size and orientation of the CSA tensor can produce significant changes in the average spectrum; ii) different types of motion can produce identical line shapes; iii) for the case of the ^{13}C NMR, which is the most common nucleus for studying organic systems, line shape analysis can only be done for systems with one or few different chemical sites, or requires ^{13}C isotopic labeling at selected molecular sites. This can actually be overcome by special 2D NMR techniques that produce 2D spectra containing high resolution MAS spectra in the first dimension and the corresponding powder line shapes in the second dimension [52, 53]; and iv) the use of high power broadband ^{1}H decoupling may also be problematic because the molecular motion can also interfere with the proton decoupling, inducing spectral line broadening and, sometimes, even precluding the acquisition of CSA powder line shapes in the intermediate motional regime.

Other NMR interactions, such as quadrupolar or dipolar, can also be used for probing molecular motions. In fact, in the case of the dipolar interaction, more traditional NMR methods, e.g., second moment calculations [6, 49] and studies of the line width as a function of temperature, allow one to obtain information about the rates and activation energies of molecular motions. However, site-specific information such as the reorientation angles and geometry of the motion cannot be obtained by these methods.

This can be done using deuterium (^{2}H) NMR. ^{2}H has nuclear spin $I = 1$, a large quadrupolar moment, and a relatively small gyromagnetic ratio. Because of these reasons, line shape and relaxation properties of ^{2}H resonances are determined almost exclusively by the first order quadrupolar interaction. Besides, X-^{2}H bonds can be arranged to replace X-^{1}H bonds (X is a low γ nucleus), and ^{2}H is a good spin label because its molecular properties are similar to ^{1}H. For ^{2}H in a X-^{2}H bond the electric field gradient tensor is almost axially symmetric, $\eta_Q = 0$, and the z-axis of its PAS points approximately along the direction of the X-^{2}H bond. Thus, the spectral shape produced by single transition between two adjacent ^{2}H energy levels is the same as for the case of spin ½ nuclei with axially symmetric chemical shift tensor. As a result, the full quadrupolar powder pattern can be calculated by superimposing two mirror-symmetric powder patterns (one for each ^{2}H transition) calculated in the same way as for the chemical shift interaction. However, the effects of the NMR interaction during the rf pulses, which may be significant for ^{2}H NMR [50], have been disregarded in equation (1.76), and some care must be taken with finite pulse effects when using this approximation. Figure 1.6d shows a set of simulated powder patterns as a function of the motional rates for two-site jump motions. In the fast exchange limit, $\tau_c << 2\pi/\delta_Q$, an averaged powder pattern characterized by $\overline{\eta}_Q$ and $\overline{\delta}_Q$ is observed. Note that the axial symmetry of the ^{2}H quadrupolar interaction was broken and a motionally averaged spectrum characteristic of a non-axial interaction was observed.

An experimental problem in the acquisition of ^{2}H line shapes is that the lines are very broad and, consequently, the decay time of the NMR signal is very short (~ μs). This makes it very difficult to acquire the first points of the FID, because it is necessary to wait a few μs to start the acquisition after the application of an intense rf pulse (dead-time problem). Thus, to acquire the ^{2}H signal, it is necessary to use a special pulse sequence, known as *solid-* or *quadrupolar-echo sequence* [50, 54]. The solid-echo pulse sequence is shown in Figure 1.7a. It produces an echo with maximum intensity at 2τ, making it possible to displace the acquisition to the echo maximum, and thus avoiding the dead time problem. Typical τ values are of the order of 10s of μs, and molecular motion occurring with correlation times within this time scale interferes with the refocusing of the echo, producing a decrease in the signal intensity. Hence, motions in the intermediate exchange regime can be studied by accompanying the τ dependence of the echo intensity. This is shown in Figure 1.7b, where the echo intensity, normalized by its rigid limit value (reduction factor), is plotted as a function of the jump frequency ($1/\tau_C$) for two-site jump motions occurring with several jump angles. Besides the reduction in the signal intensity, molecular motions in the intermediate exchange regime also affect the observed line shapes, and it is crucial to take these effects into account when performing line shape analysis. This can be done by adding two extra evolution periods τ in equation (1.76):

$$S(t) = \left\langle \vec{1} \cdot \left\{ \underbrace{e^{(\tilde{\Pi} \pm i\tilde{\omega})\tau} e^{(\tilde{\Pi} \mp i\tilde{\omega})\tau}}_{\text{solid-echo}} e^{(i\tilde{\omega} + \tilde{\Pi})t} \vec{m}(0) \right\} \right\rangle . \tag{1.77}$$

Figure 1.7c presents a set of experimental ^{2}H spectra for methyl deuterated poly(methyl acrylate) (PMA) acquired with the solid-echo pulse sequence using distinct τ values. The effect of the molecular motion in the line shape is clearly observed.

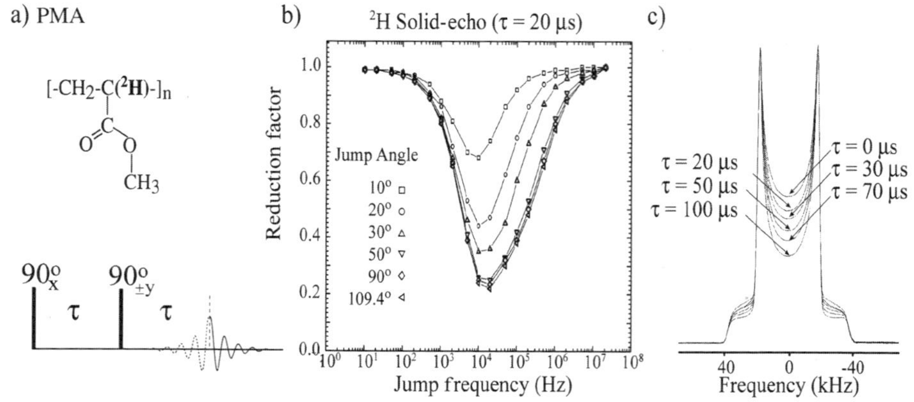

Figure 1.7 a) Molecular structure of PMA and solid-echo pulse sequence. b) Theoretical reduction factor of the solid-echo amplitude for a two-site jump between equally populated sites as a function of the jump frequency and for different jump angles. c) Effect of the solid-echo delay τ in the spectrum of deutered PMA. (Adapted with permission from references [55] and [56] . Copyright 2005 American Chemical Society)

Figure 1.8 shows an experimental example of the use of ^{2}H and CSA line shapes for obtaining information on the molecular motions in solids. ^{2}H and ^{15}N static spectra were acquired as a function of temperature to elucidate the molecular mechanism of the relaxation processes observed by dielectric relaxation in atactic poly(acrylonitrile) ($[C_3H_2N]_n$-aPAN) [57].

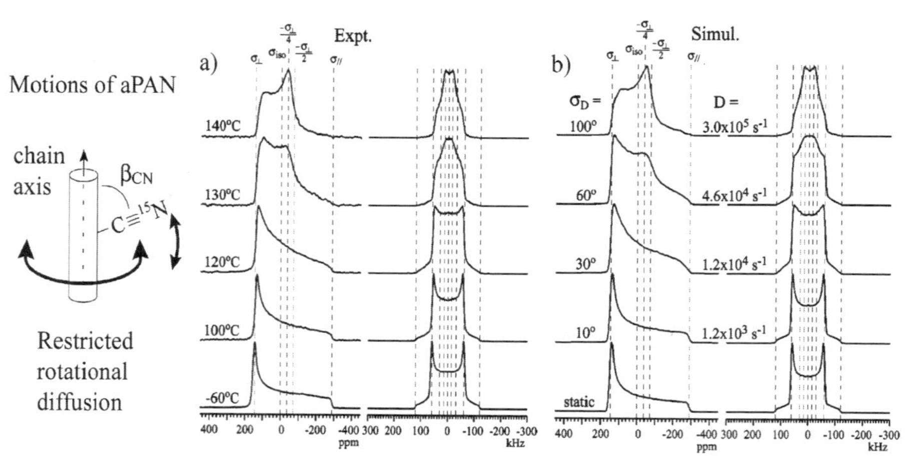

Figure 1.8 a) Experimental 1D ^{15}N CSA (left) and ^{2}H (right) spectra of aPAN as a function of temperature. b) Best-fit simulations of the experimental spectra based on the model of uniaxial rotational diffusion with distributed restricted amplitudes (see left). The width σ_D of the amplitude distribution and the diffusion coefficients D are shown for each case. (Adapted with permission from reference [57]. Copyright 2005 American Chemical Society)

The experimental spectra are well reproduced by simulations considering uniaxial rotation around the main chain with a restricted distribution of amplitudes. Below 110°C, the standard deviation σ_D of the Gaussian distribution of motional amplitudes increases with temperature up to ~30 °C. The small-angle fluctuations rapidly change to larger-amplitude motions above this temperature; σ_D increases from 30 to 100° in the temperature range of 120–140 °C, which correlates with the intensity increase observed in dielectric measurements.

Experiments involving line shape analysis under magic-angle spinning have also been performed for ^{2}H. These experiments explore the dependence of the spinning sidebands on the molecular motions. In these cases, the effects of both molecular and mechanical rotation must be put together in the simulation of the spectra, which can be done by using appropriated theories [58]. Similar approaches have also been used for spin ½ systems under the effect of the chemical shift interaction [32]. However, molecular motions can also be studied in sideband-free MAS NMR spectra, which is particularly useful for ^{13}C NMR, where multiline spectra are generally observed. This is so because molecular rotations also interfere with the MAS averaging, introducing characteristic line broadenings in the MAS line shapes [59, 60].

Figure 1.9 shows high resolution ^{13}C NMR spectra of 1,4-diphenoxy benzene (DPB) obtained at different temperatures. A clear broadening of the 2,4,1,5 lines (see assignments in the inset) as a function of temperature is observed.

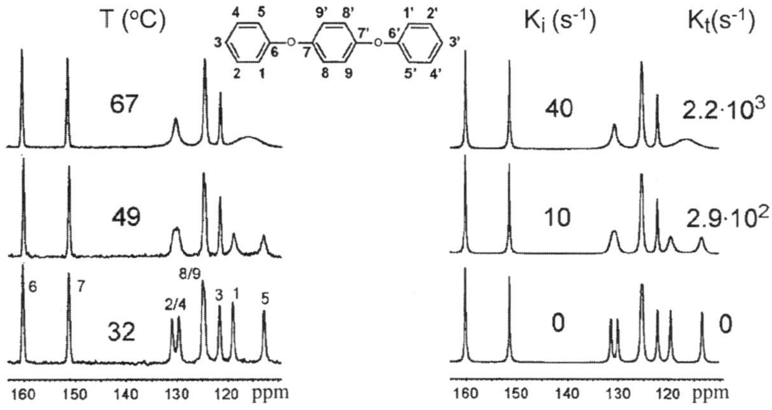

Figure 1.9 Experimental (left) and simulated (right) ^{13}C MAS spectra of DPB (see molecular structure in the inset) as a function of temperature at ν_R = 5.2 kHz. The simulated spectra were calculated considering π ring flips with the indicated rate. (Adapted from reference [60]. Copyright 2005 with permission from Elsevier)

Since carbons 2,4,1,5 belong to the terminal rings of DPB, this broadening can be associated to 180° rotations π-flips around the 1,4 axis of the rings. A slight broadening is also observed for the line attributed to carbons 8 and 9, showing that a fraction of the inner rings also executes π-flips. The temperature dependence of the experimental MAS NMR spectra is nicely reproduced considering π-flips occurring with appropriate motional rates

$k = 1/\tau_c$, k_i for inner rings and k_t for terminal rings. From these data the temperature dependence of the motional rates was obtained and the π-flips activation energies of the inner and terminal rings of DPB were estimated to be 90 and 73 kJ/mol, respectively.

The use of ^{1}H line shapes is also one of the most traditional methods for studying molecular dynamics. The ^{1}H NMR spectrum of rigid materials is characterized by broad lines, usually with Gaussian or Lorentzian shapes due to the strong coupling between the ^{1}H nuclei. As in the case of any other anisotropic interaction, molecular motion can induce the averaging of the ^{1}H-^{1}H dipolar interaction, reducing the line widths. Thus, monitoring the line width as a function of temperature and using appropriate models, information about correlation time and activation energy of molecular motions can be obtained [61, 62]. These studies, which can also be performed for other nuclei, are useful for the characterization of materials like solid polymer electrolytes, where the line widths and T_1 relaxation of the charge carriers (usually Li, Be, Ag, or Cs), provide information about their mobilities [37, 38]. However, due to the lack of spectral resolution in the static or regular MAS ^{1}H spectra, these studies have been limited to cases where no local information about molecular motions were necessary. A class of two-dimensional NMR methods that are generally called *separated-local-field* (SFL) experiments overcomes this limitation.

The simplest version of SLF experiments, the so-called *wideline separation* (WISE) [63], is derived from the conventional CP-MAS method by adding an extra time period t_1 in the CP pulse sequence, as shown in Figure 1.10a. After the ^{1}H $\pi/2$ pulse, the ^{1}H magnetization evolves under the action of the chemical shift and homonuclear dipolar interactions and then is transferred to a low natural abundance X nucleus (usually a spin 1/2 nucleus such as ^{13}C, ^{29}Si, or ^{15}N). Then, after the CP transfer, the magnetization of the X nucleus depends upon the evolution of the ^{1}H magnetization during t_1. Incrementing t_1 according to the usual rules of signal sampling, a 2D time matrix, $S(t_1,t_2)$, can be obtained and its 2D Fourier transform provides a 2D spectrum, $S(\omega_1,\omega_2)$, that contains the high resolution MAS spectra of the X nuclei along ω_2, and the corresponding ^{1}H wideline spectra along ω_1. This experiment is very convenient for distinguishing regions of the material with different mobilities. Figure 1.10b shows an example where the WISE experiment was applied to study the phase separation in the block copolymer polystyrene (PS)-*b*-poly(methyl siloxane) (PDMS). It is possible to observe that while the PS phase of the material is rigid (broad lines) the PDMS phase is mobile (narrow line), despite the high chemical connectivity and spatial proximity (~20 nm) between the phases. The WISE experiment was also modified to obtain information about the domain sizes of each phase [63].

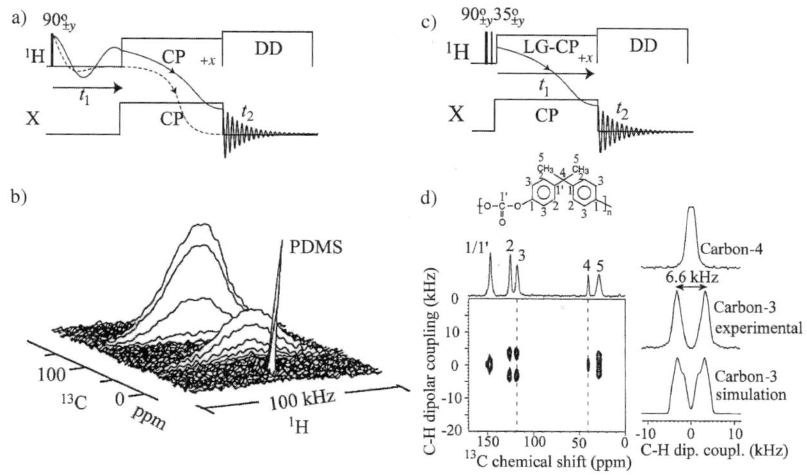

Figure 1.10 a) Pulse sequence for 2D WISE experiments. b) Application of the WISE method for identifying rigid and mobile phases in the block copolymer PS-*b*-PDMS. c) Pulse sequence for 2D LG-CP experiments. The extra 35° pulse is applied because the precession plane of the ^{1}H magnetization under LG irradiation is 35° tilted relative to the transverse plane. (Adapted with permission from references [43] and [63]. Copyright 2005 American Chemical Society)

The main problem of using WISE experiments for probing localized motions is related to the so-called *spin-diffusion phenomenom* [6, 49]. Strong homonuclear dipolar couplings induce transfer of *z* magnetization between the coupled spins (spin-diffusion or spin-exchange) with or without rf irradiation [6, 7, 49]. Hence, due to the high natural abundance and the strong ^{1}H-^{1}H dipolar coupling, the ^{1}H magnetization diffusion (^{1}H spin-diffusion) during the cross-polarization period is very effective [63, 64]. Some approaches were used to avoid this problem for the WISE experiments [65]. However, a similar experiment, but much less prone to spin-diffusion, was also proposed [43]. In this experiment, the effect of spin-diffusion is minimized by replacing the CW spin-lock pulse by the Lee-Goldburg homonuclear decoupling sequence. In addition, instead of using an extra evolution period t_1 in the pulse sequence, the CP period is incremented to produce the 2D signal (2D LG-CP experiment); see Figure 1.10c. The result is a 2D spectrum that contains the correlation between the X nucleus chemical shift, and the X-^{1}H dipolar interaction scaled by a factor of 0.577 due to the LG irradiation [66, 67]. Specifically for ^{13}C, the ω_2 spectral dimension contains the high resolution ^{13}C NMR spectrum, and the corresponding scaled ^{13}C-^{1}H dipolar powder spectrum appears in the ω_1 dimension. As in the case of ^{2}H line shape, the molecular motion breaks the axial symmetry of the dipolar interaction and in the fast exchange limit the average $\bar{\eta}_D$ and $\bar{\delta}_D$ reflect the geometry of the motion [43]. An example of local motion studied by 2D LG-CP is the π ring-flip of polycarbonate. The two phenylene rings in each repeat unit undergo π-flips superimposed with wobbling motions with a distribution of amplitudes around the equilibrium positions. Figure 1.10d shows the 2D ^{13}C-^{1}H LG-CP spectrum of polycarbonate and several cross sections along the ω_1 axis. A ^{13}C 1D LG-CP MAS spectrum is shown at the top of the 2D spectrum, where the five resolved ^{13}C signals are assigned. The C-H cross-section of the Carbon 4 signal, which results from the aliphatic quaternary carbon, does not exhibit any resolved splitting, as expected for a carbon site that is not directly bonded to any protons. In contrast, the aromatic Carbon-3 exhibits a splitting of 6.6 kHz and a smooth line shape devoid of sharp features. The spectral simulation for this group is also shown [43]. The experimental spectrum is reproduced using a model where both aromatic rings undergo π-ringflips and wobble around the equilibrium ring positions with an average amplitude of ~ 20°.

1.7.2 Two-Dimensional Exchange NMR Experiments

Exchange NMR [7, 68] has become one of the premier tools for characterizing segmental motions in organic solids. In these experiments molecular reorientations on the milliseconds-to-seconds time scale (that do not change the powder spectrum) are observed in terms of changes of orientation-dependent NMR frequencies. They permit obtaining correlation times and their distributions [69, 70], reorientation angle distributions [69,71], orienta-tional memory and rate memory [72], verifying the existence of dynamic heterogeneities [73], and determining their sizes [74]. Due to the variety of exchange experiments that can be performed [75–82], some general ideas and applications of standard 2D-exchange experiments will be discussed first and then some extensions will be addressed together with the corresponding experimental examples. For the sake of simplicity, the simpler case of spin $I = ½$ will be discussed first and then extended for the case of spin I = 1.

Figure 1.11a shows a general pulse sequence for the static 2D-exchange experiment. The first $\pi/2$ pulse (or cross-polarization) excites the transverse magnetization, which evolves under the chemical shift interaction during the first period t_1. After t_1, a second $\pi/2$ pulse stores one component of the magnetization along the z direction, where it remains during the long mixing time (t_m).

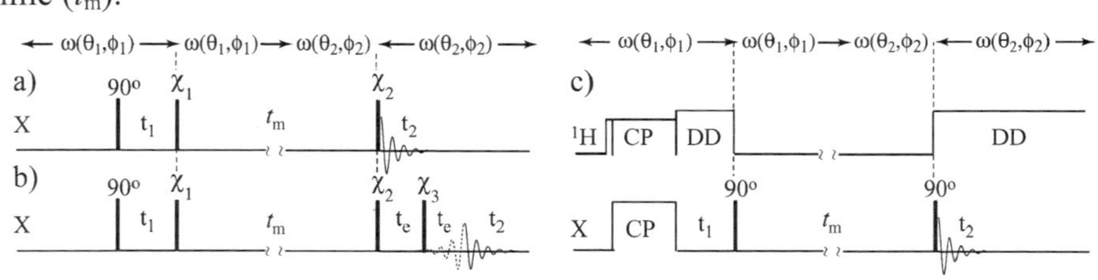

Figure 1.11 Pulse sequence used for 2D-exchange NMR experiment. a) Basic scheme using a single 90° excitation pulse. b) With single 90° excitation pulse and echo acquisition. c) Using a cross-polarization and ^{1}H decoupling. χ_1 and χ_2 represent rf pulses with general flip angles.

The remaining transverse magnetization during t_m is eliminated by cycling the phases of the $\pi/2$ pulses. A second $\pi/2$ (read-out) pulse is applied after t_m and then the magnetization evolves under the influence of the chemical shift interaction during t_2. Hence, the detected signal intensity $S(t_1,t_2,t_m)$ depends on both evolution times, t_1 and t_2, and on the mixing time t_m. The 2D Fourier transformation [28, 29] of the data produces a 2D spectrum that contains the intensity distribution as a function of the NMR frequencies experienced by the magnetization during t_1 (ω_1) and t_2 (ω_2), and also on the molecular motion that occurred during t_m, $S(\omega_1,\omega_2,t_m)$.

If no motions have occurred during t_m, the orientation-dependent frequencies are equal before and after the mixing time and the resulting 2D spectrum is purely diagonal. In contrast, if segmental reorientation occurs during t_m, the orientation-dependent frequencies change and off-diagonal intensities are observed. Moreover, the change in the NMR frequency for each molecular site is dictated by the molecular motion occurring during t_m and the spectral intensity $S(\omega_1,\omega_2,t_m)$ depends on the time scale and geometry of the molecular reorientations. While small-angle motions induce little modifications in the NMR frequencies and only near diagonal intensity is observed, high angle rotations produce large frequency changes and the corresponding 2D spectrum contains intensities in a broad spectral range. Actually, if the correlation time of the motion is much longer than the evolution times t_1 and t_2 (slow motion regime), the magnetization that gives rise to a certain intensity at a frequency point (ω_j,ω_i) depends on the fractional population of spins that precess with ω_j during t_1, $P_j(0)$, and on the conditional probability that it changes its orientation in such a way that it precesses with ω_i during t_2, $P_{ij}(t_m)$ [7, 69, 83]:

$$M_{ij}\left(t_1,t_2,t_m\right) = P_{ij}\left(t_m\right)P_j\left(0\right)e^{i\omega^j t_1}e^{i\omega^j t_2}. \tag{1.78}$$

Transverse relaxation during t_2 and longitudinal relaxation during t_1 were disregarded. Because molecular motion can be generally considered a stationary Markov process, $P_i(0)$ and $P_{ij}(t_m)$ can be properly calculated from a master equation [7, 69, 83], resulting in a 2D NMR signal given by

$$S\left(t_1,t_2,t_m\right) = \left\langle \vec{1}^T \underbrace{e^{+i\tilde{\omega}t_2}}_{\text{detect}} \underbrace{e^{\tilde{\Pi}t_m}}_{\text{mix}} \underbrace{e^{+i\tilde{\omega}t_1}}_{\text{evolve}} \vec{P}(0)\vec{1} \right\rangle. \tag{1.79}$$

The elements of $\vec{P}(0)$ represent the a priori probabilities P_j of finding a segment in a given orientation, i.e., $\vec{P}(0)$ is another representation of the initial magnetization vector $\vec{m}(0)$, which accounts for the initial distribution of magnetization isochromats. The exchange matrix $\tilde{\Pi}$ and the frequency matrix $\tilde{\omega}$ have the same definitions as before.

The 2D spectrum can be interpreted as a map of reorientation probabilities, i.e., it represents the probability density of finding a segment with frequency ω_1 before t_m and ω_2 after t_m. Thus, because these probabilities are dictated by the correlation time and the geometry of the motions, the 2D-exchange pattern is sensitive to both these features. In fact, if the molecular motion involves only one or a few reorientation angles, typical elliptical patterns are observed and the reorientation angle can be obtained directly from the experimental spectrum in a model-free approach [7, 84]. It can be accomplished by measuring the two principal axes of the observed ellipses (a and b, as indicated in Figure 1.12a), which are related to the reorientation angle β_R as [7, 84]

$$\left|\tan \beta_R\right| = b/a \,. \tag{1.80}$$

Figures 1.12a and b show a set of exchange patterns calculated for different reorientation angles, β_R. Elliptical ridges that depend on the reorientation angle can be easily observed. The 2D patterns were calculated by only considering a reorientation of β_R, i.e., the calculation does not take into account the back jumps to the original positions. However, in the limit where the correlation time of the motion is much shorter than the mixing time, many forward and back jumps may happen during this period.

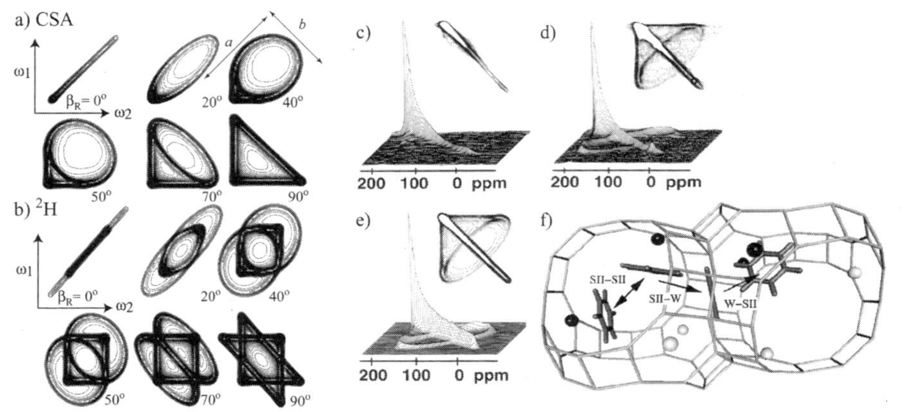

Figure 1.12 2D-exchange powder pattern for $\eta = 0$ as a function of the reorientation angle β_R. Patterns for a) $I = ½$ (CSA) and b) $I = 1$ (^{2}H quadrupolar). c) ^{13}C 2D-exchange on benzene adsorbed in the supercage of Ca-LSX zeolite at $T = 298$ K and $t_m = 1$ ms, and d) at $T = 298$ K and $t_m = 300$ ms. e) Simulated 2D-exchange spectrum based on an isotropic distribution of axially symmetric chemical shift tensors executing discrete jumps with $\beta_R = (109 \pm 3°)$. f) Diagram showing the hopping of benzene molecules adsorbed at Ca^{2+} cation sites (SII) in a Ca-LSX zeolite supercage. (Adapted with permission from reference [34]. Copyrights 2005 by American Physical Society and American Chemical Society)

As a result, even for a simple two-site jump between equally populated sites, the 2D-exchange spectrum has a diagonal pattern that arises from molecular segments that, after executing a certain number of jumps, return to their original position at the end of the mixing time. Actually, if all segments participate in the molecular reorientations with the same rate, the ratio between the integral of the diagonal pattern and the integral of the full 2D pattern gives the probability that a molecular segment has of returning to its initial orientation after t_m. In other words, it is the inverse of the number of sites accessible to the molecular motion.

An example of the application of 2D-exchange NMR for studying molecular motions can be found in the case of benzene adsorbed in nano and mesoporous materials, more specifically, benzene adsorbed in supercages of Ca-LSX zeolites [34]. Figure 1.12c shows the ^{13}C 2D-exchange NMR spectrum of benzene at T = 298 K with t_m = 1 ms. The spectrum is mostly diagonal, indicating the absence of exchange in this time scale. In contrast, a clear elliptical ridge appears when the mixing time is increased to 300 ms, Figure 1.12d. Figure 1.12e shows the simulated 2D-exchange spectrum based on an isotropic distribution of axially symmetric chemical shift tensors executing discrete jumps with $\beta_R = (109 \pm 3°)$ among tetrahedrally arranged sites (β_R = 109.5°) and correlation times of the order of the mixing times. The good agreement between the simulated and experimental spectra confirms the model for the adsorbed benzene hopping in the supercages of Ca-LSX zeolites shown in Figure 1.12f.

Another example of the application of 2D-exchange NMR in the study of discrete jump process is the case of the molecular motions responsible for the β-relaxation of poly(methyl methacrylate) (PMMA). PMMA belongs to an important class of glassy polymers, where mechanical and dielectric properties are strongly affected by the β-relaxation process, which is one of the most widely studied local relaxations in glassy polymers [85]. Because the exchange NMR experiments essentially detect reorientations of the ^{13}C chemical shift anisotropy (CSA) tensors, the molecular motion occurring as part of the β-relaxation can be well probed by the carbonyl groups, COO. The COO tensor principal values and its PAS orientation are also well known for PMMA, Figure 1.13a [86, 87]. The large CSA of COO groups and the fact that not all segments participate in the β-relaxation in the exchange time scale, make the sensitivity considerably low, which might preclude acquisition of good 2D spectra within a reasonable measuring time. To overcome this difficulty, isotopic ^{13}C labeling in the carbonyl groups was used [86]. The β-relaxation of PMMA involves rotations of the ester side group around the C-COO bond (π-ringflip) accompanied by rearrangements of the main chain (small-angle twists), which occur in several steps as shown in Figure 1.13a [86, 88, 89]: Starting from a given orientation (I), the asymmetric ester side group executes a flip coupled to a twist around the main chain axis, which avoids a steric clash with the local environment (II). Subsequently, after a new flip (flip-back), also accompanied by small-angle twist, the side group orientation is slightly different from its original position (III), which results in a small-angle reorientation of the principal axis system for the carbonyl group. The 2D ^{13}C exchange spectrum of PMMA acquired at T = 333 K is shown in Figure 1.13b. A simulation of such spectrum using only the 180° flip, Figure 1.13c, is considerably different from the experimental result, showing that another molecular motion must be involved in the β-relaxation of PMMA. Figure 1.13d shows the simulated spectrum assuming that half of the side groups reorients by 180° flip accompanied by small-angle twist with average amplitude of 20°, while the other half executes only rocking motion around the main chain with an average amplitude of 7°. The agreement between the experimental and simulated spectra is remarkable, showing that the β-relaxation of PMMA involves molecular motions with this geometry. The correlation times obtained from these studies have the same values and temperature dependence as those reported from dielectric and dynamic-mechanical studies, confirming the direct connection between the molecular motions and the β-relaxation of PMMA [85, 86].

Other studies on this topic include the merging between the α- and β-relaxation in other poly(alkyl methacrylate)s, as well as the conformational randomization that occurs in temperatures well above their glass transitions [90–92]. In all these applications, ^{13}C and ^{2}H 2D-exchange NMR were used as the main tools for elucidating the origin of the dynamic processes.

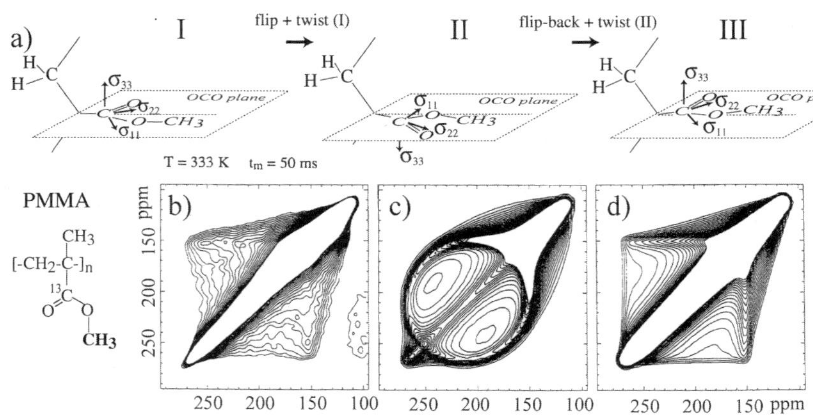

Figure 1.13 2D-exchange NMR study of β-relaxation of PMMA. a) PAS tensor orientation for the COO groups in PMMA during the molecular rotations responsible by the β-relaxation. b) 2D-exchange NMR spectrum of ^{13}COO labeled PMMA at 333 K. c) Simulation of the COO flip in PMMA. d) Simulation considering the motional process discussed in the text. (Adapted with permission from reference [86], Copyright 2005 American Chemical Society, and from reference [89], Copyright 2005 Springer)

More complex motions, such as the case of the β-relaxation in PMMA discussed above, usually lead to a distribution of reorientation angles, $R(\beta_R)$. In these cases, considering axially symmetric tensors, the 2D-exchange spectrum $S(\omega_1,\omega_2,t_m)$ can be simulated as a $R(\beta_R)$ weighted superposition of 2D spectra calculated for a specific reorientation angle, β_R, $S_{\beta_R}(\omega_1,\omega_2,t_m)$, i.e.,

$$S(\omega_1,\omega_2,t_m) = \int_0^{90^\circ} R(\beta_R) S_{\beta_R}(\omega_1,\omega_2,t_m)\, d\beta_R \,. \tag{1.81}$$

In this equation the upper limit of the integral is 90° because of the symmetry property of the 2D-exchange spectrum for axially symmetric tensors, i.e., $S_{\beta_R}(\omega_1,\omega_2,t_m) = S_{180^\circ-\beta_R}(\omega_1,\omega_2,t_m)$ [7, 84]. Thus, if the motion occurs with a distribution of reorientation angles no defined ridges appear in the 2D-exchange spectrum, but the spectral intensity is spread out. The 2D-exchange patterns can be calculated using equation (1.81) in order to obtain the distribution of reorientation angles, and, consequently, the average amplitude of the motion.

An even more complicated situation can be found in the cases where the motion not only involves a distribution of reorientation angles, but also a distribution of correlation times. This is the case of the molecular motions in many amorphous polymers at temperatures close to their glass transitions temperatures T_g. In these cases, the molecular motions have a diffusive character, which can be taken into consideration by calculating the distribution of reorientation angles from the diffusion equation that describes the rotational diffusion process [71]. In general, for rotational diffusion the reorientation angle distribution depends not only on the kind of diffusion process, but also on the ratio between the mixing time used in the experiment and the correlation time of the motion, $R(\beta_R, t_m/\tau_C)$. Consequently, for the general case where a diffusive motion occurs following a distribution of correlation times, the 2D-exchange spectrum can be calculated as:

$$S(\omega_1,\omega_2,t_m) = \int_0^{90^\circ}\int_0^{\infty} R(\beta_R, t_m/\tau_C)\, g(\tau_C)\, S_{\beta_R}(\omega_1,\omega_2,t_m)\, d\tau_C\, d\beta_R. \tag{1.82}$$

Therefore, when exchange is due to a diffusive motion a characteristic line shape that depends on $R(\beta_R, t_m/\tau_C)$ is observed.

^2H 2D-exchange NMR pattern can also be simulated by adding the two-mirror symmetric spectra, resulting in the patterns shown in Figure 1.12b. Because ^2H is a spin 1 nucleus and the quadrupolar interaction anisotropy is quite large in most cases, for obtaining absorptive undistorted ^2H 2D-exchange spectra it is necessary to select the evolution of the proper coherences in the density matrix to encode the information on the molecular motion and also to perform solid-echo acquisition to avoid the dead time problem [54]. This is accomplished by the pulse sequence shown in Figure 1.11b, where $\chi_2 = 55°$ is used in order to maximize the detected signal, together with the solid-echo detection (pulse χ_3 is set to 90° with appropriate phase). All the interpretations in terms of the probability densities also hold true for ^2H 2D-exchange, and the time signal obtained after the pulse sequence of Figure 1.11b can be written as:

$$S(t_1,t_2,t_m) = \left\langle \vec{1}^T \underbrace{e^{\pm i\tilde{\omega}t_2}}_{\text{detect}} \underbrace{e^{\pm i\tilde{\omega}\tau} e^{\mp i\tilde{\omega}\tau}}_{\text{solid-echo}} \underbrace{e^{\tilde{\Pi}t_m}}_{\text{mix}} \underbrace{e^{\pm i\tilde{\omega}t_1}}_{\text{evol}} \vec{P}(0)\vec{1} \right\rangle. \tag{1.83}$$

A more general treatment can be done considering also the cases where the correlation times of the molecular motions are comparable or smaller than the maximum evolution periods, t_1 and t_2. In these circumstances, molecular motions can also occur during the evolution periods, which can produce pronounced changes in the 2D spectral shapes. Fortunately, using the same theory presented above, these cases can also be treated. This is accomplished by adding the effect of exchange of frequencies during the evolution, detection, and echo periods to the expression for the NMR signal [69, 93],

$$S(t_1,t_2,t_m)=\left\langle \vec{1}^{T}\underbrace{e^{(\tilde{\Pi}\pm i\tilde{\omega})t_2}}_{\text{detect}}\underbrace{e^{(\tilde{\Pi}\pm i\tilde{\omega})\tau}e^{(\tilde{\Pi}\mp i\tilde{\omega})\tau}}_{\text{solid-echo}}\underbrace{e^{\tilde{\Pi}t_m}}_{\text{mix}}\underbrace{e^{(\tilde{\Pi}\pm i\tilde{\omega})t_1}}_{\text{evol}}\vec{P}(0)\vec{1}\right\rangle . \tag{1.84}$$

Figure 1.14 shows characteristic CSA ($\eta = 0$) and ^2H exchange patterns for different motional models. In Figure 1.14a the motional model is a two-site jump occurring with a distribution of reorientation angles, $R(\beta_R, t_m/\tau_C)$, as shown in the inset. The elliptical ridge characteristic of $\beta_R = 60°$ is evident in the simulations, but there is also a diagonal pattern which accounts for segments that jump forward and backward during t_m. In Figure 1.14b, the model assumed is uniaxial rotational diffusion (or diffusion in a cone) for the case where $t_m >> \tau_c$ [7, 71]. An elliptical ridge characteristic of the singularity in $R(\beta_R, t_m/\tau_c)$ is also observed in this case, but no pronounced diagonal is present. This model is typical for molecules that rotate inside restricted spaces. Figure 1.14c shows the 2D-exchange patterns obtained for isotropic rotational diffusion also in the limit $t_m >> \tau_c$ [7, 71]. The intensity is spread out over the whole frequency range and there is no diagonal pattern. This is so because this model assumes a small-step diffusion between successive sites, making the probability of returning to an original position negligible.

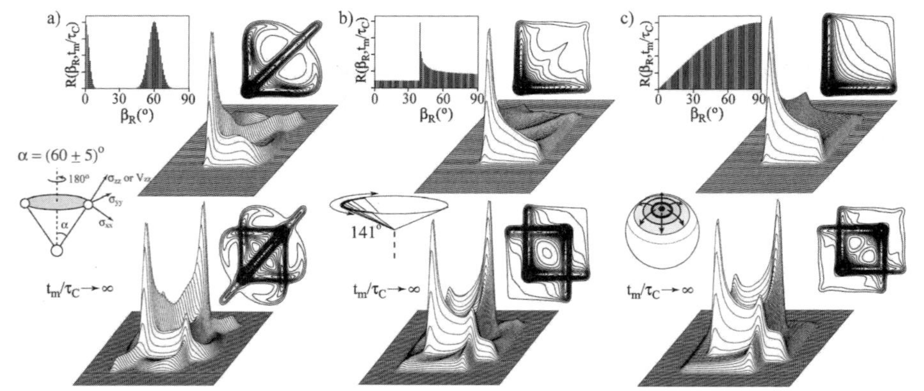

Figure 1.14 Typical 2D-exchange NMR patterns for different types of motions: a) CSA and ^2H 2D patterns for two-site jump motion occurring with a distribution of reorientation angles; b) for isotropic rotational diffusion in the limit $t_m >> \tau_C$; and c) for isotropic rotational diffusion motions in the limit $t_m >> \tau_C$.

Information about the time scale of molecular motions can also be obtained in 2D-exchange NMR experiments. A simple way of achieving that directly from the experimental spectra is accompanying the build-up of the off-diagonal exchange intensity (I_{exch}) as a function of t_m. In this case the I_{exch}

vs. t_m curve directly provides the two-time correlation function of the slow motion, allowing the estimation of the correlation time and its distribution.

Some of the most elegant examples of the application of ^2H and ^{13}C 2D-exchange NMR can be found in the studies of rotational diffusion processes [69, 71, 93, 94]. Among these examples, particular attention must be given to the rotational motions responsible by the α-relaxation (glass transition) in supercooled glass-forming systems. Amorphous systems in the glass transition range exhibit particular dynamic properties such as non-exponential α-relaxation [36, 73, 95–98]. This is attributed to the presence of a broad distribution of correlation times in the molecular motions responsible for this process. ^2H line shape analysis indicated that α-relaxation also occurs with a distribution of reorientation angles, which was further elucidated by ^2H 2D-exchange NMR performed in many of these systems. One example is the study of the α-relaxation in amorphous polystyrene ($[C_8H_8\text{-PS}]_n$) [69, 94, 99], where the hydrogen in the CH backbone groups was replaced by ^2H in order to probe the main chain motion of PS. Figure 1.15 shows ^2H 2D-exchange spectra of polystyrene at 391 K (18 K above its glass transition temperature) for mixing times of 1, 5, and 12 ms. Instead of characteristic elliptical ridges, the 2D spectrum has a broad distribution of intensities as expected for diffusive motions. In fact, the ridges parallel to the frequency axes forming prominent square ridges and the spectral intensity covering the whole 2D-plane reflect the large-angle reorientation of the chain segments.

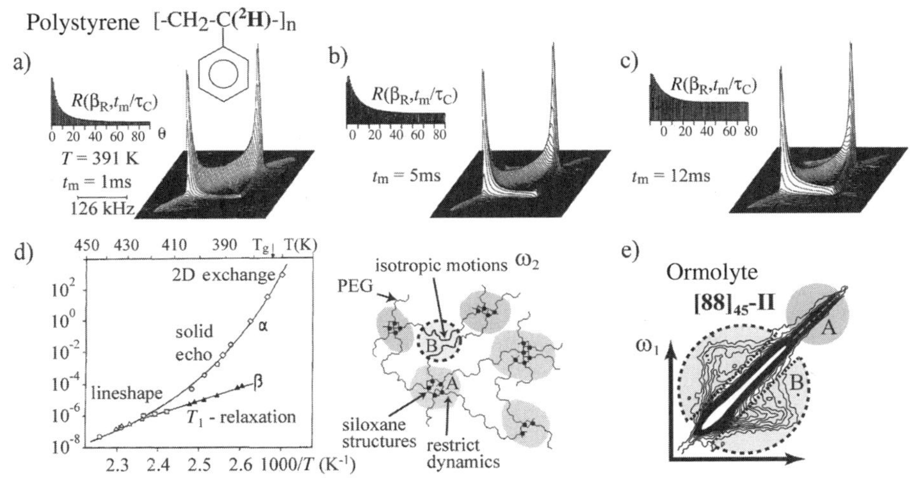

Figure 1.15 a)–c) ^2H 2D-exchange NMR spectra of amorphous PS for different mixing times with the corresponding reorientation angle distributions. d) Logarithmic plot of the correlation times vs. inverse of temperature for polystyrene obtained by different NMR methods. e) ^{13}C 2D-exchange NMR of Type II ormolytes for t_m = 200 ms. (Adapted with permission from reference [69], Copyright 2005 American Institute of Physics, and from references [79] and [99], Copyright 2005 American Chemical Society)

The fact that the exchange intensities are spread out over the whole frequency range means that a particular C-^{2}H bond has a finite probability to be found in any orientation with respect to the magnetic field after the mixing time, regardless of its starting orientation before the mixing time. Thus, the large-angle reorientations are a result of many steps of small-angle motions that occurred during t_m. This was confirmed by performing simulations based on isotropic rotational diffusion (not shown) and comparing them with the experimental spectra in order to determine the reorientation angle distribution, $R(\beta_R, t_m/\tau_C)$, shown in Figure 1.15. The behavior of $R(\beta_R, t_m/\tau_C)$ as a function of t_m was described by a model that assumes small-step isotropic rotational diffusion and a log-Gaussian distribution of correlation times with full-width-at-half-height of 3.0 decades and centered around the mean value of 6 ms [69]. Similar methods were used to study molecular dynamics in polyisopropene [93], atactic polypropylene [94], and blends of polyisoprene and poly(vinyl ethylene) [7].

Figure 1.15d also shows a plot of the correlation times as a function of the inverse of temperature (Arrhenius plot) obtained by various ^{2}H experiments (^{2}H line shape analysis or broad-line, solid-echo experiments, ^{2}H 2D-exchange NMR, and spin-lattice relaxation). The behavior of the correlation times obtained by ^{2}H line shape analysis, solid-echo experiments, and ^{2}H 2D-exchange NMR can be fitted by the Williams-Landel-Ferry equation [100], which is usual for describing the temperature dependence of the polymer structural dynamics above T_g. On the other hand, the correlation times obtained by spin-lattice relaxation measurements follow an Arrhenius behavior, characteristic of local motions in polymers (β-relaxation). Using the different NMR experiments it was possible to follow the dynamics in a broad range of frequencies and correlate the molecular dynamics with structural relaxation processes.

An example of ^{13}C 2D-exchange NMR applied to investigate molecular dynamics in a more complex system can be found in the study of the polymer molecular dynamics in siloxane/poly(ethylene glycol) nanocomposites also called *ormolytes* [79]. This hybrid system belongs to a family of versatile compounds, classified as *di-ureasils*, in which polyether-based chains are grafted on both ends to a siliceous backbone through urea functionalities (Type II ormolytes), see Figure 1.15e [38]. The polymer chain in Type II ormolytes have the following chemical structure: $\{_{3/2}OSi(CH_2)_3NHCONH(CH_2)_3[OCH_2CH_2]_nO(CH_2)_3NHCONH(CH_2)_3SiO_{3/2}\}$. The organic part of the ormolyte is composed of a linker group ($(CH_2)_3NHCONH(CH_2)_3$) and n ethylene glycol repeat units OCH_2CH_2. Figure 1.15e shows the ^{13}C 2D-exchange spectra for a Type II ormolyte sample with 45 OCH_2CH_2 repeat units and 88 wt% of PEG. The spectrum clearly shows two distinct regions. In the spectral region A the NMR signal is exclusively attributed to ^{13}C nuclei in the linkage group. The observed exchange pattern is fully diagonal, indicating that these groups do not move in the ms time scale. On the other hand, in region B a featureless off-diagonal exchange pattern is observed, indicating that the motion of the OCH_2CH_2 groups do not involve discrete jump angles. Figure 1.15e is a clear, model-free picture of the strong hindrance to the molecular motion due to the siloxane structure, since it directly reveals the distinct dynamic behavior of segments near the siloxane structures.

It is difficult to use 2D-exchange NMR in cases where only a small fraction of the segments take part in the molecular motions, because the strong and sharp diagonal ridge (due to the nonmoving segments) overwhelms the weak and broad ridges from the moving groups. This is particularly prominent for small-angle reorientations, where the exchange intensity is concentrated near the diagonal of the 2D spectrum. To overcome this difficulty, a modified version of the 2D-exchange experiment named *2D pure exchange NMR* (2D-PUREX) has been developed [101–104]. In these experiments extra pulses with adequate phase cycling, extra evolution periods, and spectral subtraction are used together to produce a 2D-exchange spectrum modulated by $\sin^2([\omega_1 - \omega_2]\tau/2)$, where τ is the extra evolution period added to the pulse sequence; see Figure 1.16 (top).

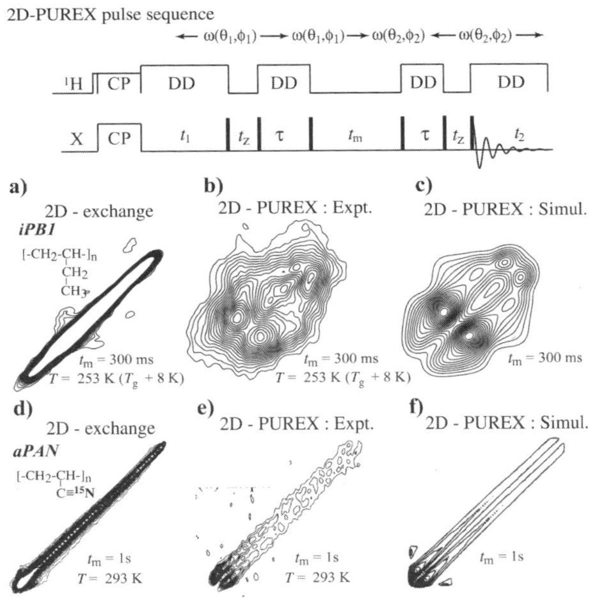

Figure 1.16 a) ^{13}C 2D-exchange NMR spectrum of iPB at T = 253 K. b) ^{13}C 2D-PUREX spectrum of iPB at T = 253 K. c) Simulation of the ^{13}C 2D-PUREX spectrum of iPB using isotropic rotational diffusion model with a log-Gaussian distribution of correlation times with width of 1.5 decades. The CSA principal values were taken from reference [53]. d) ^{15}N 2D-exchange spectrum of a PAN at T = 293 K. e) ^{15}N 2D-PUREX spectrum of aPAN at T = 293 K. f) Simulation of ^{15}N 2D-PUREX spectrum using a restricted diffusion model with average motional amplitude of 20°. (Adapted with permission from reference [57], Copyright 2005 American Chemical Society, and from reference [102], Copyright 2005 with permission from Elsevier)

Because of the $\sin^2([\omega_1 - \omega_2]\tau/2)$ spectral modulation, segments that do not reorient during t_m, $\omega_1 = \omega_2$, do not contribute to the exchange signal. Consequently, only segments reorienting on the ms to s time scale appear in the resulting 2D spectrum. One example of the performance of the 2D-PUREX method is shown in Figures 1.16a–c, where the technique was applied to the crystal form I of the semicrystalline polymer isotactic poly(1-butene), ($[CH_2\text{-}CH\text{-}R\text{-}]_n$-iPB1), at a temperature slightly above its glass transition. The regular 2D-exchange spectrum, Figure 1.16a, shows a pronounced diagonal ridge that arises from the rigid crystalline portion, and an almost imperceptible off-diagonal pattern, from the mobile amorphous region. In contrast, in the 2D-PUREX spectrum the off-diagonal intensity characteristic of diffusive motions is clearly observed, while the diagonal ridge is suppressed (Figure 1.16b). The simulation of the 2D-PUREX patterns using the isotropic rotation diffusion model reveals the good reliability of the method, Figure 1.16c [102].

The usefulness of the 2D-PUREX technique for detecting small-angle motions was also demonstrated in the amorphous polymer atactic poly(acrylonitrile) ($[C_3H_2N]_n$-aPAN), Figures 1.16d–f. At 293 K, near diagonal exchange intensity was observed in the 2D-exchange spectrum, Figure 1.16d. However, it was not possible to attribute it to small-angle motion because of the superposition of the dominant diagonal pattern. To confirm the existence of the slow small-angle motions, 2D-PUREX was carried out (Figure 1.16e). In this experiment the diagonal ridge was suppressed, and the off-diagonal signals near the diagonal were clearly detected, showing the presence of slow small-angle reorientation of the aPAN side chain. In this case the ^{15}N 2D-PUREX spectrum can be simulated by a *restrict diffusion process* with average amplitude of 20° [57].

Another important variant of the 2D-exchange experiment is the so-called *rotor synchronized 2D-MAS exchange technique*. This method was initially proposed by Veeman et al. [105], who introduced the idea of "rotor synchronization" of the pulse sequence, i.e., making sure that t_m is an integer number of rotation periods, $t_m = Nt_r$, so that the precession of the magnetization resumes at the same rotor orientation where it was at the beginning of the mixing time. Otherwise, reorientations of segments relative to the external field occur due to the macroscopic rotation of the sample (MAS) and overwhelm the more subtle effects of intrinsic segmental reorientations. In later experiments, synchronization with the beginning of the evolution period was also used, in order to achieve pure-phase 2D-MAS exchange spectra [106]. The principle of the experiment is the same as the conventional exchange NMR methods for nonrotating solids discussed above, i.e., it correlates the anisotropic NMR frequencies in two different periods of time (t_1 and t_2) separated by a long synchronized mixing time t_m [83]. The appearance of off-diagonal peaks in the two-dimensional spectra is directly associated with changes in the orientation-dependent NMR frequencies. When performed under slow-spinning frequencies relative to the anisotropy of the interaction used to probe the molecular motion, the spectrum consists of a 2D sideband pattern, which depends on the characteristics of the slow molecular reorientations. Because isotropic chemical shifts may also change according to the local conformation of a given molecular segment, MAS exchange is also a powerful tool for studying conformational dynamics. In the absence of molecular motion during t_m, a normal MAS sideband pattern along the diagonal is observed. In the case of molecular motions that only change the orientation of the chemical shift tensor principal axis, cross-peaks, labeled as *auto-cross-peaks*, appear at

$$(\omega_1, \omega_2) = \left(\omega_{iso}^{i} + M\omega_R, \omega_{iso}^{i} + N\omega_R\right), \tag{1.85}$$

where ω_{iso}^{i} is the isotropic chemical shift frequency of a nucleus at a given molecular site i; ω_R is the angular spinning frequency, and M and N are the order of the spinning sidebands. On the other hand, if the exchange process involves a change in the isotropic chemical shift, cross-peaks, labeled *hetero-cross-peaks*, appear at

$$(\omega_1, \omega_2) = \left(\omega_{iso}^{i} + M\omega_R, \omega_{iso}^{j} + N\omega_R\right), \tag{1.86}$$

where i and j indicate the change in the isotropic chemical shift of a given nucleus. Thus, if there is an exchange process that changes the isotropic chemical shift, cross-peaks linking isotropic chemical shift frequencies appear in the centerband ($M = N = 0$) even if no sidebands are observed, i.e., at high spinning frequencies.

Figure 1.17a shows a rotor synchronized 2D-MAS exchange NMR of the molecular crystal dimethylsulfone (DMS) at a spinning frequency of 1.5 kHz and mixing time of 20 ms [83]. At 293 K, DMS undergoes rotations around the main molecular axis with correlation times of ~10 ms, which lead to a reorientation of the principal axis σ_{zz} by 108° [107], resulting in the 2D sideband pattern observed in Figure 1.17a.

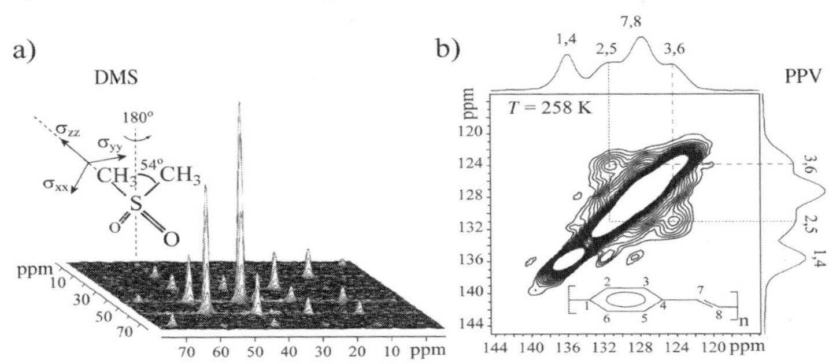

Figure 1.17 a) ^{13}C rotor synchronized 2D-MAS exchange NMR of the molecular crystal DMS at $T = 293$ K, $\nu_R = 1.5$ kHz and $t_m = 20$ ms. b) ^{13}C rotor synchronized 2D-MAS exchange of PPV at $T = 258$ K, $\nu_R = 7.5$ kHz and $t_m = 200$ ms. (Adapted from reference [83], Copyright 2005 with permission from Elsevier, and from reference [108], Copyright 2005 American Institute of Physics)

An example of motion that produces changes in the isotropic chemical shifts is the ring π-ringflips in poly(*p*-vinylene phenylene) (PPV). Figure 1.17b shows the 2D-MAS exchange of PPV acquired at 258 K with a spinning frequency of 7.5 kHz. The presence of hetero-cross-peaks linking the 2,5 and the 3,6 lines is remarkable, indicating the occurrence of an exchange process where the isotropic chemical shifts of the 2,5 and 3,6 carbons are interchanged. Since the isotropic chemical shift of the 2,5 and 3,6 carbons depends on the conformation of these segments relative to the vinylene carbon, the interchange in the isotropic chemical shift can be attributed to a 180° rotation of the phenylene rings around the 1,4 axis. In the particular case of PPV, slow conformational dynamics are directly associated with chain disorder, which affects its photoluminescence behavior [109]. Some other examples of applications of rotor synchronized 2D MAS exchange NMR are the study of conformational dynamics in amorphous and semicrystalline polymers, liquid crystals and the study of tautomerism, and molecular dynamics in molecular crystals [110–112].

2D exchange NMR experiments have also been used for studying local structure of amorphous solids. This is possible because for dipolar coupled spins 2D exchange NMR is capable of probing the relative orientation of the principal axis of the individual chemical shift tensors, which can be translated in terms of molecular torsion angles. In these cases the exchange process is not governed by molecular motions, but by local magnetization exchange, known as *spin-diffusion*, between directly bonded nuclei [113]. In fact, in studies of molecular reorientation, spin-diffusion is a limiting factor, because its effects can be confused with the exchange due to molecular motion. For this reason, low abundant and low γ nuclei such as ^{13}C or ^{2}H are preferable for probing molecular rotations, but the mixing times have to be kept short (usually smaller than 1 s) to avoid spin-diffusion effects [114]. However, in systems of highly abundant nuclei the dominant exchange process will be spin-diffusion. Since the NMR frequencies are orientation-dependent, if a nucleus in a given molecular segment exchanges its magnetization with another nucleus in a neighbor site during t_m, it will experience different frequencies before and after t_m. The shape of the resulting 2D pattern will depend on the relative orientation of the two molecular segments. However, it is very important to guarantee that spin-diffusion will be the only exchange mechanism, i.e., that no slow motion takes place during t_m. 2D exchange has been used for determining relative orientations between different carbon sites in many organic materials [115–118]. For that, the use of a ^{13}C-^{13}C or ^{15}N-^{15}N pair is necessary, which has been provided by specific ^{13}C or ^{15}N isotopic labeling at the sites of interest. Other applications of the method not requiring isotopic labeling are those where ^{31}P nuclei are used to probe relative molecular orientations. ^{31}P is a spin 1/2, 100% abundant nucleus, found in many organic and inorganic systems. An example of these studies is the measurement of relative tensor orientations of different phosphate units (Q^1

and Q^2 units in NMR terminology [9]); see molecular structure in Figure 1.18 in crystalline diphosphate $Ca_2P_2O_7$ and in binary sodium phosphate glasses [119].

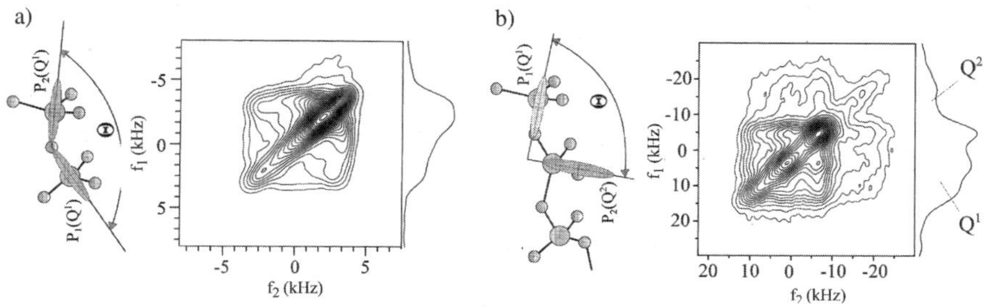

Figure 1.18 a) 2D exchange spectra using slow MAS for measuring the relative tensor orientations for a diphosphate $Ca_2P_2O_7$. b) Same as a) for a binary sodium phosphate glass with sodium to phosphorous ratio of 1:4. Molecular representation of the relative orientations of Q^1-$Q^1$and Q^1-Q^2 connectivities and their anisotropic powder patterns are also shown. (Adapted from reference [119], Copyright 2005 with permission from Elsevier)

In this case it is advantageous to use a slightly different version of the 2D exchange NMR, where slow MAS is applied together with radio frequency pulses in such a way that the local transfer of magnetization is enhanced during the mixing time (*radio frequency–driven spin-diffusion*—RFDR) [113]. For crystalline diphosphate $Ca_2P_2O_7$, which is composed of only Q^1 units, a single exchange pattern is obtained, Figure 1.18a. From this spectrum the relative orientation of two neighbor phosphate units can be estimated. In contrast, for the binary phosphate glass, where Q^1 and Q^2 units are present, the 2D exchange spectrum is a superposition of the individual spectra for Q^1- Q^1, Q^1- Q^2, Q^2- Q^2, according to their statistical weights; see Figure 1.18b. In this case, because of the spectral superposition, it is difficult to determine unambiguously the relative tensor orientation for each connectivity, but the spectrum clearly reveals the predominance of Q^1-Q^1 connectivities in the glass.

1.7.3 One-Dimensional Exchange NMR Experiments

Beyond the 2D exchange NMR methods, there is a class of one-dimensional (1D) techniques that have been used for extracting two- and multiple-time correlation functions of rotational molecular motions occurring on the millisecond-to-second time-scale [39, 75, 77, 80, 81, 102, 120, 121]. In particular, the *stimulated-echo experiment* (STE) [35, 39, 77] has been widely employed for studying time scale, amplitude, and heterogeneity of slow motions in organic and inorganic solids. The pulse sequence for the basic stimulated-echo experiment is shown in Figure 1.19a. After the excitation pulse, the quantum coherences (or magnetization components) evolve during the evolution period τ under the action of an anisotropic NMR interaction (usually chemical shift or quadrupolar). A store pulse is then applied with adequate phase and amplitude in order to convert a selected single-quantum coherence that evolved during τ into a term that does not evolve during the long mixing time, because it commutes with the spin Hamiltonian (usually it is represented by T_{20}—spin-alignment, or I_z— Zeeman order). Finally, a read-out pulse is applied to the system to reconvert the stored coherence into an observable magnetization. Using the pulse sequence shown in Figure 1.19a and disregarding relaxation effects, the NMR signal can be written as

$$S(\tau,t_m,t) = A\langle \sin(\omega(0)\tau)\sin(\omega(t_m)t)\rangle, \tag{1.87}$$

where $\omega(0)$ and $\omega(t_m)$ are the NMR frequencies before and after t_m. The amplitude A depends on the flip angles and phases of the store and read-out pulses, which are chosen according to the nuclear spin interaction used for probing the molecular motion [77]. The maximum value of $S(\tau,t_m,t)$ occurs at $t=\tau$ (echo maximum) and depends on the relation between $\omega(0)$ and $\omega(t_m)$. The decrease of the echo amplitude as a function of t_m provides the two-time correlation function of the exchange process. The number of sites involved in the exchange process also affects the stimulated-echo, in particular its intensity at long mixing times. The longitudinal relaxation, T_1, also produces echo decay as a function of the mixing time, which may happen in the same time scale of the decay due to molecular motion. Thus, to guarantee that only motion effects are encoded in the stimulated-echoes, the data must be normalized by the T_1 decay, which can be obtained by setting the evolution time $\tau = 0$ and acquiring the echo intensities as a function of t_m. Corrections for T_2 relaxation during the τ period are performed by normalizing the echo intensity by its value at short mixing times. STE experiments were extensively used to study molecular dynamics in organic systems. Most of these applications used ^{2}H STE, where the anisotropic ^{2}H quadrupolar interaction is employed to probe the molecular motion. In this case spin-alignment echoes (the T_{20} term is stored during t_m) are produced and the expression for the STE signal is

$$S(t_e,t_m,t) = \langle M_0 \tfrac{3}{4}\sin(2\chi_1)\sin(2\chi_2)\sin(\omega_Q(0)t_e)\sin(\omega_Q(t_m)t)\rangle, \tag{1.88}$$

where it becomes clear that the pulse flip angles χ_1 and χ_2 that maximize the signal are each equal to 55°.

^{2}H spin-alignment echoes have been used for many years in the study of α- and β-relaxation in supercooled liquids [35, 97, 122, 123]. Important information about the motional rates, motional geometry, and distribution of correlation times have been obtained using this technique. New experiments obtained by adding extra mixing times and evolution periods in the stimulated-echo pulse sequence have also made it possible to obtain multiple time correlation functions. These techniques were used to elucidate the origin of the non-exponential behavior of the correlation functions that characterize the molecular motions responsible for the glass transition of supercooled systems [73, 74, 78, 95, 123, 124]. The results revealed that non-exponentiality in the correlation functions are associated with dynamic heterogeneities with lifetimes comparable to that of the α-relaxation, and their sizes were estimated [74, 124]. More recently, STE experiments were also applied to study the slow exchange of the ^{7}Li and ^{29}Be cations in inorganic solid electrolytes [39, 78, 125]. In these cases the experiment was adapted to spin 3/2 systems. With the appropriate phase cycling [125], the echo intensity can be written as

$$S(\tau, t_m, t) = \left\langle M_0 \tfrac{9}{20} \sin\left(\omega_Q(0)\tau\right) \sin\left(\omega_Q(t_m)t\right)\right\rangle \tag{1.89}$$

for best pulse flip angles χ_1 and χ_2 equal to 45°. The phase cycling employed in reference [78] was used to provide pure quadrupolar order, stimulated-echoes (spin-alignment echoes) as in the case of ^{2}H. However, homonuclear dipolar coupling between the cations also contributes to the echo intensity. This is shown in Figure 1.19b, where the regular solid-echo spectra of the ionic conductor $Li_3In_2(PO_4)_3$ are compared with the spin-alignment spectra (Fourier transform of the STE signal) recorded with t_m = 50 ms. It is evident that the satellite contributions in the spin-alignment spectra are much more prominent than in the solid-echo spectra. However, the presence of significant central lines indicates that dipolar order is also produced.

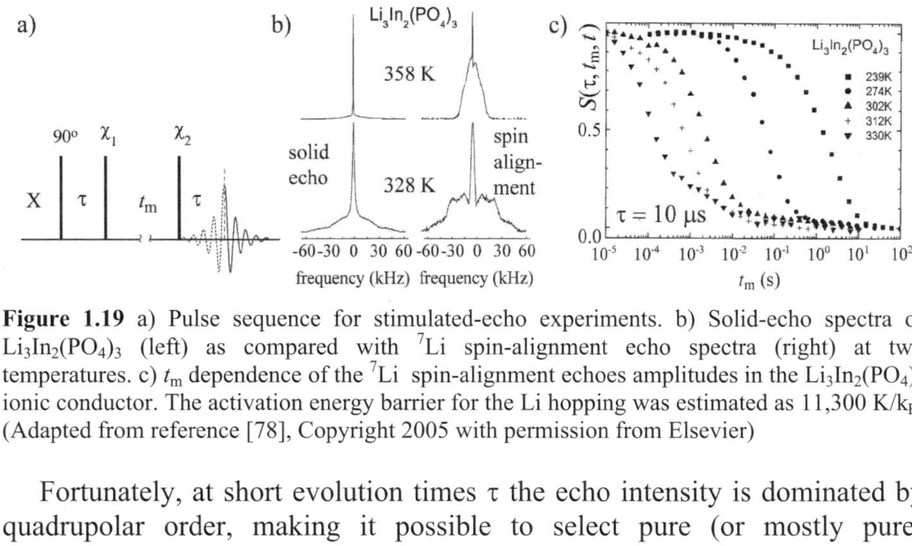

Figure 1.19 a) Pulse sequence for stimulated-echo experiments. b) Solid-echo spectra of $Li_3In_2(PO_4)_3$ (left) as compared with ^{7}Li spin-alignment echo spectra (right) at two temperatures. c) t_m dependence of the ^{7}Li spin-alignment echoes amplitudes in the $Li_3In_2(PO_4)_3$ ionic conductor. The activation energy barrier for the Li hopping was estimated as 11,300 K/k_B. (Adapted from reference [78], Copyright 2005 with permission from Elsevier)

Fortunately, at short evolution times τ the echo intensity is dominated by quadrupolar order, making it possible to select pure (or mostly pure) quadrupolar order echoes [78]. Because the quadrupolar frequency depends on the strength and orientation of the electric field gradient at the cation site, any modification in these parameters' changes ω_Q. As a consequence, jumps of the ions between electrically non-equivalent sites change the quadrupolar frequency, making it possible to detect the cation motion in STE experiments. Again, the measurement of the stimulated-echo amplitude as a function of the mixing time provides the two-time correlation function of the ionic motions, whose decay time is the average ion-hopping time. Figure 1.19c shows the mixing time dependence of the spin-alignment echo amplitudes for different temperatures in the $Li_3In_2(PO_4)_3$ ionic conductor. From these measurements the mean correlation times (decay times) as a function of the temperature and the activation energy of the ionic motion were obtained [78]. Other STE studies on ionic conductors have been carried out using ^{109}Ag (spin ½ nucleus) NMR. In this case two-time and three-time correlation functions were obtained, which allowed one to conclude that the non-exponentiality observed in the correlation function is due to a correlation time distribution, i.e., dynamic heterogeneities, rather than to an intrinsic non-exponentiality [126].

Concerning ^{13}C stimulated-echo experiments, these have been widely employed for studying not only the time scale, but also the amplitude of slow motions in organic solids. However, the use of this technique for studying complex motions suffers from the same problem as the regular 2D exchange experiment, i.e., there is no specific site resolution, and it is very difficult to characterize the exchange intensity if only a small fraction of the molecular segments takes part in the molecular motion. To overcome one of these problems a 1D pure exchange version of the stimulated-echo technique has also been developed. This technique, named *1D pure exchange NMR* (1D-PUREX), suppresses the signals of immobile molecules and thus provides spectra selectively from segments that reorient during the mixing time. Thus, the amplitude and time scale of slow motions can be determined even for a small fraction of mobile segments [101, 102]. The pulse sequence for the 1D-PUREX NMR experiment is obtained from the 2D-PUREX pulse sequence, Figure 1.16, simply by removing the t_1 evolution and the first z-period (the two 90° pulses flanking the t_z period). Hence, after the excitation the spins immediately start their evolution under the chemical shift anisotropy (CSA) during the period τ. Then, the magnetization is stored along the z-direction, so that it does not precess or dephase during a long mixing time, t_m. In the absence of slow motions during t_m, the CSA is refocused after a read-out pulse and another evolution period τ (total evolution time of $\tau_{CSA} = 2\tau$). At this point, the magnetization is stored again along the z direction during a short period t_z ($<< t_m$) and then flipped back to the transverse plane for detection. If segmental reorientations occur during t_m, the orientation-dependent frequency changes, and the CSA is not completely refocused. The resulting dephasing of the stimulated-echo is observed as a decrease in the detected spectral intensity. To remove effects of T_1 and T_2 relaxations during t_m and τ_{CSA}, respectively, a reference spectrum $S_0 = S(t_z, \delta\tau_{CSA})$—which has the same relaxation factors, but no motion effects during t_m—is measured. S_0 is obtained by simply swapping t_m and t_z. To remove the contribution of segments that do not move on the milliseconds time scale, the 1D-PUREX intensity $S = S(t_m, \delta\tau_{CSA})$ is subtracted from the reference spectrum S_0. Dividing this difference by S_0, one can obtain a normalized pure-exchange intensity $E(t_m, \delta\tau_{CSA}) = \Delta S/S_0$, which does not contain relaxation effects and just accounts for the signals from the reorienting segments. For discrete jumps with correlation times τ_C in the range of milliseconds to seconds (slow-motion regime), the normalized pure-exchange intensity $E(t_m, \delta\tau_{CSA})$ obtained by integrating the spectra yields the same information as the standard stimulated-echo experiment. In this motional regime, $E(t_m, \delta\tau_{CSA})$ can be approximated by the following expression:

$$\begin{aligned} E\left(t_m, \delta\tau_{CSA}\right) &= \frac{f_m\left(M-1\right)}{M}\left[1-\phi\left(t_m\right)\right]\left\langle 2\sin^2\left(\left[\Omega_1-\Omega_2\right]\frac{\tau_{CSA}}{4}\right)\right\rangle \\ &= E_\infty\left[\left[1-\phi\left(t_m\right)\right]\left\langle 2\sin^2\left(\left[\Omega_1-\Omega_2\right]\frac{\tau_{CSA}}{4}\right)\right\rangle\right], \end{aligned} \tag{1.90}$$

where f_m is the fraction of slow mobile segments (mobile fraction), M is the number of magnetically non-equivalent sites accessible to the moving segments, $\phi(t_m)$ is the correlation function of the molecular motion, and $\Omega_1 - \Omega_2$ is the difference between the NMR frequencies before and after t_m. Therefore, the t_m-dependence of $E(t_m, \delta\tau_{CSA})$ yields the time scale of the motion, while its τ_{CSA}-dependence provides information on the amplitude of the slow molecular reorientations that occurred during t_m.

Figures 1.20a–b show, respectively, the t_m- and the τ_{CSA}-dependence of $E(t_m, \delta\tau_{CSA})$ curves for DMS at 293 K. A correlation time of 15 ms was obtained from the t_m-dependence of the $E(t_m, \delta\tau_{CSA})$ curves. Note that the final intensity in both curves is 0.5, which is exactly what is expected for the DMS two-site jump motion. Besides, as shown in the simulation for motions with distinct reorientation angles, the 1D-PUREX experiment is very sensitive to small-angle rotations. The best fit to the DMS experimental data occurs for a reorientation angle of 70°. This is equivalent to 110° due to the symmetry property of the axially symmetric CSA tensor of DMS that makes reorientation by β_R and 180°–β_R equivalent. Figure 1.20c shows the reference S_0, the 1D-PUREX S, and the difference $S_0 - S$ spectra obtained for non-labeled PMMA at $T = 295$ K. The $S_0 - S$ spectrum shows significant intensity for all carbons confirming that the β-relaxation of PMMA involves both the side group and main chain motions. The τ_{CSA}-dependence of $E(t_m, \delta\tau_{CSA})$ for the carbonyl group in PMMA is shown in Figure 1.20d. The bimodal character of the curve is promptly recognized, showing the presence of both small (slow build-up) and high (fast build-up) angle motions. The best-fit simulation of the $E(t_m, \delta\tau_{CSA})$ 1D-PUREX intensities as a function of τ_{CSA} is also shown. In the simulations, the three-component model discussed above in the context of Figure 1.13 was used. All small-angle rotations were simulated using Gaussian distributions of reorientation angles centered at 0° with root mean square widths σ. The $E(t_m, \delta\tau_{CSA})$ vs. τ_{CSA} curves for each component and the corresponding distribution of rotation angles for the small-angle motion are shown in Figure 1.20d.

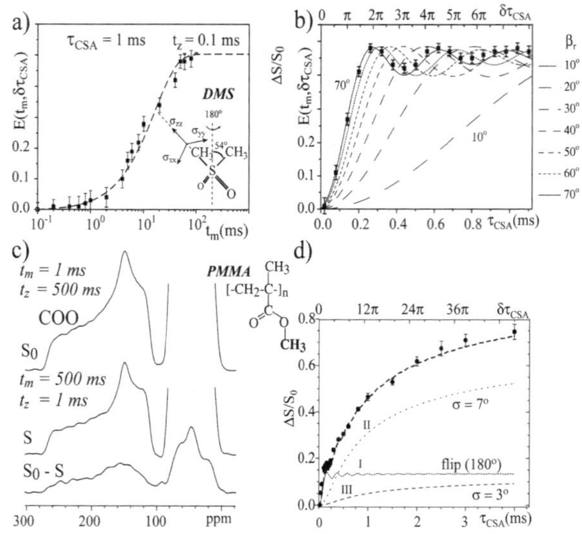

Figure 1.20 a) $E(t_m, \delta\tau_{CSA})$ vs. t_m curve of DMS at 293 K. b) $E(t_m, \delta\tau_{CSA})$ vs. τ_{CSA} curve of DMS at 293 K. c) Reference S_0, exchange S, and difference $S_0 - S$ 1D-PUREX spectra of PMMA. d) $E(t_m, \delta\tau_{CSA})$ vs. τ_{CSA} build-up curve for the carbonyl groups of PMMA together with simulated curves based on the model discussed in the text (components I, II, III). (Adapted with permission from reference [89], Copyright 2005 Springer, and from reference [101], Copyright 2005 American Institute of Physics)

Following the idea of the rotor synchronized 2D MAS exchange experiments, there is also a series of 1D MAS experiments especially designed for studying slow molecular reorientations using spinning sidebands [80, 81]. Named *one-dimensional exchange spectroscopy by sideband alternation* (ODESSA), this is an alternative version of the 2D MAS exchange. In this method, the first evolution period in Figure 1.11c is set to half of the rotor period, $t_1 = t_r/2$. Then, after the rotor synchronized mixing period, the signal is detected during the period $t = t_2$. The magnetization stored along the z direction during the mixing periods corresponds to distinct polarizations of the spinning sidebands. During the mixing period, the molecular reorientations redistribute the polarization among the various spinning sidebands, which makes their intensities dependent on the molecular motion that occurred during t_m. From a series of spectra acquired with different mixing periods, dynamic parameters such as the correlation time and, in favorable cases, the amplitude and number of accessible sites for the motion can be obtained. Other variants of these techniques were proposed in order to obtain pure absorptive MAS spectra for systems with many non-equivalent sites (time-reverse ODESSA) and to separate the sideband intensities of different chemical groups (PATROS) [80, 81, 83]. In the particular case of tr-ODESSA the correlation time of the motion can be obtained either by the increase of the spinning sidebands intensities or the decrease of the centerband intensity as a function of the mixing time.

Figure 1.21 shows an example of the use of the tr-ODESSA pulse sequence to study molecular motions responsible for the α- and β-relaxation of the glassy polymer poly(n-hexyl methacrylate) (PHMA). Using tr-ODESSA, the correlation times and their distribution were obtained as a function of temperature for the side group and main chain motion [92]. This made it possible to conclude that the alkyl side group contributes in the dynamic window of the NMR exchange experiments mainly to the β-relaxation, while the main chain and carboxyl groups contribute to both α- and β-relaxations.

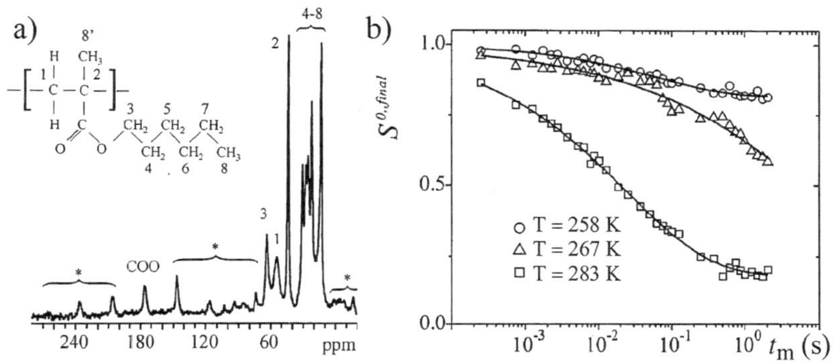

Figure 1.21 a) ^{13}C CP-MAS spectrum of poly(*n*-hexyl methacrylate) (PHMA) at $\nu_R = 3$ kHz. COO spinning sidebands are indicated with asterisks. b) t_m-dependence of tr-ODESSA experiments for PHMA. (Adapted with permission from reference [92], Copyright 2005 American Chemical Society)

The main disadvantages of the above discussed MAS exchange methods is the necessity of observing sidebands in the spectrum. This is particularly problematic for ^{13}C studies in chemically complex systems where the overlap between the sidebands of different chemical groups may preclude the accurate intensity measurement. This was overcome by a class of methods called *centerband-only detection of exchange* (CODEX), where slow molecular motions are quantified on the basis of the line intensities of a high resolution MAS spectrum, without the necessity of observing MAS sidebands. The CODEX experiment is essentially the MAS version of 1D-PUREX NMR [75, 121]. The averaging of the chemical shift anisotropy under MAS provides higher site resolution and sensitivity. However, MAS would make the pulse sequence insensitive to segmental orientations and reorientations. Then, it is necessary to reintroduce (recouple) the chemical shift anisotropy during the periods used to codify the exchange process, i.e., the evolution periods, Nt_r. This is performed by using a train of rotor synchronized 180° (π) pulses capable of reintroducing the chemical shift anisotropy even under high MAS spinning frequencies. This scheme, known as *recoupling of the CSA*, was originally developed for using in the so-called *rotational echo double resonance* (REDOR) experiments, which are a very useful class of NMR methods for measuring internuclear distances [47]. CSA recoupling during the evolution periods is achieved by replacing each evolution period τ in the 1D-PUREX pulse sequence by a train of $N/2$ 180° pulses spaced by $t_r/2$, where t_r is the sample rotation period; see Figure 1.22a. This leads to a total evolution-and-refocusing period of Nt_r for CODEX. With the CSA recoupling, the phase accumulated during the first and second evolution periods before (Φ_1) and after (Φ_2) the mixing time are given by

$$\Phi_1 = \frac{N}{2}\left(-\int_0^{t_r/2}\left[\omega_{\text{iso}} + \omega_{\text{aniso},1}(t)\right]dt + \int_{t_r/2}^{t_r}\left[\omega_{\text{iso}} + \omega_{\text{aniso},1}(t)\right]dt\right) = -N\int_0^{t_r/2}\omega_{\text{aniso},1}(t)\,dt$$

$$\Phi_2 = \frac{N}{2}\left(\int_0^{t_r/2}\left[\omega_{\text{iso}} + \omega_{\text{aniso},2}(t)\right]dt - \int_{t_r/2}^{t_r}\left[\omega_{\text{iso}} + \omega_{\text{aniso},2}(t)\right]dt\right) = N\int_0^{t_r/2}\omega_{\text{aniso},2}(t)\,dt$$

(1.91)

where $\omega_{\text{aniso},1}(t)$ and $\omega_{\text{aniso},2}(t)$ are the NMR frequencies due to the anisotropic part of the chemical shift under MAS, equation (1.71), before and after t_m, respectively. Note that equation (1.73) was used to derive the expressions for the phases. If $\omega_1(t) = \omega_2(t)$, i.e., if no exchange occurs, equation (1.91) leads to a total phase accumulation of $\Phi_1 + \Phi_2 = |\Phi_2| - |\Phi_1| = 0$; in other words, the magnetization is refocused in a stimulated-echo along its original direction.

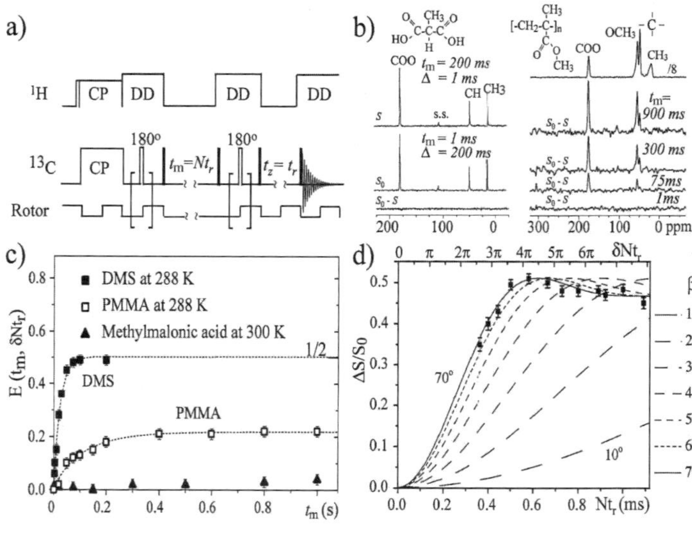

Figure 1.22 a) Pulse sequence for CODEX experiments. b) CODEX S, reference S_0, and the difference $S_0 - S$, spectra obtained for the molecular crystal methylmalonic acid (left) and difference $S_0 - S$, spectrum as a function of t_m for PMMA (right) (see chemical structures in the insets). c) t_m-dependence of the pure-exchange CODEX intensity for PMMA, methylmalonic acid, and DMS. d) Nt_r-dependence of the $E(t_m, \delta Nt_r)$ intensity obtained for DMS using $t_m = 50$ ms. (Adapted with permission from reference [75], Copyright 2005 American Chemical Society, and from reference [89], Copyright 2005 Springer)

Molecular motions during t_m reduce the stimulated-echo and, as a result, the observed line intensities. Therefore, the analysis of the motional amplitude and time scale can be performed in the same way as for the 1D-PUREX measurements. The CODEX intensity $S = S(t_m, \delta N t_r)$ and the reference intensity with no motion effects, $S_0 = S(t_z, \delta N t_r)$, also obtained after swapping t_m and t_z, are combined to produce the normalized pure exchange intensity, $E(t_m, \delta N t_r)$. $E(t_m, \delta N t_r)$ plotted as a function of t_m or $N t_r$ provides information on the motional rate and amplitude, respectively. The main advantage of CODEX is the possibility of studying the molecular motion of each chemical site independently, since a high resolution MAS spectrum is acquired. This makes possible the probing of local motions at each individual site of chemically complex systems. Figure 1.22b shows the CODEX S, the reference S_0, and the difference $S_0 - S$ spectra obtained for the molecular crystal methylmalonic acid and for the glassy polymer PMMA (see molecular structures in the insets).

The absence of lines in the difference spectrum $S_0 - S$, of methylmalonic acid indicates that the molecular segments of methylmalonic acid do not undergo slow rotations. In contrast, molecular groups in both side groups and main chain of PMMA execute molecular motions in the millisecond-to-second time scale. Figure 1.22c shows the t_m-dependence of the pure-exchange CODEX intensity for PMMA, methylmalonic acid, and DMS. The resulting curves are the correlation functions of these dynamics on the millisecond time scale. For DMS, a single exponential curve is obtained, while PMMA shows some non-exponentiality. Methylmalonic acid shows no exchange except for ^{13}C spin-diffusion (~5%). $E(t_m, \delta N t_r)$ plotted as a function of $N t_r$ for DMS is shown in Figure 1.22d. Simulations with various reorientation angles are also shown. As in the case of 1D-PUREX experiments, high sensitivity to small-angle reorientations can be observed in these curves. As expected, the best fit to the DMS experimental data is shown for a reorientation of 70°, i.e., 110°, as explained before.

Many extensions of the CODEX technique have appeared, including an extension to a four-time pulse sequence used for studying motional memory [75, 127]; a triple-resonance CODEX applied to study protein dynamics [128]; scaled-CODEX, used for measurements at low spinning rates [120]; multiple-quantum CODEX used for torsion angle determination in solids [129]; and ^{19}F CODEX, was used to study the oligomerization in lipid bilayer proteins [130]; and a CODEX variant to measure internuclear distances between a spin ½ and a quadrupolar nucleus [131].

1.7.4 Conformation-NMR

The experimental elucidation of molecular conformations in non-oriented solids, such as amorphous polymers, unoriented polypeptides, or polycrystalline proteins, is a great challenge. Solidstate NMR provides a set of techniques that show great potential for determining molecular conformations on the level of individual bonds, which it does by probing relative segmental orientations [116, 117, 132–137] or internuclear distances [138–141] through various anisotropic nuclear interactions, such as dipolar couplings and chemical shifts. In particularly, detailed information on torsion angles and their distributions can be obtained by correlating NMR interaction-tensor orientations in two-dimensional NMR spectra without magic-angle spinning. The techniques are based on 2D-exchange NMR [68, 116, 117, 132–134] or on correlation of dipolar couplings with the chemical shifts [136, 137], mostly on specifically ^{13}C-labeled samples.

Recently, one useful experiment to determine the torsion angle for ^{13}C-^{13}C pairs was proposed [142, 143], being very practical for providing information about the conformation of polymers [144–146]. This experiment, called *double-quantum NMR* (DQ-NMR), based on the INADEQUATE [147] pulse sequence (Figure 1.23a), yields two-dimensional spectra where the first spectral dimension, ω_1, reflects the torsion angle in terms of the sum of the chemical shifts of the coupled sites, without any interference from the dipolar coupling.

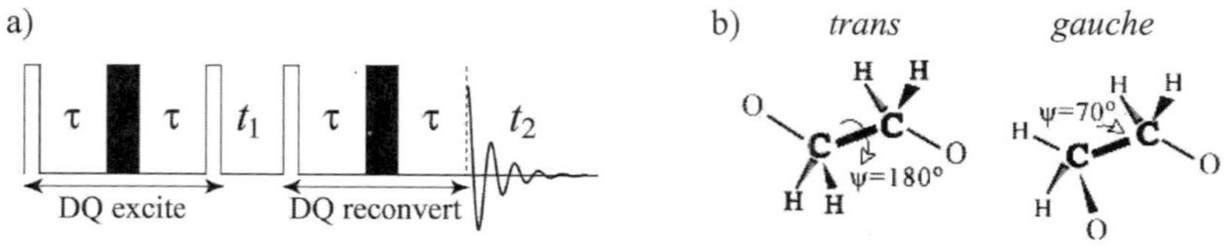

Figure 1.23 a) INADEQUATE pulse sequence. b) Trans and gauche conformations in PEO.

The conformational information is enhanced by the correlation with the individual chemical shifts in the directly detected ω_2 dimension, with or without ^{13}C-^{13}C homonuclear decoupling [143]. The application of this experiment will be demonstrated on polyethylene oxide (PEO) with dilute (~13%) ^{13}C-^{13}C spin pairs in different chemical environments in the crystalline phase [145], determining the amount of trans and gauche conformations (Figure 1.23b).

In the following, the principles of the DQ-NMR experiment are reviewed [142–144, 148]. We consider a pair of directly bonded ^{13}C nuclei with chemical shifts ω_a and ω_b, and with a dipolar coupling Hamiltonian of $\hbar\omega_D\left(3S_zL_z - \vec{S}\cdot\vec{L}\right)$. In the two-dimensional spectrum obtained, the orientation-dependent individual frequencies ω_a and ω_b in the second dimension are correlated with the torsion-angle-dependent sum of frequencies ($\omega_a + \omega_b$) in the first dimension. The resulting two-dimensional spectral pattern depends on the relative orientation and thus on the torsion angle between the ^{13}C-containing segments.

From the initial transverse magnetization $S_y + L_y$, double-quantum (DQ) coherence $2S_x L_y + 2S_y L_x$ is generated by the dipolar coupling and radiofrequency pulses of suitable phase [148, 149]:

$$\begin{aligned} \textit{DQ excitation} \quad & S_y + L_y \xrightarrow{H\tau} \xrightarrow{180^o} \xrightarrow{H\tau} -\left(2S_xL_z + 2S_zL_x\right)g\left(2\tau\right) \\ & \xrightarrow{90^o_x} -\left(2S_xL_y + 2S_yL_x\right)g\left(2\tau\right). \end{aligned} \tag{1.92}$$

The DQ excitation factor $g(2\tau)$ is given in references [144] and [148]. The DQ coherence then evolves under the sum of frequencies $\omega_a + \omega_b$ during t_1:

$$\textit{DQ evolution} \quad \xrightarrow{Ht_1} -\left(2S_xL_y + 2S_yL_x\right)g\left(2\tau\right)\cos\left[\left(\omega_a + \omega_b\right)t_1\right]. \tag{1.93}$$

Modulated by $\cos\left[\left(\omega_a + \omega_b\right)t_1\right]$, the DQ coherence is reconverted into observable magnetization $S_y + L_y$:

$$\begin{aligned} \textit{DQ reconversion} \quad & \xrightarrow{90^o_{-x}} -\left(2S_xL_z + 2S_zL_x\right)g\left(2\tau\right)\cos\left[\left(\omega_a + \omega_b\right)t_1\right] \\ & \xrightarrow{H\tau} \xrightarrow{180^o} \xrightarrow{H\tau} \left(S_y + L_y\right)g^2\left(2\tau\right)\cos\left[\left(\omega_a + \omega_b\right)t_1\right], \end{aligned} \tag{1.94}$$

which is observed in the detection period t_2. The signal acquired during t_2, under continuous-wave proton decoupling, reflects the individual chemical shifts ω_a and ω_b, and the ^{13}C-^{13}C dipole-dipole interaction. The dead-time problem is avoided by a solid-echo with Hahn-echo sandwich before detection [143]. The two-dimensional time signal is thus:

$$S\left(t_1, t_2\right) = g^2\left(2\tau\right)\cos\left[\left(\omega_a + \omega_b\right)t_1\right]f\left(t_2\right). \tag{1.95}$$

In an experiment with complete dipolar decoupling during detection [143], $f(t_2) = \exp(i\omega_a t_2) + \exp(i\omega_b t_2)$, so that in the 2D spectrum, signals of intensity $g^2(2\tau)$ are observed at $(\omega_a+\omega_b, \omega_a)$ and $(\omega_a+\omega_b, \omega_b)$.

With the DQ-NMR pulse sequence presented, let us now discuss its application to PEO conformation [145]. PEO chains in the crystalline state have been studied by infrared, Raman, and X-ray techniques. The results have shown that the chains adopt a 7_2 helix conformation with a *TTG* sequence: all OC-CO bonds are gauche and all CO-CC bonds are trans. However, the helix is distorted to improve intermolecular interactions. A detailed analysis of the X-ray diffraction peak intensities suggested a possible model for chain conformation: the helix is highly distorted with the OC-CO bonds having discrete torsion angles of 57.0°, 67.8°, 74.2°, 49.0°, 91.8°, 60.2°, and 79.1° [150].

Recently there has been much interest in the complexes of PEO with resorcinol (RES) and *p*-nitrophenol (PNP) [151]. WAXD and IR data suggest that the conformation of the PEO in the complex with RES is only slightly distorted from the original 7_2 helix. The conformation of the chains in these systems remains *TTG* but the periodicity of the helix changes slightly [152]. Contrary to what occurs in the PEO/RES complex, the conformation of the chains in PEO/PNP complexes is very different from the 7_2 helical conformation [153]. IR and WAXD analyses suggested that 1/3 of the OC-CO bonds are in the trans conformation. The hydrogen bonding between the host and guest molecules is believed to induce the chain to adopt a *TTGTTGTTT TTG'TTG'TTT* structure [154]. Another unusual property of the PEO/PNP complex is that there are two broad ^{13}C peaks at 71.4 and 69.4 ppm [155], compared with a median of 72 ppm for neat PEO. Examples of simulated DQ spectra obtained for gauche and trans torsion angles of 70° and 180°, respectively, are shown in Figures 1.24a through d [145].

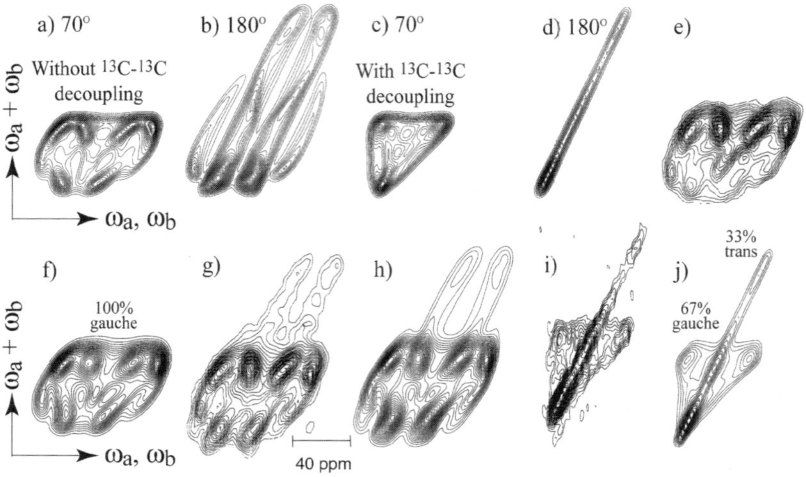

Figure 1.24 Simulated static ^{13}C DQ-NMR spectra of PEO with and without ^{13}C-^{13}C dipolar decoupling for torsion angles of (a/c) 70° and (b/d) 180°. e) ^{13}C DQ-NMR spectrum of PEO/RES without decoupling measured at 75.5 MHz and 250 K. f) The best simulation for spectrum shown in e). ^{13}C DQ-NMR spectra of PEO/PNP measured at 75.5 MHz and 293 K: g) without decoupling and i) with decoupling. The best simulations of both the undecoupled (h) and decoupled (j) spectra. (Adapted with permission from reference [145], Copyright 2005 American Chemical Society)

The observed and simulated ^{13}C-^{13}C undecoupled static DQ spectra of PEO/RES are shown in Figures 1.24e and f. The observed spectrum shows that all of the OC-CO torsion angles are gauche. The simulation suggests that the average torsion angle is $\psi = 74 \pm 9°$ with an angle distribution width of $\sigma < 10°$. The observed and simulated undecoupled and decoupled static DQ spectra of PEO/PNP are shown in Figures 1.24g through j. The spectra show a strong slope-two diagonal ridge due to trans OC-CO bonds. The simulations of the observed spectra were produced by factoring 0.33 for trans and 0.67 for gauche conformers. Both the undecoupled and decoupled static DQ spectra are consistent with one-third of OC-CO bonds of the PEO/PNP in the trans conformation ($\psi = 180°$, $\sigma < 7°$), and two-thirds in the gauche conformation ($\psi = 70 \pm 9°$, $\sigma < 10°$). The 2D ^{13}C DQ-NMR spectrum confirmed that the OC-CO torsion angles in PEO/RES are all gauche. In contrast, one-third of the OC-CO torsion angles in PEO/PNP are trans and two-thirds are gauche, confirming the previously suggested structure [154].

A MAS ^{13}C INADEQUATE NMR (MAS DQ-NMR) experiment [156] using the dipolar recoupling sequence CMR7 [157] to excite the double-quantum coherence under MAS was applied to determine the connectivities of the ^{13}C peaks in PEO/PNP at 69.4 ppm and 71.4 ppm. The MAS DQ-NMR spectrum of PEO/PNP is shown in Figure 1.25a. Three peaks, a doublet and a singlet, are observed in the spectrum. Signals at the same ω_1 value belong to a ^{13}C-^{13}C pair, and the chemical shifts of the two ^{13}Cs in the pair can be read off in the ω_2 dimension. In addition, due to special magic-angle spinning effects, the line-broadening makes it possible to identify which peaks belong to gauche or trans conformations: a broadening arises if the two dipolar-coupled ^{13}C nuclei have different CSA orientations, as is the case for a gauche conformation. In that case, the dipolar coupling is not completely removed by MAS, since the sum of the chemical shift difference tensor and the dipolar coupling tensor, which have to be considered together, does not commute at different rotor orientations. In other words, the sum of the chemical shift difference tensor and the dipolar coupling represents a homogeneous Hamiltonian, which produces broadened MAS peaks [12].

The doublet of broad peaks at (71.4, 141.3 ppm) and (69.8, 141.3 ppm) corresponds to two bonded ^{13}C nuclei with inequivalent chemical shifts. The strong broadening of the peaks shows that a gauche bond connects them. The singlet peak at (69.2, 138.4 ppm) is much sharper, which means that it must be due to a ^{13}C spin pair around a trans bond. The peak position shows that the isotropic chemical shifts of both ^{13}C nuclei of the trans bond are 69.2 ppm. This equivalence suggests that the chain structure is point-symmetric around the trans bond. The integrated area of the two gauche peaks is $(73 \pm 5)\%$ and the area of the trans peak is $(27 \pm 5)\%$.

The assignment of the peaks to gauche and trans conformations in PEO/PNP is also confirmed by the ^{13}C CODEX experiment [75, 121] with exchange by spin-diffusion. A ^{13}C CODEX experiment with ^{13}C spin-diffusion in the mixing time was performed and the normalized peak intensities in the resulting spectra were plotted as a function of the spin-diffusion time. For the two sites on a trans bond, the anisotropic chemical shift is the same; thus spin-diffusion does not result in a frequency change and does not dephase the signal. For gauche bonds, we have a two-site exchange process. Exchange in the long-time limit results in a frequency change for half the magnetization, while the other half still resides at the initial site. Therefore, dephasing reaches an asymptote of 50%. As observed in Figure 1.25b, the PEO/PNP 1D ^{13}C spectrum (reference) consists of two broad peaks. The ^{13}C CODEX spectrum obtained with a mixing time of 100 ms is presented in Figure 1.25c. The spectra are scaled to equal heights of the 69.4 ppm line to highlight the more pronounced effect of the spin-diffusion on the 71.4-ppm line. To understand this difference, it has to be noted that spin-diffusion between trans sites is not detectable because it does not introduce a change in the frequency of the ^{13}C nuclei involved; thus a spin-diffusion-induced decay is observed only for the two gauche sites. In Figure 1.25d, the mixing-time dependence of the CODEX spin-diffusion is plotted for both peaks. The relative intensity of the 71.4-ppm gauche signal for long mixing times (> 100 ms) decays to a constant value of $E_{\infty} \sim 0.5$. Meanwhile, the relative intensity for the other peak, as expected for a two-site spin-diffusion process in the absence of slow chain motion, decays to approximately 0.75. The same plot also shows the decay for the pure trans conformation, obtained by determining the difference between the intensities of the left and the right line and taking the ratios of the differences obtained from reference and spin-diffusion CODEX spectra. This ratio is very close to unity, as expected for the trans conformation. From the values of the relative intensities obtained at t_m = 200 ms for both lines, Figure 1.25d, the gauche-trans ratio is estimated to be 2, which confirms the result obtained from the double-quantum spectra, where ⅓ of the OC-CO torsion angles are trans and ⅔ are gauche.

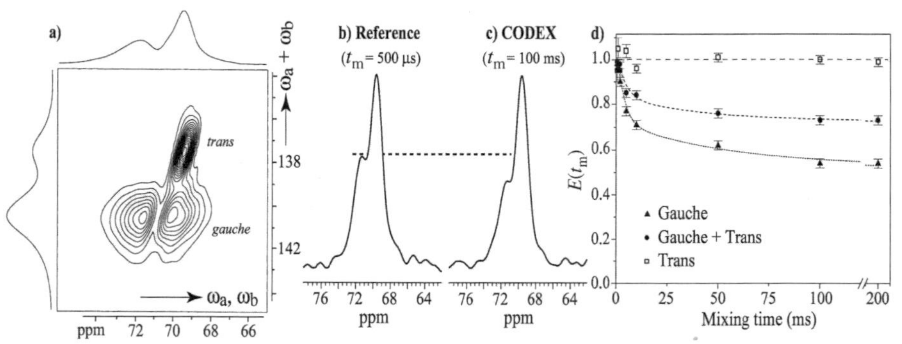

Figure 1.25 a) ^{13}C MAS DQ-NMR spectrum of PEO/PNP. The spectrum was obtained at T=293 K with ν_r=5 kHz. ^{13}C CODEX spectra used to determine the trans-gauche ratio in PEO/PNP: b) reference spectrum with t_m=0.5 ms, and c) CODEX spectrum with t_m=100 ms. d) Decay of the peak intensities due to spin-diffusion. (Adapted with permission from reference [145], Copyright 2005 American Chemical Society)

Additional static ^{13}C DQ-NMR experiments were performed for studying pure crystalline [158] and amorphous PEO intercalated in clays and MoS_2 [144]. Moreover, this method was used successfully to determine the conformation of other polymers [159–161] and biomaterials [162–164]. Similar DQ-NMR experiments are being successfully used for determining the structure of phosphate glasses, measuring the Q^n groups connectivities (where Q represents the tetrahedrally coordinated phosphate units, and n is the number of bridging oxygens to other like groups) [119]. In fact, DQ-NMR offers detailed information about the connectivities, not only in phosphate glasses, but also in binary sodium, calcium and lead phosphate glasses, sodium silicate glasses, sodium titanophosphate glasses, etc. [165–169].

ACKNOWLEDGEMENTS

The authors acknowledge Prof. A.P. Guimarães, Prof. D. Reichert, Dr. A.C. Bloise, Dr. F.A. Bonk, and J. Teles for the careful reading of this chapter and for the several comments and suggestions, and the authors and publishers for permission to adapt or reproduce figures and texts.

REFERENCES

[1] Haeberlen, U. (1976) *High Resolution NMR in Solids* (Academic Press) p 190.
[2] Mehring, M. (1983) *High Resolution NMR in Solids* (Springer-Verlag) p 342.
[3] Ernst, R.R., Bodenhausen, G. & Kokaun, A. (1987) *Principles of Nuclear Magnetic Resonance in one and Two Dimensions* (Clarendon Press) p 610.
[4] Goldman, M. (1988) *Quantum Description of High-Resolution NMR in Liquids* (Oxford University Press) p 268.
[5] Tonelli, A.E. (1989) *NMR Spectroscopy and Polymer Microstructure – The Conformational Connection* (VCH Publishers, Inc.) p 252.
[6] Slichter, C.P. (1990) *Principles of Magnetic Resonance* (Springer-Verlag) p 655.
[7] Schmidt-Rohr, K. & Spiess, H.W. (1994) *Multidimensional Solid-State NMR and Polymers* (Academic Press) p 478.
[8] Stejskal, E.O. & Memory, J.D. (1994) *High Resolution NMR in the Solid State: Fundamentals of CP/MAS* (Oxford University Press) p 189.
[9] Mackenzie, K.J.D. & Smith, M.E. (2002) *Multinuclear Solid-State NMR of Inorganic Materials* (Pergamon) 6 p 726.
[10] Cohen-Tannoudji, C., Diu, B. & Laloë, F. (1977) *Quantum Mechanics* (Hermann and John Wiley & Sons) I p 898.
[11] Heatley, F.(1993) In *NMR Spectroscopy of Polymers*, 1st ed.; Ibbett, R.N. Ed.; Blackie Academic & Professional: London, pp 1-49.
[12] Maricq, M.M. & Waugh, J.S. (1979) *J. Chem. Phys.* **70**, 3300-3316.
[13] Sarles, L.R. & Cotts, R.M. (1958) *Phys. Rev.* 111, 853-859.
[14] Bennett, A.E., Rienstra, C.M., Auger, M., Lakshmi, K.V. & Griffin, R.G. (1995) *J. Chem. Phys.* **103**, 6951-6958.
[15] Waugh, J.S., Huber, L.M. & Haeberlen, U. (1968) *Phys. Rev. Lett.* **20**, 180-182.
[16] Mansfield, P. (1971) *Journal of Physics Part C Solid State Phys.* **4**, 1444-1452.
[17] Rhim, W.K., Elleman, D.D. & Vaughan, R.W. (1973) *J. Chem. Phys.* **59**, 3740-3749.
[18] Burum, D.P. & Rhim, W.K. (1979) *J. Chem. Phys.* **71**, 944-956.
[19] Lee, M. & Goldburg, W.I. (1965) *Phys. Rev.* **140**, 1261-1271.
[20] Andrew, E.R., Bradbury, A. & Eades, R. G. (1959) *Nature* **183**, 1802-1803.
[21] Lowe, I.J. (1959) *Phys. Rev. Lett.* **2**, 285-287.
[22] Andrew, E.R. & Eades, R.G. (1962) *Discuss. Faraday Soc.* 38-42.
[23] Pines, A., Gibby, M.G. & Waugh, J.S. (1973) *J. Chem. Phys.* **59**, 569-590.
[24] Schaefer, J. & Stejskal, E.O. (1976) *J. Am. Chem. Soc.* **98**, 1031-1032.
[25] Jenner, J. (1971) In *Ampère International Summer School*: Basko Polje, Yugoslavia.
[26] Aue, W.P., Bartholdi, E. & Ernst, R.R. (1976) *J. Chem. Phys.* **64**, 2229-2246.
[27] Günther, H. (1996) NMR Spectroscopy: *Basic Principles, Concepts, and Applications in Chemistry* (John Wiley & Sons) .
[28] Redfield, A.G. & Kunz, S.D. (1975) *J. Magn. Reson.* **19**, 250-254.
[29] States, D.J., Haberkorn, R.A. & Ruben, D.J. (1982) *J. Magn. Reson.* **48,** 286-292.
[30] Chandrakumar, N. & Subramanian, S. (1987) *Modern Techniques in High-Resolution FT-NMR* (Springer-Verlag) p 388.
[31] Inigo, A.R., Chiu, H-C., Fann, W., Huang, Y-S., Jeng, U-S.T-L.L., Hsu, C-H., Peng, K-Y. & Chen, S-A. (2004) *Phys. Rev. B,* **69**.

[32] Palmer, A.G., Williams, J. & McDermott, A. (1996) *J. Phys. Chem.* **100**, 13293-13310.
[33] Andronis, V. & Zografi, G. (1998) *Pharm. Res.* **15**, 835-842.
[34] Favre, D.E., Schaefer, D.J., Auerbach, S.M. & Chmelka, B.F. (1998) *Phys. Rev. Lett.* **81**, 5852-5855; Wilhelm, M., Firouzi, A., Favre, D.E., Bull, L.M., Schaefer, D.J., Chmelka, B.F. (1995) *J. Am. Chem. Soc.* **117**, 2923-2924.
[35] Fujara, F., Petry, W., Diehl, R.M., Schnauss, W. & Sillescu, H. (1991) *Europhys. Lett.* **14**, 563-568.
[36] Sillescu, H. (1999) *J. Non-Cryst. Solids* **243**, 81-108.
[37] Donoso, J.P., Bonagamba, T.J., Panepucci, H.C., Oliveira, L.N., Gorecki, W., Berthier, C. & Armand, M. (1993) *J. Chem. Phys.* **98**, 10026-10036.
[38] Dahmouche, K., Atik, M., Mello, N.C., Bonagamba, T.J., Panepucci, H., Aegerter, M.A. & Judeinstein, P. (1997) *J. Sol.-Gel Sci.Technol.* **87**, 11-715.
[39] Bohmer, R., Jorg, T., Qi, F. & Titze, A. (2000) *Chem. Phys. Lett.* **316**, 419-424.
[40] Brow, R.K. (2000) *J. Non-Cryst. Solids* **263**, 1-28.
[41] Georgiev, D.G., Mitkova, M., Boolchand, P., Brunklaus, G., Eckert, H. & Micoulaut, M. (2001) *Phys. Rev. B,* **6413** .
[42] Schaefer, J., Stejskal, E.O., McKay, R.A. & Dixon, W.T. (1984) *J. Magn. Reson.* **57**, 85-92.
[43] Hong, M., Yao, X.L., Jakes, K. & Huster, D. (2002) *J. Phys. Chem. B,* **106**, 7355-7364.
[44] Hong, M., Gross, J.D. & Griffin, R.G. (1997) *J. Phys. Chem. B,* **101**, 5869-5874.
[45] Madhu, P.K., Goldbourt, A., Frydman, L. & Vega, S. (2000) *J. Chem. Phys.* **112**, 2377-2391.
[46] Medek, A., Harwood, J.S. & Frydman, L. (1995) *J. Am. Chem. Soc.* **117**, 12779-12787.
[47] Gullion, T. & Schaefer, J. (1989) *J. Magn. Reson.* **81**, 196-200.
[48] Schmidt-Rohr, K., Hu, W. & Zumbulyadis, N. (1998) *Science* **280**, 714-717.
[49] Abragam, A. (1994) *Principles of Nuclear Magnetism* (Oxford University Press) p 599.
[50] Vold, R.R. (1994) In *Nuclear Magnetic Resonance Probes of Molecular Dynamics*; Tycko, R., Ed.; Kluwer Academic Publishers: Dordrecht, Vol. 1, pp 27-106..
[51] Spiess, H.W. (1985) *Deuteron NMR – A new Tool for Studying Chain Mobility and orientation in Polymers* (Springer) 66 p 324.
[52] Frydman, L., Vallabhaneni, S., Lee, Y.K. & Emsley, L. (1994) *J. Chem. Phys.* **101**, 111-117.
[53] Liu, S.F., Mao, J.D. & Schmidt-Rohr, K. (2002) *J. Magn. Reson.* **155**, 15-28.
[54] Powles, J.G. & Strange, J.H. (1963) *Proc. Phys. Soc.,* London **82**, 6-15.
[55] Hirschinger, J., Miura, H., Gardner, K.H. & English, A.D. (1990) *Macromolecules* **23**, 2153-2169.
[56] Lin, W.Y. & Blum, F.D. (1997) *Macromolecules* **30**, 5331-5338.
[57] Kaji, H., Miura, N. & Schmidt-Rohr, K. (2003) *Macromolecules* **36**, 6100-6113.
[58] Schmidt, A., Smith, S.O., Raleigh, D.P., Roberts, J.E., Griffin, R.G. & Vega, S. (1986) *J. Chem. Phys.* **85,** 4248-4253.
[59] Schmidt, A. & Vega, S. (1992) *J. Chem. Phys.* **96**, 2655-2680.
[60] Reichert, D., Hempel, G., Zimmermann, H., Schneider, H. & Luz, Z. (2000) *Solid State Nucl. Magn. Reson.* **18**, 17-36.
[61] Slichter, W.P. & Mandell, E.R. (1959) *J. Appl. Phys.* **30**, 1473-1478.
[62] Slichter, W.P. & Mandell, E.R. (1958) *J. Appl. Phys.* **29**, 1438-1441.
[63] Schmidt-Rohr, K., Clauss, J. & Spiess, H.W. (1992) *Macromolecules* **25**, 3273-3277.
[64] Clauss, J., Schmidt-Rohr, K. & Spiess, H.W. (1993) *Acta Polym.* **44**, 1-17.
[65] Qiu, X.H. & Mirau, P.A. (2000) *J. Magn. Reson.* **142**, 183-189.
[66] Vinogradov, E., Madhu, P.K. & Vega, S. (1999) *Chem. Phys. Lett.* **314**, 443-450.
[67] Bielecki, A., Kolbert, A.C. & Levitt, M.H. (1989) *Chem. Phys. Lett.* **155**, 341-346.
[68] Jeener, J., Meier, B.H., Bachmann, P. & Ernst, R.R. (1979) *J. Chem. Phys.* **71**, 4546-4553.
[69] Kaufmann, S., Wefing, S., Schaefer, D. & Spiess, H.W. (1990) *J. Chem. Phys.* **93**, 197-214.
[70] Leisen, J., Schmidt-Rohr, K. & Spiess, H.W. (1993) *Physica A,* **201,** 79-87.

[71] Wefing, S., Kaufmann, S. & Spiess, H.W. (1988) *J. Chem. Phys.* **89**, 1234-1244.
[72] Heuer, A., Wilhelm, M., Zimmermann, H. & Spiess, H.W. (1995) *Phys. Rev. Lett.* **75**, 2851-2854.
[73] Schmidt-Rohr, K. & Spiess, H.W. (1991) *Phys. Rev. Lett.* **66**, 3020-3023.
[74] Tracht, U., Wilhelm, M., Heuer, A., Feng, H., Schmidt-Rohr, K. & Spiess, H.W. (1998) *Phys. Rev. Lett.* **81**, 2727-2730.
[75] DeAzevedo, E.R., Hu, W.G., Bonagamba, T.J. & Schmidt-Rohr, K. (1999) *J. Am. Chem. Soc.* **121**, 8411-8412.
[76] Bonagamba, T.J., Becker-Guedes, F., deAzevedo, E.R., Hu, W-H. & Schmidt-Rohr, K. (2000) In *Am. Phys. Soc. Meeting*: Minneapolis, USA.
[77] Rossler, E. (1986) *Chem. Phys. Lett.* **128**, 330-334.
[78] Qi, F., Jorg, T. & Bohmer, R. (2002) *Solid State Nucl. Magn. Reson.* **22**, 484-500.
[79] DeAzevedo, E.R., Reichert, D., Vidoto, E.L.G., Dahmouche, K., Judeinstein, P. & Bonagamba, T.J. (2003) *Chem. Mat.* **15**, 2070-2078.
[80] Gerardy-Montouillout, V., Malveau, C., Tekely, P., Olender, Z. & Luz, Z. (1996) *J. Magn. Reson., Ser. A,* **123**, 7-15.
[81] Reichert, D., Zimmermann, H., Tekely, P., Poupko, R. & Luz, Z. (1997) *J. Magn. Reson.* **125**, 245-258.
[82] Fujara, F., Wefing, S. & Spiess, H.W. (1986) *J. Chem. Phys.* **84**, 4579-4584.
[83] Reichert, D., Tekely, P. & Luz, Z. (2002) *Prog. Nucl. Magn. Reson. Spectrosc.* **4**, 83-113.
[84] Wefing, S. & Spiess, H.W. (1988) *J. Chem. Phys.* **89**, 1219-1233.
[85] McCrum, N.G., Read, B.E.P. & Williams, G. (1991) *Anelastic and Dielectric Effects in Polymeric Solids* (Dover) 1 p 617.
[86] Schmidt-Rohr, K., Kulik, A.S., Beckham, H.W., Ohlemacher, A., Pawelzik, U., Boeffel, C. & Spiess, H.W. (1994) *Macromolecules* **27**, 4733-4745.
[87] Schmidt-Rohr, K., Wilhelm, M., Johansson, A. & Spiess, H.W. (1993) *Magn. Reson. Chem.* **31**, 352-356.
[88] Bonagamba, T.J., Becker-Guedes, F., DeAzevedo, E.R. & Schmidt-Rohr, K. (2001) *J. Polym. Sci., Part B: Polym. Phys.* **39**, 2444-2453.
[89] Becker-Guedes, F., deAzevedo, E.R., Bonagamba, T.J. & Schmidt-Rohr, K. (2004) *Appl. Magn. Reson.* **27**, 383-400.
[90] Kulik, A.S., Beckham, H.W., Schmidt-Rohr, K., Radloff, D., Pawelzik, U., Boeffel, C. & Spiess, H.W. (1994) *Macromolecules* **27**, 4746-4754.
[91] Wind, M., Graf, R., Heuer, A. & Spiess, H.W. (2003) *Phys. Rev. Lett.* **91** .
[92] Pascui, O., Beiner, M. & Reichert, D. (2003) *Macromolecules* **36**, 3992-4003.
[93] Schaefer, D. & Spiess, H.W. (1992) *J. Chem. Phys.* **97**, 7944-7954.
[94] Schaefer, D., Spiess, H.W., Suter, U.W. & Fleming, W.W. (1990) *Macromolecules* **23**, 3431-3439.
[95] Schnauss, W., Fujara, F. & Sillescu, H. (1992) *J. Chem. Phys.* **97**, 1378-1389.
[96] Chang, I. & Sillescu, H. (1997) *J. Phys. Chem. B,* **101**, 8794-8801.
[97] Diezemann, G., Bohmer, R., Hinze, G. & Sillescu, H. (1998) *J. Non-Cryst. Solids* **235**, 121-127.
[98] Doss, A., Paluch, M., Sillescu, H. & Hinze, G. (2002) *J. Chem. Phys.* **117**, 6582-6589.
[99] Pschorn, U., Rossler, E., Sillescu, H., Kaufmann, S., Schaefer, D. & Spiess, H.W. (1991) *Macromolecules* **24**, 398-402.
[100] Williams, M.L., Landel, R.F. & Ferry, J.D. (1955) *J. Am. Chem. Soc.* **77**, 3701.
[101] DeAzevedo, E.R., Tozoni, J.R., Schmidt-Rohr, K. & Bonagamba, T. (2005) *J. Chem. Phys.* **122**, 154506.
[102] DeAzevedo, E.R., Bonagamba, T.J. & Schmidt-Rohr, K. (2000) *J. Magn. Reson.* **142**, 86-96.
[103] Vosegaard, T. & Nielsen, N.C. (2004) *Magn. Reson. Chem.* **42**, 285-290.
[104] Kaji, H. & Horii, F. (2003) *Chem. Phys. Lett.* **377**, 322-328.
[105] Dejong, A.F., Kentgens, A.P.M. & Veeman, W.S. (1984) *Chem. Phys. Lett.* **109**, 337-342.

[106] Hagemeyer, A., Schmidt-Rohr, K. & Spiess, H.W. (1989) *Adv. Magn. Reson.* **13**, 85-129.
[107] Schmidt, C., Blumich, B. & Spiess, H.W. (1988) *J. Magn. Reson.* **79**, 269-290.
[108] DeAzevedo, E.R., Franco, R.W.A., Marletta, A., Faria, R.M. & Bonagamba, T.J. (2003) *J. Chem. Phys.* **119**, 2923-2934.
[109] Collison, C.J., Rothberg, L.J., Treemaneekarn, V. & Li, Y. (2001) *Macromolecules* **34**, 2346-2352.
[110] Zemke, K., Schmidt-Rohr, K. & Spiess, H.W. (1994) *Acta Polym.* **45**, 148-159.
[111] Luz, Z., Spiess, H.W. & Titman, J.J. (1992) *Isr. J. Chem.* **32**, 145-160.
[112] Olender, Z., Reichert, D., Mueller, A., Zimmermann, H., Poupko, R. & Luz, Z. (1996) *J. Magn. Reson., Ser. A,* **120**, 31-45.
[113] Bennett, A.E., Ok, J.H., Griffin, R.G. & Vega, S. (1992) *J. Chem. Phys.* **96**, 8624-8627.
[114] Reichert, D., Bonagamba, T.J. & Schmidt-Rohr, K. (2001) *J. Magn. Reson.* **151**, 129-135.
[115] Robyr, P. & Gan, Z. (1998) *J. Magn. Reson.* **131**, 254-260.
[116] Robyr, P., Tomaselli, M., Straka, J., Grobpisano, C., Suter, U.W., Meier, B.H. & Ernst, R.R. (1995) *Mol. Phys.* **84**, 995-1020.
[117] Robyr, P., Meier, B.H. & Ernst, R.R. (1991) *Chem. Phys. Lett.* **187**, 471-478.
[118] Robyr, P., Gan, Z. & Suter, U.W. (1998) *Macromolecules* 31 .
[119] Jager, C., Hartmann, P., Witter, R. & Braun, M. (2000) *J. Non-Cryst. Solids* **263**, 61-72.
[120] Reichert, D., Pascui, O., Bonagamba, T.J., deAzevedo, E.R. & Schmidt, A. (2003) *Chem. Phys. Lett.* **380**, 583-588.
[121] DeAzevedo, E.R., Hu, W.G., Bonagamba, T.J. & Schmidt-Rohr, K. (2000) *J. Chem. Phys.* **112**, 8988-9001.
[122] Doss, A., Paluch, M., Sillescu, H. & Hinze, G. (2002) *Phys. Rev. Lett.* **88** .
[123] Hinze, G., Diezemann, G. & Sillescu, H. (1998) *Europhys. Lett.* **44**, 565-570.
[124] Tracht, U., Wilhelm, H., Heuer, A. & Spiess, H.W. (1999) *J. Magn. Reson.* **140**, 460-470.
[125] Qi, F., Diezemann, G., Bohm, H., Lambert, J. & Bohmer, R. (2004) *J. Magn. Reson.* **169**, 225-239.
[126] Vogel, M., Brinkmann, C., Eckert, H. & Heuer, A. (2004) *Phys. Rev. B,* **69**.
[127] Schmidt-Rohr, K., deAzevedo, E.R. & Bonagamba, T.J. (2002) NMR Ency. 9:Advances in NMR 633-642.
[128] DeAzevedo, E.R., Kennedy, S.B. & Hong, M. (2000) *Chem. Phys. Lett.* **321**, 43-48.
[129] Kaji, H. & Schmidt-Rohr, K. (2002) *Macromolecules* **35**, 7993-8004.
[130] Buffy, J.J., Waring, A.J. & Hong, M. (2005) *J. Am. Chem. Soc.* **127**, 4477-4483.
[131] Saalwächter, K. & Schmidt-Rohr, K. (2000) *J. Magn. Reson.* **145**, 161-172.
[132] Henrichs, P.M. & Linder, M. (1984) *J. Magn. Reson.* **58**, 458-461.
[133] Edzes, H.T. & Bernards, J.P.C. (1984) *J. Am. Chem. Soc.* **106**, 1515-1517.
[134] Dabbagh, G., Weliky, D.P. & Tycko, R. (1994) *Macromolecules* **27**, 6183-6191.
[135] Tomita, Y., Oconnor, E.J. & McDermott, A. (1994) *J. Am. Chem. Soc.* **116**, 8766-8771.
[136] Weliky, D.P., Dabbagh, G. & Tycko, R. (1993) *J. Magn. Reson. Series A,* **104**, 10-16.
[137] Nakai, T. & McDowell, C.A. (1994) *Chem. Phys. Lett.* **217**, 234-238.
[138] Raleigh, D.P., Creuzet, F., Gupta, S.K.D., Levitt, M.H. & Griffin, R.G. (1989) *J. Am. Chem. Soc.* **111**, 4502-4503.
[139] Pan, Y., Gullion, T. & Schaefer, J. (1990) *J. Magn. Reson.* **90**, 330-340.
[140] Hing, A.W., Vega, S. & Schaefer, J. (1992) *J. Magn. Reson.* **96**, 205-209.
[141] Gottwald, J., Demco, D.E., Graf, R. & Spiess, H.W. (1995) *Chem. Phys. Lett.* **243**, 314-323.
[142] Schmidt-Rohr, K. (1996) *Macromolecules* **29**, 3975-3981.
[143] Schmidt-Rohr, K. (1998) *J. Magn. Reson.* **131**, 209-217.
[144] Harris, D.J., Bonagamba, T.J. & Schmidt-Rohr, K. (1999) *Macromolecules* **32**, 6718-6724.

[145] Harris, D.J., Bonagamba, T.J., Hong, M. & Schmidt-Rohr, K. (2000) *Macromole-cules* **33**, 3375-3381.
[146] Harris, D.J., Bonagamba, T.J. & Schmidt-Rohr, K. (2002) *Macromolecules* **35**, 5724-5726.
[147] Bax, A., Freeman, R. & Kempsell, S.P. (1980) *J. Am. Chem. Soc.* **102**, 4849-4851.
[148] Nakai, T. & McDowell, C.A. (1993) *Mol. Phys.* **79**, 965-983.
[149] Nakai, T. & McDowell, C.A. (1993) *J. Magn. Reson., Ser* **A 104**, 146-153.
[150] Takahashi, Y. & Tadokoro, H. (1973) *Macromolecules* **6**, 672-675.
[151] Spevacek, J., Paternostre, L., Damman, P., Draye, A.C. & Dosiere, M. (1998) *Macromolecules* **31**, 3612-3616.
[152] Myasnikova, R.M., Titova, E.F. & Obolonkova, E.S. (1980) *Polymer* **21**, 403-407.
[153] Point, J.J. & Coutelier, C. (1985) *J. Polym. Sci., Part B: Polym. Phys.* **23**, 231-239.
[154] Damman, P. & Point, J.J. (1994) *Macromolecules* **27**, 3919-3925.
[155] Spevacek, J. & Suchoparek, M. (1997) *Macromol. Symp.* **114**, 23-34.
[156] Hong, M. (1999) *J. Magn. Reson.* **136**, 86-91.
[157] Rienstra, C.M., Hatcher, M.E., Mueller, L.J., Sun, B.Q., Fesik, S.W. & Griffin, R.G. (1998) *J. Am. Chem. Soc.* **120**, 10602-10612.
[158] Harris, D.J., Bonagamba, T.J., Hong, M. & Schmidt-Rohr, K. (2005) *Polymer* (in press).
[159] Dunbar, M.G., Novak, B.M. & Schmidt-Rohr, K. (1998) *Solid State Nucl. Magn. Reson.* **12**, 119-137.
[160] Kaji, H. & Schmidt-Rohr, K. (2000) *Macromolecules* **33**, 5169-5180.
[161] Kaji, H. & Schmidt-Rohr, K. (2001) *Macromolecules* **34**, 7368-7381.
[162] van Beek, J.D., Beaulieu, L., Schafer, H., Demura, M., Asakura, T. & Meier, B.H. (2000) *Nature* **405**, 1077-1079.
[163] Tycko, R. (2001) *Ann. Rev. Phys. Chem.* **52**, 575-606.
[164] van Beek, J.D., Meier, B.H. & Schafer, H. (2003) *J. Magn. Reson.* **162**, 141-157.
[165] Feike, M., Graf, R., Schnell, I., Jager, C. & Spiess, H.W. (1996) *J. Am. Chem. Soc.* **118**, 9631-9634.
[166] Glock, K., Hirsch, O., Rehak, P., Thomas, B. & Jager, C. (1998) *J. Non-Cryst. Solids* **234**, 113-118.
[167] Witter, R., Hartmann, P., Vogel, J. & Jager, C. (1998) *Solid State Nucl. Magn. Reson.* **13**, 189-200.
[168] Montagne, L., Palavit, G., Shaim, A., Et-Tabirou, M., Hartmann, P. & Jager, C. (2001) *J. Non-Cryst. Solids* **293**, 719-725.
[169] Olivier, L., Yuan, X., Cormack, A.N. & Jager, C. (2001) *J. Non-Cryst. Solids* **293**, 53-66.

CHAPTER 2

NUCLEAR QUADRUPOLE RESONANCE SPECTROSCOPY

Bryan H. Suits
Physics Department, Michigan Technological University, Houghton, MI, USA

2.1 INTRODUCTION

Nuclear quadrupole resonance (NQR) uses radio-frequency (RF) magnetic fields to induce and detect transitions between sublevels of a nuclear ground state, a description that also applies to nuclear magnetic resonance (NMR). NMR refers to the situation where the sublevel energy splitting is predominantly due to a nuclear interaction with an applied static magnetic field, while NQR refers to the case where the predominant splitting is due to an interaction with electric field gradients within the material. So-called "pure NQR" refers to the common case when there is no static magnetic field at all.

The beginning of NQR in solids dates back to the beginnings of NMR in the late 1940s and early 1950s [1]. The first NQR measurements reported for a solid were by Dehmelt and Kruger using signals from ^{35}Cl in trans-dichloroethylene [2]. An excellent early summary of NQR theory and technique can be found in the 1958 book by Das and Hahn [3]. Several more recent summaries can be found listed at the end of this chapter. Due to practical limitations, discussed below, NQR has not grown to be nearly as common as NMR, and is usually considered a tool for the specialist.

As is the case for NMR spectroscopy, the primary goal for NQR spectroscopy is to determine nuclear transition frequencies (i.e., energies) and/or relaxation times and then to relate those to a property of a material being studied. That property may simply be the sample temperature, for use as an NQR thermometer [4, 5], or even whether or not a sample is present when NQR is used for materials detection [6]. On the other hand, NQR is also used to obtain detailed information on crystal symmetries and bonding, on changes in lattice constants with pressure, about phase transitions in solids, and other properties of materials of interest to solid state physicists and chemists.

As will be seen in more detail below, in order to use NQR spectroscopy one must have available an isotope with a nuclear spin I > ½, which has a reasonably high isotopic abundance, and which is at a site in a solid that has symmetry lower than tetragonal. The most common NMR isotopes, ^{1}H, ^{13}C, and ^{15}N cannot be used since they have a nuclear spin ½. Of course, ^{12}C and ^{16}O cannot be used either as they have nuclear spin 0. Table 2.1 shows a selection of potential nuclei including those most commonly used for NQR, as well as a few others of possible interest.

2.2 BASIC THEORY

2.2.1 The Nuclear Electric Quadrupole Interaction

Since a nuclear wavefunction has a definite state of parity, a multipole expansion of the fields due to the nucleus yields electric 2^n-poles, where n is even (monopole, quadrupole, etc.) and magnetic 2^n-poles, where n is odd (dipoles, octupoles, etc.). In general these multipole moments become weaker very rapidly with increasing n. In a molecule or in a solid, the nucleus will be at an equilibrium position where the electric field is zero, and so in the absence of a magnetic field the first non-zero interaction is with the electric quadrupole moment of the nucleus. Higher moments, if they exist, are generally much too weak to affect NQR measurements [7–9].

A non-zero electric quadrupole moment arises for nuclei that are classically described as prolate ("stretched") or oblate ("squashed") spheroids. The nuclear charge distribution has axial symmetry and the axis of symmetry coincides with the direction of the nuclear angular momentum and the nuclear magnetic dipole moment. In general, an electric quadrupole moment is described by a 3 × 3 symmetric, traceless tensor **Q**. For a nucleus such a tensor can be determined using a single value that describes how prolate or oblate the nucleus is, plus a description of the orientation of the nucleus. Since the charge distribution for a nucleus with spin 0 or ½ is spherical, such nuclei will have no electric quadrupole moment.

If the charge distribution within the nucleus is known, the amount by which the sphere is prolate or oblate is determined by the (scalar) nuclear quadrupole moment Q, which can be calculated using

$$eQ = \int \rho \, (3z^2 - r^2)\, d\tau \tag{2.1}$$

where the z-axis is along the direction of axial symmetry, e is the magnitude of the charge on an electron, and ρ is the nuclear charge density as a function of position. While such computations may be done by a nuclear physicist to check a new model for the nucleus, the NQR spectroscopist uses values determined experimentally. Values of Q are conveniently expressed in units of 10^{-24} cm^2 = 1 barn.

Table 2.1 Selected quadrupolar nuclei.

Nucleus	Natural Isotopic Abundance %	Spin I	$\gamma/2\pi$ (kHz/G)	Q (10^{-24} cm^2)
^{2}H	0.015	1	0.654	+0.00286
^{6}Li	7.4	1	0.626	–0.0008
^{7}Li	92.6	3/2	1.655	–0.040
^{10}B	19.6	3	0.458	+0.085
^{11}B	80.4	3/2	1.366	+0.041
^{14}N	99.6	1	0.308	+0.019
^{17}O	0.048	5/2	–0.577	–0.26
^{23}Na	100	3/2	1.126	+0.10
^{27}Al	100	5/2	1.109	+0.14
^{35}Cl	75.5	3/2	0.417	–0.082
^{37}Cl	24.5	3/2	0.347	–0.064
^{50}V	0.25	6	0.425	+0.21
^{51}V	99.8	7/2	1.119	–0.05
^{55}Mn	100	5/2	1.050	+0.33
^{59}Co	100	7/2	1.005	+0.40
^{63}Cu	69.1	3/2	1.128	–0.21
^{65}Cu	30.9	3/2	1.209	–0.195
^{69}Ga	60.4	3/2	1.022	+0.17
^{71}Ga	39.6	3/2	1.298	+0.10
^{75}As	100	3/2	0.729	+0.31
^{79}Br	50.5	3/2	1.067	+0.33
^{81}Br	49.5	3/2	1.150	+0.28
^{85}Rb	72	5/2	0.411	+0.23
^{87}Rb	28	3/2	1.393	+0.13
^{93}Nb	100	9/2	1.041	–0.32
^{113}In	4.3	9/2	0.931	+0.8
^{115}In	95.7	9/2	0.933	+0.8
^{121}Sb	57.3	5/2	1.019	–0.4
^{123}Sb	42.7	7/2	0.552	–0.5
^{127}I	100	5/2	0.852	–0.7
^{138}La	0.1	5	0.564	+0.4
^{139}La	99.9	7/2	0.606	+0.2
^{181}Ta	99.99	7/2	0.510	+3.3
^{197}Au	100	3/2	0.073	+0.55
^{209}Bi	100	9/2	0.684	–0.4
^{235}U	0.72	7/2	–0.076	+5

While the electric field at the nucleus is zero, the electric field gradients (spatial derivatives of that field) may not be. Figure 2.1 is a schematic showing two orientations of a prolate nucleus ($Q > 0$) at a point where the electric field is zero in the vicinity of four fixed point charges. The configuration shown on the left will have a lower energy than that shown on the right since the positive charge of the nucleus is, on the whole, closer to the negative charges. When quantum mechanics is applied, this orientation dependence gives rise to a small splitting of the nuclear ground state.

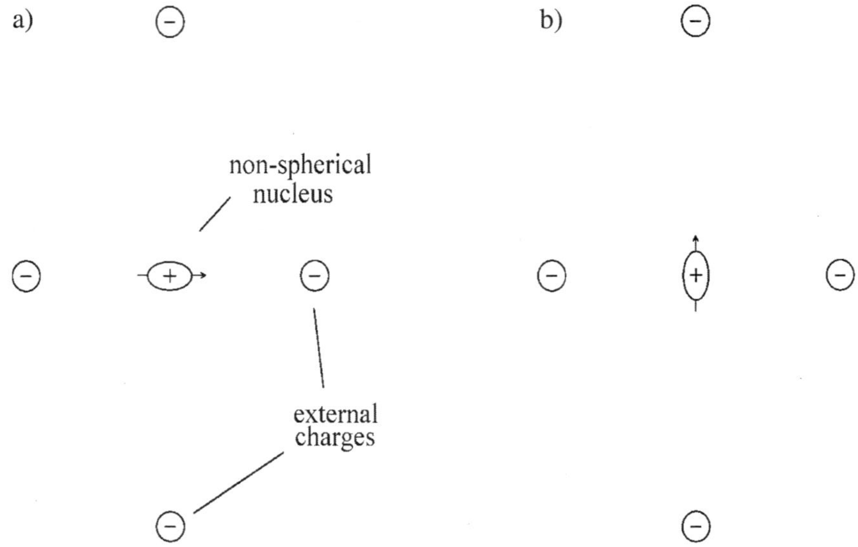

Figure 2.1 Two configurations of a non-spherical nucleus near charges external to the nucleus. The configuration at (a) has a lower energy than that shown at (b).

The electric field gradient at the nucleus due to charges external to the nucleus, $\nabla \boldsymbol{E}$, is conveniently described using spatial derivatives of the corresponding electrostatic potential, V, evaluated at the nucleus. Taking the nucleus to be at the origin of the coordinate system, the desired derivatives are

$$V_{jk} = \left. \frac{\partial^2 V}{\partial r_j \partial r_k} \right|_0 \tag{2.2}$$

where $\{r_i\} = \{x, y, z\}$. Since $V_{jk} = V_{kj}$ and, using Laplace's equation,

$$\sum_{i=x,y,z} V_{ii} = 0, \tag{2.3}$$

the field gradient can be described by a real, symmetric, traceless 3×3 tensor. Such a tensor can always be made diagonal by choosing an appropriate set of coordinate axes known as *principal axes*. Once this is done, it is conventional to define

$$eq = V_{zz} \tag{2.4}$$

$$\eta = \frac{V_{yy} - V_{xx}}{V_{yy} + V_{xx}} = \frac{V_{xx} - V_{yy}}{V_{zz}} \tag{2.5}$$

whereη is known as the *asymmetry parameter*. It is convenient to choose the principal axes such that

$$|V_{xx}| \le |V_{yy}| \le |V_{zz}| \tag{2.6}$$

giving $0 \le \eta \le 1$. Since the principal axes are determined by the environment surrounding the nucleus, those axes are sometimes also referred to as forming a "molecular" or "crystal" coordinate system. For axial symmetry $V_{xx} = V_{yy} = -V_{zz}/2$ and $\eta = 0$.

Classically, the *interaction energy* is given by the tensor scalar product

$$E_Q = \frac{1}{6} \sum_{i,j=x,y,z} V_{ij} Q_{ij} , \tag{2.7}$$

where the two tensors must be expressed in the same coordinate system. Coordinate transformations can be accomplished using well-known relations for 3 × 3 symmetric tensors.

Since the nuclear state can be described by specifying the nuclear angular momentum, the entire interaction can be written, with appropriate scale factors including the scalar quadrupole moment, in terms of the angular momentum. When written using quantum mechanical operators, the Hamiltonian $\mathcal{H}_Q$ for a nucleus of spin I expressed in the principal axis coordinate system is

$$\mathcal{H}_Q = \frac{e^2 qQ}{4I(2I-1)} \left[3I_z^2 - I^2 + \eta \left(I_x^2 - I_y^2 \right) \right]$$

where all I's in the denominator are scalar values while all I's in the square brackets are operators. The interested reader can find a detailed derivation of this result in Slichter's book [10]. In terms of the usual angular momentum raising and lowering operators, $I_\pm = I_x \pm i I_y$, the Hamiltonian can also be written

$$\mathcal{H}_Q = \frac{e^2 qQ}{4I(2I-1)} \left[3I_z^2 - I^2 + \eta \cdot \frac{I_+^2 - I_-^2}{2} \right] . \tag{2.8}$$

To represent the Hamiltonian in other coordinate systems, the appropriate angular momentum rotation operators are applied. Other forms for the operators, such as irreducible tensor operators, are also sometimes employed (see [11] for example).

One of the goals of an NQR measurement will be to determine the quadrupole coupling constant e^2qQ and the asymmetry parameter η, which contain information about the environment surrounding the nucleus.

2.2.2 Energy Levels and Transition Frequencies

In the case of axial symmetry, $\eta = 0$, the pure quadrupole Hamiltonian is easily diagonalized using eigenfunctions of the operator I_z with quantum number $m = -I, -I+1, \ldots, I-1, I$. The resulting $2I+1$ energy levels are given by

$$E_{\mathrm{m}} = \frac{e^2qQ}{4I(2I-1)}\left(3m^2 - I(I+1)\right). \tag{2.9}$$

In this case m is a good quantum number and the usual magnetic dipole transition rules apply, $\Delta m = 0, \pm 1$. Defining

$$\nu_Q = \frac{3e^2qQ}{4I(2I-1)h}, \tag{2.10}$$

where h is Plank's constant, the allowed transition frequencies are given by

$$\nu_{m,m\pm1} = \nu_Q\left|(2m\pm1)\right|;\ |m|,|m\pm1| \le I. \tag{2.11}$$

For the more general case of arbitrary η, closed form solutions are known only for $I = 1$ and $I = 3/2$. Due to the symmetry of the Hamiltonian, all the energy levels are doubly degenerate for half-integer spin nuclei. For integer spin nuclei, of which there are very few in practice, there are an odd number of levels and the degeneracy is broken. Furthermore, since the eigenfunctions of I_z are not, in general, energy eigenfunctions, additional transitions are often allowed.

2.2.2.1 Integer Spins

There are only four known stable nuclei with integer spin: ^{2}H,^{6}Li, and ^{14}N, all with $I = 1$, and ^{10}B with $I = 3$. In addition there are some very long-lived radioactive isotopes, such as ^{50}V, with $I = 6$ and ^{138}La with $I = 5$. Most of the NQR work done using integer spin nuclei is for ~100% naturally abundant ^{14}N. Deuterium (^{2}H) and ^{6}Li have very small electric quadrupole moments, making direct observation with NQR difficult. There has been some work using ^{10}B but due to its lower natural abundance compared to ^{11}B ($I = 3/2$) the latter is preferred. The long-lived radioactive isotopes also have a very low natural abundance making them quite difficult to use.

For spin 1, the three energy levels are

$$E_0 = -\frac{2}{3}h\nu_Q,\quad E_\pm = \frac{(1\pm\eta)}{3}h\nu_Q, \tag{2.12}$$

and all three possible transition frequencies

$$\nu_0 = \frac{2}{3}\eta\nu_Q,\quad \nu_\pm = \left(1\pm\frac{\eta}{3}\right)\nu_Q, \tag{2.13}$$

are allowed.

For spin 3 an exact solution is not known. Butler and Brown [12] provide a graphical representation showing 18 allowed transitions, arising from the 7 energy levels, as a function of η. Five of those 18 are forbidden when $\eta = 0$ and are somewhat inappropriately referred to as "multiple quantum transitions." They are allowed single quantum transitions, though some are quite weak, and can be useful to help unravel the wonderfully complicated ^{10}B NQR spectra [13].

2.2.2.2 Spin 3/2

Much of the NQR work in the literature is for spin 3/2 nuclei, which have two doubly degenerate energy levels,

$$E_{\pm 3/2} = h\nu_Q \left(1 + \frac{\eta^2}{3}\right)^{1/2}, \quad E_{\pm 1/2} = -h\nu_Q \left(1 + \frac{\eta^2}{3}\right)^{1/2} \tag{2.14}$$

and hence only one (non-zero frequency) transition,

$$\nu = 2\nu_Q \left(1 + \frac{\eta^2}{3}\right)^{1/2}. \tag{2.15}$$

The fact that there is only one frequency means that one cannot determine the two values ν_Q and η with a simple pure NQR measurement. The application of a small magnetic field, discussed below, is often used to separately determine the two values. For many compounds studied using spin 3/2 NQR, η has not been separately determined. Often such data are interpreted using the assumption $\eta = 0$, which can yield a maximum error of about 16% in the determination of ν_Q.

2.2.2.3 Other Half-Integer Spins

Exact solutions are not known for $I > 3/2$. Tabulated results can be used [14], or it is now quite easy to diagonalize the Hamiltonian numerically. Expansions valid for smaller values of η are also available [15, 16]. Results of numerical computations for half-integer spins 5/2, 7/2, and 9/2 are shown in Figure 2.2. As is customary, the levels are labeled according to the largest component of the wavefunction, though m is only a good quantum number when $\eta = 0$. In addition, when $\eta \neq 0$ virtually all possible transitions are allowed though many are extremely weak. This is similar to what occurs for ^{10}B, mentioned above. The dotted lines in Figure 2.2 indicate some of these weaker transitions, which are not allowed at all when $\eta = 0$ but which may be usable for large η. Those weaker transitions are rarely used in practice but can be helpful when disentangling spectra observed for samples with multiple sites having large η (for example, see [17]).

When $\eta = 1$ it is possible, with some effort, to obtain exact solutions for half-integer spins up to $I = 9/2$. The resulting energy levels for these I are

$$\begin{aligned} I = 5/2 &: \quad 0, \pm\sqrt{112/9} \\ I = 7/2 &: \quad \pm 2\left[7 \pm 4\sqrt{7/3}\right]^{1/2} \\ I = 9/2 &: \quad 0, \pm 2\left[22 \pm \sqrt{748/3}\right]^{1/2} \end{aligned} \tag{2.16}$$

in units of $h\nu_Q$.

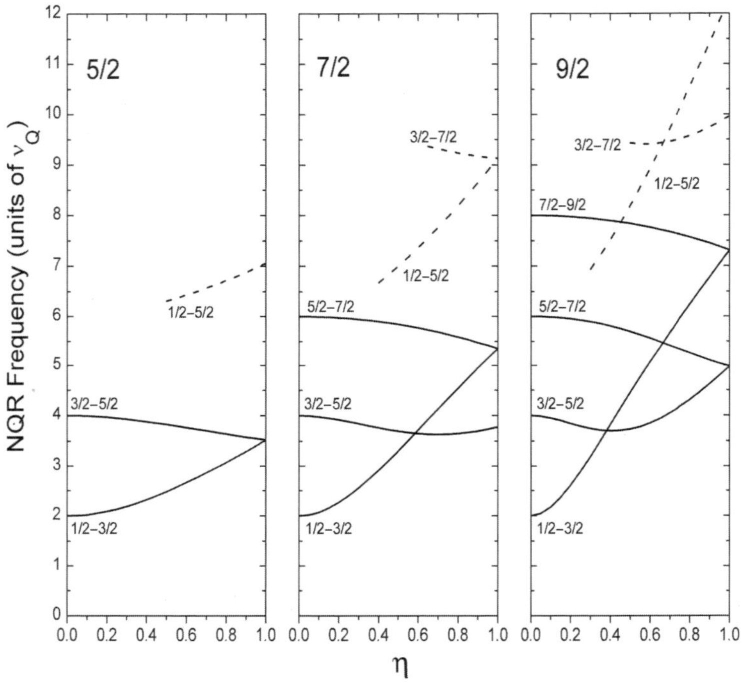

Figure 2.2 NQR transition frequencies for spins 5/2, 7/2, and 9/2. The dashed lines are weaker transitions, which are forbidden when $\eta = 0$.

2.2.3 Excitation and Detection

In a typical NQR measurement transitions are induced between energy levels via the coupling between the nuclear magnetic dipole moment and a resonant time-dependent magnetic field, as is done for NMR. One could also imagine applying a time-dependent electric field gradient, however the required field strengths are much too large to be practical in the laboratory. The required time-dependent electric field gradients can be generated indirectly by applying acoustic energy, a method important for the spectroscopic technique known as nuclear acoustic resonance (NAR) [18]. Due to practical considerations, NAR has proven to have limited utility.

To couple to the nuclear magnetic moment, the sample is placed within the RF magnetic field produced by an inductor which is carrying an alternating current (AC) of angular frequency, $\omega_r = 2\pi\nu_r$. Most commonly this is done by placing the sample within a solenoid that is part of a tuned circuit. If the AC magnetic field produced is uniform with magnitude, B_1, and is in the x'-direction in a laboratory reference frame, the *interaction Hamiltonian* is

$$\mathcal{H}_1 = \gamma B_1 I_{x'} \cos(\omega_r t + \phi)$$

where γ is the gyromagnetic (or magnetogyric) ratio for the nucleus, $I_{x'}$ is the angular momentum operator for the component along the x'-direction, and ϕ is a phase factor.

For convenience, the energy eigenfunctions in the absence of the AC field, ψ_n, with energy $E_n = \hbar\omega_n$, can be expressed in terms of the eigenfunctions of I_z, u_m, where z corresponds to the z-direction of the principal axes system. That is,

$$\psi_n = \sum_{m=-I}^{I} b_{n,m} u_m . \tag{2.17}$$

Then in turn, the total time-dependent wavefunction can be written

$$\Psi(t) = \sum_n a_n(t)\,\psi_n \exp(-i\omega_n t), \tag{2.18}$$

where the complex coefficients $a_n(t)$ are to be determined. As written, those coefficients will be time-independent when the AC magnetic field is off. Since the AC magnetic field yields a relatively weak interaction $\mathcal{H}_1$, compared to that of the static electric quadrupole field $\mathcal{H}_Q$, the coefficients a_n will vary relatively slowly with time. Placing the total wavefunction into Schrödinger's time-dependent wave equation,

$$-\frac{\hbar}{i}\frac{\partial\Psi}{\partial t} = \left(\mathcal{H}_Q + \mathcal{H}_1\right)\Psi, \tag{2.19}$$

and using the orthogonality of the eigenfunctions, the coupled equations for the coefficients, a_n(t), are obtained,

$$\frac{\partial a_j}{\partial t} = \frac{i\gamma B_1}{2}\sum_k a_k \left\langle \psi_j \middle| I_{x'} \middle| \psi_k \right\rangle \left\{ e^{-i(\omega_k - \omega_j - \omega_r)t + i\phi} + e^{-i(\omega_k - \omega_j + \omega_r)t - i\phi} \right\}, \tag{2.20}$$

where $\left\langle \psi_j \middle| I_{x'} \middle| \psi_k \right\rangle$ is a constant. Expressing $I_{x'}$ in the principal axes system as

$$I_{x'} = c_x I_x + c_y I_y + c_z I_z \tag{2.21}$$

where the c_i are a shorthand notation for the directional cosines, then

$$\begin{aligned}\langle\psi_j|I_{x'}|\psi_k\rangle = \sum_m \Big\{ & m\, c_z\, b^*_{j,m} b_{k,m} \\ & + \left(c_x - ic_y\right)\left(I(I+1) - m(m+1)\right)^{1/2} \frac{b^*_{j,m+1} b_{k,m}}{2} \\ & + \left(c_x + ic_y\right)\left(I(I+1) - m(m+1)\right)^{1/2} \frac{b^*_{j,m-1} b_{k,m}}{2} \Big\}.\end{aligned} \tag{2.22}$$

Thus far there has been no approximation.

For simplicity, assume that just two states, labeled 1 and 2, with $E_2 > E_1$, are involved. The "slowly varying" part of the solution desired occurs when the time dependence in one of the exponentials becomes small. That will occur when the frequencies in one of the exponentials nearly cancel. The other exponentials will produce rapidly oscillating terms which will tend to average to zero. Keeping only the slowly varying terms, defining $\Omega = \gamma B_1 \langle\psi_1|I_{x'}|\psi_2\rangle e^{i\phi}$ and $\Delta\omega = \omega_r - (\omega_2 - \omega_1)$, the two coupled equations that result are

$$\begin{aligned}\frac{\partial a_1}{\partial t} &= \frac{i\Omega}{2} a_2\, e^{-i\Delta\omega t} \\ \frac{\partial a_2}{\partial t} &= \frac{i\Omega^*}{2} a_1\, e^{+i\Delta\omega t}\end{aligned} \tag{2.23}$$

which have solution

$$\begin{aligned}a_1(t) &= e^{+i\Delta\omega t/2}\left[a_1(0)\cos\left(\frac{\omega_{\text{eff}} t}{2}\right) + i\,\frac{\Omega\, a_2(0) - \Delta\omega\ a_1(0)}{\omega_{\text{eff}}}\sin\left(\frac{\omega_{\text{eff}} t}{2}\right)\right] \\ a_2(t) &= e^{-i\Delta\omega t/2}\left[a_2(0)\cos\left(\frac{\omega_{\text{eff}} t}{2}\right) + i\,\frac{\Omega^* a_1(0) + \Delta\omega\ a_2(0)}{\omega_{\text{eff}}}\sin\left(\frac{\omega_{\text{eff}} t}{2}\right)\right]\end{aligned} \tag{2.24}$$

where $a_1(0)$ and $a_2(0)$ are initial values at $t = 0$, and $\omega_{\text{eff}} = \sqrt{|\Omega|^2 + (\Delta\omega)^2}$.

The detection of the signal is also done using a coupling to the nuclear magnetic dipole moment. Knowing the wavefunction, we can compute the expectation value of the nuclear magnetic moment, $\vec{\mu}$, at any time. The component which is along the direction x' is given by

$$\langle\mu_{x'}\rangle = \gamma\hbar \sum_{j,k} a^*_j a_k \langle\psi_j|I_{x'}|\psi_k\rangle e^{-i(\omega_k - \omega_j)t}. \tag{2.25}$$

The total *nuclear magnetization* from N such nuclei, $M_{x'}$, can be written in terms of the ensemble average

$$M_{x'} = \gamma\hbar N \sum_{j,k} \left\langle a_j^* a_k \right\rangle \left\langle \psi_j \middle| I_{x'} \middle| \psi_k \right\rangle e^{-i(\omega_k - \omega_j)t} .$$

The set of values $\left\langle a_j^* a_k \right\rangle$ are, of course, the elements of the density matrix.

A time-dependent magnetization will generate an *electromotive force* (EMF, a voltage) $V(t)$, in a nearby inductor by Faraday's law of induction. If this inductor is the same one used above for excitation, then by reciprocity,

$$V(t) \propto \frac{dM_{x'}}{dt} . \tag{2.26}$$

A signal measured this way must arise from terms above where $j \neq k$. In thermal equilibrium $\left\langle a_j^* a_k \right\rangle$ will be zero for all $j \neq k$; after all, one cannot expect to continually extract electrical power in equilibrium. The role of the excitation is to disturb the thermal equilibrium so that a signal can be observed.

A common method to measure NQR signals is the pulse method where a relatively large AC magnetic field ($B_1 \approx 1$ to 100 G) is applied for a short time ($\tau_p \approx 1$ to 100 μs), after which the EMF induced in the coil is detected. The other extreme is to continuously supply a low level AC magnetic field ($B_1 < 1$ G). In that case the transitions produced are balanced against those of thermal relaxation processes and a dynamic equilibrium is obtained. Then the steady state EMF induced by the nuclei which is in phase with the current in the inductor is equivalent to an electrical resistance, and that which is out of phase a reactance. Instrumentation will be discussed in more detail in the next section.

NQR is usually performed one transition at a time and hence any of these problems can be treated using the "effective spin 1/2" formalism [19, 20]. However, it is useful to consider what happens explicitly for the two common cases of $I = 1$ and $I = 3/2$ without using that formalism.

2.2.3.1 Example—Spin 1

For $I = 1$, the three energy levels and the corresponding wavefunctions are illustrated in Figure 2.3. For the sake of example, assume the transition labeled ν_+ is to be excited exactly on resonance ($\Delta\omega = 0$), and take $\phi = 0$. Then $\Omega = \gamma B_1 c_x$ and

$$\begin{aligned} a_+(t) &= a_+(0)\cos\frac{\Omega}{2}t + i\,a_0(0)\sin\frac{\Omega}{2}t \\ a_0(t) &= a_0(0)\cos\frac{\Omega}{2}t + i\,a_+(0)\sin\frac{\Omega}{2}t \\ a_-(t) &= a_-(0), \end{aligned} \tag{2.27}$$

which is exactly what one gets if one rotates the nucleus by an angle $\Omega t/2$ about the x-axis. Similar expressions are obtained for the other two transitions involving the other two axes. Such rotations are often conveniently treated using exponential operators. It is interesting that for spin 1 a simultaneous excitation of two transitions can also be described as a simple rotation of the spin [21].

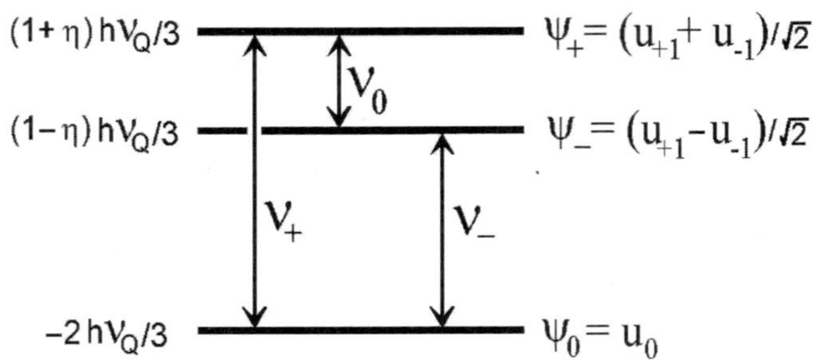

Figure 2.3 Energy levels and transitions for spin 1.

Starting from thermal equilibrium, the EMF after a time t will be

$$V(t) \propto c_x\, \omega_+ \left(\left\langle a_+^*(0)a_+(0)\right\rangle - \left\langle a_0^*(0)a_0(0)\right\rangle\right) \sin\Omega t \cos\omega_+ t\,, \quad (2.28)$$

which reaches a maximum at $\Omega t = \pi/2$ when the nucleus has been rotated by $\pi/4$. The term in parentheses is the population difference between the two levels determined by the Boltzmann distribution. For ^{14}N typical energy level splittings correspond to a temperature equivalent of about 0.1 mK and hence the thermal equilibrium population difference at room temperature is only about 1 part in 10^7.

The motion of the spin in the absence of the AC field is not easy to visualize. The simple "counter rotating" picture obtained below for $I = 3/2$ does not apply even though one can use it to some extent to form a mental picture. The classical motion of a quadrupole system is discussed by Raich and Good [22].

Note that Ω depends on sample orientation and so for powder (or polycrystalline) samples one will have a broad distribution of values. This inherently large inhomogeneity in the "effective B_1" (here $B_{1\text{eff}} = B_1 c_x$) occurs for NQR of powder samples, but does not occur for NMR.

2.2.3.2 Example—Spin 3/2

The energy levels of the spin 3/2 system are doubly degenerate. The energy levels were given above and the corresponding wavefunctions can be written

$$\begin{aligned} \psi_{\pm 3/2} &= \cos\delta\, u_{\pm 3/2} + \sin\delta\, u_{\mp 1/2} \\ \psi_{\pm 1/2} &= \cos\delta\, u_{\pm 1/2} - \sin\delta\, u_{\mp 3/2} \end{aligned} \tag{2.29}$$

where $\sin\delta = [(\rho - 1)/2\rho]^{1/2}$ with $\rho = (1 + \eta^2/3)^{1/2}$ and the only non-zero transition frequency is $\nu = \rho\nu_Q$. In general all four transitions are allowed. However, it is always possible to create a new set of wavefunctions from linear combinations of states with the same energy. In particular,

$$\psi'_{+3/2} = A\psi_{+3/2} + B\psi_{-3/2};\ \psi'_{-3/2} = A\psi_{-3/2} - B\psi_{+3/2} \tag{2.30}$$

and similarly for $\psi'_{\pm 1/2}$. Furthermore, a combination where only two of the four transitions are allowed can always be found. Once this has been done, the transformed spin 3/2 problem can be solved as two independent problems involving just two energy levels each [23]. The resulting classical picture for spin 3/2 NQR is that of an NMR experiment in an effective magnetic field involving the simultaneous measurement of two sets of (spin ½) nuclei, which are identical except for the sign of their gyromagnetic ratio. In the absence of the excitation one visualizes two sets of otherwise identical nuclei, precessing in opposite directions. Such a picture is a result of the degeneracy and generalizes to other half-integer spins.

The similarity between the classical picture used for NMR and that for spin 3/2 NQR leads to the use of "rotating reference frame" terminology for NQR even though it is perhaps not entirely appropriate.

2.2.4 The Effect of a Small Static Magnetic Field

The application of a small static magnetic field is sometimes advantageous and at other times it may simply be unavoidable. By "small" it is meant that the nuclear Zeeman interaction can be treated as a perturbation. In what follows, calculations will be carried out in the principal axes reference system. The Hamiltonian representing the Zeeman interaction can then be written

$$\mathcal{H}_z = -\gamma\hbar B_0 \left(c_x I_x + c_y I_y + c_z I_z\right), \tag{2.31}$$

where the directional cosines c_i may not be the same as those used previously. Any case can be treated numerically without difficulty, however it is worth taking a detailed look at spin 1 and spin 3/2 as representatives of what happens for integer and half-integer spins.

2.2.4.1 Spin 1

For the case of spin 1 with asymmetry parameter $\eta \neq 0$, there is no first-order shift in the energy levels. Using standard second-order perturbation theory the changes in the three transition frequencies are:

$$\Delta\nu_+ = \left(\frac{\gamma B_0}{2\pi\nu_Q}\right)^2 \left(\frac{3}{2}\frac{c_z^2}{\eta} + \frac{c_y^2}{(1-\eta/3)} + 2\frac{c_x^2}{(1+\eta/3)}\right)\nu_Q \tag{2.32}$$

$$\Delta\nu_- = \left(\frac{\gamma B_0}{2\pi\nu_Q}\right)^2 \left(-\frac{3}{2}\frac{c_z^2}{\eta} + 2\frac{c_y^2}{(1-\eta/3)} + \frac{c_x^2}{(1+\eta/3)}\right)\nu_Q \tag{2.33}$$

$$\Delta\nu_0 = \left(\frac{\gamma B_0}{2\pi\nu_Q}\right)^2 \left(3\frac{c_z^2}{\eta} - \frac{c_y^2}{(1-\eta/3)} + \frac{c_x^2}{(1+\eta/3)}\right)\nu_Q\,, \tag{2.34}$$

where ν_Q is as defined above. These are valid provided $\Delta\nu_0 << \nu_0$.

When $\eta = 0$, degenerate perturbation theory must be used and to lowest order one finds

$$\Delta\nu_{\pm} = \mp\frac{\gamma B_0}{2\pi}\,c_z \tag{2.35}$$

$$\Delta\nu_0 = 2\,\frac{\gamma B_0}{2\pi}\,c_z\,, \tag{2.36}$$

a result which can also be used when $\eta \neq 0$ if the first calculation yields $\Delta\nu_0 >> \nu_0$. The intermediate case where $\Delta\nu_0 \approx \nu_0$ is more complicated and will not be treated here.

For larger magnetic fields or when a more precise result is needed, the exact but more cumbersome solutions given by Muha can be used [24].

2.2.4.2 Spin 3/2

NQR in the presence of a small magnetic field is a principal method used to determine the asymmetry parameter η for spin 3/2.

The energy levels for spin 3/2, and for half-integer spins in general, will be degenerate regardless of the value of η and hence degenerate perturbation theory must be used. The four energy levels which result for spin 3/2, in the form presented by Brooker and Creel [25], become

$$E_{\pm 3/2} = \frac{h\nu_Q\rho}{2} \pm \frac{h\nu_0}{2\rho}\left[(\rho-1+\eta)^2 c_x^2 + (\rho-1-\eta)^2 c_y^2 + (2+\rho)^2 c_z^2\right]^{1/2} \tag{2.37}$$

$$E_{\pm 1/2} = -\frac{h\nu_Q\rho}{2} \pm \frac{h\nu_0}{2\rho}\left[(\rho+1-\eta)^2 c_x^2 + (\rho+1+\eta)^2 c_y^2 + (2-\rho)^2 c_z^2\right]^{1/2} \tag{2.38}$$

where $\nu_0 = \gamma B_0 / 2\pi$. In general all the transitions between these four levels are allowed. The four transitions shown in Figure 2.4 are of most interest. The highest and lowest frequency transitions, ν_β and $\nu_{\beta'}$, will be somewhat "weaker" than the middle two. A computation of the transition probabilities is straightforward, though complicated enough that it will not be reproduced here. The strength of each transition is a function of the relative orientations of the static magnetic field, the RF magnetic field, and the principal axes.

For multiple pulse measurements (described below) using a spin 3/2 nucleus in a small magnetic field one can observe an additional effect called "slow beats" [26] due to the fact that all four of these transitions are excited and they are not independent of each other.

To separately determine ν_Q and η the transition frequencies can be measured as a function of sample orientation for single crystals or, less accurately, the spectral line shape (i.e., the frequency distribution) can be measured for powdered or polycrystalline samples [27, 28].

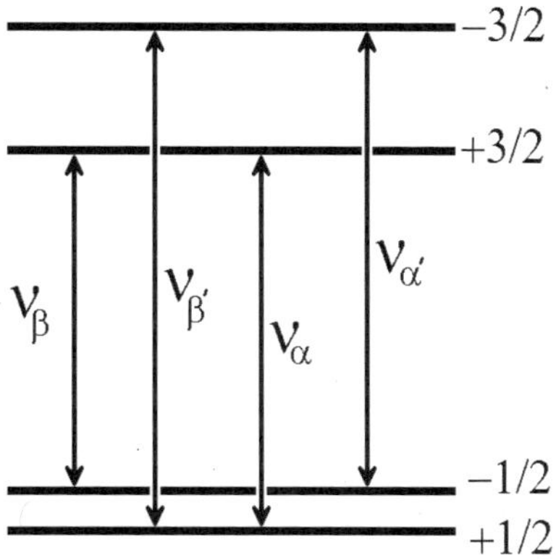

Figure 2.4 Energy levels and transitions for spin 3/2 in a magnetic field aligned along the *z*-principle axis. The energy level splitting due to the magnetic field is greatly exaggerated.

2.2.5 Line Widths and Relaxation Times

In many cases the spectral line width in an NQR measurement is due to an inhomogeneous environment. That is, there is a distribution of frequencies due to small variations in the electric field gradients within the sample. The common causes for the inhomogeneity include impurities, crystal lattice defects, and even small thermal gradients within a sample. Inhomogeneous effects can be reduced through the use of carefully prepared samples and/or can be handled with pulsed methods and the spin echo techniques.

Two other important contributors to the spectral line width are dynamic effects due to motion at the atomic scale and nuclear magnetic dipole-dipole interactions These are generally referred to as "homogeneous effects" since they apply equally to all the nuclei.

As is the case for NMR, there are several different relaxation times that can be measured. The so-called *spin-spin relaxation time,* T_2 , which may or may not actually involve spin-spin interactions, characterizes the return of the ensemble to thermal equilibrium at some "spin temperature." The *spin-lattice relaxation time,* T_1, which is the inverse of the relaxation rate, characterizes the return of the ensemble of nuclei to thermal equilibrium with the surrounding crystal, usually called the "lattice." The processes which determine T_1 involve exchange of energy between the nuclei and their surroundings—that is nuclear transitions are induced by the (random) fluctuations of the surroundings. In contrast, the interactions that are usually most effective for T_2 relaxation are those where there is no (net) gain or loss of energy from the nuclear ensemble. In general several mechanisms will contribute to the relaxation.

Typically only one of several possible transitions is excited and detected for NQR and so the phrase "the return to thermal equilibrium" is usually applied in the effective spin-½ sense. That is, the thermal ensemble includes just the two nuclear energy levels used, with the remaining levels now belonging to the lattice. This can give rise to relaxation characterized by multiple exponentials. A nice derivation of the effect for Nb ($I = 9/2$) is given by Chen and Slichter [29].

2.2.5.1 Spin-Lattice Relaxation and Temperature-Dependent Frequency Shifts

The coupling between the nuclei and the surrounding crystal will be due to time-dependent magnetic and/or electric quadrupole interactions. Magnetic interactions include those due to paramagnetic impurities and, in metals, the conduction electrons. A much weaker magnetic interaction can occur via a time-dependent spin-spin interaction. For materials with two readily available isotopes (e.g., Cl, Cu, Ga, Br, Rb, Sb, etc.), whether the relaxation is dominated by magnetic or electric quadrupole interactions can usually be determined by comparing the ratio of the relaxation times to the ratios of the magnetic dipole and electric quadrupole moments, respectively, for the two isotopes.

The contribution to the relaxation by the conduction electrons in metals is due to the so-called *Korringa result*, just as in NMR [30]. Hence, this contribution is very sensitive to changes in the electron density of states at the Fermi level, such as what one expects near superconducting phase transitions [31]. Korringa relaxation is also evident in some semiconductors [32]. Paramagnetic impurities generally contribute a constant to the rate, as they do for NMR.

In many NQR measurements of non-metals, the spin-lattice relaxation time and the temperature coefficient of the NQR frequency are both a result of lattice dynamics. This can be understood using the simple model proposed by Bayer [33]. Consider an electric quadrupole Hamiltonian that has been rotated about the y-axis by a small angle β. The new Hamiltonian (in the old principal axis coordinates) is given by

$$\mathcal{H}_Q = \frac{\nu_Q}{6}\left[\left(\frac{3}{2}\cos^2\beta - \frac{1}{2} + \frac{\eta}{2}\sin^2\beta\right)(3I_z^2 - I^2)\right.$$
$$+(3-\eta)\sin 2\beta\,(I_z(I_+ + I_-) + (I_+ + I_-)I_z)$$
$$\left.+\frac{1}{4}\left(3\sin^2\beta + \eta\,(1+\cos^2\beta)\right)(I_+^2 + I_-^2)\right] \tag{2.39}$$

and if β represents a small, rapid oscillation about an equilibrium at $\beta = 0$, then one can expand and take a time average to determine the average coupling. Keeping the lowest non-zero terms one gets an average of

$$\mathcal{H}_Q = \frac{\nu_Q}{6}\left[\left(1-\langle\beta^2\rangle\frac{3+\eta}{2}\right)\left(3I_z^2 - I^2\right) + \frac{1}{2}\left(\eta + \langle\beta^2\rangle\frac{3+\eta}{2}\right)\left(I_+^2 + I_-^2\right)\right]$$

and one can expect that $\langle\beta^2\rangle$ increases with temperature. The time-dependent terms, with RMS amplitude $\langle\beta^2\rangle^{1/2}$, can give rise to nuclear transitions and hence contribute to increase the relaxation rate $1/T_1$.

The theory of the dependence of the NQR frequency on temperature has been further developed by a number of authors [34–36]. While there are exceptions, near room temperature one can expect a temperature coefficient of order 1 kHz/K for most materials. Since NQR lines are often narrower than 1 kHz, even a small temperature change can be significant, and a small temperature gradient across the sample can significantly broaden the NQR line.

2.2.5.2 Spin-Spin Interactions

Nuclear magnetic dipole interactions between nuclei in a solid can produce some broadening of an NQR line, as happens for NMR. In an NMR experiment, the broadening due to the spin-spin interaction can be substantially reduced using a *magic angle spinning* (MAS) measurement. Spinning the sample is not effective for NQR and can make matters worse.

One way to estimate the size of the spin-spin interaction is to consider the size of the magnetic field due to a neighboring nucleus, typically of order 1 Gauss in solids, and then to treat the problem as in Section 2.2.4. It is easy to predict, quite correctly, that the effects for integer spin and half-integer spin will be very different. Except in a few isolated cases, the indirect dipole-dipole coupling (*e.g.*, J-coupling), one of the cornerstones of analysis for high resolution NMR spectroscopy, is not important for pure NQR work.

The method of moments developed by van Vleck for NMR [37], adapted to the NQR case, is often used to estimate the broadening due to the spin-spin interaction. As is the case for NMR, there are separate calculations for unlike spins and like spins. For NQR there is an additional case for spins that are otherwise alike, but for which the directions of the principal axes are different. This latter case is sometimes referred to as "semi-like." In the semi-like case the nuclear energy levels are degenerate, as is the case for like spins, however the geometry is more complicated.

One common way to compare the different situations is to compute the second moment for the somewhat artificial case of a cubic lattice of one type of nucleus. Such a comparison for several different cases is shown in Table 2.1. The characteristic decay time will be proportional to the inverse of the square root of the second moment.

Table 2.2 Results of some second moment calculations for NQR for the case of like nuclei in a cubic lattice of edge *d*. Results are in units of $\gamma^4\hbar^4 / d^6$.

Spin	Condition	Second Moment	Ref.
1	$\eta = 0$	28.2	40
1	$\eta \neq 0$	22.1	40
3/2	$\eta = 0$	60.0	38
5/2	$\eta = 0$	108.1	39

For spin 1 with $\eta \neq 0$ the first-order dipole coupling to unlike nuclei is zero and second-order calculations are necessary. For ^{14}N NQR when there are nearby hydrogen nuclei, the second-order coupling to the hydrogen nuclei can actually be larger than the coupling between (like) ^{14}N nuclei. A detailed derivation for this case is given by Vega [40].

In many cases, the addition of a small magnetic field can change "semi-like nuclei" into "unlike nuclei," resulting in a significant increase in the spin-spin relaxation time.

2.3 INSTRUMENTATION

The basic physics for NQR signal detection is the same as for NMR signal detection. Hence, NQR spectrometers are similar to NMR spectrometers in design [41]. Of course NQR does not require a large external magnetic field and associated field control circuitry. High speed sample spinning, often used for NMR, is also not appropriate for NQR and will not be present. Since magnetic field homogeneity and spinning are not at issue, sample size is limited only by the available RF power and convenience.

Early NQR and NMR instruments were largely based on relatively simple oscillator designs including the super-regenerative oscillator [42], the marginal oscillator [43], and the self-limiting "Robinson" oscillator circuit [44]. All of these are generally referred to as *continuous wave* (CW) techniques. For all of these oscillators the frequency is determined by an *LC* resonant circuit with the sample placed within (or near) the inductor *L*. The frequency is scanned by changing the capacitance *C*, either mechanically or electrically. Modern versions of these simple and inexpensive designs are still useful, particularly when searching for a signal over a broad range of frequencies and/or where small size is important [45]. Increasingly, however, computer-controlled pulsed spectrometers, similar to those now used for NMR, are employed for NQR. The sample is also within (or near) an inductor which is part of a tuned circuit, though the frequency is now determined by a separate reference oscillator.

For well-built NQR spectrometers, the principal source of electrical noise is the thermal noise from the *LC* tuned circuit. For best signal-to-noise ratios, care should be taken to minimize the resistive losses (i.e., maximize the quality factor) for that circuit. The signal-to-noise expected from an NQR measurement can be roughly estimated using expressions for an NMR measurement using the same nucleus at the same frequency [46–48].

In addition to the more common designs, several alternative techniques for NQR detection have been recently proposed, a few of which are also discussed below in Section 2.3.4.

2.3.1 CW Spectrometers

The use of a *marginal oscillator* for nuclear magnetic resonance originated with Pound and Knight [49, 50]. A number of transistorized versions of that circuit have appeared [51–54]. The marginal oscillator, as its name implies, uses just enough feedback to sustain low-level oscillations. That is, just enough energy is supplied by an active device (e.g., a transistor) as is lost and the behavior of the active device is still relatively linear. When additional energy is absorbed by the nuclei the level of oscillation can change significantly. The Robinson circuit, also now developed as a transistorized version for NQR [55], uses an additional bit of circuitry to maintain the level of feedback at a fixed value. The Robinson design is particularly useful for scans over a large frequency range.

The *super-regenerative spectrometer* is essentially a super-regenerative radio receiver but designed to detect the induced EMF from the nuclei rather than distant radio stations. An excellent explanation of how a super-regenerative receiver functions is presented by Insam [56]. In the super-regenerative circuit, the feedback condition is alternated between two states, one which maintains oscillations and one which does not. When switched from the non-oscillating or "quenched" state to the oscillating state, the time for the oscillations to build up to a predetermined level will depend on the initial signal. That is, if one assumes an exponential growth in voltage starting at a value $V(0)$ and with a time constant τ, the time t to grow to a level $V_0 > V(0)$ is given by $t = \tau \ln(V_0 / V(0))$.

Hence, in the presence of an induced EMF plus noise the oscillations will build up to the predetermined level sooner than with noise alone. The quenching signal may be generated by separate circuitry or one can design circuits which self quench. Due to the nonlinearity of the detection process (e.g., the logarithm above), one should not expect to obtain accurate line shapes with this type of spectrometer.

For the continuous wave (CW) techniques, and particularly for powder samples, additional sensitivity is often achieved using a set of external magnetic field coils which are switched on and off, combined with phase sensitive ("lock-in") detection. Typically 10–100 G fields are used at 10–100 Hz. When the magnetic field is on (with any polarity), the NQR signal is broadened sufficiently so that it is unobservable. Effectively, the magnetic field alternately turns the NQR signal on and off and only the change in the signal are recorded. Thus, all baseline errors are removed. Note that a similar technique is used for CW-NMR (usually with a sinusoidal magnetic field) resulting in a derivative signal. For NMR the magnetic field shifts the signal in frequency a bit rather than destroying it.

As an alternative to an on/off magnetic field, the frequency of an NQR oscillator circuit can be modulated electronically using a varicap diode or similar device as part of the *LC* tuned circuit. With phase sensitive detection, one obtains a derivative signal (in the limit of small modulation) though often with significant baseline problems.

2.3.2 Pulsed Spectrometers

The electronics of pulsed NQR spectrometers is virtually identical to that of broad-band NMR spectrometers except without a large magnet. In fact, many pulsed NQR spectrometers are also used (with a magnet) as broad-band NMR spectrometers and vice versa. Since the pulsed method is much more versatile than the CW techniques, the vast majority of modern NQR measurements are made using pulse methods. Many of the pulse techniques began as NMR techniques and have been adapted to the NQR environment. Since significant inhomogeneous broadening is common in NQR (See section 2.2.5), one of the most important techniques is the use of spin-echoes, and related multiple-pulse techniques, for the study of these broadened lines.

A basic computer-controlled single channel pulsed NQR spectrometer is shown schematically in Figure 2.5. As is the case for NMR, the signals are often recorded in *quadrature*. That is, signals, which are in phase (cosine-like) and 90° out of phase (sine-like) with a stable reference source, are

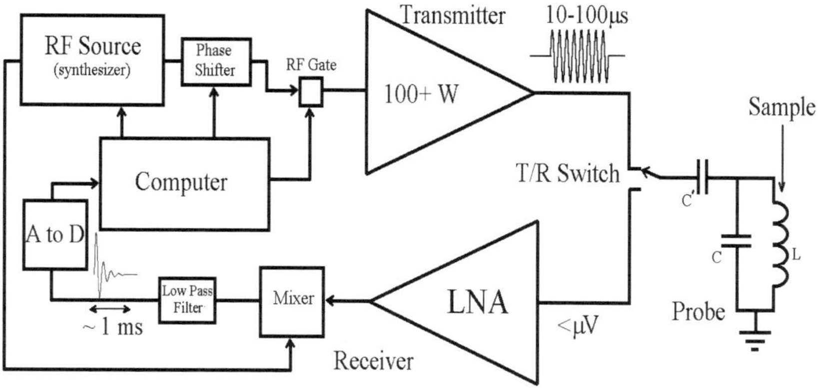

Figure 2.5 Schematic of a simple computer-controlled pulsed NQR spectrometer.

simultaneously recorded in two channels. Due to their treatment under Fourier transform, and some carryover from CW-NMR terminology, these two signals are also sometimes referred to as the "real" and "imaginary" signals and/or "absorption" and "dispersion" signals, respectively. After heterodyning and filtering, the recorded signals have a frequency which is the difference between that of the RF reference and that of the nuclear magnetization. In many modern spectrometers, the analog to digital conversion is done before heterodyning, with the mixing and filtering performed digitally.

The transmit/receive (T/R) switch shown is usually implemented using passive circuitry. One common circuit is based on the scheme developed by Low and Tarr [57], which uses semiconductor diodes and quarter wavelength transmission lines. At frequencies below about 10 MHz, common in NQR, the quarter wavelength transmission lines are replaced with lumped circuit equivalents. Other tuned circuits using diodes can also be employed [58].

While the low noise amplifiers (LNAs) are protected from damage by such a passive T/R circuit, the receiver will be overdriven and there will be some "dead time" following the pulse while the receiving circuitry recovers. At lower frequencies this is exacerbated by the ringing of the *LC* tuned circuit containing the sample. When this ringing is a bad enough, additional circuitry can be added to damp the oscillations, such as a "Q-switch," and/or one can switch the phase of the applied RF pulse by 180° for a short time just before the RF is turned off. The latter approach can be quite demanding on the high power amplifier.

The simplest pulsed experiment is the application of a single pulse with a duration τ_p set to maximize the signal (1 to 100 μs typically). This is referred to as a 90° or $\pi/2$ pulse in analogy to the NMR case, though the simple classical picture, that this corresponds to a rotation of the nucleus by 90°, is not valid. For powders, slightly longer pulses (approximately 30% longer) are used, compared to oriented single crystals, since many of the crystallites have a less than optimal orientation [59, 60]. The time-dependent signal after the pulse is referred to as the *free induction decay* (FID). If the spectral line shape is of interest, or in the rare case that there is more than one spectral line within the excitation band width (~10 kHz) then the signal will be Fourier transformed.

Simple spin-echoes are often very useful. The simplest is a two pulse measurement sometimes referred to as the *Hahn echo* with a 90° pulse, a time delay τ, a 180° pulse (twice the duration of a 90° pulse), followed by acquisition. The echo signal appears a time τ after the second pulse. All time-independent inhomogeneous interactions will be "refocused" by such a pulse sequence. Figure 2.6 illustrates FID (one pulse) and simple echo (two pulse) signals using one of the Br NQR transitions of $ZnBr_2$.

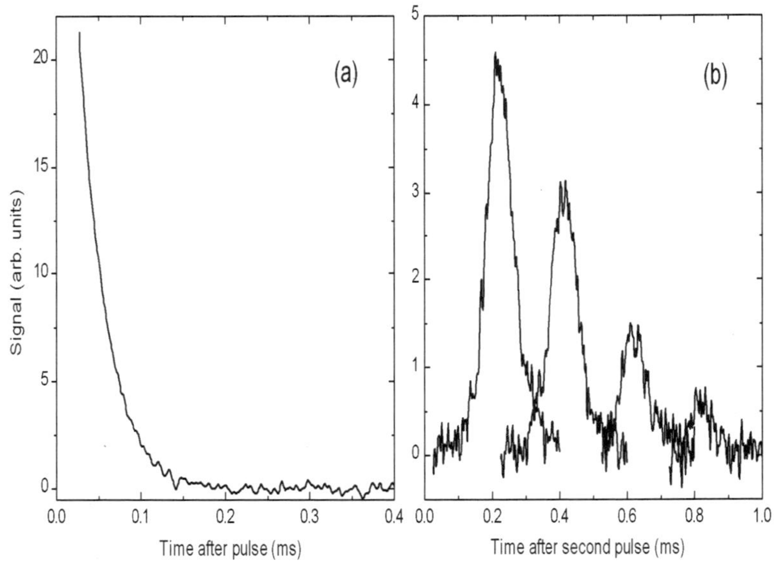

Figure 2.6 Room temperature pulse NQR signals from the 79.75 MHz ^{81}Br transition in a powder sample of $ZnBr_2$ showing (a) a free induction decay (FID) after a single pulse and (b) echoes obtained using a two-pulse sequence for four different delay times (τ = 0.2, 0.4, 0.6, and 0.8 ms).

As is the case for NMR, phase shifts are often applied to the RF pulses and they are often labeled as they are in NMR. That is, a "$\pi/2_x$" pulse and a "$\pi/2_y$" pulse are 90° out of phase with each other. Phase cycling during signal averaging, in order to remove the effects of some spectrometer imperfections, is also common [61].

In the case of weaker signals, spin-echoes may be reformed many times using *steady state free precession* (SSFP) [62, 63] or *spin-lock spin-echo*

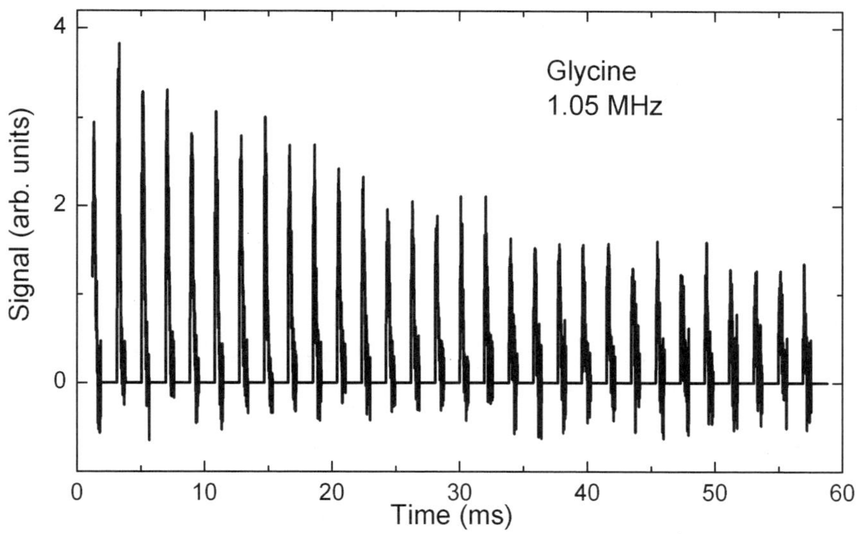

Figure 2.7 Multiple ^{14}N spin-echo signals from a spin-lock spin-echo (SLSE) pulse train. Data courtesy of J.B. Miller.

(SLSE) *pulse trains* [64, 65]. An example of such a measurement is shown in Figure 2.7.

Some more advanced work on pulse techniques includes the simultaneous excitation of two (or more) NQR transitions on the same nucleus [66–68] and various techniques to improve on the inherently inhomogeneous effective B_1 field for powder samples [69–71].

2.3.3 Field Cycling NQR Spectrometers

Field cycling spectrometers are often used to improve the sensitivity, particularly for low frequency NQR measurements and also in cases of low natural abundance. There are several different types of field cycling measurements referred to as NQR measurements – the most common are pulsed double resonance techniques where the actual measurement is an NMR measurement.

In a field cycling spectrometer the sample is alternatively subjected to a large magnetic field and a small (or zero) magnetic field. Since large magnetic fields are difficult to turn on and off rapidly, this is commonly achieved by physically moving the sample. Using pneumatics, this can be routinely accomplished over a distance of about 1 m in about 100 ms or less.

In its simplest form, the sample is placed in a very high magnetic field to obtain a large nuclear splitting and hence a large population difference. After equilibrium is reached, the magnetic field is reduced and RF is applied at the NQR frequency, the magnetic field is reintroduced, and an NMR measurement is performed. This method can be applied to half-integer spins directly [72] or for any spin using double resonance [73]. For the latter, ^{1}H is most often used as the second nucleus since it is easily observed using NMR.

For double resonance one relies on "contact" between the two nuclei some time during the measurement. Contact here refers to the case where the nuclear energy level splittings for the two nuclei match and the nuclei are physically close enough so that they can interact (e.g., via the nuclear magnetic dipole interaction). That match can occur either in the presence or absence of RF irradiation(s). When there is a match, there is efficient transfer of energy between the two types of nuclei in the same way there is transfer of energy between two weakly coupled, identical pendulums.

In an alternative form useful for compounds that contain hydrogen, the large polarizations that can be achieved for ^{1}H in a high magnetic field can be transferred, in part, to the nucleus to be measured and then a traditional NQR measurement is made. While not as sensitive, this technique has the advantage that a highly uniform magnetic field is not required for the NMR measurement. For example, initial exposure to the nonuniform magnetic field of permanent magnets can be used to obtain a large, though non-uniform, initial ^{1}H polarization, which can then be transferred to ^{14}N prior to an NQR measurement [74].

2.3.4 Some Less Common NQR Detection Schemes

The direct NQR techniques mentioned above all rely on Faraday's law of induction. Since the signal is generated by the time rate of change of the magnetic flux, these techniques lose sensitivity at low frequencies. In contrast, in a growing number of cases it is possible to detect the magnetic flux directly, rather than its time derivative. At present, these alternative detection schemes are difficult to implement on a routine basis.

A *superconducting quantum interference device*, or SQUID, is sensitive to magnetic flux and can have a very low noise level. For use as a pure NQR detector, the time-dependent nuclear magnetization is detected after a perturbation. Unlike the NMR case, however, there is no static nuclear magnetization. NQR signals have been detected using a SQUID at frequencies as low as few tens of kHz up to about 1 MHz [75–78].

A second flux detection technique recently proposed for NQR utilizes an optical transition in an alkali metal vapor in a way that is very sensitive to magnetic fields [79]. It is, in essence, a form of optically detected electron paramagnetic resonance (EPR) in a very weak static magnetic field. For NQR detection the time-dependent resonant magnetic field is supplied by the precession of the nearby nuclear magnetic moments to be detected.

A disadvantage for many pulsed NQR measurements at lower frequencies is the "dead time" after an applied pulse. Eles and Michal [80] have recently developed a method for spin 3/2 using a strong excitation at half the resonant frequency yielding a two-photon excitation, followed by traditional detection. The advantage is that the excitation frequency is well-separated from the receiver frequency, thus allowing the dead time to be virtually eliminated.

Another interesting NQR technique, which does not rely on magnetic flux at all was recently reported [81]. A close relative of *perturbed angular correlation* (PAC) techniques, the β-decay is measured from radioactive ^{8}Li nuclei after they are implanted in a material using a polarized beam. The β particles are then emitted in a direction determined by the nuclear polarization. A resonant sinusoidal magnetic field alters the nuclear polarization and thus affects the β count rate. While the number of isotopes available for such studies may be limited, it is noted that only about 10^7 nuclei are required for this technique, far fewer than are required for other NQR techniques.

2.4 INTERPRETATION OF COUPLING CONSTANTS

The contribution of the electric quadrupole field (eq in e^2qQ) can be computed for a known charge distribution surrounding a nucleus. From that the coupling constant can be estimated and compared to experiment. With modern computational techniques for materials, the largest uncertainty is often that due to the uncertainty in nuclear electric quadrupole moment, Q. For many measurements, however, of most importance are the relative changes from one situation to another, and not the absolute values. For example, one might be concerned with the differences in bonding between two similar compounds, one may be studying structural changes in the lattice during a phase transition, or one might be concerned with general correlations between NQR and other properties [82].

There are three broad situations encountered for electric quadrupole field calculations. The easiest to handle are molecular crystals where the charge distribution near the nucleus is predominantly determined by covalent bonds. For those compounds, relatively straightforward molecular computations can be made and one can expect only a small "solid effect" due to the stacking of molecules within the crystal structure. The two other cases are for ionic and metallic materials, respectively.

The electric quadrupole field at the origin due to a point charge of magnitude 1 (in cgs units) at a position $(x_1, x_2, x_3) = (x, y, z)$ relative to a nucleus at the origin is found by taking the second derivatives of the usual Coulomb potential. That computation gives

$$V_{ij} = \frac{1}{r^3}\left(\frac{3x_i x_j}{r^2} - \delta_{ij}\right),$$

where r is the distance to the origin. Since superposition applies, the field from a charge distribution is computed as the sum of the contributions from all the charges involved. Due to the $1/r^3$ dependence and charge neutrality within a unit cell, the most important contributions are those due to charges near the nucleus.

2.4.1 Molecular Crystals and Covalently Bonded Groups

Electron wavefunctions, and hence quadrupole coupling constants, can be easily computed using *ab initio* computation techniques. To understand and describe trends, however, it is often convenient to describe the electron wavefunctions in covalently bonded systems using a *linear combination of atomic orbitals* (LCAO) approach. Since filled electron (atomic) shells have a spherical charge distribution, only the outer electrons need be considered. There is also a response of the inner core electrons to the presence of an electric field gradient, which is quite important but which is ignored for the moment.

Consider an occupied electronic p_z-orbit which is (quite generally) described by a wavefunction ψ, as

$$\psi = \left(\frac{1}{2}\sqrt{\frac{3}{\pi}} \cos\theta \right) f(r) \tag{2.40}$$

where $f(r)$ is the (separately normalized) radial part of the wavefunction and, of course, $\cos\theta = z/r$. Then, for example,

$$\begin{aligned} V_{zz} &= -e \int_{r=0}^{\infty} \int_{\varphi=0}^{2\pi} \int_{\theta=0}^{\pi} \psi^* \frac{3\cos^2\theta - 1}{r^3} \psi \, r^2 \sin\theta \, d\theta \, d\varphi \, dr \\ &= -\frac{3e}{2} \int_{r=0}^{\infty} \frac{1}{r^3} r^2 |f(r)|^2 \int_{\theta=0}^{\pi} \cos^2\theta \left(3\cos^2\theta - 1\right) \sin\theta \, d\theta \, dr \\ &= -\frac{4e}{5} \int_{r=0}^{\infty} \frac{1}{r^3} r^2 |f(r)|^2 \, dr = -\frac{4e}{5} \left\langle \frac{1}{r^3} \right\rangle \equiv e q_{\mathrm{p}} \end{aligned} \tag{2.41}$$

and similarly for other terms and other orbitals. Note that $\left\langle 1/r^3 \right\rangle \approx 1/a_0^3$, where a_0 is the Bohr radius, so $V_{zz} \approx 10^{18}$ V/cm/cm . Values of $\left\langle 1/r^3 \right\rangle$ from nonrelativistic [83] and relativistic [84] atomic calculations are available.

For a more general combination of atomic orbitals the following should be noted [85, 86]. First, s-orbitals are spherical and do not contribute. Furthermore, cross-terms involving s-orbitals and p-orbits, and p-orbitals and d-orbitals are zero due to symmetry within the integral. Cross-terms between s- and d-orbitals are usually neglected. For molecular wavefunctions using the

LCAO-MO picture, the contributions from orbitals on other atoms will be small.

Hence, with a_x^2, a_y^2, and a_z^2 the respective weights for p_x, p_y, and p_z contributions to the wavefunction from orbitals on the atom in question, one gets

$$\begin{aligned} V_{zz} &= e q_p \left[3a_z^2 - \left(a_x^2 + a_y^2 + a_z^2 \right) \right] / 2 \\ V_{xx} &= e q_p \left[3a_x^2 - \left(a_x^2 + a_y^2 + a_z^2 \right) \right] / 2 \\ V_{yy} &= e q_p \left[3a_y^2 - \left(a_x^2 + a_y^2 + a_z^2 \right) \right] / 2 \end{aligned} \tag{2.42}$$

and note that $a_x^2 + a_y^2 + a_z^2 \leq 1$ though the entire wavefunction must be normalized. If there is a small s or d contribution to the wavefunction, the largest impact on the electric field gradient is likely via the reduction in the weights for the p-orbitals. The total electric field gradient is computed using the sum of the contributions from all the occupied valence wavefunctions.

2.4.2 Ionic Crystals

In principle, the computation of the electric field gradient in an ionic material is straightforward. The ions are replaced by point charges and the appropriate lattice sums are performed [87]. To achieve accurate values and reasonably fast convergence, it is necessary to take some care in the way the sum is computed.

To understand why care must be exercised, consider the following inaccurate method. First the contribution from all the negative charges is added out to a large distance R. Then, the contributions from the positive charges are added, again out to the distance R. The two sums are then subtracted from one another. Such a technique fails since the quadrupole field falls as $1/r^3$ but the number of neighboring ions in a spherical shell a distance r away grows as r^2. Hence, the total contribution falls as $1/r$, and the individual sums tend to diverge. The net result is the relatively small difference between two very large values. Furthermore, the use of an arbitrary cut-off R does not guarantee that the total calculation is charge neutral, which can lead to a significant systematic error.

To obtain accurate values and rapid convergence, terms in these lattice sums should be grouped appropriately. A simple method which has rapid convergence is to use a sum over conventional unit cells where the nucleus in question has been centered. In addition, ions of charge q which are on the boundary of n conventional unit cells are included in all of the unit cells, but with a charge q/n for each [88]. The use of a conventional unit cell ensures that charge neutrality and the symmetry of the crystal are included at every step. The field produced from each neighboring unit cell will be that of an electric dipole, or more often a higher order multipole, which will fall off much faster with distance than that of a point charge.

For numerical computations terms of comparable magnitude should be grouped together. Symbolically, the appropriate lattice sum would then be

$$V_{ij} = \sum_{\text{all } R} \left\{ \sum_{\substack{\text{unit cells at} \\ \text{distance R}}} \left[\sum_{\substack{\text{ions in conven-} \\ \text{tional unit cell}}} \frac{q}{r^3} \left(\frac{3x_i x_j}{r^2} - \delta_{ij} \right) \right] \right\} \tag{2.43}$$

where r is the total distance from the origin to the ion and q is the appropriate, possibly fractional, charge for that ion. Alternative approaches include the explicit use of multipole expansions and/or the Fourier transformation of the lattice.

Field gradients computed using this type of point charge model for the ions will certainly need to be corrected as discussed below in Section 2.4.4.

2.4.3 Metals

A very simple model for a metal is the "uniform background lattice," where the conduction electrons are considered to be uniformly distributed and the remaining positive ions are treated as point charges [89]. To achieve convergence for the electric quadrupole field, charges near the origin need to be avoided, however. Metals with narrow conduction bands can often be treated using the tight binding model, which is treated as in Section 2.4.1. To go much beyond these simple models the problem becomes very complicated very quickly and will depend on the specific metal being considered. The interested reader is referred to the review articles by Kaufmann and Vianden [90] and by Das and Schmidt [91].

For simple metals the temperature dependence of the quadrupole coupling often varies as the 3/2 power of the (absolute) temperature. That is

$$\nu_Q(T) = \nu_Q(0)\left(1 - \alpha T^{3/2}\right), \tag{2.44}$$

where α is a constant. This temperature dependence is associated with the changes in the electronic structure in the presence of thermal vibrations [92].

2.4.4 Sternheimer Shielding/Antishielding

The slight rearrangement of the core electrons in the presence of an electric field gradient has so far been neglected. A series of works by Sternheimer [93] has shown, however, that the effects on the observed quadrupole coupling constant can be far from negligible. There are two Sternheimer shielding factors which are usually considered, one associated with charges which are on the atom (or ion) in question, R, and one associated with more distant charges, γ_∞. These factors are often negative and are then referred to as "antishielding" factors. If eq_{atomic} is the magnitude of the computed electric field gradient associated with electrons on the atom (e.g., the valence electrons) and eq_{ext} is that due to charges on other atoms, then the observed value eq_{obs} is predicted to be

$$eq_{obs} = eq_{atomic} \;(1-R) \;+\; eq_{ext}(1-\gamma_\infty) \;. \qquad (2.45)$$

More rigorously the atomic and external terms should be combined as tensor, rather than scalar, quantities. Table 2.3 shows a selection of typical shielding values for γ_∞.

Table 2.3 Typical Sternheimer factors for several ions.

Atom/Ion	γ_∞
Na	–3.7
Na^+	–3.7
Al^{2+}	–1.8
Cl^-	–60
Br^-	–110
Rb^+	–50
I^-	–160

Agreement between γ_∞ values computed using different methods is no better than ±10% and in many cases much worse. It is clear, however, that the correction can be quite large. Computed values for R can be so dependent on the specific electronic configuration used that one is prone to wonder about the utility of having such values. In the majority of cases, R is of order 0.1.

Of course, a rigorous *ab initio* computation of the electronic structure of a material (including the core electrons, band structure effects, etc.) will not require these additional corrections. Computer codes available for electronic structure calculations are often optimized for computations of energies, rather than electron densities, and hence any quadrupole coupling constants produced from such programs should be used with caution.

2.5 SUMMARY

NQR is a radio frequency spectroscopy akin to wide line NMR but without a large magnet. NQR uses nuclei with spin I > ½ to probe the environment in a material. The NQR frequency will be determined by the (time-averaged) distribution of electric charge in the vicinity of a nucleus, and that distribution depends on the material being investigated. The quadrupole coupling constant, the relaxation times after an excitation, and the effects of small magnetic fields, along with modeling and comparison to other similar materials, are used to extract information about the material.

Bibliography

For additional general information, interested readers should consult one or more of the following:

Nuclear quadrupole resonance spectroscopy, T.P. Das and E.L. Hahn, Supplement 1 of *Solid State Physics* (Academic Press, NY, 1958).

Nuclear quadrupole resonance spectroscopy, E. Schempp and P.J. Bray, Chapter 11 of *Physical Chemistry, An Advanced Treatise, Vol IV, Molecular Properties*, D. Henderson (ed.), pages 521–632 (Academic Press 1970).

Nuclear quadrupole resonance spectroscopy, J.A.S. Smith, *J. Chem. Education*, Vol 48, Nos. 1–4, pages 39–49, A77–A89, A147–A148, A159–170, and A243–A252, (1971).

Nuclear quadrupole resonance in inorganic chemistry, Yu. A. Buslaev, E.A. Kravcenko, and L. Kolditz, published as Volume 82 or Coordination Chemistry Reviews (Elsevier, Amsterdam, 1987).

Nuclear Quadrupole Coupling Constants, Lucken E.A.C (Academic Press, London and New York, 1969).

Principles of Nuclear Magnetism, A. Abragam (Oxford Univ. Press, Oxford, 1961).

Principles of Magnetic Resonance, 3rd Ed., C.P. Slichter, (Springer-Verlag, Heidelberg, 1989).

Experimental Pulse NMR – A Nuts and Bolts Approach, E. Fukushima and S.B.W. Roeder (Addison-Wesley Publishing, Reading, MA, 1981).

REFERENCES

[1] Pound, R.V. (1950) *Phys. Rev.* **79**, 685-702.

[2] Dehmelt, H.G. & Kruger, H. (1950) *Naturwiss.* **37**, 111-112.

[3] Das, T.P. & Hahn, E.L. (1958) *Nuclear Quadrupole Resonance Spectroscopy, Supplement 1 of Solid State Physics,* Seitz F. & Turnbull D. (eds) (New York, Academic).

[4] Ohte, A., Iwaoka, H., Mitsui, K., Sakurai, H. & Inaba, A. (1979) *Metrologia***15**, 195-199.

[5] Huebner, M., Leib, J. & Eska, G. (1999) *J. Low Temp. Phys.* **114**, 203-230.

[6] Garroway, A.N., Buess, M.L., Miller, J.B., Suits, B.H., Hibbs, A.D., Barrall, G.A., Matthews, R. & Burnett, L.J. (2001) *IEEE Trans. Geosci. Remote Sens.* **39**, 1108-1118.

[7] Segel, S.L. (1978) *J. Chem. Phys.* **69**, 2434-2438.

[8] Thyssen, J., Schwerdtfeger, P., Bender, M., Nazarewicz, W. & Semmes, P.B. (2001) *Phys. Rev. A* **63**, 022505:1-11.

[9] Gerginov, V., Derevianko, A. & Tanner, C.E. (2003) *Phys. Rev. Lett.* **91**, 072501:1-4.

[10] Slichter, C.P. (1990) *Principles of Magnetic Resonance, 3rd Ed.,* (Heidelberg, Springer-Verlag)

[11] Bain, A.D. & Khasawneh, M. (2004) *Concepts Magn. Reson.* **22A**, 69-78.

[12] Butler, L.G. & Brown, T.L. (1981) *J. Magn. Reson.* **42**, 120-131.

[13] Hiyama, Y., Butler, L.G. & Brown, T.L. (1985) *J. Magn. Reson.* **65**, 472-480

[14] Cohen, M.H. (1954) *Phys. Rev.* **96**, 1278-1284.

[15] Bersohn, R. (1952) *J. Chem. Phys.* **20**, 1505-1509.

[16] Wang, T-C. (1955) *Phys. Rev.* **99**, 566-577.

[17] Suits, B.H. & Slichter, C.P. (1984) *Phys. Rev.* B **29**, 41-51.
[18] Sundfors, R.K., Bolef, D.I. & Fedders, P.A. (1983) *Hyperfine Interactions* **14**, 271-313.
[19] Lee, Y.K. (2002) *Concepts Mag. Reson.* **14**, 155-171.
[20] Vega, S. (1978) *J. Chem. Phys.* **68**, 5518-5527.
[21] Sauer, K.L., Suits, B.H., Garroway, A.N. & Miller, J.B. (2001) *Chem. Phys. Lett.* **342**, 362-368; (2003) *J. Chem. Phys.* **118**, 5071-5081.
[22] Raich, J.C. & Good Jr. R.H. (1963) Am. *J. Physics* **31**, 356-362.
[23] Bloom, M., Hahn, E.L. & Herzog B. (1955) *Phys. Rev.* **97,** 1699-1709.
[24] Muha, G.M. (1980) *J Chem Phys* **73**, 4139-4140; (1982) *J. Magn. Reson.* **49** 431-443.
[25] Brooker, H.R. & Creel, R.B. (1974) *J. Chem. Phys.* **61**, 3658-3664.
[26] Bloom, M. (1954) *Phys. Rev.* **94,** 1396-1397.
[27] Creel, R.B., von Meerwall, E.D. & Brooker, H.R. (1975) *J. Magn. Reson.* **20**, 328-333.
[28] Sunitha, Bai N., Reddy, N. & Ramachandran, R. (1993) *J. Magn. Reson. A* **102**, 137-143.
[29] Chen, M.C. & Slichter, C.P. (1983) *Phys. Rev. B* **27**, 278-292.
[30] Korringa, J. (1950) *Physica* **16**, 601-610.
[31] For Example see Martindale, J.A., Barrett, S.E., Durand, D.J., O'Hara, K.E., Slichter, C.P., Lee, W.C. & Ginsberg, D.M. (1994) *Phys. Rev.* B **50**, 13645-13652.
[32] Matsamura, M., Saskawa, T., Takabatake, T., Tsuji, S., Tou, H. & Sera, M. (2003) *J. Phys. Soc. Japan* **72**, 1030-1033.
[33] Bayer, H. (1951) *Z. Physik* **130**, 227-238.
[34] Kushida, T., Benedek, G.B. & Bloembergen, N. (1956) *Phys. Rev.* **104**, 1364-1377.
[35] Schempp, E. & Silva, P.R.P. (1973) *Phys. Rev. B* **7** 2983-2986; (1973) *J. Chem. Phys.* **58**, 5116-5119.
[36] Alexander, S. & Tzalmona, A. (1965) *Phys. Rev.* **138**, A845-A855.
[37] van Vleck, J.H. (1948) *Phys. Rev.* **74**, 1168-1183.
[38] Abgragam, A. & Kambe, K. (1953) *Phys. Rev.* **91**, 894-897.
[39] Nagel, O.A., Ramia, M.E. & Martin, C.A. (1996) *Appl. Magn. Reson.* **11**, 557-566.
[40] Vega, S. (1973) *Advances in Magn. Reson.* **6**, 259-302.
[41] Suits, B.H. (1994) in Trigg G. L. (ed.) *Encyclopedia of Applied Physics, Vol. 9*, 71-93 (VCH Publishers).
[42] Roberts, A. (1947) *Rev. Sci. Instrum.* **18**, 845-848.
[43] Pound, R.V. & Knight, W.D. (1950) *Rev. Sci. Instrum.* **21**, 219-225.
[44] Robinson, F.N.H. (1959) *J. Sci. Instrum.* **36,** 481-487.
[45] Kim, S.S., Mysoor, N.R., Carnes, S.R., Ulmer, C.T. & Halbach, K. (1997) In *Proceedings of the 16th Digital Avionics Systems Conference (DASC),* Irvine, CA, (IEEE, New Jersey), pp. 2.2-14-2.2-23.
[46] Hill, H.D.W. & Richards, R.E. (1968) *J. Phys. E: Sci. Inst. Series* 2, **1**, 977-983.
[47] Hoult, D.I. & Richards, R.E. (1976) *J. Magn. Reson.* **24**, 71-85.
[48] Hoult, D.I. (1973) *Prog. NMR Spec.* **12**, 41-77.
[49] Pound, R.V. & Knight, W.D. (1950) *Rev. Sci. Instrum.* **21**, 219-225.
[50] Pound, R.V. (1952) *Prog. Nucl. Phys.* **2**, 21-50.
[51] Viswanathan, T.L., Viswanathan, T.R. & Sane, K.V. (1968) *Rev. Sci. Instrum.* **39**, 472-475; (1970) *Rev. Sci. Instrum.* **41**, 477-478
[52] Zikumaru, Y. (1990) *Z. Naturforsch.* **45a**, 591-594.
[53] Sullivan, N. (1971) *Rev. Sci. Instrum.* **42**, 462-465.
[54] Offen, R.J. & Thomson, N.R. (1969) *Phys. Educ.* **4**, 264-267.
[55] Robinson, FNH. (1982) *J. Phys. E: Sci. Instrum.* **15**, 814-823.
[56] Insam, E. (2002) *Electronics World* **108**(1792) 46-53.
[57] Lowe, I.J. & Tarr, C.E. (1968) *J. Phys. E: Sci. Instrum.* **1**, 320-322.
[58] McLachlan, L.A. (1980) *J. Magn. Reson.* **39**, 11-15.
[59] Bloom, M., Hahn, E.L. & Herzog, B. (1955) *Phys. Rev.* **97**, 1699-1709.
[60] Vega, S. (1974) *J. Chem. Phys.* **61**, 1093-1100.

[61] Stejskal, E.O. & Schaefer, J. (1974) *J. Magn. Reson.* **13**, 249-251.
[62] Bradford, R., Clay, C. & Strick, E. (1951) *Phys. Rev.* **84**, 157-158.
[63] Rudakov, T.N., Mikhaltsevitch, V.T., Flexman, J.H., Hayes, P.A., & Chisholm, W.P. (2004) *Appl. Magn. Reson.* **25**, 467-474.
[64] Marino, R.A. & Klainer, S.M. (1977) *J. Chem. Phys.* **67**, 3388-3389.
[65] Cantor, R.S. & Waugh, J.S. (1980) *J. Chem. Phys.* **73**, 1054-1063.
[66] Grechishkin, V.S., Anferov, V.P. & Ja., N. (1983) *Adv. Nucl. Quadrupole Reson.***5** 1; Greschishkin, V.S. (1990) *Z. Naturforsch.* **45a**, 559-564.
[67] Sauer, K.L., Suits, B.H., Garroway, A.N. & Miller, J.B. (2001) *Chem. Phys. Lett.* **342**, 362–368; (2003) *J. Chem. Phys.* **118**, 5071-5081.
[68] Furman, G.B. & Goren, S.D. (2002) *Z. Naturforsch.* **57a** 315-319.
[69] Miller, J.B., Suits, B.H., Garroway, A.N. (2001) *J. Magn. Reson.* **151,** 228-234.
[70] Miller, J.B. & Garroway, A.N. *Appl. Magn. Reson.* **25**, 475-483.
[71] Ramamoorthy, A. & Narasimhan, P.T. (1990) *Z. Naturforsch* **45a** 581-586.
[72] Ivanov, D. & Redfield, A.G. (2004) *J. Magn. Reson.* **166**, 19-27.
[73] Slusher, R.E. & Hahn, E.L. (1968) *Phys. Rev.* **166**, 332-347.
[74] Luznik, J., Pirnat, J. & Trontelj, Z. (2002) *Sol. State Commun.* **121**, 653-656.
[75] Hürlimann, M.D., Pennington, C.H., Fan, N.Q., Clarke, J., Pines, A. & Hahn, E.L. (1992) *Phys. Rev. Lett.* **69**, 684-687.
[76] TohThat, D.M. & Clarke, J. (1996) *Rev. Sci. Instrum.* **67**, 2890-2893.
[77] Augustine, M.P., TohThat, D.M. & Clarke, J. (1998) *Solid State Nucl. Magn. Reson.* **11,** 139-156.
[78] Greenberg, Ya.S. (1998) *Rev. Mod. Phys.* **70**, 175-222; 2000 *Rev. Mod. Phys.* **72,** 329.
[79] Savukov, I.M., Seltzer, S.J., Romalis, R.V. & Sauer, K.L. (2005) *Phys. Rev. Lett.* **95**, 063004:1-4.
[80] Eles, P.T. & Michal, C.A. (2005) *J. Magn. Reson.* **175**, 201-209.
[81] Salman, Z., Reynard, E.P., MacFarlane, W.A., Chow K. H., Chakhalian, J., Kreitzman, S.R., Daviel, S., Levy, C.D.P., Poutissou, R. & Kiefl, R.F. (2004) *Phys. Rev. B* **70**, 104404:1-7.
[82] Weiss, A. & Wigand, S. (1990) *Z. Naturforsch* **45a**, 195-212.
[83] Fischer, C.F. (1977) *The Hartree-Fock Method for Atoms* (Wiley & Sons, NY).
[84] Lindgren, I. & Rosén, A. (1974) *Case Studies Atomic Phys.* **4**, 197-298.
[85] Townes, C.H. & Dailey, B.P. (1949) *J. Chem. Phys.* **17**, 782-796.
[86] Schempp, E. & Bray, P.J. (1970) Vol. IV, Chap. 11 of Henderson D (ed.) *Physical Chemistry: An Advanced Treatise*, Academic Press, New York, London.
[87] Hutchings, M.T. in Seitz, F. & Turnbull, D. (eds.) (1964) *Solid State Physics* **16** *(Academic Press, New York and London)*, 227-273.
[88] Frank, F.C. (1950) *Philos. Mag.* **41**, 1287-1289.
[89] de Wette, F.W. & Schacher, G.E. (1965) *Phys. Rev.* **137**, A92-A94.
[90] Kaufmann, E.N. & Vianden, R.J. (1979) *Rev. Mod. Phys.* **51**, 161-214.
[91] Das, T.P. & Schmidt, P.C. (1986) *Z. Naturforsch* **41a** 47-77.
[92] Jena, P. (1976) *Phys. Rev. Lett.* **36**, 418-421.
[93] Sternheimer, R.M. (1951) *Phys. Rev.* **84**, 244-253; 1951 *Phys. Rev.* **86**, 316-324; 1954 *Phys. Rev.* **95**, 736-750; 1986 *Z. Naturforsch* **41a**, 24-36.

CHAPTER 3

ELECTRON PARAMAGNETIC RESONANCE SPECTROSCOPY

[1]Sergei A. Dikanov and [2]Antony R. Crofts
[1]Department of Veterinary Clinical Medicine and [2]Department of Biochemistry, University of Illinois at Urbana–Champaign, Urbana IL 61801, USA

3.1 INTRODUCTION

This chapter is devoted to magnetic resonance spectroscopy for the investigation of unpaired electron spins. Two terms are used in the literature: electron paramagnetic resonance (EPR) and electron spin resonance (ESR). We will use the first term in this chapter. During the sixty years since its discovery in 1944 by E.K. Zavoisky [1], EPR spectroscopy has been exploited as a very sensitive and informative technique for the investigation of different kinds of paramagnetic species in solid or liquid states.

The first two parts of the chapter introduce the theoretical background and instrumentation used in two complementary approaches to EPR spectroscopy, i.e., continuous-wave and pulsed EPR. The final part describes typical areas of solid state applications.

This chapter is based on an extensive literature devoted to EPR spectroscopy. Space limitations preclude references to all original publications and the reader is therefore referred to earlier reviews for a more complete survey of earlier work. Excellent textbooks in EPR spectroscopy cover basic materials [2–5]. Many monographs are devoted to the consideration of more specific topics including theory, instrumentation, or application to selected paramagnetic species [6–19]. Handbooks of EPR spectroscopy [20, 21] are also available, as are periodic reviews on theory, experimental approaches, and different EPR applications covering the period since 1971 [22].

A comprehensive description of pulsed EPR spectroscopy is given in the recently published book of Schweiger and Jeschke [23]. Previous literature on pulsed EPR includes two monographs [24, 25], several edited books [18, 19, 26, 27], and a large number of reviews. An up-to-date list is provided in [23].

3.2 THEORETICAL BACKGROUND

3.2.1 EPR Condition

The simplest treatment of the EPR experiment is in terms of a paramagnetic center with an electron spin of S = 1/2, and this in practice covers a wide range of experiments. Representative examples of such paramagnetic species include ions of transition metals, free radicals, trapped electrons and atoms, and F-centers in alkali halides.

The magnetic moment $\boldsymbol{\mu}_e$ of the electron is given by the expression

$$\boldsymbol{\mu}_e = -g_e\beta_e\mathbf{S} \tag{3.1}$$

where β_e is the Bohr magneton equal to $e\hbar/2mc = 0.92 \cdot 10^{-23}$ J·T (e and m are charge and mass of the electron, respectively) and g_e is a dimensionless constant called the *electron g-factor,* which is equal to 2.0023 for the free electron. (In this and other equations, **bold** terms are vectorial.)

The energy of interaction of the electron magnetic moment with the magnetic field **B** is

$$E = -\boldsymbol{\mu}_e \cdot \mathbf{B} = g_e\beta_e\mathbf{S}\cdot\mathbf{B} \tag{3.2}$$

If we assume that the magnetic field is directed along the z-axis of the coordinate system, this becomes

$$E = g_e\beta_e B S_z. \tag{3.3}$$

For an electron spin $\boldsymbol{S}$ = 1/2 there are two energy states:

$$E_{\pm} = \tfrac{1}{2} g_e\beta_e B \tag{3.4}$$

variously labeled ↑ (up), +½ or α, or ↓ (down), –½, or β, corresponding to parallel and antiparallel orientations of $\boldsymbol{S}_z$ to the magnetic field (Figure 3.1).

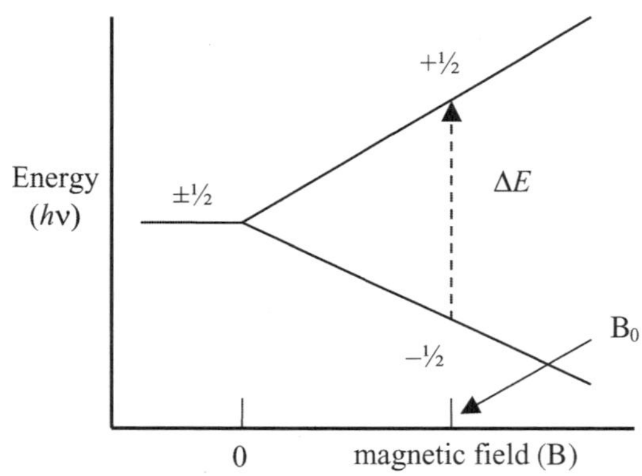

Figure 3.1. Two-level system in varying magnetic field. Resonance condition.

Application of a magnetic field $\mathbf{B_1}$ perpendicular to $\mathbf{B}$, oscillating with the frequency ν_o, induces the EPR transition if the *resonance* condition

$$\Delta E = E_+ - E_- = h\nu_o = g_e\beta_e B_o \qquad (3.5)$$

is satisfied. The transition is commonly referred to as a *spin-flip*, and ΔE is the photon energy needed to flip the spin population. The frequency, ν_o, and the field B_0, are the resonance frequency and field.

3.2.2 Continuous Wave EPR

Real samples contain a large number of spins distributed in the constant magnetic field between the two allowed energy states according to thermal equilibrium. The relative population of two levels is described by the Boltzmann distribution

$$N_+/N_- = \exp(\Delta E/kT), \qquad (3.6)$$

where k is the Boltzmann constant and T is the temperature of the sample. For the free electron with $g_e = 2.0023$, the spin population ratio is ~1.0015 for a field of 0.34 Tesla (for the most commonly used X-band EPR), or about 1 in 670 excess spins in the lower energy state. The oscillating magnetic field at the resonance frequency ν_o induces transitions from up to down and from down to up with equal probability.

However, the population of lower energy is larger than that of the upper one, leading to a larger number of transitions to the upper level, and, as a consequence, to energy absorption by the sample. This effect is the basis for the simplest and most widely used experimental approach, the so-called *continuous wave* (CW) EPR experiment, in which the sample containing paramagnetic species is irradiated by microwaves with frequency fixed by the source, and the magnetic field is varied to search for the microwave absorption. The dependence of microwave absorption on the magnetic field observed in such an experiment is the *EPR spectrum.*

3.2.3 EPR Lineshape: Relaxation Times

Further consideration of a CW-EPR experiment is simplified by introduction of a macroscopic magnetization vector. In a constant magnetic field $\mathbf{B_0}$, a population of electron spins acquires a magnetic moment $\mathbf{M_0}$, which is parallel to $\mathbf{B_0}$ under thermal equilibrium conditions. The equilibrium magnetization reflects the difference in population among the various spin states, whose degeneracy has been removed by the magnetic field, and is proportional to B_0. For the spin S = 1/2 there are only two spin levels, and the equilibrium magnetization with $\mathbf{B}_o$ applied along the z direction is:

$$\mathbf{M}_z = \frac{Ng^2\beta_e^2\mathbf{B}_o}{4kT} \qquad (3.7)$$

$N = N_+ + N_-$ is the total number of spins in the sample.

Following excitation by microwave absorption, the spin-flipped population relaxes back to the thermal equilibrium state. The time to return to the equilibrium distribution of magnetization is the *relaxation time.* Relaxation time is described by a single exponential decay. Relaxation time along the direction of a static magnetic field $\mathbf{B}_o$ is called the *longitudinal relaxation time* T_1.

The longitudinal relaxation is accompanied by a change of the energy of the spin system. The thermal motion is the source and sink of energy exchange in the relaxation processes. In solids, thermal motion is usually described by *phonons*, which are quanta (photons) with energies in the range corresponding to lattice vibrations. Longitudinal relaxation is caused by absorption or stimulated emission of phonons. Since a coupling between the spin system and the lattice is required, the longitudinal relaxation is often called *spin-lattice relaxation.* The term *lattice* comes from its use in early studies performed with ionic lattices. However, the usage has been generalized and refers to degrees of freedom other then those directly concerned with spin. Phonons have a spectrum of frequencies that range over many orders of magnitudes with varying intensities. However, only those fluctuations with frequencies that match the EPR frequency are capable of inducing transitions. There are several mechanisms by which the spin-lattice interaction can take place in condensed phases; these are called *direct, Raman* and *Orbach processes.* They all involve interaction of the spin system with phonons of the lattice.

The relaxation time in the plane perpendicular to the static magnetic field $\mathbf{B}_o$ is the *transverse relaxation time* T_2. In the absence of a microwave field, the equilibrium magnetization components $\mathbf{M}_x$ and $\mathbf{M}_y$ are zero, but these components appear in the presence of the microwaves with a magnetic field of amplitude $\mathbf{B_1}$ orthogonal to the field $\mathbf{B_o}$. Unlike longitudinal relaxation, transverse relaxation occurs without the exchange of energy with the lattice. The transverse relaxation is concerned with the mutual spin flips caused by interactions within the ensemble of spins in the sample and thus is often called *spin-spin relaxation.*

A main consequence of these relaxation processes with times T_1 and T_2 is that they determine the *lineshape* of the spectrum (the unique resonance line) of the two-level system. This can be illustrated using the *Bloch equations*, which describe the evolution of the spin system with time under the influence of magnetic fields $\mathbf{B}_o$ and $\mathbf{B}_1$. The macroscopic magnetization vector $\tilde{\mathbf{M}}$ is discussed in the context of a coordinate system (the rotating frame of reference) rotating about the z-axis with an angular velocity $\omega_o = 2\pi\nu_o = \gamma_e \mathrm{B}_\mathbf{o}$ (γ_e is gyromagnetic ratio) to match the resonance frequency.

$$\frac{d\tilde{\mathbf{M}}_x}{dt} = (\omega_o - \omega)\tilde{\mathbf{M}}_y - \frac{\tilde{\mathbf{M}}_x}{T_2}$$
$$\frac{d\tilde{\mathbf{M}}_y}{dt} = (\omega_o - \omega)\tilde{\mathbf{M}}_x + \omega_1 \mathbf{M}_z - \frac{\tilde{\mathbf{M}}_y}{T_2} \tag{3.8}$$
$$\frac{d\mathbf{M}_z}{dt} = -\omega_1 \tilde{\mathbf{M}}_y - \frac{\mathbf{M}_z - \mathbf{M}_o}{T_1}.$$

$\mathbf{M}_o$ is the equilibrium magnetization, $\tilde{\mathbf{M}}_z = \mathbf{M}_z$ in this case, i.e., transverse magnetization does not depend on the frame. After the microwave field $\mathbf{B}_1$ (directed along the x-axis in the rotating coordinate system) has been on for a sufficient time, the spin precession reaches a steady state. Then the stationary solution for the constant ω_o and ω_1, when all derivatives are equal to zero, is given by:

$$\tilde{\mathbf{M}}_x = \frac{(\omega_o - \omega)\omega_1 T_2^2}{1 + (\omega_o - \omega)^2 T_2^2 + \omega_1^2 T_1 T_2} \mathbf{M}_o$$
$$\tilde{\mathbf{M}}_y = \frac{\omega_1 T_2}{1 + (\omega_o - \omega)^2 T_2^2 + \omega_1^2 T_1 T_2} \mathbf{M}_o \tag{3.9}$$
$$\mathbf{M}_z = \frac{1 + (\omega_o - \omega)^2 T_2^2}{1 + (\omega_o - \omega)^2 T_2^2 + \omega_1^2 T_1 T_2} \mathbf{M}_o.$$

The $\tilde{\mathbf{M}}_x$ component of transverse magnetization is in phase with $\mathbf{B}_1$ and is proportional to the dispersion mode of the EPR signal, whereas $\tilde{\mathbf{M}}_y$ is 90° out of phase and is proportional to power absorption. Besides the phases, the equations describe the shape of the signal in a magnetic resonance experiment under steady state (CW) condition. The characteristic lineshapes of dispersion and absorption modes of the magnetic resonance transition as a function of the frequency ω are shown in Figure 3.2.

For small values of B_1, when $\omega_1^2 T_1 T_2 \ll 1$, the absorption is characterized by a Lorentzian lineshape:

$$g(\omega) = \frac{T_2}{\pi} \frac{1}{1 + T_2^2(\omega - \omega_o)^2} \tag{3.10}$$

with a width at the half-height equal to $2/T_2$, i.e., the width is determined by the transverse relaxation time.

With an increase of B_1, the spin-lattice relaxation at a certain power level of microwave irradiation may not be fast enough to maintain the population difference needed to measure the full amplitude of the resonance signal, so the signal level declines. This phenomenon is called *saturation*.

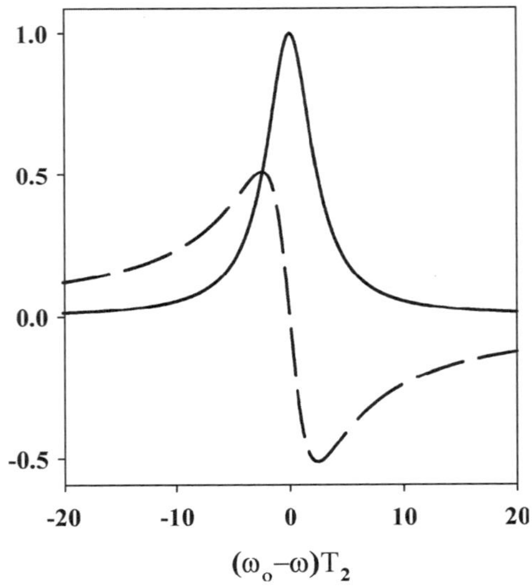

Figure 3.2 Dispersion (dashed line) and absorption (solid line) modes of the magnetic resonance transition as a function of frequency.

At resonance, where $\omega = \omega_o$, the maximal intensity of the absorption mode decreases proportionally to $\left(1+\omega_1^2 T_1 T_2\right)^{-1}$. However, the effect is smaller in the "wings" of the absorption line. As a consequence, the saturation changes the steady state resonance lineshape, suppressing the central part relative to the wings and producing an apparent broadening of the spectrum. Since the relaxation time T_1 enters into the saturation equation, EPR saturation measurements can be used in CW EPR for determination of T_1.

3.2.4 EPR Spin-Hamiltonian

3.2.4.1 Interaction of Electron Spin with the Magnetic Field: The *g*-Tensor

In different atoms and molecules the unpaired electron will occupy orbitals of different configuration and symmetry. Consequently, the interaction of electronic magnetic moment with an applied magnetic field will have a more complex character and will depend on the orientation in a molecular or atomic coordinate system (x,y,z). Let us consider this situation in greater detail.

The magnetic moment of an electron in an atom results from the orbital angular moment **L** and spin **S**. In addition, the spin and orbital moments are coupled through a so-called *spin-orbital interaction*. Thus, the complete

Hamiltonian describing the interaction of the atomic magnetic moment with the magnetic field has the form

$$H_{MZ} = \beta_e \mathbf{B}_o(\mathbf{L} + g_e\mathbf{S}) + \lambda\mathbf{LS} \tag{3.11}$$

where λ is the spin-orbit coupling constant. The value of λ increases with atomic number and, for instance, it is equal 28, 76, and 151 cm^{-1} for ^{12}C, ^{14}N, and ^{16}O, respectively [5].

The orbital angular momentum undergoes an apparent loss on going from an atomic to a molecular system and this loss is called the *quenching* of orbital angular momentum by a ligand field. Formally, the orbital angular momentum for a non-degenerate electronic ground state is equal zero (L = 0) in a molecule. However, the deviations from g_e observed experimentally imply terms with non-zero values in the Hamiltonian in addition to those with **S**, and these are produced by the spin-orbit interaction. Considered in the second order, this interaction admixes the excited states of the orbital angular momentum into the ground state. For quantitative characterization of this contribution one can use a matrix $\mathbf{\Lambda}$ of second rank (3 × 3) with elements

$$\Lambda_{ij} = \sum_{n\neq 0} \frac{\langle \psi_o | L_i | \psi_n \rangle \langle \psi_n | L_j | \psi_o \rangle}{E_o - E_n} \tag{3.12}$$

with *i, j=x,y,z* and E_o and E_n are the energies of the ground and excited orbital states. Using this matrix one can introduce the ***g*** tensor

$$\boldsymbol{g} = g_e\mathbf{U} + 2\lambda\mathbf{\Lambda} \tag{3.13}$$

which expresses the effect of spin-orbit coupling on the "true" spin (**U** is the 3 × 3 unit matrix).

With this definition, the interaction between the electron spin and the external magnetic field $\mathbf{B}_o$ in the arbitrary coordinate system is described by the electron Zeeman term

$$H_{SZ} = \beta\, \mathbf{B_o} \cdot \boldsymbol{g} \cdot \mathbf{S} \tag{3.14}$$

Usually the ***g***-tensor is considered a symmetrical tensor, i.e. $g_{ij} = g_{ji}$, with only six independent components instead of nine. In an arbitrary coordinate system the ***g***-tensor is nondiagonal. However, it is always possible to find a coordinate system for the principal axes, where the tensor is diagonal. For spin S = 1/2, the electron Zeeman interaction usually exceeds other magnetic interactions studied in EPR experiments. Therefore, the ***g*** principal axes frame is often used as a reference frame for the theoretical description and analysis of the experimental data. Three diagonal elements g_x, g_y and g_z of the ***g*** matrix are the principal elements of the ***g***-tensor. They are characteristic values for the any paramagnetic species in the solid state. They reflect the symmetry of the electron's local environment: thus, for cubic symmetry $g_x = g_y = g_z$; for axial symmetry along the *z*-axis, $g_x = g_y = g_\perp$, and $g_z = g_\parallel$; and for orthorhombic symmetry, $g_x \neq g_y \neq g_z$. For systems with low symmetry the ***g***-tensor may even become asymmetric.

3.2.4.2 Hyperfine Interaction

Interaction of an unpaired electron with the magnetic field is just one of the ways in which magnetic interaction can manifest itself in an EPR spectrum. This simplest interaction usually gives well-characterized spectra in the form of a single EPR line in liquids and up to three lines in "powder" solids corresponding to the principal directions of the ***g***-tensor. However, there are other types of magnetic interactions, usually of lower energy, which produce additional features in the EPR spectra called *hyperfine* and *fine structure.* These arise from magnetic interactions with nearby magnetic nuclei and magnetic interactions with other electron spins.

The electron-nuclear hyperfine interaction (hfi) is described by the Hamiltonian

$$H_{SI} = \mathbf{S} \cdot \mathbf{A} \cdot \mathbf{I} \tag{3.15}$$

where **I** is a nuclear spin, and **A** is a tensor of hyperfine interaction. It consists of an isotropic term or Fermi contact interaction and an anisotropic term resulting from the electron-nuclear dipolar coupling.

The Fermi contact interaction is given by

$$H_{iso} = a\mathbf{S} \cdot \mathbf{I} \tag{3.16}$$

where $a = (8/3)\pi g_e\beta_e\, g_I\beta_I\, |\psi_o(0)|^2$.

Contact interaction results from the unpaired electron spin density $|\psi_o(0)|^2$ transferred onto the nucleus. The hyperfine interaction is purely isotropic if the unpaired electron occupies an *s* orbital. A typical example is the hydrogen atom where $|\psi_o(0)|^2 = 1$ and $a = 1420$ MHz [5]. In many cases the isotropic constant *a* for a particular atom is calculated from the fraction of the *s* orbital included in the molecular wave function of the unpaired electron. The isotropic hyperfine coupling can also be significant when the unpaired electron resides in a higher *p*, *d*, or *f* orbital, and spin density in the *s* orbital is induced by configurational interactions or through a spin polarization mechanism.

The anisotropic part of the hyperfine Hamiltonian is produced by the interaction of an electron and nucleus acting as magnetic dipoles. The electron-nuclear, dipole-dipole coupling is described by the relation:

$$\hat{H} = -g\beta\, g_I\, \beta_I \left[\frac{\mathbf{S}\cdot\mathbf{I}}{r^3} - \frac{3(\mathbf{S}\cdot\mathbf{r})\ (\mathbf{I}\cdot\mathbf{r})}{r^5}\right] \tag{3.17}$$

where **r** is the vector connecting the nucleus with the different points of the electron orbital. Spatial integration over the orbital coordinates of the electron in an arbitrary coordinate system, with the components of the vector $\mathbf{r}(x,y,z)$, determines the Hamiltonian of the anisotropic hyperfine interaction in the form

$$H_{aniso} = \mathbf{S} \cdot \mathbf{T} \cdot \mathbf{I} \tag{3.18}$$

with **T** tensor components

$$T_{ij} = -g\beta g_I \beta_I \left\langle \frac{r^2 \delta_{ij} - 3r_i r_j}{r^5} \right\rangle \tag{3.19}$$

and $r_i, r_j = x, y, z$. This form of the hyperfine tensor shows that it depends on the distance between electron and nucleus and can provide information about their relative spatial location.

An important characteristic of the anisotropic hyperfine tensor is the coordinate system of the principal axes in which this tensor is diagonal. Then the three diagonal elements are called the principal values of the tensor. The determination of the hyperfine tensor in the principal coordinate system is an important venue of the magnetic resonance experiments. Because the interaction is between dipoles, the strength of the interaction shows a strong angular dependence, and hence can provide information about relative orientation of the electron and nucleus.

3.2.4.3. Nuclear Quadrupole Interaction

Nuclei with $I \geq 1$ possess an electric quadrupole moment that results from a nonspherical charge distribution. The interaction of the quadrupole moment with the gradient of the electric field due to the surrounding electrons is effectively described by the Hamiltonian of the nuclear quadrupole interaction (nqi),

$$H_Q = \mathbf{I} \cdot \mathbf{Q} \cdot \mathbf{I} \tag{3.20}$$

where **Q** is the traceless tensor of the nqi. This interaction produces additional splittings of the energy levels in the electron-nuclear spin system. In the coordinate system of the principal axes of the nqi tensor the Hamiltonian has the form

$$H_Q = Q_x I_x^2 + Q_y I_y^2 + Q_z I_z^2 = K\left[3I_z^2 - I^2 + \eta(I_x^2 - I_y^2)\right] \tag{3.21}$$

where $K = \dfrac{e^2 qQ}{4I(2I-1)h}$, eq is the electric field gradient, eQ is the nuclear quadrupole moment, and $\eta = (Q_x - Q_y)/Q_z$ is an asymmetry parameter with $0 \leq \eta \leq 1$. Because the trace of the tensor **Q** is equal to zero, its principal values are described by the two parameters K and η only. Values of quadrupole coupling constant e^2qQ/h and η are those usually used in the literature for the quantitative characterization of the quadrupole interaction of the particular nuclei in different chemical compounds.

3.2.4.4 Nuclear Zeeman Interaction

Many nuclei possess a spin I and an associated magnetic moment. The spin quantum number can be integral or half-integral between 1/2 and 6.

Characteristics of some magnetic nuclei important for EPR spectroscopy are provided in Table 3.1. The interaction of the nuclear spin with magnetic field **B** is described by the nuclear Zeeman term

$$H_{NZ} = -g_n\beta_n \mathbf{BI} = h\nu_I (\mathbf{nI}) \tag{3.22}$$

In this formula g_n is the nuclear g factor and $\beta_n = e\hbar/2Mc$ is the nuclear magneton. The term $\nu_I = -g_n\beta_n B/h$ is the nuclear Zeeman frequency, which determines the characteristic energy associated with spin-inversion of particular nuclei, as shown in Table 3.1. In contrast to the Bohr magneton, β_e, the nuclear magneton depends on the proton mass M, which is 1836 times larger then the mass of an electron. This means that the nuclear Zeeman interaction is about 10^3 times weaker than the electron Zeeman interaction in the same magnetic field. In most EPR experiments the nuclear Zeeman interaction is isotropic. Deviation from the isotropy takes place for paramagnetic centers with large hyperfine couplings or low-lying excited states.

Table 3.1. Magnetic Characteristics of Some Nuclei

Nucleus	Natural Abundance (%)	Spin	$\lvert\nu_I\rvert$, (MHz), (B_o= 0.35 T)
^1H	99.985	½	14.902
^2H	0.0148	1	2.288
^{13}C	1.11	½	3.748
^{14}N	99.63	1	1.077
^{15}N	0.366	½	1.511
^{17}O	0.038	5/2	2.021
^{19}F	100.00	½	14.027
^{31}P	100.00	½	6.038
^{57}Fe	2.15	½	0.4818

Among the nuclei shown in Table 3.1, only ^1H, ^{14}N, ^{31}P, and ^{19}F are present at a natural abundance close to 100%. Experiments with other nuclei require preparation of isotopically labeled molecules or media (for instance, by isolation of protein from bacteria grown in media with the isotope replacing the natural atom, or substitution of ^2H_2O for ^1H_2O).

3.2.5 Electron-Nuclear Interactions: Hyperfine Structure

When an $\mathbf{S} = 1/2$, $\mathbf{I} = 1/2$ spin system with an isotropic $\boldsymbol{g}$-factor is placed in a constant magnetic field $\mathbf{B_o}||z$ the complete spin-Hamiltonian (in frequency units) is:

$$H/h = \nu_o S_z + \nu_I I_z + S_z T_{zz} I_z + S_z T_{zx} I_x + S_z T_{zy} I_y \tag{3.23}$$

In deriving equation (3.23), it is generally assumed that the axis along which the quantum state of the electron spin is determined is that of $\mathbf{B_o}$, because its field greatly exceeds the local magnetic fields produced by nuclear magnetic moments. This assumption is the "high-field approximation." With this approximation, all terms proportional to S_x and S_y can be neglected, and are omitted from the equation. The Hamiltonian (equation (3.23)) is then diagonalized in the form:

$$H/h = \nu_o \mathbf{S}_z + \nu_{\alpha(\beta)} \mathbf{I}_z \tag{3.24}$$

where the following designations are used

$$A = T_{zz},\ B = \left(T_{zx}^2 + T_{zy}^2\right)^{1/2} \tag{3.25}$$

$$\nu_{\alpha(\beta)} = \left[(Am_S + \nu_I)^2 + B^2 m_S^2\right]^{1/2}$$

Here, ν_α and ν_β are frequencies of nuclear transitions between the two energy levels corresponding to m_S +1/2 and −1/2, respectively.

A simple vector model can be used to explain this presentation (Figure 3.3). The terms $(Am_S + \nu_I)$ and Bm_S correspond respectively to the local magnetic field strength along the $\mathbf{B_o}$ direction ($B_{S\alpha(\beta)}(\|)$), and in the perpendicular plane ($B_{S\alpha(\beta)}(\perp)$) at the location of the nucleus, for two m_S states. They define the direction of the quantization axis for the nuclear spin at m_S = +1/2 (for ν_α) or −1/2 (for ν_β), described by the angles φ^+ and φ^-, respectively, with:

$$\cos\varphi^{\pm} = \frac{Am_s + \nu_I}{\nu_{\alpha(\beta)}} \qquad \sin\varphi^{\pm} = \frac{Bm_s}{\nu_{\alpha(\beta)}} \tag{3.26}$$

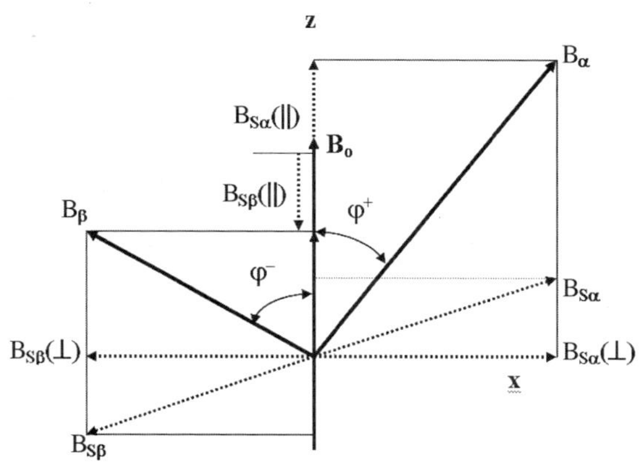

Figure 3.3 Vector addition of external and hyperfine fields for two electron spin states.

The full strength of the local magnetic field for each m_S determines two different nuclear resonant frequencies:

$$\nu_\alpha = \left[\left(\frac{A}{2}+\nu_I\right)^2 + \frac{B^2}{4}\right]^{1/2}$$
$$\nu_\beta = \left[\left(-\frac{A}{2}+\nu_I\right)^2 + \frac{B^2}{4}\right]^{1/2} \qquad (3.27)$$

The eigenstates of the Hamiltonian (equation (3.24)) are written as:

$$\frac{E_1\left(+\frac{1}{2};+\frac{1}{2}\right)}{h} = \frac{\nu_o}{2} + \frac{\nu_\alpha}{2}$$
$$\frac{E_2\left(+\frac{1}{2};-\frac{1}{2}\right)}{h} = \frac{\nu_o}{2} - \frac{\nu_\alpha}{2}$$
$$\frac{E_3\left(-\frac{1}{2};+\frac{1}{2}\right)}{h} = -\frac{\nu_o}{2} + \frac{\nu_\beta}{2} \qquad (3.28)$$
$$\frac{E_4\left(-\frac{1}{2};-\frac{1}{2}\right)}{h} = -\frac{\nu_o}{2} - \frac{\nu_\beta}{2}$$

The corresponding eigenfunctions are:

$$|1\rangle = \cos\frac{\varphi^+}{2}\left|\frac{1}{2},\frac{1}{2}\right\rangle + \sin\frac{\varphi^+}{2}\left|\frac{1}{2},-\frac{1}{2}\right\rangle$$
$$|2\rangle = -\sin\frac{\varphi^+}{2}\left|\frac{1}{2},\frac{1}{2}\right\rangle + \cos\frac{\varphi^+}{2}\left|\frac{1}{2},-\frac{1}{2}\right\rangle$$
$$|3\rangle = \cos\frac{\varphi^-}{2}\left|-\frac{1}{2},\frac{1}{2}\right\rangle + \sin\frac{\varphi^-}{2}\left|-\frac{1}{2},-\frac{1}{2}\right\rangle \qquad (3.29)$$
$$|4\rangle = -\sin\frac{\varphi^-}{2}\left|-\frac{1}{2},\frac{1}{2}\right\rangle + \cos\frac{\varphi^-}{2}\left|-\frac{1}{2},-\frac{1}{2}\right\rangle$$

The appearance of four levels, instead of two corresponding to the two different states of electron spin in the magnetic field, results from the hyperfine electron-nuclear interaction and nuclear Zeeman interaction. These interactions produce the additional splitting of each electron level. The splitting of two new levels, with $m_S = +1/2$ and equal to ν_α, differ from the splitting ν_β of the levels with $m_S = -1/2$.

Generally four EPR transitions with different frequencies can be generated between upper and lower doublets. Two of them, $1 \leftrightarrow 3$ and $2 \leftrightarrow 4$, indicated by solid lines in Figure 3.4, occur with a change of the electron spin projection (m_s changes) but without any change of the nuclear spin projection

(m_I remains the same). They are usually called the *allowed* transitions (*A*). The other two transitions, 1 ↔ 4 and 2 ↔ 3, indicated by dashed lines, involve the simultaneous change of the electron and nuclear spin projections. They are called *forbidden* transitions (*F*). The frequencies of allowed and forbidden transitions are:

$$\begin{aligned} \nu_{13}^{A} &= \nu_{o} + \frac{\nu_{\alpha} - \nu_{\beta}}{2} \\ \nu_{24}^{A} &= \nu_{o} - \frac{\nu_{\alpha} - \nu_{\beta}}{2} \\ \nu_{14}^{F} &= \nu_{o} + \frac{\nu_{\alpha} + \nu_{\beta}}{2} \\ \nu_{23}^{F} &= \nu_{o} - \frac{\nu_{\alpha} + \nu_{\beta}}{2}. \end{aligned} \tag{3.30}$$

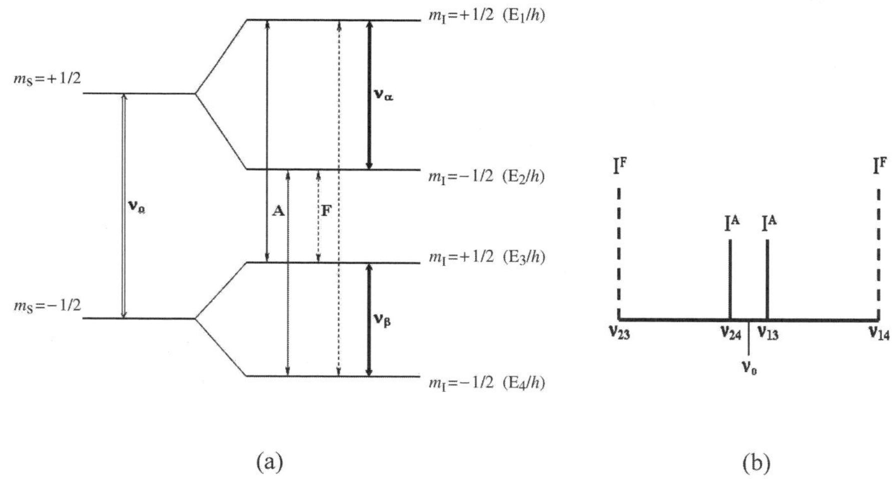

Figure 3.4 (a) Energy levels, allowed and forbidden EPR transitions, and nuclear transitions for the electron spin S = ½ interacting with a nuclear spin I = ½. (b) Schematic EPR spectrum showing allowed and forbidden transitions.

Their intensities are:

$$\begin{aligned} I^{A} &= \cos^{2}\frac{\varphi^{+} - \varphi^{-}}{2} \\ I^{F} &= \sin^{2}\frac{\varphi^{+} - \varphi^{+}}{2}. \end{aligned} \tag{3.31}$$

Taking into account equations (3.26), we can write the intensities (equations (3.31)) in the form:

$$\mathrm{I}^{A(F)} = \frac{1}{2}\left[1 \pm \frac{4\nu_{\mathrm{I}}^2 - A^2 - B^2}{4\nu_\alpha \nu_\beta}\right] \tag{3.32}$$

with I^A corresponding to the + condition. Thus, the EPR spectrum of an S = 1/2 electron spin interacting with an I = 1/2 nucleus, at an arbitrary orientation of the magnetic field in a coordinate system describing the relative location of electron and nucleus, consists of two doublets with intensities I^A and $\mathbf{I}^F$ symmetrically located relative to the center of the spectrum at ν_o. The intensities of these doublets depend on the values for A and B, and they vary when the orientation of the magnetic field is changed.

By way of illustration, we can consider several particular cases. It is obvious that when $4\nu^2_{\mathrm{I}} = A^2 + B^2$ the intensities of allowed and forbidden transitions will be equal. The intensity of the allowed transitions become larger, if $4\nu^2_{\mathrm{I}} > A^2 + B^2$, and the forbidden transitions dominate the spectrum when $4\nu^2_{\mathrm{I}} < A^2 + B^2$ (Figure 3.4). The behavior of the intensities can be understood qualitatively in terms of the variation of the local magnetic field on the nucleus. This is given by the vector sum of the external magnetic field and the effective magnetic field induced by the hyperfine interaction between electron and nucleus. Analysis of the intensities of I^A and I^F shows that, in order for intense allowed transitions to appear, it is necessary that $\varphi^+ - \varphi^- \sim 0$. If their intensity is low, then $\varphi^+ - \varphi^- \sim \pi$. The condition that $\mathrm{I}^A = \mathrm{I}^F$ occurs when $\varphi^+ - \varphi^- = \pi/2$, i.e., the local fields for two electron spin states are perpendicular to each other.

The lines produced by the interaction of the electron spin with the nucleus are called the *hyperfine structure* of the EPR spectrum. Observation of the hyperfine structure in the solid state depends on the type of sample, i.e., single crystal, powder, frozen solution, and on the value of the hyperfine coupling. Typically, the EPR spectra of a paramagnetic metal ion in the solid state will show resolved hyperfine structure from the ion's own nucleus, if this is magnetic.

Isotropic Hyperfine Interactions (hfi). In liquids with low viscosity, the molecular motion of the molecules usually leads to an averaging of all anisotropic interactions. Hyperfine interaction retains only its isotropic part when $A = a$ and $B = 0$. In this case $\mathrm{I}^A = 1$ and $\mathrm{I}^F = 0$ and only the two allowed transitions contribute to the spectrum at frequencies given by:

$$\nu_{13}^A = \nu_o + \frac{a}{2}$$
$$\nu_{24}^A = \nu_o - \frac{a}{2}. \tag{3.33}$$

The splitting between these lines $\nu_{13}^A - \nu_{24}^A$ is equal to the isotropic constant a. This basic result for the EPR spectroscopy in liquid solutions means that each nuclear spin with I = 1/2 produces a doublet of lines with a splitting equal to the isotropic hyperfine constant for this nucleus. If the electron interacts with several nuclei N with different isotropic constants a_i each of them would produce an additional splitting of the spectral lines leading to a total number of the lines in the spectrum equal to 2^N.

For a nucleus with the spin $I \geq 1$ the complete spin Hamiltonian includes an additional nqi term (equation (3.20)). In this case two levels of the electron spin doublet are split into 2I +1 additional sublevels $E_{\alpha i}$ and $E_{\beta j}$ with different projections of the nuclear spins $m_I = -I, -I + 1, \ldots, I$. Formally, an EPR experiment in this system of levels could produce $(2I + 1)^2$ transitions. Of these, $2I + 1$ are allowed with $\Delta m_I = 0$, and all others are forbidden with $|\Delta m_I| = 1, \ldots, 2I$. The intensities of these transitions depend on the orientation of the magnetic field and relative values of the different terms of the spin Hamiltonian.

3.2.6 Homogeneous and Inhomogeneous Line Broadening

The lines of the EPR spectra can be homogeneously or inhomogeneously broadened. To describe the type of broadening observed one can use the term *spin packet*. The spin packet is defined as a group of spins with the same resonance field, and its linewidth is most accurately characterized by the *phase memory* relaxation time, T_m (see Section 3.2.7.2). A homogeneous line is a superposition of spin packets with the same resonance field and the same width. This is usually the case for spectra taken in liquid media (Figure 3.5a).

An inhomogeneously broadened line consists of a superposition of spin packets with distinct, time-independent resonance fields (Figure 3.5b). The sources of the inhomogeneous broadening can include inhomogeneous external magnetic field(s), g-strain, unresolved hyperfine structure, and anisotropy of the magnetic interactions in orientation-disordered solids. This definition of lineshapes is particularly important in pulsed EPR spectroscopy because the behavior of the spin system in a pulsed experiment depends on the type of line broadening leading to the overall contour.

3.2.7 Pulsed EPR

3.2.7.1 Rotating Frame of Reference—Vector Sum of Magnetization

In the previous sections we discussed typical CW-EPR experiments in which the magnetic component $\mathbf{B}_1$ of the microwave field is constant in time, i.e., the experiment is performed under steady state conditions. We now consider a different type of EPR experiments in which the electron spins are placed in the constant magnetic field $\mathbf{B}_o$ and irradiated by microwave pulses of limited duration (time t_p), at a frequency satisfying the resonance condition for the paramagnetic species under study.

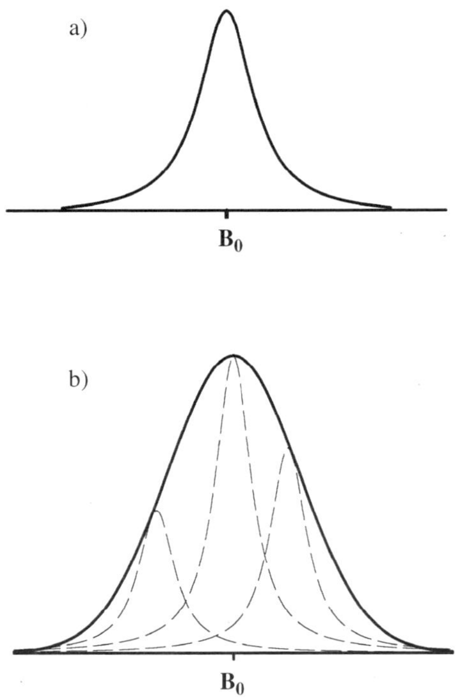

Figure 3.5 Homogeneously (a) and inhomogeneously (b) broadened lines.

To understand the influence of the pulse on the spin system, it is convenient to consider the macroscopic magnetization of the spin system, the magnetic moment resulting from the vector sum of the magnetization, referred to a coordinate system rotating around the axis of the constant magnetic field at the resonance frequency, the rotating frame discussed above. A sample with unpaired electron spins acquires a magnetic moment $\mathbf{M_o}$ parallel to $\mathbf{B_o}$ under thermal equilibrium. One can assume that in the coordinate system rotating around $\mathbf{B_o}||z$ with a resonance frequency ν_o the magnetic field $\mathbf{B_1}$ of the applied microwave pulses coincides with axis x. In the presence of the field $\mathbf{B_1}$, magnetization $\mathbf{M_o}$ rotates around the x-axis with angular velocity $\omega_1 = 2\pi\ \nu_1 = \gamma_e B_1$. For a microwave pulse of duration t_p, this rotates the vector $\mathbf{M_o}$ around an angle $\theta_p = 2\pi\nu_1 t_p$. Thus, for a pulse with duration t_p such that $\theta_p = \pi/2$ (the so-called $\pi/2$ pulse) the magnetization becomes directed along the y-axis (Figure 3.6). Application of a pulse giving $\theta_p = \pi$ (a π-pulse) inverts the magnetization vector from the z to the $-z$ direction.

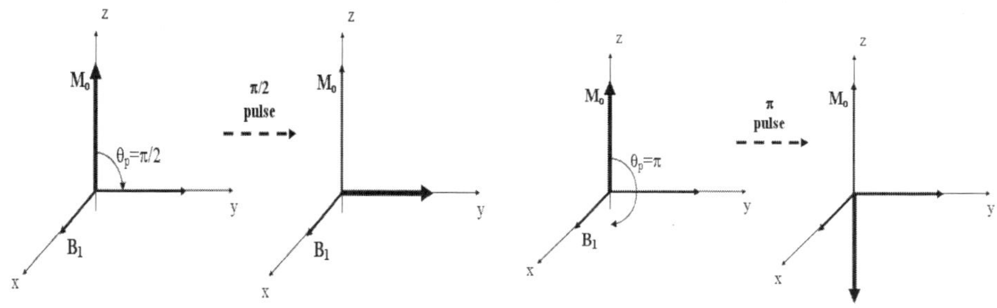

Figure 3.6 Rotation of the magnetization vector during application of a $\pi/2$-pulse (left) or a π-pulse (right).

3.2.7.2 Relaxation after a Pulse: The Free Induction Decay (FID)

The relaxation behavior of the magnetization induced by the pulse depends on the characteristics of the EPR spectrum. In solids, the EPR line is usually inhomogeneously broadened, and the distribution of intensity is described by some function $f(\nu_r)$ with a characteristic width $\Delta\nu_r$. For simplicity, one can analyze the case in which $\Delta\nu_r$ is small compared to ν_1. After the $\pi/2$-pulse (Figure 3.7a) the elementary magnetization vectors corresponding to different ν_r values will start precessing in the *xy*-plane (b) with a frequency offset from the rotational reference frequency by $(\nu_r - \nu_o)$.

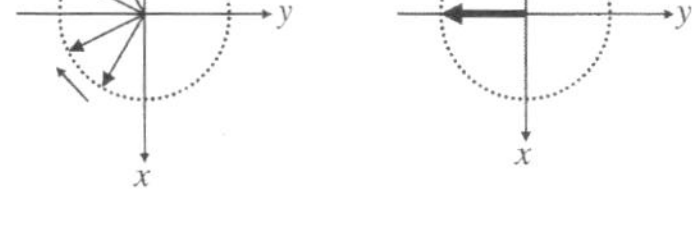

Figure 3.7 Scheme showing how the echo is generated in a 2-pulse $\pi/2$-τ-π sequence (see text for explanation).

After excitation at the resonance frequency ν_o, they defocus symmetrically from the starting point along the *y*-axis. As a result the net magnetization, always directed along *y*-axis, decreases monotonically with the characteristic time $T_2^* \sim 1/\Delta\nu_r$, the value of which can be readily estimated. For example, T_2^* will lie in the range $\sim 10^{-7}$ s for $\Delta\nu_r \sim 10$ MHz, or $\Delta B \sim 0.3$ mT. This relaxation generates microwave emission detected as a transient signal following a single microwave pulse, which is called the *free induction decay* (FID). In NMR spectroscopy, the FID following a single pulse can be readily measured because the frequencies involved are in the radio range; Fourier analysis of the FID is the basis of simple FT-NMR spectrometers. However, this simple approach is not generally practical in pulsed EPR, because typical spectrometers have electronic dead-times ~10^{-7}s after applied pulses; since the spectral features are in the same range as the response time, they cannot be quantitatively studied (discussed at greater length below). It may be noted that the behavior of the FID for broad signals, and in the case of off-resonant excitation, needs more complex consideration, but also results in an FID that always decays within the time interval ~t_p [23].

3.2.7.3 Electron Spin-Echo

Precessional evolution (*dephasing*) of the magnetization after the single π/2-pulse results in its complete defocusing over the *xy*-plane at the larger times T_2^* (Figure 3.7a,b). There is, however, a way to restore this magnetization. This is done by applying a second π-pulse after time τ, called a *refocusing* pulse. When this is applied along the same *x*-axis as the initial π/2-pulse, it rotates all elementary magnetization vectors through 180° around the *x*-axis (Figure 3.7b,c). The refocusing π-pulse does not change the direction of rotation of each elementary magnetization vector. As a consequence, as the precession continues, all these vectors meet together at the *y*-axis after the same time interval τ (Figure 3.7c,d). This refocusing produces an induced magnetization (microwave emission) called the two-pulse *electron spin-echo* (ESE).

The refocusing process produces an ESE signal with maximum amplitude at time 2τ after the initial pulse, and this is followed by a new defocusing of the magnetization vector. For this reason, the echo signal appears to consist of two FIDs, back-to-back, to generate a symmetrical pulse, which is separated in time from the initiating pulse, and therefore not masked by the instrumentation dead-time. Although in principle Fourier transformation of each FID could provide an EPR spectrum, in practice, more information at higher resolution can be obtained by the ESEEM approach detailed below.

For a pulse sequence with two pulses of arbitrary duration with rotation angles θ_{1p} and θ_{2p} the amplitude of the two-pulse or primary ESE is proportional to:

$$E \sim |\sin\theta_{1p} \;\; \sin^2(\theta_{2p}/2)| \tag{3.34}$$

The ESE amplitude is independent of the EPR lineshape for nonselective pulses with amplitudes larger than the linewidth. However, the simple considerations outlined here are not valid for broad lines, and analysis of the

FID and echo shape then requires more complex treatment. One important qualitative result following from such an analysis of broad line excitation is that only magnetization vectors in the narrow range $\pm\nu_1$ around the resonant frequency ν_o contribute to the echo formation [23].

3.2.7.4 Electron Spin Echo Envelope Modulation (ESEEM)

One of the major uses of pulsed EPR spectroscopy is to obtain information about the magnetic environment from analysis of the *echo envelope*. This generally includes important information pertaining to local structure and other properties of the solid state as discussed in Section 3.4. The echo envelope is obtained by recording the echo amplitude (or intensity) as a function of the time interval between pulses. The properties of the two-pulse spin-echo, when τ is varied, depend on the properties of the spin system and its interactions with neighboring nuclear spins. The echo intensity is influenced by two different factors (Figure 3.8).

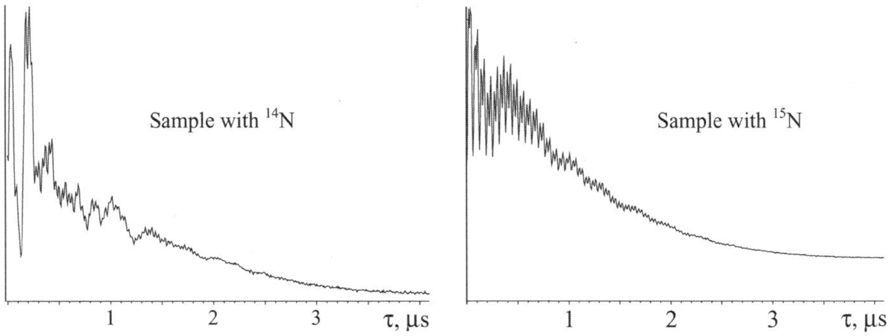

Figure 3.8 Two-pulse echo envelope showing contributions to the kinetic decay of interactions with protons, and with ^{14}N (left) and ^{15}N (right) nuclei. Experiment performed at 9.7 GHz and magnetic field of 364 mT.

The first is a magnetic relaxation processes resulting from monotonic decay of the echo signal as τ increases, mainly arising from T_2, as defined in the Bloch equations above. The second is the interaction of the unpaired electron spins with surrounding magnetic nuclei. These cause *periodic* variations of echo intensity as τ is changed. In order to quantify this effect, the intensity of the spin echo is recorded as a function of τ. The plot of this pattern is termed the *electron spin-echo envelope modulation* (*ESEEM*), and contains kinetic (time domain) information about both components contributing to the decay. After correction for monotonic decay, Fourier transformation of the echo envelope provides the ESEEM spectrum, which contains all the information about the frequencies (energies) of the nuclear interactions. Spectroscopy based on observation and analysis of echo modulation in either the time or the frequency domain is called *ESEEM spectroscopy*.

Figure 3.8 shows as an example the two-pulse ESE pattern of the reduced [2Fe-2S] Rieske cluster in the bc_1 complex, either with natural abundance of magnetic nuclei, or in a sample uniformly ^{15}N-labeled. The [2Fe-2S] cluster is coordinated by the N_δ nitrogens of two histidine ligands and by two sulfurs of cysteine ligands. The unlabeled sample shows periodical variations in two principle frequency ranges. The deep variations with lower frequencies are due to interactions with ^{14}N ($\nu_I \sim 1$ MHz, I = 1 with nuclear quadrupole moment). They disappear when the ^{14}N nuclei in the sample are replaced by ^{15}N nuclei with different magnetic properties ($\nu_I \sim 1.5$ MHz, I = 1/2). Both envelopes show also the oscillations of higher frequency ($\nu_I \sim 14.9$ MHz) produced by the protons in the environment of the cluster.

One can also see that the relaxation decay of the echo characterized by *phase memory time* T_m in both samples is the order ~1 μsec. T_m corresponds to the inverse homogeneous linewidth and is sometimes simply called T_2 in literature. Note however that there may be contributions to T_m from the other sources.

3.2.7.5 ESEEM Frequencies and Amplitudes

Simple qualitative and rigorous quantum mechanical models for ESEEM are described in detail elsewhere. In the present case, only the equations resulting from quantum mechanical analysis needed for an understanding of ESEEM frequencies and amplitudes will be discussed. In the case of S = 1/2 and I = 1/2 spins, described by the spin-Hamiltonian (3.23), the normalized expression for the two-pulse ESEEM has the form [28]

$$E(\tau) = 1-\frac{k}{2}\left[1-\cos 2\pi\nu_\alpha \tau-\cos 2\pi\nu_\beta \tau+\frac{1}{2}\cos 2\pi(\nu_\alpha+\nu_\beta)\tau+\frac{1}{2}\cos 2\pi(\nu_\alpha-\nu_\beta)\tau\right] \quad (3.35)$$

where the amplitude coefficient k is equal to

$$k = 4\mathrm{I}^A\mathrm{I}^F = \sin^2(\varphi^+ - \varphi^-) = \left(\frac{\nu_I B}{\nu_\alpha \nu_\beta}\right)^2 \quad (3.36)$$

and all other parameters are defined in Section 3.2.5.

Equation (3.35) shows that the frequencies of two-pulse ESEEM are the nuclear frequencies ν_α and ν_β at two different projections of the electron spin, together with their two combination forms $\nu_\alpha \pm \nu_\beta$. The amplitude coefficient k is proportional to the product of the probabilities of the allowed and forbidden EPR transitions in the four level system of S = 1/2 and I = 1/2 spins (Figure 3.4). This means that ESEEM appears only if the allowed and forbidden EPR transitions are simultaneously induced by the microwave pulses. The previous analysis has shown that this requires the presence of anisotropic hfi. Additional confirmation of this conclusion follows from the explicit form of the amplitude coefficient k, which contains a parameter B depending on the nondiagonal components of the anisotropic hfi tensor. Thus, ESEEM occurs only in the presence of the anisotropic hfi, and this is also a requirement for the simultaneous presence of the allowed and forbidden transitions that are usually typical for the solid state.

This result shows that ESEEM appears in relation to variation in the quantization axis of the nuclear spin accompanying a change of the electron spin orientation during the EPR transition following excitation by microwave pulses. Such a variation leads to the simultaneous reorientation of a nuclear spin. This perturbs the phase compensation in the different periods of electron spin precession and results in echo signal modulation. In pulsed experiments, ESEEM frequencies result from the EPR transitions of the spin system induced by microwave pulses and not as a result of transitions between nuclear sublevels. The cosine FT spectrum should contain lines at two nuclear frequencies of equal amplitudes, and their two combinations with the lines of opposite amplitudes and two times lesser intensity.

The result obtained when I = 1/2 can be generalized for an arbitrary nuclear spin I [29]. In this case the two-pulse ESEEM frequencies are again all possible frequencies of the nuclear transitions $\nu_{ij}^{(\alpha)}$ and $\nu_{ln}^{(\beta)}$ between the eigenstates (energy levels) with the same electron spin state (α for i, j states and β for l,n states) as well as $\nu_{ij}^{(\alpha)} \pm \nu_{ln}^{(\beta)}$ combinations. The coefficients that determine the amplitude of the harmonics depend on the probabilities of the EPR transitions between different energy levels of two manifolds reflecting the fact that ESEEM frequencies are formed as a difference of the these EPR transitions and not as a direct nuclear transition between two levels of the same manifold.

3.2.7.6 Pulse Sequences

The advantage of pulsed EPR compared to conventional CW-EPR is that different properties of the paramagnetic center, and its interactions with neighboring nuclear spins, can be explored through different pulse sequences. These can be configured so as to be suitable either for one-dimensional or for multidimensional experiments.

In addition to the two-pulse sequence $\left(\frac{\pi}{2} - \tau - \pi - \tau - echo\right)$ two other sequences are frequently used in the ESEEM experiments. One of them employs three pulses of equal length separated by the time intervals τ and T $\left(\frac{\pi}{2} - \tau - \frac{\pi}{2} - T - \frac{\pi}{2} - \tau - echo\right)$.

The echo signal appearing at the time $2\tau + T$ is formed as a result of combined action of all three pulses and is called a *three-pulse* or *stimulated-echo*. In a traditional one-dimensional (1D) three-pulse ESEEM experiment, the stimulated-echo envelope is recorded as a function of the time T at fixed time τ.

Another sequences consists of four pulses $\left(\frac{\pi}{2} - \tau - \frac{\pi}{2} - t_1 - \pi - t_2 - \frac{\pi}{2} - \tau - echo\right)$ separated by the times τ, t_1 and t_2. In the (1D) experiment with this sequence the intensity of the inverted echo after the fourth pulse is measured as a function of the times $t_1 = t_2 = T/2$, which are increased stepwise at a fixed time τ.

Differences between the three sequences introduced above are found in the relaxation time of the echo decay and in the ESEEM frequencies that they create. Three-pulse echo modulation contains the frequencies of nuclear transitions from two manifolds $\nu_{\alpha i}$ and $\nu_{\beta j}$ only but two and four pulse sequences also produce their combinations $\nu_{\alpha i} \pm \nu_{\beta j}$.

Another important factor is the time of the relaxation decay of the echo signal, which significantly affects the resolution of the ESEEM spectra. As was discussed above, the decay of the two-pulse echo is governed by the phase memory time T_m. In contrast, the relaxation decay of the three- and four-pulse echoes is controlled by the longitudinal relaxation time T_1. This is because the magnetization vector is rotated into the z-axis by the second $\pi/2$ pulse, so that relaxation of the population is determined by T_1. In solids at low temperatures the phase memory relaxation time does not usually exceed several microseconds (Figure 3.8). The relaxation time T_1 is at least one to two orders of magnitude longer and this makes possible a longer acquisition interval for echo envelope (Figure 3.9). One can also note that the selection of the time τ influences the amplitude of the different harmonics in three- and four-pulse experiments in different ways. As a consequence some frequencies can be suppressed at a particular time τ. To avoid the partial loss of the information the spectra can be recorded at several times τ (see, for example, Figure 3.20).

In addition, the amplitudes of the harmonics differ in different sequences and this allows for an enhanced contribution of relevant frequencies by selection of an appropriate experimental scheme. Therefore, these schemes do not replace but complement each other. A number of new ESEEM schemes have been introduced in recent years including dead-time and blind-spot free ESEEM methods, ESEEM with improved sensitivity, extended-time excitation experiments and others. A fairly complete description of these schemes is given in the book [23].

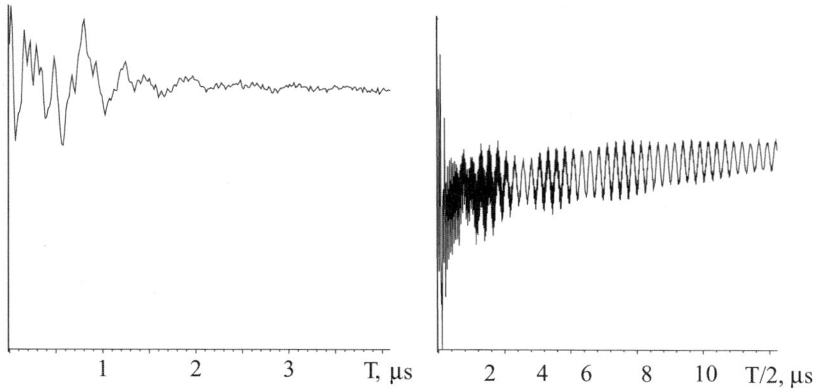

Figure 3.9 Left–Three-pulse ESE pattern of the reduced Rieske iron-sulfur protein isolated from the bc_1 complex from bacteria grown on ^{14}N (τ = 152 ns, magnetic field 365 mT). Right-Four-pulse ESE pattern from protein with ^{15}N (τ = 128 ns, magnetic field 364.7 mT). Microwave frequency was 9.7 GHz in both experiments.

3.3 EXPERIMENTAL

3.3.1 Design of CW-EPR Spectrometer

In CW-EPR, the sample is placed in a magnetic field that which can be varied in a linear field sweep while simultaneously exposing it to a fixed frequency of microwave irradiation. The type of experiment determines the design of the EPR instrument.

The standard EPR spectrometer generally consists of the following components, named according to their function: microwave source, a resonant cavity, a magnet and power supply with field control and field sweep, and the modulation and detection systems.

The source includes all components that generate the microwave waves, and transmission lines. The klystron, a vacuum tube producing microwave oscillations in a small range of frequencies, is most often used as the source of energy. Another source, used in more modern spectrometers, is a Gunn diode. The wave-guides or coaxial cables transmit microwaves to the cavity where they are concentrated at the sample. Automatic frequency control (AFC) locks the source at the cavity's resonant frequency. Interposed between the klystron and the cavity are the following components: a) an isolator, protecting the klystron from the reflected microwave energy, b) an attenuator, regulating the power input, and c) a circulator. A circulator directs the microwave power from the source to the cavity and simultaneously redirects the reflected power to the detector.

The source and detector are parts of the unit called the microwave bridge. An additional element of the microwave bridge is a reference arm, which supplies the detector with additional microwave power controlled by second attenuator. There is also a phase shifter to insure that the reference arm microwaves are in phase with the reflected microwaves when the two signals meet at the detector.

The detection system normally includes a detector, and signal preamplifier followed by phase sensitive detector (PSD), tuned at the field modulation frequency. The detector is a silicon crystal diode in contact with a tungsten wire. It converts the microwave power to an electrical current. The detector produces an inherent noise inversely proportional to the frequency of the detected signal termed $1/f$ noise. The amplifier enhances the registered signal without, however, markedly changing the signal-to-noise ratio.

A significant increase of signal-to-noise ratio is achieved by modulating the external magnetic field, generally with a 100 kHz component (Figure 3.10) in conjunction with phase-sensitive detection. The field modulator consists of small Helmholz coils placed on each side of the cavity along the direction of the field. At the resonance condition the field modulation quickly sweeps through part of the EPR signal and the reflected microwaves are amplitude-modulated at 100 kHz.

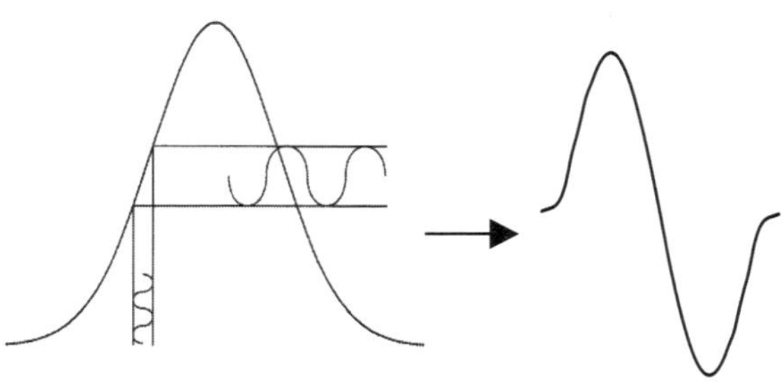

Figure 3.10 Field modulation and phase-sensitive detection.

This procedure transforms the EPR signal at the resonance field into a sine wave with an amplitude proportional to the slope of the resonance line. As a result, if the amplitude of the 100 kHz field modulation is small compared to linewidth ($\sim < 25\%$) the modulated signal after PSD appears as the first derivative of the EPR intensity vs. magnetic field strength (Figure 3.10). To improve the sensitivity further, a time constant is used to filter out more of the noise. Advantages of such an approach include reduction of $1/f$ noise, and elimination of DC drift, because the signal encoded at 100 kHz can be selectively filtered from noise at other frequencies. The processed signal is digitized, and a computer is used for recording and analyzing data and for coordinating timing functions for all components involved in acquisition of the spectrum.

3.3.2 Design of Pulsed-EPR Spectrometer

Pulsed EPR spectrometers have many of the major parts similar to those found in a CW EPR instrument including microwave source, resonant cavity, magnet and detection system. As in the CW instrument, the resonance condition is obtained by varying the field at fixed microwave frequency. The key difference in doing pulse EPR is the generation of pulse sequences with large microwave power, and detection of the weak microwave emission of the FID and echo signals. These differences determine the specific requirement for the sources, the cavities, and detectors. In a pulsed EPR instruments, a klystron or a Gunn diode generates CW microwaves in the milliwatt range at a set frequency. The source unit includes also a pulse programmer that controls generation of the high-power microwave pulse, protection of receivers, and triggering of the acquisition devices. The pulse-forming unit usually consists of several channels with individual adjustments for amplitude, phase, and length of the pulses. Several channels are usually needed for phase-cycling, which facilitates exclusion of unwanted features from the echo envelope. The low power microwave pulses are supplied to a traveling wave tube (TWT) amplifier where they are amplified to up to higher power, for example 1 kW power for $\pi/2$ pulses of ~16 ns duration.

The very low level FID and echo signals (~10^{-9} to 10^{-3} watts) emitted from the resonator are directed by the circulator for the detection. Two types of microwave amplifiers were employed in pulsed spectrometers: a low-noise GaAs field effect transistors (GaAs FET), and low noise TWT. The amplified signal is fed to the quadrature detector. The outputs from the quadrature detector provide its real and imaginary part. The quadrature detection is followed by additional amplification and filtering by a video amplifier. The bandwidth of the output signal is up to about 200 MHz. The final signal is stored directly by a fast oscilloscope or digitized, with subsequent processing by a computer. A more detailed description of CW and pulsed-ESR spectrometers and references to the original publications devoted to the EPR instrumentation can be found in several books on EPR spectroscopy [2–27].

3.3.3 Resonators

The cavity concentrates the microwaves at the sample to optimize ESR adsorption. Up until now, most CW-ESR instruments are equipped with rectangular or cylindrical cavities. Other types of the cavities include, for example, dual mode, dielectric, and loop gap resonators. Cavities are characterized by their *quality factor* Q, which indicates how efficiently the cavity stores microwave energy. As Q increases, the sensitivity of the spectrometer increases. The Q factor is defined as:

$$Q = \frac{2\pi E_{\text{stored}}}{E_{\text{dissipated}}} \tag{3.37}$$

where the dissipated energy is the amount of energy lost during one microwave period. High Q cavities are typically used in CW EPR because they are efficient at converting spin magnetization into a detectable signal.

However, a pulsed EPR experiment has different requirements of the cavity because the applied pulse causes a reflected signal called *cavity ringing,* which decays exponentially with a time constant

$$t_r = \frac{Q}{4\pi \nu_o} \tag{3.38}$$

where Q is a quality of unloaded cavity and ν_o is the resonance frequency. The power ringing after the pulse is so high that the detection system of the instrument is overloaded and fails to register a useful signal. The instrument recovers after a time (often called the *dead-time*), when the cavity ringing is reduced below the noise level. In modern spectrometers dead-time does not usually exceed a few hundred nanoseconds; the shortest reported values are less then 100 nanoseconds. An additional reason for keeping a low Q for the pulsed resonator is the need to optimize bandwidth so as not to distort broad EPR signals.

Because of the above requirements, special efforts are usually put into construction of cavities designed specifically for pulsed experiments. A variety of resonators for pulsed EPR have been described, including rectangular cavities, slotted tube resonators, strip-line cavities, bimodal cavities, several types of loop-gap resonators and dielectric resonators (see references in [23, 25, 26]).

3.3.4 EPR Bands, Multifrequency Experiments

In the basic magnetic resonance equation (3.5), two experimental characteristics ν_o and $\mathbf{B}_o$ are linearly proportional to each other and selection of the frequency uniquely defines the magnetic field necessary for resonance. The ratio between them is determined by the electronic g-factor and the Bohr magneton. Because of the fixed frequency generated by available microwave sources, EPR spectrometers are set up so that a spectrum is obtained by varying the magnetic field $\mathbf{B}_o$, at constant frequency ν_o. As the characteristic property of a paramagnetic species, the g-value of the system is a good choice because the value is independent of magnetic field strength, and therefore of the instrument. Microwave frequencies, and the corresponding fields, used in EPR spectroscopy for construction of commercial and home-built spectrometers are shown in the Table 3.2. Most commercial EPR spectrometers are operated in X-band with microwave frequency ~9.5 GHz. Resonance for species with $g \sim 2.0$ occurs at an external magnetic field ~0.35 mT. They

Table 3.2 Microwave frequency bands and magnetic fields for $g = 2$.

Band Designation	EPR Frequency (GHz)	EPR Field (mT)
L	1.5	54
S	3.0	110
X	9.5	340
K	24	900
Q	35	1250
W	95	3400
D	135	4800

represent a reasonable compromise between spectroscopic considerations such as resolution and sensitivity, and practical requirements, including equipment cost and sample handling

Spectrometers operated at lower or higher microwave frequencies are less common, but are also commercially available. Experiments at more than one microwave frequency allow the user to optimize spectral resolution and to provide additional information often not available from spectra recorded at X-band.

3.4 APPLICATIONS OF EPR SPECTROSCOPY

3.4.1 CW-EPR and Pulsed-EPR in Single Crystals

3.4.1.1 Analysis of the *g*-Tensor Anisotropy

Two major types of the solid state sample are usually studied in EPR experiments—single crystals, and orientation-disordered samples that may include frozen solutions, glasses, polymers, amorphous solids, and microcrystalline powders. The structure of single crystals consists of periodically repeated and strictly oriented units. As a result, the EPR spectrum of the paramagnetic species in the single crystal depends on the orientation of the external magnetic field relative to the crystal axes. If the magnetic field $\mathbf{B_o}$ is defined by direction cosines (l,m,n) in the reference axis system, then it can be shown that the position of the resonance is characterized by the g-factor [4] as

$$g = \left[\sum_{i=x,y,z} \left(l \cdot g_{xi} + m \cdot g_{yi} + n \cdot g_{zi} \right)^2 \right]^{1/2} \tag{3.39}$$

The usual procedure in an experiment is to orient one of the reference crystal axes perpendicular to the $\mathbf{B_o}$ and to collect the spectra at different angles obtained by rotating the crystal about this axis. For example, if $\mathbf{B_o}$ is in the xy-plane and θ is its angle with the x-axis, then the g-factor is:

$$g = \left[(g^2)_{xx} \cos^2\theta + (g^2)_{yy} \sin^2\theta + 2(g^2)_{xy} \sin\theta\cos\theta \right]^{1/2} \tag{3.40}$$

where $(g^2)_{ij}$ are elements of the $\boldsymbol{g}^2$ tensor. Thus, measurement in x, y-plane gives the three elements of this tensor. Measurements in the x,z-and$y$, z- planes give values of the other elements. The next step includes diagonalization of the $\boldsymbol{g}^2$ tensor. The square roots of its principal values give the principal values of $\boldsymbol{g}$ tensor itself and eigenvectors give the directions of the principal axes in the crystal axes system.

An important characteristic of each crystal is its symmetry. The symmetry is described by group-theoretical considerations of symmetry operations about selected point and by long-range translation order. Depending of the type of the symmetry in the crystal, several symmetry-related species may contribute to the EPR spectrum in each orientation.

A recent example of an EPR single-crystal study is the investigation of the reduced [2Fe-2S] cluster of the Rieske iron sulfur protein (ISP) of the bovine mitochondrial cytochrome (*cyt*) bc_1 complex. The inhibitor stigmatellin was present in the crystals, which binds in the ubiquinol oxidizing site (Q_o site) as an inhibitor that mimics the substrate [30]. This inhibitor forms a strong H-bond with the N_ε of His-161 of the ISP, which is also a ligand to the cluster through N_δ, and stabilizes the otherwise mobile extrinsic domain containing the cluster. The *cyt* bc_1 dimer is the asymmetric unit in the crystal, and there is one [2Fe-2S] cluster in the ISP in each monomer comprising the dimmer. In the $P2_12_12_1$ space group, there are four symmetry-related sites for each monomer, but not all four sites can be distinguished in an EPR experiment.

When the magnetic field $\mathbf{B_o}$ is perpendicular to one of the crystal symmetry axes there are degeneracies resulting in at most two distinct sites per monomer. It has already been stated that the complete determination of the $\boldsymbol{g}$-tensor normally requires rotation of the magnetic field in three orthogonal planes. However, when the crystal contains several magnetically distinct, symmetry-related sites, the measurement of each of those sites during rotation in a single plane is equivalent to a rotation in different planes for each of those sites [31]. For the bovine bc_1 crystal, rotation about one of the two-fold axes generates the same data as rotation in different planes related by those symmetry elements. Thus, a single rotation about the c-axis of the crystal has been used for the complete determination of the $\boldsymbol{g}$-tensor and its orientation without creating possible errors resulting from remounting the crystal.

The g-factors plotted as a function of the rotation angle (Figure 3.11) showed a set of data points falling along the sinusoidal path between a g of $\sim$2.03 and $\sim$1.88 assigned to monomer A. Remaining points in the rotational pattern fall along a pair of sinusoidal tracks that do not cross the track from monomer A. Each track ranges between a g of $\sim$1.94 and 1.85 but shifted in phase with respect to each other and to monomer A. They are assigned to two sets of symmetrically related sites from the other monomer B. A series of fits was made in which the principal values of the $\boldsymbol{g}$-tensor were fixed at the

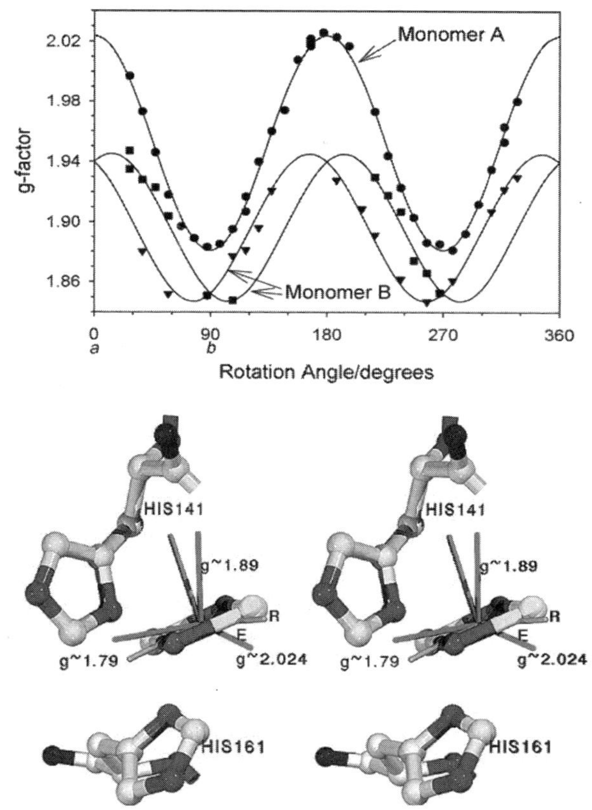

Figure 3.11 Top—The *g*-factor of the Rieske ISP as a function of rotation about the crystal *c*-axis. The solid lines are the fits to the individual monomers in the asymmetric unit. Experimental data points monomer A (•) and two distinct sets (■ and ▼) of symmetry-related sites in monomer B. The directions corresponding to the *a*- and *b*-axes are indicated below the rotation angles. Bottom—The structure of the [2Fe-2S] cluster shows the orientation of the ***g***-tensor axes of monomer B in chain R and monomer A in chain E relative to the molecular axes of the cluster. The two histidine ligands to the cluster are shown to provide its orientation but the two cysteine ligands on the other Fe are not. The figure has been redrawn for "crossed-eye" stereoviewing. (From M.K. Bowman, E.A. Berry, A.G. Roberts, and D.M. Kramer, 2004 *Biochemistry* **43** 430–6, with permission)

values (1.79, 1.89, 2.024) reported for bovine *cyt* bc_1 with stigmatellin in solution. Those yielded the cosine directions, which define the orientation of the principal axes relative to the crystal axes and made it possible to establish the orientation of the ***g***-tensor axes relative to the molecular axes of the cluster. The ***g***-tensor axes have slightly different orientations relative to the iron-sulfur cluster in the two halves of the dimer. The ***g***-tensor principal axes are skewed with respect to the Fe-Fe and S-S atom direction in the [2Fe-2S] cluster. The $g \sim 1.79$ axes makes an average angle of 30° with respect to the Fe-Fe direction and the $g \sim 2.024$ axis an average angle of 26° with respect to the S-S direction.

These data were used to test a theoretical model of the ***g***-tensor in the iron-sulfur cluster. In addition, knowledge of the ***g***-tensor orientation relative to the molecular axes is important for interpretation of magnetic resonance data in the context of the crystal structure, because hyperfine and quadrupole couplings with ligand nuclei are referred to the coordinate system of the ***g***-tensor.

The determination and analysis of the ***g***-tensors of the transition metal ions and inorganic radicals in single crystals is a significant part of EPR spectroscopy, more fully expounded in several classical monographs and textbooks [3, 4, 11, 12].

3.4.1.2 Analysis of Hyperfine Couplings

Hyperfine couplings of the central metal nuclei in transition metal complexes, and of α, β atoms in radicals, have often been measured in single crystals. From measurement of the orientation-dependence of hyperfine couplings from the splittings of spectral components, the principal values of the hyperfine tensors and their orientation relative to the crystal axes could be found. The design of the experiment and the analysis of the data are similar to those for the ***g***-tensor given in previous section. Particular examples describing the experimental determination of hyperfine tensors (and nuclear quadrupole tensors for nuclei with $I \geq 1$), and their subsequent analysis in characterization of the electronic structure, are described in detail elsewhere [2–12].

3.4.1.3 Electron Spin-Echo Envelope Modulation (ESEEM)

Since the ESEEM frequencies come from nuclear transitions and depend on the hyperfine and quadrupole interactions the study of the angular dependence of the ESEEM frequencies yields the same information as that obtained from analysis of the angular dependence of the splittings between the components in the EPR spectra. These are the principal values of the hyperfine and nuclear quadrupole tensors, and the orientation of their principal values. The difference between the pulsed and CW approaches is in the values of the interaction energies. In CW-EPR, the only couplings that can be studied are those that exceed the linewidth, which in single-crystal studies have a value of the order of several 0.1 mT. In contrast, ESEEM resolves the couplings that are masked by EPR broadening in the CW experiment. The lower boundary of the nuclear frequencies for measurement in ESEEM spectroscopy is limited by the time of the relaxation decay, or the acquisition time of the echo envelope, since the period of oscillation cannot exceed these times. This boundary can be estimated with an accuracy of several tenths of a MHz. The upper boundary of the measured ESEEM frequencies is limited by the possibility of simultaneous excitation of the allowed and forbidden transitions, separated by the splitting described by equations (3.30) (Figure 3.4). Its value reaches ~30-50 MHz in modern pulsed EPR spectrometers.

ESEEM spectroscopy has been used for investigation of radiation-induced organic free radicals in single crystals of organic acids, and in ions of transition metals, introduced as paramagnetic impurities into single crystals of inorganic compounds [25, 32, 33].

3.4.2 Orientation-Disordered Samples

3.4.2.1 CW-EPR Lineshape

In orientationally disordered solids, powders, glasses, or polymers, the paramagnetic species are randomly oriented with respect to magnetic field. The EPR spectrum of such samples is a superposition of the resonances from isotropically distributed molecular orientations with corresponding statistical weight. The formation of the powder EPR-lineshape can be easily demonstrated in the case of the axial ***g***-tensor anisotropy.

For the spin Hamiltonian written in the ***g***-tensor principal axes system the resonance *g*-factor is equal to

$$g = \left[g_{\perp}^2 \sin^2\theta + g_{\parallel}^2 \cos^2\theta\right]^{1/2} = \left[g_{\perp}^2 + (g_{\parallel}^2 - g_{\perp}^2)\cos^2\theta\right]^{1/2} \tag{3.41}$$

where θ is an angle between the symmetry axis (along $g_{\parallel}$) and the magnetic field direction. This expression leads to the conclusion that the resonance field B in the spectrum varies with the angle θ:

$$\mathrm{B} = \frac{g_e \mathrm{B}_o}{\left[g_{\perp}^2 + \left(g_{\parallel}^2 - g_{\perp}^2\right)\cos^2\theta\right]^{1/2}} \tag{3.42}$$

between the limiting values $\mathrm{B}_{\parallel}$ and $\mathrm{B}_{\perp}$ defined as:

$$g_{\parallel}\mathrm{B}_{\parallel} = g_{\perp}\mathrm{B}_{\perp} = g_e\mathrm{B}_{\mathbf{o}} \tag{3.43}$$

with $\mathrm{B} = \mathrm{B}_{\parallel}$ for $\theta = 0$ and $\mathrm{B} = \mathrm{B}_{\perp}$ for $\theta = \pi/2$.

Taking account of the fact that the fraction of the symmetry axes occurring between angles θ and $\theta + d\theta$, which is equal to $\frac{1}{2}\sin\theta d\theta = \frac{1}{2}d(\cos\theta)$, is proportional to the probability $P(B)dB$ of the resonance contribution between B and $B + dB$, one finds that

$$P(B) \sim \left(g_{\parallel}^2 \mathrm{B}_{\parallel}^2 - g_{\perp}^2 \mathrm{B}^2\right)^{-1/2}. \tag{3.44}$$

This function has the limiting value when $\mathrm{B} = \mathrm{B}_{\parallel}$ and becomes infinite at $\mathrm{B} = \mathrm{B}_{\perp}$, reflecting the contributions to the intensity of numerous orientations from the plane normal to the symmetry axis at this field. Thus, this analysis predicts the powder pattern for the axial ***g***-tensor shown in Figure 3.12. In real

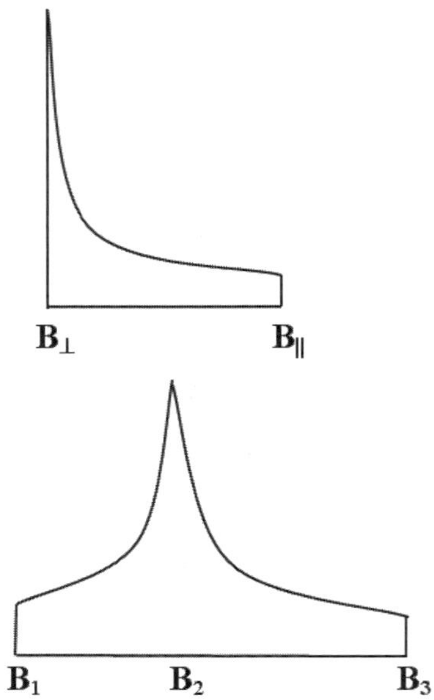

Figure 3.12 Distribution of intensity in powder EPR spectra for axial (top) and orthorhombic (bottom) systems.

samples *P*(*B*) needs to be convoluted with a suitable individual lineshape function that smoothes the step-type resonance features at $B_\perp$ and $B_{||}$. Examples of real lineshape in first derivative modes are shown (see Figures 3.13 and 3.14).

If an electron spin with an axial *g*-tensor interacts with the nuclear spin I, then the lineshape will have additional features resulting from the hyperfine structure. For the coincident principal axes of the axial ***g***-tensor and hyperfine A tensor there are two types of features, corresponding to the parallel and perpendicular orientations of magnetic field relative to the symmetry axis, located at:

$$\begin{aligned} B_{||}(m_I) &= \frac{g_e B_o}{g_{||}} - \frac{A_{||} m_I}{g_{||}\beta} \\ B_{\perp}(m_I) &= \frac{g_e B_o}{g_{\perp}} - \frac{A_{\perp} m_I}{g_{\perp}\beta} \end{aligned} \tag{3.45}$$

A schematic presentation of such a structure is shown in Figure 3.13. This type of spectrum is typical for Cu(II) and VO(II) ions, where hyperfine structure results from nuclei with the spin I = 3/2 and 7/2, respectively.

In the case of an orthorhombic ***g***-tensor, with principal values $g_1 > g_2 > g_3$, the powder EPR pattern is located between the fields B_1 and B_3 and it has a maximum at the field B_2 (Figure 3.12) where again

$$g_1 B_1 = g_2 B_2 = g_3 B_3 = g_e B_o. \tag{3.46}$$

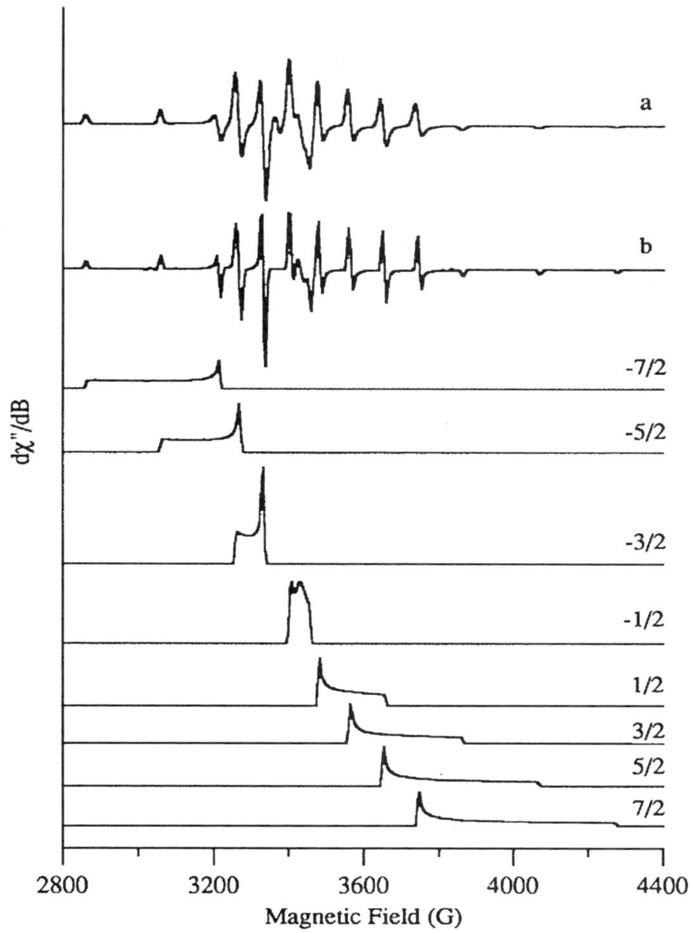

Figure 3.13 EPR spectrum (a) of $[VO(H_2O)_5]^{2+}$; trace (b) is the matrix diagonalization simulation. The additional traces show the eight individual $M_I = -7/2,..., +7/2$ powder patterns. These traces are shown in absorption mode rather than as a derivative as in the CW-EPR spectrum for simplicity. Experimental conditions are as follows: ν_{mw} = 9.68 GHz; MW power = ~25 mW; modulation amplitude = 8.0 G; time constant = 20.48 ms; conversion time = 40.96 ms; temperature = 49 K. (Taken from C.V. Grant, W. Cope, J.A. Ball, G. Maresch, B.J. Gaffney, W. Fink, R.D. Britt, 1999 *J. Phys. Chem. B* **103**: 10627–31, with permission)

In conclusion, the powder EPR spectra of paramagnetic species with anisotropic ***g***-tensor is particularly useful for determination of the principal values of the ***g***-tensor, although, in contrast to the single crystal, they cannot provide direct information about the orientation of the ***g***-tensor principal axes in the molecular coordinate system. The features corresponding to the ***g***-tensor components are easily recognizable in first-derivative spectra, where they approximate in shape to the derivative of individual components of the absorption line of the composite powder pattern (Figure 3.14).

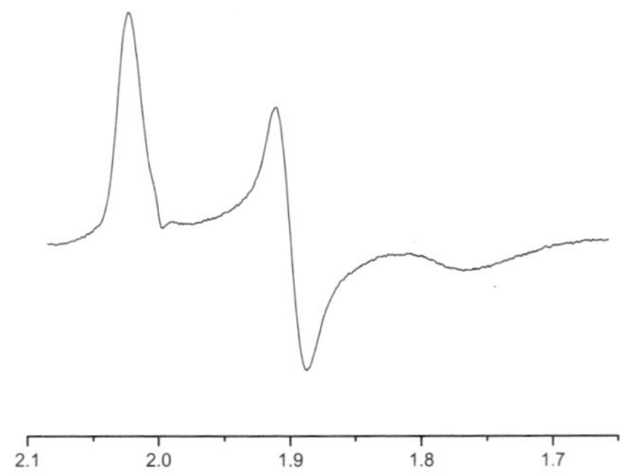

Figure 3.14 First derivative EPR spectrum of the *cyt* bc_1 complex showing the [2Fe-2S] cluster of the ISP, with an orthorhombic ***g***-tensor.

3.4.2.2 Structural Applications of the CW-EPR in Solid State: High-field EPR

EPR spectroscopy attracts the attention of researchers because it is a direct and sensitive technique for the detection and the characterization of the paramagnetic species. Usually, it requires only a small volume of material, with a concentration of the paramagnetic species from the micro- to millimolar range, depending on the width of the EPR spectrum. In any type of EPR experiment, information about the electronic and spatial structure of the paramagnetic species follows directly from the interactions of the electron magnetic moment with the applied magnetic field, and with the magnetic moments of surrounding nuclei, i.e., from the ***g***-tensor and hyperfine tensors. The principal values of the ***g***-tensor of a particular species vary within a narrow interval, and can be used for identification of that species. As an example, Table 3.3 shows principal values of the ***g***-tensor for the iron-sulfur clusters participating in the electron transport in many proteins. It is clear that the difference in the principal values makes it possible to distinguish the type of cluster and its electronic state.

Table 3.3 Principal values of the ***g***-tensor for biological iron-sulfur clusters in proteins.[a]

Cluster	g_x	g_y	g_z	g_{av}
$[2Fe\text{-}2S]^{1+}(SR)_4$	1.87–1.88	1.95–1.96	2.04–2.05	1.96
$[2Fe\text{-}2S](SR)_2\,(His)_2$	1.73–1.83	1.89–1.92	2.01–2.04	1.91
$[3Fe\text{-}4S]^{1+}(SR)_3$		~2.01		~2.01
$[4Fe\text{-}4S]^{1+}(SR)_4$	1.87–1.88	1.92	2.06	1.93
$[4Fe\text{-}4S]^{3+}(SR)_4$	2.04	2.04	2.11–2.14	2.07
[Fe-O-Fe]	1.56–1.66	1.72–1.77	1.92–1.96	

[a] Data are taken from [34–36].

Often, and this is especially true for radicals, the anisotropy of the ***g***-tensor is low, and the features corresponding to its principal values are not resolved in X-band spectra. Their resolution can be achieved with an increase of the microwave frequency (or field). This approach is illustrated in Figure 3.15, which shows several CW-EPR spectra of the cation radical of canthaxantin on a silica-alumina surface obtained at microwave frequencies between 95 and 670 GHz.

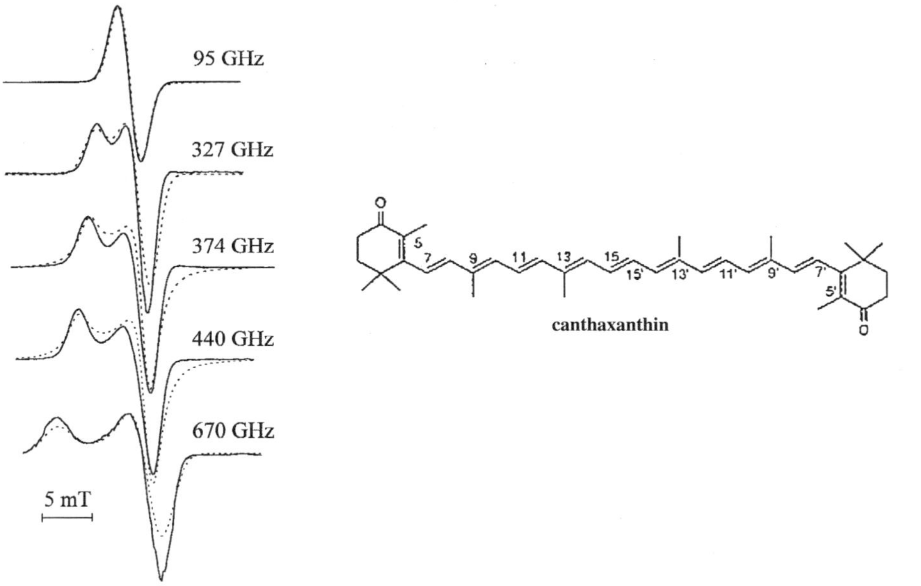

Figure 3.15 HF-EPR spectra of radical cation of the carotenoid canthaxanthin, adsorbed on silica-alumina: solid lines, experimental spectra recorded at 5 K: dotted lines, simulated spectra using ***g***-tensor values $g_{zz} = 2.0032$ and $g_{xx} = g_{yy} = 2.0023$ and line width of 1.36 mT. (Structure of canthaxanthin is shown on the right side above). (From T.A. Konovalova, J. Krzystek, P.J. Bratt, J. van Tol, L.-C. Brunel, L.D. Kispert, 1999 *J. Phys. Chem. B*, **103**: 5782–6, with permission)

One can see that the spectrum of this radical is a single isotropic line even at 95 GHz (W-band), which is ten times higher in frequency than traditional a X-band. Anisotropic splitting in the spectrum was only apparent at frequencies ~300 GHz. Complete separation of the $g_\perp$ and $g_{||}$ peaks was achieved only at frequencies higher than ~600 GHz. This measurement has provided the ***g***-tensor for this radical with $g_\perp = 2.0023$ and $g_{||} = 2.0032$ [37]. Thus, in this system, the best way to determine the ***g***-tensor was to make measurements at a sufficiently high field for B_o.

Another typical example is that of EPR of radicals in powders where the spectrum is dominated by hyperfine structure. Application of the high-field EPR provides new levels of information, allowing very precise conclusions about electronic structure of the radical. One can illustrate this by experiments with the stable tyrosyl radical (Y_{122}) (Figure 3.16) in *Escherichia coli*

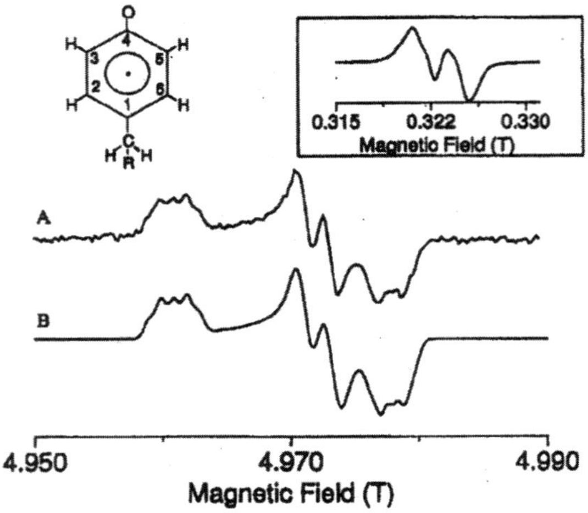

Figure 3.16 EPR spectra of the tyrosyl radical (Y_{122}) present in RDPR. (a) High frequency (139.5 GHz) spectrum of 5 μL of a 200 μM solution of the B2 subunit. (b) Simulation. The inset is a 9-GHz EPR spectrum of ~0.3 mL of the same sample. (From G.J. Gerfen, B.F. Bellew, S. Un, J.M. Bollinger Jr. J. Stubbe, R.G. Griffin, D.J. Singel, 1993 *J. Am. Chem. Soc.* **115**: 6420–1, with permission)

ribonucleoside diphosphate reductase (RDRP) [38]. The X-band EPR spectrum (~9 GHz) of shows the doublet splitting resulted from the proton hyperfine interaction. In contrast the D-band spectrum (~139.5 GHz) clearly resolves ***g***-tensor anisotropy. The principal values of the ***g***-tensor (2.00912, 2.00457, 2.00225) are readily obtained from the canonical features of the rhombic powder pattern. In addition, a doublet structure is evident at each canonical position from the β-methylene proton, with a large isotropic interaction (~2 mT) and small hyperfine anisotropy described by the tensor (2.07, 1.98, 1.88 mT). Some additional splittings are seen at the low- and high-field edges of the spectrum, assigned to the 3,5-ring protons with the couplings ~0.45–0.90 mT along different axes of the ***g***-tensor.

Multifrequency EPR, i.e., experiments at different microwave frequencies, is often employed to interpret and to perform a quantitative analysis of powder spectra with complex lineshapes resulting from ***g***-tensor anisotropy and hyperfine structure. The analysis of such spectra is often not trivial because it depends not only on the principal values of the tensors, but also on their relative orientations.

Many paramagnetic metal ions have magnetic nuclei, which produce additional hyperfine structure in the powder EPR spectra. Usually, splittings between these lines are larger than 2–3 mT. The hyperfine structure depends on nuclear spin, and has an easily recognizable typical lineform for many ions. Figure 3.13 shows examples of EPR spectra with hyperfine structure for VO(II). Table 3.4 shows general EPR characteristics for selected important transition metals.

Table 3.4 CW-ESR characteristics for selected transition metals.

Mn(II)	S = 5/2, 6 line of hyperfine structure from ^{55}Mn (I = 5/2, natural abundance 100%).
Cu(II)	S = 1/2, axial ***g***-tensor, $g_{\parallel} \sim 2.2$, $g_{\perp} \sim 2.0$, hyperfine structure from two nuclear isotopes ^{63}Cu (I = 3/2) and ^{65}Cu (I = 3/2) with natural abundance 69.2% and 30.8%, respectively.
Fe(III)	S = 5/2, large ***g***-tensor, anisotropy, $g \sim 2$–9. S = 1/2 (low spin state), rhombic ***g***-tensor: 2.4, 2.2, 1.8, stable nuclear magnetic isotope ^{57}Fe (I = 1/2, 2.3%).
VO(II)	S = 1/2, axial ***g***-tensor, $g_{\parallel} \sim 1.94$, $g_{\perp} \sim 1.98$, hyperfine structure from nuclear isotope ^{51}V (I = 7/2, 99.756%).
Ni(III)	S = 1/2, axial or rhombic ***g***-tensor, 2.02–2.30, stable nuclear magnetic isotope ^{61}Ni (I = 3/2, 1.13%).
Co(III)	S = 1/2, isotropic $g \sim 2.0$, nuclear magnetic isotope ^{59}Co (I = 7/2, 100%).
Mo(V)	S = 1/2, rhombic ***g***-tensor, 1.99, 2.00, 2.09, 6 lines of hyperfine structure from ^{95,97}Mo (I = 5/2, 15.9% and 9.6%).

Summarizing this brief consideration of CW-EPR applications in powder solids, one can conclude that this technique provides initial identification of paramagnetic species based on the ***g***-tensor anisotropy and possible hyperfine structure. However, CW-EPR spectra usually do not give any indication about the interaction with the magnetic nuclei in the ligands of the paramagnetic species and more distant environment. Interactions of the electron spin with these nuclei are weak (they produced splitting usually less than 0.3–0.5 mT or 10–15 MHz) and are not resolved in powder EPR spectra. Study of such interactions requires application of other experimental approaches, in particular pulsed EPR spectroscopy, based on observation of the electron spin echoes.

3.4.2.3 Powder-ESEEM

The ESEEM experiments in solids can usually be divided into two different cases, depending on the EPR lineshape. In one case, the EPR line has a width comparable to the excitation width of the microwave pulses, which is usually in the range <1.0–1.5 mT. This case can be considered as a complete excitation of the powder EPR spectrum because the ESEEM pattern obtained contains contribution from all possible orientations of paramagnetic species relative to the applied magnetic field. An important peculiarity of this case arises from the fact that there is no externally defined molecular coordinate system, and analysis of the data is usually performed using the coordinate system of hyperfine or quadrupole tensor principal directions. ESEEM measurement on powder systems provides information about the principal values of the tensors. However, they do not give any direct information about the orientation of these tensors.

As shown in Section 3.2.7.5, the two-pulse sequence is expected to produce four ESEEM frequencies from an I = 1/2 nucleus. A powder spectrum may have a completely different number of resolved features, because nuclear frequencies have different dependences on the orientation of magnetic field. The spectral form is determined by the relative values of nuclear Zeeman frequency, the isotropic constant *a*, and the components of anisotropic hyperfine tensor *T*.

The peculiarities of ESEEM in powder solids one can demonstrate in a simple case—the interaction of an electron spin with a remote nucleus when they are coupled via weak anisotropic coupling. This can be considered in the approximation of point dipoles [23, 25]. In this case the hyperfine tensor is purely anisotropic, with the diagonal components (–*T*, –*T*, and 2*T*), where

$$T = -\frac{g_e\beta_e g_I\beta_I}{hr^3} \tag{3.47}$$

and *r* is the distance between electron and nucleus.

If θ is the angle between the vectors of external magnetic field B_0 and symmetry axis *z* of the axial tensor, then the terms *A* and *B* in the expression (3.25) for the nuclear frequencies of I = 1/2 are

$$\begin{aligned} A &= T(3\cos^2\theta - 1) \\ B &= 3T\sin\theta\cos\theta. \end{aligned} \tag{3.48}$$

Assuming that *A* and *B* << ν_I and considering small terms only in the first order, one can rewrite the nuclear frequencies as:

$$\nu_{\alpha(\beta)} \approx \nu_I \pm \frac{A}{2} \tag{3.49}$$

and obtain the expression (3.35) for the two-pulse ESEEM in a form

$$E(\tau) = 1 - \frac{B^2}{2\nu_I^2}\left[1 - \cos 2\pi\nu_I\tau \cos 2\pi\left(\frac{A\tau}{2}\right) + \frac{1}{2}\cos 2\pi(2\nu_I\tau) + \frac{1}{2}\cos 2\pi(A\tau)\right] \tag{3.50}$$

In a disordered sample, $E(\tau)$ should be averaged over all orientations of the magnetic field:

$$\langle E(\tau)\rangle_\theta = \frac{1}{2}\int_0^\pi E(\tau)\sin\theta d\theta \tag{3.51}$$

After averaging, (3.50) becomes

$$\langle E(\tau)\rangle_\theta = 1 - \frac{3T^2}{10\nu_I^2}\left[2 - 4\cos 2\pi\nu_I\tau\, F(T\tau) + \cos 2\pi(2\nu_I\tau) + F(2T\tau)\right] \tag{3.52}$$

where $F(T\tau)$ is the normalized function decaying with the increase of $T\tau$.

This result shows that in the powder ESEEM spectrum from an I = 1/2 nucleus, there is a superposition of the two harmonics with ν_I and $2\nu_I$, and accordingly the ESEEM spectra will show the single line at ν_I as well as its double harmonic for the sequences producing combination ESEEM frequencies (Figure 3.17).

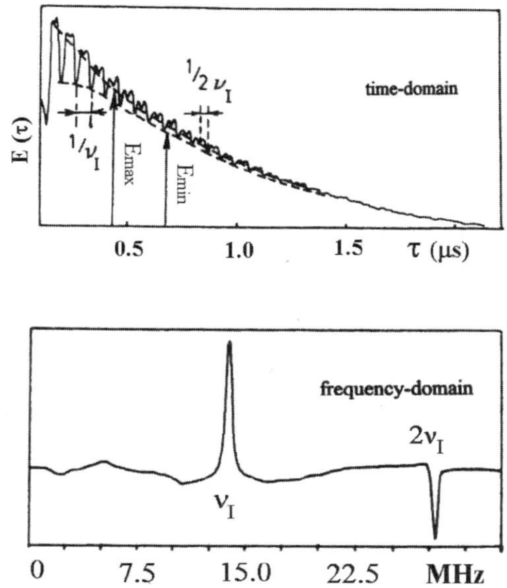

Figure 3.17 Top—An example of two-pulse ESEEM from weakly coupled protons in powder sample. Bottom—The cosine Fourier transform spectrum.

Taking into account the dependence of T on r (equation (3.47)) for point dipoles one can see that the amplitude coefficient of the ESEEM harmonics is proportional to $\left(\nu_I^2 r^6\right)^{-1}$. The environment of a paramagnetic species may involve more than one magnetic nucleus of the same type. In this case the amplitude of ESEEM harmonics is proportional to $\sum_i \left(\nu_I^2 r_i^6\right)^{-1}$. The $1/r^6$ dependence of the amplitude on distance leads to the expectation that the nearest nuclei will provide the dominant contribution. As a consequence, distances to the nearest nuclei, and their number, can be determined from analysis of the damping time of the harmonic with Zeeman frequency in the time-domain.

Theoretical analysis predicts also that the modulation amplitude from the weakly coupled nuclei depends on $\left(\nu_I^2\right)^{-1}$, i.e., on the EPR microwave frequency and resonant field. Figure 3.18 shows the increase of the proton ESEEM amplitude on the decrease of the microwave frequency. This is an

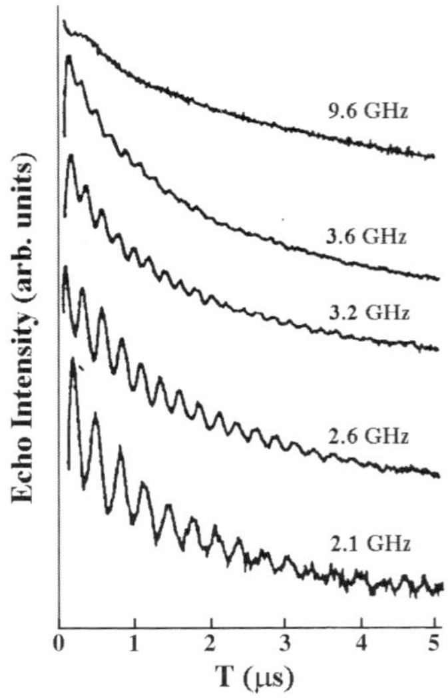

Figure 3.18 Dependence of proton ESEEM on microwave frequency (field strength for resonance) for weakly coupled protons (Pittsburgh No. 6 coal sample). (Taken from J. Hankiewicz, C. Stenland, L. Kevan, 1993 *Rev. Sci. Instr.* **64**, 2850–6, with permission.)

effective way to detect weakly coupled nuclei, using spectrometers with decreasing microwave frequency.

With increasing values of T and a compared to ν_I, the powder ESEEM becomes more complicated, representing superposition of the averaged frequencies $\nu_{\alpha(\beta)}$ and $\nu_\alpha \pm \nu_\beta$. In this case, the harmonics with low orientation dependence will produce dominating features in the frequency-domain spectra. The theoretical analysis of powder ESEEM spectra from I = 1/2 nuclei at arbitrary values of a, T and ν_I has allowed classification of all possible spectral features that may appear at different T/ν_I ratios. These features can provide in many cases an accurate initial estimate of a and T, which can be verified and improved by numerical simulation of the experimental powder ESEEM spectrum.

An important topic in ESEEM spectroscopy is the detection and characterization of the ^{14}N nitrogens involved in the ligation of metal centers or the formation of hydrogen bonds with paramagnetic species. Analysis of the literature has shown that the parameters of the ^{14}N spin Hamiltonian have the following typical values: quadrupole coupling constant ~1–5 MHz; nuclear Zeeman frequency in X-band $\nu_I \sim 1$ MHz; isotropic hyperfine constant $a \sim 1$–10 MHz. The anisotropic hyperfine coupling T is at least several times smaller than isotropic constant. Therefore, the approximation of a pure isotropic hyperfine interaction has been used for the qualitative consideration of ^{14}N powder ESEEM spectra. In this approximation the type of spectrum is governed by the ratio of effective nuclear frequencies determining the local magnetic field at two projections of the electron spin, $\nu_{\text{eff}\pm} = \left|\nu_I \pm \frac{a}{2}\right|$, to the value of $K = e^2qQ/4h$ [25].

If $\frac{\nu_{\text{eff}\pm}}{K} < 1$, then the three nuclear frequencies in a corresponding manifold will be close to the three pure nuclear quadrupole resonance (nqr) frequencies of ^{14}N, which are independent of the direction of the magnetic field, and give rise to three narrow peaks in the ESEEM spectra at

$$\nu_+ = K(3+\eta) \quad \nu_- = K(3-\eta) \quad \nu_0 = 2K\eta \tag{3.53}$$

with the property $\nu_+ = \nu_- + \nu_0$. The best condition for the observation of this triplet (equation (3.53)) is the so-called *cancellation* condition, when $\nu_{\text{eff}\pm} \approx 0$.

If $\frac{\nu_{\text{eff}\pm}}{K} > 1$, only a single narrow line from each corresponding manifold is expected, produced by a transition at the maximal frequency, which is actually a double-quantum (dq) transition between two outer states with m_I = –1 and 1. The frequency of the transition is equal to

$$\nu_{dq\pm} = 2[\nu^2_{\text{eff}\pm} + K^2(3+\eta^2)]^{1/2}. \tag{3.54}$$

Two other single-quantum transitions involving the central level with m_I = 0, have a significant orientation dependence from quadrupole interaction and do not produce informative narrow lines.

Reported ^{14}N spectra usually correspond to two typical cases: (i) $\frac{\nu_{\text{eff}}}{K} < 1$ on one manifold, $\frac{\nu_{\text{eff}}}{K} > 1$ in another manifold; and (ii) $\frac{\nu_{\text{eff}}}{K} > 1$ in both manifolds. The three-pulse ESEEM spectrum in case (i) contains four lines. They represent three frequencies (equation (3.53)) from the manifold $\frac{\nu_{\text{eff}}}{K} < 1$ and a dq line from the manifold with $\frac{\nu_{\text{eff}}}{K} > 1$. Nqr frequencies give K and η, which characterize the chemical type of the ^{14}N nitrogen and its electronic state. The dq frequency provides the hyperfine coupling.

The three-pulse ESEEM spectrum in case (ii) consists of two lines with dq frequencies (equation (3.54)) from two manifolds. Observation of these two lines allows calculation of the hyperfine coupling and $K^2(3 + \eta^2)$, which also gives an estimate of K, suggesting variation of η between 0 and 1.

Experimental spectra corresponding to the two cases considered are shown in Figure 3.19. They include the ESEEM spectra from the semiquinone (SQ) in the Q_H-site of the bo_3 oxidase and in the Q_i-site of the *cyt* bc_1 complex. The spectrum of the SQ in bo_3 consists of four lines at 0.98, 2.32, 3.30, and 5.25 MHz.

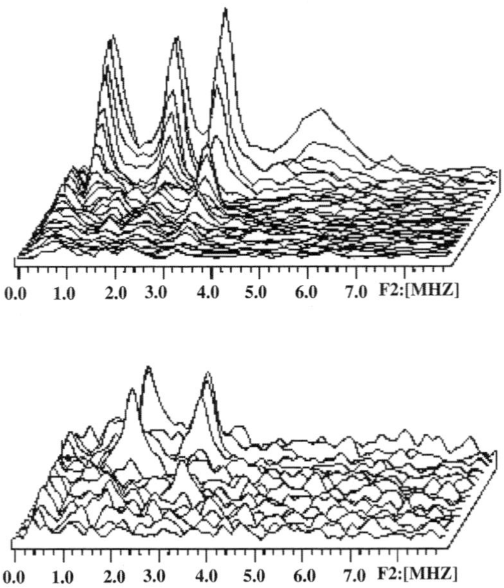

Figure 3.19 Top—Stacked plot of three-pulse ESEEM of the SQ in the Q_H-site of bo_3 oxidase. The spectra show the modulus of Fourier transform along time T-axis. The initial time is 88 ns in the farthest trace, and was increased by 16 ns in successive traces. Microwave frequency was 9.70 GHz, magnetic field was 346 mT. Bottom— Same, for the SQ in Q_i-site of bc_1 complex. Magnetic field was 354.2 mT.

Three lower frequencies satisfy the nqr property and correspond to a ^{14}N nitrogen with $K = 0.94$ MHz and $\eta = 0.52$. Hyperfine coupling of this nitrogen is 1.8 MHz. In contrast, the spectrum of the SQ in the *cyt* bc_1 complex shows only two lines at 1.7 and 3.1 MHz. Application of equation (3.53) provides values $a = 0.8$ MHz, and $K^2(3 + \eta^2) = 0.43$ MHz^2, that lead to $K = 0.35 \pm 0.03$ MHz. Thus, these spectra show the interaction of the SQ with the nitrogens in protein environment. The nitrogens are involved in hydrogen bonding with the quinone oxygens, and this is accompanied by transfer of some unpaired spin density onto the nitrogen nucleus. However, the values of the quadrupole coupling constants indicate that the SQ in bo_3 forms a hydrogen bond with a backbone peptide nitrogen [39], whereas the SQ in the bc_1 complex is involved in a hydrogen bond with the nitrogen of a histidine imidazole [40].

3.4.2.4 Orientation Selection

In the previous section we considered powder-type spectra where all orientations contribute equally to the spectrum. This type of spectrum gives a narrow single line with a width smaller than the frequency width ν_1 of the microwave pulse. Usually this condition is fulfilled only for the spectra of organic radicals like the semiquinone.

A second more typical case in solids is an anisotropic EPR spectrum with a width (usually up to several hundred Gauss) that significantly exceeds the field interval excited by the microwave pulses (~0.5–1.0 mT). Pulsed EPR measurements under such conditions are orientation-selective experiments, because the magnetic field, fixed at some point within the EPR line, selects species with different orientations of the ***g***-tensor relative to the magnetic field direction. For example, if we consider an EPR spectrum with orthorhombic ***g***-tensor (Figure 3.14), then ESEEM measurements taken at the high and low extreme edges (near the maximal and minimal g values) give "single-crystal-like" patterns from the paramagnetic species, whose principal axes correspond to the smallest and largest principal values of the ***g***-tensor and are directed along the magnetic field. The nuclear frequencies determined from such spectra can provide an estimate of the diagonal components of the hyperfine and quadrupole tensors along the g_z and g_x axes in the ***g***-tensor coordinate system. In contrast, the resonance condition at the intermediate fields is fulfilled by many different, yet well-defined, orientations. Consequently, the lineshape of nuclear frequencies measured at intermediate fields is more complex.

In many studies, the spectra measured at the field corresponding to the intermediate principal value of the ***g***-tensor (g_y) are used for estimation of the hyperfine and quadrupole splittings along the g_y direction itself. Thus, measurements at three points corresponding to the principal directions of the orthorhombic EPR spectrum provide diagonal elements of the hyperfine and quadrpole tensors. For many ESEEM spectroscopic studies this approach represents the end point, because the complete determination of the tensors (even in the simplest case of the $I = 1/2$ nucleus) would require a much more extended effort, based on the numerous simulations of the set of orientation-selected spectra collected over the entire EPR spectrum. The complexity of the analysis would increase if the tensors of the $I > 1/2$ nucleus or tensors of several nuclei needed to be determined from experimental spectra of nuclear frequencies. However, if this kind of analysis were performed, the data obtained would provide more valuable information not only about the strength of the interactions but also about their orientation in the coordinate system of the ***g***-tensor, and this would allow more detailed structural insights for the system under investigation. Examples of orientation-selected experiments and their analyses are given in the references [22, 25].

The strong influence of the magnetic field on powder ESEEM spectra via the Zeeman frequency leads to the discussion of *multifrequency* ESEEM [18, 22, 25]. In this type of experiment, ESEEM measurements are performed at several different microwave frequencies with the aim of identifying easy recognizable spectral features, which might lead to a more accurate determination of hyperfine and quadrupole parameters.

3.4.3 Two-Dimensional ESEEM

The spectra from one-dimensional spectroscopy become less useful when even a small number of magnetically nonequivalent nuclei contribute, because the congestion of lines and their overlap significantly decrease spectral resolution and create serious difficulties in interpretation. A fruitful way of simplifying the analysis and improving the spectral resolution is the use of two-dimensional (2D) techniques. In principal, the simplest 2D experiment from a technical perspective is that done by using a three-pulse sequence $\left(\frac{\pi}{2} - \tau - \frac{\pi}{2} - T - \frac{\pi}{2} - \tau - echo\right)$, when a set of stimulated ESEEM patterns is recorded as a function of time $\tau + T$ at different times τ [41]. However, the disadvantage of this experiment is that the intensity of the echo signal decays with phase memory time T_m ($\sim T_2$) along τ-axis, which is much shorter than the longitudinal relaxation time T_1 controlling the decay along $\tau + T$-axis. As a result, the linewidth along one axis of the 2D spectrum from such an experiment is significantly larger than that along the other axis. In a particular experiment with Gd^{3+}-doped $Bi_2Mg_3(NO_3)_{12}$. H_2O [41] the linewidth along one axis was about 50 times larger than that along the other. Because of this property, only a few applications using this approach have been reported, and those were mainly for the single crystals where the special techniques of spectral analysis were used for the axis with a larger linewidth [42].

The other variant of the 2D ESEEM, named HYSCORE (*hyperfine sublevel correlation*), was proposed by Höfer et al. [43], and is based on the four-pulse sequence $\left(\frac{\pi}{2} - \tau - \frac{\pi}{2} - t_1 - \pi - t_2 - \frac{\pi}{2} - \tau - echo\right)$. In this experiment the intensity of the stimulated-echo inverted by the π pulse is measured as a function of times t_1 and t_2 at constant time τ. Such a set of echo envelopes gives, after complex Fourier transform, a 2D spectrum with equal resolution in both dimensions. HYSCORE is nowadays the standard 2D sequence used in pulsed EPR applications, because its equal spectral resolution and reasonable number of pulses still allows the experiments to be performed with the samples containing the limited concentration of paramagnetic species available in biological samples.

The general advantage of two-dimensional techniques lies in the creation of nondiagonal cross-peaks, whose coordinates are nuclear frequencies from opposite electron spin manifolds. A particular property of the HYSCORE is the sensitivity to the relative signs of frequencies involved in the correlation. Due to such characteristics, experimental HYSCORE spectra are usually represented by two quadrants, (++) and (+–), of the 2D Fourier transform. For example, a nucleus with $I = 1/2$ has two hyperfine frequencies ν_α and ν_β. These may produce a pair of cross-features (ν_α, ν_β) and (ν_β, ν_α) in the (++)-quadrant, as well as another pair ($-\nu_\alpha$, ν_β) and (ν_α, $-\nu_\beta$) in the (+–)-quadrant. Peaks in the (+–)-quadrant come primarily from strong hyperfine interaction, i.e., $|T + 2a| >> 4\nu_I$, whereas the peaks in the (++)-quadrant appear predominantly from interactions with $|T + 2a| << 4\nu_I$. The appearance of cross-peaks in (++)- or (+–)-quadrants is governed by the relative values of hyperfine couplings and Zeeman frequency. Peaks may appear in both quadrants simultaneously in the intermediate case, when both parts of the inequalities are comparable.

Further development of HYSCORE has led to DONUT-HYSCORE, a 2D five-pulse experiment that correlates the nuclear transitions from the same electron spin manifold [44]. A completely new property of orientationally disordered 2D spectra is the visualization of interdependence between nuclear frequencies arising from different manifolds belonging to the same orientations in the form of cross-peak contour projection. To put it simply, the 2D spectra provide an experimental graph showing variation of one nuclear frequency from another at different orientations of the magnetic field.

For nuclear spin I = 1/2, the contour lineshape of cross-peaks at arbitrary axial hyperfine tensor and Zeeman frequency is described by simple equation [45]

$$|\nu_\alpha| = \left(Q_\alpha \nu_\beta^2 + G_\alpha\right)^{1/2} \tag{3.55}$$

where

$$Q_\alpha = \frac{\mathrm{T} + 2a - 4\nu_\mathrm{I}}{\mathrm{T} + 2a + 4\nu_\mathrm{I}}$$

$$G_\alpha = \frac{2\nu_\mathrm{I}\left(4\nu_\mathrm{I}^2 - a^2 + 2\mathrm{T}^2 - a\mathrm{T}\right)}{\mathrm{T} + 2a + 4\nu_\mathrm{I}}.$$

This equation opens a direct pathway for the analysis of powder 2D spectra from nuclei of I = 1/2 spin. The contour lineshape transforms to a straight line segment in the coordinates ν_α^2 vs. ν_β^2:

$$\nu_\alpha^2 = Q_\alpha \nu_\beta^2 + G_\alpha. \tag{3.56}$$

Here, the slope Q_α and intercept G_α of each segment determine the isotropic (a) and anisotropic (T) parts of hyperfine coupling [45].

Figure 3.20 shows the HYSCORE spectrum of reduced Rieske-type [2Fe-2S] cluster from soluble fragment of bc_1 complex, uniformly ^{15}N-labeled. Two quadrants of this spectrum contain the cross-features produced by different types of ^{15}N nitrogens. The (+–)-quadrant exhibits two pairs of cross-peaks with a contour parallel to the diagonal from two coordinated $^{15}N_\delta$ nitrogens. The location of these peaks indicates their hyperfine couplings of order 6 and 8 MHz, which significantly exceeds $\nu_I \sim 1.5$ MHz. Detailed analysis of the contour lineshape from these nitrogens in spectra recorded at different field positions provides the following hyperfine tensors in the axial approximation: $a = 6.0$, T = 1.2 MHz (N_δ 1 in figure) and $a = 7.8$ MHz, T = 1.3 MHz (N_δ 2 in figure). The other cross-features centered symmetrically around the diagonal point with ^{15}N Zeeman frequency are located in the (++)-

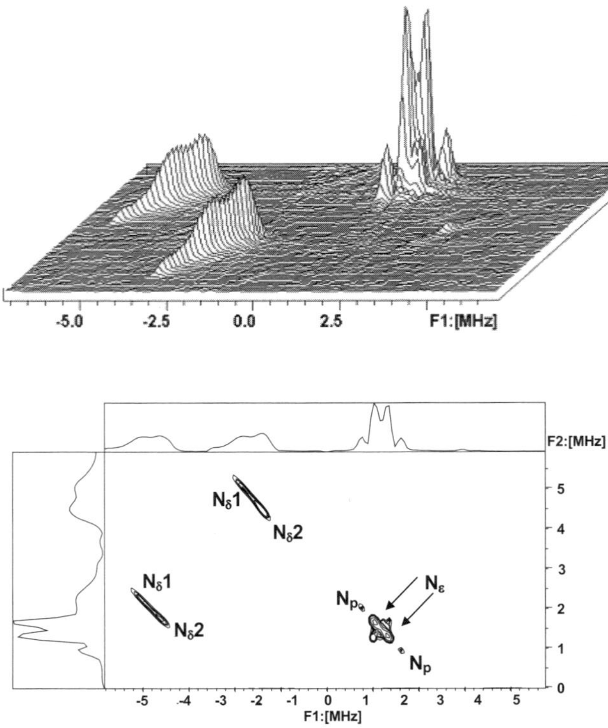

Figure 3.20 Stacked (top) and contour (bottom) presentations of the HYSCORE spectrum of the Rieske iron-sulfur fragment (ISF) from ^{15}N-labeled bc_1 complex

quadrant. These features are produced by ^{15}N nitrogens with weak couplings 0.4 and 1.1 MHz. These couplings belong to the remote N_{ε} and backbone peptide nitrogens.

At present, most reported HYSCORE studies have been of samples with ^{14}N nitrogen ligating metals in model complexes and protein active sites. The ^{14}N nucleus can produce up to 18 cross-peaks in each of the two quadrants of the HYSCORE spectra, including two [$dq_{\pm}$, $dq_{\mp}$], eight [$dq_{\pm}$, $sq^{(1,2)}{}_{\mp}$], and eight [$sq^{(1,2)}{}_{\pm}$, $sq^{(1,2)}{}_{\mp}$] correlations. However, again, only some of the possible cross-features have usually shown an observable intensity. Those cross-peaks are distributed in two quadrants, (++) and (+–), depending on the relative values of nuclear magnetic interactions and orientation dependence of nuclear frequencies. In addition, the intensity of cross-peaks is significantly influenced by the time τ between the first and second pulses, which is kept constant in each HYSCORE experiment. Due to this property, several measurements at different times τ are usually required for the detection of all cross-peaks contributing to the spectra. Moreover, the ESEEM intensity of the nuclear transition and the intensity of cross-peaks involving this transition are described by different coefficients. As a result, the absence of the line from some transitions in three-pulse ESEEM spectra does not rule out the existence of the cross-peaks. HYSCORE also possesses an enhanced capability to detect peaks of low intensity relative to the three-pulse spectra. This experiment separates overlapping peaks along a second dimension and enhances the signal-to-noise ratio by the second Fourier transform. Observation of a few cross-features from ^{14}N is often enough for the construction of its nuclear frequencies in both manifolds. These frequencies are used for calculating the nitrogen hyperfine and quadrupole couplings. The HYSCORE technique is therefore very effective for the study of the system containing a large number of ligands with nonequivalent ^{14}N. It has, for example, been used to determine hyperfine couplings from five coordinating nitrogens in the mixed-valent state of methemerythrin, with coupling values from 5 to 15 MHz, and to assign them to particular ligands [46]. A recent review of the ESEEM applications of the 1D and 2D ESEEM to the metal ions in biological systems is given by Deligiannakis et al. [47].

3.4.4 Measurement of Relaxation Times in CW- and Pulsed-EPR

The CW-EPR and ESEEM spectra discussed in previous sections allow characterization of the parameters defining the interactions of the time-independent spin-Hamiltonian. The dynamics of the electron spin system reflect interaction within the spin system itself and interactions with the environment. The characterization of the dynamic processes in the spin systems is based on the measurement of the relaxation times. There are several approaches for measuring relaxation times using CW- and pulsed-EPR experiments.

With CW-EPR techniques, progressive microwave power saturation methods can be employed. This approach can be readily pursued using a conventional spectrometer. In this experiment, the electron spins are progressively subjected to increasing microwave power. As a result, the intensity of the EPR signal also grows in proportion to the square root of the microwave power P. Microwave power saturation occurs when the rate of the microwave absorption exceeds the rate at which the system returns to equilibrium. The microwave power at half-saturation, $P_{1/2}$, is an experimentally measured parameter, which is proportional to $1/T_1T_2$. Coupled with the theoretical analysis of the relaxation processes, this value leads to an estimation of the relaxation times. A detailed description of the progressive microwave power saturation method, and examples of its application, are given in reviews [48, 49]. However, as noted by Brudvig [50], "It is difficult to extract accurate values for the spin relaxation times from progressive microwave power saturation measurements. For accurate measurements of spin relaxation times, pulsed-EPR methods are required."

Pulsed-EPR offers several different experimental approaches for the measurement of relaxation times. Different methods used for the measurement of the longitudinal relaxation time T_1 include saturation and inversion recovery, echo saturation by fast repetition, stimulated-echo decay, spectral hole-burning [23], and longitudinal detection PEPR [51]. Transverse relaxation time T_m ($\sim T_2$) can be determined using two-pulse echo decay and Curr-Parcell sequence [23].

Pulsed-EPR is still limited to those systems with relaxation times longer than the dead-time of $\sim 10^{-7}$s. This limits the temperature range for measurement of the radicals, starting at temperatures ~100–120 K and below depending on the type of the sample. Lower temperatures are usually required for most transition and rare earth ions.

A variety of pulsed-EPR experiments associated with measurement of relaxation times, and their dependence on the different characteristics of the system (e.g., concentration, temperature, matrix) have been described. These have provided a rich array of information including: (i) the irreversible relaxation time of spin magnetization associated with the homogeneous EPR line broadening and mechanism of relaxation [24, 27, 52–56]; (ii) separation of different paramagnetic species from overlapping spectra based on differences in their relaxation times [57–58]; (iii) the spatial distribution of paramagnetic centers in solids. This can be used to detect whether or not the particle distribution is homogeneous, to measure the local concentration, to restore the distribution function and its variation during the diffusion or reaction in the case of a pair distribution of paramagnetic centers, and to determine the number of spins per group in the case of a group distribution [59–64]; (iv) the times and mechanism of slow motions ($>10^{-7}$s) of paramagnetic species, paramagnetic functional groups, or molecular fragments in their environment [65–67].

Tyryshkin et al. [68] recently published an interesting study of the relaxation properties of the donor electron spins in phosphorus-doped silicon (Si:P); these spins are a candidate two-level system (*qubit*) for quantum information processing. An inversion recovery experiment and a two-pulse echo were used to measure the longitudinal T_1 and transverse T_2 relaxation times. In the inversion recovery experiment $\left(\pi - T - \frac{\pi}{2} - \tau - \pi - \tau - echo\right)$ the delay T, after the first inversion pulse was varied while time τ was kept constant, and the amplitude of the two-pulse echo formed by the second and third pulses was measured. Experiments were performed for P donors with natural Si and isotopically purified ^{28}Si.

In the inversion-recovery experiments both the Si:P and ^{28}Si:P samples showed monoexponential decays, and T_1 was obtained by fitting a simple exponential. The temperature dependence of $1/T_1$ for these two samples varies by 5 orders of magnitude (10^6 to 10^{-1} s) over the temperature interval 7–20 K. The Arrhenius dependence $\ln(1/T_1)$ vs. $(1/T)$ is linear, which is consistent with an Orbach mechanism predicting $1/T_1 \sim \exp(-\Delta E/kT)$. The energy gap to the excited state involved in the relaxation process is found to be $\Delta E = 126.1 \pm 0.5$ K. Two-pulse ESE measurements of isotopically purified ^{28}Si:P showed exceptionally long transverse relaxation times T_2. Below ~10 K the spin decoherence is controlled by instantaneous diffusion, and at higher temperatures by an Orbach process. T_2 is found to be 14 ms at 7 K for small turning angles in the samples investigated with the concentration ~10^{16} P/cm^3 that extrapolates to ~60 ms for an isolated spin. The observation of such an extremely long T_2 has important implications for efforts to use bound electron spins in Si:P as qubits for quantum information processing.

3.4.5 Interaction Between Electron Spins

Special attention in recent years has been given to the development of EPR-based techniques for the measurements of long-range distances between electron spins. Traditional CW-EPR techniques for the measurement of distances between spins that do not produce observable splittings in the EPR spectra (called *fine structure*) are based on measurements of relaxation times using a saturation approach.

More sophisticated approaches for the measurement of distances between electron spins are based on pulsed-EPR techniques. Those include pulsed electron-electron double resonance (PELDOR) or double electron-electron resonance (DEER), "2 + 1" sequence, and double-quantum EPR. All of them are described in detail in a recent book [69].

Among these techniques, increasing attention has been given to PELDOR. This technique uses pulses at two different microwave frequencies to measure the distance in a pair of paramagnetic species by exploiting the dipole-dipole spin interaction. Two different sequences were employed in the experiments.

In the PELDOR sequence originally proposed [70], two microwave pulses induce the ESE signal at frequency ν_A in a spin system. Between pulses 1 and 2 a pumping pulse is applied at frequency ν_B at time T. The pumping pulse induces transitions between the Zeeman levels of spins B and thus changes local magnetic field at spins A. This results in additional dephasing of spins A, and hence gives rise to a decrease of the PELDOR signal.

In the approximation of point dipoles, assuming that delocalization of the electrons is small on a scale compared to the distance between them, the interaction between spins is characterized by the frequency

$$\nu_{AB} = \nu_{dd}\left(1 - 3\cos^2\theta_{AB}\right) + J \tag{3.57}$$

where

$$\nu_{dd} = \frac{g_A g_B \beta_e^2}{r_{AB}^3 h}$$

is distant dependent dipole-dipole coupling, θ_{AB} is the angle between the field $\boldsymbol{B}_0$ and $\mathbf{r}_{AB}$ vectors, and J is the isotropic exchange coupling. In the powder sample, splittings for different orientations form a typical powder pattern with the two pairs of singularities corresponding to $\theta_{AB} = 0$ and 90° with the splittings $\nu_{AB}(||) = |J - 2\nu_{dd}|$ and $\nu_{AB}(\perp) = |J + \nu_{dd}|$, respectively. Fourier transformation of the echo envelope should provide a powder pattern of ν_{AB} splittings. For spin-spin distances above ~ 2 nm the exchange coupling J is usually negligible and thus observation of even one singularity $\nu_{AB}(\perp)$ with larger intensity would provide a measurement of the $\mathbf{r}_{AB}$.

Milov et al. [71] performed model PELDOR experiments with the biradicals of the rigid structure in frozen solution. They found the appearance of echo oscillations with the frequency ~7.5 MHz. These oscillations were absent in the sample with mononitroxide. Thus, assigned to $\nu_{AB}(\perp)$ it gave the distance between the nitroxide fragments equal to 1.9 ± 0.16 nm. Schienman et al. [72] in a PELDOR experiment with the RNA containing spin-labeled uridines in frozen solution detected an oscillation of the echo envelope with 1.2 ± 0.2 MHz. This was also assigned to the $\nu_{AB}(\perp)$, and gave $r_{AB} = 3.5 \pm 0.2$ nm, consistent with the distance obtained from structural modeling.

In addition to the PELDOR sequence described above, a four-pulse sequence has been proposed that avoids the problem of a large dead-time that influences the three-pulse experiment [73].

The PELDOR techniques are capable of measuring distances in the range from 1.5 up to 10 nm. Recent applications of PELDOR include measurements of the distances between spins in biradicals [74], nitroxides in spin-labeled polymers [75] and peptides [76], the redox-active components in the PSII reaction center [77], and between Mo(V) and Fe(III) centers in chicken liver sulfite oxidase [78].

ACKNOWLEDGEMENTS

Support for the authors' research by grants from the National Institute of Health (GM 62954 to S.A. Dikanov and GM 35438 to A.R. Crofts), by Fogarty NIH grant (PHS 1RO3 TW 01495 to A.R. Crofts with Dr. Rimma Samoilova, Institute of Chemical Kinetics and Combustion, Novosibirsk, Russia) and by NIH/NCRR grant S10-RR15878 for instrumentation is gratefully acknowledged.

REFERENCES

[1] Zavoisky, E. (1945) *J. Phys. USSR* **9**, 211-4.
[2] Weil, J.A., Bolton, J.R. & Wertz, J.E. (1994) *Electron Paramagnetic Resonance. Elementary Theory and Practical Applications* (New York: Wiley).

[3] Pilbrow, J.R. (1990) *Transition Ion Electron Paramagnetic Resonance* (Oxford: Clarendon).
[4] Atherton, N.M. (1993) *Principles of Electron Spin Resonance* (NY: Ellis Horwood).
[5] Carrington, A. & McLachlan, A.D. (1979) *Introduction to Magnetic Resonance* (London: Chapman and Hall).
[6] Poole, C.P. (1996) *Electron Spin Resonance: A Comprehensive Treatise on Experimental Techniques* (New York: Dover).
[7] Poole, C.P. & Farach, H.A. (1987) *Theory of Magnetic Resonance* (New York: Wiley).
[8] Abraham, A. & Bleaney, B. (1970) *Electron Paramagnetic Resonance of Transition Ions* (Oxford: Clarendon).
[9] Spaeth, J.M., Niklas, J.R. & Bartram, R.H. (1992) *Structural Analysis of Point Defects in Solids*, *Springer Series in Solid-State Sources*, *Vol. 43* (Berlin: Springer).
[10] Box, H.C. (1977) Radiation Effects: ESR and ENDOR Analysis (New York: Academic).
[11] Symons, M.C.R. & Atkins, P.W. (1967) *The structure of inorganic radicals: An application of Electron Spin Resonance to the Study of Molecular Structure* (Amsterdam: Elsevier).
[12] Mabbs, F.E. & Collison, D. (1992) *Electron Paramagnetic Resonance of d-Transition Metal Compounds, Studies in Inorganic Chemistry Series, Vol. 16* (Amsterdam: Elsevier).
[13] Slichter, C.P. (1996) *Principles of Magnetic Resonance* (Berlin: Springer).
[14] Gordy, W. (1980) *Theory and Applications of Electron Spin Resonance* (New York: Wiley).
[15] Pake, G.E. (1962) *Paramagnetic Resonance* (New York: Benjamin).
[16] Dixon, W.T. (1972) *Theory and Interpretation of Magnetic Resonance Spectra* (London: Plenum).
[17] Bencini, A. & Gatteschi, D. (1990) *EPR of Exchange Coupled System* (Berlin: Springer).
[18] Hoff, A.J. (Ed.) (1989) *Advanced EPR. Applications in Biology and Biochemistry* (Elsevier: Amsterdam).
[19] Kevan, L. & Bowman, M.K. (Eds.) (1990) *Modern Pulsed and Continuous-Wave Electron Spin Resonance* (New York: Wiley).
[20] Poole, C.P. & Farach, H.A. (Eds.) (1994) *Handbook of Electron Spin Resonance* (AIP Press).
[21] Poole, C.P. & Farach, H.A. (Eds.) (1999) *Handbook of Electron Spin Resonance Vol. 2*, (New-York: Springer).
[22] *A Specialist Periodical Reports. Electron Spin Resonance* (Cambridge: Royal Society of Chemistry).
[23] Schweiger, A. & Jeschke, G., (2001) *Principles of Pulse Electron Paramagnetic Resonance* (Oxford: Clarendon).
[24] Salikhov, K.M. Semenov, A.G. & Tsvetkov, Yu D. (1976) *Electron Spin Echo and Its Applications* (Science: Novosibirsk) (in Russian).
[25] Dikanov, S.A. & Tsvetkov, Yu D. (1992) *Electron Spin Echo Envelope Modulation (ESEEM) Spectroscopy* (Boca Raton, FL: CRC).
[26] Keijzers, C.P. Reijerse, E.J. & Schmidt, J. (Eds.) (1989) Pulsed EPR: A New Field of Applications (North Holland: Amsterdam).
[27] Kevan, L. & Schwartz, R.N. (Eds.) (1979) *Time Domain Electron Spin Resonance* (New York: Wiley).
[28] Mims, W.B. (1972) *Phys. Rev. B* **5**, 2409-19.
[29] Mims, W.B. (1972) *Phys. Rev. B* **6**, 3543-5.
[30] Bowman, M.K., Berry, E.A., Roberts, A.G. & Kramer, D.M. (2004) *Biochemistry* **43**, 430–6.

[31] Weil, J.A., Buch, T. & Clapp, J.E. (1973) *Adv. Magn. Reson.* **6**, 183-257.
[32] Shane, J.J., van der Heiden, P.A. Reijerse, A.W. & de Beer, E.J. (1994) *Appl. Magn. Reson.* **6**, 427-54.
[33] Pöpple, A., Simon, J. & Völkel, G. (1994) *Appl. Magn. Reson.* **6**, 455-70.
[34] Lippard, S.J. & Berg, J.M. (1994) *Principles of Bioinorganic Chemistry* (Mill Valley, CA: University Science Book); p. 86.
[35] Link, T.A. (1999) *Adv. Inorg. Chem.* **47**, 83-157.
[36] Guigliarelli, B. & Bertrand, P. *Adv. Inorg. Chem.* **47**, 421-497.
[37] Konovalova, T.A., Krzystek, J., Bratt, P.J., van Tol, J. Brunel, L-C. & Kispert, L.D. (1999) *J. Phys. Chem.* ***B***, **103**, 5782-6.
[38] Gerfen, G.J., Bellew, B.F. Un, S., Bollinger, Jr. J.M. Stubbe, J. Griffin, R.G. & Singel D.J. (1993) *J. Am. Chem. Soc.* **115**, 6420-1.
[39] Grimaldi, S. MacMillan, F. Ostermann, T. Ludwig, B. Michel, H. & Prisner, T. (2001) *Biochemistry* **40**, 1037-43.
[40] Kolling, D.R.J., Samoilova, R.I. Holland, J.T. Berry, E.A. Dikanov, S.A. & Crofts, A.R. (2003) *J. Biol. Chem.* **278**, 39747-54.
[41] Merks, R.P. & de Beer, R. (1979) *J. Phys. Chem.* **83**, 3319-22.
[42] Barkhuijsen, H., de Beer, R., de Wild, E.L. & van Ormondt, D. (1982) *J. Magn. Reson.* **50**, 299-315.
[43] Höfer, P., Grupp, A., Nebenführ, H. & Mehring, M. (1986) *Chem. Phys. Lett.* **132**, 279-84.
[44] Goldfarb, D., Kofman, V., Libman, J., Shanzer, A., Rahmatouline, R., Van Doorslaer, S. & Schweiger, A. (1998) *J. Am. Chem. Soc.* **120**, 7020-9.
[45] Dikanov, S.A. & Bowman, M.K. (1995) *J. Magn. Reson.* ***A*** **116**, 125-8.
[46] Dikanov, S.A., Davydov, R.M., Graslund, A. & Bowman, M.K. (1998) *J. Am. Chem. Soc.* **120**, 6797-805.
[47] Deligiannakis, Y., Louloudi, M. & Hadjiliadis, N. (2000) *Coord. Chem. Rev.* **204**, 1-112.
[48] Makinen, M.W. & Wells, G.B. (1987) *Metal Ions Biol. Systems* **22**, 129-206.
[49] Lakshmi, K.V. & Brudvig, G.W. (2000) *Biol. Magn. Reson.* **19**, 513-67.
[50] Brudvig, G. (1995) *Methods Enzymol* **246**, 536-54.
[51] Schweiger, A. & Ernst, R.R. (1988) *J. Magn. Reson.* **77**, 512-23.
[52] Bowman, M.K. (1993) *In Magnetic Resonance of Carbonaceous Solids Eds Boto*, R.E. & Sanada, Y. (Washington, DC: ACS) p. 91.
[53] Lorigan, G.A. & Britt, R.D. (1994) *Biochemistry* **33**, 12072-6.
[54] Hoffman, S.K., Hilczer, W. & Goslar, J. (1996) *J. Magn. Reson, A* **122**, 37-41.
[55] Du, J.-L., Eaton, G.R. & Eaton, S.S. (1996) *J. Magn. Reson, A* **119**, 240-6.
[56] Zhou, Y., Bowler, B.E., Eaton, G.R. & Eaton, S.S. (1999) *J. Magn. Reson* **139**, 240-6.
[57] Lawrence, C.C., Bennati, M., Obias, H.V., Bar, G. Griffin, R.G. & Stubbe, J. (1999) *Proc. Natl. Acad. Sci.* USA **96**, 8979-84.
[58] Pinhal, N.M., Vugman, N.V., Herbst, M.H. & Dias, G.H. (2000) *J. Phys. Chem.* ***A*** **104**, 1162-4.
[59] Tsvetkov, Yu D. (1983) *Russ. Chem. Rev.* **52**, 866-80.
[60] Raitsimring, A.M. & Salikhov, K.M. (1985) *Bull. Magn. Reson.* **7**, 184.
[61] Kurshev, V.V., Raitsimring, A.M. & Tsvetkov, Yu.D. (1989) *J. Magn. Reson.* **81**, 441-54.
[62] Saxena, S. & Freed, J.H. *Chem. Phys. Lett.* **251**, 102-10.
[63] Shigemori, K., Hara, H., Kawamori, A. & Akabori, K. (1998) *Bioch. Bioph. Acta* **1363**, 187-98.
[64] Ralowsky, M.H., Zecevic, A., Eaton, G.R. & Eaton, S.S. (1998) *J. Magn. Reson.* **131**, 97-110.
[65] Dzuba, S.A., Salikhov, K.M. & Tsvetkov, Yu.D. (1981) *Chem. Phys. Lett.* **79**, 568-72.
[66] Dzuba, S.A., Maryasov, A.G., Salikhov, K.M. & Tsvetkov, Yu D. (1984) *J. Magn. Reson.* **58**, 95-117.

[67] Gorcester, J., Millhauser, G.L. & Freed, J.H. (1990) In *Modern Pulsed and Continuous Wave Electron Spin Resonance* Eds. Kevan, L. & Bowman, M.K. (New York: Wiley) pp.119-94.

[68] Tyryshkin, A.M., Lyon, S.A., Astasthkin, A.V. & Raitsimring, A.M. (2003) *Phys. Rev.* **B 68**, 193207.

[69] Berliner, L.J., Eaton, G.R. & Eaton, S.S. (Eds.) (2000) *Biological Magnetic Resonance Vol. 19* (New York: Kluwer).

[70] Milov, A.D., Salikhov, K.M. & Schirov, M.D. (1981) *Solid State Physics* (Leningrad) **23**, 975-82 (in Russian).

[71] Milov, A.D., Ponomarev, A.B. & Tsvetkov, Yu D. (1985) *J. Struct. Chem.* **25**, 710-3.

[72] Schiemann, O., Weber, A., Edwards, T.E., Priesner, T.F. & Sigurdsson, S.T. (2003) *J. Am. Chem. Soc.* **125**, 3434-5.

[73] Pannier, M., Veit, S., Godt, A., Jeschke, G. & Spiess, H.W. (2000) *J. Magn. Reson.* 331-40.

[74] Jeschke, G., Bender, A., Paulsen, H., Zimmermann, H. & Godt, A. (2004) *J. Magn. Reson.* 1-12.

[75] Pannier, M., Schädler, V., Schöps, M., Wiesner, U., Jeschke, G. & Spiess, H.W. (2000) *Macromolecules* **33**, 7812-8.

[76] Milov, A.D., Tsvetkov, Yu D., Formaggio, F., Crisma, M., Toniolo, C. & Raap, J. (2001) *J. Am. Chem. Soc.* **123**, 3784-9.

[77] Kuroiwa, S., Tonaka, M., Kawamori, A. & Akabori, K. (2000) *Bioch. Bioph. Acta* **1460**, 330-7.

[78] Codd, R., Astashkin, A.V., Pacheco, A., Raitsimring, A.M. & Enemark, J.H. (2002) *J. Biol. Inorg. Chem.* **7**, 338-50.

CHAPTER 4

ENDOR SPECTROSCOPY

Lowell D. Kispert and Lidia Piekara-Sady
The University of Alabama, Chemistry Department, Tuscaloosa, Alabama, USA

4.1 INTRODUCTION

To understand what *electron nuclear double resonance* (ENDOR) transitions are, let us consider the energy levels of a system containing one unpaired electron (**S** = ½) and one nuclear spin (**I** = ½), as in a hydrogen atom. Since the electron spin and nuclear spin can each be oriented with or against the external magnetic field, there are four possible energy levels as shown in Figure 4.1. The two electron spin eigenfunctions are denoted by α_e and β_e, whereas the two nuclear spin wavefunctions are denoted α_n and β_n. In the presence of a large external applied field, the selection rule in electron paramagnetic resonance (EPR) is that the electron flips (changes directions) but the nuclear spin does not flip in an EPR transition. Thus, in Figure 4.1, an EPR transition will occur between energy levels 1 and 4 and a second EPR transitions will occur between energy levels 2 and 3. The intensity of the transition between energy levels 1 and 4 will depend on the population differences of these two energy levels. If one then applies a nuclear frequency to the system, which corresponds to either the energy difference between levels 4 and 3 or the difference between lines 2 and 1, one will induce nuclear spin flip that will change the populations in energy levels 4 and 1, respectively. This changes the intensity of the observed EPR signal. This change in intensity is referred to as the *ENDOR response*.

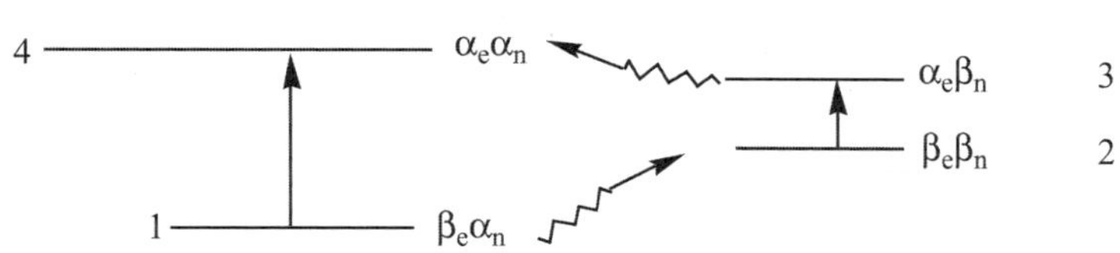

Figure 4.1 Energy diagram for **S** = 1/2 and **I** = 1/2.

Analytically, the energy levels are obtained from the spin-Hamiltonian appropriate for a system consisting of a single unpaired electron (**S** = ½) and a single magnetic nucleus (**I** = ½).

$$H = H_{EZ} + H_{NZ} + H_{HF} \tag{4.1}$$

where H_{EZ} is the *electronic Zeeman term* that describes the interaction of the electron spin (**S**) with a magnetic field **B** according to

$$H_{EZ} = \beta \mathbf{B}\cdot g\cdot \mathbf{S}. \tag{4.2}$$

g is a dimensionless number called the *g-factor* and β is the *Bohr magneton.* For a free electron, the g-factor is equal to 2.0023. However, the g value of an electron in an atomic or molecular orbital is different and may also be anisotropic, i.e., dependent on the orientation of the magnetic field **B** relative to the system.

H_{NZ} is the *nuclear Zeeman Hamiltonian* where

$$H_{NZ} = -g_n\beta_n \mathbf{B}\cdot\mathbf{I} \tag{4.3}$$

In contrast to the g value in EPR measurements, g_n is an inherent property of the nucleus and any corrections to the nuclear Zeeman energy are referred to as *screening effects* in NMR spectroscopy. The interaction of the unpaired electron (**S**) with nearby magnetic nuclei (**I**) gives rise to a hyperfine structure in the EPR spectrum. This electron-nuclear hyperfine interaction is described by the $H_{HF} = h\mathbf{S}\cdot\tilde{\mathbf{A}}\cdot\mathbf{I}$ Hamiltonian where h is the Planck constant so that the elements of the hyperfine tensor $\tilde{\mathbf{A}}$ are expressed in frequency units. This tensor can be separated into the sum of two terms giving

$$H_{HF} = ha\mathbf{S}\cdot\mathbf{I} + h\mathbf{S}\cdot\tilde{A}^{dip}\cdot\mathbf{I} \tag{4.4}$$

where the first term describes the *isotropic hyperfine interaction* (a is its constant) and the second describes the *anisotropic (dipolar) interaction* represented by the *hyperfine dipolar tensor* $\tilde{\mathbf{A}}^{dip}$.

The magnetic moments of the electron and nuclei are coupled via the contact (Fermi) interaction. The simplest system exhibiting a nuclear hyperfine interaction is the hydrogen atom, with a hyperfine constant of 1420 MHz. In more complex systems, i.e., in molecules, the general expression of the isotropic hyperfine constant is

$$a = 8/3\ \pi g\beta g_n\beta_n\rho(r_n) \tag{4.5}$$

where $\rho(r_n)$ is the *net unpaired electron-spin density at the nucleus.* The magnetic coupling also exists between the magnetic moments of an electron and a nucleus. The magnitude of this interaction depends on the orientation of the paramagnetic molecule relative to the external magnetic field **B**. In isotropic fluid solutions, the dipolar interactions are averaged out and only the orientation-independent isotropic couplings give rise to the observed hyperfine coupling in a spectrum. The complete Hamiltonian for an electron coupled to an oriented nucleus with spin **I** = ½ in the magnetic field is given by

$$H = H_{EZ} + H_{NZ} + H_{HF} = \beta\mathbf{S}\cdot\mathrm{g}\cdot\mathbf{B} - \mathrm{g}_n\beta_n\mathbf{B}\cdot\mathbf{I} + h\mathbf{S}\cdot\tilde{\mathbf{A}}\cdot\mathbf{I} \qquad (4.6)$$

A nucleus with $\mathbf{I} \geq 1$, exhibits an electric quadrupole moment **Q**, which interacts with the gradient of the electric field due to all surrounding electrons. This interaction is described by the quadrupole interaction term

$$H_Q = \mathrm{h}\mathbf{I}\cdot\mathrm{Q}\cdot\mathbf{I} \qquad (4.7)$$

where **Q** is the *quadrupole coupling tensor*. In liquids, this interaction is average out to zero. For the isotropic case of g and hyperfine coupling, as occurs for radicals in solution, the Hamiltonian given in equation (4.1) becomes:

$$H = \mathrm{g}\beta\mathbf{BS} - \mathrm{g}_n\beta_n\mathbf{BI} + ha\mathbf{S}\cdot\mathbf{I} \qquad (4.8)$$

In the high-field approximation, i.e., H_{EZ}>>H_{NZ} and H_{HF}, the Hamiltonian corresponds to a set of four energy levels:

$$E = \mathrm{m_s g}\beta\mathbf{B} - \mathrm{m_I g_n}\beta_n\mathbf{B} + ha\mathrm{m_s m_I}, \qquad (4.9)$$

where $\mathrm{g_n}$ and a may be positive or negative, thus leading to a different ordering of the levels. The EPR transitions are governed by the selection rules $\Delta \mathrm{m_S} = \pm 1$ and $\Delta \mathrm{m_I} = 0$. Adopting the notation

$$h\nu_e = \mathrm{g}\beta\mathbf{B} \qquad (4.10)$$

$$h\nu_n = \mathrm{g_n}\beta_n\mathbf{B}, \qquad (4.11)$$

two EPR transitions are obtained:

$$\nu_{EPR} = \nu_e \pm \frac{a}{2} \qquad (4.12)$$

and two NMR transitions:

$$\nu^{\pm} = \left|\nu_n \pm \frac{a}{2}\right|. \qquad (4.13)$$

The two NMR transitions defined by equation (4.13) are referred to as ENDOR *transitions*. The term "ENDOR'' is used to describe the indirect detection of the NMR transitions where the ENDOR transitions obey the selection rules $\Delta \mathrm{m_S} = 0$ and $\Delta \mathrm{m_I} = \pm 1$. The energy levels are separated into two cases: $|a|/2 > \nu_n$ and $|a|/2 < \nu_n$.

ENDOR spectra are shown schematically in Figure 4.2. For $\nu_n > |a|/2$, the two ENDOR transitions are given by $\nu_n \pm |a|/2$. Two ENDOR lines, separated by $|a|$ and centered at ν_n, are observed. The same two ENDOR transitions are observed when either EPR line is monitored, and in both cases the ENDOR lines are separated by exactly $|a|$. However, since ν_n depends on **B**, there will be a shift in the center of the spectrum as different EPR lines are monitored. For $\nu_n < |a|/2$, the two ENDOR transitions are given by $|a|/2 \pm \nu_n$. Two lines are observed separated by $2\nu_n$ and centered at $|a|/2$.

When there is more than one nucleus with $\mathbf{I} = ½$ and a particular a value, the interactions responsible for the various isotropic splittings are described by the following Hamiltonian:

$$H = g\beta \mathbf{B}\cdot\mathbf{S} - \Sigma g_n\beta_n \mathbf{B}\cdot\mathbf{I}_i + h\,\Sigma\, a_i \mathbf{S}\cdot\mathbf{I}_i. \tag{4.14}$$

For a group of equivalent nuclei having the same nuclear Larmor frequency ν_n and the same hyperfine constant $a_i = a$, the EPR transitions are

$$\nu_{EPR} = \nu_e + \mathbf{M}_I a, \tag{4.15}$$

where

$$\mathbf{M}_I = \sum_{i=1}^{n} m_{Ii}$$

and the summation is over n equivalent nuclei. The NMR transitions remain unchanged, $\nu^{\pm} = |\nu_n \pm \frac{a}{2}|$ no matter how many equivalent nuclei are coupled to the unpaired electron.

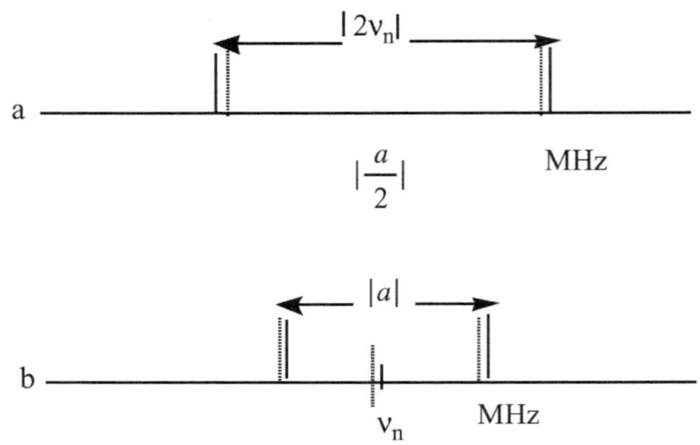

Figure 4.2 ENDOR line positions for (a) $|a|/2 > \nu_n$ and (b) $|a|/2 < \nu_n$. The solid lines give the line positions when the high field EPR line is monitored, the dashed lines when the low field EPR is monitored.

In general [1] the EPR spectrum will consist of:

(1) $(2\mathbf{I} + 1)^N$ hyperfine components, if there are N nuclei with nuclear spin $\mathbf{I}$.

(2) 2^N different hyperfine components for N nuclei with $\mathbf{I} = ½$, if all the coupling constants differ and no degeneracy occurs.

(3) $\Pi_i(2\mathbf{I}_i + 1)^{N_i}$ hyperfine components, if there are several nuclei with the individual values $\mathbf{I}_i$ and N_i; the symbol Π_i denotes the formation of a product; when some nuclei are equivalent, the number of lines is decreased but the amplitude of the components in the hyperfine pattern is increased.

In ENDOR there are always two lines separated by $|a|$ or $2\nu_n$ for any group of equivalent nuclei no matter how many nuclei there are or what their spin quantum number is. This results from the fact that each nucleus experiences the hyperfine field of only one unpaired electron, whereas an electron can couple with all nuclei.

ENDOR measurements are always useful in cases where poor resolution makes it impossible to determine the magnetic parameters from an EPR spectrum. In addition, the ENDOR lines for nuclei with different magnetic moments occur in different frequency ranges on account of their different Larmor frequencies $\nu_n = g_n\beta_n\mathbf{B}/h$ (see Table 4.1). The dramatic gain in resolution compared to EPR is presented in Figure 4.3.

From an ENDOR spectrum: (1) precise measurement of hyperfine coupling constants $|a|$ is possible when these are not obtainable from EPR spectra, and (2) the nucleus responsible for the hyperfine coupling can be identified from the relation $\nu^+ \pm \nu^- = 2\nu_n = 2g_n\beta_n\mathbf{B}/h$. Although only the absolute value of g_n (or γ_n) can be evaluated from the free nuclear frequency (ν_n), this value provides sufficient information to identify (see Table 4.1) the nucleus. Conversely, at a given experimental field value **B**, the frequencies of like nuclei can be calculated (Table 4.1) before the measurement is made, and the ENDOR spectrum obtained can be interpreted by comparison. ENDOR spectra also yield anisotropic hyperfine and quadrupole (for nuclei with **I** > 1) coupling constants in solids.

Table 4.1 Properties of selected nuclei.

Nucleus	I	Nat. abund. (%)	γ_n (MHz/T)[a]	ν_n (MHz)[b]	Q (10^{24}cm^2)
^{1}H	1/2	99.99	267.51	14.00	
^{2}H	1	0.015	41.06	2.15	0.00277
^{6}Li	1	7.42	39.37	2.06	0.007
^{7}Li	3/2	92.58	103.96	5.44	–0.03
^{10}B	3	19.78	28.75	1.50	00.74
^{11}B	3/2	80.22	85.83	4.49	0.036
^{13}C	1/2	1.11	67.26	3.52	
^{14}N	1	99.63	19.32	1.01	0.016
^{15}N	1/2	0.37	–27.11	1.42	
^{19}F	1/2	100	251.67	13.17	
^{23}Na	3/2	100	70.76	3.70	~0.14
^{25}Mg	5/2	10.13	–16.37	0.86	0.22
^{27}Al	5/2	100	69.71	3.65	0.15
^{29}Si	1/2	4.70	–53.14	2.78	
^{31}P	1/2	100	108.29	5.67	
^{39}K	3/2	93.10	12.48	0.65	0.11
^{85}Rb	5/2	72.15	25.83	1.35	0.27
^{87}Rb	3/2	27.85	87.53	4.58	0.13
^{117}Sn	1/2	7.61	–95.30	4.99	
^{119}Sn	1/2	8.58	–99.71	5.22	
^{133}Cs	7/2	100	35.09	1.84	–0.003
^{203}Tl	1/2	29.5	152.88	8.00	
^{205}Tl	1/2	70.5	154.38	8.08	

[a] 1 T (tesla) = 10^4 G.
[b] For a magnetic field value of 3288 G.

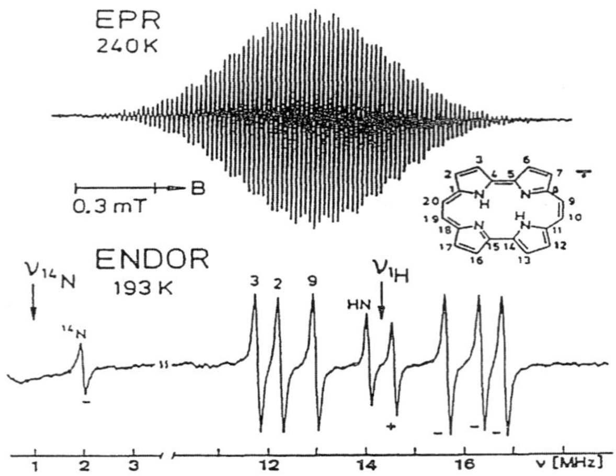

Figure 4.3 EPR, ^{1}H, and ^{14}N ENDOR spectra for free-base porphycene radical anion with the nuclear frequencies (ν_n) for ^{1}H and ^{14}N indicated. From [2].

4.2 EXPERIMENTAL CONDITIONS FOR ENDOR

The classical double resonance experiment is performed by applying microwave (mw) and radio frequency (RF) fields in a continuous fashion; this technique is called *continuous wave*, CW-ENDOR spectroscopy.

The fundamentals of the ENDOR experiment can be illustrated for a simple electron-proton system (Figure 4.1). First, the EPR spectrum is recorded. Then: (1) the magnetic field **B** is set at the resonance of one of the observed EPR transitions; (2) this transition is partially saturated by the appropriate microwave power (partial saturation allows maximum sensitivity). The saturation condition for EPR can be obtained from the condition satisfied by equation (4.16).

$$\gamma_n^2\ B_1^2\ T_{1e}T_{2e} \geq 1 \qquad (4.16)$$

where $\gamma_n = g\beta\dfrac{h}{2\pi}$ T_{1e} and T_{2e} are electron spin-lattice and relaxation times, and β_1 is the time-dependent magnetic field of the microwave radiation; (3) a second oscillating RF field $\mathbf{B}_2$ is applied perpendicular to the applied microwave $\mathbf{B}_1$ and to the externally applied field **B** to induce and saturate the NMR transition. The NMR condition for saturation has to be satisfied:

$$\gamma_n^2\ \mathbf{B}_2^2\ T_{1n}T_{2n} \geq 1 \qquad (4.17)$$

where γ_n is the gyromagnetic ratio for the nucleus, T_{1n} and T_{2n} are the nuclear spin-lattice and spin-spin relaxation times, and $\mathbf{B}_2$ is the time-dependent magnetic field at the radio frequency. The frequency of the RF field is swept and the EPR absorption is recorded as a function of NMR frequency.

When $h\nu_{RF} = h\,(\nu_n \pm |a|/2)$, the resonance condition for nuclear transition is fulfilled: saturated EPR transition can become desaturated by NMR transitions provided both transitions have energy levels in common. This desaturation of EPR transition is detected as a change (typically enhancement) in the EPR signal intensity vs. the NMR frequency.

4.2.1 Sensitivity, Magnetic Field Homogeneity, and Stability

Since the ENDOR signal-to-noise ratio is always some fraction of the ESR signal-to-noise ratio (for liquids, typically 1% of the ESR height), it is necessary to have a well-tuned ENDOR spectrometer. This requires that the ESR spectrometer be in perfect working order and ideally be capable of producing a signal-to-noise ratio of at least 100:1 or better from a standard weak pitch sample. The weak pitch (0.00033% pitch in KCl) sample contains 1×10^{13} spins cm^{-1} providing the ESR signal is recorded with a 1.7-G linewidth. Most good commercial instruments can achieve such a signal-to-noise ratio for 10^{13} spins (the effective length of a rectangular cavity is 1 cm). The magnetic field homogeneity of most commercial magnets used in ESR experiments is ±10 milligauss (mG), which is quite adequate. A homogeneity of this magnitude is usually necessary in order to properly investigate organic free radicals in solution where frequently EPR linewidths of 60 mG or smaller are encountered. Deuterium ENDOR may require slightly better homogeneity.

The most serious problem is one of magnetic field and microwave frequency stability. After setting the field to coincide with a selected position of the ESR line, the magnetic field and the microwave frequency should be invariant for the duration of the experiment. Since the experiments may take several hours, special attachments are added to the standard ESR spectrometers to provide adequate magnetic field and microwave oscillator stability required for ENDOR investigations. Small drifts in magnetic field can occur because of temperature changes in the immediate surroundings. The field and frequency can be prevented from drifting by using a field frequency lock system. An alternative method appears to work adequately whenever the cavity can be immersed in a refrigerant. In this case the cavity can be kept at constant temperature, and thus the microwave frequency remains reasonably constant for the duration of each ENDOR experiment. However, the magnetic field stability still remains a problem. ESR experiments at 35 GHz require ESR fields around 12.5 kG. Assuming 0.5 G for linewidth requires a magnetic field stability of 1 part in 10^5 or better over the duration of each ENDOR experiment. This type of stability can be obtained by locking the field to the nuclear magnetic resonance signal of deuterium (obtained by an NMR gaussmeter). Any change in the deuterium NMR signal is amplified and applied to a pair of coils on the magnet that compensate for any changes in the main magnetic field.

4.2.2 Sample Size

Sample shape and size are not really critical parameters. However, it is usually advantageous to have as large a sample size as possible to take advantage of the cavity-filling factor. The ESR signal intensity depends on the total number of spins in the active part of the cavity and not on the concentration per unit volume. Therefore signal intensity can be gained if the largest sample tube is used. However if the sample is at all polar, the microwave electric field will be absorbed by the electric dipole of the sample, and this may prevent detection of the magnetic dipoles by the microwave magnetic field. Physically this phenomenon is observed to be a loss of cavity **Q**. However, excellent solvents with low dielectric constants, such as ethyl benzene and toluene, do exist and are highly recommended for carrying out ENDOR investigations in solution. When a polar sample (i.e., ice) is frozen to a sufficiently low temperature (i.e., $T < 100$ K) the dielectric constant drops to nonpolar values and ENDOR can easily be done. Polycrystalline or glassy samples generally have broad ESR lines and thus both ESR and ENDOR sensitivity is decreased. With a large sample access cavity, it is possible to carry out successful ENDOR experiments on frozen glasses [1] or power samples. Most crystals can also be used. Thus the sample size or shape does not limit one from observing ESR and/or ENDOR spectra. However, a severe problem can arise in collecting ENDOR data from a crystal. In order to determine accurately the principal hyperfine splittings, ENDOR data must be collected from three orthogonal crystal planes. In many cases the crystal needs to be aligned to 2 minutes of an arc so very precise alignment is required.

4.2.3 Introduction of RF Power into Cavity

The method by which RF power is introduced into the cavity depends largely on the type of ENDOR experiment. For instance, ENDOR experiments on samples (transition metal complexes or semiconductors) that are carried out at 4 K normally require RF power levels on the order of 0.1 W or less because the paramagnetic relaxation times are long. As the temperature is raised above 4 K, relaxation times shorten and larger RF powers are required. Near room temperature, RF powers of 10 to 20 G are typically needed to observe free radicals in solution. When RF powers of 0.1 W or less are used, the RF coil can be wrapped around the sample and the entire sample and RF coil immersed in the liquid helium refrigerant or in cooled gas without any difficulties from RF power dissipation. At high RF power levels (1 to 20 G), it would be impossible to carry out an ENDOR experiment using this RF coil configuration as the liquid helium would boil away as the RF power is dissipated.

However this problem can be eliminated by placing the RF coil on the inside of the cavity walls, outside of a dewar insert extending through the cavity. This removes the load on the cryogenic system and a successful experiment can be carried out. The large RF coil also allows a large access area in the cavity, permitting the investigation of large amounts of liquids, frozen glasses, or large single crystals [1].

4.2.4 RF Power Level: CW versus Pulsed Schemes

Successful experiments have been carried out using a high power wideband amplifier in which no pulsed RF is used. Instead, 100 to 500 W of continuous RF power is used to irradiate the sample. Whether the 1 kW pulsed ENDOR delivering 100 W of radio frequency will be better than a CW spectrometer delivering 100 W of continuous RF power turns out to be dependent [1, 3] on how easily the nuclear signal can be saturated. If the nuclear signal is easily saturated, using continuous RF irradiation is better than a pulsed RF by a factor of $10^{1/2}$. On the other hand, if the nuclear signal saturates with difficulty, the pulsed RF will give signals that are $10^{1/2}$ times more intense than those with continuous RF.

4.2.5 Mode of Detection and Modulation Scheme

The results of several experiments show that the ENDOR lineshape and intensity depend critically on the detection and modulation scheme used [1, 3]. However, before the optimum field, RF modulation, and detection schemes can be established, it will be necessary to predetermine the ENDOR mechanism for each given free radical sample. This is unfortunately still not possible. Thus several different schemes are used, and a great deal of discussion is carried out on the merits of each.

4.2.6 ENDOR Mechanism

The diagram for a simple systems with **S** = ½ and **I** = ½ in Figure 4.4 shows the four levels involved in the relaxation [1]. All relaxation processes, both electronic and nuclear, must be considered, i.e., T_{1e} and T_{1n} are the electronic and nuclear relaxation times, respectively; T_1^{-1} describes the relaxation rates; T_{x1} and T_{x2} describe cross-relaxation processes in which electron and nuclear

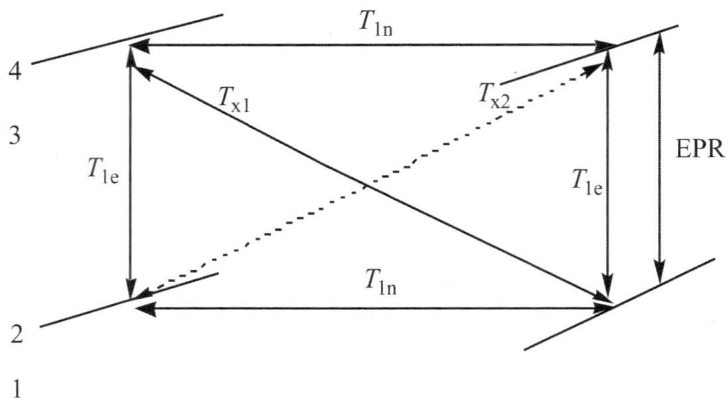

Figure 4.4 Schematic representation of the four energy levels for an **S** = ½, **I** = ½ system. The possible relaxation processes are characterized by the respective relaxation times. (The energy levels are separated horizontally to distinguish between the two electron traditions).

spin flips occur simultaneously (so-called *flip-flop* and *flop-flop processes* of electron coupled to nuclear spins). First, sufficient microwave power is applied to one of the two EPR transitions, e.g., (1) ↔ (3), to overpower the relaxation mechanism connecting the two states, so that their populations become almost equal.

Second, a radio frequency field, $\mathbf{B}_2$, at the NMR frequency is applied simultaneously. The ENDOR experiment can be described as the creation of an alternative relaxation path for electron spins which are irradiated with microwaves. The role of the RF field is to couple the nuclear and electron levels and open this alternative relaxation path. If the system is irradiated at the nuclear frequency corresponding to (3) ↔ (4) transition, e.g., then the relaxation can go via:

$$T_{x1}, \text{ i.e., } (4) \leftrightarrow (1) \tag{4.18}$$

or

$$T_{1e}, \text{ i.e., } (4) \leftrightarrow (2) \text{ and } T_{1n}, \text{ i.e., } (2) \leftrightarrow (1). \tag{4.19}$$

The extent to which the alternative relaxation paths can compete with the direct T_{1e} relaxation (3) ↔ (1) determines the degree of desaturation of the EPR, and consequently determines the ENDOR signal intensity.

After the ν^- transition is saturated, i.e., (3) ↔ (4) in Figure 4.4, the relaxation can go via T_{x1} or T_{1e} and T_{1n}, while for the ν^+ transition, i.e., (1) ↔ (2), the relaxation can occur only via T_{1e} and T_{1n} if $T_{x2} \sim 0$. In general, $T_{x1} \neq T_{x2}$ is effective in some cases.

If T_{x1} dominates relaxation, the intensity inequality of both ν^+ and ν^- ENDOR transitions is predicted because T_{x1} relaxation is effective only between the (1) and (4) levels. Their relative intensities depend on the amount by which cross-relaxation is faster than the nuclear spin-lattice relaxation process. If cross-relaxation is very slow and can be neglected, both ENDOR transitions can have equal intensity. The ENDOR effect is therefore described as the relaxation time T_{1e} effectively shortened by the application of an RF field at the NMR transition frequency.

4.2.6.1 Enhancement of Radio Frequency Field at the Nucleus

The RF enhancement effect is very important because it often allows detection of those nuclear transitions which would otherwise be difficult or even impossible to saturate with the RF fields $\mathbf{B}_2$ available in conventional ENDOR spectrometers, i.e., to meet the $\gamma_n^2\ \mathbf{B}_2^2\ T_{1n}T_{2n} \geq 1$ condition given by equation (4.17).

The effect occurs because the electric magnetic moment creates an additional magnetic field component (hyperfine field) at the nucleus. This can be described by an enhancement factor k, changing the magnetic field seen by a nucleus:

$$\mathbf{B}_2^{\text{eff}} = k\mathbf{B}_2, \tag{4.20}$$

where

$$k = |1 - \frac{\text{m}_s a}{\gamma_n \mathbf{B}}|\ | \tag{4.21}$$

The RF enhancement is particularly significant for nuclei other than protons with small magnetic moments (small γ_n), and large hyperfine constants a, e.g., ^{14}N. No enhancement exists for $\text{m}_s = 0$. Equation (4.21) holds only for the isotropic case, i.e., the liquid phase. In single crystals, the hyperfine interaction is anisotropic. Consequently, k is governed by different hyperfine components.

4.2.6.2 Signal Intensity

The absolute intensity of an ENDOR signal is not related to the fractional change in T_{1e} (desaturation) caused by the RF field, but also depends on the intensity of the EPR signal itself. The signals observed depend very critically on the balance between the various relaxation times and the magnitude of the microwave and the RF field.

In ENDOR spectra, the intensities of individual lines are not, in general, proportional to the number of nuclei contributing to the ENDOR effect, in contrast to EPR and NMR. The important criterion for assigning the hyperfine coupling constants to particular nuclei is lost, partially due to different nuclear relaxation mechanisms for different sets of nuclei. Additionally the effect of incomplete hyperfine separation of the EPR lines reduces the ENDOR signal. When microwave irradiation saturates one EPR transition, other EPR transitions may also be partially saturated, because of overlap. This effect depends on the hyperfine coupling, as well as a $T_{2e,}$ which defines the homogeneous EPR linewidth. When the experimental variables are optimized, the signal intensity in ENDOR spectroscopy is usually only a few percent of the EPR signal intensity.

The intensities of the low and high frequency lines in ENDOR have, in general, different magnitudes. One reason is the effectiveness of the cross-relaxation mentioned above. The relative intensity of the low- and high-frequency lines depends on which EPR line is monitored [Figure 4.4, (2) ↔ (4) or (1) ↔ (3)].

Another cause of different intensities is the RF enhancement effect. Frequently, the RF power available in ENDOR spectrometers is not constant over the whole frequency range; thus, line intensities are affected. Even when RF power is constant, two effects are observed as a consequence of RF enhancement: (1) for small hyperfine couplings, i.e., when the ENDOR signals are close to the free nuclear frequency, they are often very weak; (2) the high frequency line of an ENDOR line pair ($\nu^{\pm}$ pair) is usually more intense than the low frequency as the difference between the two components (ν^{+} and ν^{-}) increases. In spectra of single crystals, however, the RF enhancement factor is governed by different hyperfine tensor components. This often leads to unexpected intensity patterns of the ν^{+} and ν^{-} lines.

4.2.7 Extension of ENDOR: TRIPLE Resonance

The ENDOR technique has been expanded to include a second NMR exciting field, thus giving so-called *electron-nuclear-nuclear* (TRIPLE) *resonance*. The advantages of TRIPLE are: (1) enhanced sensitivity and resolution and (2) the possibility to directly determine relative signs of hyperfine coupling constants.

4.2.7.1 Special TRIPLE (ST)

The *special TRIPLE* (ST) technique is used for hyperfine couplings $|a|/2 < \nu_n$. Referring again to Figure 4.4, if two RF fields are tuned to simultaneously saturate both NMR transitions (ν^{+} and ν^{-}) of the same nucleus, the efficiency of the alternative relaxation path is enhanced.

Further, when both RF fields are sufficiently strong to completely saturate nuclear transitions, the EPR desaturation becomes independent of T_{1n}. Thus the line intensities are not determined by the relaxation of the various nuclei. As a consequence, an ST spectrum may reflect the number of nuclei involved in the transition [2].

4.2.7.2 General TRIPLE (GT)

In *general TRIPLE* (GT) resonance, NMR transitions of different sets of nuclei of the same kind (homonuclear TRIPLE) or of different kinds (heteronuclear TRIPLE) are saturated simultaneously. In this experiment, one EPR and one NMR transition is saturated, the NMR transition being saturated with the first RF while the second RF field is scanned over the whole range of NMR resonances. The characteristic intensity changes of the high and low frequency signals, compared with those of the ENDOR signals, give the relative signs of the hyperfine coupling constants. GT always gives larger TRIPLE line amplitudes for opposite signs of a than it does for equal signs, provided the g_n-factors of both nuclei have the same sign [4]. In general, ENDOR measurements are used to significantly improve the resolution of the EPR spectrum; directly identify the nuclei being measured; to determine the relative sign of the hyperfine coupling; to differentiate the species of overlapping spectra; and to give information obscured by inhomogeneous broadening. Steady state ENDOR requires the presence of a minimum of 10^{13} radicals in the sample, thus this technique is restricted to stable radicals or to relatively long-lived species with lifetimes of at least 10^{-3} s. ENDOR studies in liquids are confined to organic radicals including their ion pair complexes formed with alkali cations. For paramagnetic transition metal complexes in fluid solution the required microwave and RF fields cannot be achieved, so solution ENDOR spectra are almost never observed for transition metal nuclei, or any other nuclei present in a complex.

The hyperfine constants obtained from ENDOR spectra are useful in deducing the electronic structures of the radicals. Further, the temperature dependence of the spectra permits a study of the conformations and rates of intermolecular motions in radicals.

A major problem in EPR/ENDOR spectroscopy is the assignment of the observed hyperfine coupling constants to specific molecular positions in the radical. In solving this problem it is very helpful to obtain (1) intensity ratios of the signals from ST; (2) relative signs of the hyperfine constants form GT; (3) temperature dependence of the hyperfine constants; and the individual ENDOR linewidth; (4) computer simulation of the EPR spectra with ENDOR-derived coupling constants; and (5) ENDOR spectra of compounds containing specific isotropic labeling such as ^{2}H, ^{15}N, and ^{13}C.

4.3 ENDOR IN THE SOLID STATE

A variety of chemical systems and nuclei have been studied in the solid state by the ENDOR method, made possible by the wide temperature range available for solids. Since the relaxation processes affecting ENDOR signal intensity are temperature-dependent, changing the temperature can make the relaxation of the system favorable of the ENDOR experiment.

In the solid state, the hyperfine interactions usually give rise to only partially resolved EPR spectra. The unresolved couplings contribute to the EPR linewidth, giving inhomogeneously broadened lines. Linewidths in solid state EPR spectra are rarely less than a few megahertz (about 1 G), while in ENDOR spectra the linewidths range from 3 kHz to 1 MHz.

4.3.1 Single Crystals

4.3.1.1 Hyperfine Interactions

All isotropic and anisotropic data are accessible from ENDOR single-crystal studies. Since anisotropic interactions depend upon the orientation of the crystal in the applied magnetic field, the tensor elements can be determined by studying an oriented single crystal.

Let us assume, for simplicity, that g is isotropic so the spin Hamiltonian can be written as

$$H = g\beta \mathbf{B}\cdot\mathbf{S} - \Sigma\, g_n\beta_n \mathbf{B}\cdot\mathbf{I}_i + \mathrm{h}\,\Sigma\, \mathbf{S}\cdot\tilde{\mathbf{A}}_1\cdot\mathbf{I}_i, \tag{4.22}$$

where $\tilde{\mathbf{A}}_i$ is the hyperfine coupling tensor for the ith nucleus. For the strong field approximation (electron Zeeman interaction dominates, i.e., $H_{\mathrm{EZ}} >> H_{\mathrm{HF}}$ and H_{NZ}), in the principal axis system the observed splitting is

$$|\mathbf{A}_{\mathrm{obs}}| = \left[\mathbf{A}_{xx}^2 \ell_x^2 + \mathbf{A}_{yy}^2 \ell_y^2 + \mathbf{A}_{zz}^2 \ell_z^2\right]^{1/2} \tag{4.23}$$

where $\mathbf{A}_{ii}$ are the principal values of the hyperfine tensor, and the orientation of an axis system xyz with respect to the **B** field direction is specified by the ℓ_i direction cosines. If the isotropic coupling in the resonance condition for the ENDOR transition equation(4.13) is replaced by $\mathbf{A}_{\mathrm{obs}}$, the ENDOR frequencies ae given by

$$\nu^{\pm} = |\nu_n \pm m_s \mathbf{A}_{\mathrm{obs}}|$$

or, for $\mathbf{S} = \frac{1}{2}$, (4.24)

$$\nu^{\pm} = |\nu_n \pm \tfrac{1}{2}\mathbf{A}_{\mathrm{obs}}|.$$

From complete measurements of the dependence of the **A** values on the angle of rotation around the three mutually perpendicular axes, all components of the $\tilde{\mathbf{A}}$ tensor can be obtained.

The usual procedure in the EPR/ENDOR single-crystal study is to choose a set of reference axes xyz related to the crystallographic axes of the crystal. In general, the xyz-axis system will not coincide with the principal axes of any tensor. In three separate experiments, the crystal is oriented in the magnetic field with one of the reference axes perpendicular to **B**. The positions of the orientationally dependent ENDOR lines are recorded as the crystal is rotated about each axis. Since such experiments give the square of the elements of the hyperfine coupling tensor, the signs of the hyperfine coupling constants are not deduced. The tensor is then diagonalized, and the principal values of $\tilde{\mathbf{A}}$ are determined in its principal axis system. The orientation of this system relative to the reference axes is defined by means of direction cosines. Since the reference axes are related to the crystallographic axes, the location of the paramagnetic center can be identified.

In general, the total coupling tensor $\tilde{\mathbf{A}}$ is composed of isotropic and anisotropic (dipolar) components. The isotropic part is given by

$$a = \frac{1}{3}\left(\mathbf{A}_{xx} + \mathbf{A}_{yy} + \mathbf{A}_{zz}\right). \tag{4.25}$$

The $\mathbf{A}_{ii}$ principal elements of the $\tilde{\mathbf{A}}$ tensor contain both isotropic and purely dipolar components. The dipolar components of $\tilde{\mathbf{A}}$ are described by:

$$\begin{aligned} \mathbf{A}_{xx}^{\mathrm{dip}} &= \mathbf{A}_{xx} - a \\ \mathbf{A}_{yy}^{\mathrm{dip}} &= \mathbf{A}_{yy} - a \\ \mathbf{A}_{zz}^{\mathrm{dip}} &= \mathbf{A}_{zz} - a. \end{aligned} \tag{4.26}$$

For small anisotropies, i.e., when $|\mathbf{A}^{\mathrm{dip}}| << |a|$, the observed hyperfine coupling $\mathbf{A}_{\mathrm{obs}}$ can be approximated by:

$$\mathbf{A}_{\mathrm{obs}} = a + \left[\left(\mathbf{A}_{xx}^{\mathrm{dip}}\right)^2 \ell_x^2 + \left(\mathbf{A}_{yy}^{\mathrm{dip}}\right)^2 \ell_y^2 + \left(\mathbf{A}_{zz}^{\mathrm{dip}}\right)^2 \ell_z^2\right]^{1/2}. \tag{4.27}$$

In the common case of axial symmetry of $\tilde{\mathbf{A}}$, if z is a symmetry axis, the following notation is used:

$$\begin{aligned} \mathbf{A}_{xx}^{\mathrm{dip}} &= \mathbf{A}_{\parallel}^{\mathrm{dip}} \\ \mathbf{A}_{xx}^{\mathrm{dip}} &= \mathbf{A}_{yy}^{\mathrm{dip}} = \mathbf{A}_{\perp}^{\mathrm{dip}} \\ \mathbf{A}_{\parallel}^{\mathrm{dip}} &= -2\mathbf{A}_{\perp}^{\mathrm{dip}} \end{aligned}$$

and

$$\mathbf{A}_{\mathrm{obs}} = a + \frac{1}{2}\mathbf{A}_{\perp}^{\mathrm{dip}}(3\cos^2\theta - 1) \tag{4.28}$$

where θ is the angle between **B** and the symmetry axis z of the tensor.

An example of the hyperfine coupling angular dependence can be found in reference [1] pages 172–174 for a defect center (trapped) electron in a bromide ion vacancy site of a KBr crystal. Although the broad EPR signal (linewidth $\approx$ 55 G) gave no information about the structure of the defect center, the interaction of the electron with several shells of surrounding nuclei was recognized by the ENDOR method. The shells are composed of ^{39,41}K and ^{79,81}Br ions, each with spin $\mathbf{I} = 3/2$. Identification of the shell number was accomplished by a study of the angular dependence of the ENDOR lines, which were observed in the 0.5–26 MHz range. The hyperfine couplings to the nuclei in some shells have characteristically different angular dependencies. When the angular dependence of the hyperfine coupling of two shells is similar, the magnitudes of the hyperfine coupling are usually very different and the line pairs can be assigned.

i. Signs of the hyperfine coupling constants: In general, only the magnitude of the hyperfine coupling constants can be determined by EPR and ENDOR. The sign, however, is especially important, because it is related to the sign of the unpaired spin density and consequently to the electronic structure of the radical. The general TRIPLE resonance method is very helpful in determining the relative sign.

ii. Second-order shifts: Second-order effects in the hyperfine couplings measured by ENDOR can generally be detected for $\mathbf{A} \geq 30$ MHz. Although they do not contain any additional information, second order shifts must be included in the analysis in order to avoid errors in the calculated values of the elements of the hyperfine tensors.

4.3.1.2 Quadrupole Couplings

Nuclei having spin $\mathbf{I} \geq 1$ (e.g., ^{2}H, ^{14}N, ^{63}Cu, see Tables 4.1 and 4.2) possess an electric quadrupole moment in addition to a magnetic moment. If the interaction of a quadrupole moment with the electric field gradient at the nucleus is smaller than the hyperfine interaction, it will not affect the EPR spectra, to first order. However, in ENDOR spectra this interaction will lead to additional splittings. For an electron and a single nucleus with $\mathbf{I} \geq 1$, the quadrupole term, H_{Q}, should be added to the spin-Hamiltonian.

$$H = H_{\mathrm{EZ}} + H_{\mathrm{HF}} + H_{\mathrm{Z}} + h\mathbf{I} \cdot \mathbf{Q} \cdot \mathbf{I} \qquad (4.29)$$

where $\mathbf{Q}$ is the quadrupole coupling tensor. It is often assumed that the hyperfine coupling tensor $\tilde{\mathbf{A}}$ and the quadrupole coupling tensor $\mathbf{Q}$ have the same principal axis system. This coincidence of these $\tilde{\mathbf{A}}$ and $\mathbf{Q}$ tensors may be imposed by symmetry.

The energy level diagrams for an electron coupled to a nucleus with $\mathbf{I} = 1$ are given in [1] page 185. The resonance condition for the four ENDOR transitions is given by

$$\nu = \left|\tfrac{1}{2}\mathbf{A} \pm \nu_{\mathrm{n}} \pm \mathbf{Q}_{zz}\right| \qquad (4.30)$$

Table 4.2 Electronic and nuclear properties of the transition metal ions.

Ion	Electron configuration	Total spin **S**	Spin- orbit constant (cm^{-1})	Nucleus	Abundance (%)	Nuclear spin **I**	g_n	Quadruple moment (10^{24} cm^2)
Ti^{3+}	$3d^1$	1/2	154	^{47}Ti	7.32	5/2	–0.3153	...
				^{49}Ti	5.46	7/2	–0.3154	...
V^{3+}	$3d^2$	1	209	^{51}V	99.8	7/2	1.471	0.25
V^{2+}	$3d^3$	3/2	167					
Cr^{3+}	$3d^3$	3/2	273	^{53}Cr	9.55	3/2	–0.3163	...
Cr^{2+}	$3d^4$	2	230					
Mn^{3+}	$3d^4$	2	352	^{55}Mn	100	5/2	1.387	0.3
Mn^{2+}	$3d^5$	5/2	347					
Fe^{3+}	$3d^5$	5/2	...	^{57}Fe	2.21	1/2	<0.10	...
Fe^{2+}	$3d^6$	2	410					
Co^{3+}	$3d^6$	2	...	^{59}Co	100	7/2	1.328	0.5
Co^{2+}	$3d^7$	3/2	533					
Ni^{2+}	$3d^8$	1	649					
Cu^{2+}	$3d^9$	1/2	829	^{63}Cu	69.1	3/2	1.484	–0.16
				^{65}Cu	30.9	3/2	1.590	–0.15

The number of observed lines depends on which levels are common for a given EPR transition. All four lines are expected for the $m_I = 0$ EPR transition. Two ENDOR lines are observed for the $m_I = 1$ EPR (low field) transition, while for the $m_I = -1$ EPR (high field) transition another two ENDOR lines are observed. This separation of the ENDOR frequencies into two groups allows determination of the relative signs of the quadrupole and hyperfine interactions.

For **I** > 1, there are 4 × I ENDOR transitions given by

$$\nu = \frac{1}{2}\mathbf{A} \pm \nu_n \pm \mathbf{Q}(2m_I + 1) \tag{4.31}$$

Quadrupole couplings result from the electric field gradients at the nucleus and therefore give information about charge distribution in nearby chemical bonds.

4.3.2 Organic Free Radicals

Free radicals produced by ionizing radiation of organic single crystals, have been extensively studied. An ENDOR study results in the radical's identification and its orientation in the crystal. The radical formation and transformation mechanism can also be derived. The occurrence of overlapping radical spectra is more common in single crystals than in liquids, hence the ENDOR-induced EPR technique shown in [1], pages 97–98 is especially useful. Most radicals in single crystals exhibit g values with little anisotropy and range from 2.0023 to 2.0080.

4.3.2.1 Interpretation of the Hyperfine Tensors for Organic Free Radicals

The magnitudes of the hyperfine tensor components are determined from the ENDOR spectra. The relative signs can be obtained with the general TRIPLE method. The absolute sign of the principal values is often given on the basis of theoretical considerations. The isotropic and anisotropic parts of the tensor are usually separately listed. The orientation of the principal axes relative to the crystal axis is defined in terms of the direction cosines.

Because protons are the most extensively studied nuclei, the proton hyperfine tensors will be considered. Protons are labeled by Greek letters depending on how many bonds away they are from the radical center; the proton attached directly to the free radical center is labeled α. The α -proton dipolar tensors exhibit large anisotropy amounting to about 50% of the isotropic coupling for unit spin density at the radical center. The anisotropy results from electron dipole-nuclear dipole magnetic interactions. The a_{iso} of α -protons is characterized by a negative sign.

The anisotropy of protons attached to a β carbon atom is typically about 10% of the a_{iso}, for unit spin density. This small anisotropy is the result of the greater (two bonds) distance of the β proton from the atom on which the unpaired spin density is located. β-proton coupling constants are characterized by a positive sign of a_{iso}. The isotropic coupling to β-protons varies as $\cos^2\theta$ equation (4.33) where θ is the dihedral angle between two planes, both of which contain the α-carbon-β-carbon bond axis. The magnitude of a_{iso} of β-protons is related to the radical geometry. It is therefore possible to study the internal rotation of the radical fragment by analyzing the β-proton coupling constants. A typical example is the methyl group rotation. Methyl protons appear equivalent at high temperatures, i.e., their hyperfine interaction can be described by one tensor with axial symmetry around the C_3 axis of rotation. On lowering the temperature, however, it is often observed that methyl group rotation is stopped (on the EPR time scale). The hyperfine interactions for the protons are then described by their individual hyperfine tensors. The orientation of the "static" protons with respect to the rest of the radical can be determined.

γ-proton couplings are generally much smaller than those of α protons (by about two orders of magnitude), so they usually cannot be directly measured by EPR, but can be detected by ENDOR methods.

The special advantage of single-crystal ENDOR over EPR measurements is the detection of small hyperfine coupling constants (typically from 100 kHz). These small couplings are assigned to weakly coupled protons within the molecule or lattice protons that are dipolar-coupled to the free radical. From such couplings accurate dipole-dipole distances can be deduced and from these, in turn, the precise structure and location of the radical can be determined.

4.3.3 Transition Metal Ions

EPR spectra are routinely observed for paramagnetic transition metal ions in crystals, chemical complexes, and biomolecules. For ions with large spin-orbit coupling constants (see Table 4.2), the spin system is strongly coupled to the lattice vibrations so that spin relaxation is very effective. Consequently, the EPR lines are sometimes too broad to be observed at room temperature, and EPR and ENDOR studies of such compounds are usually carried out at much lower temperatures.

The central problem in EPR studies of transition metal ions is the effect of the surrounding atoms, by means of symmetry, on their electronic levels. The EPR spectra are characteristic of the transition metal ion itself rather than the surroundings. Weak hyperfine couplings to more distant nuclei are usually hidden in the inhomogeneously broadened EPR line. In contrast, ENDOR spectra provide both strong and weak hyperfine as well as quadrupole coupling tensors. The unpaired electron in transition metal complexes is mostly localized on the metal ion. The magnetic moments and the *g* factors of transition metal ions in crystals or in molecular environments are anisotropic.

The hyperfine interaction for the transition metal ion itself is described by the usual hyperfine term: $H_{HF} = h\mathbf{S}\cdot\tilde{\mathbf{A}}\cdot\mathbf{I}$, where **I** is the spin of the nucleus (see Table 4.2). The isotropic and dipolar part of the $\tilde{\mathbf{A}}$ tensor can be separated. The isotropic coupling is due to mixing of the ground and excited state configurations. In the absence of covalent bonding, the spin densities at the ion are not very sensitive to the environment of the ion. The typical situation for transition metal ion ENDOR is that $H_{HF} > H_{NZ}$, so the ENDOR lines appear asymmetrically around the value of **A**/2 (for **S** =½) because of the second-order effects. On the other hand, for ligand-ENDOR often $H_{HF} < H_{NZ}$ so the ENDOR lines are centered about the free nuclear frequency ν_n. The identity of the ligand nuclei can be determined from the number of ENDOR transitions and the value of ν_n.

As an example, the ^{63}Cu ENDOR study for a Cu mixed-ligand complex doped into the corresponding Ni complex single crystal is shown in Figure 4.5 [5]. The observed ^{63}Cu ENDOR spectrum Figure (4.5c) depends on which EPR line is saturated, provided both EPR and NMR transitions have levels in common Figure (4.5a). The ^{31}P ENDOR lines appear in a different frequency range (5.5–7 and 17–18.5 MHz).] The spectra were analyzed in terms of the

a)

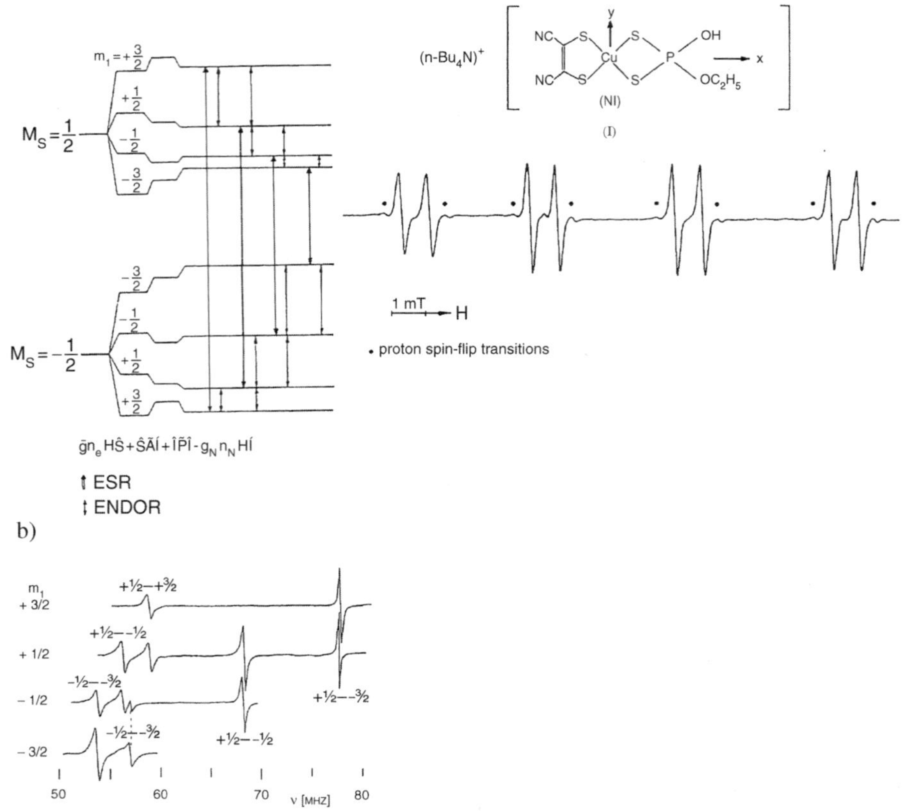

Figure 4.5 (a) ^{63}Cu energy levels and ENDOR transitions for each saturated EPR line. To the right is a typical single-crystal EPR spectrum of the Cu complex I at 27.2 K. The Cu hyperfine lines are split into doublets because of hyperfine interaction of the unpaired electron with ^{31}P. Proton spin flip transitions correspond to $\Delta m_S = 1$, and $\Delta m_I = 1$, the so-called *forbidden transitions* (b) Complete ^{63}Cu ENDOR spectrum recorded for each EPR line saturated as indicated in (a) (for $\mathbf{B} \perp g_i$, i.e., in the plane of smallest g anisotropy). From [5].

Hamiltonian expressed as the sum of the following interactions: electron Zeeman, Cu hyperfine, Cu nuclear Zeeman, Cu quadrupole, P hyperfine, and P nuclear Zeeman:

$$H = \beta \mathbf{B}\cdot\mathbf{g}\cdot\mathbf{S} + \mathbf{S}\cdot\tilde{\mathbf{A}}^{Cu}\cdot\mathbf{I}^{Cu} - g_{Cu}\beta_n \mathbf{B}\cdot\mathbf{I}^{Cu} + \mathbf{I}^{Cu}\cdot\mathbf{Q}^{Cu}\cdot\mathbf{I}^{Cu} + \mathbf{S}\cdot\tilde{\mathbf{A}}^{P}\cdot\mathbf{I}^{P} - g_p\beta_n \mathbf{B}\cdot\mathbf{I}^{P} \quad (4.32)$$

The analysis gave very precise values of the Cu hyperfine and quadrupole tensors as well as of the phosphorus hyperfine coupling tensor. The principal axes $\tilde{\mathbf{A}}^{Cu}$ tensor were found to coincide with those of **g**. Further, the relative orientations of the principal axes of $\mathbf{Q}^{Cu}$ and $\tilde{\mathbf{A}}^{P}$ with respect to those of g were determined so that the structure and symmetry of the Cu complex were established [5].

4.3.3.1 Triplet State ENDOR

An atom or a molecule with the total spin of the electrons **S** = 1 is said to be in a *triplet state*. The multiplicity of such a state is (2**S** + 1) = 3. Triplet systems occur in both excited and ground state molecules, in transition metal ions with **S** = 1, in pairs of paramagnetic centers, and in some defects in solids.

For a system with **S** = 1, there are three sublevels characterized by $m_S = \pm 1$ and $m_S = 0$. In the absence of an external magnetic field, these sublevels, in contrast to the systems with **S** = ½, may not be degenerate. The lifting of degeneracy of the spin states in the absence of a magnetic field is called *zero-field splitting*, and it is common for systems with **S** ≥ 1.

The zero-field splitting (or fine structure) term must be included in a spin-Hamiltonian when **S** ≥ 1. It can be expressed as a tensor interaction of the total effective spin with itself, i.e.,

$$H_{ss} = \mathbf{S}\cdot\mathbf{D}\cdot\mathbf{S}. \tag{4.33}$$

The spin-spin coupling (or zero-field splitting) tensor **D** is a symmetric and traceless tensor. The fine structure term is often written in terms of zero-field constants D and E in a principal axis system:

$$H_{ss} = D\left\{\mathbf{S}_z^2 - \frac{1}{3}\mathbf{S}(\mathbf{S}+1)\right\} + E\left(\mathbf{S}_x^2 - \mathbf{S}_y^2\right), \tag{4.34}$$

where D is the axial and E is the rhombic zero-field parameter. When $\mathbf{S}=1$, $H_{ss} = D(\mathbf{S}_z^2 - \frac{2}{3}) + E(\mathbf{S}_x^2 - \mathbf{S}_y^2)$. With **B** parallel to either the x, y, or z axis, a pair of EPR lines will be observed with couplings equal to ⅓ $D - E$, ⅓$D + E$, and –2/3D. For axial symmetry, $E = 0$, and two of the states will remain degenerate at zero magnetic field. The zero-field parameters D and E can, in general, be determined from the EPR spectrum (for $H_{SS} < H_{EZ}$). ENDOR studies of triplet states are especially useful because the hyperfine structure is rarely revealed in the EPR spectrum.

The total spin-Hamiltonian for the triplet system is given by:

$$H = H_{EZ} + H_{SS} + H_{HF} + H_{NZ} \tag{4.35}$$

with the H_{SS} term the second largest one because typically $H_{SS} > H_{HF}$ and H_{NZ}, and $H_{SS} < H_{EZ}$. For triplet states of aromatic molecules, g is nearly isotropic because of the small spin-orbit interaction. Thus, most of the observed anisotropy is due to spin-spin coupling as a result of magnetic dipolar interactions of the unpaired electrons in the triplet state.

In general, for metal ions with **S** ≥ 1 in crystals, g may be quite anisotropic. The spin-orbit couplings are appreciable and account for the zero-field splitting. The zero-field splittings for transition metal ions may be much larger than for organic molecules, and may even exceed the Zeeman energy.

The energy level diagram for the triplet state corresponding to the zero-field and electron Zeeman splittings is given in Figure 4.6, while the energy levels with hyperfine and nuclear Zeeman interactions, to illustrate the ENDOR spectrum, are shown in Figure 4.7. Saturation of the EPR lines will produce an ENDOR line at the free nuclear frequency, and one other ENDOR line. When the high-field EPR transitions are observed, the respective two ENDOR lines will be detected, one of them at the free nuclear frequency. Neglecting the ENDOR line at ν_n, the general expression for the ENDOR frequencies is

$$\nu_{\text{triplet ENDOR}} = |\mathbf{A} \pm \nu_n|, \tag{4.36}$$

where either the plus or the minus sign is used when a single EPR triplet transition is observed.

ENDOR studies of triplet systems must be preceded by an EPR study to determine the **D** tensor. In general, it appears that ENDOR lines of triplets are observable only when the magnetic field is almost perpendicular to a principal axis of the hyperfine tensor, often being parallel to the principal axis of the **D** tensor.

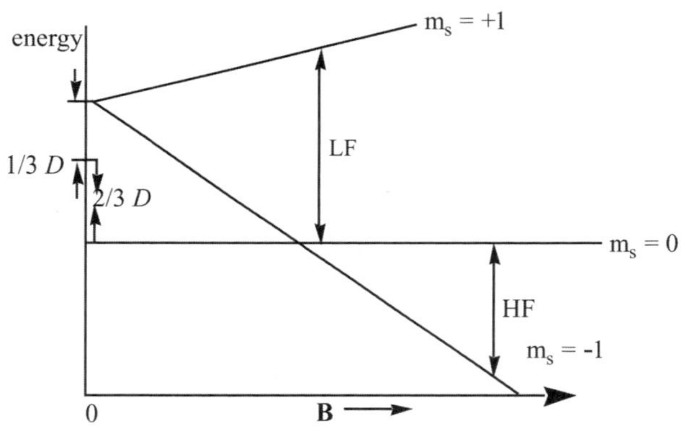

Figure 4.6 Energy of the triplet state with axial symmetry ($E = 0, D > 0$) for $\mathbf{B} \parallel z$. The EPR $\Delta m_S = \pm 1$ transitions are indicated by LF (low-field) and HF (high- field). The electron Zeeman and spin-spin interaction are taken into account, and $H_{SS} < H_{EZ}$. At zero magnetic field the triplet energy levels $m_S = 0$ and $m_S = \pm 1$ are split by D. Because the symmetry is assumed to be axial, two of the states remain degenerate.

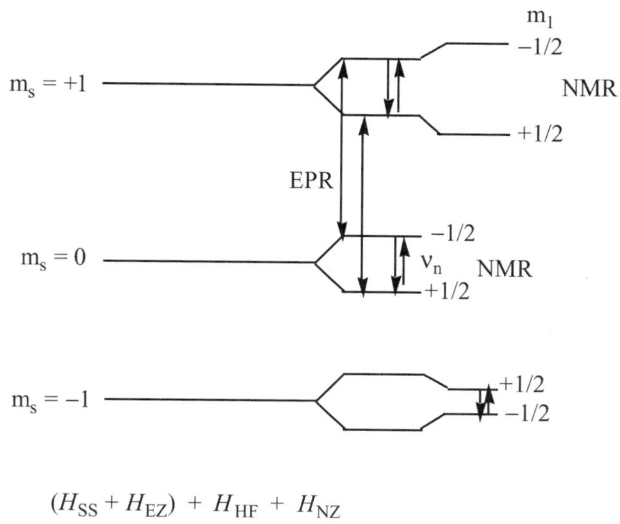

Figure 4.7 Energy levels for the triplet state at a fixed magnetic field and microwave frequency appropriate for the indicated EPR transitions. The EPR transitions correspond to the low-field (LF) EPR line in Figure 4.6. The additional splittings of the triplet state energy levels come from hyperfine (H_{HF}) and nuclear Zeeman (H_{NZ}) interactions. The two upper NMR transitions will be observed as an ENDOR spectrum.

In the ideal case, only one ENDOR line below or above the free nuclear frequency is observed for each group of equivalent protons. So if the sign of the dipolar tensor **D** is known, the sign of the hyperfine component can be determined from the ENDOR spectra. From the hyperfine tensors both spin density and geometrical information can be obtained.

4.3.4 Disordered Solids

ENDOR experiments in disordered solids are usually carried out in two cases: the paramagnetic ion or molecule can be used to probe its environment or the paramagnetic system is studied in frozen or solid solution because dilute single crystals are not available.

There are, in general, two different types of ENDOR spectra in disordered solids: (1) A powder-type spectrum is observed when all orientations contribute equally to the spectrum; the ENDOR spectrum is the same no matter which portion of the EPR line is saturated; a powder-type spectrum is also obtained when all regions of the EPR spectrum are saturated at the same time even if not all orientations contribute equally to the EPR powder pattern. (2) A crystal-like spectrum can be observed when the various orientations of paramagnetic molecules do not contribute equally to the EPR powder spectrum; the ENDOR pattern depends on the portion of the EPR spectrum being saturated; the EPR transitions can be saturated selectively.

A characteristic feature of ENDOR spectra of disordered solids is the appearance of a strong line occurring at the free nuclear frequency, called a *matrix ENDOR line*. It originates from purely dipolar coupling of the unpaired electron with magnetic nuclei of the surrounding matrix at distances within roughly 6 Å. In liquids this line is averaged to zero by the rapid tumbling of the radical. The molecular motions in disordered solids, in general, affect the intensity of the matrix ENDOR.

4.3.4.1 Powder-Type Spectra

When all orientations of the paramagnetic molecules with respect to the direction of the magnetic field are present, the ENDOR lines are broadened and may be very weak. An average of all orientations of the hyperfine coupling tensors for each individual nucleus can be measured. Thus, the ENDOR spectrum for a nucleus extends over the whole range of the hyperfine values with some build-up of intensity at the principal values of the hyperfine tensor. Narrow lines are observed if the anisotropy is small. For an α proton in organic radicals the anisotropy typically amounts to 50% of the isotropic value of a unit spin density, and hyperfine principal values are roughly $a/2$, a, and $3a/2$. Thus the α-proton powder lines are rarely observed. The anisotropy of β protons is much smaller, typically about 10% of the isotropic value.

For a system of $\mathbf{S} = \frac{1}{2}$ and one proton with axially symmetric hyperfine coupling to the electron spin, if the coupling is small, the ENDOR frequency can be expressed by

$$\nu^{\pm}(\theta) = \nu_n \pm \tfrac{1}{2}\,[\mathbf{A}_{||} \cos^2\theta + \mathbf{A}_{\perp} \sin^2\theta], \tag{4.49}$$

where θ is the angle between the magnetic field and the symmetry axis. When all orientations are equally probable, the powder-type spectrum is observed (Figure 4.8) giving the principal values of the hyperfine tensor.

This favorable situation typically occurs for the protons of a rotating methyl group. When the protons are equivalent on the EPR time scale, the hyperfine tensor is often axially symmetric with the largest component along the C-C bond direction.

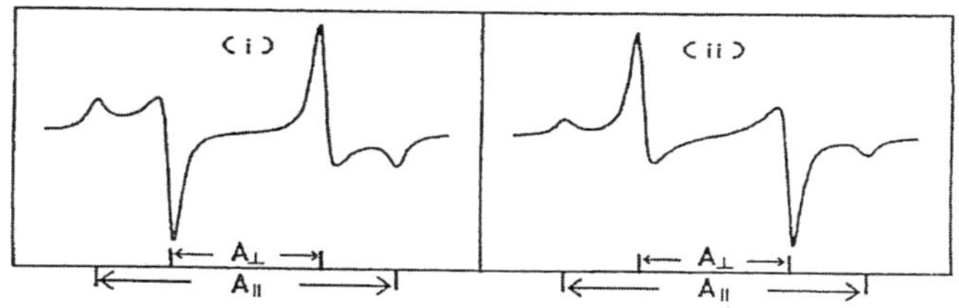

Figure 4.8 ENDOR powder lines for a proton with axially symmetric hyperfine interaction; (i) $\mathbf{A}_{\parallel}^{dip} = 2\mathbf{A}_{\perp}^{dip}$ and (ii) $\mathbf{A}_{\parallel}^{dip} = -2\mathbf{A}_{\perp}^{dip}$. From [6].

4.3.4.2 Crystal-like Spectra

Crystal-like ENDOR spectra in disordered systems can be observed when the magnetic field is set at the so-called *turning points* in the EPR powder pattern [6]. This is realized in the systems with considerable anisotropy of the magnetic interaction, i.e., in the following favorable situations: (1) the anisotropy of the **g** tensor dominates other magnetic interactions; this situation is typical for biomolecules; (2) when the anisotropy of the **g** tensor is small, but the hyperfine anisotropy is considerable (free radicals); (3) the anisotropy of the electron spin-spin interaction (described by the **D** tensor) is dominant (organic biradicals); and (4) the anisotropies of two magnetic interactions are of the same order of magnitude, but the axes with the largest anisotropy coincide (e.g., Cu^{2+} complexes).

Because of the anisotropy of the magnetic interaction, different orientations of the paramagnetic molecules with respect to the magnetic field correspond to different resonance conditions. Thus, a characteristic powder pattern, sometimes well resolved, is given by EPR spectroscopy. The EPR spectrum should be analyzed to locate the appropriate EPR field values where orientational selection of the molecules can be observed. The magnetic field set to these so-called turning points induces resonance in molecules having a unique orientation. The ENDOR spectra taken at these field settings will result in hyperfine tensor components relative to the tensor axes of the respective dominating interactions. In contrast to powder-type spectra, the crystal-like lines are characterized by linewidths close to those found in single crystal spectra. At other settings of the magnetic field, the EPR spectrum arises from molecules in many orientations and the respective ENDOR linewidths are much broader.

Figure 4.9 gives an example of how the frequency and width of the observed ENDOR lines depend on the portion of the EPR spectrum that is saturated [7]. In this system [$Cu(pic)_2$ in $Zn(pic)_2 \cdot 4H_2O$ powder] the **g** tensor and Cu hyperfine tensor are nearly coaxial. The symmetry axes of both tensors are oriented perpendicular to the plane of the complex. Because of the axial symmetry, there is only one turning point in the EPR spectrum. The EPR resonances with the magnetic field parallel to the molecular plane correspond to the high field end of the powder pattern, while the transitions with the magnetic field perpendicular to the complex plane occur at low field. By setting the magnetic field in the low field region of the EPR spectrum (indicated by "a" in Figure 4.9a), the perpendicular orientation is selected. The ENDOR spectrum taken at this field value (Figure 4.9b, line a) has a crystal-like character. It consists of two ^{14}N hyperfine lines, each split by quadrupolar coupling. Thus the nitrogen $\tilde{\mathbf{A}}$ and $\mathbf{Q}$ coupling tensor components with respect to the symmetry axis of **g** and the Cu hyperfine tensors were determined.

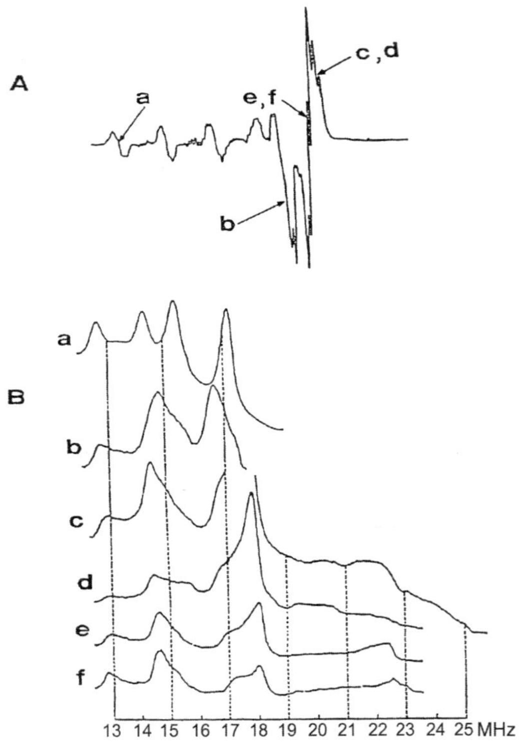

Figure 4.9 $Cu(pic)_2$ in $Zn(pic)_2{\cdot}4\ H_2O$ powder. (A) EPR spectrum. The magnetic field settings chosen for the ENDOR spectrum are indicated by arrows. (B) ^{14}N ENDOR spectra referring to the field settings indicated in the EPR spectrum. For the magnetic field set to a the ENDOR spectrum has a crystal like character. c and d as well as e and f differ only by a small shift in the field. From [7].

However, all the components of the hyperfine tensor cannot be determined form the crystal-like spectra. Additionally, when the **g** and the hyperfine tensors are not coaxial, the hyperfine component thus measured, in general, does not correspond to a principal value.

Several theoretical approaches for analyzing ENDOR spectra taken at arbitrary field values and corresponding to multiple orientations, have been reported [8, 9]. Consequently, ENDOR data can be used to characterize quite general cases of hyperfine coupling. The elements of the hyperfine tensor $\tilde{\mathbf{A}}$

and its relative orientation to the **g** tensor frame can be deduced from the variation of the hyperfine coupling value at different field settings across the EPR spectrum. These techniques are called *angle* or *orientation selection.*

In the angle selection method, ENDOR spectra are recorded at each point in the EPR powder pattern including the turning points. From crystal-like spectra approximate principal values are obtained. Simulation of selected spectra is then carried out by varying the respective parameters. For a series of metalloenzymes, ENDOR lines obtained with the angle selection method of many nuclei including ^{1}H, ^{2}H, ^{13}C, $^{14,15}N$, ^{17}O, ^{33}S, ^{57}Fe, $^{63,65}Cu$, and $^{95,97}Mo$ have given valuable information about the sites of the metal center and a bound ligand or substrate [10].

The angle selection method has been applied to many different systems: transition metal complexes [11] in which the **g** tensor anisotropy dominates; nitroxide radical [12] with strongly anisotropic nitrogen hyperfine couplings; and organic biradicals [13] in which the magnetic anisotropy results from the electron-electron coupling (**D**) tensor.

Besides the requirement of considerable anisotropy of the magnetic interaction in the system, another important condition must be fulfilled for one to observe crystal-like spectra and apply the angle selection method: Saturation of particular EPR transitions should be possible. This condition is satisfied when the process causing transfer of saturation from one orientation to another is slow compared to the spin-lattice relaxation rate. Such a process is called *spin diffusion.*

Spin diffusion is slow relative to the spin-lattice relaxation in transition metal complexes and biomolecules containing metal ions. In contrast, in the case of radicals in organic molecular crystals (at low temperature) the spin-lattice relaxation is slower than the spin-diffusion processes. Consequently, the saturation is transferred to all other EPR transitions in a time short with respect to T_{1e}, the electronic spin-lattice relaxation time. As a result, powder-type ENDOR spectra are observed [14]. At higher temperatures, however, the spin-diffusion may be slower, making selective saturation possible [12].

Regardless of the type of ENDOR spectra observed in disordered solids, they are much easier to interpret than EPR powder patterns.

4.4 PULSED-ENDOR

The conventional CW-ENDOR method has a serious disadvantage: The intensity of the ENDOR signals depends critically on the balance between the relaxation and induced transition rates (saturation conditions for both EPR and NMR transitions have to be satisfied). The experimental parameters [4], in general, cannot be independently optimized. Thus, the ENDOR transitions are often not observed because of unfavorable relaxation rates. Moreover, even when all experimental variables are optimized, the ENDOR signal intensity is only a few percent of the EPR signal. The pulsed ENDOR techniques offer greater sensitivity, higher spectral resolution, as well as reduced time of the measurement. The most important advantage of pulsed-ENDOR (or sometimes called ESE-ENDOR or *time-resolved* ENDOR) over CW is that there is no restriction to balance relaxation and induced transition rates. Consequently, the temperature is not such a critical parameter in pulsed ENDOR spectroscopy. Additionally, the pulsed technique permits a study of transient radicals.

In a CW experiment both microwave (mw) and radio frequency fields are applied continuously to the sample during the experiment. In a pulsed ENDOR experiment the pulse sequence, composed of short, intense mw (of the order of hundreds of nanoseconds) and rf (microsecond) pulses, is applied. A pulse of an oscillating field is characterized by its frequency ω as well as its time of duration t. Both of these parameters can be adjusted to invert the population of the respective levels. The general scheme (Figure 4.10) for a population transfer pulsed-ENDOR experiment (*Davies sequence*) is as follows:

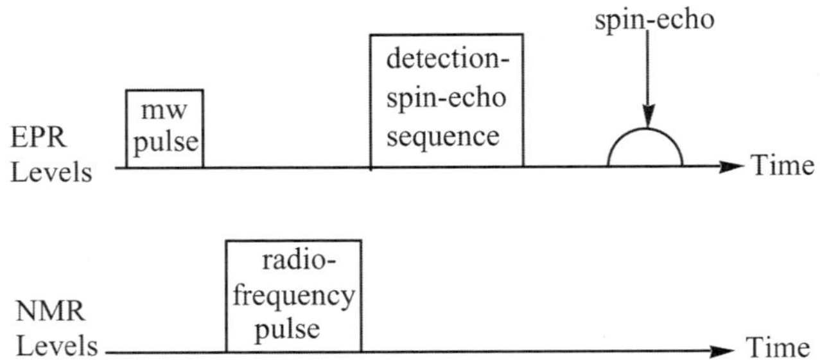

Figure 4.10 Schematic representation of the population transfer pulsed-ENDOR (Davies) method.

(1) First, a microwave π pulse is applied to invert the population of the selected EPR transition.
(2) After a time, short compared to the electron spin-lattice relaxation time, the RF π pulse is applied. Depending on the excitation frequency of this pulse (NMR frequency), the pulse will or will not change the population of the nuclear transitions.
(3) The effect of the RF pulse on the EPR transition is then detected by an electron spin-echo sequence.

While varying the NMR frequency, the whole sequence is repeated (at a rate slower than the spin-lattice relaxation rate). A plot of the amplitude of the spin-echo vs. the NMR frequency constitutes the ENDOR spectrum.

There is another commonly used pulsed sequence referred to as *Mims* ENDOR. In this technique the π pulse at the RF frequency is applied between the second $\pi/2$ and third $\pi/2$ mw pulse (two $\pi/2$ pulses to invert EPR population) in contrast to the Davies ENDOR, where the RF pulse is applied between the first π and first π/2 microwave pulses: The Mims ENDOR experiment is particularly effective for weakly coupled nuclei but some blind spots do occur where frequencies cannot be observed [15]. It is often advantageous to combine data from both the Mims and the Davies ENDOR response [16]. For nuclei with large magnetic moments such as ^1H and ^{19}F, ν_n at the X-band is often larger (14 MHz) than the hyperfine field and so the ENDOR response is centered at the Larmor frequency ν_n of the nucleus and is split by the hyperfine coupling $\mathbf{A}_n$ and by quadrupole interactions. For nuclei with smaller magnetic moments like ^2H or ^{14}N, the ENDOR response is centered at half the hyperfine coupling frequency and split into two lines that are separated by twice the Larmour frequency. Often times, ν_n = 14 MHz at the X-band is comparable to the magnitude of many nuclear hyperfine couplings, which results in substantial overlap of the proton ENDOR signals with signals from other nuclei. This problem is solved by recording ENDOR spectra at higher (fields such as 35 (Q-band) [17] or 95 (W-band) [18] GHz microwave frequencies where ν_n increases proportionally to the external field but $\mathbf{A}_n$ is independent of field, and results in a separation of overlapping ENDOR lines.

Several pulsed displays have been developed. As an example, Figure 4.11 shows (a) the overlapping ^1H and ^{14}N resonances in the Davies-type ENDOR spectrum and (b) the selected ^{14}N resonances obtained by the hyperfine-selective ENDOR techniques [19]. More complete details of the pulsed-ENDOR technique are given in [19–21]. Before attempting these books, an easier-to-understand review of the basic principles in pulsed-ENDOR and the related electron spin-echo envelope modulation (ESEEM) spectroscopy has been given [16]. An earlier review of pulsed-ENDOR techniques is given in [15].

A review of ESE-ENDOR techniques, tailored to the study of magnetic nuclei coupled to paramagnetic metals and radicals, has been published with illustrative examples from metal and radical centers in photosystem II [22]. The pulsed-Davies ENDOR experiment for S = ½ and I = ½ was developed for W band (95 GHz) ENDOR effects observed at low temperature [23].

Two-dimensional TRIPLE experiments have been carried out on orientationally disordered disordered samples and shown it can resolve overlapping powder patterns. This makes it possible to resolve and determine the relative orientation of the hyperfine tensors and the relative signs of the hyperfine couplings [24].

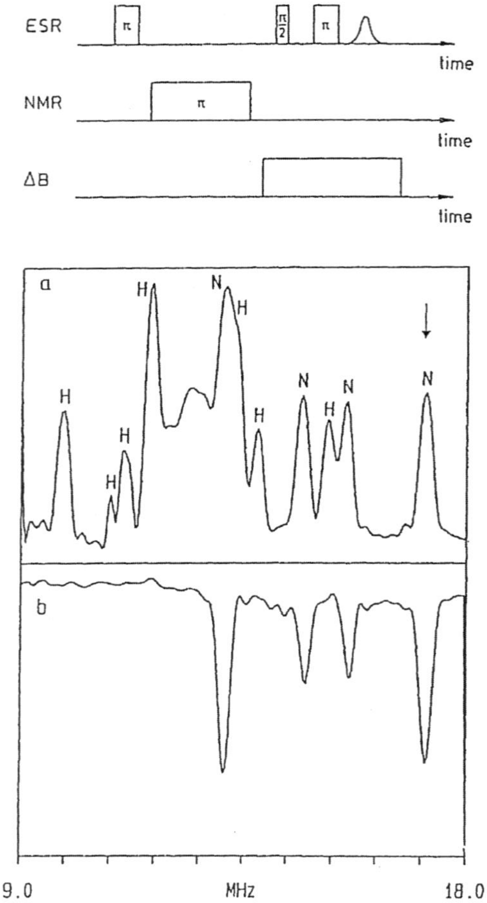

Figure 4.11 $Cu(pic)_2/Zn(pic)_2$ single crystal. (a) Conventional (Davies-types) pulsed-ENDOR spectrum with ^1H and ^{14}N lines using the pulse sequence given above. (b) Hyperfine-selective ^{14}N ENDOR spectrum. Figure 4.9 shows powder spectra. From [19].

Commercial ENDOR spectrometers at X-, Q- and W-bands are available from Bruker BioSpin Corp. in North America (epr@bruker.com) and Bruker Biospin GmbH in Europe (erp@bruker-biospin.de). The CW-ENDOR X-band spectrometer is known as an ELEXSYS E-560-D, which uses a TM cavity with vertical access. The pulsed-EPR at X-band is known as E-560-D-P using a TE cavity (cylindrical), vertical access. The pulsed-ENDOR system at Q-band (34 GHz) is the ELEXSYS E-580-Q and the pulsed-ENDOR system at W-band is known as the ELEXSYS E-680, while the CW-ENDOR at W-band is the E-600.

4.5 APPLICATIONS

ENDOR measurements can be used extensively to gain resolution in a complex spectrum, to separate overlapping EPR spectra, and to identify the nuclei responsible for the coupling. This technique has been applied to various systems. In the following section, examples are given for studying (a) organic radicals in organic host crystals, (b) radicals trapped in xenon, argon, or Freon matrices, (c) triplet state radicals in crystals or polycrystalline samples, (d) free radicals in biological systems, (e) polymeric systems, (f) inorganic radicals in irradiated inorganic single crystals, (g) paramagnetic complexes in organic single crystals, (h) F and H centers found in inorganic host lattices, (i) paramagnetic inorganic ions in organic crystals, (j) transition metal ion complexes observed in frozen solution and powders, (k) defects and complexes on surfaces such as silica alumina, zeolite, and nafion, (l) impurity centers in semiconductor host crystals, (m) spin centers in silicon and borate systems, (n) paramagnetic centers in cubic host crystal and (o) perovskite-type materials. Selected examples are given in each area with references noted.

A detailed tabulation of the published ENDOR data for single crystals, polycrystalline solids and biological materials for 58 isotropic species through 1993 has been given in the *Handbook of Electron Spin Resonance* [25]. Earlier ENDOR studies in disordered matrices, crystalline systems, hemes and hemoproteins, iron-sulfur proteins, radiation biophysics, polymer studies, and triplet state systems in solids were reviewed in Dorio and Freed (eds.) [26].

4.5.1 Organic Radicals in Organic Host Crystals

Often the crystals are irradiated at a low temperature and then the radicals identified as the temperature is raised. Electron attachment and proton abstraction occur at different temperatures so intermediate radicals can be formed. An example of the principal values of the hyperfine couplings for the α- and β-protons and the ^{14}N couplings for the radical formed [27] upon X-irradiation of deoxycytidine-5'-phosphate-hydrate at 11 K and measured by X-band ENDOR techniques at 6 K is given in Table 4.3. Two Mrad/hour of irradiation was applied for 2–4 hours. Because of the increased resolution, the couplings can be determined more precisely and it is possible to detect the presence of several radicals. Radicals I and II_a and II_b were identified at 6 K. Typically the primary radical is formed at the lowest temperature as a result of reaction with the electrons formed from the X-irradiation. However, as the temperature is raised, the unstable primary radical decays—forming other products which may or may not be radicals. An example [28] where different radicals were identified as a function of temperature occurred upon 4.2 K X-irradiation of a single crystal of citric acid (2-hydroxy-1,2,3-propane tricarboxylic acid), partially deuterated and warmed to 100 K. ENDOR analysis achieved upon cooling the radicals formed at 100 K to 4.2 K, thus quenching the reaction, are given in Table 4.4.

Radical I is formed by the loss of a carboxylic acid group from the 2-position and radical II is due to electron attachment to the carboxylic acid group located at the 3-position. Upon warming to 300 K, radical II decays and radical III is detected where loss of the α-proton from the 3-position is detected. Two nonequivalent crystal sites are detected for radical III and the α-couplings for all sites have been determined. It was also found [28] that radical I was stable at 300 K; demonstrating that upon decay of radical I, radical III is an end-product of the decay of radical II. Secondary radical products stable at room temperature have been studied to examine whether the free radical position in the crystal is the same as that of the undamaged molecules. The ENDOR resolution permits small couplings (on the order of 1 MHz due to distant protons) to be precisely measured. From the resulting proton tensor, dipolar distances can be estimated and thus proton-couplings from neighboring molecules and hydrogen-bonded protons can be assigned based on crystal structure data.

Table 4.3 ENDOR hyperfine coupling parameters at 6 K of 11 K X-irradiated deoxycytidine 5'-phosphate-H_2O (5' dCMP) single crystal) [27].

Principal Values (mT)

	Radical I		Radicals IIa and IIb
	−2.243 ± 0.008		−2.260 ± 0.005
H_α	−1.592 ± 0.005	$H_{\alpha 1}$	−0.732 ± 0.010
	−0.651 ± 0.005		−1.184 ± 0.008
	a_{iso} = −1.495		a_{iso} = − 1.392
	1.381 ± 0.004		−2.275 ± 0.003
H_β	1.439 ± 0.004	$H_{\alpha 2}$	−0.587 ± 0.003
	1.623 ± 0.004		−1.190 ± 0.010
	a_{iso} = 1.481		a_{iso} = −1.350
	1.520		
$N_1 + N_3$	0.615		
	0.011		
	a_{iso} = 0.715		

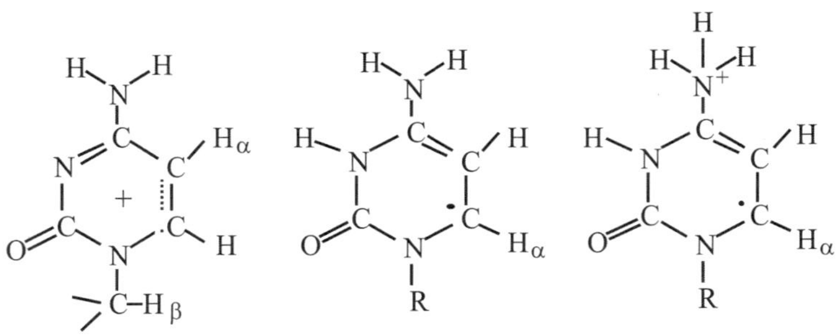

Radical I Radical IIa Radical IIb

Table 4.4 Citric acid single crystal, partially deuterated. X-irradiated at 4.2 K [28]. 4.2 K ENDOR measured β-proton couplings for radicals I and II after crystal warmed to (100 K).

	Radical I 4.2 K		Radical II 4.2 K		Radical I Stable Conformation RoomTemperature		Radical III Stable RoomTemperature
	Hyperfine Tensors (MHz)		Hyperfine Tensors (MHz)		Hyperfine Tensors (MHz)		
	113.9		90.7		105.2		
β_1	106.7	β_5	80.5	β_1	96.1	α'	–94.2
	102.3		79.4		92.0	(site 1)	–58.6
	107.5		23.1		82.0		–31.9
β_2	98.9	β_6	16.7	β_2	70.2		
	97.5		1.2		68.0		
	40.0				77.6		–92.3
β_3	26.0				70.3	α''	–58.3
	24.2			β_3	64.7	(site 2)	–31.0
	27.1				32.0		
β_4	12.7			β_4	19.3		
	11.3				17.8		

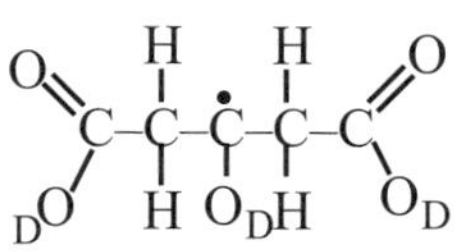

Radical I

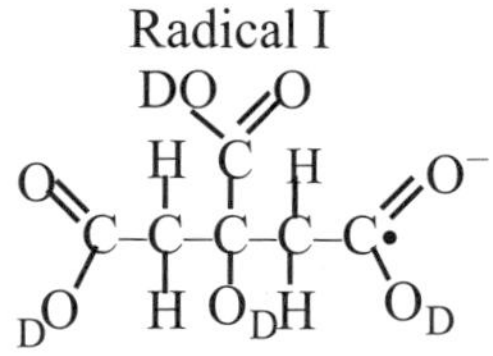

Radical II

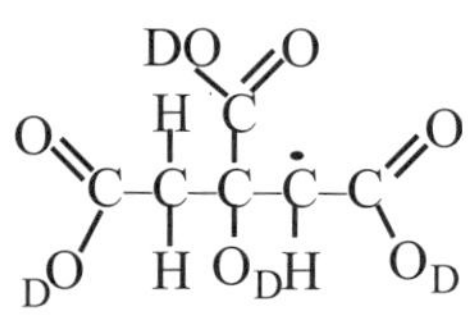

Radical III

In Table 4.5 is given an example [29] where potassium hydrogen malonate single crystals were γ-irradiated at room temperature and ^1H and ^{13}C couplings were measured for the $^-OOC^{\cdot}HCCOO^-$ radical in the host lattice.

Table 4.5 γ-irradiated potassium hydrogen malonate (KH) single crystal [29]. ^{1}H and ^{13}C hyperfine tensors (MHz) from ENDOR measurements at room temperature.

α-^{1}H	^{13}C	
–27.4	23.2	$^{-}$OOCCHCOO^{-}
–86.8	33.0	radical
–56.3	207.8	

Hyperfine tensor of the distant protons.[a]

Principal			Values		(MHz)
Proton[b]	Isotropic		Dipolar		Distance $R_{C...H}$ (Å)
A	–1.14	4.32	–2.36	–1.96	3.32
B	0.04	4.28	–2.32	–1.83	3.33
C	0.07	1.95	–0.89	–1.07	4.33
D	0.10	1.79	–1.14	0.66	4.45
E	–0.04	1.83	–1.21	0.61	4.42

[a] The signs of the couplings are not obtained from the experiment.
[b] Protons of neighboring molecules and hydrogen-bonded protons; protons assignment refers to the crystal structure data.

In this study, it was found that upon loss of an α-hydrogen, the resulting free radical retained the same position as the parent molecule. Often this is not the case, and considerable movement upon radical formation occurs, all dependent on the intermolecular packing arrangements of the original crystal. ENDOR measurements of nuclei other than ^{1}H and ^{13}C can also be carried out. Studies of deuterium [30], nitrogen [27] ^{23}Na [31], and chlorine-35 [32] hyperfine coupling constants, as well as their?coupling constants have been reported. From such data, structural relationships can be deduced. For example the relationship between deuterium quadrupole constants and the N-D and D…O bond lengths in a perdeuterates sulphamic acid (ND_3SO_3) single crystal were reported [30]. Example of the couplings for D and ^{35}Cl deduced are given in Table 4.6.

ENDOR measurements at 1.6 K have been used to deduce the exchangeable dissociation proton coupling, the α-proton, the exchangeable O-H proton in the undamaged structure and in the alkoxy radical, as well as the γ and β proton, and δ-methyl protons in a crystal of rhamnose X-irradiated at

Table 4.6 Perdeuterated sulphamic acid (ND_3SO_3) single crystal. ^2H ENDOR at 4.2 K [30]. X-irradiated phenancyl chloride: ^1H and ^{35}Cl hyperfine coupling constants and ^{35}Cl quadrupole coupling constant (MHz), ENDOR at 123 K [32].

Deuteron Quadrupole Tensors			Phenancyl chloride Single Crystal		
Deuteron				Eigenvalues of $\mathbf{Q}_{ii}$ (kHz)	
D1	z	257.4		(–)26.5	
	y	–169.6	$\mathbf{A}(^1H)$	(–)52.5	a_{iso} = –5.30
	x	–87.8		(–)80.0	
	$e^2q\mathbf{Q}/h = 171.6$				
	$\eta = 0.370$				
	z	273.0		(–)08.3	
	y	–163.9	$\mathbf{A}(^{35}Cl)$	(–)15.8	a_{iso} = 7.3
D2	x	–109.1		45.6	
	$e^2q\mathbf{Q}/h = -182.0$				
	$\eta = 0.201$				
	z	266.0		5.3	O
	y	–182.2	$\mathbf{Q}(^{35}Cl)$	5.9	||
D3	x	–83.8		–11.2	ØCCHCl
	$e^2q\mathbf{Q}/h = 177.3$				radical
	$\eta = 0.318$				

[a] Errors in tensor components are ± 2 kHz.

guanosine 5'-monophosphate single crystals have been reported [34]. 4.2 K [33]. Examples of radicals formed at 10 K and 65 K in X-irradiated

Radicals formed in γ-irradiated organic single crystals often exhibit internal motions, methyl groups rotate rapidly and t-butyl group rotate slowly on an ENDOR time scale. This motion is reflected in the temperature dependence of the electron spin relaxation times as detected by pulsed-EPR measurements [35]. Typical examples are given by the study of irradiated 4-methyl-2,6-di-t-butylphenol [35] and irradiated 10-nonadecanone in urea [36] where wobbling of the methylene groups and rotational motion of the radical in the channel were detected. ENDOR studies can be useful for determining the radiation chemistry of co-crystallized DNA base pairs [37, 38]. Varying temperature can be useful to follow the interconnecting of the planar and puckered conformers of the adduct pyrroline ring [39] by ENDOR measurements.

4.5.2 Radicals Trapped in Matrices

^{1}H ENDOR measurements have been recorded for radicals trapped in xenon, argon, or freon matrices and some typical examples can be found in references [40–48]. Typically the radicals are formed by X- or γ-irradiated frozen matrix at 77 K or 4 K. Protons from methyl groups or groups undergoing tunneling rotation give rise to narrower lines that are easy to detect, although exo- and endo-protons of a radical cation of s-trans-beta-1,3-diene has been reported [42].

4.5.3 Triplet-State Radicals in Crystal Polycrystalline Samples

Triplet-state radicals can be examined from ENDOR measurements. For example, proton couplings have been measured for the triplet state of naphthalene [49], and substituents of naphthalene [49] where intramolecular and intermolecular hyperfine interactions have been deduced for X-traps. The triplet-state of diphenyl-h_{10} in diphenyl-d_{10} single crystals indicate that the triplet-state diphenyl-h_{10} molecules have no inversional symmetry [50, 51]. The localized triplet-state (x-trap) at 4.2 K was also examined by ENDOR measurements in a charge-transfer crystal of naphthalene-tetracyano- benzene [52]. From the symmetry of the spin density distribution on the donor and acceptor, it was found that the charge-transfer (CT) sate is distributed over one donor and two acceptors. Other CT states have also been studied [53, 54]. It was deduced from a study of the triplet-state of benzil in bibenzyl that the dicarbonyl fragment has to be precisely *trans*-planar [55] while the lowest triplet state of trichlorobenzene is an excimer [56] whose structure is dimeric and located in one of the stacks of the crystal with a *mis*-orientation of 20°.

The photoexcited triplet state of free-based porphycene has been studied by time-resolved ENDOR spectroscopy at 10 K [57]. The *cis* or *trans* configuration was deduced by comparing the $\mathbf{A}_z$ coupling tensor with that predicted by density functional theory: A time-resolved ENDOR study of the photoexcited triplet state of the free-base tetraphenyl chlorine has revealed the nature of the orientation of the solute molecules [58]. Triplet energy transfer between porphyrin subunits was detected by ENDOR measurements at liquid helium temperature of the triplet state of several substituted porphyrins [59]. ^{1}H ENDOR analysis of P 700$^+$ radical cation was reported and the measurements were consistent with a slight redistribution of spin density from P_B chlorophyll to P_A as well as a slight redistribution of spin within the P_B chlorophyll [60]. Recent electron spin-cho-ENDOR was used to investigate the structure of the photosystem II oxygen-evolving complex (OEC) including the paramagnetic manganese cluster and its surroundings. The measurements suggested that the substrate waters bind within the OEC early in the S-state cycle [61].

4.5.4 Free Radicals in Biological Systems

Numerous examples exist where free radicals that occur in biological media have been examined by ENDOR measurements. For example, cytochrome oxidase has been studied by a number of authors [62–67]. ^{1}H and ^{14}N ENDOR measurements have identified two different protons with fairly large isotropic couplings of 12 and 19 MHz [62], the histidine N atoms [65] and ^{1}H couplings for protons near a heme site [65], identification of histidine ligands [64] in cytochrome a and evidence of bisimidazole coordination in cytochrome a and in cytochrome c oxidase [66]. ^{63}Cu ENDOR spectra have been observed for cytochrome c oxidase [63, 67]. This established the presence of Cu in the redox center [63] and the electronic structure of the $Cu^{2+}{}_{A}$ center was monitored by hyperfine couplings to clustering protons and the change correlated with the anti-cooperative redox interaction between cytochrome a and Cu_A [67]. High- and low-spin ferric hemoglobins and myoglobin (aquo derivatives) were studied and from the ENDOR couplings, couplings to exchangeable heme-bond water protons and the ^{13}C coupling in myoglobin and hemoglobin [68] were measured. ^{57}Fe ENDOR was recorded for azotobacter vinelandii Mo-Fe proteins in frozen solution [69], bidirectional hydrogenase [70], oxidized hydrogenase [71], costridium pasteurianum [72], escherichia coli sulfite reductase [73], hydrogenase-I [74], hydrogenase-II [71], and nitrogenase molybdenum [75].

^{17}O ENDOR has been recorded in beef heart aconetase [76] and in horseradish peroxidase [77]. In horseradish peroxidase, the O atom comes from the oxidant, the Fe portion of the system is an oxy-ferryl ($Fe^{IV} = O$) moiety. The [4Fe-4S] cluster was studied using ^{17}O ENDOR in beef heart aconetase. $^{95,97}Mo$ ENDOR was recorded for MoFe proteins [72], and ^{95}Mo protein [79], where ^{95}Mo hyperfine and quadrupole coupling constants indicated that a single Mo atom was integrated into the MoFe spin system. ^{33}S ENDOR has been recorded for MoFe protein [69] and ^{19}F ENDOR has been detected for flavin radicals [80] in flavins [81] and flavo-proteins [81] in different protonation sites. The influence of different substituents on the spin distribution within the isocelloxazine ring was deduced [81]. Proton ENDOR was recorded for photosystem-II [82], $P_{865}{}^{+}$ [83], P_{960} [84], and numerous other systems.

The effects of ionizing radiation on single crystals of quanine hydrobromide monohydrate were studied by ENDOR and field-swept-ENDOR techniques. Five different radicals were identified and comparison to other systems containing the N_7-protonated quonine bases demonstrated the effect of the environment on the primary radical formation [85]. Time-resolved ENDOR measurements were carried out on the spin-correlated radical pair $P700^{+}$... A^{-1} in frozen solution samples of PSI [86]. It was found that the mixed electronic eigenstates give rise to transitions not usually found in ENDOR and that absorptive/emissive TR-ENDOR spectra are observed.

Pulsed-ENDOR spectroscopy at X- and W-band were employed to probe the strongly coupled ^{14}N nuclei and the β-protons of the CuA center perturb by the axial methionine ligand. The high-field data resolved the Tz and CuA signals in the g perpendicular region and gave ^{1}H ENDOR spectra free of overlapping ^{14}N signals [87]. ENDOR measurements were also used to characterize the oxygen-evolving center Mn cluster [88]. Examples of Q-band ENDOR have been reported where measurements were made on the protein-bound, high-affinity ubisemiquione radical to determine its electronic spin distribution [89]. ^{1}H pulsed-ENDOR and ESEEM measurements have shown that the bis-imidazole complexes of iron(III) tetraphenylchlorine and tetraphenylporphyrin have the same g values and the same electronic ground state [90]. Several reviews are available that can provide additional information. The ENDOR crystallography of disordered systems of some Cu(II) S-containing chelate complexes and oxovanadium(II) chelate complexes in the solid state and in frozen solution has been reviewed [91]. A review has been published on the Q-band-ENDOR studies of the nitrogenase MoFe protein in which it was demonstrated how to analyze orientational selective ^{13}C and ^{57}Fe ENDOR spectra. From such analysis, the binding modes and the charge of the cofactor can be derived [92]. A briefer review details how ENDOR techniques can be used to understand metal ion functions in proteins and enzymes [93].

4.5.5 Polymeric Systems

^{1}H and ^{13}C ENDOR studies have been reported for polymeric systems [94–120]. For example studies of pristine *cis*(I) and *trans*-polyacetylene (II) have shown an isotropic coupling of 1.8 MHz for poly(methyl methacrylate) and an activation energy of 34 K. On the other hand for poly(methyl-α-chloro-acrylate, an isotropic hyperfine coupling of 1.4 MHz and 400 K was observed [95]. In stretched films of undoped polyacetylene [97], the anisotropy of the hyperfine coupling showed the π-electron character consistent with the assumption of the bond alternation kink in the π-electron system in undoped $(CH\text{-})_x$. For polyacetylene, the soliton model ENDOR data measured the size of the chain segment over which a soliton diffuses. ^{13}C-enriched undoped *cis*-rich polyacetylene was measured down to 4 K and ENDOR measurements showed that the structures observed at higher temperature [110], which were previously assigned to proton or ^{13}C hyperfine tensor components, grew into sharp single resonance lines as the temperature was lowered.

4.5.6 Inorganic Radicals in Irradiated Inorganic Single Crystals

Examples of ENDOR measurements for small inorganic radicals formed by ionizing radiation of inorganic crystals have been reported [121–129]. For example, K_2SO_4 single crystals doped with CrO_4^{2-} ions could be characterized by the location of the nearby protons [122]. In a γ-irradiated $BaClO_4{:}3H_2O$ single crystal at 120 K, the hyperfine coupling tensors for 24 protons were detected with small isotropic components and their dipolar components were found to be close to axial [124]. In γ-irradiated CsH_2AsO_4 and CsD_2AsO_4 ferroelectric crystals, ^{133}Cs and ^{1}H superhyperfine couplings for the AsO_4^- center were studied [125]. The behavior at the phase transition and in the ferroelectric phase were deduced by ENDOR measurements of the free radical in γ-irradiated single crystals of KH_2AsO_4 [126]. In X-irradiated $K_2SO_4{:}BF_4^-$ at 300 K, the FSO_2 radical was identified as trapped at the sulphate site [129].

4.5.7 Inorganic Paramagnetic Complexes in Organic Single Crystals

Numerous examples exist [130–152] where ENDOR measurements have deduced ^{1}H, ^{1}N, ^{19}F, $^{107,109}Ag$, ^{63}Cu, ^{59}Co, ^{13}C, ^{31}P couplings for inorganic complexes in a single crystal host. For example, paramagnetic Cu(II) and Ag(II) tetraphenylporphyrin (TPP) were oriented in a (H_2O) Zn(TPP) single crystal. Measurements from 20–70 K, yielded the ^{1}H, ^{63}Cu, ^{14}N, ^{107}Ag, ^{109}Ag anisotropic tensors and ^{14}N quadrupole tensors from which the electron distribution in the Cu(TPP) and Ag(TPP) could be deduced [134]. Typically a suitable diamagnetic host is selected into which paramagnetic complexes can be grown at a dopant < 1%. The effect of the axial anionic ligands on the ^{51}V nuclear quadrupolar coupling constant was deduced by ENDOR measurements [153], Q-band-ENDOR was used to detect the cation vacancies near the *cis* $[RbCl_{16}]^{4-}$ complex in NaCl [154].

4.5.8 F and H Centers in Inorganic Host Crystals

F and H color centers can be formed in inorganic host crystals by ionizing radiation [155–172]. The centers formed have been characterized by ENDOR measurements. For examples, F^+ centers [156] characterized by the parameters g = 1.9849(2), **A** = ± 49.19(2) MHz, **B** = ± 36.40(2) MHz, and **Q** = ±1.00(2) MHz can be formed by neutron irradiation of SRO while two types of F centers are formed in BaFCl [158]. They are $F(CL^-)$ and $F(F^-)$. In an irradiated $KCl{:}Li^+$ host, a new paramagnetic H center was detected [161] with a Li^+ to H^o separation of 2.17 Å. In irradiated NaCl, a $A_g^o - A_g^+$ pair center was detected [163]. In irradiated KCl and RbCl [167, 168], the superhyperfine and quadrupole interactions of the nearest Cl neighbors of atom D or cation sites in KCl and RbCl and of first- and second-shell neighbors of interstitial D in RbCl were measured. The O^- center was identified by ENDOR measurements as a self-trapped hole in X-ray irradiated $ZnWO_4$ [173].

4.5.9 Paramagnetic Inorganic Ions in Organic Host Crystals

In some systems under study by ENDOR techniques inorganic metal ions are doped into organic single crystals. Examples are Cu^{2+} [174–176], Ce^{3+} [177], Fe^{3+} [178], Nd^{3+} [179], Mn^{2+} [180, 181], VO^{2+} [182], Gd^{3+} [183, 184]. In such studies, the oriented hyperfine coupling tensor is deduced for these ions and the location of the crystal site that has been substituted is established. For instance, Cu^{2+} substituted in α-glycine crystals [174], Mn^{2+} in *tris*(sarcosine) calcium chloride [180] crystal, VO^{2+} in triglycine sulfate [182] crystal, Gd^{3+} in lanthanium nicotinate dihydrate crystal [183], Cl^{3+} in Cs_2NaYCl_6 [177] crystal, Fe^{3+} in $RbAl(SO_4)_2 \cdot 12H_2O$ [178] crystal, Gd^{3+} in Cs_2NaYCl_6 [184] crystal, and Nd^{3+} in YCl_3 crystal [179]. Metal-quinone high spin complexes were studied at X- and W-band by ENDOR measurements [185]. The VO^{2+} complex was examined by ENDOR techniques [186] as well as Fe^{3+} in a heme protein [187].

4.5.10 Transition Metal Ion Complexes in Frozen Solutions and Powders

Often it is not possible to grow single host crystals, so the transition metal ion complex is incorporated in a powder or a frozen solution. Although it may not always be possible to obtain a complete ENDOR hyperfine tensor from a powder spectrum, numerous examples exist [188–205] where ^{1}H [188–205], ^{14}N [192, 205] $^{47,49}Ti$ [201], ^{31}P [195], ^{13}C [202] tensors have been deduced from ENDOR measurements for VO^{2+} [188, 191, 192, 197, 198, 202, 204], Cu^{2+} [190, 193, 195], Co^{2+} [199, 200], Gd^{3+} [196], Ti^{+3} [201], and Mo^{+5} [203]. More recently ^{61}Ni hyperfine and nuclear quadrupole tensors, the ^{77}Si hyperfine tensor for four nearest neighbors, and the $^{69,71}Ga$ hyperfine and nuclear quadrupole matrixes for four neighbors, have been detected [206]. The water coordination to Gd^{3+} has been deduced by pulsed-ENDOR measurements [201].

4.5.11 Defects and Complexes on Surfaces

Radicals can be formed by ionizing irradiation of surfaces [208] such as silica, alumina, zeolite, and nafion. For example, F^+ centers are formed at O(1), O(2), O(3), O(4) and O(5) sites [209] in irradiated sodium β-alumina single crystals at room temperatures. ^{27}Al and ^{23}Na ENDOR measurements at 60 K provided the Al and Na hyperfine coupling tensors that, when combined with the *g* and quadrupole tensors, made it possible to identify the site. F^{+} were also identified as being found in α-alumina upon fast-neutron irradiation [210], and in Na, K, and Li β-alumina single crystals upon irradiation with 1.5 meV electrons [211]. ENDOR measurements were carried out at 2 K. From ^{1}H and ^{23}Na ENDOR measurements at 120 K, the VO^{2+} species was identified as undergoing adsorption on Y zeolite [211] and on the perfluorinated ionomer nafion [213] and from ^{27}Al ENDOR at 8 K as being adsorbed on δ-alumina [214]. ^{1}H ENDOR measurements at 110 K [215] showed that perylene radicals could be adsorbed on alumina and silica-alumina powders. Copper(II) complexes were identified on δ-alumina from ^{1}H ENDOR measurements at 6 K [216]. Examples have been reported where the carotenoid radical cation can be formed and stabilized on a silica-alumina matrix at room temperature as shown by ^{1}H ENDOR measurements [217]. The methyl groups on the carotenoid radical cation rapidly rotate from 5 K up to 300 K, which results in the appearance of narrow detectable ENDOR lines. Radical cations of benzene-d_6 and benzene-d_1 were generated by ionizing radiation and stabilized on silica gel and HY molecular sieves [218]. The electronic structure and dynamic features were different from those stabilized on a freon matrix ($CFCl_3$) as deduced from ENDOR and ESEEM measurements between 3.5 and 105 K.

4.5.12 Impurity Centers in Semiconductor Host Crystals

Inorganic ions of ^{57}Fe [219] ^{55}Mn [220, 221], Te [221], ^{53}Cr [219, 223], Ge [224], Ni [225], Co [226], Zn [226], were identified by ENDOR measurements of these nuclei as being impurity centers in semiconductor GaP host crystals. The V^{3+} ion was detected in the GaAs semiconductor host [227] where hyperfine interactions with the V nucleus and the first As and Ga neighbor shells were detected. The V resides in a Ga substitutional or in a high symmetry interstitial site tetrahedrally surrounded by four nearest As neighbors. ENDOR measurements made it possible to determine the symmetry and interaction with typically three or four neighboring shells by detecting the ^{31}P nucleus [224, 228]. The use of advanced ENDOR techniques such as pulsed- and CW-ENDOR optically detected ENDOR and field-stepped ENDOR techniques in the study of boron-vacancy and thermal donors in silicon and the gallium vacancy in gallium phosphide has been reviewed [229]. A general review can be found in reference [230].

4.5.13 Spin Centers in Silicon and Borate Systems

Silicon has been shown to be a convenient host for trapping various interstitial impurities. Ions of $^{47,49}Ti$, ^{61}Ni, Si, ^{105}Pb, ^{29}Si, ^{31}P, ^{31}Al, ^{10}B, ^{17}O, ^{33}S have been detected by ENDOR measurements, usually at low temperatures < 20 K. Some general review papers discussing the various vacancies that can be formed upon ionizing irradiation at low temperature and subsequent annealing have been published [231]. More recently, pulsed-ENDOR measurements have been carried out at 20 K on room temperature γ-irradiated glasses consisting of (100 – x) mol % B_2O_3·x mol % Li_2O. CW-EPR, ID-ESEEM, 2D-HYSCORE results, and SCF-HF calculations have shown the formation of various boroxal ring structures, boron oxygen hole centers, pentaborate, diborate, tetraborate and borate networks with nonbridging atoms [232].

4.5.14 Paramagnetic Centers in Cubic Host Crystals

Cubic crystal hosts of PbF_2 [233], PbF_2 [244, 245, 247] CdF_2 [233], $SrCl_2$ [234], CdF_2 [235], CaF_2, CaF_2 [236–346], BaF_2 [243, 248], and SrF_2 [238, 243, 246] have been used to trap the ions of Gd^3 [233, 234, 236, 240, 244, 245, 247], Eu^{2+} [233, 241, 244, 245, 248], Cu^{2+} [235], Nb^{3+} [237], Tm^{2+} [238], Yb^{3+} [243], Cr^+ [246]. ENDOR measurements provide the site symmetry of the ion and the nearest neighbor interactions up to four shells away from the trapped ion.

4.5.15 Perovskite-Type Materials

Inorganic ions can also be doped in perovskite-type materials. A perovskite crystal structure is cubic with ions at the cube corners, face centers and the body center. A typical example is barium titanate where Ba^{2+} is located at the cube corners, O^{2-} ions at the face center, and a Ti^{4+} ion at the body center. ENDOR measurements permit a measure of the site symmetry and the interaction with the nearest neighbors surrounding the paramagnetic impurity that are located in the first or second or third nearest neighbors. The ^{19}F nuclei have been often detected surrounding the ion. ENDOR spectra due to the doped inorganic ion have been recorded for ^{181}Ta [264], ^{57}Fe [265, 266], ^{7}Li [250], ^{51}V [267], Mn [268] in various lattices, and ^{133}Cs [269] in hosts as varied as $KMgF_3$ [266, 253], $LiBaF_3$ [250], $KmgF_3$ [251, 260–262], $CsCdF_3$ [252, 254, 268, 258, 259], $RbCdF_3$ [255–257], $KZnF_3$ [263], $KTaO_3$ [264, 265] and $LiTaO_3$ [266], and $PbTiO_3$ [268]. The magnetic properties of double perovskites of $A_2B'NbO_6$ where A = Ba, Sr, or BaSr, and B' = Fe or Mn have been examined [270].

REFERENCES

[1] Kevan, L. & Kispert, L.D. (1976) *Electron Spin Double Resonance* (Wiley, New York).

[2] Möbius, K., Lubitz, W. & Plato, M. (1989) *in Advanced EPR*, edited by Hoff, A.J., (Elsevier Amsterdam), Chap. XIII.

[3] Hyde, J.S., Astlend, T., Eriksson, L.E.G. & Ehrenber, A. (1970) *Rev. Sci. Instr.* **41**, 1598.

[4] Kurreck, H., Kirste, B. & Lubitz, W. (1988) *Electron Nuclear Double Resonance Spectroscopy of Radicals in Solution* (VCH, New York).

[5] Kirmse, R., Böttcher, R. & Keijzers, C.P. (1982) *Chem. Phys. Lett.* **87**, 467.

[6] Brustolon, M. & Segre, U., in Ref. 2 Chap. XVI.

[7] Rist, G.H. & Hyde, J.S. (1970) *J. Chem. Phys.* **52** 4633.

[8] Hoffman, B.M., Martinsen, J. & Venters, R.A. (1984) *J. Magn. Res.* **59**, 110; Hoffman, B.M., Venters, R.A. & Martinsen, J. (1985) *ibid.* **62** 537.

[9] Hurst, G.C., Henderson, T.A. & Kreilick, R.W. (1985) *J. Am. Chem. Soc.* **107**, 7294.

[10] Hoffman, B.M., Grubiel, R.J., Werst, M.M. & Sivaraja, M., in Ref. 2 Chap. XV.

[11] Henderson, T.A., Hurst, G.C. & Krelick, R.W. (1985) *J. Am. Chem. Soc.* **107,** 7299.

[12] Brustolon, M., Maniero, A.L., Segre, V. & Greci, L. (1987) *J. Chem. Soc. Faraday Trans.1* **83** 69; Brustolon, M., Maniero, A.L. & Segre, U. (1985) *Mol. Phys.* **55,** 713.

[13] Van Willigen, H. & Mulks, C.F. (1981) *J. Chem. Phys.* **75,** 2135.

[14] Dalton, L.R. & Kwiram, A.L. (1972) *J. Chem. Phys.* **57,** 1132.

[15] Thomann, H. & Bernardo, M. (1993) *Biol. Magn. Resonance* **13,** 275-322.

[16] Eaton, G.R. & Eaton, S.S. (2004) *Comprehensive Coordination Chemistry* II Eds. McCleverty, J.A. & Meyer, T., **2**, 49-55; Chastein, N.D. & Svetsinger, P.A. (2000) *Phys. Methods in Bioinorganic Chem.* Ed. Que, L. Jr. (Sausalito, CA: University Science Books) **187** 231.

[17] Hoffman, B.M., DeRose, V.J., Doan, P.E., Gurbiel, R.J., Houseman, A.L.P. & Telser, J. (1993) *Biol. Magn. Resonance* **13**, 151-218.

[18] Slutter, C.E., Gromov, I., Epel, B., Peckht, I., Richards, J.H. & Goldfarb, D. (2001) *J. Am. Chem. Soc.* **123**, 5325-5336.

[19] Grupp, A. & Mehring, M. (1990) in Modern Pulsed and Continuous-*Wave Electron Spin Resonance,* edited by Kevan, L. & Bowman, M.K., (Wiley, New York) Chap. IV.

[20] Borbat, P.P. & Freed, J.J. (2000) *Biological Magnetic Resonance* Berliner, L.J., Eaton, S.S. & Eaton, G.R., eds. (New York: Kluwer Academic/Plenum) **19**, 405-441.

[21] Schweiger, A. & Jeschke, G. (2001) *Principles of Pulse Electron Paramagnetic Resonance* (New York: Oxford) pp. 359-405.

[22] Britt, R.D. (2003) ACS Symposium Series: *Paramagnetic Resonance of Metallobiomolecules* **858**, 16-54.

[23] Epel, B., Poppl, A., Manikandan, P., Vega, S. & Goldfarb, D. (1997) *J. Magn. Reson.* **148**, 388-97.

[24] Goldfarb, D., Epel, B., Zimmermann & Jeschke, G. (2004) *J. Magn. Res.* **168**, 75-87.

[25] Poole, C.P. & Farach, H.A. (1994) *Handbook of Electron Spin Resonance* (New York: American Institute of Physics) pp. 451-626.

[26] Dorio, M. & Freed, J. (1979) *Multiple Electron Resonance Spectroscopy* (New York: Plenum Press.

[27] Close, D.M. & Bernhard, W.A. (1979) *J. Chem. Phys.* **70**, 210.

[28] Finch, L.L., Johnson, J.E. & Moulton, G.C. (1979) *J. Chem. Phys.* **70**, 3662.

[29] Colliziani, A., Pinzino, C., Brustolon, M. & Corvaja, C. (1978) *J. Magn. Reson.* **32**, 419.

[30] Reuveni, A., Marcellus, D., Parker, R.S. & Kwiram, A.L. (1981) **74**, 179.

[31] Westerling, J. & Lund, A. (1988) *Chem. Phys. Lett.* **147**, 111.

[32] Hwang, J.S., Wang, H.C., Anderson, B., Kispert, L.D. & Geoffroy, M. (1981) *J. Chem. Phys.* **75**, 3758.

[33] Budzinski, E.E. & Box, H.C. (1985) *J. Chem. Phys.* **82,** 3487.

[34] Sagystuen, E., Hole, E.O., Nelson, W.H. & Close, D.M. (1988) *Radiat. Res.* **116**, 196.

[35] Brustolon, M., Maniero, A. L., Bonora, M. & Segre, U. (1996) *App. Magn. Reson.* **11**, 88-113.
[36] Brustolon, M. (1995) *Rad. Phys. Chem.* **45**, 953-4.
[37] Nelson, W.H., Nimmala, S., Close, D.M., Hole, E.O. & Sagstuen, E. (1995) *Radiation Damage in DNA* Ed. Fuciarelli, A.F. & Zimbrick, J.D., (Columbus, Ohio: Battelle Press) 97-103.
[38] Close, D.M. (1993) *Radiation Research* **135**, 1-15.
[39] Honeywill, J.D. & Mile, B. (2000) *Magn. Reson. in Chem.* **38**, 423-428.
[40] Toriyama, K., Iwasaki, M., Nunome, K. & Muto, H. (1981) *J. Chem. Phys.* **75**, 1633.
[41] Cirelli, G., Russu, A., Wolf, R., Rudin, M., Schweiger, A. & Guenthard, H.H. (1982) *Chem. Phys. Lett.* **92**, 223.
[42] Gerson, F. & Qin, X. (1988) *Helv. Chim. Acta.* **71**, 1065.
[43] Gerson, F., Qin, X.Z., Bally, T. & Aebischet, J.N. (1988) *Helv. Chim. Acta.* **71**, 1069.
[44] Gerson, F., DeMeijere, A. & Qin, X.Z. (1989) *J. Am. Chem. Soc.* **111**, 1135.
[45] Gerson, F. & Qin, X.Z. (1988) *Helv. Chim. Acta.* **71**, 1498.
[46] Gerson, F. & Qin, X.Z. (1988) *Chem. Phys. Lett.* **153**, 546.
[47] Gerson, F. & Qin, X.Z. (1989) *Chim. Acta.* **72**, 383.
[48] Weltner, W., Jr. & Zee, R.J.V. (1986) *Rev. Sci. Instrum.* **57,** 2763.
[49] Doerner, H. & Schmid, D. (1979) *Chem. Phys. Lett.* **61**, 171.
[50] Hutchinson, C.A., Jr. & Kemple, M.D. (1979) *J. Chem. Phys.* **71**, 866.
[51] Hutchinson, C.A., Jr. & Kemple, M.D. (1981) *J. Chem. Phys.* **74**, 192.
[52] Grupp, A., Wolf, H.C. & Schmid, D. (1982) *Chem. Phys. Lett.* **85**, 330.
[53] Corvaja, C., Maniero, A.L. & Pasimeni, L. (1985) *Chem. Phys.* **100**, 265.
[54] Maniero, A.L., Corvaja, C. & Pasimeni, L. (1988) *Chem. Phys.* **122**, 327.
[55] Chan, I.Y. & Sandroff, C.J. (1984) *J. Chem. Phys.* **80**, 5916.
[56] Waldmann, J., Von Schuetz, J.U. & Wolf, H.C. (1985) *Chem. Phys.* **92**, 1.
[57] Kay, C.W.M., Gromadecki, U., Torring, J.T. & Weber, S. (2001) *Mol. Physics* **99**, 1413-1420.
[58] Kay, C.W.M., DeValentine, M. & Mobius, K. (1997) *J. Chem. Soc., Perkin Trans.* **2**: *Phys. Org. Chem.* 2563-2568.
[59] Hamacher, V., Wrachtrap, J., vonMaltzan, B., Plato, M. & Moebius, K. (1993) *App. Magn. Reson.* **4**, 297-319.
[60] Li, Y., Lucas M-G Konovalova, T., Abbott, B., MacMillan, F., Petrenko, A., Sivokumar, V., Wang, R., Hasting, G., Gu, F., Yantol, J., Brunel L-C Timkovich, R., Rappaport, F. & Redding, K. (2004) *Biochemistry* **43**, 12634-12647.
[61] Britt, R.D., Campbell, K.A., Peloquin, J.M., Gilchrist, M.L., Aznar, C.P., Dicus, M.M., Robbles, J. & Messinger, J. (2004) *Biochim. Biophys. Acta*, **1655**, 158-171.
[62] Camp, H.L.V., Wei, Y.H., Scholes, C.P. & King, T.E. (1978) *Biochim. Biophys. Acta.* **53**, 238.
[63] Hoffman, B.M., Roberts, J.E., Swanson, M., Speck, S.H. & Margoliash, E. (1980) *Proc. Natl. Acad. Sci.* USA **77**, 1452.
[64] Stevens, T.H., Martin, C.T., Wang, H., Brudvig, G.W., Scholes, C.P. & Cahn, S.I. (1982) *J. Biol. Chem.* **12**, 206.
[65] LoBrutto, R., Wei, Y.H., Mascarenhas, R., Scholes, C.P. & King, T.E. (1983) *J. Biol. Chem.* **258**, 7437.
[66] Martin, C.T., Scholes, C.P. & Chan, S.I. (1985) *J. Biol. Chem.* **260**, 2857.
[67] Fan, C., Bank, J.F., Dorr, R.G. & Scholes, C.P. (1988) *J. Biol. Chem.* **263**, 3588.
[68] Mulks, C.F., Scholes, C.P., Dickinson, L.C. & Lapidot, A. (1979) *J. Am. Chem. Soc.* **101**, 1645.
[69] True, A.E., Nelson, M.J., Venters, R.A., Orme-Johnson, W.H. & Hoffman, B.M. (1988) *J. Am. Chem. Soc.* **110**, 1935.
[70] Wang, G., Benecky, M.J., Huynh, B.H., Cline, J.F., Adams, M.W.W., Mortenson, L.E., Hoffman, B.M. & Muenck, E. (1984) *J. Biol. Chem.* **14**, 328.
[71] Telser, J., Benecky, M.J., Adams, M.W., W Mortenson, L.E. & Hoffman, B.M. (1987) *J. Biol. Chem.* **262**, 6589.

[72] Telser, J., Benecky, M.J., Adams, M.W.W., Mortenson, L.E. & Hoffman, B.M. (1986) *J. Biol. Chem.* **261**, 13536.
[73] Cline, J.F., Janick, P.A., Siegel, L.M. & Hoffman, B.M. (1985) *Biochemistry* **24**, 7942.
[74] Cline, J.F., Janick, P.A., Siegel, & Hoffman, B.M. (1986) *Biochemistry* **25**, 4647.
[75] Hoffmann, B.M., Venters, R.A., Roberts, J.E., Nelson, M. & Orme-Johnson, W.H. (1982) *J. Am. Chem. Soc.* **104**, 4711.
[76] Telser, J., Emptage, M.H., Merkle, H., Kennedy, M.C., Beinert, H. & Hoffman, B.M. (1986) *J. Biol. Chem.* **261**, 4840.
[77] Roberts, J.E., Hoffman, B.M., Rutter, R. & Hager, L.P. (1981) **103**, 7654.
[78] Venters, R.A., Nelson, M.J., McLean, P.A., True, A.E., Levy, M.A., Hoffman, B.M. & Orme-Johnson, W.H. (1986) *J. Am. Chem. Soc.* **108**, 3487.
[79] Hoffman, B.M., Roberts, J.E. & Orme-Johnson, W.H. (1982) *J. Am. Chem. Soc.* **104**, 860.
[80] Weilbacher, E., Helle, N., Elsner, M., Kureck, H., Mueller, F. & Allendoerfer, R.D. (1988) *Magn. Reson. Chem.* **26**, 64.
[81] Kurreck, H., Bretz, N.H., Helle, N., Henzel, N. & Weilbacher, E. (1988) *J. Chem. Soc. Faraday Trans.* **184**, 3293.
[82] Chandrashekar, T.K., O'Malley, P.J., Rodriguez, I. & Babcock, G.T. (1986) *Photosynth. Res.* **10** 423.
[83] Lubitz, W., Isaacson, R.A., Abresch, E.C. & Feher, G. (1984) *Proc. Natl. Acad. Sci.* USA **81**, 7792.
[84] Lendzian, F., Lubitz, W., Scheer, H., Hoff, A.J., Plato, M., Traenkle, E. & Moebuis, K. (1988) *Chem. Phys. Lett.* **148**, 377.
[85] Hole, E.O., Sagstuen, E., William, H., Close, D.M. (1991) *Radiation Research* **125**, 119-28.
[86] Fursman, C.E., Teutloff & Bittl (2002) *J. Phys. Chem. B* **106**, 9679-9686.
[87] Slutter, C.E., Gromov, I., Epel, B., Pecht, I., Richards, J.H. & Goldfarb, D. (2001) *J. Am. Chem. Soc.* **123**, 5325-5336.
[88] Peloquin, J.M. & Britt, R.D. (2001) *Biochim. Biophys. Acta.* **1503** 96-111.
[89] Veselov, A.V., Osborne, J.P., Gennis, R.B. & Scholes, C.P. (2000) *Biochemistry* **39**, 3169-75.
[90] Astashkin, A.V., Raitsimring, A.M. & Walker, F.A. (2001) *J. Am. Chem. Soc.* **123**, 1905-13.
[91] Zdravkova, M. & Yordanov, N.D. (1994) *App. Magn. Reson.* **6**, 83-105.
[92] Lee, H.I., Cameron, L.M., Christiansen, J., Christie, P.D., Pollock, R.C., Song, R., Sorlie, M., Orme-Johnson, W.H., Dean, D.R., Holes, B.J. & Hoffman, B.M. (2003) *ACS Symposium Series: Paramagnetic Resonance of Metallobiomolecules* **858**, 150-178.
[93] Hoffman, B.M. (2003) *Proc. Nat. Acad. Sci.* **100**, 3575-3578.
[94] Schlick, S. & Kevan, L. (1982) *J. Appl. Polym. Sci.* **27**, 319.
[95] Thomann, H., Dalton, L.R., Tomkiewicz, Y., Shiren, N.S. & Clark, T.C. (1982) *Mol. Cryst. Liq. Cryst.* **83**, 1065.
[96] Hartl, W. & Schwoerer, M. (1982) *Chem. Phys.* **69**, 443.
[97] Kuroda, S. & Shirakawa, H. (1982) *Solid State Commun.* **43**, 591.
[98] Thomann, H., Dalton, L.R., Tomkiewicz, Y., Shiver, N.S., Clarke, T.C. (1983) *Phys. Rev. Lett.* **50**, 533.
[99] Kuroda, S., Tokumoto, M., Kinoshita, N., Ishiguro, T. & Shirakawa, H. (1983) *J. Magn. Mater.* **31-34**, 1149.
[100] Dalton, L.R., Thoman, H., Morrobel-Sosa, A., Chiu, C., Galvin, M.E., Wnek, G.E., Tgomkiewica, Y., Shirer, N.S., Robinson, B.H. & Kwiram, A.L. (1983) *J. Appl. Phys.* **54,** 5583.
[101] Heeger, A.J. & Schrieffer, J.F. (1983) *Solid State Commun* **48**, 207.
[102] Dalton, L.R., Thomann, H., Tomkiewicz, Y., Shiren, N.S. & Clarke, T.C. (1982) *Polym. Prep.* **23**, 86.
[103] Baker, G.J., Raynor, J.B. & Pron, A. (1984) *J. Chem. Phys.* **80**, 5250.
[104] Kuroda, S., Barido, H. & Shivakawa, H. (1984) *Solid State Commun.* **52**, 893.

[105] Cline, J.F., Thomann, H., Kim, H., Morrobel-Sosa, A., Dalton, L.R. & Hoffman, B.M. (1985) *Phys. Rev. B* **31**, 1605.

[106] Kuroda, S., Barrdo, H. & Shirakawa, H. (1985) *J. Phys. Soc. Japan.* **54**, 3956.

[107] Bartl, A. & Stasko, A. (1986) *Phys. Status. Solidi. A* **95**, 659.

[108] Kahol, P.K. & Mehring, M. (1986) *J. Phys. C* **19**, 1045.

[109] Kuroda, S. & Shirakawa, H. (1986) *Solid State Commun.* **59**, 261.

[110] Kuroda, S. & Shirakawa, H. (1987) *Synth. Met.* **17**, 423.

[111] Kuroda, S., Murase, I., Onishi, T. & Noguchi, T. (1987) *Synth. Met.* **17**, 663.

[112] Veregin, R.P., Harbour, J.R., Kotake, Y. & Jansen, E.G. (1987) *Carbon* **25**, 541.

[113] Kuroda, S. & Shirakawa, H. (1987) *Phys. Rev. B* **35**, 9380.

[114] Baker, G.J., Holloway, J.H., Raynor, J.B. & Selig, H. (1987) *Synth. Met.* **20**, 323.

[115] Raynor, J.B. (1987) *Mater. Sci. Forum* **21**, 111.

[116] Kuachadre, N. & Tomashevskii, E.E. (1988) *Fiz Tverd. Tela.* **30**, 1859.

[117] Dorio, M.M. (1979) in Multiple Electron Resonance Spectroscopy, edited by Dorio, M.M. & Freed, J.H., (Plenum, New York) p. 393-407.

[118] Clarkson, R.B. (1980) *MMI Press Symp. Ser.* **1** 263.

[119] Mehring, M. & Kahol, P.K. (1985) *Springer Ser. Solid State Sci.* (Electron, Prop. Polym. Relat. Compd.) (Springer, New York) **63**, 264.

[120] Young, C.L., Whitney, D., Vistnes, A.I. & Dalton, L.R. (1986) *Ann. Rev. Phys. Chem.* **37**, 459.

[121] Brustolon, M., Maniero, A.I. & Segre, U. (1988) *Mol. Phys.* **65**, 447.

[122] Ravi, S.Y. & Bill, H. (1982) *Chem. Phys. Lett.* **92**, 349.

[123] Wells, J.W., Budzinski, E. & Box, H.C. (1987) *J. Chem. Phys.* **85**, 6340.

[124] Atherton, N.M. & Blackford, R.D.S. (1987) *J. Chem. Soc. Faraday Trans.* **283**, 569.

[125] Kahol, P.K. & Dalal, N.S. (1988) *Solid State Commun.* **65**, 823.

[126] Gaillard, J. & Gloux, P. (1980) *J. Phys.* **41**, 647.

[127] Reddy, M.V., V, S., Lingam, K.V. & Rao, T.K.G. (1981) *J. Mol. Struct.* **73**, 105.

[128] Angelone, R., Colligiani, A., Pinzino, C. & Brustolon, M. (1981) *Mol. Phys.* **44**, 997.

[129] Sekhar, Y.R., Bill, H. & Lovy, D. (1987) *Chem. Phys. Lett.* **136**, 57.

[130] Schweiger, A., Rudin, M. & Guenthard, H.H. (1979) *Mol. Phys.* **37**, 1573.

[131] Kita, S., Hashimoto, M. & Iwaizumi, M. (1979) *Inorg. Chem.* **18**, 3432.

[132] Rudin, M., Joerin, E., Schweiger, A. & Guenthard, H.H. (1979) *Chem. Phys. Lett.* **67**, 134.

[133] Joerin, E., Rudin, M., Schweiger, A. & Guenthard, H.H. (1980) *Chem. Phys. Lett.* **69**, 193.

[134] Brown, T.G. & Hoffman, B.M. (1980) *Mol. Phys.* **39**, 1073.

[135] Snaathorst, D., Keijzers, C.P., Klaassen, A.A., K DeBoer, E., Chacko, V.P. & Gomperts, R. (1980) *Mol. Phys.* **40**, 585.

[136] Kirmse, R., Stach, J. & Boettcher, R. (1980) *Chem. Phys. Lett.* **75**, 565.

[137] Rudin, M., Schweiger, A., Berchten, N. & Guenthard, H.H. (1980) *Mol. Phys.* **41**, 1317

[138] Butler, J.E. & Hutchinson, C.A. Jr. (1981) *J. Chem. Phys.* **74**, 3102.

[139] Kita, S., Hashimoto, M. & Iwaizumi, M. (1982) *J. Magn. Reson.* **46**, 361.

[140] Kirmse, R., Abram, U. & Boettcher, R. (1982) *Chem. Phys. Lett.* **88**, 98.

[141] Kirmse, R., Abram, U. & Boettcher, R. (1982) *Chem. Phys. Lett.* **90**, 9.

[142] Kirmse, R., Boettcher, R. & Keijzers, C.P. (1982) *Chem. Phys. Lett.* **87**, 467.

[143] Rudin, M., Schweiger, A. & Guenthard, H.H. (1982) *Mol. Phys.* **47**, 171; **46**, 1027.

[144] Rudin, M., Fauth, J.M., Schweiger, A., Ernst, R.R., Zoller, L. & Ammeter, J.H. (1983) *Mol. Phys.* **49**, 1257.

[145] Iordanov, N. (1984) *Chem. Phys. Lett.* **107**, 457.

[146] Kirmse, R., Stach, J., Abram, U., Dietzsch, W., Boettcher, R., Gribnau, M.C., M. & Keijzers, C.P. (1984) *Inorg. Chem.* **23**, 3333.

[147] Boettcher, R. & Kirmse R. Stach, J. (1984) *Chem. Phys. Lett.* **112**, 460.

[148] Wolf, R., Schweiger, A. & Guenthard, H.H. (1984) *Mol. Phys.* **53**, 567-585.

[149] Kirmse, R., Dietzsch, W., Stach, J., Golic, L., Boetteher, R., Brunner, W., Gribnau, M.C.M. & Keijzers, C.P. (1986) *Mol. Phys.* **57**, 1139.

[150] Baker, G.J., Raynor, J.B., Smits, J.M.M., Beurskens, P.T., Vergooseen, H. & Keijzers, C.P. (1986) *J. Chem. Soc. Dalton Trans.* 2655.

[151] Baker, G.J. & Raynor, J.B. (1986) *J. Chem. Soc. Dalton Trans.* 2663.

[152] Boettcher, R., Kirmse, R., Stach, J., Reijerse, E.J. & Keijzers, C.P. (1986) *Chem. Phys.* **107**, 145.

[153] Brynda, M., Aznar, C.P., Britt, D.R., Kabanos, T., Tolis, E.J. & Deligiannakis, Y. (2004) *Abstracts of Papers, 22th ACS National Meeting March* 28 - April 1.

[154] Vrielinck, H., Callens, F. & Matthys, P. (2001) *Phys. Rev. B: Condensed Matter and Materials Physics* **64**, 214105/6-2141036.

[155] DuVarney, R.R. & Spaet, J.M. (1979) *Solid State Commun.* **32**, 1237.

[156] Abraham, M.M., Chen, Y., Peters, W.C., Rubio, O.J. & Unruh, W.P. (1979) *J. Chem. Phys.* **71**, 3658.

[157] Abraham, M.M., Chen, Y., Olson, D.N., Orera, V.M., Wilson, T.M. & Wood, R.F. (1981) *Phys. Rev. B* **23**, 51.

[158] Bauer, R., Niklas, J.R. & Spaeth, J.M. (1983) *Radiat. Effects* **72**, 287.

[159] Bauer, R.U., Niklas, J.R. & Spaeth, J.M. (1983) *Phys. Status Solidi B* **118**, 557.

[160] Niklas, J.R., Bauer, R.U. & Spaeth, J.M. (1983) *Phys. Status Solidi. B* **119**, 171.

[161] Heder, G. & Spaeth, J.M. (1984) *Phys. Status Solidi. B* **125,** 523.

[162] Soethe, H., Studizinski, P. & Spaeth, J.M. (1985) *Phys. Status Solidi. B* **130,** 339.

[163] Muessing, T., Niklas, J.R., Spaeth, J.M. & Granzer, F. (1987) *Cryst. Lattice Defects Amorphous Mater.* **16** 169.

[164] Kalinowski, H.J. (1988) *J. Magn. Reson.* **77**, 562.

[165] Olim, M.T., Niklas, J.R., Spaeth, J.M. & Symons, M.C.R. (1988) *Phys. Rev. B* **38**, 4343.

[166] Studzinski, P., Niclas, J.R. & Spaeth, J.M. (1980) *Phys. Status Solidi. B* **10**, 673.

[167] Hoentzsch, C. & Spaeth, J.M. (1979) *Phys. Status Solidi. B* **94,** 497.

[168] Heder, G., Niklas, J.R. & Spaeth, J.M. (1980) *Phys. Status Solidi. B* **100,** 567.

[169] Heder, G., Niklas, J.R. & Spaeth, J.M. (1983) *Radiat. Effects* **73,** 61.

[170] Spaeth, J.M. (1985) *Cryst. Lattice Defects Amorphous Mater.* **12**, 381.

[171] Spaeth, J.M. (1986) *NATO ASI Ser., Ser. B* **147,** 205.

[172] Spaeth, J.M. (1989) *Electron Spin Reson.* **11**, 89.

[173] Watterich, A., Kovacs, L., Wurz, R., Schon, F., Hofstaetter, A. & Scharmann, A. (2001) *J. Phys.: Condensed Matter* **13**, 1595-1607.

[174] Fujimoto, M., McDowell, C.A. & Takui, T. (1979) *J. Chem. Phys.* **70**, 3694.

[175] Atherton, N.M. & Horsewill, A.J. (1979) *Mol. Phys.* **37**, 1349.

[176] McDowell, C.A. & Sastry, D.L. (1988) *J. Chem. Phys.* **89**, 2697.

[177] Fish, G. & Stapleton, H.J. (1978) *J. Chem. Phys.* **69**, 4068.

[178] Brisson, J.R. & Manoogian, A. (1978) *Phys. Rev. B* **18**, 4576.

[179] Hutchison, C.A. Jr. & Kim, S.S. (1979) *Phys. Rev. B* **19**, 4454.

[180] Metz, H., Kuechler, J., Boettcher, R. & Windsch, W. (1982) *Chem. Phys. Lett.* **89**, 351.

[181] Metz, H., Boettcher, R. & Windsch, W. (1986) *Chem. Phsy. Lett.* **127**, 278.

[182] Boettcher, R., Heinhold, D. & Windsch, W. (1979) *Chem. Phys. Lett.* **65**, 452.

[183] Fields, R.A. & Hutchison, C.A. Jr. (1985) *J. Chem. Phys.* **82**, 1711.

[184] Bill, H., Magne, G., Guedel, H.U. & Neuenschwander, K. (1984) *Chem. Phys. Lett.* **104,** 258.

[185] Chemerisov, S.D., Perekkodtsev, G.D., Tipikin, D.S., Lebedev, Y.S., Prokofev, A.I., Aleksandrov, A.I., Dubinskii, A.A., Moebius, K., Poluektov, O.G. & Schmidt, J.A. (1996) *J. Chem. Soc. Faraday Trans.* **92**, 1959-68.

[186] Aznar, C.P., Deligeonnakis, Y., Toles, E.J., Kabanos, T., Brynda, M. & Britt, R.D. (2004) *J. Phys. Chem. A* **108**, 4310-21.

[187] Fahnenschmidt, M., Bittl, R., Rau, H.K., Haehnel, W. & Lubitz, W. (2000) *Chem. Physics Lett.* **323**, 329-339.

[188] Van Willigen, H. (1979) *Chem. Phys. Lett.* **65**, 490.

[189] Van Willigen, H. (1980) *J. Magn. Reson.* **39**, 37.

[190] Van Camp, H.L., Sands, R.H. & Fee, J.A. (1881) *J. Chem. Phys.* **75**, 2098.

[191] Van Willigen, H., Mulks, C.F. & Atherton, N.M. (1982) *Inorg. Chem. Soc.* **21**, 1708.

[192] Mulks, C.F., Kirste, B. & Van Willigen (1982) *J. Am. Chem. Soc.* **104**, 5906.
[193] Kirste, B. & Van Willigen, H. (1983) *J. Phys. Chem.* **87**, 781.
[194] Henderson, T.A., Hurst, G.C. & Kreilick, R.W. (1985) *J. Am. Chem. Soc.* **107**, 7299.
[195] Yordanov, N. & Zdravoka M. (1986) *Chem. Phys. Lett.* **127**, 487.
[196] Yin, M.B. & Makinen, M.W.A. (1986) *J. Magn. Reson.* **70**, 89.
[197] Attanasio, D. (1986) *J. Phys. Chem.* **90**, 4952.
[198] Mulks, C.F. & Van Willigen, H. (1981) *J. Phys. Chem.* **85**, 1220.
[199] Babonneau, F. & Livage, J. (1986) *Inorg. Chem.* **25**, 2741.
[200] Baumgarten, M., Lubitz, W. & Winscom, C.J. (1987) *Chem. Phys. Lett.* **133**, 102.
[201] Gourier, D. & Samuel, E. (1987) *J. Am. Chem. Soc.* **109**, 4571.
[202] Mustafi, D. & Makinen, M.W. (1988) *Inorg. Chem.* **27**, 3360.
[203] Attanasio, D., Funicello, M. & Suber, L. (1988) *Chem. Phys. Lett.* **147**, 273.
[204] Branca, M., Micera, G. & Dessi, A. (1989) *J. Chem. Soc. Dalton Trans.*1289.
[205] Buchler, J.W., Drecher, C., Lay K-L Raap, A. & Gersonde, K. (1983) *Inorg. Chem.* **22**, 879.
[206] Stevens, K.T., Garces, N.Y., Bai, L., Giles, N.C., Halliburton, L.E., Stezler, S.D., Schunemann, P.G., Pollak, T.M., Route, R.K. & Fergelson, R.S. (2004) *J. Phys: Condensed Matters* **16**, 2593-2607.
[207] Astashkin, A.V., Raitsimring, A.M. & Caravan, P. (2004) *J. Phys. Chem. A* **108**, 1990-2001.
[208] Mehring, M. (1987) *Z. Phys. Chem.* **151,** 1.
[209] Barklie, R.C., Niklas, J.R. & Spaeth, J.M. (1980) *J. Phys. C* **13,** 1745-1757.
[210] DuVarney, R.Ç., Garrison, A.K., Niklas, J.R. & Spaeth, J.M. (1981) *Phys. Rev. B* **24,** 3693.
[211] Kurtz, S.R., Stinson, D.G., Stapleton, H.J. & Abraham, M.M. (1981) *Phys. Rev.B* **24**, 4983.
[212] Van Willigen, H. & Chandrashekar, T.K. (1983) *J. Am. Chem. Soc.* **105**, 4232.
[213] Martini, G., Ottaviani, M.F., Pedocchi, L. & Ristori, S. (1989) *Macromolecules* **22**, 1743.
[214] Motschi, H. & Rudin, M. (1984) *Colloid Polym. Sci.* **262**, 579.
[215] Clarkson, R.B., Belford, R.L., Rothenberger, K.S. & Crookham, H.C. (1987) *J. Catal.* **106**, 500.
[216] Rudin, M. & Motschi, H. (1984) *J. Colloid Interface Sci.* **98**, 385.
[217] Jeevarajan, A.S., Kispert, L.D. & Piekara-Sady, L. (1993) *Chem. Phys. Lett.* **209**, 269; Konovalova, T.A., Dekanov, S.A., Bowman, M.K. & Kispert, L.D. (2001) *J. Phys. Chem. B* **105**, 8361.
[218] Erickson, R., Lindgren, M., Lund, A. & Sjoeqvist, L. (1993) *Colloids and Surfaces A: Physicochem. Eng. Aspects* **72**, 207-16.
[219] Kirillov, V.I. & Teslenko, V.V. (1979) *Fiz. Tverd. Tela* **21**, 3209.
[220] Van Engelen, P. & Sie, S.G. (1979) *Solid State Commun.* **30**, 515.
[221] Engelen, P.V. (1980) *Phys. Rev.* **B 22**, 3144.
[222] Igelmund, A., Hausmann, A. & Utsch, B. (1979) *Z. Phys.* **32**, 265.
[223] Teslenko, V.V. (1984) *Fiz. Tekh. Poluprovodn.* **18**, 1536
[224] Hausmann, A., Igelmund, A. & Utsch, B. (1980) *Z. Phys. B* **39**, 33.
[225] Ueda, Y., Niklas, J.R., Spaeth, J.M., Kaufmann, U. & Schneider, J. (1983) *Solid State Commun.* **46**, 127.
[226] Kirillov, V.I., Pribylov, N.N., Rembeza, S.I., Spirin, A.I. & Teslenko, V.V. (1983) *Fiz. Tekh. Poluprovodn.* **17**, 1149.
[227] Hage, J., Niklas, J.R. & Spaeth, J.M. (1984) International Conference on Defects in Semiconductors, Proceedings of the 13th, Meeting, edited by, L.C., Kimerling and, J.M., Parsey Jr. (Metall. Soc. AIME Warrendale, PA) 1051.
[228] Gurbiel, R., Hoffmann, B.M. & Wessels, B.W. (1986) *Mater. Sci. Forum 10* (*Proc. Int. Conf. Defects Semicond.*, 14th 3) 1063.
[229] Gregorkiewicz, T., Altink, H.E., Ammerlaan, C.A.J. (1991) *Acta Physics Polonica A* **80**, 161-70.

[230] Meyer, B.K. (1988) *Rev. Phys. Appl.* **23**, 809.
[231] Sano, Y., Morigaki, K. & Hirabayashi, I. (1983) *Physica* B + C 117-118 923; Ammerlaan, C.A.J., Sprenger, M., Van Kemp, R., Van Wezep, D.A. & Sieverts, E.G. (1985) *Res. Soc. Symp. Proc.* **46**, 227; Defects Crystals, Proceedings of the International Sch., 7th Meeting (World Scientific, Signapore) 1973.
[232] Kordas, G. (2000) *Phys. and Chem. of Glasses* **41**, 325-329.
[233] Baker, J.M. & Wood, R.L. (1979) *J. Phys. C* **12**, 4033.
[234] Van Ormondt, D., Reddy, K.V., Van Ast, M.A., Den Hartog H. & Bijvank, E.J. (1980) *J. Magn. Reson.* **37**, 195.
[235] Chan, I.Y. & Mushlin, R.A. (1980) *Chem. Phys. Lett.* **70**, 138.
[236] Bijvank, E.J. & Den Hartog, H.W. (1980) *Phys. Rev. B* **22**, 4121.
[237] Kon'kin, A.L., Meiklyar, V.P. & Falin, M.L. (1980) *Fiz. Tverd. Tela* **22**, 2816.
[238] Fainstein, C., Tovar, M. & Ramos, C. (1982) *Phys. Rev. B* **25**, 3039.
[239] Vainshtein, D.I., Shcherpkin, V.D., Safin, V.A. & Vinokurov, V.M. (1982) *Fiz. Tverd. Tela* **24**, 3480.
[240] Nakata, R. & Den Hartog, H.W. (1984) *J. Magn. Reson.* **56**, 385.
[241] Anikeenok, O.A., Eremin, M.V., Falin, M.L., Konkin, A.L. & Meiklyar, V.P. (1984) *J. Phys. C* **17**, 2813.
[242] Studzinski, P., Casas-Gonzales, J. & Spaeth, J.M. (1984) *J. Phys. C* **17**, 5411.
[243] Ramos, C.A., Fainstein, C. & Tovar, M. (1985) *Phys. Rev. B* **32**, 64.
[244] Baker, J.M. & Bluck, L.J.C. (1985) *Phys. Rev. C* **18** 6051.
[245] Eremin, M.V. & Khutsishvii, O.G. (1987) *Fiz. Tverd. Tela.* **29**, 2687.
[246] Cases, R., Alonso, P.J., Alcala, R. & Spaeth, J.M. (1987) *Cryst. Lattice Defects Amorphous Mater.* **16**, 289.
[247] Kuriata, J., Rewaj, T. & Guskos, N. (1989) *Acta Phys. Pol. A* **75** 427.
[248] Kiknadze, M., Mirianashvili, R. & Sanadze, T. (1989) *Phys. Status Solidi* **152**, K13.
[249] Falin, M.L., Meiklyar, V.P. & Konkin, A.L. (1980) *J. Phys. C* **13**, 1299.
[250] Yosida, T. (1980) *J. Phys. Soc. Japan* **49**, 127.
[251] DuVarney, R.C., Niklas, J.R. & Spaeth, J.M. (1981) *Phys. Status Solidi* **B 103**, 329.
[252] Ziaei, M.E. (1981) *Can. J. Phys.* **59**, 298.
[253] Anikeenok, O.A., Eremin, M.V., Falin, M.L. & Meiklyar, V.P. (1982) *J. Phys. C* **15** 1557.
[254] Aoki, H., Arakawa, M. & Yosida, T. (1983) *J. Phys. Soc. Japan* **52**, 2216.
[255] Studzinski, P. & Spaeth, J.M. (1983) *Radiat. Effects* **73**, 207.
[256] Studzinski, P. & Spaeth, J.M. (1986) *Phys. Status Solidi B* **136**, 735.
[257] Studzinski, P. & Spaeth, J.M. (1986) *J. Phys. C* **19**, 6441.
[258] Casas-Gonzalez, J., Studzinski, P., Andriessen, J., Buzare, J.Y., Fayet, J.C. & Spaeth, J.M. (1986) *J. Phys.* **C 19**, 6767.
[259] Allsopp, R.A., Baker, J.M., Bluck, L.J.C. & Buzare, J.Y. (1987) *J. Phys. C* **20**, 451.
[260] Grachev, V.G., Ibragimov, I.R. & Falin, M.L. (1987) *Fiz. Tverd. Tela* **29**, 721.
[261] Zaripov, M.M., Ibragimov, I.R. & Falin, M.L. (1987) *Fiz. Tverd. Tela.* **29**, 1546.
[262] Zolrita, E., Alonso, P.J., Alcala, R., Spaeth, J.M. & Soethe, H. (1988) *Solid State Commun.* **66**, 7773.
[263] Falin, M.L., Eremin, M.V., Zaripov, M.M., Ibragimov, I.R., Leushin, A.M. & Yu, R. (1989) *J. Phys. B* **1**, 2331.
[264] Laguta, V.V., Glinchuk, M.D., Bykov, I.P., Karmazin, A.A., Grachev, V.G. & Troitskii, V.V. (1987) *Fiz. Tverd. Tela* **29**, 2473.
[265] Glinchuk, M.D. (1987) *Ser. Fiz.* **51**, 2124.
[266] Soethe, H., Rowan, L.G. & Spaeth, J.M. (1989) *J. Phys. B* **1**, 3591.
[267] Takeuchi, H., Horai, K. & Arakawa, M. (1979) *J. Phys. Soc. Japan* **46**, 18.
[268] Voelkel, G., Boettcher, R., Brunner, W., Milsch, B., Poeppl, A., Windsch, W. (1988) *Exp. Tech. Phys.* **35**, 359.
[269] Meiklyar, V.P., Usachev, A.E., Yablokov, Y.V. & Shuston, V.A. (1985) *Phys. Status Solidi* **B 132**, K73.
[270] Gross, R. & Rao, M.S.R. (2004) *J. Applied Physics* **95**, 7528-7530.

CHAPTER 5

MÖSSBAUER SPECTROSCOPY

J.M. Cadogan[1] and D.H. Ryan[2]
[1]School of Physics, The University of New South Wales, Sydney, NSW 2052, Australia
[2]Centre for the Physics of Materials, Department of Physics, Rutherford Building, McGill University, 3600 University Street, Montreal, Quebec, Canada H3A 2T8

5.1 INTRODUCTION

The Mössbauer effect was discovered in 1957 by R.L. Mössbauer [1] and involves the recoilless emission of a gamma ray (γ-ray) by an excited nucleus with a subsequent recoilless absorption by another nucleus. Mössbauer spectroscopy has found numerous applications in physics, chemistry and biology and one notable early success was the measurement of the gravitational redshift [2]. This spectroscopic technique continues to make significant contributions, such as 2004/5's analysis of soil on the surface of Mars [3], with spectra collected in situ. Rudolf Mössbauer's work was awarded the 1961 Nobel Prize for Physics.

In this chapter we present an introduction to Mössbauer spectroscopy. The article is divided into three main sections. We begin with an introduction to the Mössbauer effect and the nuclear hyperfine interactions underlying the use of this spectroscopy. In section two, we give a detailed discussion of the various methodologies of Mössbauer spectroscopy, encompassing all aspects of this technique from the standard transmission experiment to more exotic techniques. Finally, in section three we illustrate the use of Mössbauer spectroscopy with a variety of spectra taken from our work, including studies of crystalline, nanocrystalline, and amorphous materials.

5.1.1 Recoilless Processes

To begin, we consider the recoil-free nature of Mössbauer transitions, which is a consequence of the nuclei being embedded in a solid matrix, which takes up the momentum associated with the γ-ray emission and absorption processes. The lattice's mass is clearly much greater than that of a single nucleus and the energy lost to nuclear recoil is therefore negligible. As such, the transition may be thought of as a *zero-phonon process*. Naturally, not all such γ-ray transitions in a material are recoil-free and the recoil-free fraction *f* varies markedly with isotope, material matrix, and temperature. The most commonly used Mössbauer isotope, ^{57}Fe, has an *f*-factor ~0.7 at room temperature. The dependence of the *f*-factor on the host material through the lattice's Debye temperature has implications for quantitative phase analysis since the common assumption that all phases in a mixed-phase sample have the same *f*-factor is almost certainly merely a useful simplification.

The recoil-free fraction *f* is related to the average squared displacement $\langle x^2 \rangle$ of the probe nucleus from its equilibrium position, and for the one-dimensional case, $f = \exp\left(-k_\gamma^2 \langle x^2 \rangle\right)$ where k_γ is the wavevector of the incident γ-ray. At temperatures much higher than the Debye temperature (Θ_D) of the host lattice, $\langle x^2 \rangle$ is given by

$$\left\langle x^2 \right\rangle = \frac{3\hbar^2}{Mk_B \Theta_D}\left(\frac{T}{\Theta_D}\right), \tag{5.1}$$

i.e., it is linear in *T*, whereas at low temperatures one finds

$$\left\langle x^2 \right\rangle = \frac{3\hbar^2}{Mk_B \Theta_D}\left[\frac{1}{4} + \frac{\pi^2}{6}\left(\frac{T}{\Theta_D}\right)^2\right], \tag{5.2}$$

i.e., it is quadratic in temperature. Within the Debye model the latter expression can be rewritten in terms of Θ_D to give the *f*-factor as

$$\ln f = \left\{\frac{-\hbar^2 k_\lambda^2}{2M}\frac{3}{2k_B \Theta_D}\right\}\left[1 + \frac{2\pi^2}{3}\left(\frac{T}{\Theta_D}\right)^2\right], \tag{5.3}$$

where *M* is the mass of the probe atom.

This expression for the *Debye-Waller factor* contains important consequences for the observation of the Mössbauer effect. First, *f* decreases with increasing recoil energy so the observation of a Mössbauer transition is limited to γ-ray energies below about 100 keV. Second, lower temperatures are favored for the experiments (notice the T^2 dependence in $\ln f$). There is also a temperature-dependent *second-order Doppler shift* due to the average squared velocity of the probe atom

$$SOD = \frac{\left\langle v^2 \right\rangle}{2c^2} E_\gamma . \tag{5.4}$$

5.1.2 Doppler Velocity

The emission and absorption γ-ray distributions most likely will not overlap due to such effects as hyperfine interactions. However, they may be brought into coincidence by Doppler shifting of the γ-ray source. The resulting change in energy of the emitted γ-rays, to first-order, is $\Delta E = E_\gamma \frac{v}{c}$, where v is the velocity of the γ-ray source relative to the absorber, c is the speed of light, and E_γ is the unperturbed γ-ray transition energy. To give some idea of the remarkable energy resolution provided by the Mössbauer effect, a typical Doppler shift of 1 mm/s in the most commonly used isotope ^{57}Fe, for which E_γ = 14.4 keV, yields ΔE = 48 neV and thus $\Delta E/E_\gamma = 3.3 \times 10^{-12}$.

5.1.3 Lineshape

The energy uncertainty associated with a nuclear excited state of lifetime τ is given by Heisenberg's relationship $\Gamma = \frac{\hbar}{\tau}$, where Γ is the natural linewidth. The energy distribution of the emitted γ-rays has a Lorentzian lineshape in which the γ-ray intensity is given by

$$\mathrm{I}(v) = \left[1 + \left(\frac{v - v_0}{\Gamma} \right)^2 \right]^{-1}, \tag{5.5}$$

where the line is centered at a relative source-absorber velocity v_0 and the *half-width at half-maximum intensity* (HWHM) is Γ. The lower limit on Γ is due to the Heisenberg uncertainty principle and experimental considerations such as sample thickness increase Γ.

5.1.4 Hyperfine Interactions

The hyperfine coupling of the atomic electron charge density with the nuclear charge density shifts the nuclear energy levels and can remove their degeneracy either fully, partially, or not at all. The removal of nuclear level degeneracy results in multiple γ-ray transitions. There are three main types of hyperfine interaction, namely (i) the electric monopole term, leading to spectral displacements from zero Doppler velocity commonly referred to as the *isomer* or *chemical shift*; (ii) the electric quadrupole term, and (iii) the magnetic dipole term.

5.1.4.1 Electric Monopole Term

The monopole term is electrostatic and arises from the overlap of the atomic s-electron density with the nuclear charge distribution. This shifts the various nuclear energy levels by different amounts, due to their different spatial extents (hence different degrees of overlap), without affecting their degeneracies, and the resulting single-transition Mössbauer spectrum is referred to as a "singlet." The energy of a nuclear state of mean square radius $\langle r^2 \rangle$ is shifted by $E_o = \frac{Ze^2}{6\varepsilon_o}|\psi(0)|^2 \langle r^2 \rangle$ relative to a point nucleus, where Z is the atomic number and $|\psi(0)|^2$ is the atomic electron density at the nucleus. The isomer shift is normally quoted relative to a standard, such as the source or an α-Fe calibration foil. It is easy to show that resonant absorption between an absorber (A) and a source (S) occurs at a relative Doppler velocity of:

$$v_{\text{res}} = \frac{Ze^2c}{6\varepsilon_o E_\gamma}\left[\psi_A(0)^2 - \psi_S(0)^2\right]\left[\langle r_e^2 \rangle - \langle r_g^2 \rangle\right]. \tag{5.6}$$

This relative Doppler velocity is called the isomer shift (or sometimes the chemical shift). Due to the presence of the overlap term $|\psi(0)|^2$ in the expression for the isomer shift, this measurement can often unambiguously determine the electronic valence of the parent ion. In Figure 5.1 we show, schematically, the effect of the electric monopole interaction on the nuclear energy levels of ^{57}Fe and we illustrate the ensuing isomer shift with the ^{57}Fe room temperature (RT) Mössbauer spectrum of a stainless-steel foil that shows an isomer shift of ~ – 0.4 mm/s relative to the ^{57}CoRh source.

5.1.4.2 Electric Quadrupole Interaction

A nuclear state with a spin I greater than 1/2 has an aspherical charge density characterized by its electric quadrupole moment Q, which is a measure of the

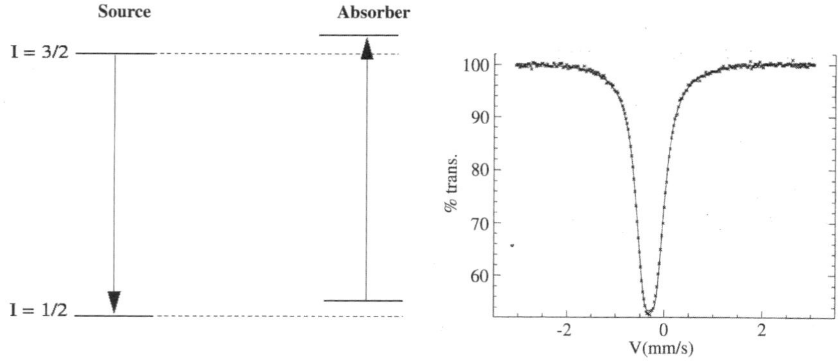

Figure 5.1 Left—Effect of electric monopole interaction on the nuclear energy levels in the source and absorber, giving rise to the isomer shift. Right—^{57}Fe Mössbauer spectrum at RT of a stainless-steel foil [Cadogan, unpublished].

degree of asphericity; (most readers will be familiar with the terms "oblate" and "prolate" when referring to the sign of Q). This energy state splits in the presence of an electric field gradient (EFG). The EFG is a tensor that has contributions from the unfilled atomic electron shells, valence electrons, surrounding lattice charges, amongst other sources. The components V_{ij} of the EFG tensor are defined as the second derivatives of the electrostatic potential $V_{ij} = \frac{\partial^2 V}{\partial x_i \partial x_j}$ acting at the nucleus. The principal axis frame of the EFG tensor is such that the off-diagonal V_{ij} terms disappear and the tensor is traceless, as a consequence of Laplace's equation. Thus $V_{xx} + V_{yy} + V_{zz} = 0$ and it is conventional to define the axes such that $|V_{zz}| \geq |V_{yy}| \geq |V_{xx}|$. The electric quadrupole Hamiltonian may be written in the form

$$H_Q = \frac{eQV_{ZZ}}{4I(2I-1)}\left[3\hat{I}_z^2 - \hat{I} + \eta\left(\hat{I}_x^2 - \hat{I}_y^2\right)\right], \qquad (5.7)$$

where the asymmetry parameter is defined as $\eta = \frac{V_{xx} - V_{yy}}{V_{zz}}$ and is a measure of the deviation of the EFG from uniaxial symmetry (in the presence of uniaxial point symmetry at the site in question, $\eta = 0$). The dependence of the quadrupole Hamiltonian on the squares of the nuclear spin operators means that nuclear substates such as $\left|I = \tfrac{3}{2}, m_I = +\tfrac{3}{2}\right\rangle$ and $\left|I = \tfrac{3}{2}, m_I = -\tfrac{3}{2}\right\rangle$ remain degenerate. In ^{57}Fe, the ground state is unaffected by EFG effects since I_g = 1/2 and thus Q = 0. The excited state I_e = 3/2 splits into two levels $\left|I = \tfrac{3}{2}, m_I = \pm\tfrac{3}{2}\right\rangle$ and $\left|I = \tfrac{3}{2}, m_I = \pm\tfrac{1}{2}\right\rangle$ with the corresponding energy shifts $E\left(\pm\frac{3}{2}\right) = \frac{eQV_{zz}}{4}\sqrt{\left(1+\frac{\eta^2}{3}\right)}$ and $E\left(\pm\frac{1}{2}\right) = -\frac{eQV_{zz}}{4}\sqrt{\left(1+\frac{\eta^2}{3}\right)}$. So, for a 3/2 ↔ 1/2 system such as in ^{57}Fe, ^{119}Sn, or ^{169}Tm one observes two Mössbauer transitions (a "doublet") with a *quadrupole energy splitting* (QS) of $\Delta = \left|\frac{eQV_{zz}}{2}\sqrt{\left(1+\frac{\eta^2}{3}\right)}\right|$. We note that the sign of the EFG component V_{zz} is indeterminate in this case. If the excited state spin is greater than 3/2, such as is the case of ^{166}Er and ^{170}Yb (I_e= 2 and I_g = 0), one can determine the sign of V_{zz} [4].

In Figure 5.2 we show the splitting of the nuclear energy levels of ^{57}Fe due to EFG effects and we illustrate this with the ^{57}Fe RT Mössbauer spectrum of $FeSO_4 \cdot 7H_2O$, which is a sharp quadrupole doublet with a large quadrupole splitting $\Delta \sim 3.4$ mm/s, typical of Fe^{2+}.

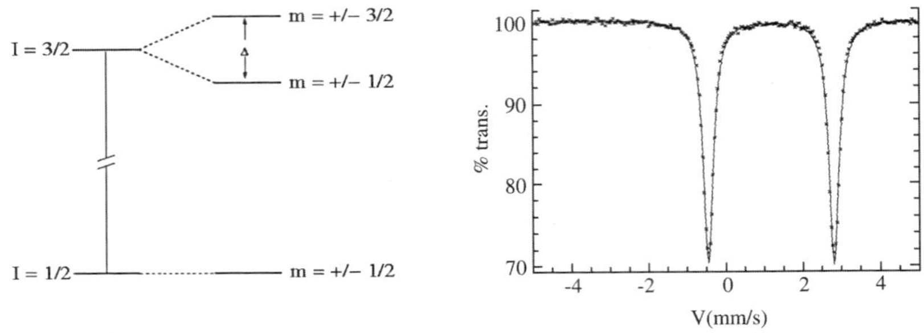

Figure 5.2 Left—Effect of the electric quadrupole interaction on the nuclear energy levels in ^{57}Fe. Right—^{57}Fe Mössbauer spectrum at RT of $FeSO_4 \cdot 7H_2O$ [Cadogan, unpublished].

5.1.4.3 Magnetic Dipole Interaction

A nuclear state with a non-zero spin has a magnetic moment and is therefore affected by magnetic fields acting at the nucleus via the magnetic dipole interaction $H_M = -\boldsymbol{\mu} \cdot \mathbf{B}_{hf} = -g_n \mu_N \hat{\mathbf{I}} \cdot \mathbf{B}_{hf}$ where μ_N is the nuclear magneton, B_{hf} is the hyperfine magnetic field, and g_n is the g-factor of the nuclear state in question (g_n can take either a + or – sign, depending on the state). The eigenvalues of the magnetic dipole Hamiltonian are $E_m = -m_I g_n \mu_N B_{hf}$ and thus the degeneracy of this nuclear state is completely removed. The selection rule for a magnetic dipole M1 transition (excited ↔ ground) is $\Delta m_I = 0, \pm 1$, which yields six possible transitions (a "sextet") for a 3/2 ↔ 1/2 transition, as in ^{57}Fe, ^{119}Sn, or ^{169}Tm. The relative intensities of the six Mössbauer transitions are given by the well-known Clebsch-Gordan coefficients and depend on the angle ϑ between the incident γ-ray direction and the nuclear magnetic moment. From lowest energy to highest one observes relative line intensities in the ratio 3:R:1:1:R:3 where $R = \dfrac{4\sin^2\vartheta}{1+\cos^2\vartheta}$ and, for a "perfect" powder averaged sample, $R = 2$. This angular dependence allows one to study magnetic texturing of samples; e.g., if the sample magnetization is completely parallel to the incident γ-ray direction, $R = 0$ and lines 2 and 5 of the sextet disappear. One can also use this approach to study spin-reorientations of the atomic magnetization by using magnetically aligned powder samples.

It is clear that the sign of the hyperfine magnetic field, usually referred to the direction of the atomic magnetic moment giving rise to the field, is indeterminate in a simple experiment. However, the application of a relatively small external field (a few tesla) is enough to determine the sign of the hyperfine field via its effect on the atomic magnetization. (We also note at this point that the sign of the EFG term V_{zz} in the magnetic state can only be determined relative to the sign of the magnetic hyperfine field.)

In general, both magnetic dipole and electric quadrupole effects are present and this results in an asymmetric sextet. The situation is further complicated by the fact that the quantization z-axes of the magnetic dipole and electric quadrupole Hamiltonians often do not coincide. Thus, one must transform one of the axial frames onto the other in order to obtain the eigenvalues and eigenvectors of the combined nuclear Hamiltonian, and therefore the transition energies and relative line intensities. Obviously, the eigenvalues cannot depend on the choice of reference frame. This problem was treated in detail in a classic paper by Kündig [5]. The reader is also referred to papers by Häggström [6] and Szymanski [7] for detailed treatments of the ^{57}Fe hyperfine Hamiltonian problem.

In a coordinate system coincident with the principal axis frame of the EFG, the hyperfine Hamiltonian may be written

$$H = H_{\mathrm{M}} + H_{Q} = -g_n \mu_N B_{hf}\left[\hat{\mathrm{I}}_Z \cos\theta + \left(\hat{\mathrm{I}}_X \cos\phi + \hat{\mathrm{I}}_Y \sin\phi\right)\sin\theta\right] + \frac{eQV_{zz}}{4\mathrm{I}(2\mathrm{I}-1)}\left[3\hat{\mathrm{I}}_Z^2 - \hat{\mathrm{I}} + \eta\left(\hat{\mathrm{I}}_X^2 - \hat{\mathrm{I}}_Y^2\right)\right], \tag{5.8}$$

where (θ, ϕ) are the polar angles of the hyperfine magnetic field in the EFG principal axis frame. Alternatively, one could rotate the EFG frame onto the magnetic frame. In the simple case where the nuclear hyperfine Hamiltonian is diagonal and $\eta = \theta = 0$ the nuclear energy levels are given by

$$E_{\mathrm{m}} = -m_I g_n \mu_N B_{hf} + \frac{eQV_{zz}}{4\mathrm{I}(2\mathrm{I}-1)}\left[3m_I^2 - \mathrm{I}(\mathrm{I}+1)\right] \tag{5.9}$$

In the simple situation where this EFG can be treated as a perturbation on the magnetic effect, the eigenvalues are given by

$$E_{\mathrm{m}} = -m_{\mathrm{I}} g_n \mu_N B_{hf} + \frac{eQV_{zz}}{4\mathrm{I}(2\mathrm{I}-1)}\left[3m_{\mathrm{I}}^2 - \mathrm{I}(\mathrm{I}+1)\right]\frac{1}{2}\left(3\cos^2\theta - 1\right). \tag{5.10}$$

The usefulness of this approach is that the angular term in the quadrupole contribution allows one to investigate spin-reorientations, i.e., temperature-driven changes in the atomic magnetization direction. As an example, one can easily see that if the principal axis of the EFG is along the crystal c-axis, a simple change from an easy magnetization axis along the crystal c-axis to one in the basal ab-plane leads to a relative change in effective quadrupole splitting of –1/2, which is fairly straightforward to observe in a Mössbauer experiment. In the presence of an asymmetry parameter, the preceding eigenvalue expression becomes

$$E_{\mathrm{m}} = -m_{\mathrm{I}} g_n \mu_N B_{hf} + \frac{eQV_{zz}}{4\mathrm{I}(2\mathrm{I}-1)}\left[3m_{\mathrm{I}}^2 - \mathrm{I}(\mathrm{I}+1)\right] \times \frac{1}{2}\left(3\cos^2\theta - 1 + \eta\sin^2\theta\cos 2\phi\right). \tag{5.11}$$

The hyperfine magnetic field contains numerous contributions, including the Fermi contact term due to the s-electron overlap with the nucleus, conduction electron polarization (in metals), dipole and transferred fields from the surrounding lattice, and externally applied fields. In ^{57}Fe the hyperfine field in α-Fe is 33.0 T at room temperature and α-Fe foils are the standard method for calibrating the Doppler energy-velocity scale (see Sections 5.2 and 5.3). Mössbauer experiments are commonly used to "determine" the atomic magnetic moment from the nuclear hyperfine magnetic field, which raises the problem of the conversion factor between the two. In ^{57}Fe, a commonly accepted conversion factor is 15 T/μ_B where μ_B is the Bohr magneton. However, large variations in the conversion factor can occur between different compounds and even between different crystal sites in the same compound.

In Figure 5.3 we show the 3/2 ↔ 1/2 energy level splitting in the presence of a magnetic hyperfine field along with the effect of an additional quadrupole splitting.

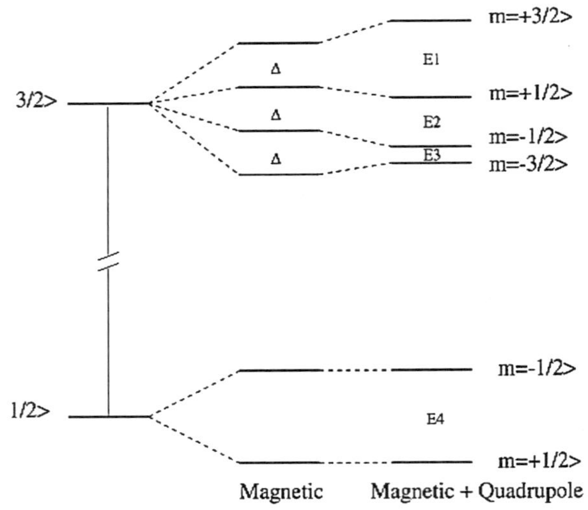

Figure 5.3 Nuclear energy level splittings for a 3/2 ↔ 1/2 Mössbauer transition (appropriate to ^{57}Fe, ^{119}Sn, or ^{169}Tm) in the presence of magnetic and electric quadrupole interactions.

To illustrate the effect of a magnetic hyperfine field, in Figure 5.4 we show the ^{57}Fe RT spectrum of an α-Fe foil, used to calibrate the Doppler velocity scale in a Mössbauer experiment. The high crystal symmetry of the single Fe crystallographic site in body-centered cubic α-Fe results in a zero EFG and the spectrum is a symmetric magnetic sextet. The velocity splitting between the outside peaks is 10.6246 mm/s, which corresponds to a magnetic hyperfine field of 33.0 T.

In Figure 5.5 we show the corresponding energy level splittings for 2 ↔ 0 transitions such as ^{166}Er and ^{170}Yb, and in Figure 5.6 we illustrate this with the ^{166}Er spectrum (4.5 K) of $Er_2Ge_2O_7$, which shows a magnetic splitting corresponding to a hyperfine field of 663.7(5) T [8]. For such a transition, the relative intensities of the five lines are 1:1:1:1:1.

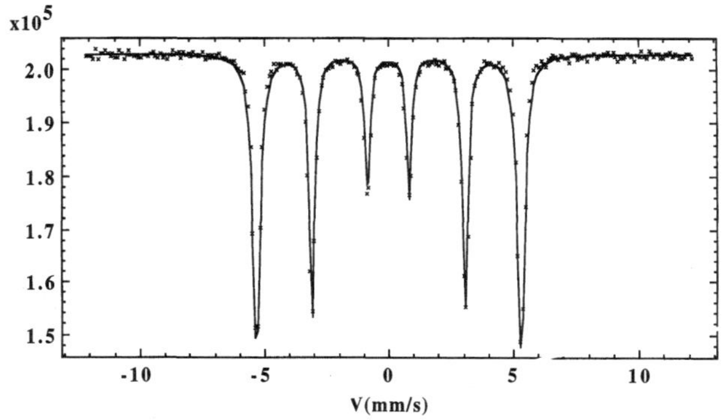

Figure 5.4 ^{57}Fe RT spectrum of a 25 μm α-Fe foil [Cadogan, unpublished].

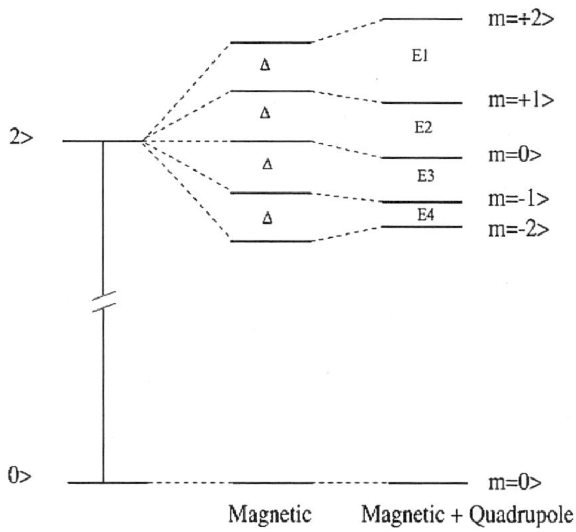

Figure 5.5 Nuclear energy level splittings for a 2 ↔ 0 Mössbauer transition (appropriate to ^{166}Er and ^{170}Yb) in the presence of magnetic and electric quadrupole interactions.

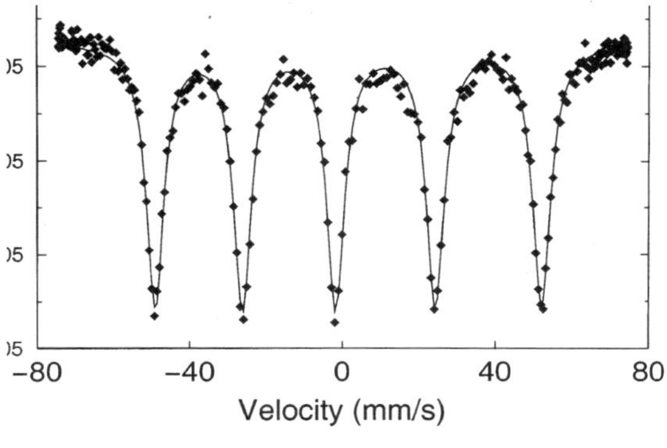

Figure 5.6 ^{166}Er spectrum of $Er_2Ge_2O_7$ obtained at 4.5 K [8].

5.1.4.4 Electronic Relaxation

The hyperfine magnetic field seen at a nucleus is predominantly determined by the electronic configuration of the parent atom, i.e., the unfilled 3d or 4f shells for the transition metals or "rare earths," respectively. For example, to a first approximation, the hyperfine field acting on a rare earth nucleus is proportional to the 4f electronic ensemble thermal average $\langle J_Z \rangle$ whereas the nuclear quadrupole splitting is proportional to $\langle 3J_Z^2 - J(J+1) \rangle$. It is implicit here that the 4f electronic configuration is static within the Mössbauer measuring time frame. However, one often observes the effects of "electronic relaxation" in which the nucleus senses the time-dependent fluctuations of the surrounding electronic configuration and the resulting Mössbauer spectrum reflects this behavior. There are two important timescales involved here: the lifetime of the nuclear excited state and the Larmor precession time. Basically, if the nucleus is to "see" a well-resolved magnetic splitting, unaffected by relaxation broadening, there must be a sufficient sensing time, i.e., at least one complete Larmor precession must occur before the nuclear state will decay. In the case of very fast electronic relaxation, the electronic configuration changes at a rate significantly faster than that corresponding to the nuclear Larmor precession rate. Hence, the nucleus senses a time-averaged hyperfine environment, manifest as the collapse of the well-defined magnetic sextet. Finally, we have a restriction on the usefulness of Mössbauer spectroscopy to study time-dependent effects in that if the characteristic time associated with the phenomenon under study is extremely slow and is greater than the nuclear excited state lifetime (and hence the Larmor precession time) then the process cannot be studied effectively by Mössbauer spectroscopy.

Time-dependent effects are observed in (i) isomer shift, (ii) magnetic field, and (iii) EFG effects, but are most often observed via the magnetic field. A number of approaches have been used to give a good account of the resulting broadening and/or collapse of the magnetic spectrum. In particular we mention the classic works of Blume and Tjon [9] and Wickman and Wertheim [10], both of which basically treat the relaxation within a two-state electronic model and then investigate the ensuing nuclear consequences on the Mössbauer spectrum. In Figure 5.7 we show simulated ^{57}Fe spectra illustrating the collapse of the magnetic sextet with increasing electronic relaxation rate.

There are, of course, other time-dependent effects which are amenable to study by Mössbauer spectroscopy. For example, lattice dynamics in which the Mössbauer "atom" diffuses, i.e., moves, during the Mössbauer process have been studied, once again given the aforementioned time considerations.

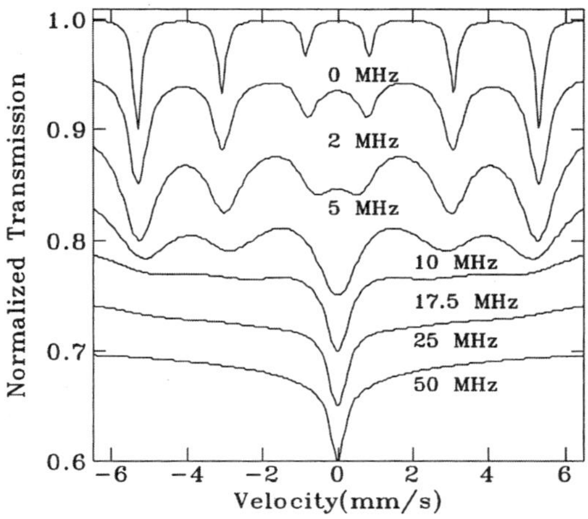

Figure 5.7 Simulated ^{57}Fe spectra as a function of increasing electronic relaxation rate (Ryan and van Lierop, unpublished).

5.1.4.5 Crystal-Field Effects

As mentioned above, it is a good approximation in rare earth magnetic materials that proportionality exists between the magnetic hyperfine field at the rare earth nucleus and $\langle J_Z \rangle$, and the 4f contribution to the EFG is proportional to $\langle 3J_Z^2 - J(J+1) \rangle$. The temperature dependences of these 4f electronic averages are determined by the interplay of magnetic exchange and electrostatic crystal field effects, and Mössbauer spectroscopy has played an important role over the years in the study of crystal fields acting of the 4f shell [11]. The 4f elements are quite similar chemically but exhibit a wealth of varied magnetic behaviour due to the progressive filling of the 4f shell. As we shall demonstrate later in this article, the combination of rare earth isotope Mössbauer spectroscopy and neutron diffraction has proved itself to be a formidable means of determining the magnetic behavior of intermetallic compounds.

5.1.4.6 Amorphous Materials

The case of disordered or amorphous materials is complicated by the fact that the probe atoms are subject to a distribution of environments, which leads to distributions of hyperfine parameters such as hyperfine field and electric field gradient. In such cases, Mössbauer spectroscopy is predominantly used to determine parameters such as the average hyperfine field (and hence the average atomic magnetic moment) or average quadrupole splitting in the case of a paramagnetic material, both of which can be reliably determined despite the presence of a hyperfine parameter distribution. It is known that the form of the hyperfine parameter distribution is strongly dependent on the method used to calculate it. However, the average parameter values deduced are quite robust. The calculation proceeds as follows: One selects a histogram [12] or Fourier series [13] approach to the form of the hyperfine parameter distribution, field, or EFG; simulates appropriate doublets or sextets with a common linewidth; and finally builds up the form of the distribution by varying the relative intensities of the individual components to obtain the best fit to the experimental spectrum. Asymmetries in the spectra are accounted for by introducing correlations between parameters such as isomer shift and hyperfine field in a somewhat ad hoc manner. Deviations from simple Lorentzian lineshapes can also arise, particularly for low hyperfine field values where the commonly employed field-EFG perturbation approach is questionable. One must also be aware of the significant effects of histogram size, i.e., how many steps there are in the histogram. This can lead to severe overlap effects between "adjacent" hyperfine field sextets.

At low field values, the electric quadrupole interaction can no longer be treated as a perturbation on the magnetic hyperfine field. This can give rise to a peak in the lower part of the hyperfine field distribution and some authors have considered the resulting bimodality of the distribution as a signal of two distinct environments in amorphous materials. While this might be true, one must always bear in mind the somewhat limited validity of the approach employed. As shown by Keller [14], for example, one can simulate a spectrum with a single Gaussian distribution of hyperfine field and subsequently obtain a bimodal distribution using the commonly employed fitting methods. Le Caër and coworkers [15] published a validity diagram that discusses this important aspect of Mössbauer data analysis in amorphous materials.

Mössbauer spectroscopy has also been used to great advantage to determine the relative volume fractions of a number of commercially important soft-magnetic materials such as Vitroperm and Finemet, where one has an intimate mixture of nanocrystalline materials (usually bcc-α-Fe) embedded in an Fe-rich amorphous matrix [16]. Such information is vital to the development of a model for the random anisotropy observed in these commercially viable soft-magnetic materials. We shall illustrate this work in Section 5.3.

In the following Table 5.1 we present a compendium of information relevant to the more commonly used Mössbauer transitions. This information is taken from publications by the Mössbauer Effect Data Centre [17]. At the end of this article we list a number of books and reviews that deal with Mössbauer spectroscopy and hyperfine interactions in more detail than is possible here [18–26].

5.2 METHODOLOGY

To obtain a Mössbauer spectrum one needs a suitable source of γ-photons, a detector capable of isolating the desired γs from other radiations emitted by

Table 5.1. Mössbauer parameters for a number of important Mössbauer transitions [17].

	^{57}Fe	^{119}Sn	^{151}Eu	^{155}Gd	^{166}Er	^{169}Tm	^{170}Yb	^{197}Au
Isotopic Abundance (%)	2.14	8.58	47.82	14.73	33.41	100	3.03	100
E_γ(keV)	14.41	23.87	21.53	86.55	80.56	8.40	84.25	77.345
I_e	3/2	3/2	7/2	5/2	2	3/2	2	1/2
I_g	1/2	1/2	5/2	3/2	0	1/2	0	3/2
Excited state half-life (ns)	97.8	17.75	9.7	6.33	1.87	4.00	1.608	1.88
Γ (mm/s) FWHM	0.194	0.647	1.31	0.499	1.82	8.1	2.02	1.882
Excited state magnetic moment (μ_N)	– 0.1553	+ 0.633	+2.587	–0.529	+0.629	+0.534	+0.669	+0.416
Excited state gyromagnetic ratio (mm/s /T)	–0.067897	0.167	0.3244	–0.0231	0.0369	0.400	0.0375	+0.1017
Ground state magnetic moment (μ_N)	+0.090604	–1.0461	+3.465	–0.2584	0	–0.231	0	+0.1448
Ground state gyromagnetic ratio (mm/s /T)	–0.118821	–0.8283	+0.6083	–0.02713	0	–0.520	0	+0.0118
Excited state electric quadrupole moment (barns)	0.21	– 0.06	1.50	0.32	–1.59	–1.20	–2.11	0
Ground state electric quadrupole moment (barns)	0	0	1.14	1.59	0	0	0	0.594
Energy equivalent of 1 mm/s in joules ($\times 10^{-27}$)	7.70278	12.757	11.507	46.253	43.052	4.490	45.0275	41.336
Energy equivalent of 1 mm/s in MHz	11.62477	19.253	17.367	69.803	64.973	6.776	67.954	62.382
Source half-life	271 d	293 d	87 y	4.8 y	26 h	9.4 d	130 d	19 h

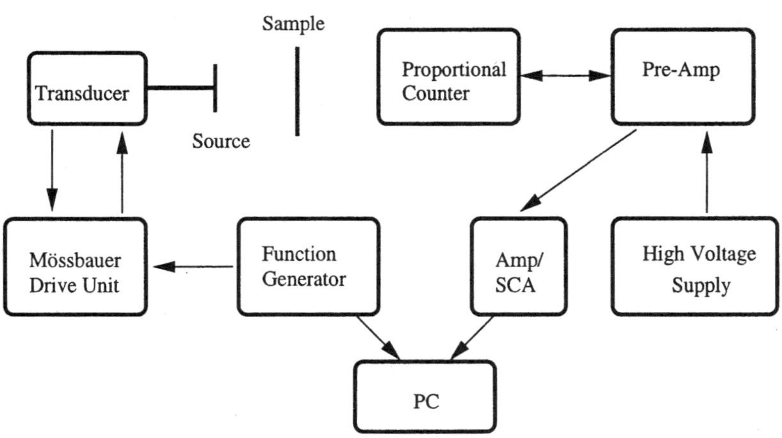

Figure 5.8 Typical layout of a transmission Mössbauer spectrometer (Ryan, unpublished).

the source and, most critically, a means to modulate the energy of the Mössbauer γ-photons. In Figure 5.8 we show a schematic of a standard Mössbauer spectrometer and in the following sections we describe the various aspects of spectral acquisition.

5.2.1 Drives

The velocity range demanded by most commonly used Mössbauer transitions lies below ±50 mm/s and is easily achieved mechanically. Therefore, most modern Mössbauer spectrometers employ a commercial electromechanical drive. Such a drive is basically a pair of rigidly coupled loudspeakers designed to provide long-term stability and extremely linear motion. The power section of the pair is used to drive the motion, while the pick-up section is used to measure the instantaneous velocity. A digital function generator provides the desired velocity waveform, and feedback from the pick-up coils is used to slave the drive motion to the velocity waveform and so control the source velocity. Accuracies and long-term stabilities of better than 0.1% are routinely exceeded.

Drives are normally run at or close to their natural frequency (typically 30 to 40 Hz) in order to minimize the errors that have to be corrected by the velocity feedback loop. Two velocity waveforms are commonly used: triangular or "constant acceleration," and sinusoidal. The linear mapping between time and velocity provided by constant acceleration mode simplifies data visualization and analysis, and also yields spectra with a constant velocity resolution and time weighting. The instantaneous change in acceleration demanded of the drive when the "constant" acceleration changes sign at maximum velocity puts significant demands on the mechanical and electronic designs and some ringing in the velocity error is inevitable. However, the error occurs at the extremal velocities where only background data should be collected; its effects can be minimized in a properly designed system. Constant acceleration offers many advantages and is widely used for the modest velocity ranges needed for ^{57}Fe, ^{119}Sn, and ^{151}Eu Mössbauer spectroscopy.

Any drive that operates well in constant acceleration mode will give better performance when operated with a sinusoidal velocity waveform. However, the improvements are rarely significant and generally lie well below other mechanical or resolution limitations. Disadvantages of sine-mode operation include a nonlinear mapping between time and velocity that adds some minor complexity to the data analysis, and a severely nonuniform velocity coverage. A sine wave spends most of its time out near its extrema, where we would expect only background contributions, and much less time traversing the region around zero velocity where most of the Mössbauer lines are expected to occur. This decreased counting time and lower central resolution need to be carefully compensated for during data acquisition.

There are, however, two situations in which sinusoidal velocity operation is unavoidable: (i) high velocities, (ii) long drive couplings. When velocities of hundreds of mm/s are needed, e.g., for ^{161}Dy or ^{169}Tm, the drive must be operated sinusoidally as the demands placed on the mechanical system preclude operation in constant acceleration mode: the ringing at the acceleration reversals becomes more severe and persistent. Furthermore, the harmonic content of the triangular waveform becomes increasingly problematic at higher velocities; achieving adequate vibration isolation between the drive/source assembly and the sample is more difficult. Dedicated drives with much stiffer internal springs are generally used for high velocity applications, as this moves the natural frequency up above about 100 Hz and reduces the vibration amplitude. At 30 Hz, a maximum velocity of 1000 mm/s yields an amplitude of about 5 mm and requires an extremely linear velocity pick-up over a very wide position range to ensure accurate velocity control. Raising the operating frequency to 100 Hz reduces the amplitude to around 1.6 mm.

When a source is operated within a cryostat, either to apply a large magnetic field or to keep the source at a low temperature, the source is often placed on the end of a long (~1 m) extension rod, well away from the actual drive. This situation puts severe demands on the drive performance, adding the natural vibration modes of the long rod to the mix of problems. Again, sinusoidal drive operation is generally the only viable option, and careful choice of operating frequency is essential so that driving natural modes in the extension is avoided. Proper design of the drive-extension-source assembly is important to ensure that transverse motions are prevented and the rod is stiff enough to provide proper coupling of the drive motion to the source.

5.2.2 Detectors

The Mössbauer γ originates from a nucleus in an excited state. This state is generally populated by the prior decay of a radioactive parent nucleus. The path to the Mössbauer level often involves one or more intermediate steps, each leading to the emission of a γ. As most Mössbauer transitions involve quite low energy γ-photons, the process is often quite heavily internally converted (an s-electron is emitted instead of the γ), which creates a hole in an inner electron shell, which if filled radiatively rather than by an Auger process, leads to the further emission of characteristic K or L X-rays. Finally, with many energetic γ-photons and X-rays being emitted directly by the source, other materials in the source and shielding environment may fluoresce, adding their own characteristic K or L X-rays to the emission spectrum.

For example, the standard Mössbauer source for ^{57}Fe is ^{57}Co, diffused into a metallic matrix, commonly rhodium. As shown in Figure 5.9, ^{57}Co decays by electron capture, so that a significant emission of Co-K_α X-rays at ~6.9 keV is inevitable. The ^{57}Co decay populates the 136 keV level of ^{57}Fe which decays directly to the ground state in ~10% of cases yielding a 136-keV γ. The remaining 90% of the nuclei emit a 122-keV γ and populate the 14.4-keV Mössbauer level. Ninety percent of the time, the 14.4-keV level decays by internal conversion, emitting a low energy (~7 keV) electron and

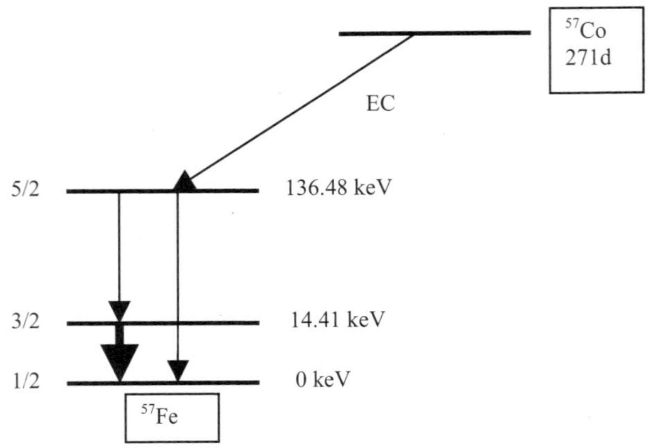

Figure 5.9 Decay scheme associated with ^{57}Fe Mössbauer spectroscopy (the thicker arrow represents the Mössbauer γ-ray).

leading subsequently to the emission of a Fe-K_α X-ray at ~6.4 keV. As a result, only about 9% of the original $^{57}Co \rightarrow ^{57}Fe$ decays lead to the emission of the 14.4 keV Mössbauer γ, the rest yield other γ-photons and X-rays. There is also a significant emission of Rh-K_α X-rays at ~20 keV, and likely also at ~75 keV if lead is used in the shielding. The job of the detector and associated nuclear electronics is to efficiently detect and isolate the 14.4-keV Mössbauer γ from the six or more other photon energies emitted with similar or substantially higher intensity.

The choice of detector will be determined in part by budget, but primarily by the characteristics of the source being used. None of the detectors discussed here, or in fact no available detector, comes close to matching the $1{:}10^{12}$ energy resolution of a typical Mössbauer transition. The detectors used in a Mössbauer spectrometer serve only to isolate the Mössbauer γ from other radiations emitted by the source; the fundamental resolution of the Mössbauer transition is only observed in the spectrum obtained by velocity modulation of the source energy. One therefore needs only modest energy resolution, and the simplest and most rugged detector that will isolate the γ of interest is usually the best choice. Indeed, a "better", i.e., more expensive, detector may actually yield much poorer performance.

5.2.2.1 Proportional Counters

Noble gas-filled proportional counters achieve relatively poor energy resolution (~12% at 14-keV), yet this resolution is more than sufficient to separate the 14.4-keV Mössbauer γ from the neighboring X-rays at 7 keV and 20-keV. They are also comparably cheap, robust, and have an excellent operational lifetime. They are the detector of choice for ^{57}Fe Mössbauer spectroscopy.

A typical configuration would be a 50-mm diameter cylinder with 0.25-mm thick beryllium side entry and exit windows, 25 mm in diameter, and a Xe+3% CO_2 fill at one atmosphere. The thin Be windows admit the 14.4-keV γ without loss and the 1 atm Xe fill has enough stopping power to efficiently collect them without providing too much efficiency for the much more intense 122-keV γ-photons, which mostly exit undetected through the second Be window. A thin (~1 mm) plexiglass filter over the detector window can also reduce detection of the Co and Fe K_α and count rates on the 14.4-keV γ of 5 to $10 \times 10^3\ s^{-1}$ can be obtained without saturating the electronics.

A Kr+3% CO_2 fill is also very popular as it yields even lower sensitivity to the higher energy γ-photons and so reduces background rates. However, the 14.4-keV γ is above the Kr K-edge and a significant fraction of the Mössbauer γ-photons appear in the 1.8 keV "escape" peak, necessitating the use of an additional single channel analyzer if good efficiency is to be obtained. The fluorescence yield of Kr is also poor, with only about 50% of the absorbed γ-photons yielding a photoelectron that can be detected.

The increased high energy sensitivity of NaI(Tl) or HPGe detectors means that their response is dominated by the 136-keV and 122-keV γ-photons. Counting rates are more limited and increased backgrounds from Compton scattering of higher energy γ lead to very poor performance in ^{57}Fe Mössbauer spectroscopy.

Proportional counters with a 1 atm Ar+3% CO_2 fill are commonly used to detect the 8.4-keV ^{169}Tm Mössbauer γ, and can also be used for *X-ray backscatter* (XBS, see below) work with the ~6.4-keV Fe-K_α X-ray. In principle, a 2-atm Xe+10% CO_2 fill can be used for either ^{151}Eu or ^{119}Sn Mössbauer spectroscopy, but freedom from intense higher energy γ contamination and the availability of more efficient detectors for the 21-keV and 24-keV γ make NaI(Tl) or HPGe better choices for these Mössbauer resonances.

5.2.2.2 NaI(Tl) Scintillation Detectors

For many applications in the 20–100 keV range, a thallium-activated sodium iodide scintillation detector is a good choice. A 0.25-mm thick beryllium or 25-μm thick aluminium entry window minimizes attenuation losses above 20 keV, and a crystal 6–8 mm thick is sufficient to give over 95% efficiency out to 100 keV (there are no useful Mössbauer resonances beyond this energy). These detectors actually give slightly poorer energy resolution than proportional counters (~50% at 6 keV and ~20% at 90 keV) but can be used at similar count rates. The presence of a photomultiplier tube behind the crystal does make these detectors significantly heavier and more fragile than the gas-filled proportional counter. The gain of the photomultiplier can also be affected by stray magnetic fields and while they are typically manufactured with μ-metal shielding, they should be protected from external magnetic fields.

5.2.2.3 High Purity Germanium (HPGe) Detectors

HPGe solid state detectors exhibit energy resolutions of < 0.5%, with better than 95% efficiency from 6 keV to 200 keV. When equipped with a pulsed optical feedback network to stabilize the detector bias voltage, count rates in excess of 10^5 s^{-1} are readily achievable. This performance comes at a significant price, with a good HPGe detector costing 10–20 times that of a NaI(Tl) scintillation detector. The detector must also be continuously cooled with liquid nitrogen during operation, adding a bulky Dewar assembly to the unit. However, the detector can be allowed to warm up, and may be stored at room temperature when not in use, unlike Si(Li) detectors, which would be damaged by such treatment. The comparable performance and increased convenience has led to HPGe detectors replacing Si(Li) units for most applications.

Few Mössbauer resonances demand the performance of a HPGe detector and for the commonly used isotopes such a detector is either inappropriate (^{57}Fe) or expensive overkill (^{151}Eu and ^{119}Sn). Where HPGe detectors are essential is in work with noncommercial Mössbauer sources. Neutron activation or proton irradiation, even of sources prepared with isotopically separated starting materials, inevitably leads to the presence of one or more unwanted radioisotopes. Even when the source is effectively pure, γ-energies tend to be higher than the 24-keV ^{119}Sn resonance, and copious amounts of fluorescence X-rays can accompany the Mössbauer γ. In this situation, good resolution is needed to isolate the Mössbauer energy from surrounding X-rays, and excellent rate capability permits operation with high-activity sources to compensate for low *f*-factors without saturating the detector. Count rates of 5×10^4 s^{-1} on the Mössbauer γ in the presence of comparable rates in non-Mössbauer X-rays are routinely encountered.

5.2.3 Data Collection

Collecting a Mössbauer spectrum requires the proper synchronization of two sets of signals: the motion of the drive and the detection of γ-photons. The charge pulses generated in the detector by the stopping of a γ are shaped and amplified in a pre-amplifier and amplifier. The voltage pulses are then passed through a single-channel analyzer (SCA), where those within a certain amplitude range, chosen to correspond to the γ-energy of interest, generate a TTL pulse, while those outside the selected range are rejected. The pre-amp/amp/SCA chain is available as a dedicated commercial Mössbauer system, or can be assembled from individual nuclear instrumentation modules (NIM). Costs are comparable and both approaches have advantages. A dedicated unit is simpler to operate, but more difficult to troubleshoot when problems develop, whereas a NIM-based system requires a larger initial investment of time to set up, but it is more flexible and far simpler to troubleshoot.

The stream of TTL pulses being generated by the SCA must now be synchronized with the motion of the Mössbauer drive. A typical arrangement is as follows. The velocity waveform to which the drive's motion is slaved is produced digitally by a function generator that also provides two additional data streams: *sweep START* and *channel advance* (CHA). These are used to control an array of counters in a multichannel scaler (MCS), typically an add-on card for a PC. The START pulse resets the multichannel scaler and opens the first data register to receive counts. Any TTL pulses that arrive from the SCA while the register is open are added to the current contents. When each successive CHA pulse arrives, the current register is closed, and the next one is opened. In this way, the arriving TTL pulses are placed in a register that corresponds to a specific velocity of the drive. When the complete waveform has been generated (commonly 512 or 1024 channels), a new START is issued and the cycle repeats. At velocities corresponding to absorption lines in the sample, the count rate will be slightly lower and those data channels will increase more slowly. Over time, a Mössbauer spectrum develops.

Once the spectrum is of sufficient quality (determined by visual inspection, predetermined time, or by evaluation of intermediate fits carried out during the data collection) collection is halted and the spectrum is transferred and folded ready for analysis. Since the motion of the drive is necessarily cyclic, every velocity occurs twice during the waveform, but the two copies are not identical. The source is slightly closer to the detector during one half of the cycle than the other, leading to a higher average count rate. Indeed, since the source position changes continuously during the cycle, so does the count rate, and the background in the absence of a sample exhibits an approximately sinusoidal modulation with an amplitude that can exceed 1% for set ups with short source-detector distances. A second problem arises through mechanical lags in the drive motion, which can be as much as 1%, and lead to the center of symmetry of the actual source motion being offset from the center of the intended velocity waveform. Folding the data greatly reduces the effects of position-induced count rate variations, but this must be done with care to avoid distortions due to mechanical phase shifts. There are three common strategies:

Auto-fold. It is possible to arrange for some multichannel scalers to count down through their registers as well as up. Therefore, once the drive reaches its maximum velocity and starts back down, the MCS can reverse its sweep through the registers. This approach can mask any number of failures in the instrumentation and runs the risk of destroying the spectrum as it is collected while only avoiding a simple arithmetic manipulation that is trivially carried out after the collection is complete.

Fold before analysis. This is the simplest method. The center of symmetry (of the *spectrum*) can be determined either by fitting the two halves, seeking the maximum correlation between them, or by simply observing the folding point at which the maximum signal is obtained. Some care needs to be exercised when folding at noninteger points as some form of interpolation of the data will be needed, and this will make the actual data distribution non-Poisson and can affect subsequent data analysis. Folding the data after collection greatly reduces the background variations due to source motion and has the added benefit that it is nondestructive: the unfolded data are still available for refolding. Unexpected variations in folding point or poor overlap between the two halves are immediate indications that all is not well with the spectrometer.

Fit the unfolded halves together with a background term. A final option is to not fold at all, but simply fit the duplicated half and include a symmetry-center variable explicitly in the fit. Some form of background function must also be included in the fit. The data can be folded for presentation and interpolation-included distortions of the counting statistics are irrelevant as they do not affect the analysis.

5.2.4 Calibration

Accurate calibration of the spectrometer is critical and it should provide three pieces of information: (i) velocity-zero (ii) the velocity scale, either V_{max} or the velocity step per channel, and (iii) confirmation of drive linearity and performance. Ideally, calibration is achieved by measuring a standard sample, but various forms of laser interferometers are also popular.

A thin foil (< 25 μm) of iron metal provides an excellent calibration standard for ^{57}Fe Mössbauer spectroscopy as the line positions have been determined absolutely (see Section 5.1) and the material is extremely stable. The 6-line magnetic pattern also provides a convenient drive linearity test. All ^{57}Fe isomer shifts are quoted relative to the center of the iron metal spectrum at room temperature. Hematite (α-Fe_2O_3) can be used to extend the velocity range out to ±10 mm/s, and with confirmed linearity, velocity ranges two or three times larger than this can be safely used. While calibrating with a standard sample does use up counting time that could be spent measuring "real" samples, it is an essential step and also provides a complete check of the spectrometer performance in the actual mode that it used to acquire data. Every component of the spectrometer is tested together, and problems that might be missed on isolated tests can be caught on a clean standard long before they distort the results on a more complex material.

For those not working with ^{57}Fe, calibration standards are less clear. The simplest approach is to mount a ^{57}Co source on the back of the drive and use iron metal as a calibrant, with a second standard being used to set velocity-zero for the isotope being studied (e.g., $BaSnO_3$ is a common choice for ^{119}Sn work). This procedure works well if ^{57}Fe Mössbauer is also being used in the same laboratory and if the velocity scales being used are compatible. Other standards that have been used include $ErFe_2$ (± 60 mm/s, for ^{166}Er), Dy metal (± 200 mm/s, for ^{161}Dy) and $TmFe_2$ (± 800 mm/s, for ^{169}Tm; see Figure 5.18).

Laser interferometers are a popular and versatile alternative to calibration that uses a standard sample. They can be run in parallel and so do not cost counting time, and they have a much wider velocity range. However, they only provide information on the motion of the reflecting surface mounted on the back of the drive, and strictly speaking do not measure the actual relative motion of the source-sample pair. Furthermore, laser interferometers cannot be used to check source linewidths, nor do they provide any information on the performance of the nuclear counting electronics. Laser interferometers employ a Michelson interferometer to count the number of fringes that pass during the opening of each velocity channel, and knowing the time each channel is open, the average velocity of each channel is readily calculated. However, for almost all Mössbauer applications, the motion of the drive during any given step is comparable to the wavelength of the laser light. For example, at a V_{max} of 10 mm/s working at 30 Hz and 512 data channels, the drive moves a maximum of about 650 nm over one channel. Coupled with the highly stable, deterministic motion of the drive, a He-Ne laser (λ = 633 nm) signal shows large blocks of channels with zero, one, and two counts. Various schemes for introducing position noise have been employed to smooth out this behavior, but low velocity calibrations are not ideal, and they are best at higher velocities (> 50 mm/s).

Mechanical recoil can also affect calibrations. While the mass of the moving coil and source assembly is generally kept as small as possible, it cannot be made zero and its motion necessarily induces some recoil of the drive. For conventional transmission work, the effects of this recoil can be minimised by rigid coupling of the drive to a heavy mount; however when one works with a cold-source system, or with the source and sample in a magnet cryostat, the driven mass increases and rigid mounting becomes more problematic. Elasticity in the long coupling rod between the drive and source can also affect the actual motion of the source. In such situations, an interferometric measure of the drive motion may not provide an accurate measure of the source-sample relative motion. Deviations of 5% are not uncommon and cross-checking with one or more standard samples is essential.

5.2.5 Sources

Commercial sources are currently available for three of the most commonly used Mössbauer resonances: ^{57}Fe, ^{119}Sn, and ^{151}Eu. ^{57}Fe Mössbauer sources are prepared by diffusing ^{57}Co into a cubic metal matrix (commonly rhodium) to give a single-line source with a good recoil-free fraction. The half-life is 271 days and initial activities of 10–50 mCi (300–2000 MBq) provide useful count rates for two or more years (depending on the application). One side benefit is that the source can be retired to an undergraduate teaching lab and be used for teaching experiments for another year of use. Observed linewidths for a thin iron metal absorber are around 0.113 mm/s (half-width at half maximum–HWHM). ^{119}Sn is unusual in that the parent for the decay is a long-lived excited state of ^{119}Sn, usually designated ^{119m}Sn, as shown in Figure 5.10.

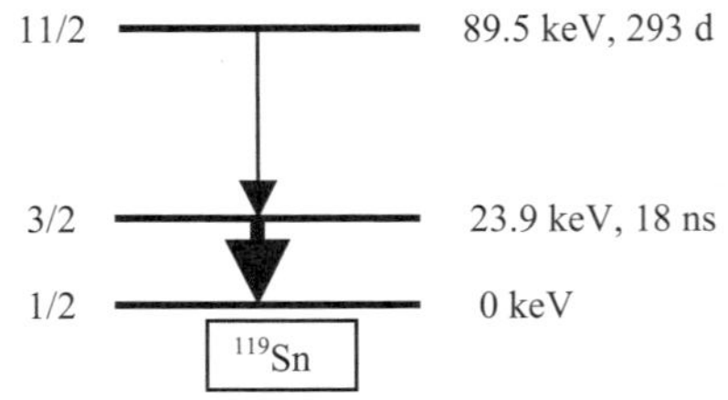

Figure 5.10 Decay scheme associated with ^{119}Sn Mössbauer spectroscopy.

As the parent and daughter are chemically identical they cannot be separated. It is essential therefore that sources be prepared from fresh ^{119m}Sn, as a build-up of ^{119}Sn in the source leads to resonant self-absorption, which can cause a significant reduction in source *f*-factor (recoil-free γ are preferentially reabsorbed in the source, while those from recoil events escape) and an increase in linewidth (γ further from the center of the Mössbauer line are less likely to be reabsorbed in the source and so are over-represented in the emission spectrum). A source that starts out bad, can only get worse as ground state ^{119}Sn continues to accumulate. The standard matrices are $CaSnO_3$ and $BaSnO_3$. The parent half-life is ~290 days so the working lifetime of a source is comparable to that of ^{57}Co sources. The γ is emitted quite efficiently, so activities in the 2–10 mCi (70–400 MBq) range yield acceptable to high counting rates. Typical source linewidths observed with a $BaSnO_3$ absorber are ~0.42 mm/s (HWHM).

Sources for ^{151}Eu Mössbauer spectroscopy usually employ ^{151}Sm in a SmF_3 matrix (Figure 5.11). Linewidths of 0.12 mm/s are typical. The parent half-life is 87 years, so apart from the inevitable line broadening and decline in *f*-factor associated with the build-up of ^{151}Eu in the source, a good ^{151}Eu Mössbauer source will last a researcher's career. Unfortunately, the γ emission is extremely inefficient. Only about 1% of the electron capture decays of the parent ^{151}Sm actually lead to a ^{151}Eu nucleus in the 21.5-keV Mössbauer state.

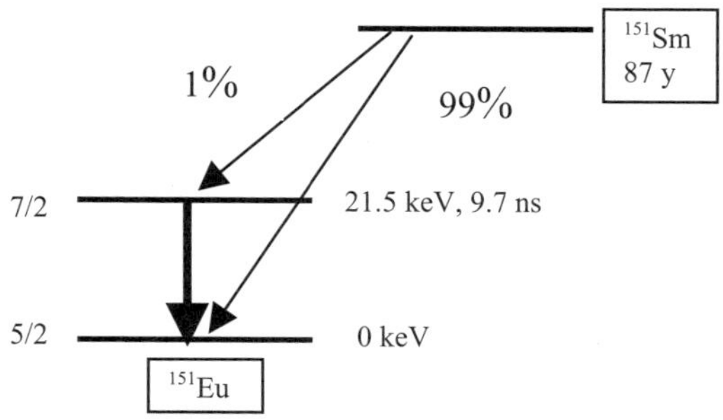

Figure 5.11 Decay scheme associated with ^{151}Eu Mössbauer spectroscopy.

Internal conversion dominates the subsequent decay process so that only about 1% of the transitions are radiative. The inefficiency of the decay process means that significant source activities are essential to attain useful count rates. Even with 100 mCi (4 GBq) counting rates are of order 200 s^{-1}. High activity sources are not currently available, with 30 mCi (1 GBq) being the limit. Such a source would not lend itself to high throughput experiments.

5.2.5.1 Nonstandard Resonances

Many Mössbauer resonances are underused because sources are not readily available. The parent isotope is often short-lived and some investment in source materials preparation is needed and isotopically separated starting materials must be used. In addition, higher γ energies mean that useful recoil-free fractions are only obtained with both the source and the sample at low temperatures (typically below 50 K). We consider several examples here.

^{197}Au: The principal problems with this transition are the short half-life of the parent ^{197}Pt (19 hours) and interference between the Mössbauer γ at 77.35 keV and the $K_{\beta 1}$ X-rays from Au (77.984 keV). The former demands nearby access to a nuclear reactor while the latter must be accepted. Source preparation is relatively easy as ^{196}Pt is available as a metal and can be irradiated directly. Neutron fluxes of 3×10^{11} cm^{-2} s^{-1} are sufficient to obtain useful source activities.

^{170}Yb: The radioactive parent (^{170}Tm) is obtained by neutron activation of ^{169}Tm which is, conveniently, the only naturally occurring isotope of thulium (Figure 5.12). A 10 wt.% alloy of thulium in aluminium (a mixture of $TmAl_2$ in an Al matrix) makes a convenient single-line source. The 130-day half-life makes it a relatively convenient source to work with, but it demands higher neutron fluxes so it can be activated in a reasonable time (15 h at 3×10^{13} cm^{-2}s^{-1} works well). The 84.2-keV Mössbauer γ is well separated from the many X-rays emitted by the source.

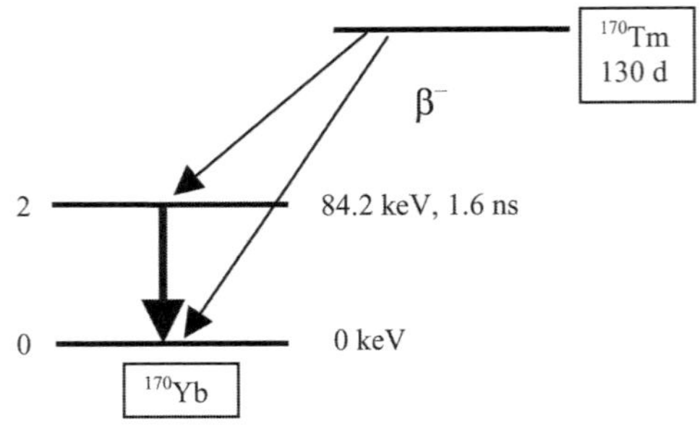

Figure 5.12 Decay scheme associated with ^{170}Yb Mössbauer spectroscopy.

^{166}Er: This resonance is also accessed by activation of a single-isotope precursor, in this case ^{165}Ho (Figure 5.13). The short half-life (~27h) and large capture cross-section make high activity sources possible. Overnight irradiation at 3×10^{11} cm^{-2} s^{-1} works well, but the reactor facility needs to be fairly close by. A standard source matrix is $Ho_{0.4}Y_{0.6}H_2$ which remains unsplit down to about 5 K. At lower temperatures, the source needs to be heated to prevent magnetic relaxation broadening the emission line.

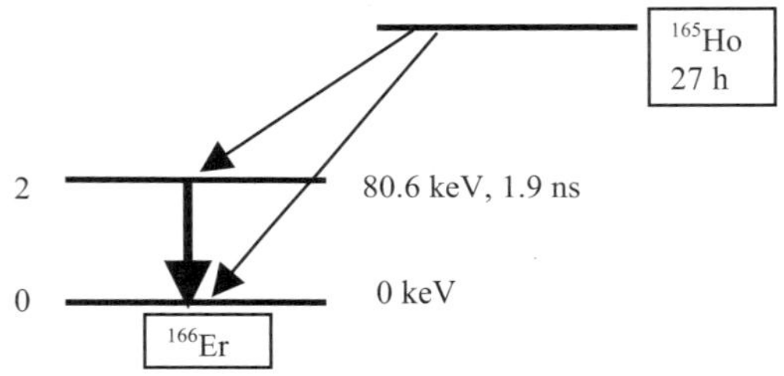

Figure 5.13 Decay scheme associated with ^{166}Er Mössbauer spectroscopy.

^{169}Tm: There are many problems limiting access to this very useful Mössbauer resonance. The source, ^{169}Er, must be prepared by neutron activation of isotopically pure ^{168}Er in order to avoid the presence of interfering radiations (Figure 5.14). As with ^{170}Yb, the best source matrix is Al (+10 wt.% ^{168}Er). Unfortunately ^{168}Er is only available as an oxide, and the reduction of $^{168}Er_2O_3$ and subsequent alloying with aluminium is an interesting challenge. The capture cross-section is small and the source must be extremely thin to permit the 8.4-keV Mössbauer γ to escape, thus necessitating long irradiations in order to attain useful source activities.

5.2.5.2 Environments

Mössbauer spectroscopy is a well-established nuclear technique in solid state physics, therefore it is not the technique itself that is central to Mössbauer work, but rather the samples and the environments to which they are exposed.

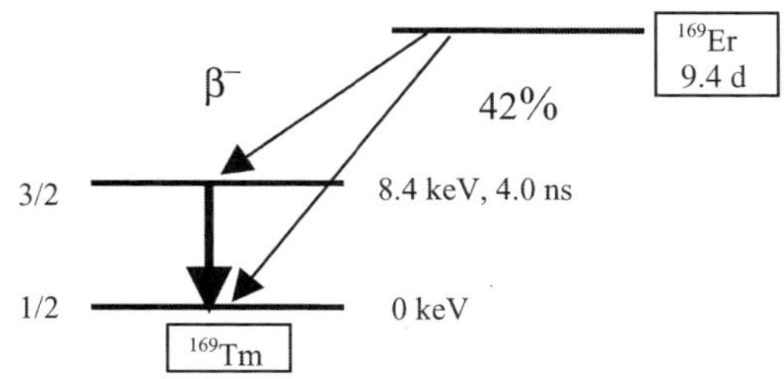

Figure 5.14 Decay scheme associated with ^{169}Tm Mössbauer spectroscopy.

A significant amount of work is carried out under ambient conditions, varying the composition or pretreatment of the material under study; however, the majority of Mössbauer studies involve reduced temperatures or applied magnetic fields. For many of the more popular resonances (^{57}Fe, ^{119}Sn, ^{151}Eu, and ^{169}Tm) it is also possible to work at elevated temperatures (metallic iron has been studied well above its Curie temperature).

5.2.6 Cryostats

Helium-flow cryostats have largely replaced older cold-finger and exchange-gas systems, as the former are far more efficient. Current designs use either a transfer line to a large transport dewar, or have an integrated liquid helium reservoir. The latter designs will generally also have a liquid nitrogen-cooled radiation shield and can exhibit consumption rates as low as 100 cm^3/hour for operation from 5–300 K. Liquid helium is drawn through a capillary from the storage vessel and vaporized as it passes through a small sintered copper plug fitted with a resistive heater. Flow rates are metered using a needle valve and temperature control is achieved by balancing cooling derived from the flow of helium vapor with resistive heating applied to the copper plug. Silicon diodes are popular as thermometers in this temperature range; however, for work below 50 K, departures from standard curves can become significant and they should either be individually calibrated, or replaced with a calibrated ruthenium oxide "cernox" thermometer. Temperatures below 5 K can be achieved by evacuating the sample space with a reasonably large roughing pump and using the needle valve as a throttle valve. Work down to 1.3 K is possible with some care. Much lower temperatures demand a dedicated ^{3}He system or a ^{3}He/^{4}He dilution refrigerator.

5.2.6.1 Closed-Cycle Refrigerators

Substantial improvements in closed-cycle refrigerator technology have led to the widespread availability of simple two-stage units that can cool below 4 K without demanding liquid cryogens. While they represent a larger capital investment than a simple helium-flow cryostat, they are extremely reliable and can quickly save on liquid helium costs. Unfortunately, the reciprocating action of the expander head and the action of the large compressor combine to make the cold stage a mechanically noisy environment. Making a Mössbauer measurement while the sample is mounted on a vibrating stage is impossible. Isolating the sample stage from the mechanical noise of the cold head, while still coupling it thermally, adds some complexity to the system and usually raises the base temperature of the sample. However, with proper designs, it is possible to reduce the vibration levels to the point where they do not lead to a measurable broadening of the Mössbauer lines of an ^{57}Fe spectrum. Current best designs place the cold head under the sample where it is used to cool helium gas at atmospheric pressure, which in turn cools the sample. The sample mounting stage is mechanically isolated from the cold head, and only coupled through soft (often rubber) bellows. The top-loading nature of this design is a major advantage as samples are surrounded by cold helium gas and so are uniformly cooled. They can be removed and remounted without shutting down the system, Turn-around times of about an hour are normal. Older, bottom-loading designs place the sample in vacuum, greatly increasing the equilibration times and opening up the possibility of temperature gradients between the sample and the thermometers. The refrigerator must be shut down and warmed to ambient temperature before the sample can be changed. Turn-around times for these systems can be as long as a day. With samples being studied at many temperatures over several days, the sample change delay is a nuisance but not a major problem.

5.2.6.2 Magnetic Fields

The magnetic ordering of materials can be strongly affected by an applied magnetic field, and Mössbauer spectroscopy is widely used to study spin-flop transitions, field-induced reorientations, and magnetostructural transitions. Such work brings its own set of restrictions and problems. The large fields inevitably demand a superconducting solenoid and so a liquid helium cryostat is needed. High-field magnets are physically large so a large source-sample-detector path is unavoidable. High-activity sources and efficient detection become important. Detector choice becomes more limited as the stray fields from a large solenoid are significant and these can affect the photomultiplier used with NaI(Tl) scintillation detectors and also damage HPGe detectors through penning discharges in the vacuum space around the biased detector. Gas-filled proportional counters are popular choices, however it is possible to use scintillation detectors if a long light pipe is placed behind the NaI(Tl) crystal so that the photomultiplier can be located well away from the magnet.

The simplest cryostat geometry has a vertical field with the sample placed in the center of a superconducting solenoid. This arrangement has the field and γ directions parallel to each other (axial geometry). The detector is placed below the cryostat and the source is above the sample, located at a field-null point created by a bucking coil. A zero-field region is essential if one is to avoid magnetic splitting of the source, which will not only broaden the line, but can also lead to significant polarization corrections to the observed line intensities on absorption by the sample. The source is then connected to the Mössbauer drive by a long rigid rod. Source-detector separations of 20–30 cm are common with the γ passing through a series of aluminized mylar or kapton windows to exit through the base of the cryostat. It is essential to ensure that the long drive rod is braced against transverse motions. Allowing the rod to slide through several guide rings works, but this practice is prone to jamming as the various ices build up. A more satisfactory arrangement uses two or three disc-springs of the same kind used in the construction of Mössbauer drives. These are flexible to longitudinal motions but extremely stiff in the transverse direction and, as they do not involve a sliding contact, they do not jam.

If it is necessary to apply the field perpendicular to the γ direction (transverse geometry), then some form of split solenoid is needed (two superconducting coils with a small, ~3-cm gap to provide optical access). Such systems come in two forms. In the first, the solenoid is mounted horizontally and the source-sample path is vertical as in the design discussed above. A bucking coil may be provided to create a zero-field region at the source, but such an arrangement would be unusual. Measurements can also be made in an axial field by providing optical access across the cryostat body, and hence along the bore of the solenoid, through windows on the cryostat walls. In the second, the field is vertical and a bucking coil again provides a zero-field point for axial work, while side windows on the cryostat permit transverse field measurements. While significantly more expensive, split-coil magnet systems are far more versatile as they make both axial and transverse measurements possible. The choice of magnet orientation will be dictated in part by the expected frequency of use for the two field directions, but the vertically oriented split-coil design is more common.

Temperature control is typically through a helium-flow system, with thermometry provided by resistance devices such as carbon-glass or ruthenium oxide. Si diode thermometers are unsuitable as they exhibit severe field shifts.

5.2.6.3 Cold-Source Spectrometers

As the Mössbauer γ energy exceeds about 30 keV, the recoil-free fraction becomes progressively smaller. Mössbauer spectra cannot be recorded at ambient temperatures and both the source and sample must be cooled. A standard arrangement is to place the source and sample inside a conventional helium-flow cryostat, with the source coupled to an externally mounted drive using a long connecting rod in much the same arrangement as that used for systems with superconducting solenoids. Issues with the calibration of these long-rod spectrometers were discussed earlier. One key difference from the magnet systems is that the source generally needs its own thermometer and heater, as it may be necessary to maintain the source at a slightly elevated (with respect to the sample) temperature to avoid ordering or relaxation effects broadening the source linewidth.

Cold-source Mössbauer isotopes often have many interfering γ-photons and X-rays, so high-rate optimized HPGe detectors would be the best choice. There are many useful Mössbauer isotopes with γ energies in the 60–80-keV range, with the 103-keV transition in ^{153}Eu being about the highest practical Mössbauer energy in use.

5.2.6.4 High Temperatures

Many Mössbauer isotopes, notably ^{57}Fe, ^{119}Sn, ^{151}Eu, and ^{169}Tm, exhibit useful recoil-free fractions well above ambient temperatures, and so furnaces can be used to heat samples and measurements can be made as high as 1000°C in some cases. Heating is typically achieved resistively, with the sample and hot stage in vacuum, and thermocouples are used for thermometry. The polymer windows used in cryostats are replaced by thin beryllium sheets (typically 0.1 mm thick). The outer, vacuum-supporting, windows can be made from polymers as long as they are adequately protected from the radiated heat from the hot sample stage. A number of commercial designs are available, but with access to a reasonable workshop, it is possible to build a good Mössbauer furnace.

5.2.6.5 High Pressure

The effects of pressure on structural and magnetic phase transitions are of fundamental interest. Furthermore, many iron-containing minerals undergo geologically relevant structural changes under pressure, making high pressure Mössbauer studies an area of active interest. The detailed methodologies are quite specialized so only basic ideas are outlined here.

Achieving geologically relevant pressures of 100 GPa places a researcher in the realm of diamond-anvil pressure cells. Diamonds make perfect windows for ^{57}Fe Mössbauer spectroscopy but they are rather expensive, and reaching such high pressures with accessible forces means that the actual contact area is limited. As a result, samples are extremely small (less than 1 mm in diameter) and conventional sources are impractical. Special high specific activity "point" sources ~0.5 mm in diameter are used so that they can be placed quite close to the sample. Collimation is provided by the diamond support structure, which is far less transparent to the 14.4-keV γ.

Conventional drives and detectors can be used. The real experimental challenges center on the loading and operation of the diamond-anvil pressure cell.

5.2.7 Emission-Based Techniques

The decay from the Mössbauer level in many Mössbauer isotopes is heavily internally converted, leading to the emission of a low energy K-shell or L-shell conversion electron rather than the Mössbauer γ. While this is often a disadvantage from the standpoint of source preparation, it does open some important windows onto the sample. In a conventional transmission Mössbauer experiment, the signal appears as resonant absorption dips: When the modulated source energy matches an allowed Mössbauer transition in the sample, some γ are absorbed and the count rate at the detector is reduced. However, each of these resonant absorption events creates a nucleus in the appropriate excited state. In isotopes where internal conversion is significant (e.g., ^{57}Fe) most of these excited nuclei will return to the ground state through the emission of one or more conversion electrons, often followed by characteristic X-rays. If these electrons or X-rays are detected, then an emission Mössbauer spectrum can be obtained.

As the most common isotope used for emission studies is ^{57}Fe, most of what follows deals specifically with Fe-Mössbauer work. However, with suitable minor modifications, ^{119}Sn and ^{151}Eu work can be performed with essentially the same methods.

5.2.7.1 Conversion Electron Mössbauer Spectroscopy (CEMS)

In Figure 5.15 we show the possible de-excitation processes after irradiation with the 14.4-keV ^{57}Fe Mössbauer γ-ray.

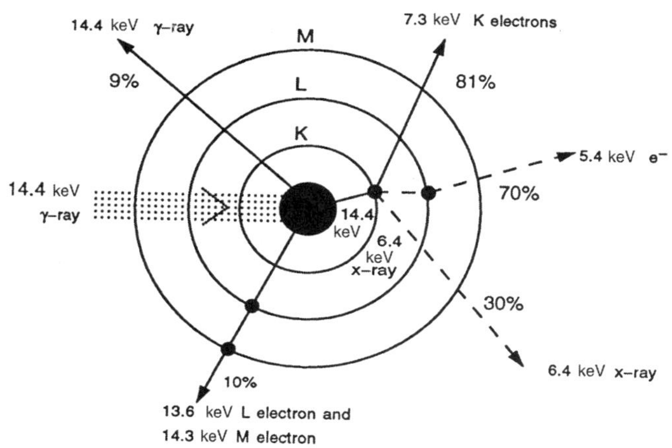

Figure 5.15 De-excitation processes in ^{57}Fe (Ryan and van Lierop, unpublished).

The conversion electrons typically have energies of order 10 keV and have penetration depths of only ~100 nm, so if the electron is detected, it must have originated within the top 100 nm of the sample surface. *Conversion electron Mössbauer spectroscopy* (CEMS) is therefore a surface-sensitive technique. The conversion electrons are emitted following a resonant absorption event in the sample so that the signal appears as a series of emission peaks rather than absorption dips; however the Mössbauer pattern is unchanged and the exact same information can be obtained from the spectrum.

The extremely low penetrating power of the conversion electrons rules out the use of windows between the sample and detector. The sample must be placed within the detector and this detector must be made sensitive to the backscattered electrons but insensitive to the substantial flux of γ- and X-rays incident from the source. Gas-filled proportional detectors are the norm, with He+10% CH_4 at atmospheric pressure being the standard fill gas. Detector construction must employ low atomic number materials throughout in order to minimize fluorescence contributions to the background. Polymers overcoated with thin aluminium layers for conduction are common. The thickness of the active volume is kept to a few mm to minimize γ- and X-ray sensitivity and a thin polymer filter in front of the source can be used to minimize incident X-ray flux. These detectors provide essentially no useful energy resolution and a simple lower-level discriminator after the amplifier is generally used just to cut off electronic noise.

The choice of materials, thin windows and the presence of the sample within the active volume preclude the evacuation and aggressive bake-out normally used to out-gas a proportional counter prior to sealing, therefore a continuous flow of counter gas is used to flush out contaminants before they can accumulate and affect detector operation. Flow rates of a few cc/min are typical.

With no windows between the sample and detector, non-ambient work requires both sample and detector to be at the same temperature and the sample is necessarily exposed to the counter gas at that temperature. Operating temperatures are limited above by failure of the polymers used in the construction, and below by condensation of the quench gas (CH_4). Flow systems down to 90 K are quite convenient, but the increasing gas density at one atmosphere affects the detector efficiency so that sealed units (with limited operating times) are preferred for lower temperatures. Alternatively, the sample can be mounted in vacuum and electron multipliers such as channeltrons or microchannel plates can be used.

Count rates tend to be much lower than those normally encountered in conventional transmission experiments. However, the signal appears as emission over a small background so that good signal-to-noise ratios can be achieved in comparable times with a well-designed detector. CEMS is used to study very thin samples (detection limits for a usable Mössbauer spectrum in a few days can be as low as one monolayer of ^{57}Fe on an area of 1 cm^2) such as magnetic multilayers or spin-valve systems. It also finds applications where only the surface of the sample is of interest, such as in corrosion or surface damage studies, or where the sample is simply too thick for conventional transmission work and cannot be thinned.

5.2.7.2 X-Ray Backscattering (XBS) Spectroscopy

The emission of a K- or L-conversion electron leaves a hole in the corresponding atomic level. This hole can be filled radiatively through the emission of a K_α or L_α X-ray. These X-rays are typically more penetrating than the conversion electrons and have ranges of a few microns. The X-ray backscatter (XBS) Mössbauer spectrum takes the form of a series of emission peaks, but the information contained in the spectrum comes from a thicker surface region than with CEMS.

Applications include deeper surface examination and comparison of immediate and deeper surface layers. As with CEMS, XBS is also employed where the sample cannot be made thin enough for transmission work.

An effective XBS detector can be made by simply changing the fill gas in a CEMS detector from helium to Ar+10% CH_4, also at atmospheric pressure. This mix gives sufficient sensitivity to X-rays without appreciable sensitivity to the 14.4-keV γ-rays from the source. It is critical to attenuate fully all X-rays from the source as these will be detected as a background and greatly reduce the observed signal-to-noise ratio. If the sample is mounted within the detector (as it would be for CEMS work) then about 20% of the detected signal will be due to conversion electrons, which will introduce a slight surface bias to the spectrum. This electron contribution can be eliminated by replacing the back plate of the CEMS detector with a thin aluminized mylar window, which is transparent to the X-rays, but will block electrons. The sample is then placed outside the detector.

With the sample mounted outside the detector it is, in principle, possible to cool or heat the sample without affecting the detector. However, the very limited penetrating power of the X-rays puts severe constraints on the number of windows that can be placed between the sample and the detector.

5.2.7.3 Selective Excitation Double Mössbauer (SEDM) Spectroscopy

Some fraction of the nuclei excited by a resonant absorption event will return to the ground state through the re-emission of a γ at the original Mössbauer energy. If this re-emission is also a Mössbauer event, i.e., it is recoil-free, then the γ can be Mössbauer analyzed in a second Mössbauer experiment. This effectively turns the sample into a Mössbauer source. Selective excitation double Mössbauer (SEDM) spectroscopy is one of the most difficult Mössbauer techniques, but it provides unique and unambiguous information on dynamics.

If the original source is driven in conventional constant-acceleration (or sine) mode, then the sample is exposed to all possible γ energies and the re-emission spectrum is identical to that which would be obtained by transmission, CEMS, or XBS. No new information is available. However, if the source is driven at constant velocity, and this velocity is chosen to coincide with a specific hyperfine level in the sample, then only a single state in the sample is populated, and all re-emissions are produced by decays from this selectively excited state. Even in the simplest case of a well-defined static hyperfine environment, selection rules governing the decay process can lead to multiple energies in the re-emitted γ spectrum. Indeed, the strongest emission line need not even correspond to the initial drive energy.

The key application of SEDM lies in the study of dynamics. In a static system, the re-emitted spectrum of energies is controlled entirely by quantum mechanical selection rules and no new information about the sample is obtained (except that the hyperfine environment is in fact static). However, dynamics such as superparamagnetic relaxation or valence fluctuations can mix the nuclear levels and allow new transitions to occur that would otherwise be forbidden. In many cases, systems that exhibit dynamic behavior also contain some level of static disorder, and both dynamics and disorder can lead to very similar changes in the Mössbauer spectrum. Unambiguous separation is extremely difficult and often subject to bias. SEDM provides for a complete and totally model-independent determination of both contributions.

Disorder leads to broad lines in the transmission Mössbauer spectrum as a wide range of hyperfine environments are present. This broad distribution of environments has no effect on the SEDM spectrum as only a narrow subset of environments is probed at any given drive energy, and the SEDM pattern is very slightly broadened, with line intensities given by selection rules. By contrast, a dynamic hyperfine environment mixes the nuclear levels and "forbidden" transitions appear in the SEDM spectrum. The existence of such lines can *only* be due to dynamics, while the identity and intensity of the new lines can be used to determine the nature and rate of the fluctuations present.

SEDM requires two Mössbauer drives. The first is used to drive the source at a constant velocity chosen to correspond to a selected hyperfine transition (absorption line) in the sample. Since we are working with 14.4-keV γ-rays, we do not have significant window problems, and the sample can be cooled or heated in conventional furnaces or cryostats. Straight-through windows are needed to obtain a transmission spectrum of the sample and to allow for tuning of the source drive to the required velocity. The emission signal is detected at 90° to the source to minimize backgrounds from the source. Detection is best carried out using a resonant CEMS detector containing an ^{57}Fe-enriched stainless steel sample. This resonant single-line Mössbauer detector is driven at constant acceleration and Mössbauer analyzes the re-emitted 14.4-keV γ-photons. The detector is gated so that it is active only when the source is being driven at the required constant velocity, and disabled during the fly-back and acceleration periods. The detector drive is also synchronized with the source drive so that it sweeps through its full velocity range while the source is being held at constant velocity. Even with a 50-mCi source and a large sample data rates are only a few counts per second, so effective shielding is critical to the success of the measurement.

5.2.7.4 Absorber Thickness

One of the most important considerations one must make when undertaking a Mössbauer experiment is thickness of the absorbing sample. The spectrum registered during an experiment comprises both resonant and non-resonant absorption events, and it is vital that an optimal sample thickness be chosen to maximize the signal-to-noise of the desired Mössbauer process [27–29]. An excellent resource is the article by Long et al. [28] in which they demonstrate the calculation of optimal absorber thicknesses.

Basically, the optimal sample thickness lies between $1/\mu_e$ and $2/\mu_e$ in g/cm^2, where μ_e is the electronic mass absorption coefficient appropriate to the radiation and sample being used. These authors provide an excellent didactic example in which they consider two samples: $FeWO_4$ and $Na_2[Fe(CN)_5(NO)] \cdot 2H_2O$. Both samples contain virtually the same amount (by weight) of Fe (18.4 wt% and 18.7 wt%, respectively) however, the presence of the heavy element W in the former leads to significant non-resonant absorption of the 14.4-keV Mössbauer γ-ray and demands a much thinner sample than the latter, which only contains light elements and consequently experiences much less non-resonant absorption. In fact, the optimal $FeWO_4$ sample is an order of magnitude lighter than the $Na_2[Fe(CN)_5(NO)]{\cdot}2H_2O$ sample, despite the fact that the samples contain the same wt% of Fe!

5.2.7.5 Data Analysis

Unlike the situation in crystallography (both X-ray and neutron) where a veritable cornucopia of programs exists in the public domain for the analysis of diffraction data, the field of Mössbauer data analysis seems to have developed somewhat haphazardly, with a general trend of individual groups writing their own programs. Some commercial programs for Mössbauer spectroscopy exist and we mention here the programs NORMOS [30], RECOIL/MOSMOD [31], and MOSSWIN [32]. A number of free programs are available through the program repository associated with the journal Computer Physics Communications. However, the task of writing a basic least-squares fitting program to fit Lorentzian lines (or other lineshapes) to a Mössbauer spectrum is not difficult.

The fitting of Mössbauer spectra involves the setting up of the respective nuclear Hamiltonians for the excited and ground states and diagonalizing, something which can be accomplished using a variety of free routines such as those found in CERNLIB or "Numerical Recipes" [33]. The transition energies, and hence the Mössbauer line positions, can be found from the energy eigenvalues while the relative line intensities can be calculated from the eigenfunctions using the well-known Clebsch-Gordan coefficients. However, many cases are well treated within the framework of the simpler perturbation methods described in Sections 5.1.

The fitting of distributions of hyperfine parameters observed in the spectra of amorphous materials, for example, can be carried out using methods such as the Hesse/Rübartsch and Le Caër/Dubois method [12] or the Window method [13] mentioned earlier. The fitting of spectra showing the effects of electronic relaxation is most usefully carried out using the two-state models of Blume and Tjon [9] or Wickman and Wertheim [10].

In reference [34] we list a number of papers that describe some of the processes involved in fitting Mössbauer spectra. This list is by no means exhaustive but should provide the interested reader with a good overview of the fitting process.

5.3 APPLICATIONS

Mössbauer spectroscopy has been widely applied in physics, chemistry, materials science, geology, and biology. In this section we will illustrate briefly the use of Mössbauer spectroscopy, and the information it provides, with a selection of examples taken from our own work.

5.3.1 Magnetism

The spectrum of α-Fe shown earlier in Figure 5.4 is quite simple, due to the single, high-symmetry, Fe crystallographic site. In Figure 5.16 we show a more complicated example of magnetic splitting. Here, we present RT (room temperature) spectra of the permanent-magnet material $Nd_2Fe_{14}B$ [35]. This intermetallic compound has a tetragonal crystal structure with six distinct Fe crystallographic sites, in the population ratio 4:4:8:8:16:16. Each site gives rise to a magnetically split sextet, including EFG effects due to the lower crystal symmetry. The subspectral areas are proportional to the number of Fe atoms in the respective crystal sites and this provides a useful constraint in the fitting process, thereby reducing the number of free parameters. In the example shown, we compare the ^{57}Fe spectra of a commercial Nd-Fe-B material (upper spectrum) with that of single-phase $Nd_2Fe_{14}B$ (lower spectrum). Commercial permanent-magnet materials comprise the pure $Nd_2Fe_{14}B$ phase and smaller amounts of other phases. The upper spectrum

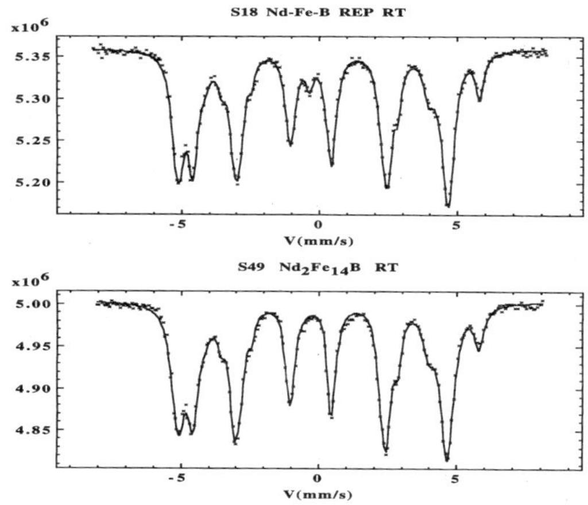

Figure 5.16 RT ^{57}Fe spectra of commercial Nd Fe-B (top) and single-phase $Nd_2Fe_{14}B$ (bottom) [Cadogan, unpublished].

clearly shows additional lines due to these other phases; particularly noticeable is a nonmagnetic doublet, which produces a distinct line around −0.5 mm/s. This is the incommensurate $Nd_{1.1}Fe_4B_4$ minor phase. The relative spectral areas of the various components allow us to carry out reliable phase analysis.

$Nd_2Fe_{14}B$ is the basis for the world's strongest permanent-magnets but it is slightly limited by its modest Curie temperature (~585 K). Much subsequent work focused on either identifying intermetallic compounds with higher Curie temperatures or enhancing known intermetallics, mostly by absorption of small elements such as H, N, and C. In Figure 5.17 we show RT ^{57}Fe spectra of the hexagonal intermetallic compound Y_2Fe_{17} and interstitially modified $Y_2Fe_{17}N_3$. The parent compound Y_2Fe_{17} contains 4 Fe sites in the population ratio 4:6:12:12 and has a Curie temperature of ~325 K, only slightly above room temperature. Thus, its Mössbauer spectrum has almost collapsed into a superposition of nonmagnetic doublets, with only a small broadening due to the remaining weak magnetism. By comparison, the spectrum of $Y_2Fe_{17}N_3$ shows a large, well-established magnetic splitting. As shown by Coey and Sun [36], the absorption of interstitials such as N or C leads to a dramatic increase in Curie temperature and in $Y_2Fe_{17}N_3$ the enhanced Curie temperature is ~700 K. Thus, the room temperature of $Y_2Fe_{17}N_3$ is well split magnetically.

As mentioned in Section 5.2 and illustrated in the preceding two figures, the typical ^{57}Fe magnetic spectrum is fully contained in a velocity range of ±10 mm/s, with typical hyperfine fields up to a maximum of ~52T in hematite. Rare earth nuclei generally experience much larger hyperfine fields and in Figure 5.18 we show ^{169}Tm spectra of the cubic intermetallic compound $TmFe_2$ at 1.3 K and 295 K [11]. This cubic Laves phase compound has a single Tm crystallographic site and the spectra show a large magnetic splitting with a magnetic hyperfine field at the ^{169}Tm nuclei of around 700 T.

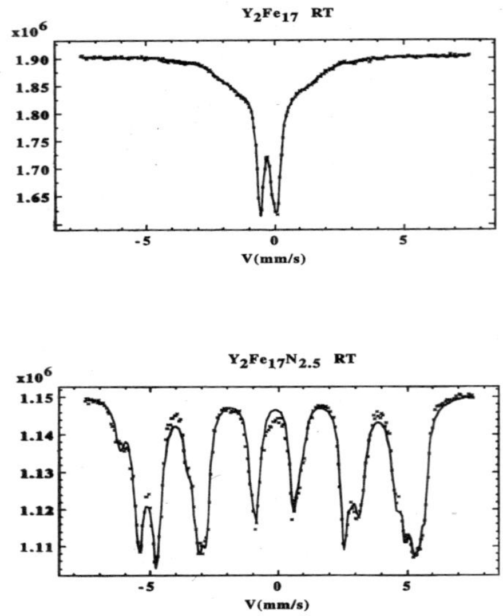

Figure 5.17 RT ^{57}Fe spectra of Y_2Fe_{17} (top) and $Y_2Fe_{17}N_3$ (bottom) [Cadogan, unpublished].

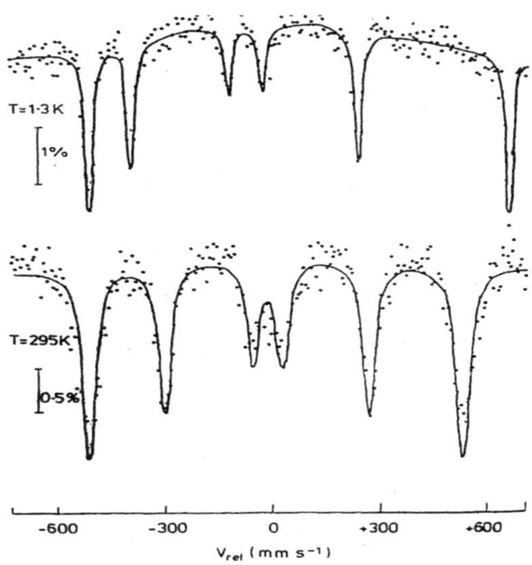

Figure 5.18 ^{169}Tm spectra of $TmFe_2$ obtained at 1.3 K (top) and 295 K (bottom) [Cadogan, unpublished]; see also [11].

The asymmetry of the spectra is a clear indication of a large EFG, in addition to the magnetic splitting, and the temperature dependences of both hyperfine parameters were used to derive magnetic exchange and crystal field parameters. $TmFe_2$ is an ideal compound for calibration of ^{169}Tm spectrometers. It is interesting to note the velocity scale here; on this scale of ±700 mm/s the ^{57}Fe calibration spectrum shown above in Figure 5.4 would occupy only 2 to 3 channels.

5.3.2 Magnetic Reorientations

As explained in Section 5.1, the Mössbauer spectrum in the presence of both magnetic and quadrupole effect depends critically on the angle between the quantization axes of the respective Hamiltonians, thereby allowing one to investigate reorientations of the magnetic structures. The ^{119}Sn spectra of the hexagonal intermetallic compound $TbMn_6Sn_{5.6}Ga_{0.4}$, obtained over the temperature range 12 K to 300 K and shown in Figure 5.19 [37], illustrate clearly the effect of magnetic reorientation. At low temperatures the easy direction of magnetization is along the hexagonal *c*-axis. This compound undergoes a spin-reorientation to the hexagonal basal plane at 244 K. An axis ($\theta = 0°$) to plane ($\theta = 90°$) reorientation leads to a change in the measured quadrupole splitting by a factor of $-1/2$, easily observed by Mössbauer spectroscopy. The temperature dependences of the hyperfine fields and quadrupole splittings of the three Sn sites in this structure are also shown in Figure 5.19. At each site we see the expected change in quadrupole splitting. The hyperfine field "seen" by the ^{119}Sn nuclei is a transferred field from its magnetic neighbors (Sn is nonmagnetic) and clear changes occur at the spin-reorientation transition, which are related to orbital magnetism effects.

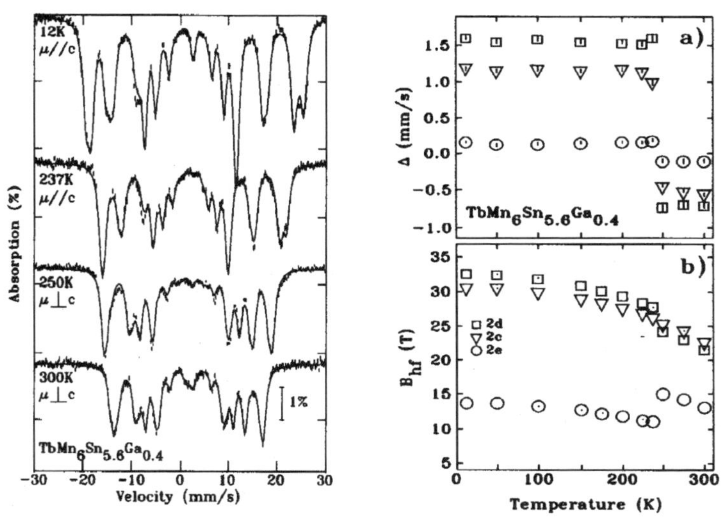

Figure 5.19 Left — ^{119}Sn spectra of hexagonal $TbMn_6Sn_{5.6}Ga_{0.4}$. Right — Temperature dependences of the hyperfine field and the quadrupole splitting at each of the Sn sites [37].

Continuing this theme of transferred hyperfine fields, in Figure 5.20 we show the ^{119}Sn RT spectrum of the intermetallic compound YFe_6Sn_6. There are eight Sn crystallographic sites in this orthorhombic structure and the spectrum shows that 1/3 of the Sn nuclei experience quite a large hyperfine field of 24.6 T, transferred from the surrounding magnetic Fe sites,

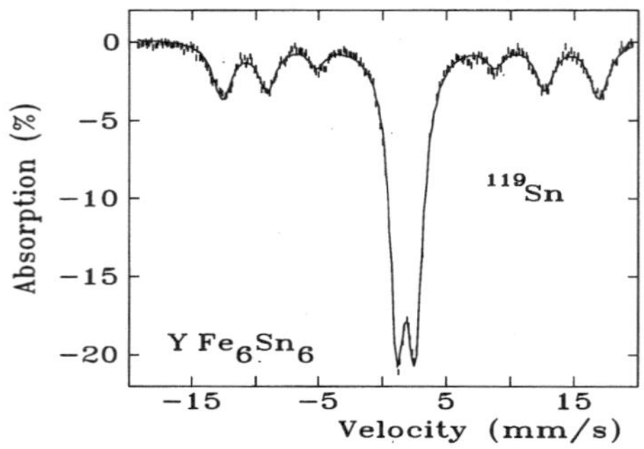

Figure 5.20 ^{119}Sn spectrum (RT) of YFe_6Sn_6 [38].

while the remaining 2/3 experience no field. This finding provided confirmation of the suggested antiferromagnetic structure of the Fe sublattice in YFe_6Sn_6, deduced by neutron diffraction [38].

The temperature dependences of the hyperfine parameters can also be used to great effect to study the nature of magnetic transitions, including the ordering transition. In Figure 5.21 we show a series of ^{119}Sn spectra of the intermetallic compound Gd_5Sn_4 obtained at various temperatures [39]. We also show the temperature dependences of the ^{119}Sn hyperfine magnetic field and the relative fraction of the magnetic component subspectra. This compound, and related compounds, are currently of interest due to their "giant magnetocaloric" behavior, which opens up the possibility of future applications as refrigerant materials [40]. The ^{119}Sn spectra clearly demonstrate the first-order magnetic ordering at 82 K, in addition to magnetic inhomogeneity over a wide range of temperature.

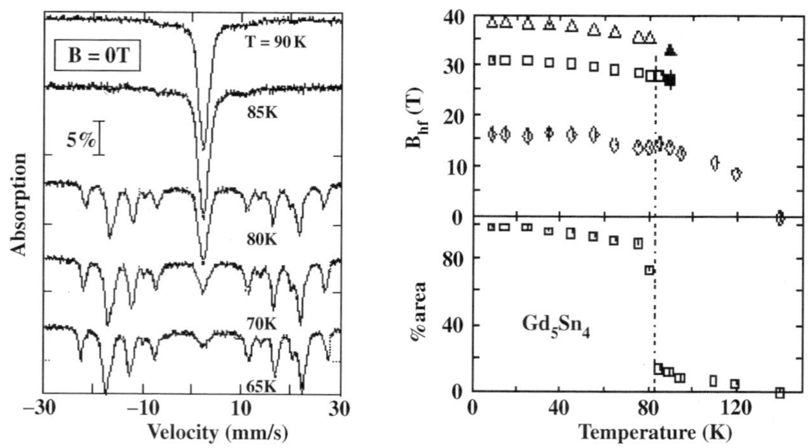

Figure 5.21 ^{119}Sn spectra of Gd_5Sn_4 together with the temperature dependences of the hyperfine field and the quadrupole splittings at the three Sn sites [39].

5.3.3 Crystal Fields

We have already mentioned the use of Mössbauer spectroscopy to study crystal-field effects. The form of the crystal-field Hamiltonian of a rare earth ion in a compound is determined by the point-symmetry at the rare earth crystallographic site. Thus, a study of the temperature-dependences of hyperfine parameters, not only allows one to deduce crystal-field parameters but can also distinguish clearly between different crystallographic models. To illustrate this, in Figure 5.22 we show ^{169}Tm spectra of $Tm_2Ge_2O_7$ [41]. This structure had been suggested to exhibit five-fold crystal symmetry, forbidden in extended crystal structures [42]. The ^{169}Tm spectra allowed us to test this claim by measuring the electric quadrupole splitting at the ^{169}Tm nucleus and analyzing these data with a model for the crystal field at the parent Tm^{3+} site. Our crystal field fit to the quadrupole splitting clearly shows that the proposed 5-fold symmetry is incorrect and the site symmetry is in fact triclinic.

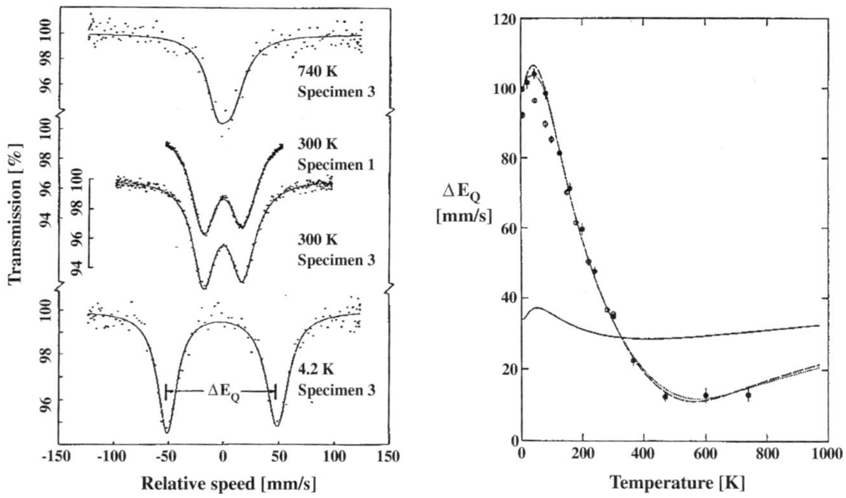

Figure 5.22 Left —^{169}Tm spectra of $Tm_2Ge_2O_7$. Right— temperature dependence of the ^{169}Tm quadrupole splitting, with a triclinic crystal-field fit. Also shown is the temperature-dependence predicted by the 5-fold symmetry model [41].

5.3.4 Phase Analysis

Mössbauer spectroscopy is well suited to the problem of phase analysis, in particular in Fe-bearing materials. Not only can individual phases present in a sample be identified, but a quantitative analysis can often be achieved, keeping in mind the caveat related to the recoil-free fractions of the individual phases, mentioned in Section 5.1. As an example of the use of ^{57}Fe Mössbauer spectroscopy for phase analysis, in Figure 5.23 we show a series of RT ^{57}Fe spectra of Nd-Fe-B, with the commercial composition $Nd_{15}Fe_{77}B_8$, after various stages in the HDDR process (*hydrogen disproportionation desorption*

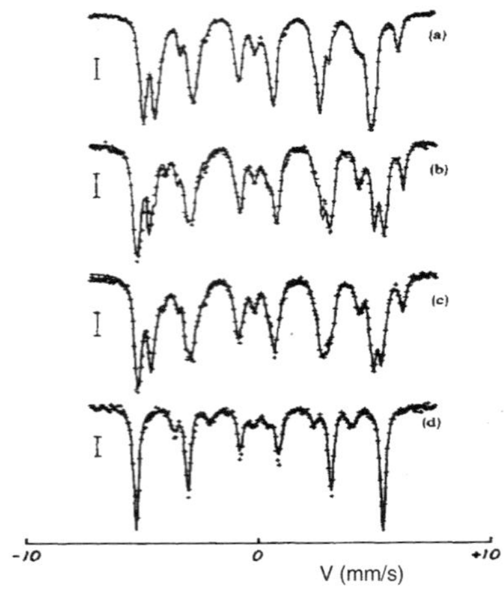

Figure 5.23 RT ^{57}Fe Mössbauer spectra of $Nd_{15}Fe_{77}B_8$ in the virgin, unhydrogenated state (spectrum (a)) and following hydrogenation up to (b) 320°C, (c) 650°C and (d) 1000°C [43].

recombination) used in the preparation of permanent-magnet materials. Heating the starting Nd-Fe-B material to about 200°C in hydrogen gas leads to absorption, with a resulting slight increase in magnetic ordering temperature. Further heating to above 700°C leads to the complete disproportionation of the $Nd_2Fe_{14}B$ phase into α-Fe, Fe_2B, and NdH_2 (identified by X-ray diffraction) [43]. The presence of α-Fe completely dominates the disproportionated spectrum.

As an example of quantitative phase analysis, we show in Figure 5.24 the RT spectrum of the Allegan meteorite which was found in Allegan County, Michigan, US [44]. Four distinct Fe-bearing phases are observed (Olivine 38.9 ± 1.0 area-%, Pyroxene 34.1 ± 0.8 area-%, Troilite 20.3 ± 1.3 area-%, and Fe-Ni metal 7.0 ± 2.4 area-%).

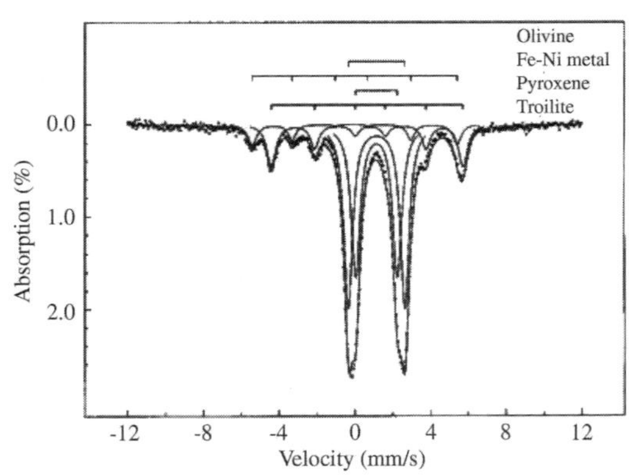

Figure 5.24 RT ^{57}Fe spectrum of the Allegan meteorite with the individual spectral components identified [44].

5.3.5 Amorphous Materials

As mentioned in Section 5.1, the Mössbauer spectra of amorphous materials are broad due to the distribution of hyperfine parameters such as hyperfine magnetic field or quadrupole splitting. A typical hyperfine field distribution fit involves 15–30 subspectral components and the shape of the fitted distribution depends on numerous factors, not only related to the actual sample. However, the average field of such a distribution is fairly robust.

In Figure 5.25 we show a series of room temperature Mössbauer spectra of amorphous $Fe_{86-x}Nb_xB_{14}$ ribbons, prepared by the melt-spinning technique [45]. These materials are precursors for the preparation, by controlled crystallization, of soft-magnetic materials similar to the FINEMET alloys. These spectra are broad (refer to the crystalline α-Fe spectrum in Figure 5.4) and their respective hyperfine field distributions clearly show the effect of the increasing Nb content in reducing the Curie temperature and hence the average hyperfine field.

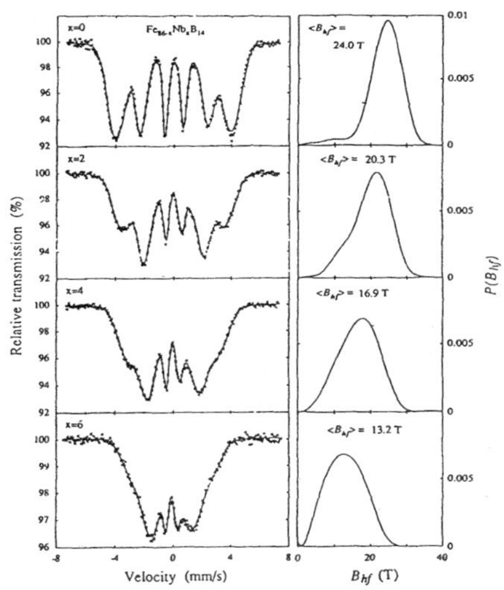

Figure 5.25 RT ^{57}Fe spectra of amorphous $Fe_{86-x}Nb_xB_{14}$ ribbons [45].

The preceding set of room temperature spectra showed the effect of reducing the Curie temperature in a series of Fe-Nb-B amorphous alloys. We have also followed the collapse of the magnetic spectrum of amorphous $Fe_{92}Zr_8$ by obtaining a series of spectra at various temperatures above room temperature [46] (see Figure 5.26). The hyperfine field distributions shown in Figure 5.26 provide a dramatic representation of this collapse. Often, such temperature dependences are analyzed in terms of a modified Brillouin function, taking into account the effects of the distribution of environments in the amorphous state, which manifest themselves as a distribution of the magnetic exchange [47].

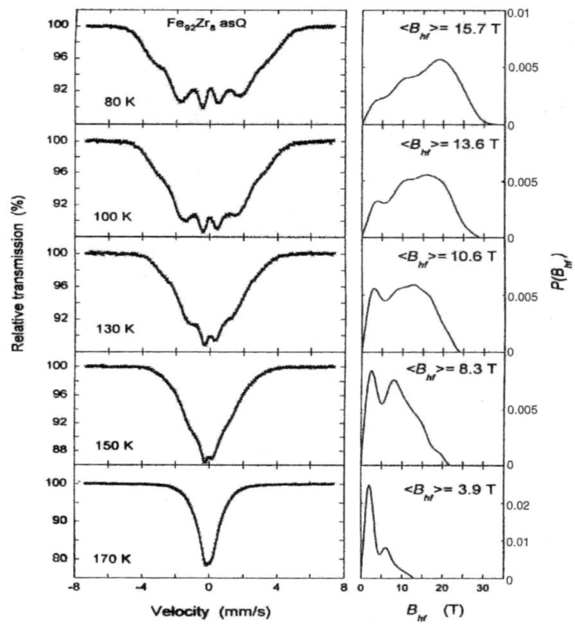

Figure 5.26 ^{57}Fe spectra of amorphous $Fe_{92}Zr_8$ ribbons over the temperature range 80 K to 170 K [46].

It is clear the ^{57}Fe Mössbauer spectroscopy is ideally suited to the comparative study of crystalline and amorphous materials and mixtures of the two. In Figure 5.27 we show a series of room temperature Mössbauer spectra of the permanent-magnet material Nd-Fe-B, as previously illustrated in Figure 5.23 [48]. Here, we prepared a series of samples by melt-spinning at various wheel speeds, and hence quench rates, and the transition from crystalline through microcrystalline to amorphous with increasing quench rate is clearly reflected in the broadening of the spectrum.

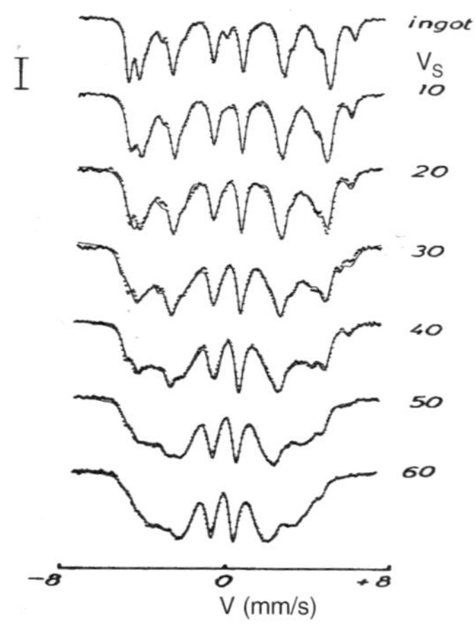

Figure 5.27 ^{57}Fe spectra of Nd-Fe-B prepared at various melt-spinning wheel speeds V_S (m/s) [48].

To illustrate the quantitative fitting of spectra comprising both broad amorphous and sharper nano-crystalline components, we show in Figure 5.28 the room temperature Mössbauer spectra of (i) $Fe_{79}V_7B_{14}$ and (ii) $Fe_{80}Nb_6B_{14}$ ribbons, prepared fully amorphous by the melt-spinning technique and subsequently partially crystallized to produce bcc-Fe(V,Nb,B) nanocrystallites, embedded in an amorphous matrix [49]. These spectra show the combined features of the broad amorphous phase and the sharper nano-crystalline precipitates. Fitting these combinations allow one to deduce the volume fractions of the amorphous and nano-crystalline phases in these materials. Such data are important in the development of a comprehensive model for the random anisotropy in such commercially important soft-magnetic materials [16, 50].

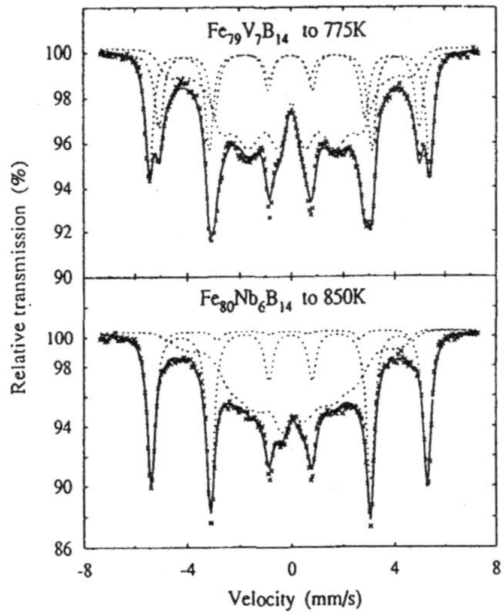

Figure 5.28 ^{57}Fe spectra of partially crystallized (top) $Fe_{79}V_7B_{14}$ and (bottom) $Fe_{80}Nb_6B_{14}$ ribbons [49].

Mössbauer spectroscopy also provides a simple method for estimating the magnetic ordering temperature of a material. The "thermal scan" technique consists of switching off the velocity drive unit and counting for fixed periods of time in the zero-velocity channel at various temperatures. As the temperature increases, the magnetic spectrum collapses towards a quadrupole doublet or a singlet, depending on the crystallographic site symmetry. Thus, the γ absorption increases at zero-velocity and the number of counts registered in the given time period falls. This is illustrated in Figure 5.29 where we show spectra of amorphous $Fe_{90}Sc_{10}$ obtained at 4.2 K and 300 K [51]. The temperature dependence of the average hyperfine field yields a Curie temperature of 98 K. The spectrum at 300 K is a paramagnetic, broad doublet due to the distribution of the electric quadrupole splitting. This figure also shows the Curie temperature determination using the thermal scan technique,

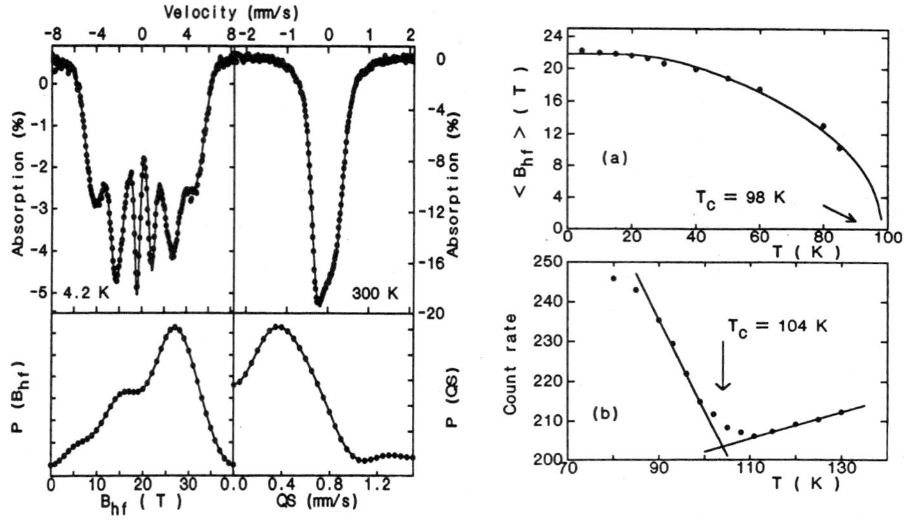

Figure 5.29 Left—^{57}Fe spectra of a-$Fe_{90}Sc_{10}$ at 4 K and 300 K, together with their respective hyperfine field or quadrupole splitting distributions. Top right—Temperature dependence of the ^{57}Fe average hyperfine field. Bottom right—Thermal scan determination of the magnetic ordering temperature [51].

yielding a value of 104 K, slightly higher than the previous value but nonetheless useful as a quick method of obtaining a reasonable estimate of the magnetic ordering temperature.

A further example of quantitative phase analysis in a mixed amorphous/crystalline sample is shown in Figure 5.30, where we present a series of RT (and 4.2 K) ^{57}Fe Mössbauer spectra obtained on amorphous and crystallized $Sc_2Fe_{14}B$ [52]. This experiment was an attempt to prepare crystalline $Sc_2Fe_{14}B$ by crystallizing an amorphous precursor, since $Sc_2Fe_{14}B$ does not form by traditional processing routes. Differential scanning calorimetry had shown that the crystallization of amorphous $Sc_2Fe_{14}B$ is a two-stage process with crystallization temperatures of 600°C and 685°C. Spectrum (a) was obtained with the amorphous, as-quenched sample at 4.2 K and shows a well-split broad magnetic sextet, indicative of magnetic order with a distribution of hyperfine field. Spectrum (b) is the RT spectrum of the as-quenched amorphous sample and is a broad, very weakly magnetically split spectrum, obtained just below the Curie temperature of 328 K. Spectrum (c) was obtained after annealing the sample at the first crystallization stage and clearly shows the growth of crystalline α-Fe, superimposed on a residual amorphous component. The final spectrum (d) shows the sample after annealing at 700°C. This fully crystallized material comprises α-Fe, $ScFe_2$, and ScB_6 (the latter phase being identified by X-ray diffraction).

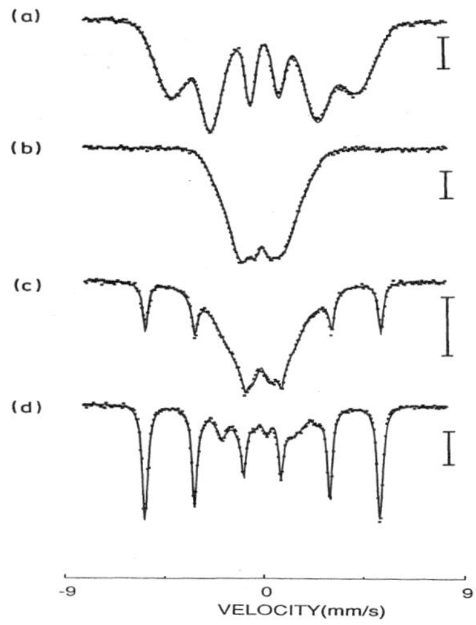

Figure 5.30 ^{57}Fe spectra of $Sc_2Fe_{14}B$: (a) as-quenched amorphous ribbons at 4 K, (b) as-quenched amorphous ribbons at 295 K, (c) partially crystallized, and (d) fully crystallized [52].

5.3.6 Electronic Relaxation

In Figure 5.31 we show the ^{166}Er spectra obtained in the temperature range 2–20 K of the intermetallic phase $ErFe_6Sn_6$ [53]. The magnetic ordering temperature of the Er sublattice, deduced by neutron diffraction, is 4.5 K. The ^{166}Er spectra show the collapse of the spectrum with increasing temperature, but more noticeable is the fact that virtually the full magnetic splitting is seen well above the magnetic ordering temperature. Er^{3+} is prone to slow electronic relaxation effects, yielding the hyperfine splitting observed in the paramagnetic regime. Similar effects have been observed by ^{169}Tm spectroscopy in Tm_3Al_2 [54] and Tm_2Al [55]. $ErFe_6Sn_6$ provided a rare opportunity to study all crystallographic sites in a compound by Mössbauer spectroscopy, and in Figure 5.31 we also show the ^{57}Fe, ^{119}Sn, and ^{166}Er spectra of this intermetallic.

5.3.7 Electronic Valence

As mentioned in Section 5.1, hyperfine parameters such as quadrupole splitting and isomer shift can be quite sensitive to the electronic valence of the parent ion.

To illustrate this sensitivity to electronic valence, we show in Figure 5.32 a series of ^{57}Fe Mössbauer spectra obtained at room temperature on a number of compounds in the pseudobrookite series $Fe_{1+x}Ti_{2-x}O_5$ [56]. In the Fe_2TiO_5 end-member, Fe is ferric, i.e., Fe^{3+}, and a broad, nonmagnetic doublet is

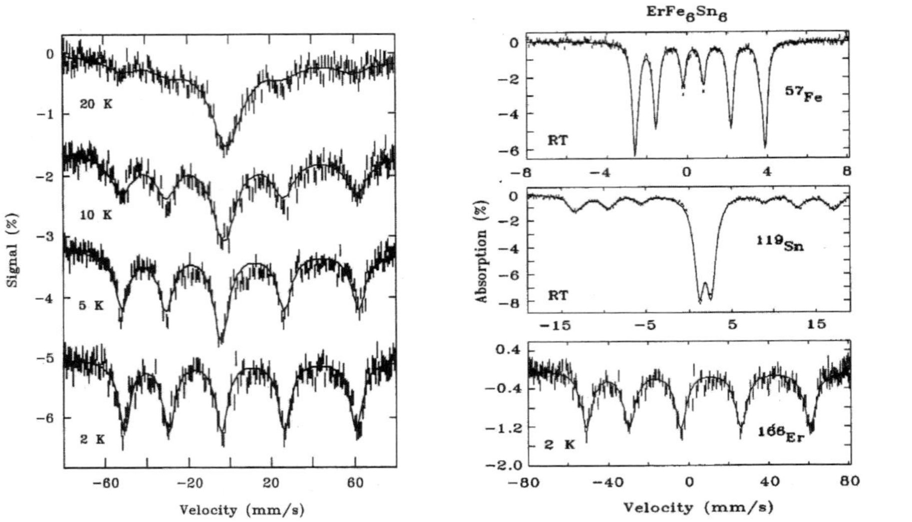

Figure 5.31 Left— ^{166}Er spectra of $ErFe_6Sn_6$. Right— ^{57}Fe, ^{119}Sn, and ^{166}Er spectra of $ErFe_6Sn_6$ (note the different velocity scales) [53].

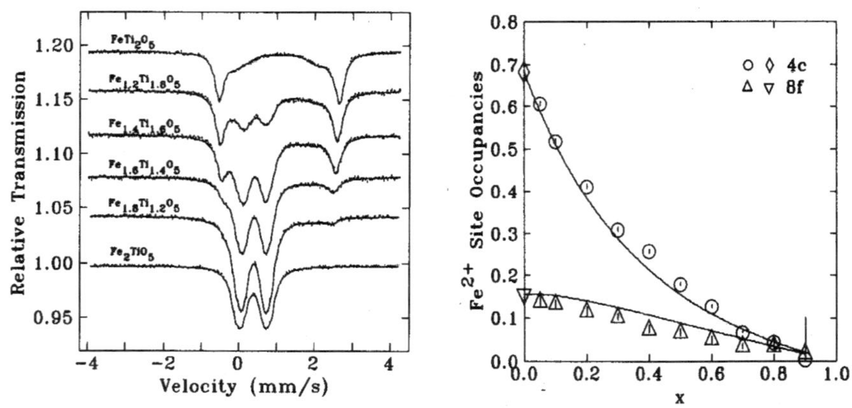

Figure 5.32 Left— RT ^{57}Fe spectra of the pseudobrookite series $Fe_{1-x}Ti_{2-x}O_5$. Right—Relative Fe^{2+} site occupancies as a function of x [56].

observed with a quadrupole splitting of ~ 0.55 mm/s. The other end-member $FeTi_2O_5$ comprises Fe^{2+}, and produces a sharper, nonmagnetic doublet with a much larger quadrupole splitting of ~ 2.16 mm/s. Pseudobrookite contains two crystallographic sites available to the Fe ions and we also show the site occupancies of Fe^{2+} deduced from the Mössbauer spectra. These site occupancies are in excellent agreement with those derived from neutron diffraction measurements.

A powerful demonstration of the use of Mössbauer spectroscopy to study electronic valence is provided by ytterbium. The typical 4f electronic configuration for rare earth ions in compounds is 3+. Yb occurs at the end of the 4f series and Yb^{3+} has the $4f^{13}$ configuration, which is magnetic. However, by gaining an extra electron to form Yb^{2+}, a full 4f electronic shell is obtained and Yb^{2+} is nonmagnetic. In Figure 5.33 we show the ^{170}Yb spectra of intermetallic $YbMn_2Ge_2$ and $YbMn_2Si_2$ [57]. The Yb ion in $YbMn_2Ge_2$ is divalent and therefore nonmagnetic. The spectrum of $YbMn_2Ge_2$ at 4.5 K is a typical nonmagnetic triplet. $YbMn_2Si_2$ has trivalent Yb and is consequently magnetic. The spectrum of $YbMn_2Si_2$ at 40 K is also a quadrupole-split triplet, being obtained well above the magnetic ordering temperature of the Yb^{3+} sublattice. However, the quadrupole splitting in the silicide is much larger than in the germanide and is typical of a Yb^{3+} state. The series of $YbMn_2(GeSi)_2$ spectra at 4.5 K allows one to determine accurately the relative amounts of Yb^{2+} and Yb^{3+} across this series.

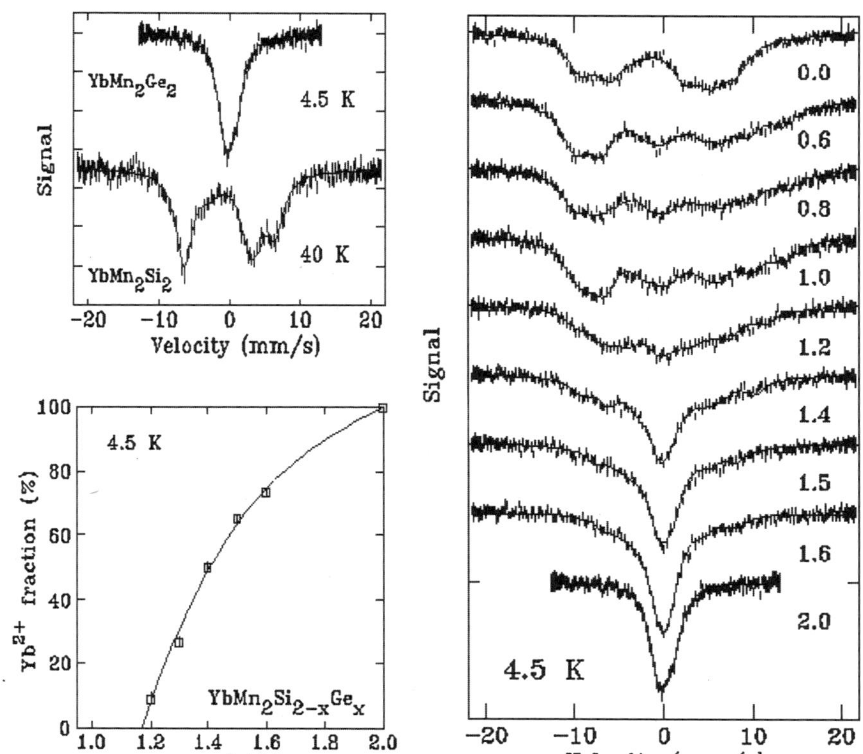

Figure 5.33 Top Left— ^{170}Yb spectra of $YbMn_2Ge_2$ (4 K) and $YbMn_2Si_2$ (45 K). Right—4.5 K spectra of the $YbMn_2Ge_xSi_{1-x}$ series. Bottom Left—Fraction of Yb^{2+} in the $YbMn_2Si_xGe_{1-x}$ series [57].

5.3.8 Industrial Applications

^{57}Fe Mössbauer spectroscopy is used in numerous industrial applications. We have used ^{57}Fe to study the direct reduction process in steel-making [58, 59]. In Figure 5.34 we show the RT ^{57}Fe Mössbauer spectrum of an Australian iron ore raw material, used as the starting material in our study. About 95% of the spectral area is a magnetically split sextet with a hyperfine field of 51.5 T. This corresponds to hematite (α-Fe_2O_3). The remaining 5% of the spectrum is a paramagnetic, quadrupole-split doublet due to wüstite (Fe_xO, with x slightly less than 1). We also show the spectrum of this iron ore material after it has been reduced in H_2 at 600°C for 13 minutes. The spectrum of the reduced material contains about 90 wt% α-Fe, with traces of magnetite (Fe_3O_4).

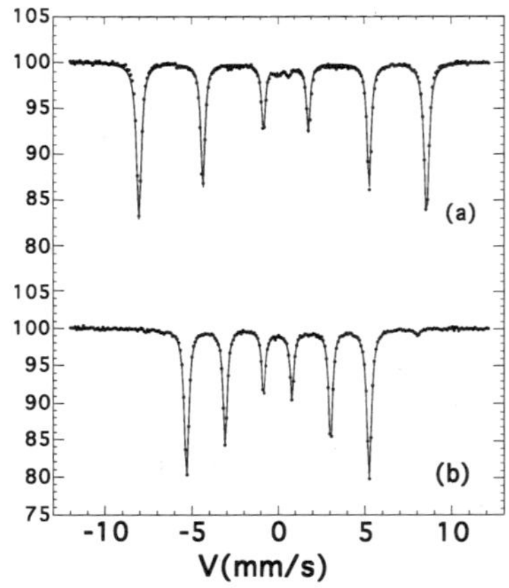

Figure 5.34 RT ^{57}Fe spectra of raw iron ore (a) before and (b) after reduction in hydrogen [58].

^{57}Fe Mössbauer spectroscopy also allowed us to follow the reduction process quantitatively. In Figure 5.35 we show the RT spectra of the iron ore after various reduction times [58, 59]. After two minutes, the hematite is gone and the sample comprises α-Fe, magnetite and wüstite. In Figure 5.35 we also show the relative amounts of these three phases as a function of reducing time.

5.3.8.1 CEMS (Conversion Electron Mössbauer Spectroscopy)

As mentioned in Section 5.2, conversion electron Mössbauer spectroscopy is ideally suited to the study of surface phenomena. It is well known that many amorphous materials, which are malleable in the as-quenched state, become quite brittle upon heating to a critical temperature below the crystallization temperature. Thus, the embrittled materials remain amorphous. It is believed that the embrittlement phenomenon is related to changes in the surface of the amorphous ribbons, possibly involving the propagation of surface micro-cracks into the bulk material. In an attempt to investigate the surface

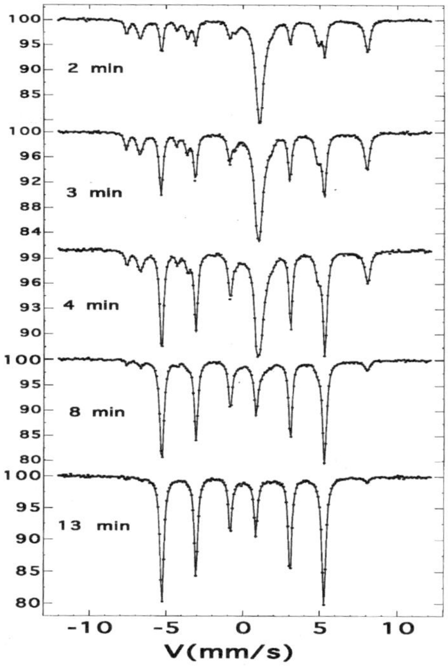

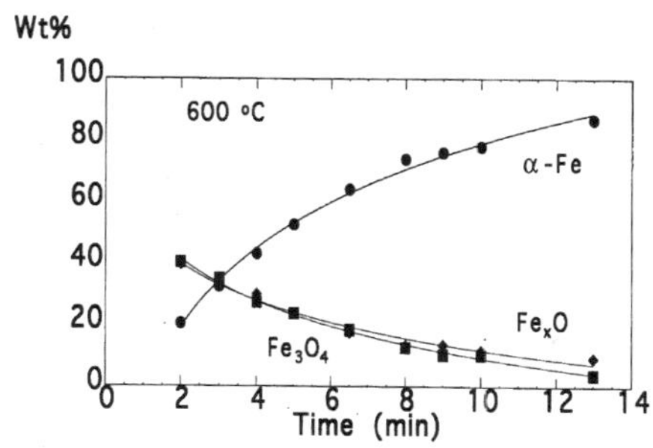

Figure 5.35 Left—Reduction time-dependence of the RT spectra. Right—Relative amounts of α-Fe, Fe_3O_4, and Fe_xO in reduced iron ore as a function of reducing time [58, 59].

magnetism of an embrittled amorphous material we carried out both transmission and CEMS Mössbauer spectroscopy on amorphous Fe-Si-B. In Figure 5.36a and 5.36b we show ^{57}Fe transmission and CEMS Mössbauer spectra of as-quenched amorphous Fe-Si-B [60]. In both cases, the spectra are broad, magnetically-split sextets. The most significant difference between the two spectra is the relative intensity of lines 2 and 5.

As discussed in Section 5.1, the intensity of these two lines, relative to the other four, provides a measure of the magnetic texture of the sample. The transmission spectrum samples the bulk material and the relative line intensities indicate a random arrangement of magnetic moments. In contrast, the enhanced intensities of lines 2 and 5 in the CEMS spectrum, which samples the surface of the material, indicates significant magnetic texturing of the surface with the Fe magnetic moments lying preferentially along the plane of the amorphous ribbon samples (see Section 5.1). In Figure 5.36c and 5.36d we show the spectra of an annealed sample of Fe-Si-B (360°C for 5 min). The annealed sample is still amorphous, as indicated by the broad magnetic spectra, but is now brittle. Changes in the CEMS spectra indicate structural relaxation on the surface of the annealed sample, which probably causes the embrittlement.

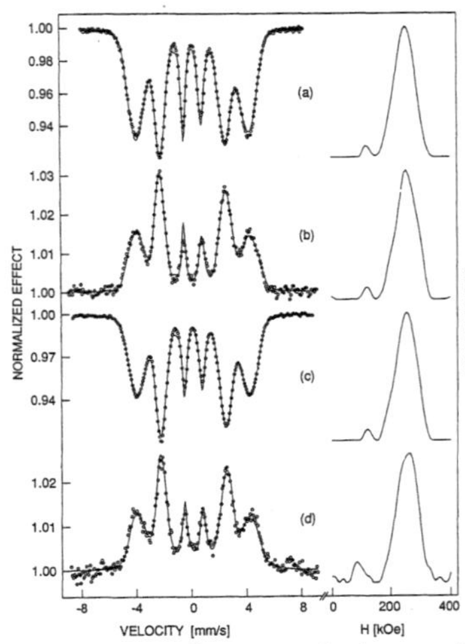

Figure 5.36 RT ^{57}Fe Mössbauer spectra of as-quenched amorphous Fe-Si-B (spectra a and b) and embrittled Fe-Si-B (spectra c and d), obtained in transmission and CEMS modes [60].

5.3.8.2 CEMS—Thin Films

Figure 5.37 shows room temperature CEMS spectra for a 316 stainless steel sheet covered by increasing thicknesses of α-Fe. As the iron overlayer gets thicker, fewer conversion electrons from the stainless steel substrate are able to escape, and the spectrum is dominated by the α-Fe sextet. A plot of the relative area of the 316 stainless steel component as a function of iron overlayer thickness shows that the CEMS signal derives from two distinct conversion electron populations. The softer K-electrons (7.3 keV) have a penetration depth of 52(8) nm and are lost quite rapidly, while the L-electrons

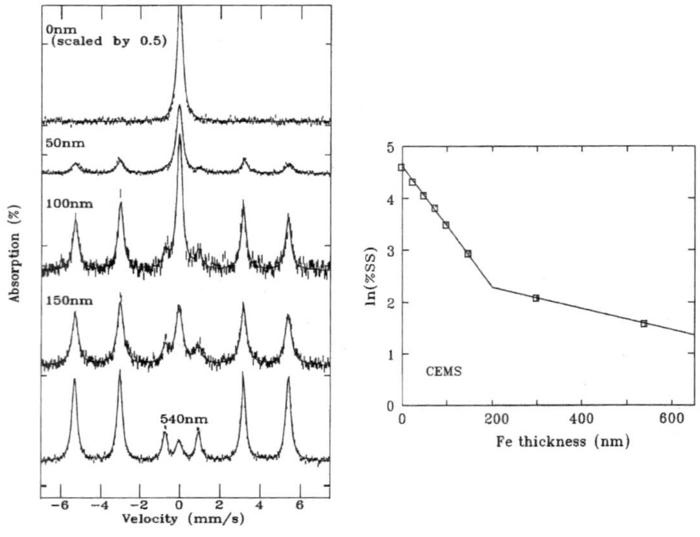

Figure 5.37 Left—RT ^{57}Fe Mössbauer CEMS spectra of a 316 stainless steel sheet covered by increasing thicknesses of α-Fe. Right—Relative area of the stainless steel spectral component as a function of α-Fe layer thickness [Perry and Ryan, unpublished].

(13.6 keV) are more penetrating (~ 300 nm) and yield a signal from quite deep below the surface.

A comparison of XBS and CEMS data for a 540-nm thick iron overlayer (Figure 5.38) shows that the 6.4-keV Fe-K_α X-rays are far more penetrating (penetration depth = 3.5(2) μm) and provide information from quite deep into the sample. The combination of CEMS and XBS measurements allows surface and bulk contributions to be distinguished and is particularly valuable in corrosion and other surface-damage studies where the samples are too thick for conventional transmission work.

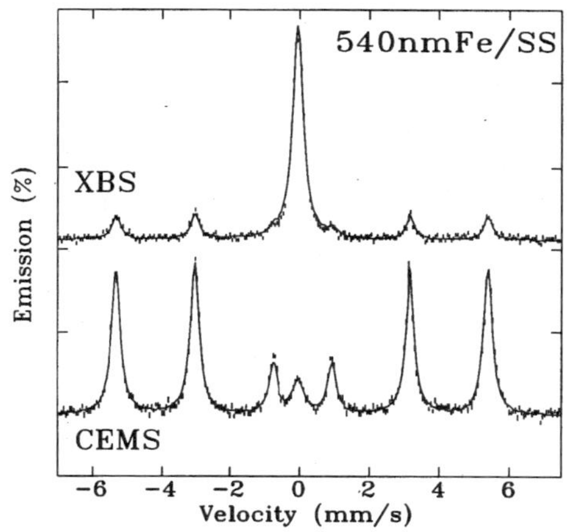

Figure 5.38 ^{57}Fe Mössbauer XBS (top) and CEMS (bottom) spectra of a 316 stainless steel sheet covered by a 540-nm layer of α-Fe [Perry and Ryan, unpublished].

5.3.8.3 SEDM (Selective Excitation Double Mössbauer) Spectra

Figure 5.39 shows a conventional Fe transmission spectrum (top) followed by SEDM spectra obtained by setting the constant velocity drive (source) on each of lines 1, 2, and 3 in succession. This illustrates that SEDM actually leads to the population of a specific hyperfine state in the target nucleus and is not simply a scattering or fluorescence technique. Driving line 1 yields emission only at line 1 as the $m_I = +\frac{3}{2}$ state has one allowed transition. Driving line 2 populates the $m_I = +\frac{1}{2}$ excited state and there are two allowed decays (and hence two lines in the SEDM spectrum) to the $m_I = \pm\frac{1}{2}$ ground states, yielding lines 2 and 4. Driving line 3 leads to a more striking demonstration that the nuclear selection rules play a major role in determining the SEDM pattern as the emission at the line 5 position is actually stronger than that at line 3 where the source energy is centered. Note also that no "forbidden" transitions are observed for any of the SEDM patterns.

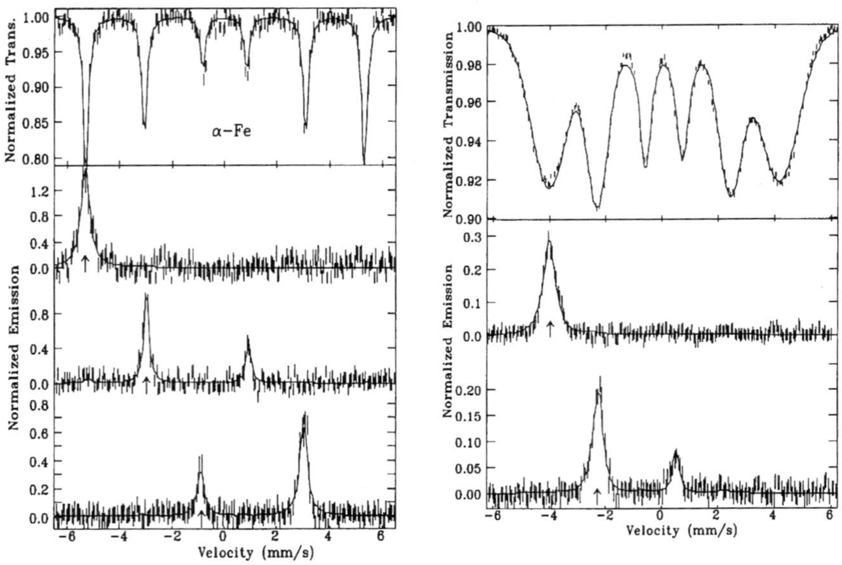

Figure 5.39 Left—RT ^{57}Fe Mössbauer transmission and SEDM spectra of α-Fe. Right—RT ^{57}Fe Mössbauer transmission and SEDM spectra of $Fe_{80}B_{20}$ metallic glass [61].

Figure 5.39 shows the same experiment carried out on a chemically and structurally disordered sample ($Fe_{80}B_{20}$ metallic glass). The SEDM emission lines follow the same pattern seen for α-Fe and the severe disorder associated with the glass simply leads to some line broadening. It is interesting to note that the SEDM lines are slightly sharper than those observed by transmission as only a limited subset of the iron environments present in the glass are excited by the source at any given constant velocity.

Figure 5.40 shows the SEDM patterns expected if the nuclei in the sample are subject to a rapidly fluctuating magnetic field, due for example to superparamagnetic moment fluctuations [61]. As the fluctuation rate increases, a new line appears at position 6 even though the sample is being driven only at position 1. This arises because if the field reverses rapidly enough, the excited nucleus is unable to follow the change and is now in a spin-up orientation, but in a spin-down field. This effectively converts the created $m_I = +\frac{3}{2}$ state into a $m_I = -\frac{3}{2}$ state, and the subsequent decay appears at the line 6 position. More rapid fluctuations increase the intensity of the line 6 emission at the expense of the line 1 signal and also lead to a broadening of both lines.

Figure 5.41 shows precisely this behavior for a sample of fine particle magnetite in oil (a ferrofluid) [61]. As the temperature is increased, superparamagnetic fluctuations appear and the field reversals lead to an emission signal at the line 6 position. Figure 5.41 also shows the fitted relaxation rates derived from these spectra.

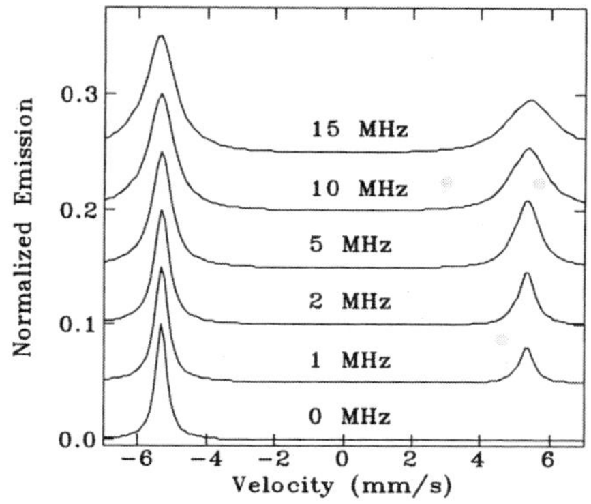

Figure 5.40 Calculated ^{57}Fe Mössbauer SEDM spectra in the case of a rapidly fluctuating magnetic field [61].

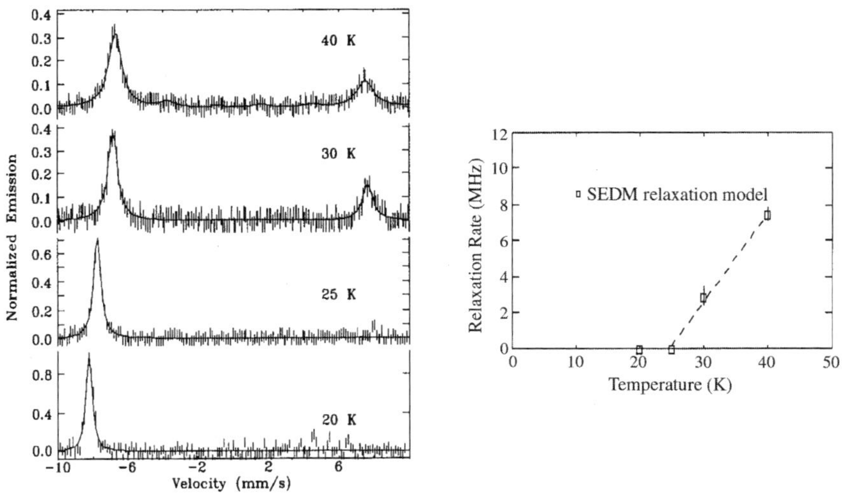

Figure 5.41 Left—^{57}Fe Mössbauer SEDM spectra of a fine-particle suspension of magnetite in oil. Right—Relaxation rates derived from the SEDM spectra [61].

Figure 5.42 shows a particularly clear demonstration of the power of SEDM. The metallic glass, a-$Fe_{92}Zr_8$, undergoes two magnetic ordering transitions as a result of exchange frustration [62]. The upper one establishes long-range ferromagnetic order, while the lower transition (at T_{xy}) marks the ordering of transverse spin components. Conventional Mössbauer measurements cannot easily detect any changes at T_{xy} due to the pre-existing magnetic order combined with the inevitable chemical disorder from the glassy matrix. However, SEDM has been used to detect the fluctuations that occur in the vicinity of T_{xy} and SEDM patterns obtained at T_{xy} while driving at the line 1 (top) and line 2 (bottom) positions are shown in Figure 5.42. The derived relaxation rates are in perfect accord with values obtained independently on the same sample using μSR.

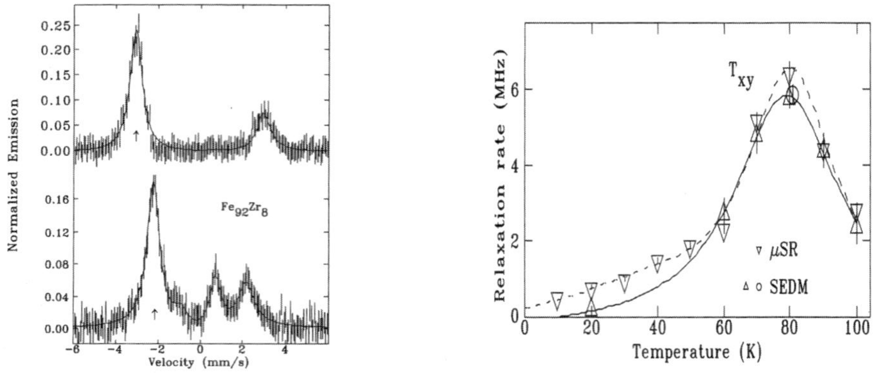

Figure 5.42 Left—^{57}Fe Mössbauer SEDM spectra of amorphous $Fe_{92}Zr_8$. Right—A comparison of relaxation rates derived from SEDM and μSR experiments [62].

5.4 CONCLUDING REMARKS

Mössbauer spectroscopy is now a well-established analytical technique and transmission ^{57}Fe spectroscopy is routine. Despite its age (nearly 50 years) significant advances are still being made in this research area. Selective excitation double Mössbauer spectroscopy, for example, is more difficult to perform than the standard transmission technique but provides a unique insight into dynamic processes. Another recent development in this technique involves the use of synchrotron radiation. Clearly, Mössbauer spectroscopy will continue to play a vital role in the study of materials.

REFERENCES

[1] Mössbauer, R.L. (1958) *Z. Physik* **151,** 124-143.
[2] Pound, R.V.& Rebka, G.A. (1960) *Phys. Rev. Lett.* **4,** 337-341.
[3] Klingelhöfer, G. et al., http://akguetlich.chemie.unimainz.de/klingelhoefer/rahmen/eframrows.htm (see also http://athena.cornell.edu/ the_mission/ins_moss.html).
[4] e.g. Cadogan, J.M. & Ryan, D.H. (2004) *Hyp. Int.* **153,** 25-41.
[5] Kündig, W. (1967) *Nucl. Instr. Meth.* **48,** 219-228.
[6] Häggström, L. (1974) Report UUIP-851 *Inst. of Physics*, University of Uppsala.
[7] Szymanski, K. (2000) *J. Phys.: Condensed Matter* **12,** 7495-7507.
[8] Cadogan, J.M., Ryan, D.H., Stewart, G.A. & Gagnon, R. (2003) *J. Magn. Magn. Mater.* **265,** 199-203.
[9] Blume, M. & Tjon, J.A. (1968) *Phys. Rev.* **165,** 446-456; Tjon, J.A., Blume, M., (1968) ibid. pp 456-461.
[10] Wickman, H.H. & Wertheim, G.K. (1968) Ch 11 pp 548-621 in 'Chemical Applications of Mössbauer Spectroscopy' Eds Goldanskii, V.I., Herber, R.H., Academic Press (London).
[11] Bleaney, B., Bowden, G.J., Cadogan, J.M., Day, R.K. & Dunlop, J.B. (1982) *J. Phys. F: Metal Phys.* **12,** 795-811.
[12] Hesse, J. & Rübartsch, A. (1974) *J.Phys. E: Sci. Instrum.* **7,** 526-532; Le Caër, G., Dubois, J.M. (1979) *J. Phys. E: Sci. Instrum.* **12,** 1083-90.

[13] Window, B. (1971) *J. Phys. E: Sci. Instrum.* **4,** 401-402.
[14] Keller, H. (1981) *J. Appl. Phys.* **52,** 5268-5273.
[15] Le Caër, G., Dubois, J.M., Fischer, H., Gonser, U. & Wagner, H.G. (1984) *Nucl. Instr. Meth. In Phys. Research B* **233,** 25-33.
[16] For a review see Herzer, G. (1997) Ch. 3 in *Handbook of Magnetic Materials* vol 10 ed K.H.J. Buschow, Elsevier Science BV.
[17] Mössbauer Effect Data Centre http://orgs.unca.edu/medc/.
[18] Frauenfelder, H. (1963) The Mössbauer Effect, W.A. Benjamin, New York.
[19] Wertheim, G.K. (1964) Mössbauer Effect Principles and Applications, Academic Press, New York.
[20] Greenwood, N.N. & Gibb, T.C. (1971) Mössbauer Spectroscopy, Chapman and Hall, London.
[21] Gonser, U., (ed) (1975) *Topics in Applied Physics* Vol. V: Mössbauer Spectroscopy, Springer-Verlag, Berlin.
[22] Gibb, T.C. (1977) Principles of Mössbauer Spectroscopy, Chapman and Hall, London.
[23] Long, G.J. (ed) (1984) *Mössbauer Spectroscopy Applied to Inorganic Chemistry Vol. 1*, Plenum Press, New York.
[24] Dickson, D.P.E. & Berry, F.J. (eds) (1986) Mössbauer Spectroscopy, Cambridge University Press, Cambridge.
[25] Schatz, G. & Weidinger, A. (1996) *Nuclear Condensed Matter Physics*, Wiley, Chichester.
[26] Cadogan, J.M. (1996) *J. Phys. D: Applied Physics* **29,** 2246-2254.
[27] Margulies, S. & Ehrman, J.R. (1961) *Nucl. Instr. Meth.* **12,** 131-137.
[28] Long, G.J., Cranshaw, T.E. & Longworth, G. (1983) Mössbauer Effect Reference and Data Journal **6(2),** 42-49.
[29] Sarma, P.R., Prakash, V. & Tripathi, K.C. (1980) *Nucl. Instr. Meth.* **178,** 167-171.
[30] Brand R.A. 'NORMOS' – available from http://www.wissel-instruments.de/produkte/software.html.
[31] Rancourt, D. & Lagarec, K. 'RECOIL' – Available from http://www.isapps.ca/recoil/.
[32] Klencsar, Z., 'MOSSWIN' – Available from http://www.mosswinn.com. See also Klencsar Z, Kuzmann E and Vertes A 1998 *Hyp. Int.* **112,** 269-273.
[33] Press, W.H., Flannery, B.P., Teukolsky, S.A. & Vetterling, W.T. (1986) Numerical Recipes: The Art of Scientific Computing, Cambridge University Press, Cambridge.
[34] Barb, F.D., Netoiu, O., Sorescu, M. & Weiss, M. (1992) *Comp. Phys. Commun.* **69,** 182-186; Verbiest, E., (1983) ibid **29,** 131-154; Chipaux, R., (1990) ibid **60,** 405-415; Kündig, W. (1969) *Nucl. Instr. Meth.* **75,** 336-340; Mei, Z. & Morris, J.W. (1990) *Nucl. Instr. Meth. in Phys. Research* B **47,** 181-186; Davidson, G.R. (1973) *Nucl. Instr. Meth.* **107,** 557-561; Ruebenbauer, K. & Birchall, T. (1979) *Hyp. Int.* **7,** 125-133 .
[35] Herbst, J.F. (1991) *Rev. Mod. Phys.* **63,** 819-898.
[36] Coey, J.M.D. & Sun, H. (1990) *J. Magn. Magn. Mater.* **87,** L251-L254.
[37] Perry, L.K., Ryan, D.H., Venturini, G. & Cadogan, J.M. (2006) *J. Appl. Phys.* **99,** 08J302(1-3).
[38] Cadogan, J.M., Suharyana, Ryan D.H., Moze, O. & Kockelmann, W. (2000) *J. Appl. Phys.* **87,** 6046-6048.
[39] Ryan, D.H., Elouneg-Jamroz, M., van Lierop, J., Altounian, Z. & Wang, H.B. (2003) *Phys. Rev. Lett.* **90,** 117202(1-4).
[40] Gschneidner, Jr K.A. & Pecharsky, V.K. (2000) *Ann. Rev. Mater. Sci.* **30,** 387-429.
[41] Stewart, G.A., Cadogan, J.M. & Edge, A.V.J. (1992) *J. Phys.: Condensed Matter* **4,** 1849-1858.
[42] Sengupta, A., Bhattacharyya, S. & Ghosh, D. (1989) *Phys. Lett.* **A 140,** 261-264.
[43] Cadogan, J.M. & Coey, J.M.D. (1986) *Appl. Phys. Lett.* **48,** 442-444.
[44] Bland, P.A., Kelley, S.P., Berry, F.J., Cadogan, J.M. & Pillinger, C.T. (1997) American Mineralogist **82,** 1187-1197.
[45] Suzuki, K., Sahajwalla, V., Cadogan, J.M., Inoue, A. & Masumoto, T. (1996) *Mater. Sci. Forum* **225-7**, 707-709.

[46] Suzuki, K. & Cadogan, J.M. (2000) *J. Appl. Phys.* **87,** 7097-7099.
[47] Handrich, K. (1969) *Phys. Stat. Sol.* **32,** K55-K58.
[48] Cadogan, J.M., Ryan, D.H. & Coey, J.M.D. (1988) *Mat. Sci. Eng.* **99,** 143-146.
[49] Suzuki, K., Sahajwalla, V., Cadogan, J.M., Inoue, A. & Masumoto, T. (1996) *Mater. Sci. Forum* **225-7**, 665-668.
[50] Suzuki, K. & Cadogan, J.M. (1998) *Phys. Rev. B* **58,** 2730-2739.
[51] Ryan, D.H., Strom-Olsen, J.O., Muir, W.B., Cadogan, J.M. & Coey, J.M.D. (1989) *Phys. Rev. B* **40,** 11208-11214.
[52] Cadogan, J.M., Day, R.K., Dunlop, J.B. & Foley, C.P. (1990) *J. Less-Common Metals* **158,** L45-L49.
[53] Cadogan, J.M., Ryan, D.H., Moze, O., Suharyana & Hofmann, M. (2003) *J. Phys. Condensed Matter* **15,** 1757-1771.
[54] Bowden, G.J., Cadogan, J.M., Day, R.K. & Dunlop, J.B. (1981) *J. Phys. F: Metal Physics* **11,** 503-510.
[55] Bowden, G.J., Cadogan, J.M., Day, R.K. & Dunlop, J.B. (1988) *Hyp. Int.* **39,** 359-367.
[56] Guo, W., Malus, S., Ryan, D.H. & Altounian, Z. (1999) *J. Phys. Condensed Matter* **11,** 6337-6346.
[57] Ryan, D.H., Cadogan, J.M. & Edge, A.V.J. (2004) *J. Phys. Condensed Matter* **16,** 6129-6138.
[58] Sahajwalla, V., Cadogan, J.M. & Zhao, A. (1996) *Proc. Annual Meeting of the Australian Inst. of Mining and Metallurgy* **153,** 267-272.
[59] Zhao, A., Suzuki, K., Sahajwalla, V. & Cadogan, J.M. (1999) *Scand. J. Metallurgy* **28,** 65-69.
[60] Jing, J., Campbell, S.J. & Cadogan, J.M. (1993) *Non-Destructive Testing – Australia* **30,** 6-8.
[61] van Lierop, J. & Ryan, D.H. (2002) *Phys. Rev. B* **65,** 104402 (1-9).
[62] van Lierop, J. & Ryan, D.H. (2001) *Phys. Rev. Lett.* **86,** 4390-4393.

CHAPTER 6

CRYSTAL FIELD SPECTROSCOPY

Albert Furrer and Andrew Podlesnyak
Laboratory for Neutron Scattering, ETH Zurich & PSI Villigen, CH-5232 Villigen PSI, Switzerland

6.1 INTRODUCTION

The crystal field interaction is an essential ingredient in a discussion of the magnetic properties of materials. It is therefore important to be able to describe and characterize the bonding between the central magnetic ion and its (nonmagnetic) ligand ions in terms of some electronic theory. For such a procedure it is obvious to classify the central magnetic ions according to the character of the partly filled electronic shells giving rise to a permanent magnetic dipole moment due to the orbital motion of the electrons, or to their intrinsic spin, or to both.

Partly filled shells occur naturally in the transition metal groups which comprise the 3d iron group, the 4d palladium group, the 5d platinum group, the 4f lanthanide or rare earth group, and the 5f actinide group. In a crystal, however, magnetic ions (rather than atoms) are by no means free, but they are surrounded by a cage of (diamagnetic) ligand ions, the complex-forming part of an extended lattice. The charged ligand ions have a strong interaction with the magnetic ions, producing an electrostatic field (the crystal field or the ligand field). The energy associated with this interaction varies roughly from 1 to 1000 meV; it may thus exceed the spin-orbit interaction, and in some cases it may exceed the electrostatic interaction with other electrons on the central ion responsible for the LS-coupling. Thus the crystal field is an additional interaction in the already complicated problem of a free many-electron atom. A perturbation approach is necessary in which terms must be considered in order of decreasing importance (i.e., decreasing interaction energy), and the crystal field interaction must be introduced at the appropriate point relative to the sequence of interactions internal to the magnetic ion. In order of decreasing interaction energy, these are [1]:

(i) The interaction of an electron with the Coulomb field of the nucleus, modified by the repulsion field of the other electrons. With a suitably averaged electronic field with central symmetry this interaction results in the electronic levels being grouped into *configurations*. The ground configuration of Cr^{3+} is $3d^3$, which lies some 10,000 meV below the first excited configuration $3d^24s$.

(ii) The residual mutual electrostatic repulsion of the electrons, not represented by a central field, gives rise to Russell-Saunders or LS-coupling, in which the orbital angular momenta and the spins of the electrons are vectorially coupled to give a resultant total angular momentum L and a total spin S, respectively. This coupling produces energy splittings of order 1000 meV between *terms* of different quantum numbers L and S associated with a particular configuration. e.g., the ground term of $3d^3$ is by Hund's rule ^4F, with $S = 3/2$ and $L = 3$, some 1000 meV below the ^4P term, with $S = 3/2$ and $L = 1$, belonging to the same $3d^3$ configuration.

(iii) The spin-orbit coupling $H_{so} = \lambda L{\cdot}S$, with $J = L \pm S$ being a good quantum number (upper and lower sign appropriate to electron shells being more or less than half-filled, respectively), splits a given term into a *multiplet* of levels with different values of J. The components of the multiplets are split by about 10 meV for 3d electrons, and by larger amounts for ions with larger atomic number (typically 1000 meV for 4f electrons).

The importance of the crystal field interaction relative to the interactions just enumerated is different for the various transition groups. They may conveniently be grouped as follows:

(i) Strong crystal field, i.e., crystal field interaction > LS-coupling. Typical examples are the 4d and 5d electron systems.

(ii) Intermediate crystal field, i.e., crystal field interaction ≈ LS-coupling > spin-orbit coupling. Typical examples are 3d electron systems and to a lesser extent the 5f actinide group.

(iii) Weak crystal field, i.e., crystal field interaction < spin-orbit coupling. This group comprises the 4f electron systems.

In Figure 6.1 we give an example of the effect of the crystal field interaction on the level scheme of an ion with $L = 3$ and $S = 1$. The left hand part belongs to case (ii) and the right hand part to case (iii).

The present chapter is organized as follows: Section 6.2 summarizes the basic formalism and some commonly used models of the crystal field interaction for the case (iii) mentioned above, i.e., for weak crystal fields as realized for the 4f lanthanides. Some experimental techniques are described in Section 6.3, with emphasis on neutron spectroscopy. Section 6.4 shows how crystal field parameters can be determined from spectroscopic data. The effect of interactions of crystal field split ions with phonons and conduction electrons as well as exchange coupling effects are discussed in Section 6.5. Section 6.6 demonstrates the relevance of crystal field studies towards understanding the mechanism of high temperature superconductivity. Some final conclusions are given in Section 6.7.

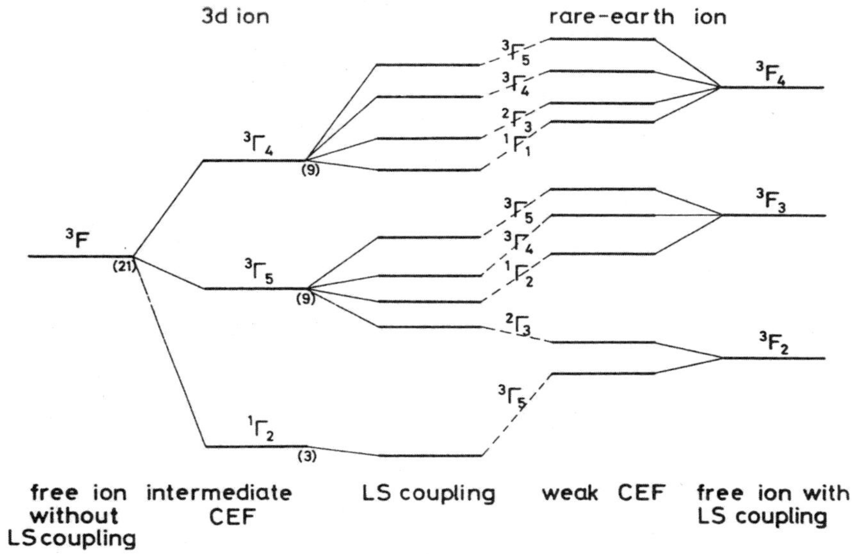

Figure 6.1 Crystal field interaction of an ion with L = 3 and S = 1 for intermediate and weak crystal field strength (after [2]).

6.2 THE CRYSTAL FIELD INTERACTION

6.2.1 Basic Formalism

The theory of crystal fields originates from the work of Bethe [3]. He showed that the eigenvalues of a Hamiltonian, describing an open shell of electrons in a crystal, can be classified according to the irreducible representations of the point group of the site. Since that time the theory has made substantial progress toward becoming elegant and simple especially applied to the 4f lanthanide group as demonstrated by the work of Stevens [4, 5], Elliott and Stevens [6], Judd [7], Hutchings [8], Newman [9], and Fulde [10]. Although the description of crystal fields is rather well known and standard by now, their computation from first principles is far from being satisfactory. Nevertheless, the available models serve as a convenient parameterization scheme to describe the crystal field interaction and the related magnetic properties, but the model parameters have to be considered phenomenological parameters rather than parameters reflecting the detailed electronic properties.

In the following paragraphs, and in the remainder of this article, we will restrict our considerations to the crystal field interaction associated with the 4f lanthanide group. This choice is motivated by the fact that most of the crystal field studies reported in the literature have been devoted to rare earth compounds for which the crystal field is usually the dominant interaction looked at to understand the magnetic properties. However, the formalism can be applied to the 5f actinide group as well as to the d-electron groups with some extensions and modifications.

The mathematical problem, which has to be solved, is the calculation of the electrostatic potential around the rare earth ion site. In elementary crystal field theory the electrostatic potential arises from the electric charge distribution on the surrounding ions. If overlap with the 4f electrons is neglected, the electrostatic potential V_{CF} is a solution of Laplace's equation

$$\Delta V_{CF}(r) = 0. \tag{6.1}$$

The solutions of this equation can be expanded in terms of multipoles of the 4f electrons as:

$$V_{CF}(r) = \sum_{n=0}^{\infty} \sum_{m=-n}^{n} a_n^m r^n Y_n^m(\Theta, \Phi). \tag{6.2}$$

The spherical harmonics $Y_n^m(\Theta,\Phi)$ are normalized to $\int d\cos\Theta d\Phi \,|\, Y_n^m(\Theta,\Phi)|^2 = 1$. The summation is restricted to $n \leq 6$, since a 4f electron with orbital quantum number $l = 3$ does not have multipoles with $n > 6$. Alternatively one may expand V_{CF} in terms of *tesseral harmonics*, which has the advantage that the expansion is with respect to real functions:

$$V_{CF}(r) = \sum_{n=0}^{6} \sum_{\alpha} a_n^\alpha r^n Z_n^\alpha(\Theta, \Phi), \tag{6.3}$$

where the tesseral harmonics $Z_n^\alpha(\Theta,\Phi)$ are defined by [8]:

$$\begin{aligned} Z_n^0 &= Y_n^0 \\ Z_n^m(c) &= \tfrac{1}{\sqrt{2}}[Y_n^{-m} + (-1)^m Y_n^m] \\ Z_n^m(s) &= \tfrac{i}{\sqrt{2}}[Y_n^{-m} - (-1)^m Y_n^m] \end{aligned} \tag{6.4}$$

The index α in equation (6.3) is understood such that for a given value of n, Z_n^α runs over $Z_n^0 = Y_n^0$, $Z_n^m(c)$, $Z_n^m(s)$. The spherical and tesseral harmonics are listed in [8]. The expansion coefficients a_n^m are calculated from the charge distribution $\rho(r)$ as

$$a_n^\alpha = \frac{4\pi}{2n+1} \int d^3r \frac{\rho(r)}{r^{n+1}} Z_n^\alpha(\Theta, \Phi). \tag{6.5}$$

Until now we have only considered a single 4f electron. For more than one 4f electron the potential energy of the rare earth ion in the crystal field V_{CF} is given by

$$E_{CF} = e \sum_i V_{CF}(r_i), \tag{6.6}$$

where e is the electron charge and the index i runs over all the 4f electrons. For rare earth compounds the spin-orbit interaction usually exceeds the crystal field interaction by at least an order of magnitude, thus one can treat the crystal field interaction as a perturbation of the lowest J-multiplet alone. The restriction to the lowest J-multiplet simplifies the calculations considerably as demonstrated by Stevens [5]. He showed that the crystal field $V_{CF}(r_i)$ being a sum of polynomials in $r_i = (x_i, y_i, z_i)$ can be replaced by a sum of polynomials of the total angular momentum operators $J = (J_x, J_y, J_z)$, which have the same transformation properties as the original expression. They act on the unfilled 4f shell as a whole and therefore are much more convenient than the (x_i, y_i, z_i) polynomials that act on an individual 4f electron. The rules for transforming an expression in the r_i space to the J space are such that any products of (x_i, y_i, z_i) are replaced by the corresponding products of (J_x, J_y, J_z) but written in a symmetrized form [e.g., $xy \rightarrow \frac{1}{2}(J_xJ_y + J_yJ_x)$]. Constants of proportionality χ_n have to be introduced, which depend on the order n as well as on the quantum numbers L, S, and J. Furthermore, the dimensionality of the r_i^n operators has to be retained by introducing the averages $\langle r^n \rangle$ over the 4f wave-functions. Some simple examples are [8]:

$$\begin{aligned}
&\sum_i (3z_i^2 - r_i^2) \equiv \chi_2 \langle r^2 \rangle [3J_z^2 - J(J+1)] = \chi_2 \langle r^2 \rangle O_2^0 \\
&\sum_i (x_i^2 - y_i^2) \equiv \chi_2 \langle r^2 \rangle [J_x^2 - J_y^2] = \chi_2 \langle r^2 \rangle O_2^2 \\
&\sum_i (x_i^4 - 6x_i^2 y_i^2 + y_i^4) = \sum_i \tfrac{1}{2}[(x_i + iy_i)^4 + (x_i - iy_i)^4] \\
&\qquad \equiv \chi_4 \langle r^4 \rangle \tfrac{1}{2}[J_+^4 + J_-^4] = \chi_4 \langle r^4 \rangle O_4^4
\end{aligned} \tag{6.7}$$

where $J_\pm = J_x \pm iJ_y$. A complete list of the operators O_n^m which are called *Stevens operators* [4] as well as of the reduced matrix elements χ_n can be found in [8]. A list of the radial integrals $\langle r^n \rangle$ can be found in Freeman and Desclaux [11]. In the Stevens notation the crystal field Hamiltonian reads

$$H_{CF} = \sum_n \chi_n \langle r^n \rangle \sum_m A_n^m O_n^m = \sum_{n,m} B_n^m O_n^m, \tag{6.8}$$

where the A_n^m and B_n^m are the crystal field parameters. We immediately recognize that the crystal field Hamiltonian consists of a large number foparameters that however, can be drastically reduced due to the point symmetry at the rare earth site. In particular, a center of inversion cancels all the odd n terms, and a p-fold axis of symmetry when chosen as polar axis reduces the Hamiltonian to terms containing O_n^p. e.g., for cubic point symmetry (with the four-fold symmetry axis taken as polar axis) the crystal field Hamiltonian reads

$$H_{CF} = B_4^0[O_4^0 + 5O_4^4] + B_6^0[O_6^0 - 21O_6^4], \tag{6.9}$$

i.e., the number of independent crystal field parameters is reduced to two. Lea et al. [12] rewrote equation (6.9) in the form

$$\mathrm{H}_{\mathrm{CF}} = \mathrm{W}(\tfrac{x}{F(4)})[O_4^0 + 5O_4^4] + \tfrac{1-|x|}{F(6)}\mathrm{B}_6^0[O_6^0 - 21O_6^4] \quad (6.10)$$

where $\mathrm{B}_4^0\mathrm{F}(4) = \mathrm{W}x$ and $\mathrm{B}_6^0\mathrm{F}(6) = \mathrm{W}(1-|x|)$ with $-1 \le x \le 1$ and W being an energy scale factor. The eigenfunctions and eigenvalues of equation 6.10 are tabulated in [12] as a function of the parameter x. The factors F(4) and F(6) depend only on J and are also listed there. For hexagonal point symmetry, the crystal field Hamiltonian has the form:

$$\mathrm{H}_{\mathrm{CF}} = \mathrm{B}_2^0 O_2^0 + \mathrm{B}_4^0 O_4^0 + \mathrm{B}_6^0 O_6^0 + \mathrm{B}_6^6 O_6^6 \quad (6.11)$$

with four independent parameters. For orthorhombic symmetry, there are as much as nine adjustable crystal field parameters:

$$\mathrm{H}_{\mathrm{CF}} = \mathrm{B}_2^0 O_2^0 + \mathrm{B}_4^0 O_4^0 + \mathrm{B}_4^2 O_4^2 + \mathrm{B}_4^4 O_4^4 + \mathrm{B}_6^0 O_6^0 + \mathrm{B}_6^2 O_6^2 + \mathrm{B}_6^4 O_6^4 + \mathrm{B}_6^6 O_6^6 . \quad (6.12)$$

When the overall crystal field splitting is large and comparable to the intermultiplet splittings, the Stevens formalism described above is no longer appropriate. This situation was discussed in detail, e.g., by Wybourne [13]. In this case the crystal field interaction leads to a mixing of the different J-multiplets (J-mixing), and furthermore the J-multiplets are contaminated by states of different quantum numbers L and S (intermediate coupling). Therefore, the electrostatic, spin-orbit and crystal field interactions have to be diagonalized simultaneously; the corresponding Hamiltonian reads

$$\mathrm{H} = \mathrm{H}_{\mathrm{EL}} + \mathrm{H}_{\mathrm{SO}} + \mathrm{H}_{\mathrm{CF}}, \text{ with } \mathrm{H}_{\mathrm{EL}} = \sum_{k=0}^{3} \mathrm{E}^k e_k \text{ and } \mathrm{H}_{\mathrm{SO}} = \xi \mathrm{A}_{\mathrm{SO}} . \quad (6.13)$$

The adjustable free ion parameters E^k and ξ correspond to Slater electrostatic and spin-orbit integrals, respectively, which can be taken, e.g., from [13]. E^k and A_{so} represent matrix elements for the angular parts of the electrostatic and spin-orbit interactions, respectively, which have been tabulated by Nielson and Koster [14]. The crystal field Hamiltonian takes then the form [13]

$$\mathrm{H}_{\mathrm{CF}} = \sum_{m,n,i} \mathrm{D}_m^n (\mathrm{C}_m^n + \mathrm{C}_{-m}^n)_i , \quad (6.14)$$

where C_m^n are one-electron tensor operators and the summation involving i runs over all the electrons of the ion of interest. The crystal field parameters D_m^n of the tensor formalism are related to the crystal field parameters B_n^m in the Stevens notation through

$$\mathrm{D}_m^n = \frac{\mathrm{F}_n^m}{\chi_n} \mathrm{B}_n^m , \quad (6.15)$$

where the reduced matrix elements χ_n and the numerical factors F_n^m are tabulated by Hutchings [8] and Wybourne [13], respectively.

6.2.2 Model Calculations of the Crystal Field Interaction

6.2.2.1 Point-Charge Model

The point-charge model is characterized by ascribing effective charges q_i to the neighboring sites of the rare earth ion under consideration. In that case the charge distribution $\rho(\mathbf{r})$ is given by

$$\rho(\mathbf{r}) = \sum_i q_i \delta(\mathbf{R}_i - \mathbf{r}) \quad , \tag{6.16}$$

where the $\mathbf{R}_i$ denote the positions of the neighboring ions. If we only consider the nearest-neighboring ligand ions carrying an effective charge q and being placed at a distance d from the rare earth ion, then the expansion coefficients of equation (6.5) can be expressed as

$$a_n^m = \frac{4\pi}{2n+1} q\langle r^n\rangle f_n^m \text{, with } f_n^m = \sum_j \frac{(-1)^m}{d^{n+1}} \mathrm{Y}_n^{-m}(\vartheta_j, \varphi_j), \tag{6.17}$$

where the sum runs over all the nearest neighbors located at angular coordinates ϑ_j, φ_j . The f_n^m are called *geometrical coordination factors*, which are discussed in detail in [8]. Thus, the crystal field parameters can be decomposed in a charge and a geometrical term:

$$\mathrm{B}_n^m = \mathrm{A}_n^m \chi_n \langle \mathrm{r}^n \rangle = |e| q \chi_n \langle r^n \rangle f_n^m . \tag{6.18}$$

For a crystal field of cubic symmetry as defined by equation (6.9), one finds the crystal field parameters for eight fold coordination (cube):

$$\mathrm{B}_4^0 = \frac{7}{18}\frac{|e|q}{d^5}\chi_4\langle r^4\rangle,\ \mathrm{B}_6^0 = -\frac{1}{9}\frac{|e|q}{d^7}\chi_6\langle r^6\rangle, \tag{6.19}$$

and for six-fold-coordination (octahedron):

$$\mathrm{B}_4^0 = -\frac{7}{16}\frac{|e|q}{d^5}\chi_4\langle r^4\rangle,\ \mathrm{B}_6^0 = -\frac{3}{64}\frac{|e|q}{d^7}\chi_6\langle r^6\rangle. \tag{6.20}$$

6.2.2.2 Extended Point-Charge Model

In general the simple point-charge model is not able to reproduce observed crystal field spectra in a satisfactory manner. In particular, the second- and sixth-order crystal field parameters are usually found to be an order of magnitude too large and too small, respectively, thus several extensions have been introduced to improve the point-charge model:

(i) Sternheimer [15] showed that due to the screening of the 4f electrons by the outer shells, the $\langle r^n\rangle$ terms should be replaced by $\langle r^n\rangle$ $(1 - \sigma_n)$. The shielding factors σ_n are given as a function of the number of electrons N of the R ion as [16]:

$$\begin{aligned}\sigma_2 &= 0.6846 - 0.00854 \cdot N \\ \sigma_4 &= 0.02356 + 0.00182 \cdot N \\ \sigma_6 &= -0.04238 + 0.00014 \cdot N\end{aligned} \tag{6.21}$$

(ii) Morrison [16] showed that the nth moment of the radial wavefunctions $\langle r^n \rangle$ for the free ion have to be replaced by $\langle r^n \rangle / \tau^n$ when it is placed into a crystal. τ as a function of N is given for the system R:$CaWO_4$ as

$$\tau = 0.75 \cdot (1.0387 - 0.0129 \cdot N) \tag{6.22}$$

(iii) For metallic compounds a further correction is necessary due to the screening effect of the conduction electrons which can be taken into account by a Yukawa-type potential:

$$V(\mathbf{r}) = \sum_j \frac{|e| q_j}{|\mathbf{r} - \mathbf{R}_j|} \exp\left(-\kappa |\mathbf{r} - \mathbf{R}_j|\right) \tag{6.23}$$

where κ is the inverse screening length. As a consequence, equation (6.17) has to be modified in the following way [17]:

$$a_n^m = \frac{4\pi}{2n+1} q K_{n+1/2}(\kappa d) \langle r^n \rangle f_n^m . \tag{6.24}$$

The $K_{n+1/2}(x)$ are the modified Bessel functions, normalized to obtain $K_{n+1/2}(0) = 1$. Assuming nominal charges for the nearest neighboring ions, we are left with a model containing only a single parameter κ. As we will see in Section 6.4.2, this model works well for metallic rare earth compounds.

6.2.2.3 Superposition Model

In its most general form the superposition model corresponds to the assumption that the total crystal field can be built up from separate contributions from each of the ions in the crystal [9]. The spherical symmetry of the ions then ensures that each contribution can be represented by a cylindrically symmetrical field, which is described by just three parameters $A_n^0 \langle r^n \rangle$ ($n = 2, 4, 6$) in the Stevens notation if the z-axis is the polar axis. The functional dependence of the field of a single ligand ion on its distance R from the central ion is made explicit by introducing the "intrinsic" parameters $\bar{A}_n(R)$:

$$A_n^m \left\langle r^n \right\rangle = \sum_i \kappa_{nm}(\mathrm{i}) \bar{A}_n(R_i), \tag{6.25}$$

where the coordination factors $\kappa_{nm}(i)$ depend on the angular positions of all ligand ions at a given distance R_i from the central ion. Explicit formulae for the coordination factors $\kappa_{nm}(i)$ have been given by Newman [9]. It is often convenient to decompose the function $\bar{A}_n(R)$ into a value $\bar{A}_n(R_0)$ at the mean distance R_0 of the coordinated ligands and a power law t_n over a restricted range of R:

$$\bar{A}_n(R) = \bar{A}_n(R_0)\cdot(R_0/R)^{t_n}. \tag{6.26}$$

In practice the superposition model works well for $n = 4$ and $n = 6$, but not for $n = 2$, thus the fourth- and sixth-order crystal field terms can be represented by the four quantities $\bar{A}_4(R)$, $\bar{A}_6(R)$, t_4, t_6, provided that all the ligands are of the same type.

6.2.2.4 Angular Overlap Model

The angular overlap model (AOM) may be seen as a first-order perturbation approach to covalent bonding [18]. Being based on covalency, the AOM is thus in contrast to the point-charge model resting on the electrostatic interaction. With the assumption of weak covalency the perturbation energies of the central orbitals are proportional to the squares of the overlap integrals of the central ion with the ligand. Like in the electrostatic theory the AOM assumes the effects of several ligands bonded to the central ion to be additive, which is equivalent to the assumption of no ligand-ligand overlap. The antibonding energy E^* of a given f-orbital is given as [18]

$$E^* = e_\lambda\left(F_\lambda^{(\ell)}\right)^2, \tag{6.27}$$

where F_λ^ℓ, the angular overlap integral, is a property of the orientations of the overlapping orbitals and applies to equivalent λ-bonding. The superscript ℓ labels the quantum number of the wavefunction of the central ion. e_λ is the angular overlap parameter for a particular bonding type.

Let us consider a specific example, namely the crystal field potential of an octahedron formed by six identical, axially symmetric ligands with the quantization axis along the three-fold axis. This leads to a Hamiltonian with five crystal field parameters, which are related to four independent angular overlap parameters e_λ [18]:

$$\begin{aligned}
B_4^0 &= \frac{3}{16\chi_4}\left(-2e_\sigma - \frac{2}{3}e_\pi + \frac{14}{3}e_\delta - 2e_\varphi\right)\\
B_4^3 &= \frac{15\sqrt{2}}{4\chi_4}\left(-2e_\sigma - \frac{2}{3}e_\pi + \frac{14}{3}e_\delta - 2e_\varphi\right)\\
B_6^0 &= \frac{13}{224\chi_6}\left(\frac{8}{3}e_\sigma - 4e_\pi + \frac{8}{5}e_\delta - \frac{4}{15}e_\varphi\right)\\
B_6^3 &= -\frac{65}{128\chi_6}\left(\frac{8}{3}e_\sigma - 4e_\pi + \frac{8}{5}e_\delta - \frac{4}{15}e_\varphi\right)\\
B_6^6 &= \frac{143}{256\chi_6}\left(\frac{8}{3}e_\sigma - 4e_\pi + \frac{8}{5}e_\delta - \frac{4}{15}e_\varphi\right)
\end{aligned} \tag{6.28}$$

Confining ourselves to σ bonding only [19], we reduce the number of ligand-field parameters to one, which is e_σ. This can be done for all kinds of co-ordination symmetries, assuming identical ligands and equal bond lengths. The reduction of the number of independent ligand-field parameters is a particular strength of the AOM that will simplify the interpretation of the magnetic properties and energy spectra of compounds with low symmetry.

6.2.2.5 First Principles Density Functional Theory

The *density functional theory* (DFT) is a first principles method to calculate the crystal field parameters B_n^m. Within this method the electronic structure and the corresponding distribution of the ground state charge density is obtained using the full potential *linearized augmented plane wave method* (LAPW) which is implemented in the latest version of the WIEN97 code [20]. This was demonstrated, e.g., for $ErNi_2B_2C$ [21]. The crystal field parameters originating from the aspherical part of the total single-particle DFT potential in the crystal can be obtained from:

$$B_n^m = \frac{f_n^m}{\chi_n}\left(\int_0^{R_{MT}} |R_{4f}(r)|^2 V_1^m(r) r^2 dr + \int_{R_{MT}}^{R_c} |R_{4f}(r)|^2 W_1^m(r) r^2 dr \right), \tag{6.29}$$

where $V_1^m(r)$ and $W_1^m(r)$ are the components of the total (Coulomb and exchange correlation) potential inside the atomic sphere with radius R_{MT} and in the interstitial region, respectively. The term $V_1^m(r)$ is readily obtained with the WIEN97 code. The term $W_1^m(r)$ is calculated using an exact transformation of the Fourier representation of the LAPW potential into a spherical Bessel function expansion between the radius R_{MT} and an upper radial limit R_c beyond which the 4f charge density can be neglected. The term $R_{4f}(r)$ describes the radial shape of the localized 4f charge density of the R^{3+} ion. The conversion factors f_n^m establish the relation between the symmetrized spherical harmonics used within LAPW and the real tesseral harmonics, which transform in the same way as the tensor operators O_n^m [22]. The crystal field parameters for $ErNi_2B_2C$ obtained from equation (6.29) were found to be in reasonable agreement with those determined by neutron spectroscopy [23], which suggests that the DFT crystal field parameters can be used as good starting values in the analysis of experimental data. A more recent example concerns the intermetallic compounds $PrAl_3$ and $NdAl_3$ for which the substantial pressure-induced changes of the crystal field parameters could be well reproduced by DFT calculations [24].

6.2.3 Parameterization of the Crystal Field Interaction

The determination of the crystal field parameters from experimental data is not an easy task, except perhaps for the case of cubic systems with only two adjustable parameters. For low symmetry sites such as orthorhombic symmetry with nine independent crystal field parameters (see equation (6.12)), it is a highly nontrivial task in the least squares fitting procedure to the experimental data to find a correct set of starting parameters. However, we can make use of the fact that the crystal field parameters B_n^m for a particular degree n can reasonably be determined by taking account of the geometry of the nearest neighboring coordination polyhedra (see equation (6.18)):

$$B_n^m = B_n^0 f_n^m / f_n^0 \tag{6.30}$$

i.e., we are left with only three independent crystal field parameters B_n^0 ($n = 2, 4, 6$). Following [12] we introduce a parameterization which covers all possible ratios of B_2^0 / B_4^0 and B_2^0 / B_6^0:

$$\begin{aligned} B_2^0 F_2 &= W(1-|y|) \\ B_4^0 F_4 &= Wxy \\ B_6^0 F_6 &= W(1-|x|)y \end{aligned} \tag{6.31}$$

with $-1 \leq x,\ y \leq 1$. Equation (6.31) corresponds to the most general combination of second-, fourth- and sixth-order diagonal crystal field parameters. Usually only one set of x, y parameters gives a reasonable agreement with the spectroscopic data, thus we have solved the problem of assigning the leading diagonal start parameters of the least squares fitting procedure in a reasonable manner. In a second step, we can then try to achieve a fine-tuning of the remaining nondiagonal crystal field parameters.

An alternative way to derive a starting set of crystal field parameters is to consider the extended point-charge model (see Section 6.2.2.2), provided that the nearest-neighbor point charges, as well as the screening length (in metallic systems), are known.

6.2.4 Extrapolation Schemes

In many models the structure and charge aspects of the crystal field interaction can be separated as exemplified for the point-charge model by equation (6.18) and for the superposition model by equation (6.25). This offers a convenient extrapolation scheme for the crystal field potential along a series of rare earth compounds provided that the charge distribution of the coordinating ligand ions is not affected by the particular rare earth ions. The procedure for extrapolating the crystal field parameters from one compound (R_1) to another compound (R_2) based on equation (6.18) yields:

$$B_n^m(R_2) = B_n^m(R_1) \frac{\left(\chi_n \langle r^n \rangle f_n^m\right)(R_2)}{\left(\chi_n \langle r^n \rangle f_n^m\right)(R_1)}. \tag{6.32}$$

This procedure was successfully applied to a large number of both insulating and metallic rare earth (R) compounds, e.g., for Cs_2NaRBr_6 [25] and for RPd_3 [26], respectively; see Section 4.1 for Cs_2NaRBr_6.

6.2.5 Calculation of Thermodynamic Magnetic Properties

In order to check the reliability of the crystal field parameters it is important to calculate various thermodynamic magnetic properties and to compare the results with experimental data. These properties depend explicitly upon both the crystal field energies E_i and the crystal field wavefunctions $|\Gamma_i\rangle$. Based on general expressions of statistical mechanics for the Gibbs free energy $F = -k_B T \ln Z$ and the internal energy $U = F - T(\partial F/\partial T)_V$, where Z is the partition function, we obtain for the magnetization:

$$\mathrm{M}_\alpha = \frac{1}{k_B T}\frac{\partial(\ln Z)}{\partial \mathrm{H}_\alpha} = g\mu_B \sum_i p_i \langle \Gamma_i | \mathrm{J}_\alpha | \Gamma_i \rangle, \tag{6.33}$$

for the single-ion susceptibility:

$$\chi_{\alpha\alpha} = \frac{\partial \mathrm{M}_\alpha}{\partial \mathrm{H}_\alpha} = (g\mu_B)^2 \left[\sum_i \frac{|\langle \Gamma_i | \mathrm{J}_\alpha | \Gamma_i \rangle|^2}{k_B T} p_i + \sum_{i \neq j} \frac{|\langle \Gamma_j | \mathrm{J}_\alpha | \Gamma_i \rangle|^2}{E_i - E_j} (p_j - p_i) \right], \tag{6.34}$$

and for the Schottky heat capacity:

$$c_V = \left(\frac{\partial U}{\partial T}\right)_V = k_B \left[\sum_i \left(\frac{E_i}{k_B T}\right)^2 p_i - \sum_i \left(\frac{E_i}{k_B T} p_i\right)^2 \right], \tag{6.35}$$

with the Boltzmann population factor

$$p_i = \frac{1}{Z} \exp\left\{ -\frac{E_i}{k_B T} \right\}. \tag{6.36}$$

6.3 EXPERIMENTAL TECHNIQUES

6.3.1 Introductory Remarks

The crystal field interaction discussed in the preceding section gives rise to discrete energy levels E_i and wavefunctions $|T_i\rangle$ which can be determined by a variety of experimental methods. The most powerful techniques are certainly spectroscopic methods such as inelastic neutron scattering, light scattering, and point-contact spectroscopy, which allow a direct determination of the crystal field states. Light scattering offers several methods working with different modes of operation at different wavelengths, e.g., Raman, infrared, photoemission, X-ray absorption, and luminescence spectroscopy. Some information about the crystal field states can also be obtained by resonance techniques, e.g., in a paramagnetic resonance experiment one is measuring the Zeeman splitting of a given ground state crystal field level, which has seen extensive applications in rare earth salts [1]. Sometimes it is possible to use the Mössbauer effect to observe a zero-field magnetic hyperfine splitting associated with the rare earth ions yielding valuable information about the crystal field parameter B_2^0 [27], which is difficult to determine by model calculations due to the long-range nature of the 2nd order crystal field potential. Information on the crystal field states is also contained intrinsically in the thermodynamic magnetic properties (see equations (6.33)–(6.36)); however, an extraction of crystal field parameters is likely to fail (as demonstrated in Section 6.4.1) due to the integral nature of these properties.

In this work we will describe only a few spectroscopic techniques that have been applied to the study of crystal field effects. These include primarily inelastic neutron scattering and to a lesser extent Raman scattering and point-contact spectroscopy. Both inelastic neutron and Raman scattering experiments have their merits and should be considered as complementary methods. Raman scattering, on the one hand, can be applied to very small samples of the order of 10 μm^3; it provides highly resolved spectra so that small line shifts and splittings can be detected [28], and it covers a large energy range so that intermultiplet transitions can easily be observed [29]. Neutron scattering, on the other hand, is not restricted to particular points in reciprocal space, i.e., interactions between the rare earth ions can be observed through the wavevector dependence [30], the intensities of crystal field transitions can easily be interpreted on the basis of the wavefunctions of the crystal field states, and data can be taken over a wide temperature range, which is important when studying linewidths of crystal field transitions. Inelastic neutron scattering as the most widely used spectroscopic technique is described below in detail, followed by short descriptions of Raman and point-contact spectroscopy.

6.3.2 Neutron Spectroscopy

The principal aim of a neutron scattering experiment is the determination of the probability that a neutron incident on the sample with wavevector **k** is scattered into the state with wavevector **k'**. The intensity of the scattered neutrons is thus measured as a function of the momentum transfer

$$\hbar\mathbf{Q} = \hbar(\mathbf{k} - \mathbf{k}'), \tag{6.37}$$

where **Q** is known as the *scattering vector*, and the corresponding energy transfer is given by

$$\hbar\omega = \frac{\hbar^2}{2\mathrm{m}}(k^2 - k'^2). \tag{6.38}$$

Equations (6.37) and (6.38) describe the momentum and energy conservation of the neutron scattering process, respectively. The momentum conservation is schematically sketched in Figure 6.2. For $k =$ k' we have from equation (6.38) $\hbar\omega = 0$, i.e., elastic scattering. Figure 6.2a shows the particular situation $\mathbf{Q} = \mathbf{k}-\mathbf{k}' = \tau$, which is just the condition known as *Bragg's law*. Inelastic scattering is visualized in Figure 6.2b where the scattering vector can be decomposed according to $\mathbf{Q} = \tau + \mathbf{q}$, with $\mathbf{q}$ being the wavevector of an elementary excitation to be specified. Neutron scattering turns out to be the only experimental technique that is able to measure the dispersion relation $\hbar\omega(\mathbf{q})$ at any predetermined point in reciprocal space.

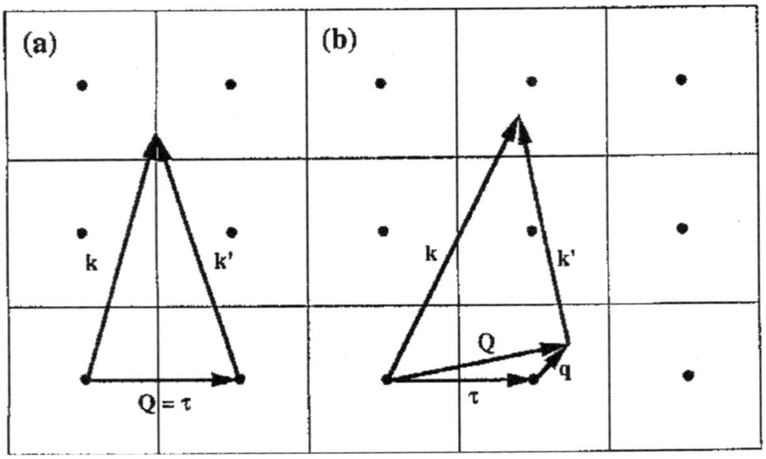

Figure 6.2 Visualization of equation (6.37) in reciprocal space for elastic (a) and inelastic (b) neutron scattering. The lines indicate the boundaries of the Brillouin zone, and the full circles denote the zone centers.

The determination of $\hbar\omega(\mathbf{q})$ by neutron scattering techniques requires a controlled access to the variables $\mathbf{Q}$ and ω. This can be done in various ways, but by far the most effective experimental method is the triple-axis crystal spectrometry developed by Brockhouse [31] who received the 1994 physics Nobel Prize for this achievement. The principle of this method is sketched in Figure 6.3. An incident beam of neutrons with a well-defined wavevector **k** is selected from the white spectrum of the neutron source by the monochromator crystal and scattered from the sample. The intensity of the scattered beam is measured as a function of **k'** by the analyzer crystal and the neutron detector.

The outstanding advantage of the triple-axis spectrometer is that data can be taken at predetermined points in reciprocal space (which is known as the "constant-**Q**" or "constant-ω" method), so that single-crystal measurements of the dispersion relation $\hbar\omega(\mathbf{Q})$ can be performed in a controlled manner. Of course, general scans in **Q** and ω are also possible.

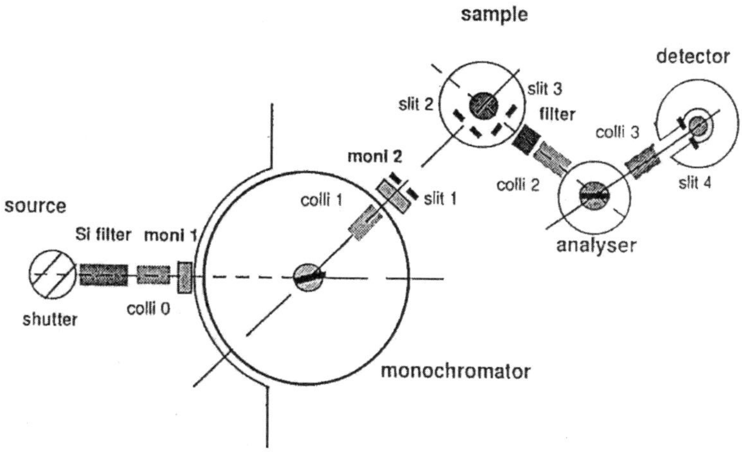

Figure 6.3 Basic layout of a triple-axis spectrometer. The three axes around which the respective rest of the spectrometer is rotated are: the monochromator axis (variation of **k**), the sample axis (variation of the scattering angle), and the analyzer axis (variation of **k**').

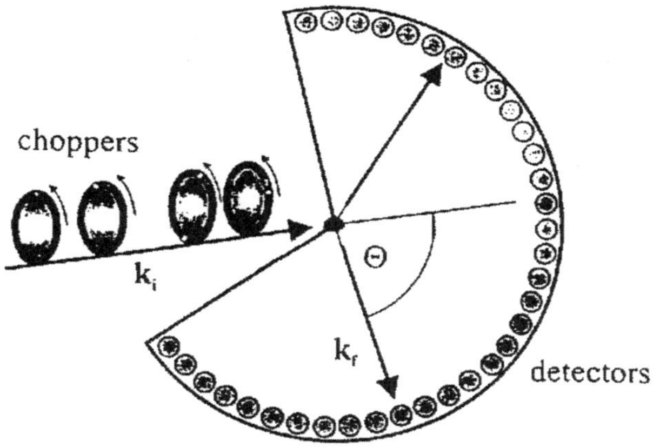

Figure 6.4 Schematic of a direct time-of-flight spectrometer. The four choppers define the energy and the pulse width of the incident neutrons. In addition, they suppress higher order neutrons and frame overlap.

For neutron scattering experiments on polycrystalline, liquid, and amorphous materials, various types of time-of-flight spectrometers are usually more appropriate. In the time-of-flight method the neutron beam is monochromatized by a series of choppers that produce pulses of neutrons with the desired wavelength and that eliminate higher order neutrons and frame overlap of pulses from different repetition periods as well (see Figure 6.4). The monochromatic neutron pulses are then scattered from the sample and are detected by arrays of neutron counters covering a large solid angle. The energy transfer $\hbar\omega$ and the modulus of the scattering vector $\mathbf{Q}$ are then determined by the flight time of the neutrons from the sample to the detector and by the scattering angle at which the detector is positioned, respectively.

The neutron scattering cross-section corresponds to the number of neutrons scattered persecondintoa(small) solid angle $d\Omega$ with energy transfers between $\hbar\omega$ and $\hbar(\omega + d\omega)$, divided by the flux of the incident neutrons. Theoretical expressions for the cross-section usually start from Fermi's "golden rule":

$$\frac{d^2\sigma}{d\Omega d\omega} = \left(\frac{m}{2\pi\hbar^2}\right)^2 \frac{k'}{k} \sum_{\lambda} p_{\lambda} \sum_{\lambda'} |\langle k', \lambda' | \hat{U} | k, \lambda\rangle|^2 \delta(\hbar\omega + E_{\lambda} - E_{\lambda'}). \qquad (6.39)$$

Here, $|\lambda\rangle$ denotes the initial state of the scatterer, with energy E_λ and thermal population factor P_λ, and its final state is $|\lambda'\rangle$. $\hat{U}$ is the interaction operator of the neutron with the sample, which depends on the specific scattering process. For magnetic scattering the interaction operator $\hat{U}_m$ of the neutron with the sample may be described by the interaction of a neutron with a magnetic field **H**:

$$\hat{U}_m = \hat{\mu} \cdot \mathbf{H} = -\gamma\mu_N \hat{\sigma} \cdot \mathbf{H}, \qquad (6.40)$$

where $\hat{\mu}$ is the magnetic moment operator of the neutron, $\gamma = -1.91$ the gyromagnetic ratio, $\mu_N = 0.505 \cdot 10^{-19}$ erg·T^{-1} the nuclear magneton, and $\hat{\sigma}$ a Pauli spin operator. For a large class of magnetic compounds the magnetic field **H** used in equation (6.40) is generated by unpaired electrons. The magnetic field due to a single electron moving with velocity $\mathbf{v}_e$ is given by:

$$\mathbf{H} = \text{curl}\left\{\frac{\pi_e \times \mathbf{R}}{|\mathbf{R}|^3}\right\} - \frac{e}{c}\frac{\mathbf{v}_e \times \mathbf{R}}{|\mathbf{R}|^3}, \qquad (6.41)$$

where **R** is the distance from the electron to the point at which the field is measured, $e = 1.602 \cdot 10^{-19}$ C the elementary charge, and $c = 3 \cdot 10^{10}$ cm·s^{-1} the velocity of light. The magnetic moment operator of an electron is

$$\hat{\mu}_e = -2\mu_B \hat{\mathbf{s}}, \qquad (6.42)$$

where $\mu_B = 0.927 \cdot 10^{-16}$ erg·T^{-1} is the Bohr magneton and $\hat{\mathbf{s}}$ the spin operator of the electron. The first term of equation (6.41) arises from the spin of the electron, and the second from its orbital motion.

The major task in the evaluation of the magnetic neutron cross-section is the calculation of the transition matrix element in equation (6.39). This was first done by Halpern and Johnson [32], and more recently in many excellent text books on neutron scattering [33, 34]. For unpolarized neutrons, identical magnetic ions with localized electrons, and spin-only scattering the following "master formula" is obtained:

$$\frac{d^2\sigma}{d\Omega d\omega} = (\gamma r_o)^2 \frac{k'}{k} F^2(\mathbf{Q}) \exp\{-2W(\mathbf{Q})\} \sum_{\alpha,\beta} \left(\delta_{\alpha\beta} - \frac{Q_\alpha Q_\beta}{Q^2}\right) S^{\alpha\beta}(\mathbf{Q}, \omega), \qquad (6.43a)$$

where $S^{\alpha\beta}(\mathbf{Q}, \omega)$ is the magnetic scattering function:

$$S^{\alpha\beta}(\mathbf{Q}, \omega) = \sum_{i,j} \exp\{i\mathbf{Q} \cdot (\mathbf{R}_i - \mathbf{R}_j)\} \sum_{\lambda,\lambda} p_\lambda \left\langle \lambda | \hat{S}_i^\alpha | \lambda' \right\rangle \left\langle \lambda' | \hat{S}_j^\beta | \lambda \right\rangle \qquad (6.43b)$$

$$\cdot \delta(\hbar\omega + E_\lambda - E_{\lambda'}).$$

$r_0 = 0.282 \cdot 10^{-12}$ cm is the classical electron radius, $F(\mathbf{Q})$ the dimensionless magnetic form factor defined as the Fourier transform of the normalized spin density associated with the magnetic ions, $\exp\{-2W(\mathbf{Q})\}$ the Debye-Waller factor, and $\hat{S}_i^\alpha$ ($\alpha = x,y,z$) the spin operator of the ith ion at site $\mathbf{R}_i$. From the magnitude of r_0 we expect the magnetic neutron cross-section to be of the order 10^{-24} cm^{-2}.

The essential factor in equation (6.43) is the magnetic scattering function $S^{\alpha\beta}(\mathbf{Q},\omega)$ which will be discussed in more detail below. There are two further factors that govern the cross-section for magnetic neutron scattering in a characteristic way: First, the magnetic form factor $F(\mathbf{Q})$ which usually falls off with increasing modulus of the scattering vector $\mathbf{Q}$. Second, the polarization factor $(\delta_{\alpha\beta} - Q_\alpha Q_\beta / Q^2)$ tells us that neutrons can only couple to magnetic moments or spin fluctuations perpendicular to $\mathbf{Q}$, which unambiguously allows us to determine moment directions or to distinguish between different polarizations of spin fluctuations.

Equation (6.43) strictly applies to cases where the orbital angular momentum of the magnetic ions is either zero or quenched by the crystal field. A theoretical treatment of the scattering by ions with unquenched orbital moment was given by Johnston [35], however, the calculation is complicated, and we simply quote the result for $\mathbf{Q} \to 0$. In this case the cross-section measures the magnetization, $\boldsymbol{\mu} = -\mu_B(\mathbf{L} + 2\mathbf{S})$, i.e., a combination of spin and orbital moments that does not allow their separation. This clearly contrasts to magnetic scattering by X-rays. For magnetic neutron scattering an approximate result can be obtained for modest values of $\mathbf{Q}$. We replace the spin operator $\hat{S}_i^\alpha$ in equation (6.43) by:

$$\hat{S}_i^\alpha = \frac{1}{2} g \hat{J}_i^\alpha , \tag{6.44}$$

where

$$g = 1 + \frac{J(J+1) - L(L+1) + S(S+1)}{2J(J+1)} \tag{6.45}$$

is the Landé splitting factor and $\hat{J}_i^\alpha$ is an effective angular momentum operator (e.g., for rare earth ions J is the total angular momentum quantum number resulting from the spin-orbit coupling which combines the spin and orbital angular momentum L and S, respectively).

If ω is a positive quantity in the scattering function $S^{\alpha\beta}(\mathbf{Q},\omega)$ of equation (6.43), the neutron loses energy in the scattering process and the system is excited from the initial state $|\lambda\rangle$ which has energy $\hbar\omega$ less than the final state $|\lambda'\rangle$. Consider now the function $S^{\alpha\beta}(\mathbf{Q},-\omega)$ where ω is the same positive quantity. This represents a process in which the neutron gains energy. The transitions of the system are between the same states as for the previous process, but now $|\lambda'\rangle$ is the initial state and $|\lambda\rangle$ is the final state. The probability of the system being initially in the higher state is lower by the factor $\exp\{-h\omega/k_BT\}$ than its probability of being in the lower energy state, hence

$$S^{\alpha\beta}(\mathbf{Q},-\omega)=\exp\left\{-\frac{\hbar\omega}{k_BT}\right\}S^{\alpha\beta}(\mathbf{Q},\omega), \tag{6.46}$$

which is known as the principle of detailed balance. Equation (6.46) must be applied in the analysis of experimental data taken in both energy-gain and energy-loss configurations which correspond to the so-called Stokes and anti-Stokes processes, respectively.

Using the integral representation of the δ-function,

$$\delta(\hbar\omega+E_\lambda-E_{\lambda'})=\frac{1}{2\pi\hbar}\int_{-\infty}^{+\infty}\exp\left\{i\frac{(E_{\lambda'}-E_\lambda)t}{\hbar}\right\}\exp\{-i\omega t\}dt, \tag{6.47}$$

the scattering function $S^{\alpha\beta}(\mathbf{Q},\omega)$ (equation (6.43)) transforms into a physically transparent form:

$$S^{\alpha\beta}(\mathbf{Q},\omega)=\frac{1}{2\pi\hbar}\sum_{i,j}\int_{-\infty}^{+\infty}\exp\{i\mathbf{Q}\cdot(\mathbf{R}_i-\mathbf{R}_j)\}\left\langle\hat{S}_i^\alpha(0)\hat{S}_j^\beta(t)\right\rangle\exp\{-i\omega t\}dt. \tag{6.48}$$

$\left\langle S_i^\alpha(0)S_j^\beta(t)\right\rangle$ is the thermal average of the time-dependent spin operators. It corresponds to the van Hove pair correlation function [36] and gives essentially the probability that, if the magnetic moment of the ith ion at site $\mathbf{R}_i$ has some specified (vector) value at time zero, then the moment of the jth ion at site $\mathbf{R}_j$ has some other specified value at time t. A neutron scattering experiment measures the Fourier transform of the pair correlation function in space and time, which is clearly just what is needed to describe a magnetic system on an atomic scale.

The van Hove representation of the cross-section in terms of pair correlation functions is related to the *fluctuation-dissipation theorem* [33, 34]:

$$S^{\alpha\beta}(\mathbf{Q},\omega)=\frac{N\hbar}{\pi}\left(1-\exp\left\{-\frac{\hbar\omega}{k_BT}\right\}\right)^{-1}\operatorname{Im}\chi^{\alpha\beta}(\mathbf{Q},\omega), \tag{6.49}$$

where N is the total number of magnetic ions. Physically speaking, the neutron may be considered a magnetic probe, which effectively establishes a frequency- and wavevector-dependent magnetic field $H^\beta(\mathbf{Q},\omega)$ in the scattering sample, and detects its response, $M^\alpha(\mathbf{Q},\omega)$, to this field by

$$M^\alpha(\mathbf{Q},\omega) = \chi^{\alpha\beta}(\mathbf{Q},\omega)H^\beta(\mathbf{Q},\omega), \tag{6.50}$$

where $\chi^{\alpha\beta}(\mathbf{Q},\omega)$ is the generalized magnetic susceptibility tensor. This is really the outstanding property of the neutron in a magnetic scattering measurement, and no other experimental technique is able to provide such detailed microscopic information about magnetic compounds.

If the coupling between the magnetic ions is weak, we are left with a single-ion problem, thus the excitation energies will be independent of the scattering vector $\mathbf{Q}$. Typical examples are rare earth compounds that exhibit very low magnetic-ordering temperatures or do not order at all. In this case the dominant mechanism is the crystal field interaction.

The effect of the crystal field on a rare earth ion is to partially or totally remove the (2J + 1)-fold degeneracy of the ground state J-multiplet. This is exemplified at the top of Figure 6.5 for Pr^{3+} ions in the hexagonal compound $PrBr_3$ [37]. The crystal field levels resulting from the Hamiltonian (11) are denoted by the irreducible representations Γ_n, and the corresponding wave-functions are:

$$|\Gamma_n\rangle = \sum_{M=-j}^{J} a_n(M)\,|M\rangle. \tag{6.51}$$

From the sequence of the energy levels, properly identified by their irreducible representations Γ_n, the crystal field potential can be unambiguously determined.

In evaluating the cross-section for the crystal field transition $\Gamma_n \rightarrow \Gamma_m$ we start from the scattering law $S^{\alpha\beta}(\mathbf{Q},\omega)$ defined by equation (6.43). Since we are dealing with single-ion excitations, we have $i=j$, so that the **Q**-dependence vanishes. For N identical magnetic ions we can even drop the index i. $S^{\alpha\beta}(\mathbf{Q},\omega)$ then reduces to:

$$S^{\alpha\beta}(\omega) = Np_{\Gamma_n}\,\langle\Gamma_n|\,\hat{J}^{\alpha}|\Gamma_m\rangle\,\langle\Gamma_m|\,\hat{J}^{\beta}|\Gamma_n\rangle\,\delta\!\left(\hbar\omega + E_{\Gamma_n} - E_{\Gamma_m}\right), \tag{6.52}$$

where N is the total number of magnetic ions and p_{Γ_n} the Boltzmann population factor. From the symmetry relations associated with the matrix elements we find the cross-section

$$\frac{d^2\sigma}{d\Omega d\omega} = N\left(\frac{1}{2}g\gamma r_o\right)^2 \frac{k'}{k} F^2(\mathbf{Q})\exp\{-2W(\mathbf{Q})\}\,p_{\Gamma_n} \sum_{\alpha}\left(1-\frac{\mathbf{Q}_{\alpha}^{\;2}}{\mathbf{Q}^2}\right)|\langle\Gamma_m|\,\hat{\mathbf{J}}^{\alpha}\,|\Gamma_n\rangle|^2\,\delta\!\left(\hbar\omega + E_{\Gamma_n} - E_{\Gamma_m}\right). \tag{6.53}$$

The polarization factor permits discrimination between transverse ($\alpha = x,y$) and longitudinal ($\alpha = z$) crystal field transitions by measuring at different **Q**. This is nicely demonstrated in Figure 6.5. For $\mathbf{Q}\|c$ only transverse transitions are observed, whereas for $\mathbf{Q}\perp c$ the transverse transitions lose half their intensities, and in addition longitudinal transitions appear. The lines in Figure 6.5 were calculated without any disposable parameters, i.e., the intensities of the crystal field transitions are excellently described by equation (6.53).

For experiments on polycrystalline material, equation (6.53) has to be averaged in **Q** space:

$$\frac{d^2\omega}{d\Omega d\omega} = N\left(\frac{1}{2}g\gamma r_o\right)^2 \frac{k'}{k} F^2(\mathbf{Q})\,\exp\{-2W(\mathbf{Q})\}\,p_{\Gamma_n}\,|\langle\Gamma_m|\,\hat{\mathbf{J}}_{\perp}\,|\Gamma_n\rangle|^2 \times \delta\!\left(\hbar\omega + E_{\Gamma_n} - E_{\Gamma_m}\right), \tag{6.54}$$

where $\mathbf{J}_\perp = \mathbf{J}-(\mathbf{J}\cdot\mathbf{Q})\mathbf{Q}/Q^2$ is the component of the total angular momentum perpendicular to the scattering vector **Q**, and

$$\left\langle \mid \Gamma_m \mid \hat{\mathbf{J}}_\perp \mid \Gamma_n \right\rangle \mid^2 = \frac{2}{3} \sum_\alpha \mid \left\langle \Gamma_m \mid \hat{\mathbf{J}}^\alpha \mid \Gamma_n \right\rangle \mid^2 . \tag{6.55}$$

These matrix elements have been tabulated by Birgeneau [38] for cubic crystal fields.

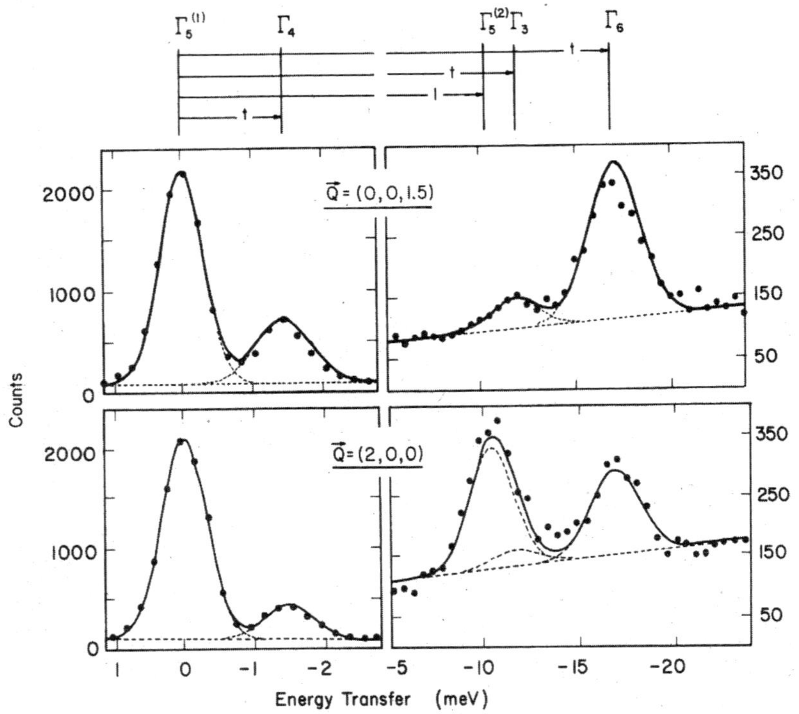

Figure 6.5 Energy spectra of neutrons scattered from single-crystalline $PrBr_3$ at T=1.5 K for **Q** parallel and perpendicular to the *c*-axis. The resulting crystal field level scheme and the observed transverse (t) and longitudinal (l) ground state transitions are indicated at the top (after [37]).

6.3.3 Raman Spectroscopy

Raman spectroscopy utilizes light scattering to gain information about elementary excitations of matter such as phonons, magnons, and electronic single particle or collective excitations within a sample (for a review and references see, e.g., [39]). When a photon (usually produced from a laser in the visible, near infrared, or near ultraviolet range) interacts with a solid it can be: i) elastically scattered and thus retain its incident energy (Rayleigh scattering); ii) inelastically scattered by quasi-particle excitations (Raman process), thereby either giving energy to the medium (Stokes scattering) or removing energy from it (anti-Stokes scattering). A Raman process arises when the photon causes an induced transition from state $|b\rangle$ to an excited state $|c\rangle$ whose energy is higher than the Debye energy $k_B T_D$. This excitation

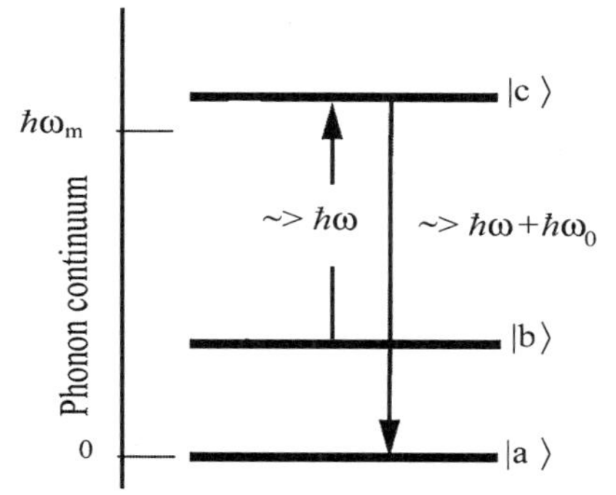

Figure 6.6 A schematic diagram for the anti-Stokes Raman process.

is followed by a second virtual transition $|c\rangle \rightarrow |a\rangle$, as indicated in Figure 6.6. This gives an indirect two-stage transfer of an ion from state $|b\rangle$ to state $|a\rangle$, which could become an important relaxation process if the direct process between the states $|a\rangle$ and $|b\rangle$ is negligible.

Raman processes are governed by two conservation laws as in neutron spectroscopy (see equations (6.37) and (6.38)): energy conservation and momentum conservation, i.e., the energy and momentum of the scattered radiation and the elementary excitation created (absorbed) in the interaction process equals the energy and momentum of the exciting photon. As the photon wavevector is about 10^3 times smaller than a typical reciprocal lattice vector, only excitations at the center of the Brillouin zone are observed by Raman scattering.

The Raman spectroscopy obeys characteristic selection rules, which reflect the symmetry properties of the elementary excitations. Crystal field excitations are forbidden if the rare earth ions occupy a center of inversion. This explains the difficulties of detecting crystal field excitations in hole-doped, high-temperature superconductors of type $RBa_2Cu_3O_{7-x}$. On the other hand, this drawback does not exist for electron-doped, high-temperature superconductors as reported, e.g., for Raman spectroscopy studies carried out for the compounds R_2CuO_4 (R = Pr, Nd, Sm) [28, 29] and $R_{2-x}Ce_xCuO_4$ [40]. Other interesting examples are the elpasolites Cs_2NaRX_6 (R = rare earth; X = Cl, Br), which will be highlighted in Section 6.4.1. For $Cs_2NaTbCl_6$ a most comprehensive data set was collected with a total of 125 levels, which include states originating from 10 of the 17 terms and 36 of the 87 multiplets [41]. Figure 6.7 displays transitions in the $^5D_4$-multiplet obtained by luminescence spectroscopy, emphasizing the ability to resolve Zeeman splittings as well as the effect of linear polarizations.

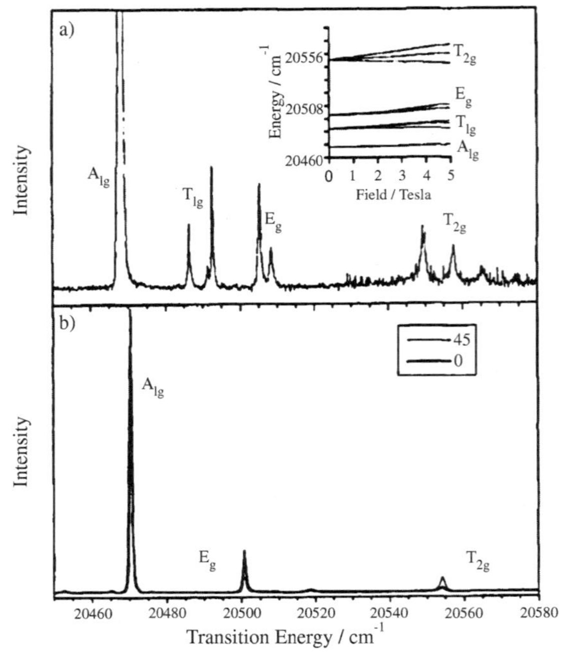

Figure 6.7 Two-photon excitation spectra of a single crystal of $Cs_2NaTbCl_6$, at ~ 6 K (a) in the presence of a 4.5 T field along [110] and (b) in zero-field but in two linear polarizations making the angle indicated to [001]. The inset shows the experimental field dependence of the Zeeman energies (after [41]).

The calculation of intensities of the observed Raman-active crystal field transitions is a nontrivial task. This is in contrast to neutron spectroscopy where the intensities of crystal field transitions are directly proportional to the square of the magnetic dipole matrix elements (see equations (6.53) and (6.54)). Since Raman scattering is always at least a two-step process, it involves intermediate states that are often not known, thus approximate models have to be employed for the calculation of transition matrix elements [42]. On the other hand, as a laser is usually used for Raman experiments, the specific polarization of the exciting source can be exploited to obtain extra information from the spectra as exemplified in [28].

6.3.4 Point-Contact Spectroscopy

Electrical *point-contact spectroscopy* (PCS) is a technique to study the interaction mechanism of electrons with all kinds of elementary excitations in metals. Small constrictions between two metals show deviations from the linearity governed by Ohm's law. The nonlinearity is a measure of the inelastic scattering of the conduction electrons, where the applied voltage defines the energy scale for the interaction process. Yanson [43] discovered these nonlinearities in the *I–V* characteristics of microcontacts between metals and used them for a direct microscopic measurement of the energy dependence of the interaction between conduction electrons and quasiparticle excitations such as phonons, magnons, paramagnetic impurities, crystal field excitations, and so on. For the ballistic electron transport (i.e., both the inelastic and the elastic mean free paths are larger than the contact size) the second derivative d^2V/dI^2 of the voltage with respect to the current turns out to be proportional to the spectral function of the interaction process. For instance, in the case of the electron-phonon interaction, the spectral function is the well known Eliashberg function α^2F [44]. In case of crystal field interactions the point-contact spectra show additional maxima associated with the allowed crystal field transitions.

As an example we quote the splitting of the Pr^{3+} ions which were directly observed in the hexagonal van Vleck paramagnet $PrNi_5$ [45]. Figure 6.8 displays the corresponding PCS spectra, which exhibit several maxima and kinks marked by arrows, however, only the peaks at 4.6, 13.6, and 29.4 meV can be associated with crystal field transitions (see [46] and Section 6.5.1), whereas the peaks at 9.2 and 18.5 meV are due to electron-phonon interaction. The PCS was subsequently used to measure the Zeeman splitting of the crystal field excitations in external magnetic fields up to 20 T for the same compound [47].

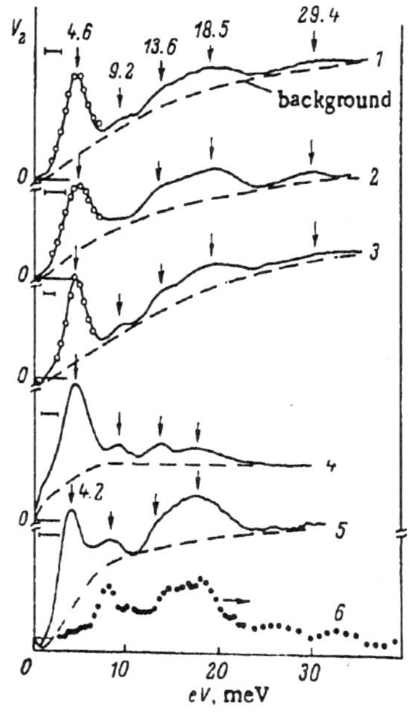

Figure 6.8 Microcontact $PrNi_5$ spectra (after [45]).

Nowadays the PCS has become a standard technique for the study of the electron-phonon interaction in conductors (for a review see [48]), but at present it seems to be limited to simple metals and intermetallic compounds.

The limitation of the PCS as a spectroscopic tool is due to the fact that the contact diameter has to be smaller than the inelastic electron mean free path, which is hardly fulfilled in most of the strongly correlated electron systems, so that the data analysis becomes ambiguous and complicated. Moreover, the PCS technique cannot discriminate a priori between crystal field and electron-phonon interactions, i.e., a proper identification of the observed PCS spectra strongly relies on the availability of other spectroscopic data.

6.4 DETERMINATION OF CRYSTAL FIELD PARAMETERS FROM EXPERIMENTAL DATA

6.4.1 A Simple Two-Parameter Crystal Field Problem

In this section we present a simple crystal field problem in order to exemplify the way in which the crystal field parameters can be determined from spectroscopic data. We choose the insulating compound series Cs_2NaRBr_6 (R = rare earth) called (bromo-)elpasolites, which were investigated by several experimental techniques, so that we can compare the results and thereby obtain information about the performance of the different experimental methods.

The crystal field is the dominant interaction for the rare earth elpasolites, since no magnetic ordering is observed down to the lowest temperatures, thus exchange effects can be neglected. Moreover, the elpasolites have a cubic crystal structure with the R^{3+} ions situated at the sites of octahedral symmetry, i.e., each R^{3+} ion is hexacoordinated by six nearest Br^- ions. Thus the crystal field problem is rather simple with only two independent crystal field parameters (see equation (6.9)).

Figures 6.9 and 6.10 show the results of neutron spectroscopic investigations for $Cs_2NaErBr_6$ ($J_{Er} = {}^{15}/_2$) [25]. The cubic crystal field splits the 16-fold degenerate ground state ${}^4I_{15/2}$-multiplet of the Er^{3+} ion into two doublets Γ_6 and Γ_7 and three quartets $\Gamma_8^{(1)}$, $\Gamma_8^{(2)}$, and $\Gamma_8^{(3)}$. The low-temperature spectra of Figure 6.9 give evidence for two inelastic lines A and B at 3.0 and 5.8 meV, respectively, that can be attributed to crystal field transitions out of the ground state. Upon raising the temperature the intensity of line B decreases as expected (because of the decreasing Boltzmann population factor), whereas the intensity of line A increases slightly because of an excited crystal field transition that occurs accidentally at the same energy. For Er^{3+} the higher lying crystal field states usually have very small matrix elements for transitions out of the ground state, thus their energies can be determined only in connection with excited-state transitions. This is demonstrated in Figure 6.10 by lines C and D at 21 and 25 meV, respectively, whose intensities clearly increase with increasing temperature.

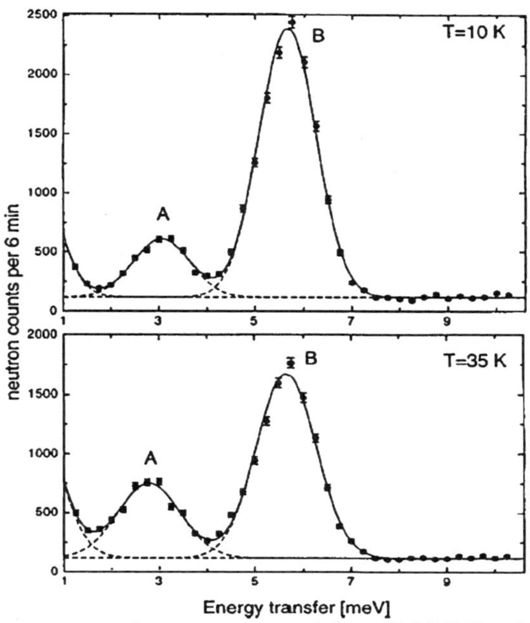

Figure 6.9 Low-energy spectra of neutrons scattered from $Cs_2NaErBr_6$ observed with use of a triple-axis spectrometer. The analyzer energy and the modulus of the scattering vector were set to E_A=13.7 meV and Q =1.25 Å^{-1}, respectively (after [25]).

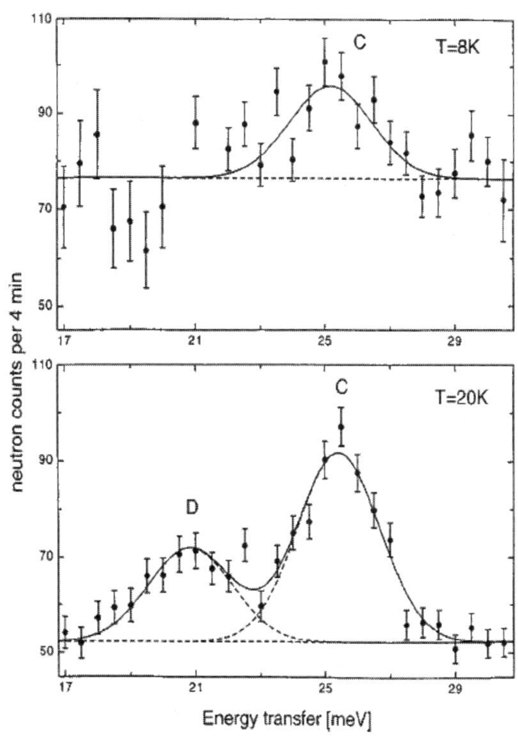

Figure 6.10 Medium-energy spectra of neutrons scattered from $Cs_2NaErBr_6$ with E_A=13.7 meV and Q = 2.4 Å^{-1} (after [25]).

These experiments tell us immediately that the crystal field splitting comprises three low-lying states in the energy range $0 < \hbar\omega < 6$ meV, separated from two high-lying states at $\hbar\omega > 20$ meV. In order to identify the crystal field states by their correct representations, an inspection of the energy level sequence in terms of the parameters W and x (see equation (6.10)) is most useful. Figure 6.11 offers at least two sets of parameters that are compatible with the spectroscopic data, namely $x \approx -0.6$ with a Γ_6 ground state and $x \approx 0.65$ with a $\Gamma_8^{(1)}$ ground state, both with $W > 0$. This means that a unique identification of the crystal field states is not possible by consideration of the energy splittings alone. In addition, we have to consider the intensities of the observed crystal field transitions (as tabulated in [38]), which in the present case unambiguously favor the parameter set $\{x \approx 0.65,\ W > 0\}$. By doing this, we can completely reconstruct the crystal field level scheme of $Cs_2NaErBr_6$ from the spectroscopic data (see Figure 6.12). The resulting crystal field parameters, mainly determined from neutron (N) and optical (O) spectro-scopic measurements, which are presently available for the rare earth bromo-elpasolites, are listed in Table 6.1. The agreement between N and O is rather good for R = Ho, R = Tm, and R = Yb, with a maximum deviation of 0.5 meV for all crystal field levels (which can readily be calculated from equation (6.9)). For R = Pr there is a discrepancy for the highest crystal field level Γ_5, which is located at 55 meV in the neutron-scattering experiments, whereas an energy of 77 meV was derived from the luminescence spectrum. The discrepancies for R = Tb can be explained by the fact that neutron spectroscopy was able to establish only the low-energy part of the crystal field level structure.

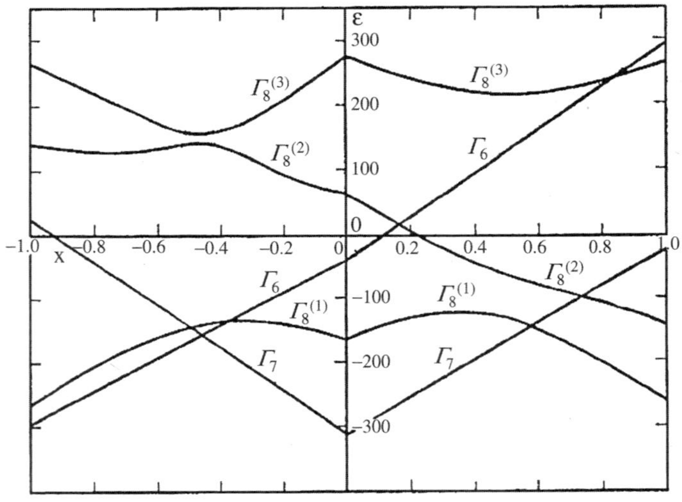

Figure 6.11 Eigenvalues of Eq. (10) (after [12]) for Er^{3+}.

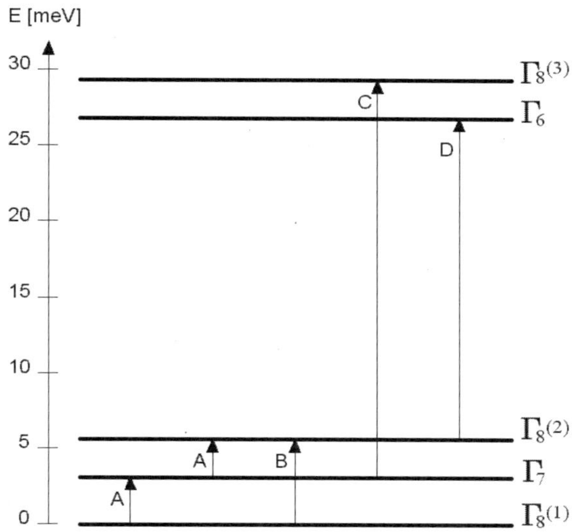

Figure 6.12 Crystal field level scheme of $Cs_2NaErBr_6$. The arrows and corresponding labels denote the transitions observed by neutron spectroscopy.

We may conclude that both N and O spectroscopies have their merits for the determination of crystal field splittings in the ground state J-multiplet of rare earth compounds; however, the correct identification of the observed crystal field states is more reliable in neutron spectroscopic experiments, since no approximations have to be applied to calculate the transition matrix elements. On the other hand, optical spectroscopies have distinct advantages for measurements of intermultiplet crystal field excitations. Finally, the data for R = Yb show that a reliable determination of crystal field parameters from thermodynamic magnetic data is usually not possible.

The crystal field parameters listed in Table 6.1 are found to follow nicely the extrapolation scheme of equation (6.32). However, there is a slight increase of the parameters from the light to the heavy rare earth ions. This was explained in terms of dipolar polarizability and covalency [49], which seem to have an increased relative importance for the heavy rare earth compounds.

6.4.2 A Complicated Many-Parameter Crystal Field Problem

In this section we exemplify the crystal field procedure for a metallic compound with low symmetry at the rare earth site, for which the use of neutron spectroscopy is indispensable. This is particularly true for the rare earth-based high-temperature superconductors for which early attempts to unravel and to parametrize the crystal field interaction by other techniques failed completely. We choose as an example the compound $ErBa_2Cu_3O_7$

Table 6.1 Crystal field parameters B_4^0 and B_6^0 of the elpasolites Cs_2NaRBr_6 determined by different experimental techniques (N: neutron spectroscopy; O: optical spectroscopies; M: magnetic susceptibility).

R	B_4^0 (meV)	B_6^0 (meV)	Technique	Ref.
Ce	0.162 ± 0.002	0	N	A
Pr	$(-1.32 \pm 0.06)\cdot 10^{-2}$	$(1.39 \pm 0.13)\cdot 10^{-4}$	N	A
	$(-1.93 \pm 0.10)\cdot 10^{-3}$	$(1.20 \pm 0.29)\cdot 10^{-4}$	O	B
Nd	$(-7.6 \pm 0.2)\cdot 10^{-3}$	$(-7.2 \pm 0.3)\cdot 10^{-5}$	N	A
Tb	$(2.1 \pm 0.2)\cdot 10^{-3}$	$(-1.5 \pm 0.4)\cdot 10^{-6}$	N	A
	$(2.96 \pm 0.05)\cdot 10^{-3}$	$(-1.35 \pm 0.17)\cdot 10^{-6}$	O	C
Dy	$(-1.3 \pm 0.1)\cdot 10^{-3}$	$(1.5 \pm 0.3)\cdot 10^{-6}$	N	A
Ho	$(-6.2 \pm 0.5)\cdot 10^{-4}$	$(-1.4 \pm 0.2)\cdot 10^{-6}$	N	A
	$(-6.56 \pm 0.08)\cdot 10^{-4}$	$(-1.58 \pm 0.10)\cdot 10^{-6}$	O	B
Er	$(8.1 \pm 0.3)\cdot 10^{-4}$	$(1.8 \pm 0.1)\cdot 10^{-6}$	N	A
Tm	$(3.0 \pm 0.4)\cdot 10^{-3}$	$(-4.2 \pm 0.8)\cdot 10^{-6}$	N	A
	$(3.32 \pm 0.18)\cdot 10^{-3}$	$(-4.06 \pm 2.18)\cdot 10^{-6}$	O	B
Yb	$(-3.1 \pm 0.1)\cdot 10^{-2}$	$(1.1 \pm 0.3)\cdot 10^{-4}$	N	A
	$(-3.53 \pm 0.31)\cdot 10^{-2}$	0	O	B
	$-2.13\cdot 10^{-2}$	$1.38\cdot 10^{-4}$	M	D

A: Furrer A and Güdel H U 1997 *Phys. Rev. B* 56 15062–72
B: Tanner P A, Ravi Kanth Kumar V V, Jayasankar C K, and Reid M F 1994 *J. Alloys Compd.* 215 349–70
C: Berry A J, McCaw C S, Morrison I D, and Denning R G 1996 *J. Lumin.* 66 & 67 272–7
D: Urland W 1979 *Z. Naturforsch.* 34a 1507–11

($J_{Er} = {}^{15}/_2$) which has orthorhombic symmetry at the Er^{3+} site, i.e., the crystal field Hamiltonian is defined by equation (6.12) with nine independent parameters. The crystal field splits the 16-fold degenerate ground state $^4I_{15/2}$-multiplet of Er^{3+} into eight Kramers doublets. All seven crystal field transitions out of the ground state have been observed by neutron spectroscopy as shown in Figure 6.13 [50, 51], thus the crystal field level splitting is completely established. Yet the crystal field states have to be characterized by their proper representations, which is a highly nontrivial task in view of the large number of adjustable crystal field parameters. This is due to the difficult problem of any nonlinear least squares fitting procedure of choosing reasonable start values of the fitting parameters, which, however, can be achieved as described below.

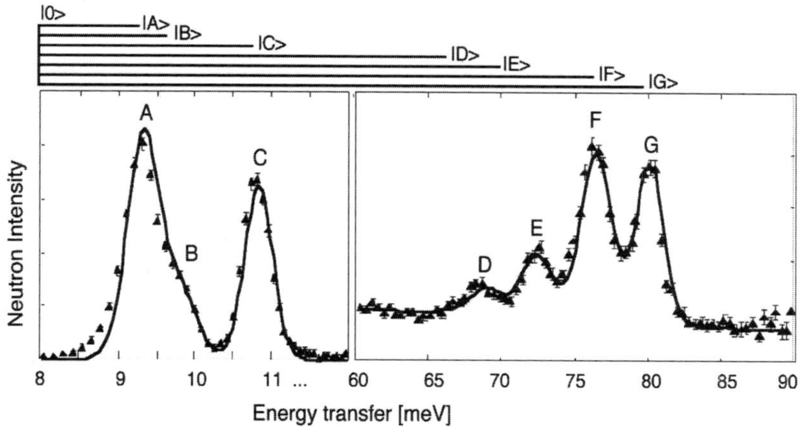

Figure 6.13 Energy spectra of neutrons scattered from $ErBa_2Cu_3O_7$ at T=10 K. The lines represent the calculated spectra. The arrows on top of the figure denote the observed ground state transitions (after [50]).

The coordination polyhedra of Er^{3+} in $ErBa_2Cu_3O_7$ formed by the nearest neighbors (eight oxygen ions) as well as by the next-nearest neighbors (eight copper ions) are slightly distorted cubes. Thus we expect the overall features of the crystal field potential to be reasonably described in a first approximation by cubic point symmetry, which limits the number of disposable parameters to two as for $Cs_2NaErBr_6$ discussed in the preceding section. Now we make use of the fact that the crystal field transitions occur in a low-energy window ($\hbar\omega < 12$ meV, lines A–C) and in a high-energy window ($65 < \hbar\omega < 82$ meV, lines D–G), i.e., the crystal field splitting is characterized by four low-lying doublets and four doublets at much higher energies. Inspection of Figure 6.11 tells us that this particular situation can only be realized for the parameter $x \approx -0.45$ with either $W > 0$ or $W < 0$ (note that in cubic symmetry six doublets merge into three quartets). When considering the intensities, the parameter set with $W < 0$ has to be discarded, thus the set $\{x \approx -0.45,\ W > 0\}$ gives us a good estimate of the dominant crystal field parameters B_4^0 and B_6^0. In a second step we make use of equation (6.30) in order to arrive at reasonable values for the remaining "orthorhombic" crystal field parameters. After having determined reasonable start values of the crystal field parameters as outlined above, a least squares fitting procedure is then applied to the observed energy spectra, which usually converges successfully to a final, appropriate set of crystal field parameters (see [51]).

Another useful scheme to predict crystal field parameters in metallic systems is based on the extended point-charge model (see Section 6.2.2.2). As shown by Mesot and Furrer [52], the crystal field parameters of $ErBa_2Cu_3O_7$ calculated from equation (6.24), with the inverse screening length adjusted to $\kappa = 1.25$ Å^{-1}, provide values very close to the fitted ones.

In order to check the reliability of the derived crystal field parameters it is important to calculate various thermodynamic properties with the help of equations (6.33)–(6.34) and to compare the results with experimental data. Figures 6.14 and 6.15 [51] show the good agreement between high-field magnetization [53] and specific heat [54, 55] measurements of powder samples and calculations using the crystal field parameters determined by neutron spectroscopy [51]. Moreover, the crystal field interaction reduces the zero-field moment of Er^{3+} in $ErBa_2Cu_3O_7$ drastically below the free ion value of 9 μ_B. A mean-field calculation based on equation (6.33) yields 4.2 μ_B, in agreement with the value derived from Mössbauer experiments [56].

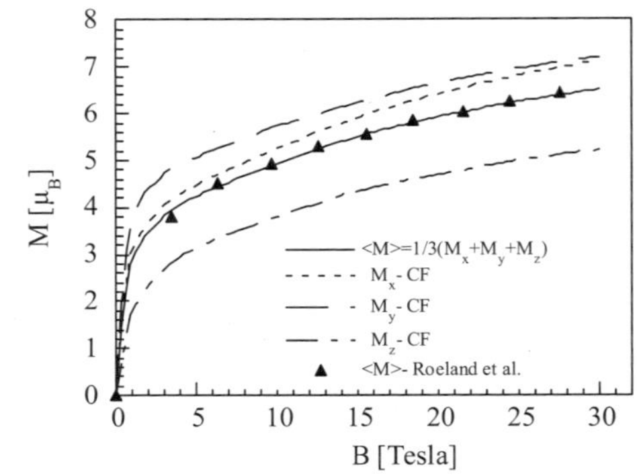

Figure 6.14 High-field magnetization observed and calculated for $ErBa_2Cu_3O_7$ (after [51]).

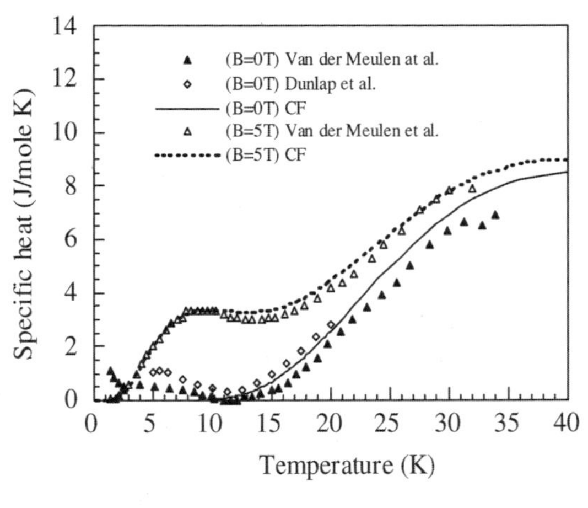

Figure 6.15 Schottky anomaly of the heat capacity observed and calculated for $ErBa_2Cu_3O_7$ (after [51]).

6.5 INTERACTIONS OF CRYSTAL FIELD SPLIT IONS

6.5.1. Introductory Remarks

In the preceding sections, we considered crystal field effects for non-interacting rare earth ions. However, the crystal field states are usually subject to interactions with phonons, spin fluctuations, conduction electrons (or more generally charge carriers), etc., which limit the lifetime of the crystal field states, thus the observed crystal field transitions exhibit line broadening. This has major consequences for the cross-section formulae, which have been derived to interpret spectroscopic experiments. More specifically, in the neutron cross-section formulae (6.53) and (6.54) the δ-function describing the energy conservation has to be replaced by a Lorentzian with a finite linewidth whose temperature dependence is characteristic of the particular type of interaction. Moreover, if the rare earth ions are coupled by the exchange interaction, the crystal field excitations can propagate through the lattice. All these effects will be discussed in the following subsections.

6.5.2 Interaction with Phonons

Theoretical models of the effects of lattice vibrations in distorting the crystal field have been suggested since the early 1930s by Heitler and Teller [57], Kronig [58], and van Vleck [59]. A more detailed and general approach was developed later by Orbach [60]. His main idea was to treat the spin-lattice interaction as a dynamic crystal field effect, which produces deformations of the equilibrium charge configuration of the surrounding ions. Orbach proposed to expand the crystal field potential V in powers of the fluctuating strain ε caused by the lattice vibrations [60]:

$$V = V^{(0)} + \varepsilon V^{(1)} + \varepsilon^2 V^{(2)} + \cdots, \tag{6.56}$$

where $V^{(0)}$ is the static term, $V^{(1)}$ and $V^{(2)}$ represent the dynamical part of the magnetic ion-lattice interaction generated by the strain in first- and second-order, respectively. In these calculations the spin-orbit coupling has to be included. The terms $V^{(n)}$ in equation (6.56) turned out to be of the same order as the static $V^{(0)}$. These calculations made it possible to evaluate the spin-lattice relaxation time in terms of a few constants. However, the additional distinction of whether the rare earth ion has an odd (Kramers ion) or an even number of electrons must be made.

Orbach considered in detail one-phonon (or direct) and two-phonon processes. Suppose that the ground state J-multiplet of the rare earth ion is split into two low-lying states $|a\rangle$ and $|b\rangle$, energetically separated from the excited state $|c\rangle$. In a direct process a single transition Δ is made from $|b\rangle$ to $|a\rangle$ and a phonon of energy $\hbar\omega_0$ is emitted (see Figure 6.16a). This process will be dominant only at the lowest temperatures, for most rare earth compounds typically around 1 K. The spin-lattice relaxation time τ_1 for the direct process in an external field H exhibits the typical behavior $\tau_1 \sim T^{-1}H^{-4}$ and $\tau_1 \sim T^{-1}H^{-2}$ for the Kramers and non-Kramers ions, respectively.

In the two-phonon process a phonon of energy $\hbar\omega$ is absorbed by the spin system and an intermediate transition from $|b\rangle$ to $|c\rangle$ occurs. To complete the process a transition from $|c\rangle$ to $|a\rangle$ occurs and a phonon of energy $\hbar\omega + \hbar\omega_0$ is then emitted as shown in Figure 6.16b. In this case the spin-lattice relaxation time τ_1 turns out to be $\tau_1 \sim \exp(\Delta_1/k_B T)$.

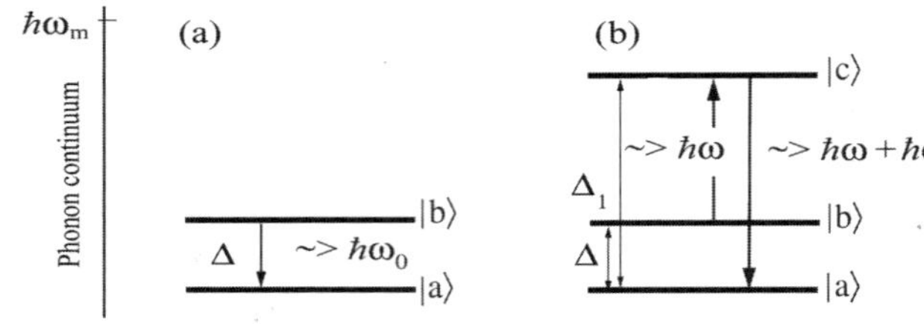

Figure 6.16 Schematic diagram for the direct (a) and the two-phonon (b) process. Only the anti-Stokes process is shown. $\hbar\omega_m$ is the maximum allowed phonon energy in the phonon continuum.

The basic difference between the Raman (see Section 6.3.3) and the Orbach processes is that in the former process "virtual" transitions are involved, i.e., the crystal field state $|c\rangle$ has an energy $\Delta_1 > k_B T_D$ that lies outside the continuum of allowed phonon frequencies (compare Figures 6.6 and 6.16b. For this case the relaxation times turn out to be $\tau_1{\sim}T^{-9}$ and $\tau_1{\sim}T^{-7}$ for the Kramers and non-Kramers ions, respectively. The relaxation time of the Raman process is independent of a magnetic field. Note that the Raman process dominates the Orbach process at high temperatures. For further discussion and bibliography the reader is referred to some classical textbooks (see, e.g., [1]).

6.5.3 Interaction with Conduction Electrons

The interaction with the charge carriers is by far the dominating relaxation mechanism in metallic rare earth compounds. The corresponding linewidth $\Gamma_n(T)$ increases almost linearly with temperature according to the well-known Korringa law [61]:

$$\Gamma_n(T) = 4\pi(g-1)^2 J(J+1)\left[N(E_F)\cdot j_{ex}\right]^2 \cdot T\ , \qquad (6.57)$$

where g denotes the Landé factor, $N(E_F)$ the density-of-states of the charge carriers at the Fermi energy E_F, and j_{ex} the exchange integral between the charge carriers and the 4f electrons of the rare earth ions. The inclusion of crystal field effects slightly modifies the low-temperature limit of equation (6.57) according to the theory of Becker, Fulde, and Keller [62]:

$$\Gamma_n^{ij}(T) = 2j_{ex}^2 \left[M_{ij}^2 \coth\left(\frac{\hbar\omega_{ij}}{2k_B T}\right)\chi''(\hbar\omega_{ij}) + \sum_{i \neq k} M_{ik}^2 \frac{\chi''(\hbar\omega_{ik})}{\left(\exp\left\{\frac{\hbar\omega_{ik}}{k_B T}\right\}-1\right)} + \sum_{k \neq j} M_{kj}^2 \frac{\chi''(\hbar\omega_{kj})}{\left(\exp\left\{\frac{\hbar\omega_{kj}}{k_B T}\right\}-1\right)} \right], \tag{6.58}$$

where $\hbar\omega_{ij} = E_i - E_j$, and $\chi''(\hbar\omega)$ is the imaginary part of the susceptibility summed over the Brillouin zone. For a noninteracting Fermi liquid we have

$$\chi''(\hbar\omega_{ij}) = \pi N^2(E_F)\hbar\omega_{ij}. \tag{6.59}$$

The situation is anomalous for superconducting compounds because the Cooper pairing of the charge carriers creates an energy gap Δ below the superconducting transition temperature T_c, thus crystal field excitations with energy $\hbar\omega < 2\Delta$ do not have enough energy to span the gap, and consequently there is no interaction with the charge carriers. For an isotropic gap, the intrinsic linewidth in the superconducting state is then given by

$$\Gamma_s(T) = \Gamma_n(T) \cdot \exp\left\{-\frac{\Delta}{k_B T}\right\}. \tag{6.60}$$

This means that $\Gamma_s(T \ll T_c) \approx 0$, and line broadening sets in just below T_c where the superconducting gap opens. The exponential temperature dependence of $\Gamma_s(T)$ was nicely demonstrated in the first neutron spectroscopic study on the classical superconductor $La_{1-x}Tb_xAl_2$ [63] (see Figure 6.17).

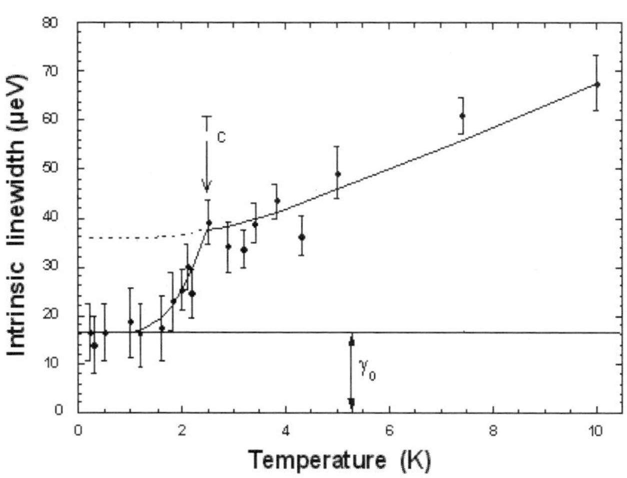

Figure 6.17 Observed linewidth of the crystal field transition at 0.68 meV in $La_{0.997}Tb_{0.003}Al_2$ vs. temperature. The superconducting transition temperature T_c is indicated by an arrow. The solid curve results from a fit to eqs. (58) and (60) (after [63]) for $T > T_c$ and $T < T_c$, respectively. The calculations assume a constant additive contribution γ_0 due to local structural distortions.

6.5.4. Magnetic Exchange Interaction

In extended, undiluted systems the exchange coupling between the rare earth ions is often of the same order of magnitude as the crystal field interaction, thus it has to be considered in the total spin Hamiltonian: $\mathrm{H} = \mathrm{H}_{\mathrm{CF}} + \mathrm{H}_{\mathrm{EX}}$. The most widely used approximation for the exchange interaction is the Heisenberg model:

$$\mathrm{H}_{\mathrm{EX}} = -\sum_{ij} \mathrm{J}_{ij} \; \mathbf{J}_i \cdot \mathbf{J}_j \, . \tag{6.61}$$

The random phase approximation is then used to express the generalized dynamical susceptibility tensor in the following form [64]:

$$\chi^{\alpha\beta}(\mathbf{q},\omega) = \frac{\chi_0^{\alpha\beta}(\omega)}{1 - 2\mathrm{J}(\mathbf{q})\chi_0^{\alpha\beta}(\omega)} \, , \tag{6.62}$$

where $\mathrm{J}(\mathbf{q})$ is the Fourier transform of the exchange couplings J_{ij}, $\mathbf{q}$ the wavevector of the propagating magnetic excitation, and $\chi_0^{\alpha\beta}(\omega)$ the single-ion susceptibility:

$$\chi_0^{\alpha\beta}(\omega) = (g\mu_B)^2 \sum_{n,m} \frac{\langle \Gamma_n | \mathrm{J}_\alpha | \Gamma_m \rangle \langle \Gamma_m | \mathrm{J}_\beta | \Gamma_n \rangle}{E_n - E_m + \hbar\omega} \left(p_{\Gamma_n} - p_{\Gamma_m} \right) \, . \tag{6.63}$$

Here E_n and Γ_n denote the energy and the wavefunction of the crystal field states, respectively, and p_{Γ_n} the Boltzmann population factor. The magnetic excitation energies are determined by the poles of equation (6.62) which yields the dispersion relation

$$\hbar\omega(\mathbf{q}) = \sqrt{\Delta^2 - 2\Delta \mathrm{M}^2 \mathrm{J}(\mathbf{q}) \left[p_{\Gamma_n} - p_{\Gamma_m} \right]} \, , \tag{6.64}$$

where $\Delta = E_m - E_n$ and M is the transition matrix element between the two crystal field states.

Figure 6.18 shows neutron spectroscopy data observed for the $\Gamma_4 \rightarrow \Gamma_5^{(1)}$ crystal field transition in $PrNi_5$ [46] which gave rise to a well-defined peak in the PCS spectra of Figure 6.8. The full dispersion measured for different wavevectors $\mathbf{q}$ allowed to determine the exchange coupling parameters parallel and perpendicular to the hexagonal *c*-axis. Figure 6.18 nicely shows the flattening of the dispersion curve with increasing temperature which is due to the increasing thermal population of the excited state (see equation (6.64)).

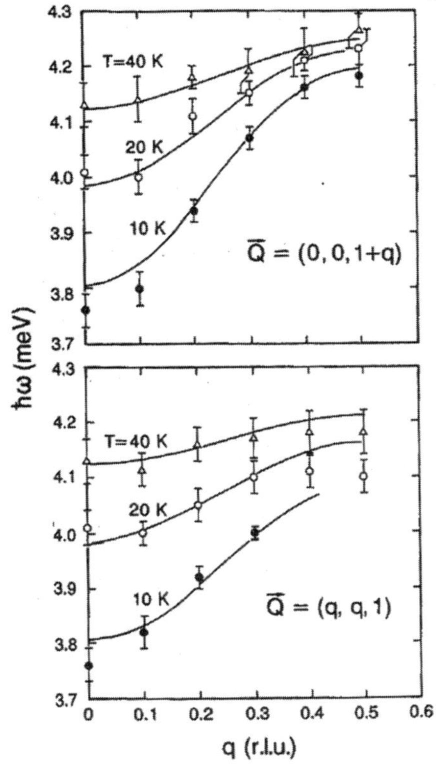

Figure 6.18 Dispersion of the $\Gamma_4 \rightarrow \Gamma_5^{(1)}$ crystal field transition in $PrNi_5$ as a function of **Q** scanned parallel and perpendicular to the hexagonal *c*-axis for different temperatures. The lines represent fits to eq. (64) (after [46]).

6.6 CRYSTAL FIELD EFFECTS RELATED TO HIGH-TEMPERATURE SUPERCONDUCTIVITY

6.6.1 Introductory Remarks

The discovery of high-temperature superconductivity in the copper-oxide perovskites [65] has given rise to a huge amount of materials research. In particular, there have been myriad publications reporting on the change of the superconducting properties upon ion substitution at various sites in these compounds. For the compounds $YBa_2Cu_3O_x$ the surprising observation was made that the superconducting transition temperature T_c is essentially unchanged upon replacing the Y^{3+} ions by paramagnetic rare earth (R^{3+}) ions, in contrast to conventional superconductors, for which paramagnetic ions usually have a large detrimental effect on superconductivity. It is therefore important to achieve a detailed understanding of the low-energy electronic properties which define the magnetic ground state of the R^{3+} ions. In particular, information on the crystal field interaction at the R site is highly desirable, for the following reasons:

(i) For many high-T_c compounds, superconductivity and long-range three-dimensional magnetic ordering of the R ion sublattice coexist at low temperatures (with the exception of R = Ce and R = Pr which have a drastic detrimental effect on superconductivity). An understanding of both the nature of the magnetic ordering and its apparent lack of influence on T_c requires a detailed knowledge of the crystal field states of the R ions.

(ii) In most high-T_c compounds the R ions are situated close to the superconducting copper-oxide planes where the superconducting carriers are located, thus the crystal field interaction at the R sites constitutes an ideal probe of the local symmetry as well as the local charge distribution of the superconducting CuO_2 planes and thereby monitors directly changes of the carrier concentration induced by doping. This property results from the short-range nature of the leading fourth- and sixth-order terms in the crystal field potential.

(iii) The temperature dependence of the intrinsic linewidths of crystal field transitions reflects essentially the density-of-states associated with the charge carriers at the Fermi energy. Linewidth studies can therefore reveal information about the opening as well as the symmetry of the energy gap.

In the following subsections we focus on aspects (ii) and (iii), which provide crucial information on two important issues of the phenomenon of high-T_c superconductivity, namely the inhomogeneous evolution of the superconducting state upon doping and the existence of the so-called *pseudogap*, respectively.

6.6.2 The Crystal Field as a Local Probe: Evidence for Materials Inhomogeneities

One of the most interesting aspects of the superconductivity in the compounds of type $YBa_2Cu_3O_x$ is the relation between the superconducting transition temperature T_c and the oxygen stoichiometry x [66]. Annealed systems exhibit a two-plateau structure of T_c as visualized in Figure 6.19f. Charge transfer from the CuO chains to the CuO_2 planes certainly plays a crucial role in the oxygen vacancy-induced suppression of superconductivity in $YBa_2Cu_3O_x$, and this should manifest itself in the oxygen stoichiometry dependence of the observed crystal field energy spectra. This is exemplified for $ErBa_2Cu_3O_x$ in Figure 6.20 [51]. When going from $x = 6$ to $x = 7$, the transitions B, D, E shift slightly to lower energies and the transitions A, F, G move up to higher energies, whereas the energy of the transition C remains unchanged. By analyzing the crystal field parameters on the basis of equation (6.18) it could be shown that the observed shifts of the crystal field transitions are due to both structural changes and charge transfer effects [51]. In particular, it was found that a charge of 0.07 $|e|$/O is transferred into the planes when going from $x = 6$ to $x = 7$, which means that about 28% of the created holes go into the planes. Similar conclusions have been obtained for the related compound $HoBa_2Cu_3O_x$ [67]. These results show an almost linear dependence of the charge transferred as a function of x. This is in agreement with the results of neutron diffraction experiments on $ErBa_2Cu_3O_x$, which give evidence for a linear decrease of the c-axis upon hole doping [68], but it is in contrast to the conclusions of bond valence sum arguments derived in similar experiments on $YBa_2Cu_3O_x$ where the two-plateau structure of T_c has been suggested to be due to the nonlinearity of the hole transfer into the planes [66].

Figure 6.21 displays the low-energy part of the energy spectra of $ErBa_2Cu_3O_x$ upon increased resolution conditions [69], which show that the lowest crystal field transition A is actually built up of three different components, A_1, A_2, and A_3, whose main characteristics can be summarized as follows:

(i) *Intensities*: The individual components A_i have maximum weight close to $x = 7.0$, $x = 6.5$, and $x = 6.0$, respectively. With the crystal field interaction being a local probe, there is no doubt that these substructures originate from different local environments of the Er^{3+} ions which obviously coexist in the compound $ErBa_2Cu_3O_x$.

(ii) *Energies*: Whereas all the crystal field transitions are independent of energy for oxygen content $x \geq 6.5$ within experimental error, they shift slightly when going from $x \approx 6.5$ to $x \approx 6.0$. This may be due to the structural discontinuities at the orthorhombic to tetragonal phase transition at $x \approx 6.4$.

(iii) *Linewidth*: As visualized in Figure 6.19d, the intrinsic linewidths of the transitions A_i are much smaller for oxygen contents where these transitions individually reach their maximum weight, namely for $x \approx 6.0$, 6.5, and 7.0.

All these facts and their interpretation provide clear experimental evidence for cluster formation. It is tempting to identify the three clusters associated with the transitions A_1, A_2, and A_3 by two local regions of metallic ($T_c \approx 90$ K, $T_c \approx 60$ K) and a local region of semiconducting character, respectively [69]. Figures 19a,b,c show the fractional proportions of the three cluster types that exhibit a continuous behavior versus oxygen content x, consistent with the earlier conclusion that the transfer of holes into the CuO_2 planes is linearly related to the oxygenation process [51]. Furthermore, the continuous increase of the metallic states A_1 and A_2 can explain the increase of the superconducting volume fraction as observed by magnetic susceptibility measurements [68] when the oxygen content is raised from $x = 6$ to $x = 7$. It should be mentioned that these results have been nicely confirmed by μSR experiments [70] showing the coexistence of superconductivity and antiferromagnetism at low doping.

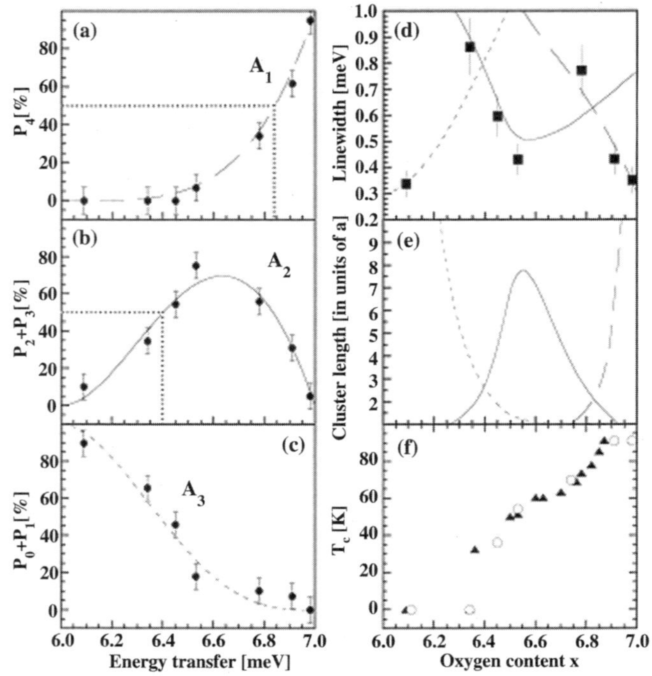

Figure 6.19 (a,b,c) Proportions (expressed as a percentage) of the lowest-lying crystal field transitions A_i of $ErBa_2Cu_3O_x$ as a function of the oxygen content x. The full, dashed, and broken lines refer to geometrical probability functions as explained in the text. The dotted lines mark the critical concentration for bond percolation. d) Intrinsic linewidths of the crystal field transitions A_i at T=10 K. e) Mean cluster length versus oxygen content x determined by a Monte Carlo simulation [75]. f) Superconducting transition temperature T_c versus oxygen content x for $ErBa_2Cu_3O_x$ (after [50]).

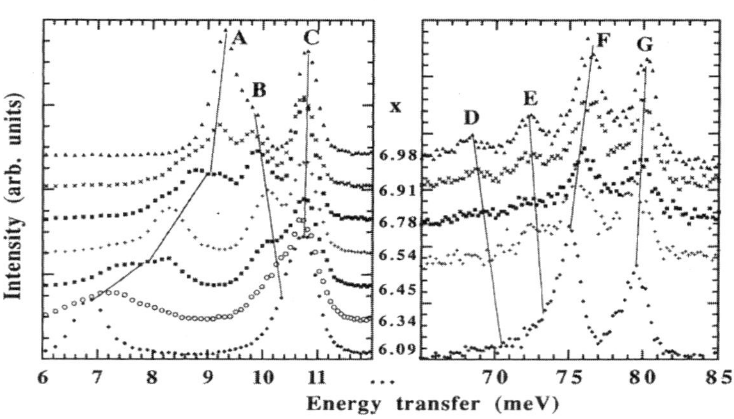

Figure 6.20 Energy spectra of neutrons scattered from $ErBa_2Cu_3O_x$ at T=10 K. The lines indicate the x dependence of the observed crystal field transitions (after [51]).

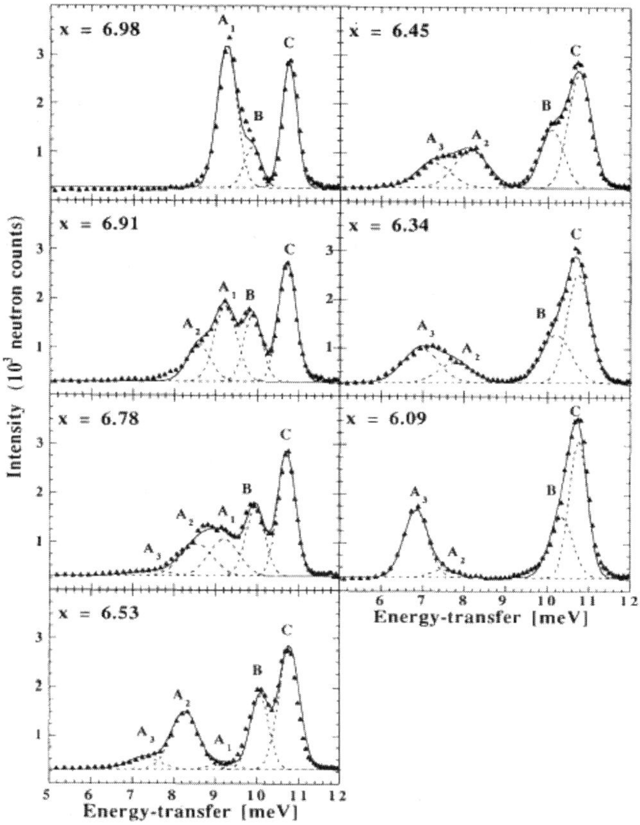

Figure 6.21 Energy spectra of neutrons scattered from $ErBa_2Cu_3O_x$ at $T = 10$ K. The full lines are the result of a least squares fitting procedure. The broken lines indicate the subdivision into individual crystal field transitions (after [69]).

Our current understanding of the superconducting properties of $ErBa_2Cu_3O_x$ (and more generally all the $RBa_2Cu_3O_x$ compounds) involves a percolation mechanism of electric conductivity as discussed in both theoretical [71] and experimental [72] work. For $x = 6$ the system is a perfect semiconductor. When adding oxygen ions into the chains, holes are continuously transferred into the CuO_2 planes. By this mechanism the number of local regions with metallic character (associated with the crystal field transition A_2) rises, which can partially combine to form larger regions. For some critical concentration a percolative network is built up, and the system undergoes a transition from the semiconducting to the conducting state (with $T_c \approx 60$ K). Upon further increasing of hole concentration a second (different) type of metallic cluster (associated with the crystal field transition A_1) is formed, which start to attach to each other and induce a transition into another conducting state (with $T_c \approx 90$ K) at the percolation limit.

For a two-dimensional square structure the critical concentration for bond percolation is $p_c = 50\%$ [73]. From the fractional proportions of A_2 and A_1

displayed in Figures 6.19a,b,c we can then immediately determine the critical oxygen concentrations for the transitions from the semiconducting to the $T_c \approx 60$ K superconducting state and to the $T_c \approx 90$ K superconducting state to be $x_2 = 6.40$ and $x_1 = 6.84$, respectively, which is in excellent agreement with the observed two-plateau structure of T_c (see Figures 6.19a,b,f). For three-dimensional structures, on the other hand, the critical concentration for bond percolation is 20% (face-centered cubic) $< p_c <$ 30% (simple cubic) [73], resulting in $6.21 < x_2 < 6.31$ and $6.64 < x_1 < 6.73$, which is inconsistent with the observed two-plateau structure of T_c. This reinforces the well known fact that the superconductivity in the perovskite-type compounds has indeed a two-dimensional character.

Combined statistical and geometrical considerations may be useful to understand the x-dependent profiles of the fractional proportions of the three cluster types visualized in Figures 19a,b,c. A local symmetry model [74] was developed that defines specific probabilities $P_k(y)$ to have, for a given oxygen content $x = 6 + y$, k of the four oxygen chain sites (0,1/2,0), (1,1/2,0), (0,1/2,1), and (1,1/2,1) nearest to the R^{3+} ion occupied:

$$P_k(y) = \begin{pmatrix} 4 \\ k \end{pmatrix} y^k (1-y)^{4-k} \quad , \quad 0 \le k \le 4 \ . \tag{6.65}$$

The fractional proportion of the cluster type A_1 exhibits the behavior predicted by the probability function $P_4(y)$ (i.e., all the oxygen chain sites being occupied). Similarly, the fractional proportions of the cluster types A_2 and A_3 follow the sum of the probability functions $P_3(y) + P_2(y)$ (i.e., one or two empty oxygen chain sites) and $P_1(y) + P_0(y)$ (i.e., one or no oxygen chain site occupied), respectively. The above probability functions are shown in Figures 6.19a,b,c by lines which excellently reproduce the experimental data.

The x-dependent linewidth of the crystal field transitions allows one to estimate the size of the clusters, since a line broadening is due to structural inhomogeneities and gradients in the charge distribution, which occur predominantly at the border lines between different cluster types. For a cluster of mean length L (in units of the lattice parameter a) in the (a,b)-plane the fraction of unit cells at the border of the cluster and inside the cluster is $4(L-1)/L^2$ and $[(L-2)/L]^2$, respectively. In a first approximation the total linewidth is then given by [75]

$$\Gamma = \frac{1}{L}\sqrt{(L-2)^2\Gamma_0^2 + 4(L-1)\left[\left(\frac{\Delta E}{2}\right)^2 + \Gamma_0^2\right]} \quad , \tag{6.66}$$

where Γ_0 is the intrinsic linewidth for infinite cluster size (i.e., for $x = 6$ and $x = 7$ exactly realized), and ΔE corresponds to the energetic separation of the crystal field transitions A_i for two coexisting cluster types. More specifically, from Figure 6.21 we derive $\Delta E = E(A_1) - E(A_2) \approx 1.0$ meV and $\Delta E = E(A_2) -$

$E(A_3) \approx 1.5$ meV at high and low oxygen concentrations x, respectively, and $0.3 \leq \Gamma_0 \leq 0.4$ meV.

The lines of Figure 6.19d are the results of Monte Carlo simulations of the doping mechanism [75]. These simulations start from the ortho-II structure at $x = 6.5$ with some degree of disorder and show that the sizes of the clusters at percolation (Figure 6.19e) are of the order of $L = 2 - 3$ (7 – 10Å), which is of the order of the superconducting coherence length in these materials. These numbers compare favorably with theoretical estimates [71].

Evidence for the strongly inhomogeneous distribution of holes in the CuO_2 planes also results from other techniques such as μsR [70], magnetic susceptibility [72], NMR [76], electron paramagnetic resonance [77], Mössbauer [27], Raman [78], or even neutron diffraction [68] experiments. Various theoretical models have also predicted that under doping these systems become highly unstable and tend to phase separate into hole-rich and hole-poor regions [71, 79, 80].

6.6.3 Relaxation Phenomena to Probe the Pseudogap

Superconductivity is the result of two distinct quantum phenomena, pairing of the charge carriers at a characteristic temperature T^* and long-range phase coherence at the superconducting transition temperature T_c. In conventional superconductors these two phenomena occur simultaneously, i.e., $T^*=T_c$. In contrast, for high-temperature superconductors we have $T^* > T_c$ over a large doping range, thus the so-called *pseudogap region* ($T_c < T < T^*$) is clearly the most challenging part of the phase diagram. The experimental discovery of the pseudogap region gave rise to an impressive number of models for the mechanism of high-temperature superconductivity [81]. Consequently, experiments that produce changes of the pseudogap temperature T^* are of crucial importance for discriminating between the different pairing scenarios developed for the cuprate superconductors. The experimental situation, however, is still rather controversial. Different techniques gave different results concerning the shift of T^* upon doping, isotope substitution, and pressure application, thus particular care has to be taken in choosing adequate experimental tools. Ideally, the experimental method should be (i) truly bulk sensitive, (ii) fast in order to detect dynamical effects with a timescale of $\tau \leq 10^{-13}$ s (typical of spin and lattice fluctuations), and (iii) direct in order to reveal a distinct crossover temperature without involving a complicated machinery to derive T^* from the raw data. Among the frequently used methods—ARPES and tunneling spectroscopy, NMR and NQR experiments— bulk transport and thermodynamic measurements have severe deficiencies concerning the criteria (i), (ii), and (iii), respectively. Neutron crystal field spectroscopy is probably the only experimental tool that fulfills all three criteria simultaneously. The technique was summarized in a recent review article [82] and is exemplified below for the case of isotope substitution.

The $HoBa_2Cu_4O_8$ compound is excellently suited for studying oxygen isotope effects because it is highly stoichiometric and the stoichiometry does not change upon both $^{16}O \rightarrow ^{18}O$ and $^{18}O \rightarrow ^{16}O$ substitution. The critical temperatures T_c of the ^{16}O and ^{18}O compounds were determined by magnetometry to be (79.0 ± 0.1) K and (78.5 ± 0.1) K, respectively. The observed oxygen isotope shift $\Delta T_c \approx (-0.5 \pm 0.2)$ K can be described by the isotope coefficient α defined by the relation $T_c \propto M^{-\alpha}$, where M is the mass of the oxygen ion. For $HoBa_2Cu_4O_8$ we deduce $\alpha = 0.05 \pm 0.02$, in agreement with literature data on the isostructural compound $YBa_2Cu_4O_8$ [83]. Figure 6.22 shows energy spectra obtained for $HoBa_2Cu_4{}^{18}O_8$ ($T_c = 79$ K) by Rubio Temprano et al. [84]. There are two strong ground state crystal field transitions at $\hbar\omega_1 = 0.6$ meV and $\hbar\omega_2 = 1.3$ meV. With increasing temperature the crystal field transitions exhibit line broadening. In addition, excited crystal field states become increasingly populated giving rise to excited crystal field transitions. The energy spectra were fitted according to the neutron cross-section for crystal field transitions defined by equation (6.54). The final spectra were obtained by convoluting the instrumental resolution function with the crystal field intrinsic Lorentzian function, whose linewidth is given in the normal state by the modified Korringa law (equation (6.58)). The only free parameters in the fitting procedure were an overall scale factor for the intensities and a temperature-dependent linewidth $\Gamma(T) \propto [N(E_F)j_{ex}]^2$. The results of the fitting procedure are shown by solid lines in Figure 6.22.

Figure 6.23 (upper panel) shows the temperature dependence of the intrinsic linewidth Γ (HWHM) corresponding to the crystal field transition from the ground state to the first excited state at $\hbar\omega_1 = 0.6$ meV for the ^{18}O compound. The linewidth is zero below 60 K, then it increases almost linearly up to 210 K, without any anomaly at $T_c = 79$ K. The linewidth is step-like enhanced between 210 and 220 K. Above 220 K it increases again linearly as expected for the normal state. We therefore identify the temperature where the step-like enhancement occurs with the temperature where the pseudogap opens, i.e., we set $T^* \approx 220$ K. The lower panel of Figure 6.23 shows the temperature dependence of the intrinsic linewidth derived for the same crystal field transition in the ^{16}O compound. We recognize that the relaxation behavior of the ^{16}O sample has a similar shape as for the isotope substituted ^{18}O compound, but the step-like enhancement of the linewidth occurs at much lower temperatures between 160 and 170 K, i.e., $T^* \approx 170$ K. These relaxation data therefore give evidence for a large isotope shift $\Delta T^*(O) \approx 50$ K, which, in analogy to the isotope effect on T_c, gives rise to a huge oxygen isotope coefficient $\alpha^*(O) \approx -2.2$. Similar experiments performed for the copper isotope effect in $HoBa_2Cu_4O_8$ gave $T^*(^{63}Cu) \approx 160$ K and $T^*(^{65}Cu) \approx 185$ K, i.e., $\Delta T^*(Cu) \approx 25$ K and a giant copper isotope coefficient $\alpha^*(Cu) \approx -4.9$ [85]. Neutron spectroscopic experiments performed for underdoped, optimally doped and overdoped $La_{2-x-y}Sr_xHo_yCuO_4$ with $x = 0.10$, 0.15, 0.20, 0.25 and $y = 0.04$ yielded oxygen isotope shifts of $\Delta T^* \approx 20$, 10, 5, and 2 K, respectively [86, 87, 88].

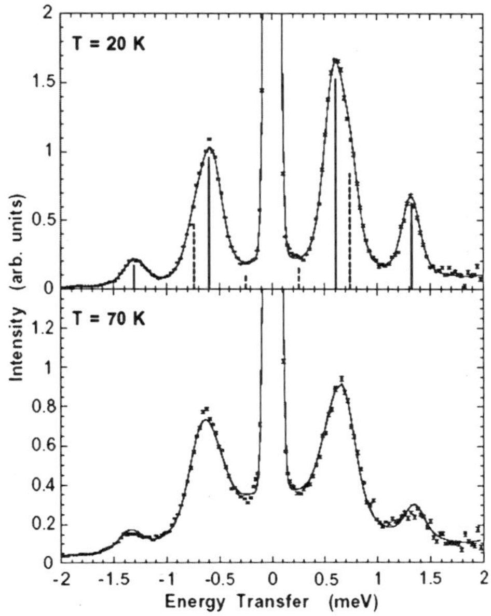

Figure 6.22 Energy spectra of neutrons scattered from $HoBa_2Cu_4{}^{18}O_8$. The solid lines are the results of least squares fitting procedures as explained in the text. The solid and dashed vertical bars indicate the energy and intensity of the ground state and excited state crystal field transitions, respectively (taken from [84]).

The application of hydrostatic pressure affects the lattice degrees of freedom in a manner different from isotope substitution. Recently, neutron spectroscopy experiments under hydrostatic pressure were performed for the optimally doped compound $La_{2-x-y}Sr_xHo_yCuO_4$ ($x = 0.15$, $y = 0.04$) [87, 89]. It was found that the pseudogap temperature decreases from $T^* \approx 61$ K at ambient pressure to $T^* \approx 55$ K for $p = 0.8$ GPa, i.e., $\Delta T^* \approx -6$ K. The opposite effects of oxygen isotope substitution and pressure can be qualitatively understood by simple models based on phonon-mediated mechanisms. In a first approximation the phonon energy $h\nu$ is inversely proportional to the square root of the oxygen mass, thus the substitution of ^{16}O by ^{18}O leads to a decrease of the phonon energy by 5.7%. On the other hand, the application of pressure yields an enhancement of the phonon energy according to $\partial h\nu/(h\nu)\partial p = -\gamma\kappa$, where $\gamma \approx 2$ is the Grüneisen parameter and $\kappa = -6.8 \cdot 10^{-3}$ Gpa^{-1} the volume compressibility, which results in an increase of the phonon energy by 1.4%. The combined observation of oxygen isotope and pressure effects on T^* provides strong support for phonon-mediated pairing mechanisms in high-T_c superconductors, in agreement with theoretical models based on dynamical charge ordering [90], phonon-induced stripe formation [91] and bipolaron formation associated with Jahn-Teller oxygen vibrations [92].

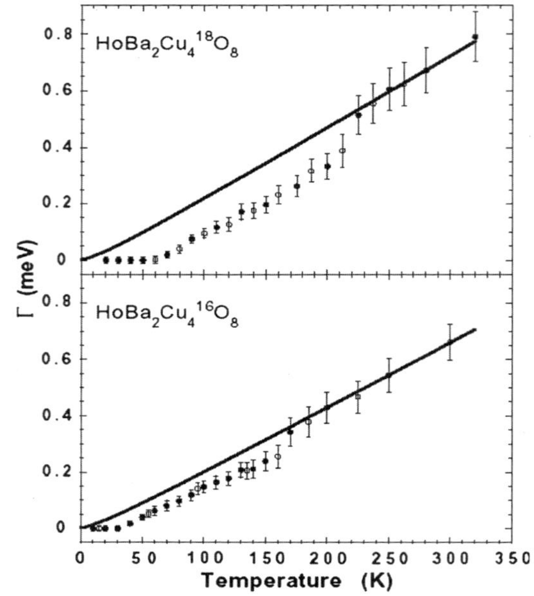

Figure 6.23 Temperature dependence of the intrinsic linewidth Γ (HWHM) corresponding to the $\Gamma_3 \rightarrow \Gamma_4$ ground state crystal field transition in $HoBa_2Cu_4{}^{16}O_8$ and $HoBa_2Cu_4{}^{18}O_8$. The lines denote the linewidth in the normal state calculated from eq. (58) (after [84]).

6.7 CONCLUDING REMARKS

The crystal field interaction in transition metal compounds and, in particular, in rare earth compounds, is responsible for a variety of magnetic phenomena. A detailed insight into the crystal field interaction accompanied by a convenient set of crystal field parameters is therefore indispensable for the understanding and numerical analysis of magnetic properties. The calculation of crystal field parameters from first principles is a difficult task, thus experimental techniques and in particular spectroscopic methods are needed. Fortunately the experimental determination of crystal field parameters is now extremely well established as demonstrated in the present work. Special emphasis was put on the inelastic neutron scattering technique, which has been used as a unique method to study the crystal field interaction in a large number of both metallic and insulating rare earth compounds. Our work had to focus on a few specific topics, thus we refer finally to the excellent and comprehensive review article by Moze [93] about neutron spectroscopy studies of crystal field effects in intermetallic compounds for further reading and for an almost complete compilation of the experimental data.

ACKNOWLEDGMENTS

Financial support by the Swiss National Science Foundation (project No. 20-66948.01) and by the NCCR MaNEP project is gratefully acknowledged.

REFERENCES

[1] Abragam, A. & Bleaney, B. (1970) *Electron Paramagnetic Resonance* (Oxford: Clarendon).
[2] Purwins, H.G. & Leson A. (1990) *Adv. Phys.* **39,** 309-405.
[3] Bethe, H.A. (1929) *Ann. Phys.* **3,** 133-208.
[4] Stevens, K.W.H. (1952) *Proc. Phys. Soc. A*, **65**, 209-15.
[5] Stevens, K.W.H. (1967) *Rep. Progr. Phys.* **30**, 189-226.
[6] Elliott, R.J. & Stevens, K.W.H. (1953) *Proc. R. Soc. A*, **219**, 387-404.
[7] Judd, B.R. (1963) *Operator Techniques in Atomic Spectroscopy* (New York: McGraw-Hill).
[8] Hutchings M.T. (1964) *Solid State Physics* **16**, 227-73.
[9] Newman, D.J. (1971) *Adv. Phys.* **20**, 197-255.
[10] Fulde, P. (1978) *Handbook on the Physics and Chemistry of Rare Earths*, Chapter 17 (Amsterdam: North-Holland).
[11] Freeman, A.J. & Desclaux, J.P. (1979) *J. Magn. Magn. Mater.* **12,** 11-21.
[12] Lea, K.R., Leask, M.J.M. & Wolf, W.P. (1962) *J. Phys. Chem. Solids* **23**, 1381-405.
[13] Wybourne, B.G. (1965) *Spectroscopic Properties of Rare Earths* (New York: John Wiley).
[14] Nielson C.W. & Koster, G.F. (1964) *Spectroscopic Coefficients for the* p^n, d^n, *and* f^n *configurations* (Cambridge: M.I.T. Press).
[15] Sternheimer, R.M. (1966) *Phys. Rev.* **146**, 140-60.
[16] Morrison, C.A. (1988) *Angular Momentum Theory Applied to Interactions in Solids* (Berlin: Springer).
[17] Tellenbach, U. (1974) Report AF-SSP-75 (ETH Zurich, unpublished).
[18] Urland, W. (1976) *Chem. Phys.* **14,** 393-401.
[19] Jörgensen C.K., Pappalardo, R. & Schmidtke, H.H. (1963) *Chem. Phys.* **39**, 1422-30.
[20] Blaha, P., Schwarz, K. & Luitz (1997) WIEN91 (Vienna Technical University, unpublished).
[21] Divis, M., Rusz, J., Hilscher, G., Michor, H., Blaha, P. & Schwarz, K. (2002) *Czech. J. Phys.* **52**, 283-6.
[22] Novak, P. (1996) *Phys. Stat. Sol. B*, **198**, 729-40.
[23] Gasser, U., Allenspach, P., Fauth, F., Henggeler, W., Mesot, J., Furrer, A., Rosenkranz, S., Vorderwisch, P. & Buchgeister, M. (1996) *Z. Phys. B,* **101**, 345-52.
[24] Strässle, Th., Divis, M., Rusz, J., Janssen, S., Juarnyi, F., Sadikov, R. & Furrer, A. (2003) *J. Phys. Condens. Matter* **15**, 3257-66.
[25] Furrer, A. & Güdel, H.U. (1997) *Phys. Rev. B*, **56**, 15062-72.
[26] Elsenhans, O., Furrer, A., Purwins, H.G. & Hulliger, F. (1990) *Z. Phys. B*, **80**, 281-6.
[27] Hodges, J.A., Bonville, P., Imbert, P., Jéhanno, G. & Debray, P. (1991) *Physica C*, **184**, 270-82.
[28] Jandl, S., Dufour, P., Strach, T., Ruf, T., Cardona, M., Nekvasil, V., Chen, C., Wanklyn, B.M. & Pinol, S. (1996) *Phys. Rev B*, **53**, 8632-7.
[29] Strach, T., Ruf T., Cardona, M., Lin, C.T., Jandl. S., Nekvasil, V., Zhigunov, D.I., Barilo, S.N. & Shiryaev, S.V. (1996) *Phys. Rev. B*, **54**, 4276-82.
[30] Henggeler W., Chattopadhyay, T., Roessli, B., Vorderwisch, P., Thalmeier, P., Zhigunov, D.I., Barilo, S.N. & Furrer, A. (1997) *Phys. Rev. B*, **55**, 1269-79.

[31] Brockhouse, B.N. (1955) *Can. J. Phys.* **33**, 889-91.
[32] Halpern, O. & Johnson, M.H. (1939) *Phys. Rev.* **55**, 898-923.
[33] Squires, G.L. (1996) *Thermal Neutron Scattering* (New York: Dover Publications).
[34] Lovesey, S.W. (1987) *Theory of Neutron Scattering from Condensed Matter* (Oxford: Oxford Science Publishers).
[35] Johnston, D.F. (1966) *Proc. Phys. Soc.* **88**, 37-52.
[36] Van Hove, L. (1954) *Phys. Rev.* **93**, 268-9.
[37] Schmid, B., Hälg, B., Furrer, A., Urland, W. & Kremer, R. (1987) *J. Appl. Phys.* **61**, 3426-8.
[38] Birgeneau, R.J. (1972) *J. Phys. Chem. Solids* **33**, 59-68.
[39] Lewis Ian, R. & Edwards Howell, G.M. (2001) *Handbook of Raman Spectroscopy* (New York: Marcel Dekker).
[40] Sanjurjo, J.A., Martins, G.B., Pagliuso, P.G., Granado, E., Torriani, I., Rettori, C., Oseroff, S. & Fisk, Z. (1995) *Phys. Rev. B*, **51**, 1185-9.
[41] Berry, A.J., McCaw, C.S., Morrison, I.D. & Denning, R.G. (1996) *J. Lumin.* **66**, & **67**, 272-7.
[42] Cardona, M. (1982) *Light Scattering in Solids* II (Berlin: Springer).
[43] Yanson, I.K. (1974) *Sov. Phys. JETP* **39**, 506-13.
[44] Jansen, A.G.M., van Gelder, A.P. & Wyder, P. (1980) *J. Phys. C*, **13**, 6073-118.
[45] Akimenko, A.I., Ponomarenko, N.M., Yanson, I.K., Janos, S. & Reiffers, M. (1984) *Sov. Phys. Solid State* **26**, 1374-9.
[46] Amato, A., Bührer, W., Grayevski, A., Gygax, F.N., Furrer, A., Kaplan, N. & Schenck, A. (1992) *Solid State Comm.* **82**, 767-71.
[47] Reiffers, M., Naidyuk, Yu.G., Janseen, A.G.M., Wyder, P., Yanson, I.K., Gignoux, D. & Schmitt, D.P. (1989) *Phys. Rev. Lett.* **62**, 1560-3.
[48] Naidyuk, Yu.G. & Yanson, I.K. (2004) *Point-Contact Spectroscopy* (Berlin: Springer).
[49] Richardson, F.S., Reid, M.F., Dallara, J.J. & Smith, R.D. (1985) *J. Chem. Phys.* **83**, 3813-30.
[50] Mesot, J., Rubio, Temprano, D. & Furrer, A. (2002) *Trends in Applied Spectroscopy, Vol. 4* (Trivandrum: Research Trends) p 75.
[51] Mesot, J., Allenspach, P., Staub, U., Furrer, A., Mutka, H., Osborn, R. & Taylor, A. (1993) *Phys. Rev. B*, **47**, 6027-36.
[52] Mesot, J. & Furrer, A. (1998) *Neutron Scattering in Layered Copper-Oxide Superconductors* p 335 (Dordrecht: Kluwer).
[53] Roeland, L.W., de Boer, F.R., Huang, Y.K., Menovsky, A.A. & Kadowaki, K. (1988) *Physica C*, **152**, 72-6.
[54] Dunlap, B.D., Slaski, M., Hinks, D.G., Soderholm, L., Beno, M., Zhang, K., Segre, C., Crabtree, G.W., Kwok, W.K., Malik, S.K., Schuller, I.K., Jorgensen, J.D. & Sungaila, Z. (1987) *J. Magn. Magn. Mater.* **68**, L139-44.
[55] Van der Meulen, H.P., Franse, J.J.M., Tarnawski, Z., Kadowaki, K., Klasse, J.C.P. & Menovsky, A.A. (1988) *Physica C*, **152**, 65-71.
[56] Hodges, J.A., Imbert, P., Marimon de Cunha, J.B. & Sanchez, J.P. (1989) *Physica C*, **160**, 49-54.
[57] Heitler, W. & Teller, E. (1936) *Proc. Roy. Soc. A*, **155**, 629-39.
[58] Kronig, R. de L. (1939) *Physica* **6**, 33-43.
[59] van Vleck, J.H. (1939) *J. Chem. Phys.* **7**, 72-84; 1940 Phys. Rev. **57**, 426-47.
[60] Orbach, R. (1961) *Proc. Roy. Soc. A*, **264**, 458-84.
[61] Korringa, J. (1950) *Physica* (Utrecht) **16**, 601-10.
[62] Becker, K.W., Fulde, P. & Keller, J. (1977) *Z. Phys. B*, **28**, 9-18.
[63] Feile R., Loewenhaupt M., Kjems J.K. & Hoenig H.E. (1987) *Phys. Rev. Lett.* **47**, 610-3.
[64] Buyers, W.J.L., Holden, T.M., Svenson, E.C., Cowley, R.A. & Hutchings, M.T. (1971) *J. Phys. C*, **4**, 2139-59.
[65] Bednorz, J.G. & Müller, K.A. (1986) *Z. Phys. B*, **64**, 189-93.
[66] Cava, R.J., Hewat, A.W., Hewat, E.A., Batlogg, B., Marezio, M., Rabe, K.M., Krajevski, J.J., Peck, W.F.Jr. & Rupp, L. W. (1990) *Physica C*, **165**, 419-33.

[67] Staub, U., Mesot, J., Guillaume, M., Allenspach, P., Furrer, A., Mutka, H., Bowden, Z. & Taylor, A. (1994) *Phys. Rev. B*, **50**, 4068-74.
[68] Radaelli, P.G., Segre, C.U., Hinks, D.G. & Joergensen, J.D. (1992) *Phys. Rev. B*, **45**, 4923-9.
[69] Mesot, J., Allenspach, P., Staub, U., Furrer, A. & Mutka, H. (1993) *Phys. Rev. Lett.* **70**, 865-8.
[70] Niedermayer, C., Bernhard, C., Blasius, T., Golnik, A., Moodenbaugh, A. & Budnick, J.I. (1998) *Phys. Rev. Lett.* **80**, 3843-6.
[71] Hizhnyakov, V. & Sigmund, E. (1988) *Physica C*, **156**, 655-66.
[72] Kremer, R., Sigmund, E., Hizhnyakov, V., Hentsch, F., Simon, A., Müller, K.A. & Mehring, M. (1992) *Z Phys. B*, **86**, 319-24.
[73] Kirkpatrick, S. (1973) *Rev. Mod. Phys.* **45**, 574-88.
[74] Allenspach, P., Furrer, A. & Rupp, B. (1990) *Progress in High-Temperature Superconductivity* (Singapore: World Scientific) p 318.
[75] Furrer, A., Allenspach, P., Fauth, F., Guillaume, M., Henggeler, W., Mesot, J. & Rosenkranz, S. (1994) *Physica C*, 235-240 261-4.
[76] Hammel, P.C., Reyes, A.P., Fisk, Z., Takigawa, M., Thompson, J.D., Heffner, K.H. & Cheong, S.W. (1990) *Phys. Rev. B*, **42**, 6781-3.
[77] Wübbeler, G. & Schirmer, O. F. (1990) *Phys. Stat. Sol.* (b) **174** K21-5.
[78] Poulakis, N., Palles, D., Liarokapis, E., Conder, K. & Kaldis, E. (1996) *Phys. Rev. B*, **53**, R534-7.
[79] Emery, V.J., Kivelson, S.A. & Lin, H.Q. (1990) *Phys. Rev. Lett.* **64**, 475-8.
[80] Grilli, M., Raimondi, R., Castellani, C., di Castro, C. & Kotliar, G. (1991) *Phys. Rev. Lett.* **67**, 259-62.
[81] Batlogg, B. & Varma, C. (2000) *Physics World* 13 (No 2) 33-7.
[82] Furrer, A. (2005) *Structure and Bonding* **114**, 171-204.
[83] Raffa, F., Ohno, T., Mali, M., Roos, J., Brinkmann, D., Conder, K. & Eremin, M. (1998) *Phys. Rev. Lett.* **81**, 5912-5.
[84] Rubio Temprano, D., Mesot, J., Janssen, S., Conder, K., Furrer, A., Mutka, H. & Müller, K.A. (2000) *Phys. Rev. Lett.* **84**, 1990-3.
[85] Rubio Temprano, D., Mesot, J., Janssen, S., Conder, K., Furrer, A., Sokolov, A., Trounov, V., Kazakov, S.M., Karpinski, J. & Müller, K.A. (2001) *Eur. Phys. J. B*, **19**, 5-8.
[86] Rubio Temprano, D., Conder, K., Furrer, A., Mutka, H., Trounov, V. & Müller, K.A. (2002) *Phys. Rev. B*, **66**, 184506.
[87] Furrer, A., Conder, K., Häfliger, P.S. & Podlesnyak, A. (2004) *Physica C*, 408-410 773-4.
[88] Häfliger, P.S., Podlesnyak, A., Conder, K., Pomjakushina, E. & Furrer, A. (2004) *Europhys. Lett.* **67**, 1018-23.
[89] Podlesnyak, A., Häfliger, P.S., Conder, K. & Furrer, A. (2005) *J. Phys.: Condens. Matter* **17**, S801-6.
[90] Andergassen, S., Caprara, S., di Castro, C. & Grilli, M. (2001) *Phys. Rev. Lett.* **87**, 056401.
[91] Bussmann-Holder, A. (2000) *J. Supercond.* **13**, 773-6.
[92] Gor'kov, L.P. (2000) *J. Supercond.* **13**, 765-9.
[93] Moze, O. (1998) *Handbook of Magnetic Materials, Vol. 11* (Amsterdam: Elsevier) p 493.

CHAPTER 7

SCANNING TUNNELING SPECTROSCOPY (STS)

K.W. Hipps
Department of Chemistry and Materials Science Program, Washington State University, Pullman, WA, USA

7.1 INTRODUCTION

The nanoscale world is exciting because it is governed by rules different than those in the macroscopic, or even microscopic, realm. It is a world where quantum mechanics dominates the scene and events on the single molecule—or even single atom—scale are critical. What we know about the behavior of material on our scale is no longer true on the nanometer scale and our formularies must be rewritten. In order to study this quantum world, a quantum mechanical probe is essential. Electron tunneling provides such a quantum mechanical tool.

In the Newtonian world, a particle can never be in a region where its potential energy is greater than its total energy. Being there would require a negative kinetic energy—a clear impossibility since $mv^2/2 \geq 0$. As the scale shrinks to molecular dimensions, of the order of 1 nanometer (nm), classical concepts fail and the correct mechanics is quantum mechanics. Thus, it is possible for a particle to move from one classically allowed region to another through a region where its potential energy is greater than its total energy—this is the phenomenon of *tunneling*. While it can occur for relatively heavy particles such as protons, it is far more probable for light particles such as electrons. Electron tunneling is a particularly useful probe because it is easy to control the flow and energy of electrons and to set up precisely controlled regions through which the electron must tunnel. An early example of an electron-tunneling device was the *metal-insulator-metal* (M-I-M) *tunnel diode*, Figure 7.1 [1–7]. Also shown in Figure 7.1 are the corresponding features of a *scanning tunneling microscope* (STM) [8–11]. Both devices rely on exactly the same physics. Within the conductors (metal electrodes in the M-I-M case, substrate and atomically sharp tip in the STM case) the electrons are in classically allowed regions I and III. Within these regions, their total energy E is greater than their potential energy. In the gap between conductors (the insulator in the M-I-M case, the vacuum or solvent gap in the STM case) however, the potential is greater than E, region II. This region is classically forbidden but quantum mechanically allowed. A simple quantum mechanical calculation quickly demonstrates that the probability of transmission through the barrier decreases exponentially with the thickness of the barrier and the square root of the potential (barrier height) relative to the electron energy. If distance d is measured in Angstroms (1Å = 0.1 nm) and energies (E and U) are measured in electron volts, then the constant A in Figure 7.1 is approximately 1 [6–8]. If one assumes that the bias voltage is small compared to the barrier height, $U - E$ is approximately equal to the work function Φ, and the tunneling current is roughly given by:

$$\mathrm{I} = CV\exp(-d\sqrt{\Phi}). \tag{7.1}$$

A more sophisticated treatment of the tunneling problem based on the WKB method can be used and one then finds that the tunneling current is given by the expression [10]:

$$I = \int_0^{eV} \rho_s(r,E)\rho_t(r+eV-E)T(E,eV,r)dE \tag{7.2}$$

where the density of states of the tip (ρ_t) and sample (ρ_s) as a function of position along the surface plane must be considered along with the exponential transmission probability (T).

The exponential dependence of tunneling current on electrode separation is the essential element of the STM, a device that can produce exquisitely well resolved images of molecules and atoms. Modern scanning tunneling microscopes are capable of resolving single atoms at temperatures ranging from near 0 K to above 600 K. STM images have been acquired in ultra high vacuum (UHV), in air, even in electrochemical cells. The STM has allowed us to visualize the nanoscale world in a way that is essential for understanding processes on that scale. But there is more information in the tunneling current than just the surface geometry. If a structured barrier is considered, one in which there is a nonuniform distribution of states, the tunneling current can also reflect this energetic variation. Changes in current with applied bias voltage at constant electrode separation provide spectroscopic information about the surface of the electrodes and any material located in the barrier region. Moreover, there are several different types of interactions that can lead to distinctly different spectroscopic methods. These electron tunneling spectroscopies can be generally classed based upon either inelastic or elastic electron tunneling processes. Our primary interest in this article is on the spectroscopic information that can be obtained with the STM. This encompasses both inelastic and elastic processes. On the other hand, some understanding of the experimental requirements and limitations on the imaging technique are essential for a real appreciation of the range of applications of the spectroscopic methods. Thus, we will begin this article with an overview of STM imaging and then proceed to an introduction of the basic spectroscopic methods. Along the way we will provide examples from

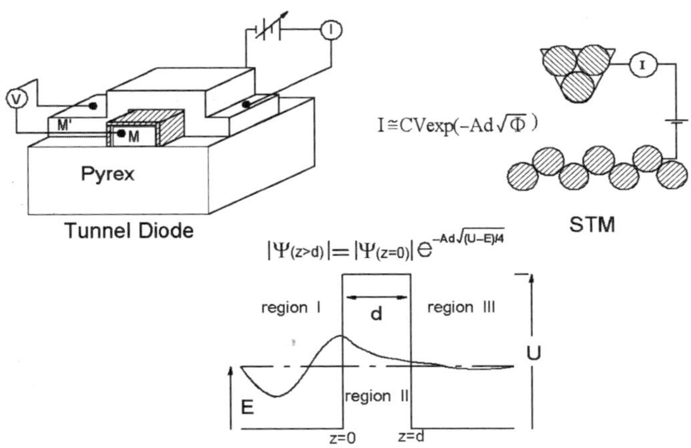

Figure 7.1 Schematic drawings of a tunnel diode, an STM, and the electronic energy diagram appropriate for both. U is the height of the potential barrier, E is the energy of the incident electron, d is the thickness of the barrier, A is approximately 1.02 Angstrom/(eV)$^{1/2}$ if U and E are in electron volts and d is in Angstroms, ψ_0 is the wavefunction of the incident electron and ψ_d is the wavefunction after transmission through the barrier. I is the measured tunneling current, V is the applied bias, and M and M' are the electrode metals.

the literature of applications of tunneling spectroscopy. Because STM imaging has been extensively reviewed, we will limit our presentation on this topic to a few basic points and refer the reader to the literature for a deeper understanding of the method [8–19]. We will also only give a cursory treatment of STM-based spectroscopy of semiconductors and metals. Understanding of the spectra of these crystalline surfaces has been well documented in a number of outstanding reviews and books [8, 12, 13, 18], with our personal favorite being the chapter by Hamers [9].

7.2 THE SCANNING TUNNELING MICROSCOPE (STM)

The forebearer of the STM was the *topographiner*. The topographiner was a device demonstrated in the early 1970s that utilized field emission rather than tunneling as the surface sensing technique [20]. Like the modern STM, the device consisted of a sharp metallic tip mounted on a piezoelectric element for positioning and an electronic feedback system that maintained the tip-sample spacing during a raster scan of the surface. The resolution was limited to about 3 nm in the vertical direction and 400 nm in the horizontal due to the weak distance dependence of the field emission current. In the early 1980s, Gerd Binnig and Heinrich Rohrer began investigating the possibility of using tunneling electrons to probe the surface of conductors and semiconductors [21–23]. In 1983, Binnig and Rohrer succeeded in producing an atomic resolution image of two unit cells of the 7×7 reconstruction surface of silicon(111). It was this image that captured the attention of the surface science world and resulted in their receiving the Nobel Prize in 1986 [23].

From the STM has evolved an entire menagerie of *scanning probe microscopy* (SPM) techniques. While this chapter is solely concerned with STM, it is useful to introduce it by way of the more general SPM approach. A schematic representation of the essential elements for any scanning probe microscope is presented in Figure 7.2. The tip-sample interaction defines the type of SPM and controls the spatial resolution possible with the particular technique. For example, if mechanical forces are measured by physical contact between tip and sample, then the radius of curvature of the tip and the elastic compliance of the substrate limit the possible spatial resolution in the (x, y)-plane.

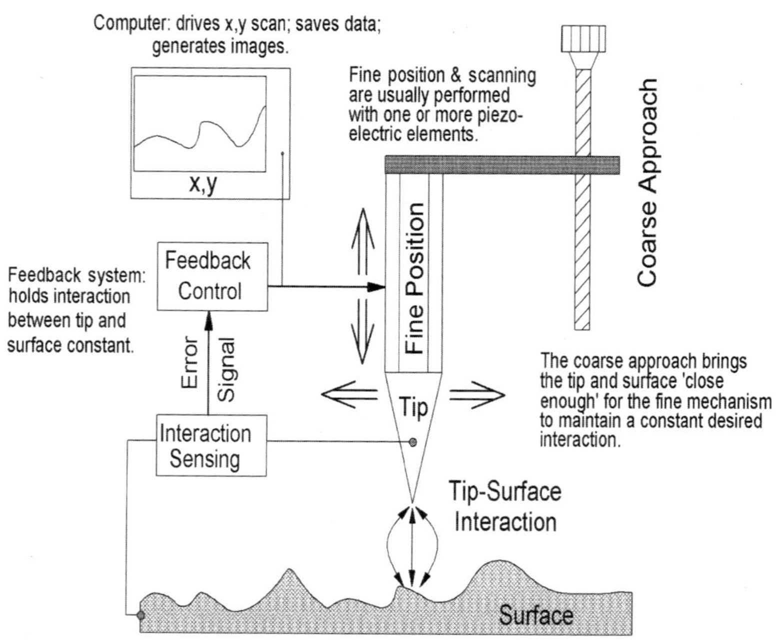

Figure 7.2 Schematic representation of a scanning probe microscope. The critical element in differentiating the different probe techniques and their relative resolutions is the tip-surface interaction and the method of sensing that interaction. In STM it is the tunneling current that is sensed.

In STM it is the tunneling current that is probed. The exponential dependence of the current on tip-sample separation results in the possibility of single atom resolution. This is generally explained by considering a tip formed from a single atom sitting in a three-fold hollow site. For conditions normally used in high resolution STM (1 nA set point and 300 mV bias), about 90% of the current is carried by the apical atom because of the difference in distance between it and the atoms at its base. For the general

SPM technique, the finer one can make the tip, the higher the x, y spatial resolution. Good STM tips are generally atomically sharp and methods for fabricating them will be discussed later. On the other hand, the z (or normal direction) resolution depends solely upon the z dependence of the interaction between tip and sample. Because of its exponential form, STM is the most sensitive SPM technique, easily achieving 0.005-nm sensitivity.

Another essential feature of all SPMs is the feedback control loop. This electronic system maintains the tip-sample interaction at a preset value by controlling the z position (or deflection) of the tip relative to the surface (or to an undeflected position). In the case of the STM, the absolute distance of the tip from the surface is difficult to impossible to determine, and the relative height is controlled through setting fixed values of current and voltage ($I \sim cV \exp(-d\ \Phi^{1/2})$). Typical current values range from picoamps to nanoamps, while the bias voltage can vary from millivolts to volts. Commercial feedback loops generally incorporate both integral and proportional gain stages. As in all feedback loop applications, it is desirable to set the gain as high as possible but not so high as to drive the system into oscillation. Oscillation of the feedback loop is especially bad in STM where repeated contacts between tip and surface (called "crashes") can destroy the usefulness of the tip.

The SPM image is generated by performing a raster scan of the tip over the surface while recording the z deflection required to maintain the set-point. Almost all commercial instruments use one or more piezoelectric elements to perform the fine motion required to generate images with sub Angstrom data intervals. Instruments designed for high resolution are usually limited to piezoelectric elements having 1 to 10 micron maximum scan widths, while 100 micron scan widths are available for lower resolution SPM studies. It is important to note that in the feedback controlled scan regime, one is almost never measuring a true height. In STM, for example, the changes in tip height with position under feedback control reflect both the tip-sample separation and the spatial variation of the *local density of surface states* (LDOS) of the sample ($\Phi(x, y)$ in the simple model). Thus, the constant current image only reflects true height changes if the LDOS of the surface (the local work function) is constant across the surface. This would be the case for atomic steps on clean metal surfaces, but would NOT be the case for adsorbates on surfaces.

The range of z motion possible for the fine position control is generally only a few microns. Thus, some method of bringing the tip within a few microns of the surface without crashing it is required. This is identified as the coarse approach mechanism in Figure 7.2. Often, this course approach is accomplished by a fine pitch screw (as in the Veeco electrochemical STM [24]), or a piezoelectric driven slip-stick (inertial approach) mechanism (as in the McAllister [25]). Some designs incorporate x, y coarse motion into their coarse approach mechanism, as does the RHK design [26].

The computer control system does more than integrate the feedback loop with the scanning mechanism while saving piezoelectric position coordinates.

Another essential feature is in displaying the data in a manner easily interpreted by the human eye. The left hand image of Figure 7.3 shows the STM constant current height versus position of a large area of a Au-111 surface. While the data were actually acquired with a 512 × 512 grid, only a small sampling of the grid is shown for clarity. The change in tip height relative to the initial set-point at each grid crossing is mapped both as a 3D projection and by brightness. The lighter the area, the higher the feature.

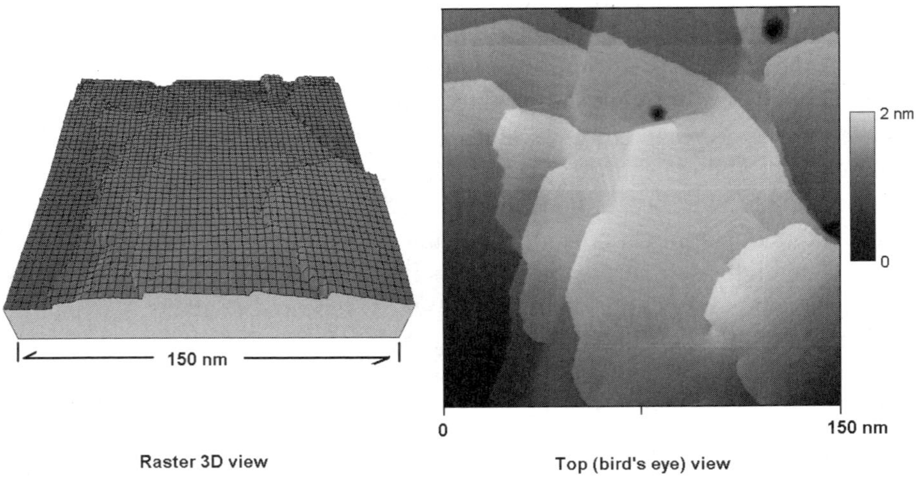

Figure 7.3 Comparison of different display methods for the same data set of constant current height versus position for the same large area of a Au-111 surface. The left hand image shows a portion of the 512 x 512 grid used for data collection with the heights mapped both as a 3D projection and by brightness. The right hand image is a standard top view where all the height information is contained in the gray scale shown to the far right.

While these 3D projection images are lovely, a real appreciation of the data requires the ability to rotate and tilt the image interactively relative to the observer. Almost all modern SPM software is capable of doing this real time, but this interactive image is not something that can be placed in a publication. Instead, a top view, sometimes called a "bird's eye view," is most frequently published. The right hand image is a standard top view of the same area as shown in the left image. Here, all the height information is contained in the color scale, which is displayed to the far right.

While virtually any commercial STM will do a good job of taking pictures of surfaces, spectroscopic applications place extra requirements on the instrumentation that must be met if even low quality data are desired.

1) The control and data acquisition system must be able to ramp the bias voltage or the tip sample separation while acquiring the tunneling current or some other voltage signal supplied by an external device. The rate at which data are acquired must be variable over a wide range in time (from about 5 microseconds to about 10 milliseconds per data point), in voltage (from about

±0.5 volt to ±10.0 volts, with the ability to set asymmetrical limits being highly desirable), or in z span (from about –2 nm to +5 nm, where negative distance moves the tip closer to the substrate than at the set-point. The most commonly used auxiliary input is the output of a lock-in amplifier that detects a modulated signal in the tip current.

2) It must be possible to add modulation (typically sine or square wave) to the bias or tip position voltage. An old version of the Digital Instruments (now Veeco) software, version 3.2x, had built in square wave modulation and the software could display directly dI/dV, determined from the resulting modulation in the current. This system worked very well, but was dropped from, or incorrectly implemented in, later versions of the software. To our knowledge, none of today's manufacturers offer such complete "built-in" spectroscopic capability.

3) The control system must be able to shut off the feedback loop during data acquisition, when desired.

4) The drift (both x, y and z) must be low. The extent of acceptable x, y drift is determined by the amount of spatial resolution desired for the spectroscopic data. If one is acquiring spatially averaged molecular spectra from a dense monolayer with a large curvature tip, drifts of the order of 1 nm/s can be tolerated. On the other hand, if one is trying to locate particular atoms through their vibrational signature in STM-based IETS, drifts of less than 0.05 nm/min are required. The amount of acceptable drift in the z direction is determined by the intensity of the spectral feature to be studied relative to the background tunneling current in the absence of that feature. Since this background increases (approximately) as I and $I = CV\exp(-Az\sqrt{\Phi})$, a small change in z, dz, results in an increase in relative current (dI/I) of $-A\sqrt{\Phi}dz$. Or, using z in nm and Φ in volts, the background change in relative current with a small z drift of dz is given by $dI/I \sim -0.1\ dz$. For true vibrational IETS (with no resonance enhancements), dI/I for the spectral transition, $dI/I \sim 0.002$. Thus, less than 1×10^{-2} nm of z drift is allowed during the time required to scan a particular spectral band. For STM-OMTS, on the other hand, $dI/I \sim 0.1$, and 100 times as much z drift is allowed. Moreover, because the OMTS signal is so much stronger than that of IETS, equal signal-to-noise can be obtained about $(0.1/0.002)^{1/2} \sim 7$ times faster. Overall, STM-OMTS is expected to be about 1000 times less z drift-sensitive than STM-IETS of non-resonance enhanced transitions.

5) Sample and tip geometry and shielding: Because of the drift constraints discussed above, one needs to take spectra as fast as the electronic bandwidth allows. With the feedback loop turned off, the limiting term is the capacitance in the tip-sample assembly and the wires leading to the preamplifier circuit. Thus, the preamplifier needs to be close to the tip, and the tip and sample need to be electrically isolated and as small as possible.

6) An essential requirement for STM-IETS is that the working parts of the STM be at or below 10 K. Otherwise, the thermal line-width destroys the information inherent in identifying vibrational peak positions. It is not enough

to cool the sample; the sample, tip, and all parts physically close must be cooled to this temperature. In the case of STM-based OMTS or spectroscopic studies of density of states in metals or semiconductors, room temperature measurements are usually satisfactory, and many measurements can be made at significantly elevated temperatures.

7.2.1 Commercial Instruments

There are currently available a wide range of SPM instruments that incorporate all of the features discussed above (except for tips, which we will discuss later). A number of commercial suppliers, many of whom currently only provide atomic force microscopy (AFM) instruments and not STM, are listed below. It may be noted, however, that this is a rapidly evolving business and companies may broaden or narrow their offerings on short notice. This list is not exhaustive and the absence of a manufacturer from the list does not indicate any preference. In the United States, Veeco Metrology [24] is the major supplier of ambient and solution phase SPMs, followed by Molecular Imaging Corporation [27]. Asylum Research is an offshoot the original Digital Instruments (now owned by Veeco) and specializes AFM and pico-force measurements [28]. Novascan provides ambient scanning force microscopes, AFM tips, and chemically modified tips and samples [29]. Quesant, in partnership with Novascan, provides a full range of ambient and liquid scanning probe microscopes [30]. For UHV systems made in the United States, one must turn either to McAllister Technical [25] or RHK Technology [26]. The McAllister system is very inexpensive and has provided very high resolution images in the hands of a number of research groups, but it is a room temperature-only STM. RHK now offers both STM and AFM in UHV with variable temperature achieved by cooling the sample (only). Note, as indicated in a previous section, that for the purposes of measuring vibrational inelastic spectroscopy the entire microscope (sample and tip) must be cooled. To our knowledge there are currently no U.S. suppliers of such microscopes. While commenting on spectroscopy, it is useful to note that STM heads where the sample and tip are electrically isolated from the rest of the microscope (including the sample holder) are especially desirable for spectroscopic purposes. These designs (like that of the RHK) minimize the capacitive coupling that can limit the sampling speed.

The best-known UHV STM and AFM system provider is the German company, Omicron [31]. Omicron offers a range of UHV systems with cooled sample STM and AFM capability, and the multiprobe LT which is a dedicated low temperature STM wherein both the tip and sample are cooled. Witech is another German company that makes commercial AFM and SNOM equipment, but appears to have dropped their STM line [32]. Nanosurf is a Swiss company that manufactures extremely compact and inexpensive STM and AFM systems suitable for use in undergraduate laboratories [33]. Both

Witech and Nanosurf are respresented in the United States by Nanoscience Instruments [34]. Nanotech Electronica is a Spanish company specializing in scanning force microscopy and distributing a free SPM (including STM) data analysis program WSxM [35]. In Russia, NT-MDT is a comprehensive supplier of STM and AFM instruments and supplies, and it also offers a small education-oriented multipurpose scanning probe microscope [36] Other SPM manufacturers include Danish Micro Engineering [37], JPK [38] and Attocube [39] instruments specializing in AFM and SNOM, PSIA in Korea specializing in AFM [40], and in Germany, SIS offering ambient AFM systems [41] and Triple-O offering AFM and SNOM systems [42].

7.2.3 Tips

Once an appropriate SPM has been purchased or built from scratch, one must obtain appropriate tips and samples. Since the emphasis here is on STM, suffice it to say that there are a number of commercial sources for silicon, silicon nitride, and variously coated tips appropriate for scanning force microscopy [24, 29, 34, 36, 43]. While research grade AFM tips are commercially available at reasonable prices, this is not the situation for STM.While a few companies offer etched STM tips, they are expensive and generally unsuitable for spectroscopic applications because of their long exposure to various ambient environments. Thus, the STM practitioner makes his own tips. These fall generally into two classes, mechanically formed and etched tips.

The first STM tips were mechanically formed by Binnig et al. [44, 45]. These were formed by mechanical grinding (at 90°!) 1-mm diameter tungsten wire. Today, by far the most common mechanically formed tips are cut wire tips. While Au, $Pt_{0.9}Ir_{0.1}$, $Pt_{1-x}Rh_x$, and similar metal wires have been used, the most popular cut tip is made from 0.25-mm diameter $Pt_{0.8}Ir_{0.2}$ wire. Almost every laboratory has its own preferred cutting method. Some anneal the wire before cutting with dull scissors using a pulling motion. Others prefer to use very sharp dykes. Some suggest that a 60° angle cut is best, while others prefer larger or smaller angles. Since the "tip" is really an atomic asperity at the end of a rather rough mass of metal, it is not too surprising that the macroscopic and microscopic (as in microns) morphology have little relevance to the quality of the tip. In fact, the tips that give the best images do not always appear sharp under an optical microscope (200 to 400 power). There are almost always a number of tips like fingers of a hand extending towards the surface. See, for example, Figure 6.4a in reference [46]. Because of the strong exponential dependence of the tunneling current, only the longest "finger" is important when measuring flat surfaces. When imaging rough surfaces, or intermediate to large features (a nm or more tall) on a flat surface, one may see "ghost images" resulting from tunneling through secondary tips of nearly the same length as the longest. The offset between the ghost and the primary image is related to the offset between the longest and second longest asperity. The cure for ghost images is to re-cut the tip until the ghosts disappear. Cut $Pt_{0.8}Ir_{0.2}$ tips are inexpensive, quickly made, and have a fairly good yield (usually 1 in 2 will show atomic reconstruction lines on Au).

Electrochemically etched tips offer a well-defined geometry near the atomic asperity that functions as the tip. Methods for preparing atomically sharp tips from a number of different metals were first developed primarily for field ion microscopy [47, 48]. Etched tips are commonly made from Pt, Ir, Au, W, Pd, Ni, and Ag. Prescriptions for a number of these metals were given by Nam et al. [49]. For solution, and especially electrochemical applications, etched $Pt_{0.8}Ir_{0.2}$ tips are preferred. Various methods for making these are described in the literature [49–52], including descriptions of how to limit Faradaic currents through tip coating [53, 54]. Au tips, both etched [55, 56] and coated [57] have been reported in the literature, as have etched silver tips [58]. For spin-polarized tunneling studies, Cavallini reports that etched nickel tips work well and have better oxidation resistance than tungsten tips [59]. By far the most commonly used tip material for UHV studies is tungsten, and these tips are almost always etched tips.

One has a great variety of methods to choose from when etching W tips [49, 60–62]. One may either follow the prescriptions and designs in the literature, or purchase commercial tip etching stands such as those offered by W-tech (through Omicron) or Shrodinger's Sharpener from Obbligato Objectives [63]. Electrochemically etched W tips have an oxide layer on the surface. This oxide layer can be up to 20 nm thick [64]. Dipping the W tips in 47% HF prior to loading in the UHV chamber has been reported to improve tips [65–67], but an insulating layer is still apparently left on the tip surface [65, 67]. For simple imaging, the oxide layer on freshly prepared tips is usually not a problem. For spectroscopic studies, however, it is a major impediment since the spectrum can be dominated by the density of states of the oxide layer. The stable form of the oxide is WO_3, and can be removed by heating above 800 C [68]. At this temperature WO_3 reacts with W to form WO_2, which then sublimes. W melts at 3410 C, so a wide window exists to remove the oxide without deforming the tip. A convenient method of heating the tip to remove contaminants is to use electron bombardment. However care still needs to be taken since the local temperature at the tip can easily reach the melting point – even to the extent of forming an obvious round ball at the end of the tip wire.

Some crude in situ methods for tip cleaning have also been successful. In the earliest days of STM, it was found that applying a 10 kHz 2 nm peak to peak (vertical) oscillation to a tip initially in contact with a platinum plate could produce clean sharp tips [44]. Binnig thought that this procedure might clean the tip through some type of ultrasonic interaction. The application of a large voltage pulse (from 3 volts to hundreds of volts) has also been used for tip cleaning. Field emission cleaning is proposed to account for why application of about 100 V between tip and sample at a distance suitable to produce nA to mA currents can result in clean tips. In the above two cases, one had best use either a clean portion of the sample (that is sacrificed), or another clean metal sample. Another common method for eliminating the oxide involves 'controlled' crashing of the tip on a clean gold surface. This method does work on occasion, but it can also lead to dull (poor resolution) or even bent (unrealistic surface images) tips. An extreme example of this is to simply let the tip scan over a large area of the surface for an extended time. For all the methods of this paragraph, it is not clear whether one cleans the W, or coats the tip with the metal counter surface.

7.3 STS OF SEMICONDUCTORS & METALS

The scanning tunneling spectroscopy of metals and semiconductors primarily focuses on elastic tunneling current changes associated with the local density of states (LDOS), $\rho_s(r, E)$, which appears in equation (7.2). To a good approximation, $\rho_s(r, E)$ is proportional to dI/dV when the tip is far from the substrate and the density of states of the tip is reasonably smooth. These are elastic tunneling spectra, as will be described in a later section.

Jacklevic and coworkers first demonstrated imaging of metal surface states when they identified the Au(111) surface state in its dI/dV spectrum [69, 70]. They found that the surface state peak was centered just above –500 mV sample bias with a full width at half height of about 300 mV. They observed changes in the peak intensity and position that correlated with surface features. The surface state intensity was found to be substantially reduced at step edges as compared to values observed for large terraces. A change in the intensity by a factor of 2 over the $23 \times \sqrt{3}$ reconstruction unit cell was also observed. These effects were attributed to a spatial variation of the surface state intensity with the local potential. Upward shifts of the surface state energy were also observed on narrow terraces. Kuk and Silverman performed tunneling spectroscopy of Au(100)-(5 × 20) and Fe on Au(100) surfaces [71]. Using a well-defined tunneling tip, $I - d$, $I - V$, and $d\ln I/d\ln V - V$ spectra were obtained (d is the tip – Au separation). The results confirmed that the characteristics of the spectra resemble those of previously reported semiconductors. From $I - d$ relations, they found that the tunneling barrier decreased abruptly when the tunneling gap was < 0.6 nm. Later, Kuk reviewed the elastic scanning tunneling spectroscopy of a number of clean metal surfaces [72]. He presented the operating principles of the scanning tunneling microscope as applicable to the problem of the small corrugations seen on metallic samples. Various spectroscopies were described and compared with theory. Some examples of past accomplishments on metal surfaces were given.

Fonden and coworkers investigated unoccupied surface resonances seen in a plot of *dI/dV* versus *V* [73]. The authors extend and detail a previously developed model for formation of electronic resonances at free electron-like metal surfaces to calculate scanning tunneling spectra. The effect of the tip is mimicked by inclusion of an external field, self-consistently, in a jellium description of the surface potential. The lattice-induced corrugation of the potential is included perturbatively via a pseudopotential. The authors compare the calculated spectra for Al(111) with experimental results and conclude that a peak occurring below the metal vacuum level is a "crystal-derived" resonance, in the sense that lattice effects are crucial for its manifestation. Hoermandinger [74] and Doyen and Drakova [75] also investigated the theoretical underpinnings for the observation of metallic surface states by *dI/dV* spectroscopy.

Bischoff and coworkers examined the role of impurities on the surface state of V(001) [76]. They reported the first scanning tunneling spectroscopy measurements on V(001). A strong surface state was detected that was very sensitive to the presence of segregated carbon impurities. This surface state energy shifted from 0.03 eV below the Fermi level in clean regions, up to as much as +0.2 eV above the Fermi level in contaminated areas. Because of the negative dispersion of this state, the upward shift could not be described in a simple confinement picture. Rather, Bischoff concluded that the surface state energy was governed by vanadium surface s-d interactions, which are altered by carbon coverage. Differences in the tunneling spectra of metals have been used to provide chemical selectivity for one metal on another [77, 78]. Himpsel and coworkers studied the growth of copper stripes on stepped W(110) and Mo(110) surfaces. Contrast between copper and the substrate metal was achieved by resonant tunneling via surface states and image states.These states are characterized independently by inverse photoemission. Image states provide elemental identification via the work function, since their energy is correlated with the local work function. Element-specific surface states produce contrast at higher spatial resolution, but the contrast is smaller than that for image states [77]. Weisendanger and coworkers studied the topography and chemical surface structure of a submonolayer Fe film on a W(110) substrate by combined STM and spectroscopy [78]. Local tunneling spectra revealed a pronounced difference in the electronic structure between nanometer-scale Fe islands of monolayer height and the bare W(110) substrate. In particular, a pronounced empty-state peak at 0.2 eV above the Fermi level was identified for the Fe islands. Based on the pronounced difference in the local tunneling spectra measured above the Fe islands and the W substrate, element-specific imaging was achieved. Scanning tunneling spectroscopy has also permitted real-space observation of one-dimensional electronic states on an Fe(100) surface alloyed with Si [79]. These states are localized along chains of Fe atoms in domain boundaries of the Fe(100) c(2 × 2)Si surface alloy. The calculated spin charge densities illustrate the d-like orbital character of the one-dimensional state and show its relationship to a two-dimensional state existing on the pure Fe(100) surface.

Scanning tunneling spectroscopy can also be used to study small metallic structures on surfaces. Scanning tunneling spectroscopy and microscopy show that the empty states of linear Au clusters supported on a metal surface behave as if they are the states of an electron in an empty one-dimensional box [80]. It was suggested that certain difficulties of this description are removed by a particle-in-a-cylinder model. Their interpretation was supported by density functional calculations. Crommie et al. [81] studied the local properties of low-dimensional electrons (properties such as standing wave patterns in the surface local density of states, which is due to the quantum mechanical interference of surface state electrons scattering off of step edges and adsorbates). The authors found that Fe adatoms strongly scatter the surface state and, as a result, are good building blocks for constructing atomic-scale barriers ("quantum corrals") to confine the surface state electrons. Tunneling spectroscopy performed inside of the corrals reveals discrete resonances, consistent with size quantization [81].

The potential of *spin-polarized scanning tunneling microscopy and spectroscopy* (SP-STM/S) was by demonstrated Wiesendanger and coworkers [82, 83] on antiferromagnetic and ferromagnetic transition metals and on rare earth metals. Data measured on the antiferromagnetic Cr(001) surface revealed that screw dislocations cause topology-induced spin frustrations leading to the formation of domain walls with a width of about 120 nm [82]. On another antiferromagnetic surface a pseudomorphic monolayer film of chemically identical manganese atoms on W(110), they showed that SP-STM provides the surface magnetic structure with atomic resolution. SP-STS also allows the imaging of the domain structure of self-organized Fe nanostructures, which are antiferromagnetically coupled due to dipolar interaction. Using spin-polarized scanning tunneling microscopy in an external magnetic field, the Wiesendanger's group observed magnetic hysteresis on a nm scale in an ultrathin ferromagnetic film [83]. The film was an array of Fe nanowires two atomic layers thick grown on a stepped W(110) substrate. The microscopic sources of hysteresis in this system (domain wall motion, domain creation, and domain annihilation) were observed with nm spatial resolution [83]. A saturation field stable residual domain was found measuring 6.5 nm by 5 nm. Its stability was ascribed to the consequences of a 360° spin rotation.

In the early days of STM it was noted (first with dismay, but then with excitement) that the image obtained from a semiconductor surface depended significantly on the polarity and magnitude of the applied bias [6, 7, 10, 12–15]. A wonderful example of this is provided by the Si(001) reconstructed surface. Figure 7.4 shows the constant current image (a) at –2 V bias while (b) was taken at +2 V sample bias [8]. As will be discussed in detail in subsequent sections, the negative bias image reflects occupied regions of the LDOS while the positive bias image probes unoccupied regions. In the case of this reconstructed surface, two Si atoms form strong and relatively localized double bonds. These pairs of atoms have occupied π and unoccupied π* orbitals near the Fermi surface. All the other occupied states lie deeper than the π and the other unoccupied states lie above the π*.

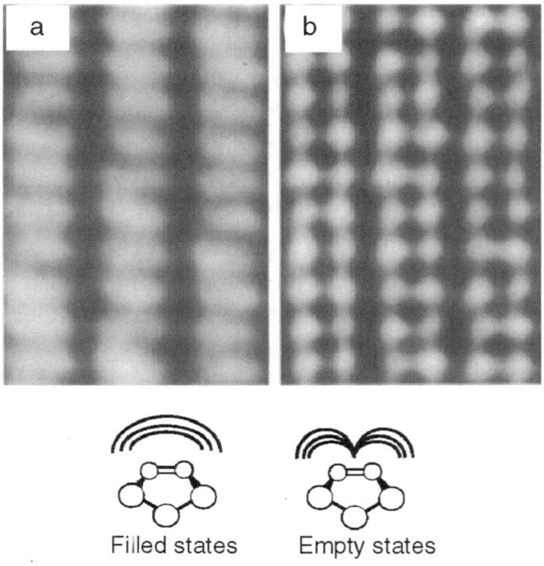

Figure 7.4 a) Negative bias, and b) positive bias constant current STM images of the Si(001)-(2 × 1) reconstructed surface. The cartoons at the bottom depict side views of the surface dimer structure giving rise to filled and empty π and π* type orbitals. Reprinted with permission from [8]. Copyright 1996 The American Chemical Society.

When the sample bias voltage is near –2 V most of the tunneling is going through the localized π orbital (the max in the LDOS at that bias), and the image is essentially a picture of the π bonding orbitals of the Si-Si dimers (Figure 7.4a). If the bias is set to +2 V, most of the tunneling current is carried through the π*-like unoccupied portion of the LDOS. Because there is a node in the spatial distribution of the antibonding dimer wavefunction, the tip must push in close to the surface to maintain constant current. Similarly, the tip can pull back as it moves over the regions where the antibonding wavefunction (and therefore available electron density) is large.

There are several extremely important general messages here. First, the images seen in constant current STM reflect the density of states at the bias energy and tip position ($\rho_s(r, E)$), and are generally not what one identifies as the "structure." The structure is generally derived from scattering experiments in which all the electrons contribute. Second, the STM is actually providing a surface site-dependent spectroscopy. In the example above, the occupied π states were probed when the bias was near –2 V and the unoccupied π* were probed with the bias set at +2 V. Moreover, we only "saw" them when the tip was in a region of high electron density for that particular state. Finally, the data in Figure 7.4 suggest that there should be a way to literally measure the surface state spectra as a function of position.

The first site-resolved STM-based elastic tunneling spectrum was reported by Hamers, Tromp, and DeMuth [84] and is partially reproduced in Figure 7.5. *I*/*V* and *dI*/*dV* curves were taken at fixed sample-tip height at different points in the 7 × 7 unit cell [8, 10, 84]. The unit cell contains 12 adatoms and 6 rest atoms (positions indicated by squares and dots in Figure 7.5). The *I*/*V* curves show that the adatom sites contribute significantly to the LDOS near +0.7 V while the rest atom sites do not. On the other hand, the occupied states near –1 V bias are predominantly located on the rest atom sites. *Ultraviolet photoelectron spectroscopy* (UPS) and *inverse photoemission spectroscopy* (IPS) are techniques that provide the spatially averaged occupied and empty (respectively) state densities, and these spectral results are shown in the middle portion of Figure 7.5. Finally, the spatially averaged tunneling spectrum (reported as a normalized intensity — *dI*/*dV*/(*I*/*V*)) is presented in the lower third of Figure 7.5. As stated above, the *dI*/*dV* plot is expected to be proportional to the local density of states, and the spatially averaged spectrum should approximate the density of states for the entire surface. It is gratifying, therefore, to see that the UPS, IPS, and tunneling spectra are all in close agreement.

7.4 ELECTRON TUNNELING SPECTROSCOPY OF ADSORBED MOLECULES

Instead of the quantum mechanical conceptual structure depicted in Figure 7.1, let us now consider a real tunneling device — either a metal-insulator-metal (M-I-M) tunnel diode, or a substrate and STM tip. In a real device, there are many electrons and the Pauli principle plays a key role. A simple and useful model for the conduction electrons in a metal assumes that one begins by removing the valence electrons, then spreads the remaining positive charge into a uniform distribution (jelly) producing a simple constant potential box. Into this box the valence electrons are returned two at a time into each energy level until the metal is just neutrally charged. The energy of the last electron to go in is the Fermi energy, E_F, and the energy required to just remove it from the metal is the work function, Φ. If there are no molecules in the barrier region (tunneling gap), the current is approximately proportional to the voltage difference between the two metals (called the *bias*) and $\exp(-Ad\sqrt{\Phi})$. This current is said to be due to elastic tunneling since the electron loses no energy to the barrier.

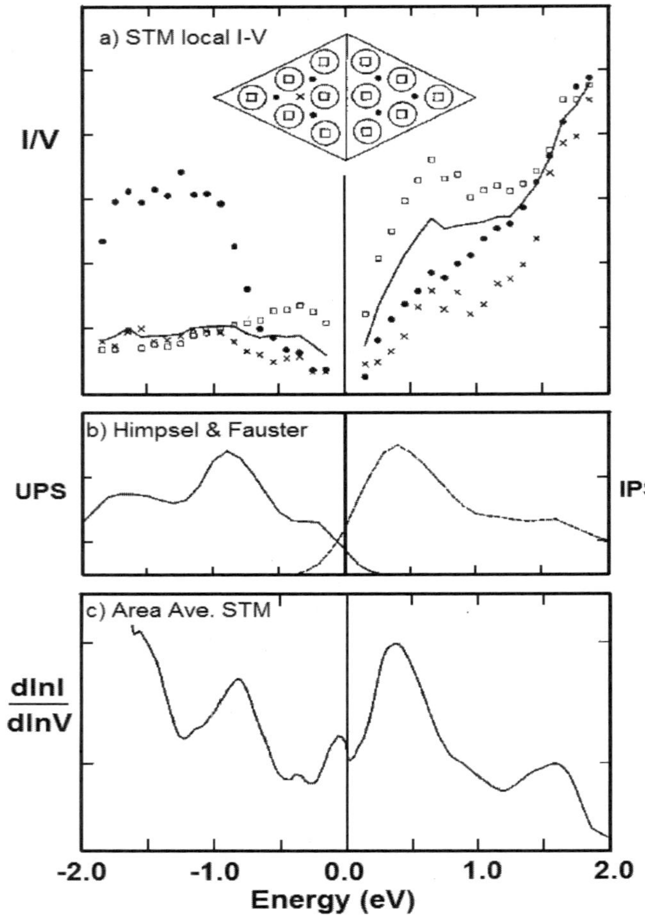

Figure 7.5 First atomically resolved tunneling spectrum obtained on Si(111)-(7 × 7). (a) Conductance as a function of position, (b) ultraviolet photoelectron spectrum and inverse photoemission spectrum, and (c) area averaged normalized tunneling spectrum of the reconstructed surface. Reprinted from [8]. Copyright 1996 The American Chemical Society.

If the gap (barrier) between the electrodes is not a vacuum, equation (7.1) must be modified in several ways. The simplest effect is a reduction in the effective barrier height. For an insulator or semiconductor, it may only require a volt or two of energy above E_F for the electron to reach the conduction band in the barrier, while the work function may be 4 to 6 V. In these cases, the work function in equation (7.1) is replaced by barrier height, Φ_b, where Φ_b is approximately the difference in energy between the bottom of the conduction band (in the insulator) and the Fermi energy in the electrodes at zero applied bias. If individual molecules are present in the barrier, several new interaction mechanisms can affect the tunneling current. The best known of these is inelastic electron tunneling and is the basis for *inelastic electron tunneling spectroscopy* (IETS) [4–7]. In IETS the moving electronic charge interacts with the time-varying molecular dipoles (electronic or vibrational) to induce excitation of the molecule in the barrier with concomitant loss of energy by the electron. This process is similar to a Raman photon process. Consider a vibrational motion with frequency ν and energy spacing *h*ν. This is shown in Figure 7.6 as an excitation from the ground vibrational state to the first excited vibrational state with a corresponding loss of energy by the tunneling electron.

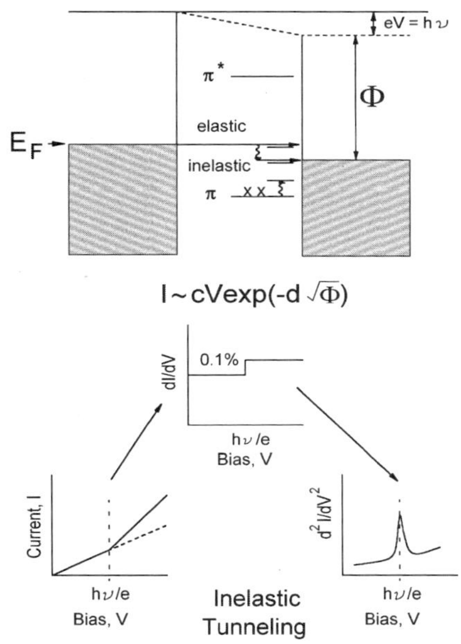

Figure 7.6 Inelastic tunneling process and associated spectral peaks. *h*ν is the molecular energy level spacing, ϕ is the barrier height in eV (approximately the metal work function for a symmetrical vacuum barrier), *d* is the barrier width in Angstroms, *I* is the tunneling current, and *V* is the applied bias voltage. The reader is encouraged to consult references [5] and [27] for a detailed description of this diagram and its interpretation. Reprinted with permission from [85]. Copyright 2005 the Journal of Chemical Education.

If the applied voltage is less than *h*ν/*e*, the inelastic channel is closed because the final states for the tunneling electron, the electronic levels in the metal electrode of the appropriate energy, are already filled. At $V = h\nu/e$ the inelastic channel opens. Further increases in *V* result in additional possible final states with an associated increase in current due to this channel. As is depicted in Figure 7.6, this opening of an inelastic conductance channel results in a break in the *I*(*V*) curve at $V = h\nu/e$. Note that the size of the break is exaggerated. In the conductance *dI*/*dV*, the opening of the inelastic channel is signaled by a step — typically 0.1% of the tunneling electrons utilizes a vibrational inelastic channel and 5% utilize electronic inelastic channels. To obtain an IETS spectrum we can plot d^2I/dv^2 versus V and expect to see peaks whenever the energy difference between the ground and excited state (electronic or vibrational) just matches the applied bias voltage. The width of IETS bands depends upon the sharpness of the thermal distribution of electron energies. Thus, the IETS linewidth (full width at half height) is 5.4 kT (3.5 T cm^{-1} or about 0.5 T mV, where temperature is in Kelvin) [86] and vibrational IETS is most often performed below 10 K. As the bias voltage, V_{bias} or V_b, increases (in either sign!) higher vibrational excitation (as seen in Figure 7.7) or even inelastic electronic excitation can occur. As shown in Figure 7.7 for a tunnel diode containing a submonolayer of VOPc (vanadyl phthalocyanine), both electronic and vibrational inelastic transition can be seen [87]. It is important to note that *IETS bands appear at the same bias magnitude independent of sign, although the intensities may differ* [5–7]. This is a diagnostic feature for non-resonant IETS. For many electronic transitions, the intensity and bandwidth are orders of magnitude greater than for vibrational IETS, and cooling to 100 K, or even higher temperatures, is sufficient.

In its simplest form, an IET spectrum is a plot of d^2I/dV^2 versus V. It turns out that using $d^2I/dV^2/(dI/dV)$ as the y-axis provides spectra having flatter baselines and is most appropriate for high bias work [6, 7, 88–90]. These are called *normalized tunneling intensities* (NTI) or constant modulation spectroscopy. Tunneling spectra are measured by applying both a variable bias V and a small modulation component V_f at frequency f. A lock-in amplifier is used to detect the $2f$ signal, which is proportional to d^2I/dV^2. The instrumentation required for obtaining normalized intensities NTI is a bit more complex [88–90]. In general, the bias voltage may be converted to the more conventional wavenumbers through the factor of 8066 cm^{-1}/volt. The amplitude of the modulation affects both the observed signal strength and resolution. The signal increases as V_f^2 but the experimental linewidth is proportional to V_f [5, 7, 86].

Until about 1988 essentially all tunneling spectroscopy was carried out in tunnel diodes and almost all of it was IETS. In 1989 Hipps and Mazur began observing strange vibrational line shapes and huge new signals that were as big or greater in intensity than electronic IETS but that could not be explained by a simple molecular excitation process [91]. These new transitions produce peaks in dI/dV (rather than d^2I/dV^2) and are due to direct tunneling via either unoccupied or occupied molecular orbitals and were thus termed *orbital mediated tunneling* (OMT) bands. The very intense but weirdly shaped band *seen only in positive bias* (near +0.3 volt) in Figure 7.7 is an example of one of these bands. These transitions are due to what is (approximately) an *elastic tunneling mechanism* in that the energy of the tunneling electron that causes the excitation is the absolute energy (not an energy difference) of a molecular state (see Figure 7.7). The temperature dependence of a delta function line is only 3.5 kT (0.3 T mV) and amounts to 90 mV at 300 K. In fact, the orbital mediated tunneling (OMT) bands are usually not temperature-dependent until well in excess of 500 K. This is due to the large intrinsic width of OMT bands (usually more than 0.25 V).

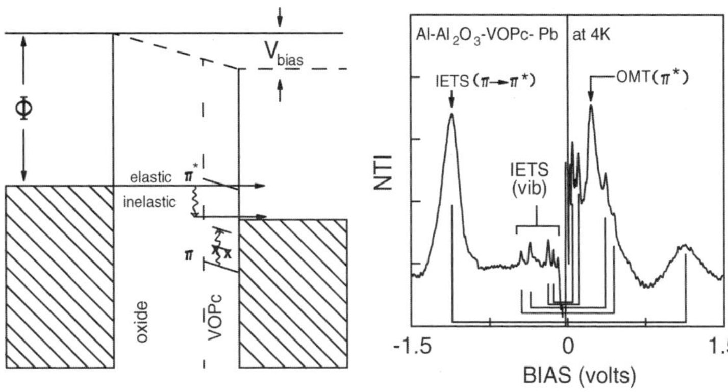

Figure 7.7 Schematic diagram of inelastic and resonant elastic tunneling processes (left). Tunneling spectroscopy data obtained from an Al-Al_2O_3-VOPc-Pb tunnel junction at 4 K with 10 mV rms modulation (VOPc is vanadyl phthalocyanine). Vertical lines guide the eye to the inelastic excitations which occur in both bias directions. The orbital mediated tunneling spectroscopy (OMTS) band associated with transient reduction of the Pc ring, on the other hand, appears only on the Pb+ bias side. Reproduced by permission from [87]. Copyright (2000) The American Chemical Society.

The exact mechanism of the OMT process can vary from case to case. It might be true resonance tunneling where the effective residence time of the tunneling electron on the molecule is negligible compared to nuclear motion. It might be a real oxidation or reduction of the molecule followed by thermally induced return to the original charge state (electron hopping), or it might be a redox that occurs too rapidly for thermal relaxation, such as in ultraviolet photoelectron spectroscopy (UPS) or inverse photoemission spectroscopy (IPS). Because there are a number of different physical processes that can give rise to these bands, the spectroscopy associated with measuring these transitions is called *orbital mediated tunneling spectroscopy* (OMTS) [92–94]. Because the timescales are usually unknown, we often refer to these transitions "transient" redox processes. The technique might equally well be called ionization and affinity level spectroscopy. Ionization spectroscopy is the measurement of the energy required to remove electrons from a filled (or partially filled) orbital. Affinity level spectroscopy measures the energy released when an electron is captured by an atom or molecule. Since there are generally several vacant orbitals that may be occupied, there is a spectrum of affinity levels associated with the addition of a single electron.

A qualitative understanding of OMTS may be obtained with reference to Figure 7.8. When the sample is biased positively ($V_{bias} > 0$) with respect to the tip, and assuming that the molecular potential is essentially that of the substrate [95], only the normal elastic current flows at low bias (≤ 1.0 V in Figure 7.8). As the bias increases electrons at the Fermi surface of the tip approach, and eventually surpass, the absolute energy of an unoccupied molecular orbital (the LUMO at +1.7 V). OMT through the LUMO at $\Phi - 1.7$ V below the vacuum level produces a peak in *dI*/*dV* seen in the actual STM-based OMTS data for cobalt (II) tetraphenylporphyrin (CoTPP). If the bias is increased further, higher unoccupied orbitals produce additional peaks in the OMTS. Thus, the positive sample bias portion of the OMTS is associated with electron affinity levels (transient reductions). In reverse (opposite) bias as in the lower part of Figure 7.8, the LUMO never comes into resonance with the Fermi energy and no peak due to unoccupied orbitals is seen. However, occupied orbitals are probed in reverse bias. In the CoTPP case, there are two occupied orbitals near the Fermi energy. The half-filled $d_z{}^2$ orbital lies only –0.1 volt below the Fermi energy while the highest occupied porphyrin ring orbital is found at –1.20 V bias. The fully occupied ring MO, therefore, is located at $\Phi + 1.20$ V below the vacuum level and produces a peak in *dI*/*dV* at –1.20 V sample bias [96, 97]. It is also clear from Figure 7.8 that there are deeper occupied orbitals, for example, there is one near –1.5 V bias giving a well-defined shoulder.

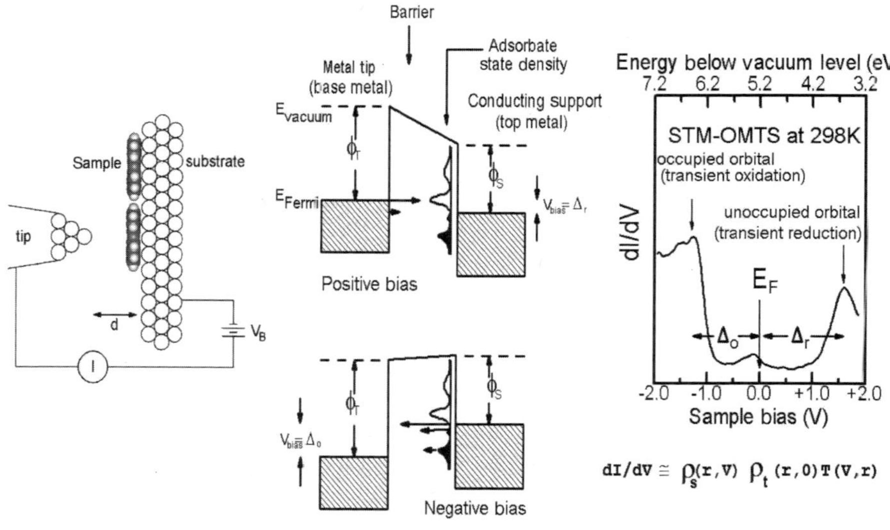

Figure 7.8 Schematic diagram of orbital mediated tunneling spectroscopy and representative spectrum obtained from cobalt (II) tetraphenylporphyrin in an STM under UHV conditions at room temperature. The central diagram shows resonant tunneling through unoccupied (upper) and occupied (lower) orbitals in positive and negative bias, respectively. This diagram works equally well for an M-I-M junction (base and top metal labels) and for an STM (tip and substrate labels).

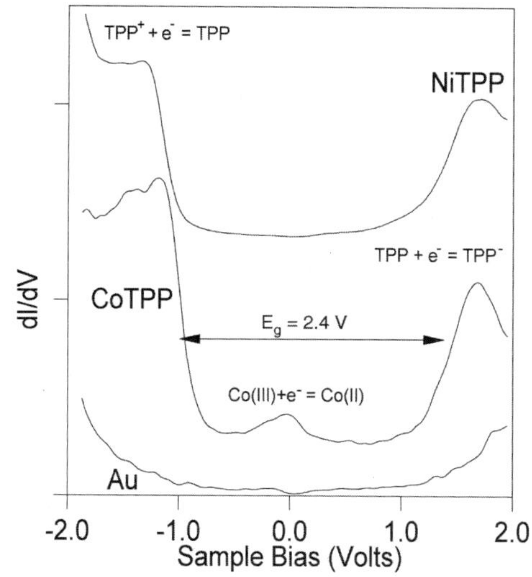

Figure 7.9 Elastic tunneling spectra of a Au substrate, CoTPP on Au, and of NiTPP on Au taken with a clean tungsten tip in UHV at room temperature. The orbital mediated tunneling peaks obtained from the metallorganic adlayers are clearly distinguished from the contributions of tip (W) and substrate (Au). Data obtained by K.W. Hipps, D.E. Barlow, L. Scudieo, and U. Mazur during the study reported in [96, 97].

Note that peaks are observed in *dI*/*dV* (and not *I*). This is because once current starts to flow through orbital-mediated channels, increasing bias doesn't turn it off. On the other hand, the probability of tunneling is greatest for electrons near the Fermi surface; so, as the Fermi surface passes the appropriate orbitals, *dI*/*dV* is maximized. Another way of seeing this is through differentiation of equation (7.2). This derivative is given to good approximation by the formula for *dI*/*dV* shown in Figure 7.8. Note that the density of states of the tip contributes equally with that of the sample. Thus, any contamination (including surface molecules picked up during scanning) of the tip will lead to contributions in the OMTS. It is very important, therefore, to ensure that the tip and substrate density of states peaks are not confused with those of the adsorbate. Figure 7.9 provides an example of how the tip and substrate density of states contribute to the overall OMTS in the case of two different tetraphenylporphyrin complexes. In the case of the data presented in Figure 7.9, the surface coverage was about 0.7 of a monolayer and regions of clean gold, separated, well-defined islands of either CoTPP or NiTPP [96, 97]. By alternately acquiring spectra over the molecular islands and the clean substrate regions, it is possible to precisely identify the OMTS of the adsorbate.

7.5 PRACTICAL CONSIDERATIONS RELATING TO STM-IETS AND STM-OMTS

The first molecular excitations seen in the STM were vibrational IET bands [98–100], followed closely by OMT spectra [87, 96, 101]. One might think that vibrational IETS in the STM should be the Holy Grail of surface analysis. It offers the exquisite selectivity of vibrational spectroscopy combined with the possibility of submolecular spatial resolution. However, in actual practice, it has proven to be less valuable than expected.

Consider, for example, the STM-based IETS obtained from C_{60} on Ag(110) at 4.5 K by Pascual and coworkers (Figure 7.10) [102]. This is an interesting case because both the tunnel junction and STM-based IETS are available. Pascual's data are as clean as any STM-IETS reported, and his paper provides the STM-OMTS as well as the STM-IETS. Thus, he could correlate the active vibration with the change in geometry induced by the electronic excitation. As in tunnel diode IETS, one expects that inelastic transitions should appear symmetrically in either bias direction. This is observed. The number and intensity of bands is not as expected from junction-based IETS.

C_{60} has 176 possible vibrational eigenvalues, but the high symmetry and associated degeneracy reduce the number of unique frequencies to 46. These modes have one-, three-, four- and five-fold degeneracy. Four of these modes (the F_{1u}) are IR active, and 10 are Raman active ($2A_g + 8H_g$). While conventional M-I-M spin-doped tunnel junctions provided only 24 of the 46 modes, these were only partially overlapped on the 24 modes seen in inelastic neutron scattering and the 4 IR and 10 Raman modes [103]. Thus, all but 9 of the 46 fundamental bands could be assigned through a combination of all four data sets. Very recently, Nolen and Ruggiero used an interesting composite barrier design to make tunnel junctions that provided IETS wherein 26 of the silent modes of C_{60} were identified [104]. Thus, all but 6 of the 46 fundamentals have been directly observed through the concerted use of IR, Raman, INS, and M-I-M-based IETS. In the tunnel junction IETS, more than 14 vibrational bands are observed in the same region (0 to 100 mV) where only one is clearly seen in STM-IETS (Figure 7.10). Moreover, the one band seen in Figure 7.10a is about 100 times the intensity of the bands observed in tunnel junction experiments [102–104].

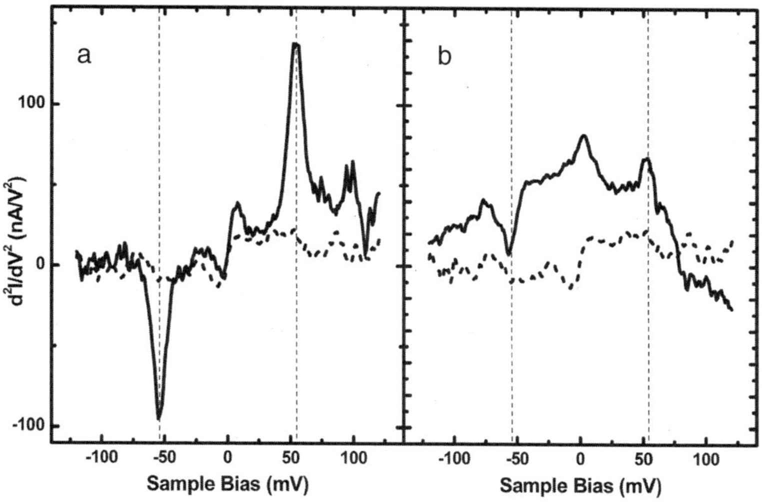

Figure 7.10 Comparison of consecutive STM-IETS obtained from two neighboring C_{60} molecules adsorbed on Ag(110). The dashed spectra were taken over bare silver. The normalized intensity for the 54-mV peaks in spectrum (a) is about 9%. Set-point (I = 1.6 nA, V = 0.5 V), modualtion = 5 mV, temperature = 4.5 K, modulation frequency = 341 Hz. Reprinted with permission from [102]. Copyright 2002 American Institute of Physics.

Observing vibrational IETS in an STM is experimentally demanding, and interpretation of the data is challenging. The 5 kT linewidth associated with IETS bands requires that STM-IETS experiments be performed at cryogenic temperature (usually near or below 10 K) and in ultrahigh vacuum (UHV). The signals are generally weak and require extremely stable instruments [98–100, 102, 105]. Unlike M-I-M' diode IETS where data are easily correlated with IR and Raman peak positions and lineshapes [5, 6, 7], surprisingly few vibrational modes are seen in STM-IETS and their lineshapes are generally not easy to predict [99, 105–107]. The STM-IETS bands are often derivatives or even inverted relative to diode-IETS bands. This behavior is reminiscent of the vibrational peaks seen in diode-IETS when there are OMT bands near the vibrational bands [108]. It is very likely that a resonance mechanism is essential for producing sufficient vibrational IETS intensity in the STM environment [102, 106]. While the M-I-M diode spectroscopist has long taken advantage of the "fingerprint" bands familiar to the IR chemical analyst, they are not accessible in STM-IETS because the resolution required (~0.5 mV) is beyond the current sensitivity limit for STM-IETS and because most are strangely silent.

In sharp contrast to STM-IETS, STM-OMTS can be performed at room temperature, and most commercial UHV scanning tunneling microscopes have sufficient mechanical and electronic stability to allow spectra to be acquired. Moreover, the location and nature of bands observed can be easily interpreted in terms of the electronic orbitals (both occupied and unoccupied) of the molecular system of interest. This latter strength is sometimes a weakness, since molecular systems having large band gaps (> 5 eV) have OMTS bands that can only be observed at very large applied voltages. At large bias voltage the elastic tunneling current and the potential for current-induced instabilities in the tip and substrate can effectively mask the OMTS. Fortunately, a very large percentage of all molecules have either occupied or unoccupied states within a few volts of the Fermi energy of a typical metal substrate, and many have both occupied and unoccupied states within this range. The large background elastic currents at high bias can be reduced in significance by reporting the normalized OMTS, $(dI/dV)/(I/V)$ (vide infra), thereby canceling out much of the exponential dependence of $T(E, eV, r)$ on V. We will place our primary focus on STM-OMTS in the examples that follow.

7.5.1 STM-Based Orbital-Mediated Tunneling Spectra and Electrochemistry

One view of the OMT process is that the molecule M is reduced, M^-, or oxidized, M^+, during the tunneling process [92, 93, 96, 101, 109–112, 115]. In this picture a fully relaxed ion is formed in the barrier. The absorption of a phonon (the creation of a vibrational excitation) then induces the ion to decay back to the neutral molecule with emission of the electron—which then completes tunneling through the barrier. For simplicity, the reduction case will be discussed in detail; but the oxidation arguments are similar. A transition of the type $M + e^- \Rightarrow M^-$ is conventionally described as formation of an electron affinity level. The most commonly used measure of condensed phase electron affinity is the half-wave reduction potential measured in non-aqueous solvents, $E_{1/2}$. Often these values are tabulated relative to the saturated calomel electrode (sce). In order to correlate OMTS data with electrochemical potentials, we need the data referenced to an electron in the vacuum state. That is, we need the potential for the half-reaction:

$$M(\text{solution}) + e^-(\text{vac}) \Rightarrow M^-(\text{solution})$$

These values can be closely approximated from those referenced to sce by adding 4.70 V to $E_{1/2}$ (sce) [111–114]. That is, $E_{1/2}(\text{vacuum}) = E_{1/2}(\text{sce}) + 4.71$ V. This connection between solution phase electrochemical potentials and vacuum level-based spectroscopic values such as OMTS and UPS is extremely useful, but the derivation is rather complex. For example, the difference in electron affinity in the gas phase and in solution is primarily due to solvation stabilizing the reduced form. The reader wishing to better understand its origins is encouraged to consult [113] and [114]. The various energy conventions are depicted in Figure 7.11, where the connection between OMTS bands and electron affinities is made within the context of the model. This diagram is based on one presented by Loutfy et al. [113] and expanded to OMTS by Mazur and Hipps [7, 92, 93, 96, 101, 115–118]. By using the measured value of E_F (from UPS), the OMTS bands can be located both relative to the vacuum level and also to electrochemical potentials. The details of the procedure for locating the vacuum level will be presented in a later section.

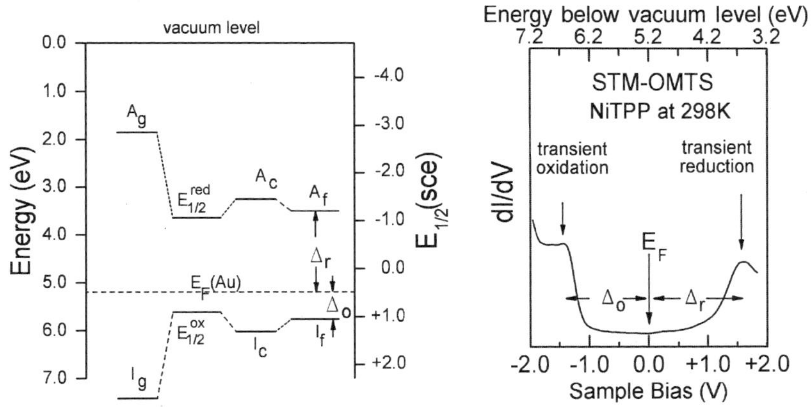

Figure 7.11 Electrochemical energy level model for orbital-mediated tunneling. A_g and A_c are the gas and crystalline phase electron affinities, $E_{1/2}$(sce) is the electrochemical potential referenced to the saturated calomel electrode, and provides the solution phase electron affinity. E_F is the Fermi level of the substrate (Au here). The corresponding positions in the OMT spectrum are shown by Δ_r and Δ_o and correspond to the electron affinity and ionization potential of the adsorbate film modified by interaction with the supporting metal, A_f. The spectrum is that of nickel (II) tetraphenylporphyrin on Au(111). Reprinted with permission from [85]. Copyright (2005) The Journal of Chemical Education.

Because the dielectric constant of most organic solids is less than that of common solvents, redox potentials in the solid state are expected to differ from those in solution as shown (qualitatively) in Figure 7.11 [113]. There will also be shifts associated with intermolecular interaction that are very difficult to predict and that vary considerably in different types of molecules that are not depicted. Moreover, there will be shifts in the potentials of a thin film relative to that of a solid due to interactions with the metal support and counter electrode, including image charge effects. These all tend to stabilize ion formation. Thus, they act to return the ionization and reduction potentials for the species adsorbed on a metal surface to those for the species in solution, (as shown in Figure 7.11). There also may be an opposite signed shift due to the absence of a covering layer of solvent or adsorbed molecules in the case of a monolayer (or less) in UHV [116]. Another complication is the fact that electrochemical potentials are equilibrium values and therefore reflect the energy associated with the formation of an ion in its equilibrium state. OMTS transitions, as discussed in the next section, may occur so rapidly that the ion is formed in an excited state—a vertical transition in the Frank-Condon sense.

For a wide range of materials and film thickness (sub—monolayer to about 0.5 nm) studied to date, a fortuitous cancellation of polarization terms and differences between vertical and equilibrium affinities has resulted in many OMTS bands lieing close to the positions predicted from electrochemistry (see Figure 7.12). This correlation is especially good for *unoccupied* orbitals. Unfortunately, OMTS bands associated with *occupied* orbitals generally lie *deeper* than predicted by solution phase electrochemical oxidation potentials.

The transferability of electrochemical values to thin film band positions for affinity levels but not ionization levels indicates that the polarization energy terms differ for these processes. This is a failure in the simple model used to generate Figure 7.11, where it was assumed that only the sign of the polarization energy changed [101]. This failure is particularly large for porphyrins. Given the trends in stabilization of ion energies by the surrounding molecules and image charges induced in the metal substrate, we would expect the ionization potential of thin film NiOEP to be about 0.5 to 1.0 eV less than for the gas phase. Instead, the ionization energies measured from a thin film are nearly identical to those reported from the gas phase [101, 119]. As we shall see in the next section, the STM-OMTS band positions are consistent with UPS observations, suggesting that the problem is in the model, not the technique.

While the first electrochemical reduction potential provides an estimate for A_c (assuming it is a reversible nonreaction process), the second and higher reduction potentials do not provide the spectrum of single electron affinity levels. Rather, they provide information about 2-electron, 3-electron, and higher electron reduction processes and, therefore, depend on electron pairing energy. Thus, the utility of solution phase reduction potentials for estimating solid state affinity levels is limited to the lowest affinity level. The same argument applies to oxidation potentials beyond the first. OMTS, on the other hand, probes the single electron reduction energies for the spectrum of states of the negative ion and the single electron ionization energies for the spectrum of states of the positive ion. Thus, OMTS can be used to determine ionization spectra and affinity levels beyond the first transitions of each type [7, 115].

Molecular ball and stick models of a metal tetraphenylporphyrin (MTPP) and of a metal phthalocyanine (MPc) are presented in Figure 7.13. Metal ions are in the +2 oxidation state while the rings are in the –2 state. Note that there are 4 nitrogen atoms coordinating each metal. The Pc ring has a total of eight nitrogen atoms while the TPP ring has only four.

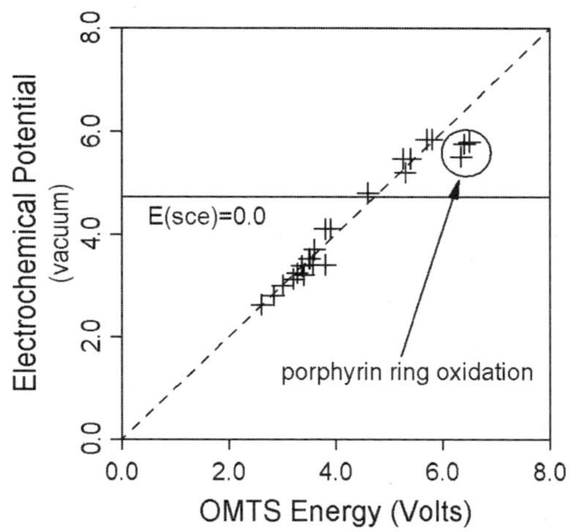

Figure 7.12 Correlation between electrochemical potentials and OMTS bands for more than 10 compounds including polyacenes, phthalocyanines, and porphyrins. OMTS data were acquired both from tunnel junctions and STM measurements, and individual points have been reported in [92–94, 96, 101, 115,117, 118]. The potential on the left is that associated with the half-reaction $M(\text{solution}) + e^{-}(\text{vac}) \Rightarrow M^{-}(\text{solution})$.

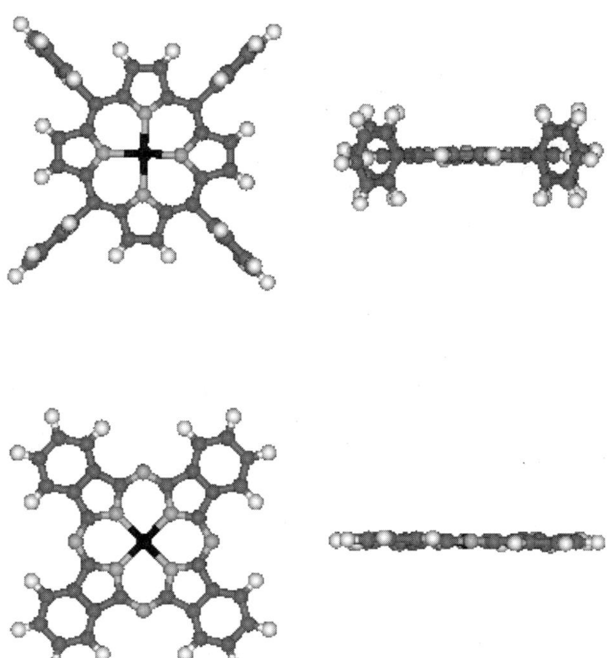

Figure 7.13 Molecular models of MPc and MTPP.

7.5.2 STM-Based OMTS and Ultraviolet Photoemission Spectroscopy

An alternative mechanism for OMTS is one in which the electron residence time is long enough to cause electronic excitation of the molecule, but not so long as to allow vibrational relaxation to occur. Thus, the electron capture or emission is a vertical process in the Frank-Condon sense. For the reverse bias region of the OMTS, this is essentially the condition for ultraviolet photoemission. In general, ultraviolet photoemission spectroscopy (UPS) is described by the expression

$$M_0 + h\nu \text{ ---->}\ M_j^+ + e^-$$

The ionization energy to produce state j of the positive ion M^+ is expressed by: $\varepsilon^F_j = h\nu - KE_j$, where ε^F is the ionization energy measured relative to the Fermi energy (E_F), KE is the kinetic energy of the ejected photoelectron, and $h\nu$ is the energy of the photon. KE and ν are the directly measured quantities. M_0 is the ground state molecule of interest and M_j^+ is the ionized molecule in its jth excited electronic and vibrational state reflecting a vertical Frank-Condon transition.

To acquire a UPS spectrum, one irradiates the sample with UV light, causing electrons to be ejected from the higher bands. The kinetic energy of these ejected electrons is then measured. UPS is an extremely surface-sensitive tool, probing only about 1 nm into the surface of the sample. Because of the photoionization cross-section differences at the relatively low energies typically used for UPS, the technique is much more sensitive to p-type valence orbitals than d-type. Thus, valence shell UPS is the technique of choice for studying the highest energy π orbitals of a molecular system. Figure 7.14 shows schematic band diagrams for a metal, and for a molecular film coated on a metal, that illustrate the important energetic parameters derivable from UPS data [17, 120–122]. The energy of the vacuum level over the molecular coating relative to that of the clean metal is given by the quantity Δ (shown for the case where Δ is a negative quantity: This is *not* Δ_o or Δ_r). The work function of the clean metal is given by Φ_m, and can be calculated from the relationship $h\nu = W_m + \Phi_m$, where W_m is the width of the photoelectron spectrum of the clean metal expressed in eV. In order to compare ionization energies to other quantities, one needs the ionization energy of an occupied orbital relative to the vacuum level. This is given by

$$\varepsilon_j = \varepsilon_j^F + \Phi_m + \Delta \tag{7.3}$$

where ε_j is the energy of the jth occupied state relative to the vacuum level and ε_j^F is the energy of the jth occupied state relative to the Fermi level of the metal contact.

Both the UPS and the STM-OMTS obtained from a nickel(II) octaethlyporphyrin (NiOEP) film on Au (111) are depicted in Figure 7.15. The UPS sample was about 1monolayer (~10^{12} molecules) while the STM-OMTS was taken from a single molecule. In order to place both spectra on the same scale, one correction and one assumption was required. The assumption was that the potential at the molecule was essentially the same potential as the substrate (gold electrode). We will address this assumption in more detail later.

The correction required first that the vacuum level over the NiOEP film on gold be correctly located. This was done using equation (7.3), where Φ_m and Δ were obtained form clean gold and NiOEP-covered gold spectra. The OMTS band energies were also referenced to the vacuum level through the measured work function. The vacuum level-referenced OMTS energy ε_{OMTS} is then calculated from the bias voltage of the transition peak V_{peak} using equation (7.4):

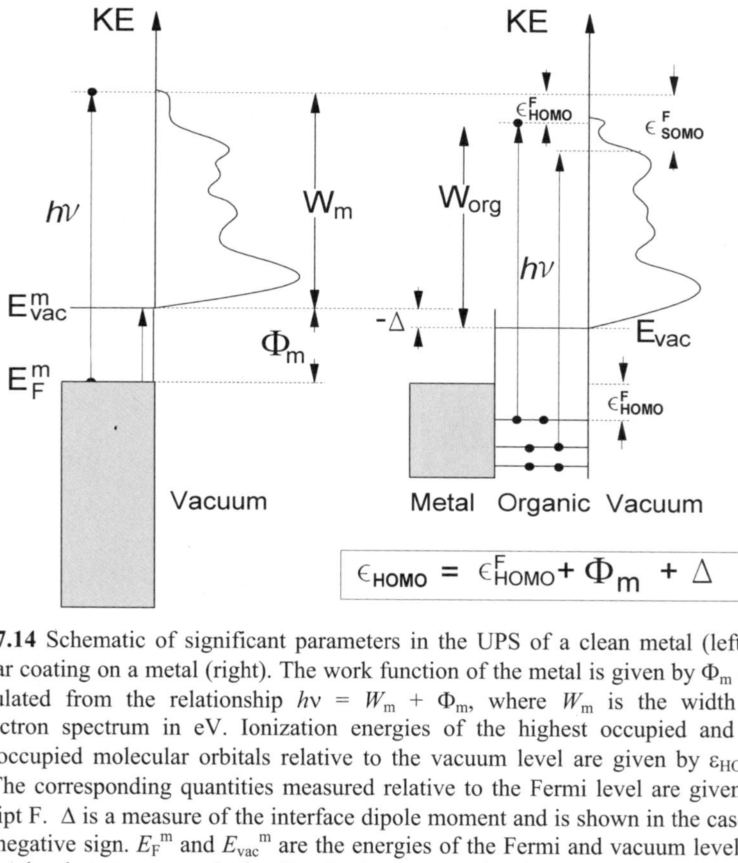

Figure 7.14 Schematic of significant parameters in the UPS of a clean metal (left) and a molecular coating on a metal (right). The work function of the metal is given by Φ_m and can be calculated from the relationship $h\nu = W_m + \Phi_m$, where W_m is the width of the photoelectron spectrum in eV. Ionization energies of the highest occupied and second highest occupied molecular orbitals relative to the vacuum level are given by ε_{HOMO} and ε_{SOMO}. The corresponding quantities measured relative to the Fermi level are given with a superscript F. Δ is a measure of the interface dipole moment and is shown in the case where it has a negative sign. $E_F{}^m$ and $E_{vac}{}^m$ are the energies of the Fermi and vacuum levels of the clean metal substrate, respectively. E_{vac} is the vacuum level energy over the adsorbate-covered metal surface. Note that these are cartoons. In real cases, one sometimes sees the Fermi edge of the metal superimposed on that of the adlayer (high *KE* end of the spectrum). Reprinted with permission from [85]. Copyright (2005) the Journal of Chemical Education.

$$\varepsilon_{\mathrm{OMTS}} = -e\,(V_{\mathrm{peak}}) + \Phi_{\mathrm{m}} + \Delta\,. \tag{7.4}$$

The values of Φ_m and Δ were the same as used for the UPS spectrum. The derived values of energy relative to the vacuum level are shown in the right panel of Figure 7.15 as the top scale.

In some cases the UPS spectrum is taken with the adsorbate on a different crystal face or metal than is used for the OMTS. In these cases caution is required. Different values of Φ_m must be used for UPS and OMTS, and the values of Δ may also vary.

Thus, it is possible to make a direct comparison of occupied orbital energies as derived from UPS and those observed by OMTS. We have made these measurements for a number of phthalocyanines and porphyrins [85, 96, 101, 118] and find good agreement between the UPS positions of the occupied π-type orbitals and the band positions seen in OMTS. In cases where there are occupied d orbitals near the Fermi surface, the OMTS spectrum is richer than that of the UPS. Because of the difference in cross-section for electron emission by tunneling and electron ejection by photon absorption, d-type orbitals are seen with much greater intensity in OMTS than in UPS. For example, the UPS spectra of nickel (II) tetraphenylporphyrine (NiTPP) and cobalt (II) tetraphenylporphyrine (CoTPP) near E_F are almost the same, but the STM-OMTS of CoTPP has a well-defined band due to the partially occupied d_{z^2} orbital that is absent in the OMTS of NiTPP (see Figure 7.9) [96].

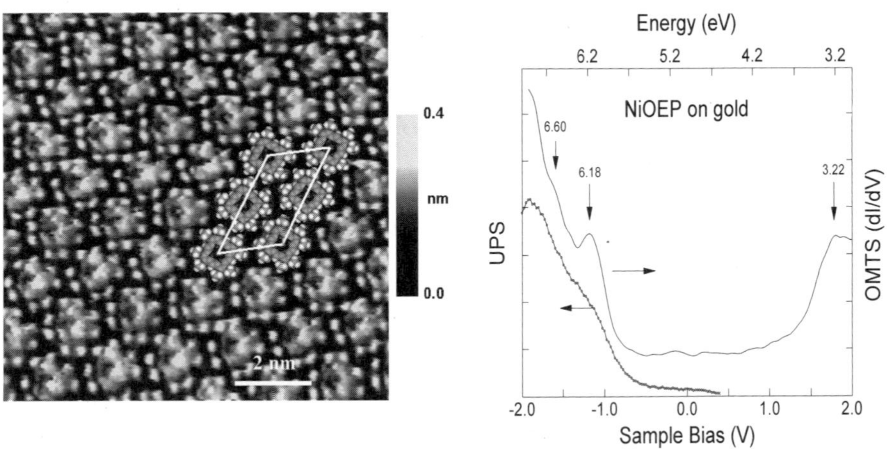

Figure 7.15 Constant current STM image, OMTS, and UPS of a monolayer of nickel (II) octaethylporphyrin (NiOEP) on Au (111). The absolute energy scale was derived by taking Φm = 5.20 V and Δ = – 0.2 V. The STM image was acquired at a sample bias of – 0.6 V and a set-point voltage of 0.30 nA. The STM image is reprinted with permission from [101]. Copyright (2002) American Chemical Society.

Because of the relatively relaxed selection rules associated with near-resonant elastic electron scattering, OMTS provides more information about occupied orbitals near E_F than does UPS. But OMTS offers much more than that. The ability to observe unoccupied orbitals (affinity levels) gives OMTS the capabilities normally associated with inverse photoemission without the problems created when organic molecules are subjected to the intense electron beams essential to the inverse photoemission (IPS) process. The practical resolution of IPS when studying organic adlayers is generally of the order of 0.5 volt [123]. OMTS, on the other hand, has significantly greater resolution capabilities. Even at room temperature, an instrumental resolution of less than 0.1 volt is easily obtained. By cooling the sample to 100 K, the resolution can be improved below 0.03 volt (240 cm^{-1}). While this has little effect on the overall band intensity (as discussed above), it does provide sufficient resolving power to make possible the observation of vibronic structuring on the OMTS bands.

In the previous section we stated that the agreement between occupied orbital positions as predicted by oxidation potentials did not compare well with the OMTS band positions for the ring oxidations of metal porphyrins. Armstrong and coworkers have also noticed this difficulty in reconciling solution phase oxidation potentials with thin film UPS data [97, 124]. Rather than using the simple formula that results from Figure 7.11 ($I_1 = 4.71$ eV + $E^{ox}(sce)_{1/2}$) they find that multiplication of the oxidation potential by a factor of about 1.7 is necessary to bring UPS HOMO peaks and solution phase electrochemical first oxidation potentials into agreement. The close agreement between UPS and OMTS positions, and the disagreement with electrochemical predictions, suggests that the OMT process is a vertical process in the Frank-Condon sense. As shown in Figure 7.16, the center of the UPS band is the vertical transition labeled ΔE_p. The width of the band arises from vibrational (molecular and lattice) vibrations associated with the change in geometry inherent in $M \Rightarrow M^+$ transition. The electrochemical potential, on the other hand, is given by ΔE_0. The larger the difference in geometry between the parent and the ion, the greater the difference between ΔE_0 and ΔE_p. If the OMTS in the porphyrins case is a vertical excitation process, then the OMTS energy should agree well with that from UPS. Based on our results, it would appear that the two processes are very similar.

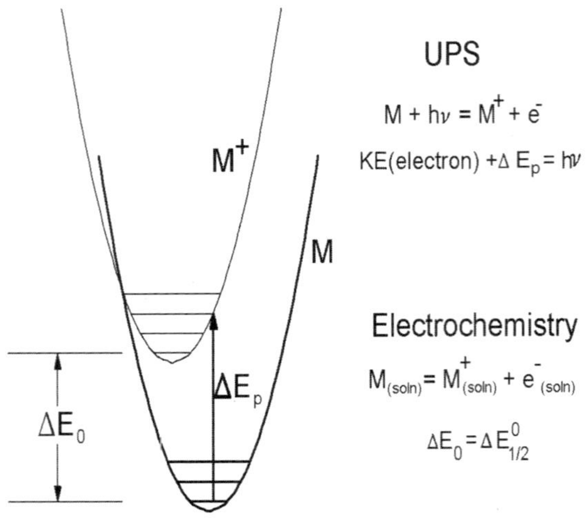

Figure 7.16 Schematic potential energy surfaces for a parent molecule M and its most stable positive ion M^+.

This conclusion should be drawn with care. First, the bandwidths seen in both UPS and OMTS of thin films are of the order of 0.5 eV, or 4000 cm^{-1}. Since a typical C-C stretch or C-H bend is less than half this value, there could be some significant vibronic differences in the transition types that are hidden within the room temperature peak shapes. Low temperature measurements are essential in resolving this issue. The second caution is that each molecular type will interact in a complex way with the tunneling environment. There is no single mechanism that describes OMTS for all molecules.

Why, one might ask, is it only the porphyrins that show the large disparity? One likely answer within the context of Figure 7.16 is that the porphyrin ring is smaller and less rigid than that of phthalocyanine. Moreover, all of the porphyrins studied have peripheral substituents that are subject to low frequency vibrational modes. Thus, one would expect a larger geometry change upon removing an electron from a porphyrin relative to a phthalocyanine. A larger geometry change translates into a greater difference between ΔE_0 and ΔE_p. If this argument is accepted, one must then ask why the other OMTS bands of the porphyrins agree with the electrochemical model. The answer here is more complex and ultimately will require detailed quantum mechanical calculations. Consider the cobalt ion. The molecular geometry change most likely to play a role is the change in force constant between cobalt and nitrogen—not a change in position since the entire ring tends to restrict the possibilities. A force constant change with no change in equilibrium geometry leads to no difference in ΔE_0 and ΔE_p.

Tsiper, Soos, Gao, and Kahn have performed STM-OMTS, UPS, and IPS measurements of PTCDA (perylenetetracarboxylic acid dianhydride) films as a function of thickness [125]. They also find good agreement between the three techniques. In this context it is interesting to note that PTCDA is a large rigid planar molecule and that the first ionization and affinity levels involve electrons delocalized over much of the molecule.

7.5.3 OMTS as a Chemical Analysis Tool: Direct Spectral Characterization

Based upon its ability to determine the location of HOMOs and LUMOs, OMTS offers *at least* as much chemical selectivity as does electrochemistry. The limiting factor in the selectivity of OMTS is the thermal line width—of the order of 3 kT or about 0.09 eV at room temperature. Because typical UV-Vis bands observed in molecular species are of the order of 0.25 eV, the thermal broadening is not significant until about 700°C. Thus the selectivity of OMTS is better than, or of the order of that afforded by, UV-Vis spectra. To quantify these ideas, consider Figure 7.17.

The room temperature STM-OMTS obtained from single molecules of four chemically similar species are shown in Figure 7.17. All have a central +2 transition metal ion, all have four coordinating nitrogens, and all have large π systems. Nevertheless, each is clearly distinguishable from the others by its OMTS spectrum in a 4-eV window centered about the Fermi energy of gold (the substrate used). In the cases where the π systems are different, the positions of the π HOMO and LUMO distinguish the compounds. For those complexes having the same ligand but different metals, the metal d-orbital occupancy provides clear selectivity.

What about other types of molecules? Are the porphyrin and phthalocyanine cases really representative, or do they constitute some kind of special (best) case? To answer that question we have prepared Table 7.1. This table gives a selection of electrochemical potentials for the first reduction or oxidation of a wide range of molecules. Polyacenes, hetero-organics, metalorganics, and transition metal complexes all appear on the list. We have then used the correlation between first reduction and oxidation potentials discussed in a previous section, and exemplified in Figure 7.12, to estimate the position of the first ionization and affinity levels in OMTS. We used the formula

$$E_{\mathrm{OMTS}} \approx E_{1/2}(\mathrm{sce}) + 4.71\ \mathrm{V}$$

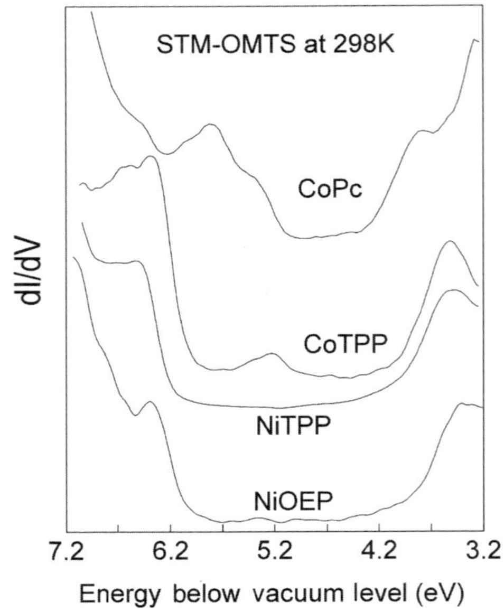

Figure 7.17 STM-OMTS spectra of four compounds on Au (111) at room temperature. Zero bias ~5 eV. Reprinted with permission from [85]. Copyright (2005) American Chemical Society.

to produce the values in the far right column of Table 7.1. Note first that these values span a range of about 5 volts and that including second ionizations and or affinity levels would expand this range even further. Thus, even with a line-width (determined by the intrinsic width) of 0.25 volt, there is enough spread in values to make identification from even a single ionization or affinity level possible in a suitable mixture of molecular species. When one takes the pair of energies formed by the first ionization and affinity levels, it is clear that almost any pair of molecular species can be identified at a particular site on a surface using STM-based OMTS. In fact, the situation here is nearly ideal for the analytical chemist wishing to identify individual molecules on a surface even if there are a large number of components. *This is NOT like conventional spectroscopic analysis of mixtures where the individual spectra must be sufficiently unique to allow their presence to be identified in the complex spectrum that results from a multicomponent mixture.* Because STM-OMTS provides the spectra of one molecule at a time, it is only necessary to match this *pure spectrum* to that of the suspected component. Thus, not only is the sensitivity of STM-OMTS exceptional (one molecule at a time), its selectivity is also significantly better than conventional UV-Vis absorption and fluorescence techniques.

Another interesting issue to the analyst, is the range of energies that can be measured in any one experiment. At very high (positive or negative) bias, field emission will occur and OMTS will no longer be available in the current-voltage curve. If we take this range (somewhat arbitrarily) as ±2 volts, we will have a 4-volt span over which to measure spectral bands of interest.

The central location of that span will be determined by the Fermi energy of the substrate-adsorbate system. For gold, values near 5 eV are expected so one has a 3 to 7 volt "window" for spectral observation. If silver is used as the substrate instead, the window is shifted up by about 0.5 volt, while choosing a platinum substrate shifts the window down by about 0.4 volt. This assumes, of course, that the adsorption mechanism is physisorption. By judicious choice of substrate(s), the entire range of affinity and ionization levels shown in Table 7.1 is accessible by STM-OMTS. In fact, OMTS transitions beyond the 2 to 7 volt range are accessible.

In this section we have assumed that the potential at the molecule is essentially that of the substrate. One can (and should) question the validity of this assumption [95, 109, 112, 126–133]. Consider STM-based spectroscopy of quantum dots [132]. Bakkers et al. found that by varying the tip-dot distance they could change the relative rate of tunneling into, versus tunneling out of, the dot. If the tip is retracted relatively far from the dot, tunneling into the dot was much slower than tunneling out, resulting in single electrons tunnel through the dot. This is similar to OMTS where the resonances in the conductance spectrum correspond to the single-particle energy levels of the CdSe quantum dot. When the tip was brought closer to the dot, tunneling into the dot became as fast as tunneling out. In this regime they found that up to three (extra) electrons could be present in the particle. The resulting electron–electron Coulomb interactions lead to a much more complex conductance spectrum. This is Coulomb *blockade behavior*. In order to see this, Bakkers et al. had to electronically isolate the nanoparticle from the substrate by using a self-assembled monolayer (SAM) as a spacer. Katz and coworkers performed a similar experiment with InAs nanocrystals [133]. While they were never able to enter the Coulomb blockade regime, they did see the positions of the resonant tunneling bands shift with dot-tip distance. They attributed this change as due to the local potential at the dot being intermediate between that of the substrate and tip, with the SAM and tip-dot gaps acting as two parts of a voltage divider. As the size of adsorbed molecules increases, especially in cases like proteins, which have an electroactive portion buried inside a relatively electronically inert sheath, the molecule might become indistinguishable from a small quantum dot. In this case one would expect that the potential at which oxidation or reduction occurred would begin to depend on tip-molecule separation and not be simply related to spectroscopic or electrochemical values. Lindsey and coworkers attempted to measure experimentally the ratio of the local potential (at a molecule experiencing OMTS) to the applied potential (between tip and substrate) in the case of porphyrins, but were unable to demonstrate a significant dependence [112].

Deng and Hipps set out to test the tip-molecule dependence of the OMTS in the case of nickel (II) tetraphenylporphyrin (NiTPP) adsorbed on Au-111 under UHV conditions [95]. Adjusting the set-point prior to spectral measurement allowed for the control of the tip-sample distance. A sequence of $dI/dV(V)$, $I(V)$, and $I(z)$ curves were acquired over a wide range of set-point currents and bias voltages. The $I(z)$ data were measured to provide a means of converting set-point values to relative tip displacements. Determining the peak positions and peak shapes from the dI/dV curves was difficult because of the strong variation in both the resonant and elastic (background) intensities with tip-sample separation. Stroscio and Feenstra [134–136] considered this problem several years ago and determined that this difficulty could often be eliminated by using the logarithmic derivative, $d\ln I/d\ln V(V)$, as the spectral intensity function. Ukrainstev discussed some problems with this method [137], but they were not relevant in the NiTPP case. The results obtained by Deng and Hipps are presented in Figure 7.18. They found that changes in tip-sample distance over several Angstroms and a factor of 20 in set-point current produce no measurable changes in orbital energy splitting [95]. Thus, the ionization and affinity spectra obtained by STM-OMTS for molecules of the order of the size of porphyrins can be used reliably. It was suggested that it may be possible to reliably measure STM-OMTS on molecules larger than tetraphenylporphyrin without concern about the tip-molecule separation distance provided that the effective gap impedance (between electroactive moiety in the molecule and the substrate) does not drop below about 500 MΩ. An unexpected result of this study was the small but persistent shift of all OMTS bands to energies deeper than the vacuum level.

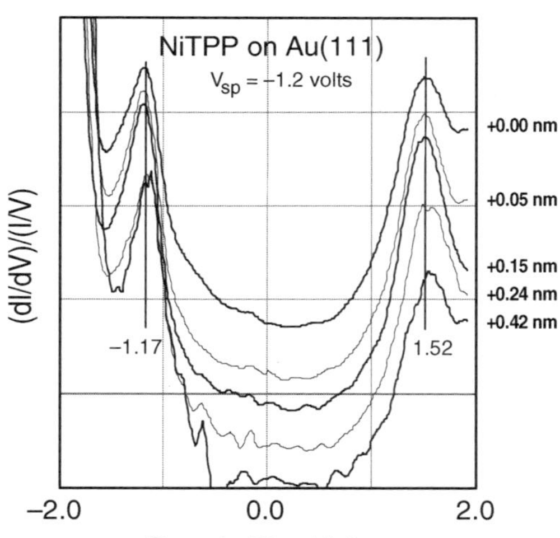

Figure 7.18 Orbital mediated tunneling spectra of NiTPP on Au (111) in UHV at room temperature. Relative tip-molecule distances are given on the far right. Reprinted with permission from [95]. Copyright (2002) American Chemical Society

Table 7.1 Energy estimate for LUMO and HOMO OMTS energies (relative to the vacuum level) based on electrochemical potentials.

Molecule	$E°_{1/2}$ (sce)	E_{OMTS} (volts)
REDUCTION (M^+ electron ⇒ M^-)		
Pyridine	–2.72[c]	1.99
Pyrimidene	–2.34[c]	2.37
$Ni(acac)_2$	–2.24[a]	2.47
quinoline	–2.12[c]	2.59
s-triazine	–2.05[c]	2.66
anthracene	–1.91[d]	2.80
perylene	–1.70[d]	3.00
tetracene	–1.59[d]	3.12
Zinc (II) tetrabenzoporphine	–1.47[a]	3.24
pentacene	–1.31[d]	3.40
$Co(acac)_2$	–1.24[a]	3.47
cobalt (II) etioporphyrin I	–1.04[a]	3.67
$Cr(C_6H_6)_2^{+2}$	–0.88[a]	3.83
copper (II) tetrasulfophtalocyanine	–0.73[a]	3.98
$Ru(acac)_2$	–0.54	4.16
cobalt (III) ethylenediimine cation	–0.43[a]	4.28
itc-TPP-FeCl	–0.34[b]	4.37
itc-TPP-MnBr	–0.26[b]	4.45
tetracyanoethylene	–0.20[a]	4.50
tetracyanoquinodimethane	–0.15[a]	4.56
ferricyanide	+0.09	4.80
OXIDATION (M ⇒ M^+ + electron)		
cobalt (II) tetrasulfophtalocyanine	+0.45[a]	5.16
chlorophyll b	+0.65[a]	5.36
diphenypycrilhydrazyl	+0.73[a]	5.44
perylene	+0.85[a]	5.56
tetraethylthioethylene	+0.90[a]	5.61
2,4,6-tri-*t*-butylnitrosobenzene	+1.0[a]	5.71
anthracene	+1.09[a]	5.80
coronene	+1.23[a]	5.94
dibenzylthiophene	+1.35[a]	6.06
1-nitronapthalene	+1.62[a]	6.36
propylene sulfide	+1.69[a]	6.40
methoxybenzene	+1.76[a]	6.47
thiophene	+1.84[a]	6.55
biphenyl	+1.91[a]	6.62
nitrosobenzene	+2.0[a]	6.71
methyliodide	+2.12[a]	6.83
benzene	+2.30[a]	7.01

[a] [138]; [b] [112]; [c] [139]; [d] [140]. All measurements in nonaqueous solvents. Oxidation and reduction processes relative to a substrate work function of 5 eV.

7.5.4 OMTS as a Chemical Analysis Tool: Bias-Dependent Imaging

While the selectivity offered by measuring the full OMTS for each molecule on the surface is extremely good, it can be experimentally demanding in terms of data acquisition time, the need for very low drift rates, and when large areas of a sample must be considered. It is possible to use this high selectivity to do a much faster and easily visualized chemical map of the surface. If one monitors dI/dV (with a lock-in amplifier) at the peak bias voltage for a particular band as a function of position on the sample surface, a spatial distribution of that species can be obtained. The downside of this method is that we forgo the excellent selectivity inherent in collecting one full spectrum per molecule, and return to the equivalent of the mixed sample problem. Since spectral data are collected at only one energy (bias voltage) per component to be identified, that component must have a strong feature in OMTS that is absent or weak for all the other components. Consideration of Figure 7.17, for example, suggests that setting the bias at about –0.7 volt (near 5.7 eV relative to the vacuum level) would allow the CoPc complex to be easily distinguished from the nickel complex, and would even allow discrimination between the two cobalt complexes (CoPc and CoTPP).

The difference in the cobalt and nickel centered porphyrins and phthalocyanines is so large, in fact, that an even simpler method of chemical mapping is possible. Namely, one need only take two constant current images with the bias set at appropriate values. Figure 7.17 provides two such constant current STM images of a mixed layer of NiTPP and CoTPP on Au (111) at room temperature. These were taken within minutes of each other on the same area of the same sample. The only difference was the bias voltage setting. In the –1.4 V image, the NiTPP and CoTPP molecules cannot be distinguished. In the –1.0 V image, however, the CoTPP molecules become very apparent by "lighting up." At –1.0 V a significant fraction of the tunneling current is orbitally mediated through the half-filled d_z^2 orbital on the cobalt (II) center, a current pathway not available to the nickel (II) ion. Thus, by adjusting the bias at which the constant current images are taken, one can often rapidly map out chemical composition.

Some points to note about Figure 7.19. It is a very large area scan and it would be possible (by counting bright and dark molecules) to arrive at a statistically significant estimate of the relative concentrations of the two species. Second, while there may be some drift, it has no significant effect on molecular identification because the pictures are so big that reference points can be tracked from one picture to the next. Finally, note that there appears to be more CoTPP near the edges of the islands than in the centers. This is in fact the case since the NiTPP was deposited first and then followed by the CoTPP. The fact that we do not see a core of NiTPP surrounded by a shell of CoTPP allows us to learn about surface diffusion.

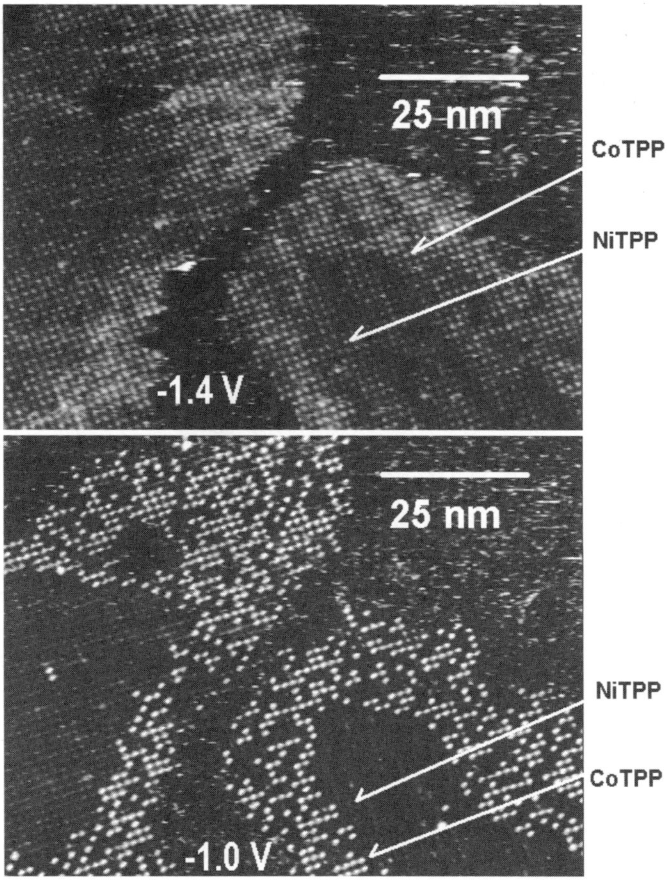

Figure 7.19 Constant current tunneling images of mixed composition islands of NiTPP and CoTPP at different bias voltages. The image at –1.4 V was acquired with a set-point of 300 pA and that at –1 V with a set-point of 200 pA. Reprinted with permission from [97]. Copyright (2000) American Chemical Society.

7.5.5 OMTS as a Submolecular Electron Transport Mapping Tool

We can use spectroscopic mapping to image the electron transport pathways on a submolecular scale. Since the resolution of STM-OMTS can be 0.1 nm, it is conceivable that one may map electron transport at the single atom level within a given molecule. A tantalizing first look at how this might be done is provided in Figure 7.20 [85]. Shown are three points of a map that should eventually include a number of different ionization and affinity levels. A map of the electron transport path from the molecule to the tip via the HOMO is provided by the image taken at –1.0 volt bias (6.3 eV below the vacuum level). Clearly, most of the current is flowing through the carbon p_z

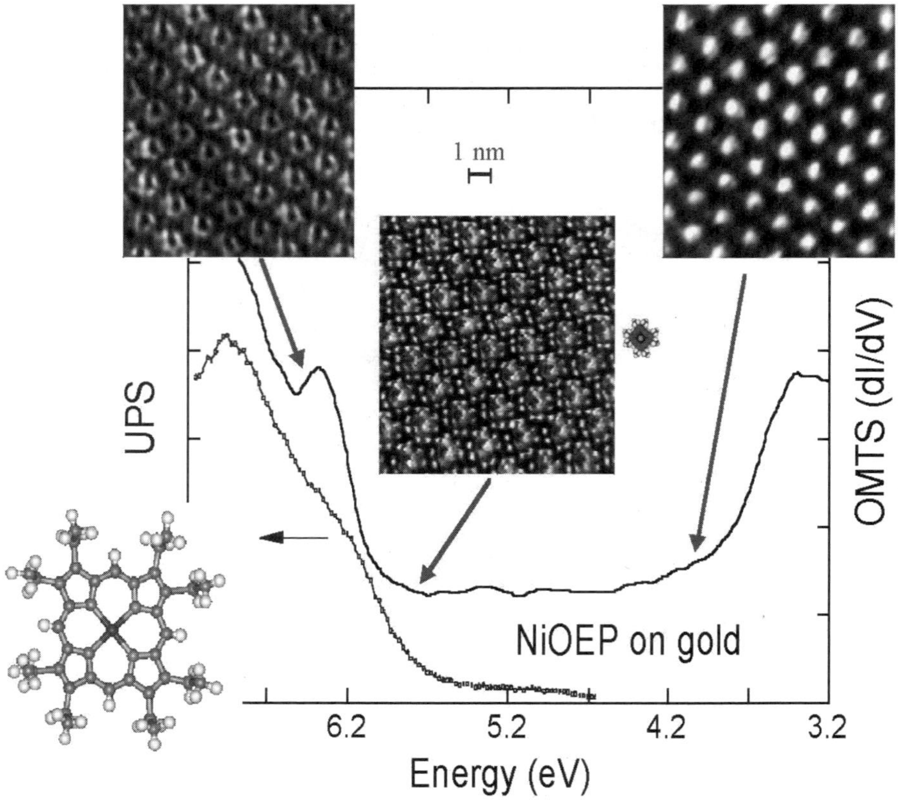

Figure 7.20 Bias-dependent constant current STM imaging of a mixed composition adlayer of NiTPP and CoTPP. The bright spots in the right hand image (bias = –1.0 V) are due to enhanced tunneling through the cobalt (II) d_z^2 orbital in CoTPP. Reprinted with permission from [85]. Copyright 2005, the American Chemical Society.

orbitals of the porphyrin ring. The ethyl groups, central nitrogen atoms, and the nickel (II) ion are not conducting at this bias. In the intermediate bias region the tunneling pathway is not strongly affected by any particular orbital and an image similar to the actual molecular structure is observed. Note the eight individual ethyl groups per molecule are clearly seen (at about 5.7 eV below the vacuum level). Compare the space-filling CPK model of NiOEP (shown to the center right) with the actual STM image (center). As the bias is increased positively and approaches the first affinity level, a new image appears. This image is a map of the electron-accepting regions of the NiOEP molecule. At 4 eV below the vacuum level most of the current from the tip to the molecule flows through the central portion of the molecule. The central area is so bright that it seems likely that there are contributions from all the atoms in the porphyrin ring, with the greatest current being carried through the nitrogen and nickel (II) orbitals.

7.6 SOME CONCLUDING POINTS

The future of STM-based spectroscopy is a bright one. The advantages offered are great and can be grouped into those based on the particular physical advantages inherent in STM-based spectroscopy:

1) As a single-molecule spectroscopy it can:
    a) Locate and identify species at particular sites on surfaces.
    b) Perform highly selective surface analysis even where there are several different components present.
    c) Map electron transport through different portions of a particular molecule or nanostructure.
    d) Map molecular force fields (through local vibrational frequencies) on an atomic level.
2) As an elastic electron tunneling spectroscopy:
    a) It can locate both electron affinity levels and ionization states simultaneously and without utilizing energy probes that can decompose the molecule.
    b) The selection rules are more relaxed than for UPS or IPS, thereby allowing a more complete description of the electronic states.
    c) It can be applied to samples over a range of temperature (0 to ≥ 700 K).
    d) Conduction pathways both in space and in energy can be mapped and used in the design of complex molecular electronic devices.
3) As an inelastic tunneling spectroscopy:
    a) It can utilize the higher intensity and wider acceptable temperature range of inelastic electronic state tunneling as a surface analytical tool when OMTS is not appropriate.
    b) In resonant IETS, knowing which vibrational mode is activated by excitation through a particular resonant electronic state can lead to better assignments of the nature of the electronic states involved (similar to resonance Raman analysis).
    c) (For non-resonance enhanced IETS) the selection rules are more relaxed than for infrared or Raman spectroscopy, thereby allowing a more complete description of the vibrational states.
4) Combined advantages:
    a) It can identify chemical species formed through surface or *in situ* manipulation by the tip or tip current.
    b) It can utilize the voltage dependence of resonant tunneling-induced structural and physical changes (either through inelastic or quasi-elastic processes) to understand the mechanism of molecular change.
    c) By comparing the electronic spectra of single molecules on surfaces to those of aggregates (multiples of the same molecules or different geometry configurations of different molecules)

precise measurements of the role of intermolecular interaction on electronic states can be made.

Some of these advantages will only be realized when we have improved the stability of the STM to the point where non-resonance enhanced IETS can be observed. This is a point we have not yet reached, but which should be attainable. The only reason STM-IETS on electronic states has not yet been demonstrated is that the materials studied to date have stronger OMTS bands masking them. They will be seen once the appropriate materials are examined. Most of the advantages enumerated above, however, have been demonstrated and only require that they be applied to a wider range of problems.

ACKNOWLEDGEMENT

We thank the NSF for support in terms of grants CHE–0138409 and CHE–0234726. Acknowledgement is made also to the donors of the Petroleum Research Fund for direct support of part of this research and for their funding of a Summer School on Physical Chemistry at the Nanometer Scale. Thanks go to my students and colleagues who participated in this research, with special appreciation directed to Professors Ursula Mazur and Louis Scudiero.

REFERENCES

[1] Hurst, H.G. & Ruppel, W. (1964) *Zeitschrift fuer Naturforschung* **19a**, 573-579.

[2] Lambe, J. & Jaklevic, R.C. 1968 *Physical Review* **165**, 821-832. (These authors are the fathers of all tunneling spectroscopy.)

[3] Thomas, D.E. & Klein, J.M. (1963) *Rev. Sci. Instrum.* **34**, 920-924.

[4] Jaklevic, Robert C., Lambe, John. (1966) *Physical Review Let.* **17**, 1139-1140.

[5] Hansma, P.K., Ed. (1982) *Tunneling Spectroscopy*, Plenum: New York.

[6] Hipps, K.W. & Mazur, U. (1993) *J. Phys. Chem.* **97**, 7803-7814.

[7] Mazur U. & Hipps K.W. (2001) Inelastic electron tunneling spectroscopy, in *Handbook of Vibrational Spectroscopy*, John Chalmers & Peter Griffiths, Volume 1. John Wiley and Sons 812-829.

[8] Hamers, R.J. (1996) *J. Phys. Chem.* **100**, 13103-13120.

[9] Giancarlo, L.C., Fang, H., Avila, L., Fine, L.W., Flynn & George, W. (2000) *J. Chem. Educ.* **77**, 66-71.

[10] Hamers, R.J. (1989) *Ann. Rev. Phys. Chem.* **40**, 531-59.

[11] Lehmpuhl, D.W. (2003) *J. Chem. Educ.* **80**, 478.

[12] Guntherodt, H.-J. & Wiesendanger, R. (1992) *Scanning Tunneling Microscopy* I. Springer Verlag, Berlin.

[13] Guntherodt, H.-J. & Wiesendanger, R. (1992) *Scanning Tunneling Microscopy* II, Springer Verlag, Berlin.

[14] Magonov, S.N. & Whangbo, M.H. (1996) *Surface Analysis with STM and AFM.* VCH, New York.

[15] Stroscio, J.A. & Kaiser, W.J. (1993) *Scanning Tunneling Microscopy.* Academic Press, New York.

[16] Meyer, E., Jarvis, S.P. & Spencer, N.D. (2004) *Scanning probe microscopy in materials science*. In *MRS Bull.* 29(7).

[17] Bhushan, B., Fuchs, H. & Hosaka, S. (2004) *Applied Scanning Probe Methods*. Springer-Verlag, Berlin.
[18] Bonnel, D. (2000) John Wiley and Sons, *Scanning Probe Microscopy and Spectroscopy*. New York.
[19] Bard, A.J. & Mirkin, M.V. (2001) *Scanning Electrochemical Microscopy*. Marcel Dekker New York.
[20] Young, R., Ward, J. & Scire, F. (1971) *Phys. Rev. Lett.* **27**, 922.
[21] Binnig, G. & Rohrer, H. (1982) *Hen. Phys. Acta* **55**, 726.
[22] Binnig, G., Rohrer, H., Gerber, C. & Weibel, E. (1983) *Phys. Rev. Lett.* **50**, 120.
[23] Binnig, G. & Rohrer, H. (1987) *Rev. Mod. Phys.* **59,** 615.
[24] Veeco Metrology (formerly Digital Instruments), 112 Robin Hill Road, Santa Barbara, CA 93117. http://www.veeco.com/.
[25] McAllister Technical Services, West 280 Prairie Avenue, Coeur d'Alene, ID 83815. http://www.mcallister.com/.
[26] RHK Technology, 1050 East Maple Road, Troy, MI 48083.http://www.rhk-tech.com/.
[27] Molecular Imaging, 4666 S. Ash Avenue, Tempe, Arizona 85282. http://www.molec.com/.
[28] Asylum Research, 6310 Hollister Ave, Santa Barbara, CA 93117. www.Asylum Research.com.
[29] Novascan Technologies, Inc. 131 Main Street, Ames, IA 50010. http://www.novascan.com/.
[30] Quesant. 29397 Agoura Road, Suite 104, Agoura Hills, Ca 91301. http://www.quesant.com/.
[31] Omicron, NanoTechnology GmbH, Limburger Str. 75, 65232 Taunusstein, Germany. http://www.omicron.de/.
[32] WITec GmbH. Hoervelsinger Weg 6, 89081 Ulm, Germany. http://www.witec.de/.
[33] Nanosurf AG, Grammetstrasse 14, CH-4410 Liestal, Switzerland. http://www.nanosurf.com/.
[34] Nanoscience Instruments, Inc. 9831 South 51st Street, Suite C119, Phoenix, AZ 85044. http://www.nanoscience.com/.
[35] Nanotech Electronica, Parque Cientifico de Madrid. Pabellon C, UAM, Cantoblanco E-28049 Madrid Spain http://www.nanotec.es/.
[36] NT-MDT Co., Zelenograd, Moscow, 124482, Russia korp 317 A, P O 158. http://www.ntmdt.ru/.
[37] Danish Micro Engineering A/S. Transformervej 12, DK-2730 Herlev, Denmark. http://www.dme-spm.dk/.
[38] JPK Instruments AG. Bouchéstrasse 12, 12435 Berlin, Germany. http://www.jpk-instruments.de/.
[39] Attocube systems AG. Viktualienmarkt 3, D-80331 Munchen, Germany. www.attocube.com.
[40] PSIA, Induspia 5F, Sang-Daewon-Dong 517-13, Sungnam 462-120, Korea. http://www.advancedspm.com/.
[41] Surface Imaging Systems (S.I.S.), Rastersonden- und Sensormesstechnik GmbH, Kaiserstrasse 100 (Technologiepark Herzogenrath, TPH), D-52134 Herzogenrath, Germany. http://www.sis-gmbh.com/.
[42] Triple-O Microscopy GmbH. Behlertstrasse 26, D-14469 Potsdam, Germany. http://www.triple-o.de/.
[43] Budget Sensors, 6, Kestenova Gora Str., 1404 Sofia, Bulgaria. http://www.budgetsensors.com/.
[44] Binnig, G., Rohrer, H., Gerber, C. & Weibel, E. (1982) *Appl. Phys. Lett.* **40**, 178.
[45] Binnig, G., Rohrer, H., Gerber, C. & Weibel, E. (1982) *Appl. Phys. Lett.* **49**, 57.
[46] Smith, R.L. & Rohrer, G.S. (2001) In *Scanning Probe Microscopy and Spectroscopy*, Ed: Bonnell, D. Chapter 6 John Wiley and Sons, New York.
[47] Muller, W.E. & Tsong, T.T. (1969) *Field Ion Microscopy*, American Elsevier: New York.

[48] Bowkett, K.M. & Smith, D.A. (1970) In *Field Ion Microscopy* Vol. 2, Amelinckx, S., Gevers, R., Nihoul, J., Ed., Defects in cxrystalline solids. North Holland: Amsterdam.
[49] Nam, A.J., Teren, A., Lusby, T.A. & Melmed, A.J. (1995) *Journal of Vacuum Science & Technology B*, **13**, 1556-9.
[50] Weinstein, V., Slutzky, M., Arenshtam, A. & Ben-Jacob, E. (1995) *Review of Scientific Instruments* **66**, 3075-6.
[51] Gueell, A.G., Diez-Perez, I., Gorostiza, P. & Sanz, F. (2004) *Analytical Chemistry* **76**, 5218-5222.
[52] Fainchtein, R., Zarriello, P.R. (1992) *Ultramicroscopy* **42-44**, 1533-7.
[53] Heben, M.J., Dovek, M.M., Lewis, N.S., Penner, R.M. & Quate, C.F. (1988) *Journal of Microscopy* **152**(3), 651-61.
[54] Nagahara, L.A., Thundat, T. & Lindsay, S.M. (1989) *Review of Scientific Instruments* **60**, 3128-30.
[55] Baykul, M.C. (2000) *Materials Science & Engineering B*, **74**, 229-233.
[56] Ren, B., Picardi, G. & Pettinger, Bruno. (2004) *Review of Scientific Instruments* **75**, 837-841.
[57] Fried, G.A., Wang, X.D. & Hipps, K.W. (1993) *Rev. Sci. Instrum.* **64**, 1495-1501.
[58] Iwami, M., Uehara, Y. & Ushioda, S. (1998) *Review of Scientific Instruments* **69**, 4010- 4011.
[59] Cavallini, M. & Biscarini, F. (2000) *Review of Scientific Instruments* **71**, 4457-4460.
[60] Mendez, J., Luna, M. & Baro, A.M. (1992) *Surface Science* **266**, 1-3, 294-8.
[61] Zhang, R. & Ivey, D.G. (1996) *Journal of Vacuum Science & Technology B*, **14**, 1-10.
[62] Oliva, A.I., Romero G.A., Pena, J.L., Anguiano, E. & Aguilar, M. (1996) *Review of Scientific Instruments* **67**, 1917-1921.
[63] Obbligato Objectives, 68 Corporate Drive Suite 2025, Toronto, Ontario, Canada M1H 3H3. http://www.obbligato.com/
[64] Ekvall, I., Wahlstrom, E., Claesson, D., Hakan, O., Olsson, E. (1999) *Meas. Sci. Technol.* **10**, 11-18.
[65] Hockett, L.A., Creager, S.E. (1993) *Rev. Sci. Instrum.* **64**, 263-264.
[66] Paparazzo, E., Moretto, L., Selci, S., Righini, M. & Farne, I. (1999) *Vacuum* **52**, 421-426.
[67] Ottaviano, L., Lozzi, L. & Santucci, S. (2003) *Review of Scientific Instruments* **74**, 3368- 3378.
[68] Chen, C.J. (1993) *Introduction to Scanning Tunneling Microscopy*, Oxford University: New York.
[69] Kaiser, W.J. & Jaklevic, R.C. (1987) *Surface Science* **181**, 55-68.
[70] Everson, M.P., Jaklevic, R.C. & Shen, W. (1990) *J. Vac. Sci. Tech. A*, **8**, 3662-5.
[71] Kuk, Y., Silverman & P.J. *(*1990) *J. Vac. Sci. Tech. A*, **8**, 289-92.
[72] Kuk, Y. (1992) *Springer Series in Surface Sciences* **20**, 17-37.
[73] Fonden, T., Papadia, S. & Persson, M. (1995) *Journal of Physics: Condensed Matter* **7**, 2697-716.
[74] Hoermandinger, G. (1994) *Physical Review B*, **49**, 13897-905.
[75] Doyen, G. & Drakova, D. (1997) *Progress in Surface Science* **54**, 249-276.
[76] Bischoff, M.M.J., Konvicka, C., Quinn, A.J., Schmid, M., Redinger, J., Podloucky, R. Varga, P. & van Kempen, H. (2001) *Physical Review Letters* **86**, 2396-2399.
[77] Himpsel, F.J., Jung, T., Schlittler, R., Gimzewski, Jim K. (1996) *Japanese Journal of Applied Physics, Part 1* **35**, 3695-3699.
[78] Bode, M., Pascal, R., Wiesendanger, R. (1996) *Applied Physics A*, **62**, 571-573.
[79] Biedermann, A., Genser, O., Hebenstreit, W., Schmid, M., Redinger, J., Podloucky, R. & Varga, P. (1996) *Physical Review Letters* **76**, 4179-4182.
[80] Mills, G., Wang, B., Ho, W. & Metiu H. (2004) *J. Chem. Phys.* **120**, 7738-40.
[81] Crommie, M.F., Lutz, C.P., Eigler, D.M. & Heller, E.J. (1995) *Surface Review and Letters* **2**, 127-37.
[82] Wiesendanger, R. & Bode, M. (2001) *Solid State Communications* **119**, 341-355.

[83] Pietzsch, O., Kubetzka, A., Bode, M. & Wiesendanger, R. (2001) *Science* **292**, 2053-2056.
[84] Hamers, R.J., Tromp, R.M. & Demuth, J.E. (1986) *Surf. Sci.* **56**, 1972-75.
[85] Hipps, K.W. & Scudiero, L. (2005) *J. Chem. Ed.* **82**, 704-711.
[86] Hipps, K.W. & Peter, S.L. (1989) *J. Phys. Chem.* **93**, 5717-5722.
[87] Hipps, K.W., Barlow, D.E. & Mazur, U. (2000) *J. Phys. Chem. B* **104**, 2444-2447.
[88] Hipps, K.W. & Mazur, U. (1988) *Rev. Sci. Instrum.* **59**, 1903-1905.
[89] Hipps, K.W. & Mazur, U. (1987) *Rev. Sci. Instrum.* **58**, 265-268.
[90] Seman, T.R. & Mallik, R.R. (1999) *Rev. Sci. Instrum.* **70**, 2808-2814.
[91] Hipps, K.W. (1988) *J. Phys. Chem.* **93**, 5958-5960.
[92] Mazur, U. & Hipps, K.W. (1994) *J. Phys. Chem.* **98**, 5824-5829.
[93] Mazur, U. & Hipps, K.W. (1995) *J. Phys. Chem.* **99**, 6684-6688.
[94] Mazur, U., Hipps, K.W. (1999) *J. Phys. Chem. B*, **103**, 9721-9727.
[95] Deng, W. & Hipps, K.W. (2003) *J. Phys. Chem. B*, **107**, 10736-10740.
[96] Scudiero, L., Barlow, D.E., Mazur, U. & Hipps, K.W. (2001) *JACS*, **123**, 4073-4080.
[97] Scudiero, L., Barlow, D.E. & Hipps, K.W. (2000) *J. Phys. Chem. B*, **104**, 11899-11905.
[98] Stipe, B.C., Rezaei, M. A. & Ho, W. (1998) *Science* **280**, 1732-1735.
[99] Gaudioso, J. & Ho, W. (2001) *J. Amer. Chem. Soc.* **123**, 10095-10098.
Ho, W.; Qiu X.; Nazin, G.V. (2004) **92**, 206102.
[100] Moresco, F., Meyer, G. & Rieder, K.H. (1999) *Mod. Phys. Lett. B*, **13**, 709-715.
[101] Scudiero, L., Barlow, D.E. & Hipps, K.W. (2002) *J. Phys. Chem. B*, **106**, 996-1003.
[102] Pascual, J.I., Gomez-Herrero, J., Sanchez-Portal, D. & Rust, H.P. (2002) *J. Chem. Phys.* **117**, 9531-9534.
[103] Brousseau, J.L., Tian, K., Gauvin, S., Leblanc, R.M. & Delhaes, P. (1993) *Chem. Phys. Lett.* **202**, 521-7.
[104] Nolen, S. & Ruggiero, S.T. (1999) *Chem. Phys. Lett.* **300**, 656-660.
[105] Komeda, T., Kim, Y., Fujita, Y., Sainoo, Y. & Kawai, M. (2004) *J. Chem. Phys.* **120**, 5347-5352.
[106] Pascual, J.I., Lorente, N., Song, Z., Conrad, H. & Rust, H.-P. (2003) *Nature* **423**, 525-528.
[107] Komeda, T., Kim, Y. & Kawai, M. (2002) *Surf. Sci.* **502-503**, 12-17.
[108] Hipps, K.W. & Hoagland, J.J. (1991) *Langmuir* **7**, 2180-2186.
[109] Sumi, H., (1998) *J. Phys. Chem. B*, **102**, 1833-1844.
[110] Kuznetsov, A.M. & Ulstrup, J. (2000) *J. Phys. Chem. A*, **104**, 11531-11540.
[111] Schmickler, W. & Tao, N. (1997) *Electrochimica Acta* **42**, 2809-2815.
[112] Han, W., Durantini, E.N., Moore, T.A., Moore, A.L., Gust, D., Rez, P., Letherman, G., Seely, G., Tao, N. & Lindsay, S.M. (1997) *J. Phys. Chem. B*, **100**, 10719-10725.
[113] Loutfy, R.O., Hsiao, C.K., Ong, B.S. & Keoshkerian, B. (1984) *Can. J. Chem.* **62**, 1877- 86.
[114] Richardson, D.E. (1990) *Inorg. Chem.* **29**, 3213-23.
[115] Mazur, U. & Hipps, K.W. (1994) *J. Phys. Chem.* **98**, 8169-8172.
[116] Hill, I.G., Kahn, A., Soos, Z.G. & Pascal, Jr. R.A. (2000) *Chem. Phys. Lett.* **327**, 181-188.
[117] Scudiero, L., Hipps, K.W. & Barlow, D.E. (2003) *J. Phys. Chem. B*, **107**, 2903-2909.
[118] Barlow, D.E., Scudiero, L. & Hipps, K.W. (2004) *Langmuir* **20**, 4413-21.
[119] Westcott, B.L., Gruhn, N., Michelsen, L. & Lichtenberger, D. (2000) *J. Am. Chem. Soc.* **122**, 8083-89.
[120] Ishii, H. & Seki, K. (1997) *Trans. Electr. Dev.* **44**, 1295-1301.
[121] Ishii, H., Sugiyama, K., Ito, E. & Seki, K. (1999) *Advanced Materials* **11**, 605-625.
[122] Gao, W. & Kahn, A. (2003) *J. Appl. Phys.* **94**, 359-366.
[123] Yan, L. & Gao, Y. (2002), *Thin Solid Films* **417**, 101-106.
[124] Schmidt, A., Armstrong, N.R., Goeltner, C. & Mullen, K. (1994) *J. Phys. Chem.* **98**, 11780.
[125] Tsiper, E.V., Soos, Z.G., Gao, W. & Kahn, A. (2002) *Chemical Physics Letters* **360**, 47–52.

[126] Plihal, M. & Gadzuk, J.W. (2001) *Phys. Rev. B: Condens. Matter Mater. Phys.* **63**, 085404/1.

[127] Onipko, A.I., Berggren, K.-F., Klymenko, Yu O., Malysheva, L.I., Rosink, J.J.W.M., Geerligs, L.J., van der Drift, E. & Radelaar, S. (2000) *Phys. Rev. B: Condens. Matter Mater. Phys.* **61**, 11118.

[128] Snyder, S.R. & White, H.S. (1995) *J. Electroanyl. Chem.* **394**, 177.

[129] Muller, A.-D., Muller, F. & Hietschold, M. (1999) *Appl. Phys. Lett.* **74**, 2963.

[130] Gasparov, V., Riehl-Chudoba, M., Schroter, M. & Richter, W. (2000) *Europhys. Lett.* **51**, 527.

[131] Schmickler, W.J., (1990) *Electroanal. Chem.* **296**, 283.

[132] Bakkers, E.P., Hens, Z., Kouwenhoven, L.P. Gurevich, L. & Vanmaekelbergh, D. (2002) *Nanotechnology* **13**, 258-262.

[133] Katz, D., Millo, O., Kan, S. & Banin, U. (2001) *Appl. Phys. Lett.* **79** 117-119.

[134] Stroscio, J.A., Feenstra, R.M. & Fein, A.P. (1986) *Phys. Rev. Lett.* **57**, 2579.

[135] Feenstra, R.M. (1994) *Phys. Rev. B*, **50**, 4561.

[136] Stroscio, J.A., Feenstra, R.M., Stroscio, J.A. & Kaiser, W.J. (1993) Scanning tunneling microscopy; *Methods Expermental Physics. Vol.27* (Academic, New York).

[137] Ukraintsev, V.A. (1996) *Phys. Rev. B*, **53**, 11176.

[138] Mann, C. & Barnes, K. (1970) Electrochemical Reactions in Nonaqueous Systems. Marcel Dekker, New York.

[139] Nenner, I. & Schulz, G.J. (1975) *Chem. Phys.* **62**, 1747. (Hg pool taken to be +.52 vs SCE.).

[140] Bergman, I. (1954) *Trans. Faraday Soc.* **50**, 829. (Corrected to SCE by subtracting 0.45V.).

CHAPTER 8

RESONANCE ACOUSTIC SPECTROSCOPY

Farhang Honarvar and Esmaeil Enjilela
K.N. Toosi University of Technology, Faculty of Mechanical Engineering
P.O. Box 16765-3381, Tehran, Iran.

8.1 INTRODUCTION

Resonance acoustic spectroscopy (RAS) is the study of resonance effects present in acoustic echoes of an elastic target. These resonance effects are caused by the excitation of eigenvibrations of the target by an incident acoustic wave. Such use of resonance effects for the purpose of characterization of an elastic object (shape, size, composition, and integrity) is similar to the way chemical elements are identified from their optical properties. RAS is a powerful tool for monitoring various characteristics of elastic components and in many engineering applications for quality control purposes.

Resonances are intrinsic characteristics of the elastic object, characteristics that depend only on its physical properties and are completely independent of the source of excitation. The resonances of an elastic component are its *fingerprints*. For a given geometry, resonances are indicators of the elastic properties of the material such as density and elastic constants. Any defects, inclusions, or inhomogeneities in the material will also affect its resonance frequencies. Appropriate exploitation of the information contained within the vibratory signals and identification of the resonance frequencies of the elastic object can serve as a powerful tool for material characterization purposes.

In resonance acoustic spectroscopy, the target is insonified by a harmonic acoustic wave. If the frequency of the wave coincides with one of the resonance frequencies of the target, the corresponding eigenmode of the target is excited and the target starts to vibrate, emitting energy into the surrounding fluid. The vibrations of the target can then be measured by a sensing device. By changing the carrier frequency of the incident wave, various eigen-frequencies of the scatterer can be identified and measured. These measurements can either be compared to predetermined reference values or can be matched to a theoretical model. In the latter case, the physical properties used in the mathematical model will correspond to those of the elastic scatterer.

Since RAS is based on the scattering of waves from elastic targets, in this chapter we will first review the principles of wave scattering (Section 8.2). This will be followed by a discussion of the mathematical models for acoustic wave scattering from cylindrical components (Section 8.3). Experimental methods developed for measurement of the scattered field, used for "identification" and "isolation" of resonances observed in the backscattered echoes of elastic targets, are also discussed (Section 8.4). Combination of the theoretical capabilities of resonance scattering theory (RST) with these experimental tools makes the application of RAS practically viable. Sample applications of RAS for characterization of the shape, composition, integrity, and health conditions of engineering components are presented.

8.2 SCATTERING OF WAVES

The scattering and propagation of waves is a classical area that has engaged the interest of many scientists in various disciplines. Extensive research in the areas of acoustics, elastodynamics, and electromagnetism has resulted in many mathematical, experimental, and computational techniques.

In acoustics, the medium in which the wave propagates is a fluid. Since ideal fluids cannot support transverse stresses, the wavefield consists of compression waves only. In the more general case of elastic waves, the medium can be an elastic solid, which can support both normal and shear stresses. Hence, the wavefield is composed of both compression and transverse components. Moreover, there is also the possibility of "mode conversion," i.e., conversion between compression waves and shear waves during the process of scattering.

The target can be either penetrable (transparent) or impenetrable (opaque). For impenetrable targets, no internal fields can exist inside the body of the scatterer. The impenetrable targets can be either soft (e.g., a bubble in water) or rigid (e.g., an aluminum cylinder in air). On the other hand, in the case of penetrable targets, the wavefield can penetrate into the body of the target and generate different types of waves on the surface and inside the body of the target.

In this section, the scattering of acoustic waves from elastic targets is considered. The geometry of interest is a cylinder; however, similar theoretical and experimental procedures are available for spheres and cylindrical shells as well.

8.2.1 Physics of Acoustic Resonance Scattering

The physics of the scattering of an acoustic wave from an elastic target is a complicated phenomenon that involves the generation of different types of surface waves outside and inside the elastic target. For example, in the case of an elastic cylindrical target, when a normally incident acoustic wave strikes the boundary of the cylinder, in addition to the reflected specular echo, surface waves are generated on the boundary of the cylinder. If the wave is normally incident on the cylinder as shown in Figure 8.1, surface waves circumnavigate the body in spiral trajectories, and if the wave is obliquely incident, the travel paths are helical.

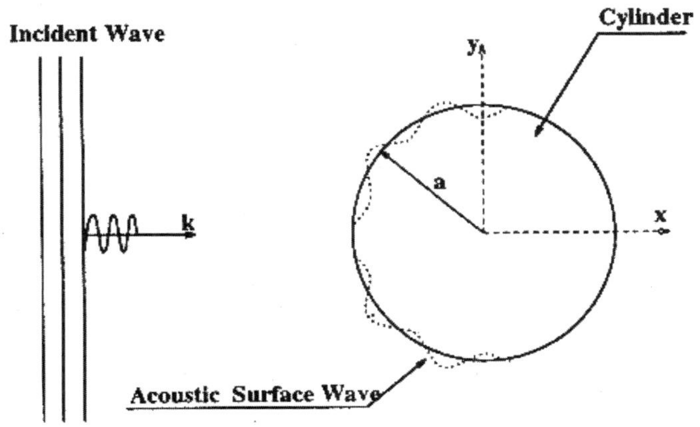

Figure 8.1 Insonification of an elastic cylinder by a normally incident plane acoustic wave.

These surface waves propagate near the boundary, either in the outer medium, or in the body of the scatterer [1]. The waves that propagate on the fluid side of the fluid-solid interface are Franz and Scholte-Stoneley waves (also called *creeping waves*) [2, 3]. These waves are primarily related to the geometry of the interface along which they travel and are present in the extreme cases of rigid and soft cylinders. By contrast, at normal incidence, two distinct families of waves can propagate inside the elastic cylinder. The first group comprises *Rayleigh-type* (or, more correctly, *pseudo-Rayleigh-type*) waves and the other group comprises *whispering gallery-type* (WG) waves [1]. Both of these two types of surface waves were first explained by Rayleigh [4, 5], and were further studied by many other researchers [3, 6–10]. In the case of scattering from an elastic cylinder in vacuum, a real Rayleigh-type wave exists, which becomes a true Rayleigh wave in the limit when the cylinder radius tends to infinity. Rayleigh and WG waves have been experimentally photographed inside submerged glass cylinders [11, 12].

When the wave is obliquely incident on the cylinder, additional types of waves are generated in the cylinder, and correspondingly additional resonance modes of the cylinder are excited. These additional (surface) waves which travel on helical paths are called (axially) *guided waves* (GW) [13–15] and are mostly of a transverse nature. The angle of each helix depends on the phase velocity of the corresponding surface wave [16].

A close relationship exists between the resonances of an elastic scatterer and the circumferential waves that travel along its periphery [17]. These surface waves and their relationship with the natural resonances of the scatterer have been extensively studied for the case of elastic cylinders. If the frequency of the incident wave coincides with one of the vibrational eigenfrequencies of the cylinder, surface waves match phases upon circumnavigation so as to build up to the resonance by constructive interference. It was the resonance scattering theory (RST) formulated by Uberall and his coworkers [18, 19] that physically described this phenomenon for the first time. RST will be discussed in more detail in further sections.

8.2.2 Acoustic Wave-Scattering from Elastic Targets

The dynamic scattering response of impenetrable scatterers, such as soft or rigid cylinders, to an incident wavefield has been studied since the pioneering works of Rayleigh [4]. In scalar form, the wave equation can be separated and solved in terms of infinite Rayleigh series in eleven coordinate systems [20]. Due to slow convergence of Rayleigh series, this is usually is a time-consuming process. The Watson-Summerfeld method was suggested as a better solution technique in 1968 [21]. In this method, the standard contour of integration for the slowly convergent Rayleigh series is transformed such that the new contour results in a quickly convergent series. The new path encloses new poles, which are associated with a quickly convergent residue series. The Watson-Summerfeld method has been most successful in explaining the physical behavior of the waves generated around impenetrable targets (called *Franz waves*).

In the case of elastic targets, the scattering equations are more complicated, and even in the case of isotropic materials they can be solved in a closed form for only a few geometries. If either the scatterer or the surrounding medium (or both) is elastic, then the wave equations can only be solved in three coordinate systems; viz. in spherical, in two-dimensional cylindrical, and in Cartesian (for an infinite plane fluid layer) coordinate systems [20]. The only true 3D target with closed form solution is the elastic sphere. Numerical schemes have been used for treating elastic scatterers of nonspherical geometry [22, 23].

The first pioneering study of acoustic wave scattering from submerged solid elastic circular cylinders and spheres based on normal-mode expansion is due to Faran [24]. In the case of an elastic cylinder insonified by a normally incident, plane acoustic wave, with the assumption that both the incident plane wave and the cylinder be infinite, the problem has no dependence on the axial coordinate of the cylinder (2D problem). Hickling [25] formulated the problem of acoustic wave scattering from elastic spheres and carried out experiments on different spherical targets.

Although White [26] formulated the problem of the scattering of an obliquely incident wave from an infinite elastic cylinder embedded in another elastic medium in 1958, it was not until 1980 that the problem of scattering of an obliquely incident acoustic wave from a solid elastic cylinder was specifically addressed. Flax et al. [27] formulated this problem and numerically evaluated the far-field backscattered amplitude spectrum for an isotropic aluminum cylinder at various angles of incidence. They noted the appearance of the resonance effects in the numerically calculated spectra. The same problem was also studied by Li and Ueda [28] who carried out experimental measurements on aluminum and steel rods at small angles of incidence. Bao et al. [29] studied the relationship between resonances and surface waves in the scattering of an obliquely incident wave from an infinite elastic cylinder. They also evaluated the dispersion curves of Rayleigh-type helical surface waves.

Development of the resonance scattering theory by Uberall and his coworkers [18, 19] provided a physical explanation for the phenomena observed in scattering of waves from elastic targets.

Surface waves generated by a plane acoustic wave obliquely incident on a cylinder were studied by Nagl et al. [30], basing their work on RST for elastic cylinders and the Watson-Summerfeld transformation for impenetrable cylinders. Uberall [22] studied the helical surface waves on cylinders and cylindrical cavities of finite length and found that a refraction effect takes place between incident and helical wave directions. His calculations for finite length cylinders were based on a transition matrix (or T-matrix) approximation [16]. Veksler [31] computed the modal resonances for an aluminum cylinder insonified by an obliquely incident wave and showed their correspondence to the extrema of the backscattered spectrum. A geometrical interpretation of the phase matching of the helical surface waves was given by Conoir et al. [32, 33], who also showed that the resonances shift towards higher frequencies when the angle of incidence is increased. Further explanations on this phenomena were presented by Ying et al. [34].

The theoretical explanations of RST were later complemented by experimental procedures of the *method of isolation and identification of resonances* (MIIR) proposed by French researchers [35–44]. RST and MIIR provide the necessary theoretical and experimental tools for resonance acoustic spectroscopy applications.

The mathematical models developed for scattering of waves from isotropic cylinders for normally [24] and obliquely [27] incident waves were improved by Honarvar and Sinclair [45] to include both isotropic and transversely isotropic cylinders. The mathematical model developed by Honarvar and Sinclair was the first attempt in considering anisotropic scatterers. Ahmad and Rahman suggested an alternative solution for the problem of scattering of waves from transversely isotropic cylinders [46]. Since the mathematical model for wave scattering from anisotropic cylinders is more general and includes the simpler case of isotropic cylinders, this model will be discussed in the next section.

8.3 MATHEMATICAL MODELS

The mathematical model for the scattering of a plane harmonic acoustic wave obliquely incident on a transversely isotropic cylinder will be discussed in this section. Transversely isotropic materials have hexagonal symmetry, and in the case of a cylinder, it is assumed that the axial elastic properties are different from elastic properties in the transverse plane.

Figure 8.2 shows an infinite monochromatic plane acoustic wave of frequency $\omega/2\pi$ incident at an angle α on a submerged circular cylinder of infinite length and outer radius *a,* and density ρ_c. The cylinder is submerged in a fluid of density ρ_w.

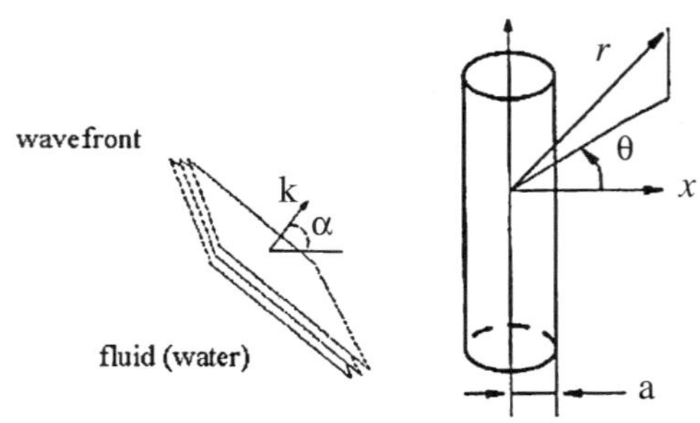

Figure 8.2 Geometry used for formulating the problem.

A cylindrical coordinate system (r,θ,z) is chosen with the *z*-direction coincident with the axis of the cylinder. The pressure p_i of the incident plane compression wave, external to the cylinder at an arbitrary point $M(r,\theta,z)$ is defined by,

$$p_i = p_0 e^{i(\mathbf{k}.\mathbf{r}-\omega t)}, \tag{8.1}$$

where p_0 is the incident wave amplitude, **k** is the wave vector, and **r** is the position vector of the point *M*. Moreover, $|\mathrm{k}| = \omega/c$, where ω is the angular frequency and c is the compression wave velocity in the fluid medium surrounding the cylinder. In a Cartesian coordinate system, one can write:

$$\mathbf{r} = r\cos\theta\hat{\mathbf{e}}_x + r\sin\theta\hat{\mathbf{e}}_y + z\hat{\mathbf{e}}_z, \tag{8.2}$$

and

$$\mathbf{k} = k(\cos\alpha\hat{\mathbf{e}}_x + 0\hat{\mathbf{e}}_y + \sin\alpha\hat{\mathbf{e}}_z). \tag{8.3}$$

Therefore,

$$\mathbf{k}\cdot\mathbf{r} = kr\cos\alpha\cos\theta + kz\sin\alpha. \tag{8.4}$$

After introducing the notation

$$k_z = k\sin\alpha; \quad k_\perp = k\cos\alpha, \tag{8.5}$$

one can write,

$$p_i = p_0 e^{ik_\perp r\cos\theta} e^{i(k_z z-\omega t)}. \tag{8.6}$$

The first exponential term in equation (8.6) can be written in terms of cylindrical functions [47, (Formula 10.1.47)],

$$e^{ik_\perp r\cos\theta} = \sum_{n=0}^{\infty} \varepsilon_n i^n J_n (k_\perp r)\cos(n\theta), \tag{8.7}$$

where ε_n is the *Neumann factor* ($\varepsilon_0 = 1$ and $\varepsilon_n = 2$ for $n \geq 1$) and J_n is the *Bessel function of the first kind* of order n. Thus, equation (8.6) can be written as,

$$p_i = p_0 \sum_{n=0}^{\infty} \varepsilon_n i^n J_n(k_\perp r)\cos(n\theta)\, e^{i(k_z z-\omega t)}. \tag{8.8}$$

We consider the steady state situation and drop the time dependence term from now on. The outgoing scattered wave pressure p_s at point M satisfies the wave equation,

$$\left(\nabla^2 - \frac{1}{c^2}\frac{\partial^2}{\partial t^2}\right) p_s = 0, \tag{8.9}$$

where ∇^2 is the Laplace operator. In a cylindrical coordinate system, equation (8.9) can be expanded to

$$\left(\frac{\partial^2}{\partial r^2} + \frac{1}{r}\frac{\partial}{\partial r} + \frac{1}{r^2}\frac{\partial^2}{\partial \theta^2} + \frac{\partial^2}{\partial z^2} - \frac{1}{c^2}\frac{\partial^2}{\partial t^2}\right) p_s = 0. \tag{8.10}$$

By using the method of separation of variables, the scattered pressure field will be found as follows,

$$p_s = p_0 \sum_{n=0}^{\infty} \varepsilon_n i^n A_n H_n^{(1)}(k_\perp r)\cos(n\theta)\, e^{k_z z i}. \tag{8.11}$$

where A_n are the unknown scattering coefficients and $\mathrm{H}_n^{(1)}$ is the Hankel Function of the First kind of order n. The general *Hooke's law* for a transversely isotropic material is,

$$\begin{Bmatrix} \sigma_{rr} \\ \sigma_{\theta\theta} \\ \sigma_{zz} \\ \sigma_{z\theta} \\ \sigma_{rz} \\ \sigma_{r\theta} \end{Bmatrix} = \begin{bmatrix} c_{11} & c_{12} & c_{13} & 0 & 0 & 0 \\ c_{12} & c_{11} & c_{13} & 0 & 0 & 0 \\ c_{13} & c_{13} & c_{33} & 0 & 0 & 0 \\ 0 & 0 & 0 & c_{44} & 0 & 0 \\ 0 & 0 & 0 & 0 & c_{44} & 0 \\ 0 & 0 & 0 & 0 & 0 & (c_{11}-c_{12})/2 \end{bmatrix} \begin{Bmatrix} \varepsilon_{rr} \\ \varepsilon_{\theta\theta} \\ \varepsilon_{zz} \\ 2\varepsilon_{z\theta} \\ 2\varepsilon_{rz} \\ 2\varepsilon_{r\theta} \end{Bmatrix}, \tag{8.12}$$

where σ_{ij} are stress components, ε_{ij} are strain components, and c_{ij} are the elastic constants. In a continuum, the equations of motion in the absence of body forces are [48],

$$\frac{\partial \sigma_{rr}}{\partial r}+\frac{1}{r}\frac{\partial \sigma_{r\theta}}{\partial \theta}+\frac{\partial \sigma_{rz}}{\partial z}+\frac{1}{r}(\sigma_{rr}-\sigma_{\theta\theta})=\rho_c\frac{\partial^2 u_r}{\partial t^2} \tag{8.13}$$

$$\frac{\partial \sigma_{r\theta}}{\partial r}+\frac{1}{r}\frac{\partial \sigma_{\theta\theta}}{\partial \theta}+\frac{2}{r}\sigma_{r\theta}+\frac{\partial \sigma_{z\theta}}{\partial z}=\rho_c\frac{\partial^2 u_\theta}{\partial t^2} \tag{8.14}$$

$$\frac{\partial \sigma_{rz}}{\partial r}+\frac{1}{r}\frac{\partial \sigma_{z\theta}}{\partial \theta}+\frac{1}{r}\sigma_{rz}+\frac{\partial \sigma_{zz}}{\partial z}=\rho_c\frac{\partial^2 u_z}{\partial t^2} \tag{8.15}$$

where ρ_c is the cylinder density and u_r, u_θ, and u_z are the displacements in the r, θ, and z directions, respectively. Substituting equation (8.12) in equations (8.13) – (8.15) gives,

$$\begin{aligned}
&c_{44}\left(\frac{\partial^2 u_z}{\partial z\partial r}+\frac{\partial^2 u_r}{\partial z^2}\right)\\
&+c_{11}\left(\frac{\partial^2 u_r}{\partial r^2}+\frac{1}{r}\frac{\partial u_r}{\partial r}-\frac{3}{2r^2}\frac{\partial u_\theta}{\partial \theta}+\frac{1}{2r^2}\frac{\partial^2 u_r}{\partial \theta^2}-\frac{u_r}{r^2}+\frac{1}{2r}\frac{\partial^2 u_\theta}{\partial r\partial \theta}\right)\\
&+\frac{1}{2r}c_{12}\left[\frac{\partial}{\partial \theta}\left(\frac{\partial u_\theta}{\partial r}-\frac{1}{r}\frac{\partial u_r}{\partial \theta}+\frac{u_\theta}{r}\right)\right]+c_{13}\frac{\partial^2 u_z}{\partial z\partial r}=\rho_c\frac{\partial^2 u_r}{\partial t^2},
\end{aligned} \tag{8.16}$$

$$\begin{aligned}
&c_{44}\left[\frac{\partial}{\partial z}\left(\frac{\partial u_\theta}{\partial z}+\frac{1}{r}\frac{\partial u_z}{\partial \theta}\right)\right]\\
&\quad+c_{11}\left(\frac{1}{r^2}\frac{\partial^2 u_\theta}{\partial \theta^2}+\frac{1}{2r}\frac{\partial u_\theta}{\partial r}+\frac{1}{2}\frac{\partial^2 u_\theta}{\partial r^2}+\frac{3}{2r^2}\frac{\partial u_r}{\partial \theta}+\frac{1}{2r}\frac{\partial^2 u_r}{\partial r\partial \theta}-\frac{1}{2}\frac{u_\theta}{r^2}\right)\\
&\quad+\frac{c_{13}}{r}\frac{\partial^2 u_z}{\partial z\partial \theta}=\rho_c\frac{\partial^2 u_\theta}{\partial t^2},
\end{aligned} \tag{8.17}$$

$$\begin{aligned}
&c_{44}\left(\frac{1}{r^2}\frac{\partial^2 u_z}{\partial \theta^2}+\frac{\partial^2 u_z}{\partial r^2}+\frac{1}{r}\frac{\partial^2 u_\theta}{\partial z\partial \theta}+\frac{1}{r}\frac{\partial u_z}{\partial r}+\frac{1}{r}\frac{\partial u_r}{\partial z}+\frac{\partial^2 u_r}{\partial z\partial r}\right)\\
&\qquad+c_{13}\left[\frac{\partial}{\partial z}\left(\frac{u_r}{r}+\frac{1}{r}\frac{\partial u_\theta}{\partial \theta}+\frac{\partial u_r}{\partial r}\right)\right]+c_{33}\frac{\partial^2 u_z}{\partial z^2}=\rho_c\frac{\partial^2 u_z}{\partial t^2}.
\end{aligned} \tag{8.18}$$

The displacement vector can be written in terms of three scalar potentials, ϕ, χ, and ψ [49],

$$\mathbf{U}=\nabla\phi+\nabla\times(\chi\hat{e}_z)+a\nabla\times\nabla\times(\psi\hat{e}_z), \tag{8.19}$$

where a is the radius of the cylinder. Substituting equation (8.19) in equations (8.16) – (8.18) results in the following set of equations:

$$\left(\nabla^2 - \frac{\partial^2}{\partial z^2}\right)\left\{c_{11}\nabla^2\phi + (c_{13}+2c_{44}-c_{11})\frac{\partial^2\phi}{\partial z^2} - \rho_c\frac{\partial^2\phi}{\partial t^2}\right.$$

$$\left. + a\frac{\partial}{\partial z}\left[(c_{11}-c_{13}-c_{44})\nabla^2\psi + (c_{13}+2c_{44}-c_{11})\frac{\partial^2\psi}{\partial z^2} - \rho_c\frac{\partial^2\psi}{\partial t^2}\right]\right\} = 0 \quad (8.20)$$

$$\frac{\partial}{\partial z}\left[(c_{13}+2c_{44})\nabla^2\phi + (c_{33}-c_{13}-2c_{44})\frac{\partial^2\phi}{\partial z^2} - \rho_c\frac{\partial^2\phi}{\partial t^2}\right]$$

$$+a\left(\frac{\partial^2}{\partial z^2} - \nabla^2\right)\left[c_{44}\nabla^2\psi + (c_{33}-c_{13}-2c_{44})\frac{\partial^2\psi}{\partial z^2} - \rho_c\frac{\partial^2\psi}{\partial t^2}\right] = 0 \quad (8.21)$$

$$\left(\nabla^2 - \frac{\partial^2}{\partial z^2}\right)\left[\frac{(c_{11}-c_{12})}{2}\nabla^2\chi + \left(c_{44} - \frac{(c_{11}-c_{12})}{2}\right)\frac{\partial^2\chi}{\partial z^2} - \rho_c\frac{\partial^2\chi}{\partial t^2}\right] = 0 \quad (8.22)$$

The first two equations, (8.20) and (8.21) indicate that the compression wave P (represented by ϕ) and the vertically polarized shear wave SV (represented by ψ) are coupled. Equation (8.22) shows that the horizontally polarized shear wave SH (represented by χ) is decoupled. Equations (8.20) – (8.22) should be solved for ϕ, ψ, and χ. Applying the separation of variables technique and taking advantage of irrotational and solenoidal properties of the longitudinal and transverse fields, the solution to equation (8.20) can be obtained by setting the terms in curly brackets equal to zero [50],

$$\left[\overbrace{c_{11}\nabla^2\phi + (c_{13}+2c_{44}-c_{11})\frac{\partial^2\phi}{\partial z^2} - \rho_c\frac{\partial^2\phi}{\partial t^2}}^{dilational}\right]$$

$$+\left[a\frac{\partial}{\partial z}\left[\overbrace{(c_{11}-c_{13}-c_{44})\nabla^2\psi + (c_{13}+2c_{44}-c_{11})\frac{\partial^2\psi}{\partial z^2} - \rho_c\frac{\partial^2\psi}{\partial t^2}}^{rotational}\right]\right] = 0, \quad (8.23)$$

that is,

$$c_{11}\nabla^2\phi + (c_{13}+2c_{44}-c_{11})\frac{\partial^2\phi}{\partial z^2} - \rho_c\frac{\partial^2\phi}{\partial t^2} = 0, \quad (8.24)$$

and,

$$(c_{11}-c_{13}-c_{44})\nabla^2\psi+(c_{13}+2c_{44}-c_{11})\frac{\partial^2\psi}{\partial z^2}-\rho_c\frac{\partial^2\psi}{\partial t^2}=0. \tag{8.25}$$

By solving the resulting separated partial differential equations, the scalar potential functions for an infinite cylinder are found to be of the form [50],

$$\phi(r,\theta,z,t)=\sum_{n=0}^{\infty}B_nJ_n(sr)\cos(n\theta)e^{k_zzi}, \tag{8.26}$$

$$\psi(r,\theta,z,t)=\sum_{n=0}^{\infty}C_nJ_n(sr)\cos(n\theta)e^{k_zzi}, \tag{8.27}$$

$$\chi(r,\theta,z,t)=\sum_{n=0}^{\infty}D_nJ_n(sr)\sin(n\theta)e^{k_zzi}. \tag{8.28}$$

where s is an unknown parameter that must be determined. Substituting equations (8.26)–(8.28) in equations (8.20) – (8.22) gives the $A_{1\times3}\cdot B_{1\times3}=O_{1\times3}$ matrix equation:

$$\begin{bmatrix} -s^2(c_{11}s^2-(\rho_c\omega^2-(c_{13}+2c_{44})k_z^2)) & -aik_zs^2((c_{11}-c_{13}-c_{44})s^2-(\rho_c\omega^2-(c_{44}k_z^2)) & 0 \\ ik_z((c_{13}+2c_{44})s^2-(\rho_c\omega^2-c_{33}k_z^2)) & as^2(c_{44}s^2-(\rho_c\omega^2-(c_{33}-c_{13}-c_{44})k_z^2)) & 0 \\ 0 & 0 & -s^2((c_{11}-c_{12})s^2-2(\rho_c\omega^2-c_{44}k_z^2)) \end{bmatrix}\begin{bmatrix} B_n \\ C_n \\ D_n \end{bmatrix}=\begin{bmatrix} 0 \\ 0 \\ 0 \end{bmatrix}. \tag{8.29}$$

For a nontrivial solution, the coefficient determinant in equation (8.29) must be zero. This yields the following characteristic equation:

$$(c_{44}c_{11}s^4-\xi s^2+\zeta)\left(\left(\frac{c_{11}-c_{12}}{2}\right)s^2-\rho_c\omega^2+c_{44}k_z^2\right)=0, \tag{8.30}$$

where

$$\xi=(c_{13}+c_{44})^2k_z^2+c_{11}(\rho_c\omega^2-c_{33}k_z^2)+c_{44}(\rho_c\omega^2-c_{44}k_z^2), \tag{8.31}$$

$$\zeta=(\rho_c\omega^2-c_{44}k_z^2)(\rho_c\omega^2-c_{33}k_z^2). \tag{8.32}$$

Three roots s_1, s_2, and s_3 of equation (8.30) correspond to the wave numbers of P, SV, and SH waves, respectively, and can be written as,

$$s_1^2=\frac{\xi-\sqrt{\xi^2-4c_{11}c_{44}\zeta}}{2c_{11}c_{44}}, \tag{8.33}$$

$$s_2^2 = \frac{\xi + \sqrt{\xi^2 - 4c_{11}c_{44}\zeta}}{2c_{11}c_{44}}, \tag{8.34}$$

$$s_3^2 = \frac{2(\rho_c\omega^2 - c_{44}k_z^2)}{c_{11} - c_{12}}, \tag{8.35}$$

which implies that the potential functions should be of the form [45, 50]:

$$\phi(r,\theta,z,t) = \sum_{n=0}^{\infty}\left[B_n J_n(s_1 r) + q_2 C_n J_n(s_2 r)\right]\cos(n\theta)e^{k_z zi}, \tag{8.36}$$

$$\psi(r,\theta,z,t) = \sum_{n=0}^{\infty}\left[q_1 B_n J_n(s_1 r) + C_n J_n(s_2 r)\right]\cos(n\theta)e^{k_z zi}, \tag{8.37}$$

$$\chi(r,\theta,z,t) = \sum_{n=0}^{\infty} D_n J_n(s_3 r)\sin(n\theta)e^{k_z zi}. \tag{8.38}$$

and the coupling coefficients are

$$q_1 = -\frac{(c_{11}s_1^2 - (\rho_c\omega^2 - (c_{13} + 2c_{44})k_z^2))}{aik_z((c_{11} - c_{13} - c_{44})s_1^2 - (\rho_c\omega^2 - (c_{44}k_z^2))}, \tag{8.39}$$

$$q_2 = -\frac{aik_z((c_{11} - c_{13} - c_{44})s_2^2 - (\rho_c\omega^2 - (c_{44}k_z^2))}{(c_{11}s_2^2 - (\rho_c\omega^2 - (c_{13} + 2c_{44})k_z^2))}. \tag{8.40}$$

For the submerged cylinder, the boundary conditions at $r = a$ are [51]

$$\frac{1}{\rho_w\omega^2}\frac{\partial}{\partial r}(p_i + p_s) = u_r;\ \sigma_{ij} = -(p_i + p_s);\ \sigma_{r\theta} = 0;\ \sigma_{rz} = 0. \tag{8.41}$$

Expanded expressions for displacements and stresses in terms of potential functions can be found in [45]. By inserting the potential functions from equations (8.36) – (8.38) into equation (8.41), a system of four linear algebraic equations is obtained for each value of n:

$$\begin{bmatrix} a_{11} & a_{12} & a_{13} & a_{14} \\ a_{21} & a_{22} & a_{23} & a_{24} \\ a_{31} & a_{32} & a_{33} & a_{34} \\ a_{41} & a_{42} & a_{43} & a_{44} \end{bmatrix} \begin{bmatrix} A_n \\ B_n \\ C_n \\ D_n \end{bmatrix} = \begin{bmatrix} b_1 \\ b_2 \\ 0 \\ 0 \end{bmatrix}. \tag{8.42}$$

Elements a_{ij} and b_i of the matrices in equation (8.42) are given in [45]. Equation (8.42) can be solved for A_n for any specified frequency using Cramer's rule.

$$A_n(ka) = \frac{\begin{vmatrix} b_1 & a_{12} & a_{13} & a_{14} \\ b_2 & a_{22} & a_{23} & a_{24} \\ 0 & a_{32} & a_{33} & a_{34} \\ 0 & a_{42} & a_{43} & a_{44} \end{vmatrix}}{\begin{vmatrix} a_{11} & a_{12} & a_{13} & a_{14} \\ a_{21} & a_{22} & a_{23} & a_{24} \\ 0 & a_{32} & a_{33} & a_{34} \\ 0 & a_{42} & a_{43} & a_{44} \end{vmatrix}}. \tag{8.43}$$

The scattered pressure field is usually evaluated in the far-field ($r >> a$) at a fixed angle θ for a range of frequencies. The resulting far-field amplitude spectrum, which is called a *form function*, is obtained from the equation [19],

$$f_\infty(\theta, ka) = \left(\frac{2r}{a}\right)^{1/2} \left(\frac{\mathrm{p_s}}{\mathrm{p_o}}\right) e^{-k_\perp r i}. \tag{8.44}$$

The total form function can be written as the sum of individual normal modes,

$$f_\infty(\theta, ka) = \sum_{n=0}^{\infty} f_n(\theta, ka), \tag{8.45}$$

where the individual normal modes are defined as,

$$f_n(\theta, ka) = \frac{2}{\sqrt{i\pi\, k_\perp a}} \varepsilon_n A_n \cos(n\theta). \tag{8.46}$$

For solving the corresponding problem for isotropic cylinders, it is a common practice to use the Helmholtz decomposition method in which the displacement field is written in terms of two distinct components [49], i.e.,

$$\mathbf{U} = \nabla \cdot \phi + \nabla \times \psi, \tag{8.47}$$

where ϕ is the rotation-free component corresponding to the longitudinal wave and ψ is the solenoidal term, which represents the S-wave. This approach is easier but cannot be used in the case of anisotropic materials since in these materials the wave modes are not of a purely compression or shear nature any more.

It can be shown that the model developed in the previous section can also be applied to isotropic cylinders [45]. For an isotropic cylinder, the values of the elastic constants c_{ij} can be written in terms of the *Lame constants* λ and μ such that,

$$c_{11} = c_{33} = \lambda + 2\mu;\ c_{12} = c_{13} = \lambda;\ c_{44} = \mu. \tag{8.48}$$

By substituting the above values into equations (8.36), (8.37) and (8.38) these equations are reduced to

$$\phi(r,\theta,z,t) = \sum_{n=0}^{\infty} B_n J_n(k_{\mathrm{L}} r)\cos(n\theta)e^{k_z zi}, \quad (8.49)$$

$$\psi(r,\theta,z,t) = \sum_{n=0}^{\infty} C_n J_n(k_{\mathrm{T}} r)\cos(n\theta)e^{k_z zi}, \quad (8.50)$$

$$\chi(r,\theta,z,t) = \sum_{n=0}^{\infty} D_n J_n(k_{\mathrm{T}} r)\sin(n\theta)e^{k_z zi}. \quad (8.51)$$

where,

$$k_{\mathrm{L}} = \left(\frac{\omega^2}{c_{\mathrm{L}}^2} - k_z^2\right)^{1/2}; \quad k_{\mathrm{T}} = \left(\frac{\omega^2}{c_{\mathrm{T}}^2} - k_z^2\right)^{1/2}. \quad (8.52)$$

Equations (8.49) to (8.51) are the corresponding potential functions for an isotropic cylinder.

8.3.1 Resonance Scattering Theory (RST)

The formalism of *resonance scattering theory* (RST) for acoustic resonance scattering is based on the corresponding theory in scattering [52]. RST provides a physical explanation of the wave phenomena that are observed in the acoustic-vibrational interaction. RST states that in acoustic scattering, the spectrum of the returned echo from a submerged elastic cylinder consists of two distinct parts: a smooth background plus a resonant part. The background varies smoothly with frequency and is present even for impenetrable scatterers and the resonant part consists of a number of resonance peaks that coincide with the eigenfrequencies of the vibrations of the cylinder. As mentioned by its originators, RST carries out two major functions:

(1) It separates the resonance components from the background signal.
(2) It mathematically expresses the resonance amplitude in a form where its resonance nature can be observed explicitly.

RST has been discussed in detail in many references [18–19, 39, 53–54]. In this section, principles of RST are briefly discussed. Let's start with considering an infinite plane acoustic wave

$$\mathrm{p_i} = \mathrm{p_o} e^{i(kx-\omega t)}, \quad (8.53)$$

with a propagation constant $k = \omega / c$ incident along the x-axis ($\alpha = 0^{\circ}$) on a solid elastic cylinder (see Figure 8.1).

We follow closely the derivation presented by Flax et al. [19]. At a point $M(r,\theta)$ the incident wave produces the following scattered field p_s,

$$p_s = p_o \sum_{n=0}^{\infty} \varepsilon_n i^n A_n(x) H_n^{(1)}(kr)\cos(n\theta), \tag{8.54}$$

where

$$A_n(x) = -\frac{J_n(x)F_n - xJ_n'(x)}{H_n^{(1)}(x)F_n - xH_n^{(1)}(x)}. \tag{8.55}$$

The quantities F_n, related to the modal mechanical impedance of the cylinder [55], are the quotient of two 2×2 determinants,

$$F_n(x) = \frac{\rho_w}{\rho_c} x_T^2 \frac{D_n^{(1)}(x)}{D_n^{(2)}(x)}, \tag{8.56}$$

where

$$D_n^{(1)} = \begin{vmatrix} a_{22} & a_{23} \\ a_{32} & a_{33} \end{vmatrix} \text{ and } D_n^{(2)} = \begin{vmatrix} a_{12} & a_{13} \\ a_{32} & a_{33} \end{vmatrix}. \tag{8.57}$$

Expressions for a_{ij} are given in [19]. It should be noted that the matrices given in equation (8.57) are similar to those given in equation (8.43), however, isotropy of the material and normalcy of the incident wave has resulted in much smaller matrices. The argument x of the Bessel and Hankel functions in equation (8.56) is $x = ka = \omega a / c_w$, where c_w is the speed of sound in the surrounding fluid. The prime denotes differentiation with respect to the argument. In the far-field where $r >> a, H_n^{(1)}(x)$ may be written in its asymptotic form [47],

$$H_n^{(1)}(x) \approx (2/\pi x)^{1/2} i^{-n} e^{ix} . \tag{8.58}$$

Then the far-field scattered pressure becomes

$$p_s(\theta) = p_o e^{ikr} (2/i\pi\, kr)^{1/2} \sum_{n=0}^{\infty} \varepsilon_n A_n(x)\cos(n\theta), \tag{8.59}$$

and the form function f_∞ of the scattered pressure is

$$f_\infty(\theta) = \sqrt{\frac{2r}{a}} \frac{p_s}{p_o} e^{ikr}. \tag{8.60}$$

The form function can also be represented by using equation (1.59),

$$f_\infty(\theta) = \frac{2}{\sqrt{\pi i x}} \sum_{n=0}^{\infty} \varepsilon_n A_n(x)\cos(n\theta). \tag{8.61}$$

where individual normal modes or partial waves which make up the form functions are defined as

$$f_n(\theta) = \frac{2}{\sqrt{\pi i x}} \varepsilon_n A_n(x)\cos(n\theta), \tag{8.62}$$

so that, $f_\infty(\theta) = \sum_{n=0}^{\infty} f_n(\theta)$. (8.63)

There are two limiting cases of these results. If $\rho_c \to \infty$, the solution represents the scattering by a rigid cylinder:

$$A_n(x)^r = -J_n'(x)/H_n^{(1)'}(x), \tag{8.64}$$

and if $\rho_c \to 0$, the solution represents the scattering by a soft cylinder:

$$A_n(x)^s = -J_n(x)/H_n^{(1)}(x). \tag{8.65}$$

The scattered pressure of equation (8.54) may be written as a normal mode series (setting $p_0 = 1$),

$$p_s(r,\ \theta) = \frac{1}{2}\sum_{n=0}^{\infty} \varepsilon_n i^n (S_n - 1) H_n^{(1)}(kr)\cos(n\theta) \tag{8.66}$$

where the scattering function of the nth mode with a constant unit magnitude containing the scattering phase shifts δ_n is introduced as follows:

$$S_n = e^{2i\delta_n}. \tag{8.67}$$

In the present case,

$$S_n - 1 = 2A_n(x). \tag{8.68}$$

For rigid and soft cylinders, the scattering functions are, respectively,

$$S_n^r = -H_n^{(2)'}(x)/H_n^{(1)'}(x) = e^{2i\delta_n^r} \tag{8.69}$$

and

$$S_n^s = -H_n^{(2)}(x)/H_n^{(1)}(x) = e^{2i\delta_n^s} \tag{8.70}$$

Superscripts r and s have been used for terms corresponding to rigid and soft cylinders, respectively. The corresponding phase shifts can be shown to be the real quantities,

$$\tan\delta_n^r = J_n'(x)/Y_n'(x), \text{ and } \tan\delta_n^s = J_n(x)/Y_n(x). \tag{8.71}$$

A rigid or soft scattering function may be factored out from the elastic scattering function as follows:

$$S_n = S_n^r(F_n^{-1} - z_n^{(2)-1})/(F_n^{-1} - z_n^{(1)-1}) \tag{8.72}$$

or

$$S_n = S_n^s(F_n - z_n^{(2)})/(F_n - z_n^{(1)}), \tag{8.73}$$

where the quantities

$$z_n^{(i)} = xH_n^{(i)'}(x)/H_n^{(i)}(x), \qquad i = 1,\ 2 \tag{8.74}$$

are related to the modal specific acoustic impedances [55].

After some mathematical manipulations [18–19], it can be shown that equation (8.73) can be approximated as,

$$S_n - 1 = 2i\,e^{2i\,\delta_n^{\mathrm{r}}}\left[\sum_{l=1}^{\infty}\frac{\frac{1}{2}\Gamma_{nl}^{\mathrm{r}}}{x_{nl}^{\mathrm{r}} - x - \frac{1}{2}i\Gamma_{nl}^{\mathrm{r}}} + e^{-i\delta_n^{\mathrm{r}}}\sin\delta_n^{\mathrm{r}}\right], \tag{8.75}$$

or

$$S_n - 1 = 2i\,e^{2i\,\delta_n^{\mathrm{s}}}\left[\sum_{l=1}^{\infty}\frac{\frac{1}{2}\Gamma_{nl}^{\mathrm{s}}}{x_{nl}^{\mathrm{s}} - x - \frac{1}{2}i\Gamma_{nl}^{\mathrm{s}}} + e^{-i\delta_n^{\mathrm{s}}}\sin\delta_n^{\mathrm{s}}\right], \tag{8.76}$$

where x_{nl}^{r} and x_{nl}^{s} are the lth resonance frequencies of the nth mode and Γ_{nl}^{r} and Γ_{nl}^{r} are the resonance widths related to radiation damping. Based on expressions (8.75) and (8.76), resonance scattering theory argues that the scattered field is a summation of two components, i.e., a resonance component, which is the first term of equations (8.75) and (8.76), and a smooth background component, which is the second term of these equations.

Following the approach proposed by Rhee and Park [56], the scattering function S_n can be written as,

$$S_n = S_n^{\mathrm{r}} S_n^{*}, \tag{8.77}$$

where

$$S_n^{*} = (F_n^{-1} - z_n^{(2)-1})/(F_n^{-1} - z_n^{(1)-1}). \tag{8.78}$$

Equation (8.77) states that S_n is the product of a rigid background S_n^{r} and the remaining term S_n^{*}, which includes resonances. However, S_n^{*} is not a pure resonance term. It has a constant term, which hides resonances unless it is removed. To see this, S_n^{*} can be written as

$$\begin{aligned} S_n^{*} &= (F_n^{-1} - z_n^{(2)-1})/(F_n^{-1} - z_n^{(1)-1}) \\ &= (z_n^{(1)-1} - z_n^{(2)-1})/(F_n^{-1} - z_n^{(1)-1}) + 1 = S_n^{\mathrm{res}} + 1, \end{aligned} \tag{8.79}$$

where S_n^{res} is defined as the resonance scattering function that consists purely of resonance information. From equation (8.77), the resonance scattering function S_n^{res} can be expressed as

$$S_n^{\mathrm{res}} = \frac{S_n}{S_n^{\mathrm{r}}} - 1 = \frac{z_n^{(1)-1} - z_n^{(2)-1}}{F_n^{-1} - z_n^{(1)-1}} = 2\frac{A_n - A_n^{r}}{1 + 2A_n^{r}}, \tag{8.80}$$

where $A_n = (1/2)(S_n - 1)$ and $A_n^{\mathrm{r}} = (1/2)(S_n^{\mathrm{r}} - 1)$.

In equation (8.66), $S_n - 1$ contributes to both the background and resonance parts of the scattered field. S_n^{res} is the component that only contains resonance effects of S_n. Therefore, from equation (8.66), the scattered wave due to resonance can be obtained by replacing $S_n - 1$ with S_n^{res}:

$$p_{\text{s}}^{\text{res}}(r,\ \theta) = \frac{1}{2}\sum_{n=0}^{\infty} \varepsilon_n i^n (S_n^{\text{res}}) H_n^{(1)}(kr)\cos(n\theta)$$
$$= \sum_{n=0}^{\infty} \varepsilon_n i^n \frac{A_n - A_n^{\text{r}}}{1+2A_n^{\text{r}}} H_n^{(1)}(kr)\cos(n\theta) \tag{8.81}$$

and the individual normal mode for resonances in the far-field can be written as

$$f_n^{\text{res}}(\theta) = \frac{2}{\sqrt{\pi i x}} \varepsilon_n \frac{A_n - A_n^{\text{r}}}{1+2A_n^{\text{r}}} \cos(n\theta). \tag{8.82}$$

The expressions containing the corresponding soft background parameters are analogous,

$$S_n^* = (F_n - z_n^{(2)})/(F_n - z_n^{(1)})$$
$$= (z_n^{(1)} - z_n^{(2)})/(F_n - z_n^{(1)}) + 1 = S_n^{\text{res}} + 1 \tag{8.83}$$

$$S_n^{\text{res}} = \frac{S_n}{S_n^{\text{s}}} - 1 = \frac{z_n^{(1)} - z_n^{(2)}}{F_n - z_n^{(1)}} = 2\frac{A_n - A_n^{\text{s}}}{1+2A_n^{\text{s}}} \tag{8.84}$$

$$f_n^{\text{res}}(\theta) = \frac{2}{\sqrt{\pi i x}} \varepsilon_n \frac{A_n - A_n^{\text{s}}}{1+2A_n^{\text{s}}} \cos(n\theta). \tag{8.85}$$

The expressions in equations (8.82) and (8.85) are also applicable to acoustic wave scattering from shell structures or fluid cylinders with appropriate changes in F_n. Using equations (8.82) and (8.85), the resonance part of the backscattered field can be completely separated from the background. A typical extraction of the resonance spectrum from the form function based on RST is shown in Figure 8.3.

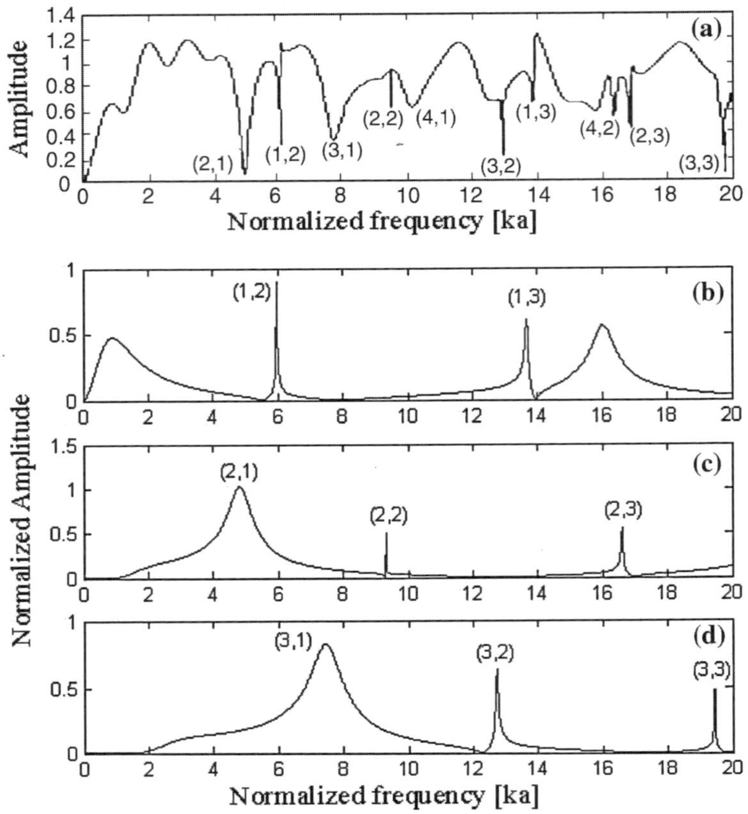

Figure 8.3 a) The form function for backscattering by a solid aluminum cylinder in water. b, c, d) Isolated modal resonances of the first, second, and third (n = 1, 2, 3) modes for backscattering by an aluminum cylinder in water [76].

8.3.2 Resonance Acoustic Spectroscopy

Figure 8.4a shows the form function of an aluminum cylinder insonified by a normally incident plane acoustic wave. The backscattered field is composed of a series of dips, which are due to resonances, superimposed on a smooth background. Each dip in Figure 8.4a is due to a certain resonance frequency identified by the integers (n,l). The first of these two integers defines the mode number and the second one indicates the eigenfrequency label for a given mode. $l=1$ corresponds to a Rayleigh-type wave and $l=2,3,...$ to a whispering gallery-type wave. The mode number n identifies the shape of the pressure contours around the cylinder at a resonance frequency. For the breathing mode $(n=0)$, the pressure contours around the cylinder move radially in and out. For the dipole mode $(n=2)$, the pressure contour motion is rigidly back and forth, and for the quadrupole mode $(n=2)$, the contours oscillate between prolate and oblate ellipses. For higher modes, the pressure contours around the cylinder have rosette patterns with $2n$ lobes.

The values of the resonance frequencies are dependent on the elastic wave velocities (or elastic constants) of the solid cylinder. The way these resonances can be used for material characterization purposes, resonance acoustic spectroscopy (RAS) [57–59], resembles the way chemical elements are identified from their optical spectra. The so-called *Regge pole trajectories* [59] are a plot of resonance frequencies in the (ka, n)-plane where ka is the normalized frequency and n is the mode number. The Regge trajectories for an aluminum cylinder are shown in Figure 8.4b. Each resonance frequency in Figure 8.4a corresponds to a pole on the Regge trajectories plot highlighted by a "+" in Figure 8.4b. The poles belonging to the same family of surface waves are connected by a line in Figure 8.4b. Regge poles are obtained by finding the roots of the denominator of equation (8.43).

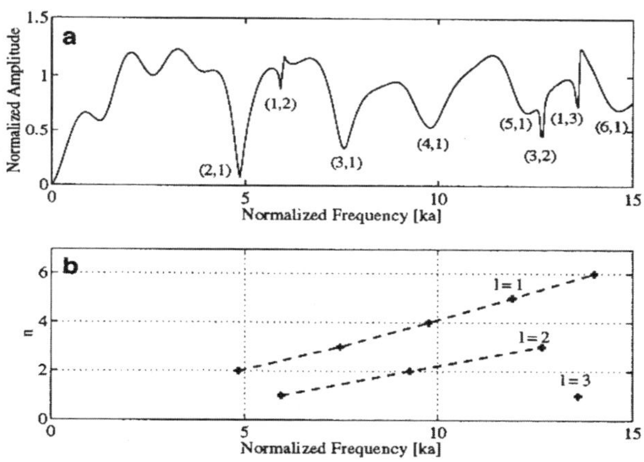

Figure 8.4 a) Form function for an aluminum cylinder. b) Regge trajectories for an aluminum cylinder [76].

Figure 8.5a shows the numerically calculated form function of the same aluminum cylinder used to generate Figure 8.4 when it is insonified by a plane acoustic wave at an incident angle of $\alpha = 3^\circ$. It can be observed that extra families of resonances have been generated. These resonances are due to axially guided waves. Resonances associated with axially guided waves are designated by $\langle n, p \rangle$ where, similar to resonances associated with Rayleigh and whispering gallery waves, n is the mode number and p is the frequency label of the resonance. Regge trajectories of the aluminum cylinder are also shown in Figure 8.5b.

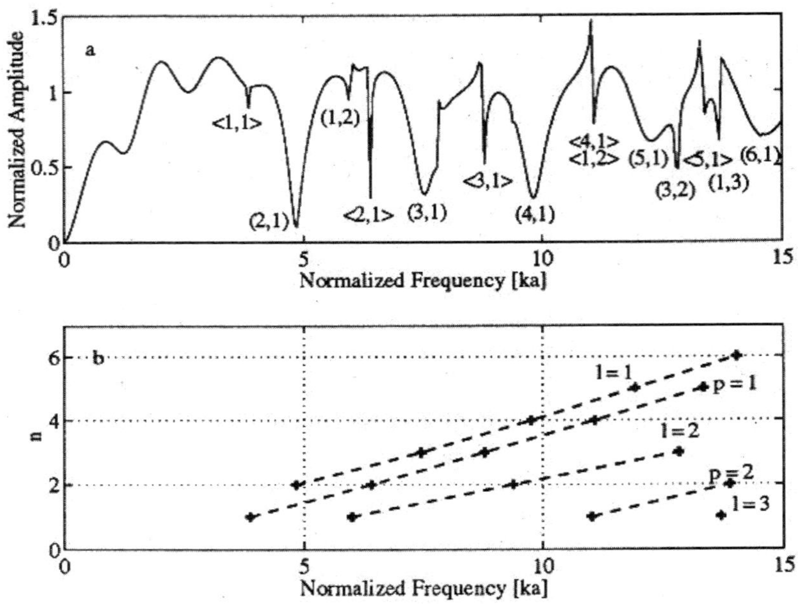

Figure 8.5 a) Form function for aluminum cylinder. b) Regge trajectories for aluminum cylinder. Resonances associated with Rayleigh and WG waves are designated by (n, l) and resonances associated with axially guided waves by $\langle n, p \rangle$ [76].

Any component has its own resonances, which depend on the geometry and material constituents of the component. Presence of any flaws in the component also affects the resonance frequencies. Each resonance has its own characteristics and consequently shows sensitivity to certain variations in the component. One of the objectives of resonance acoustic spectroscopy is to identify the resonances and their relative sensitivity to variations in properties of the component. After identifying the right resonances, one can use them to verify or monitor certain properties of the component.

Resonance acoustic spectroscopy has been used by a number of researchers for characterization of various materials. Brill et al. [57, 58] used RAS for characterization of shapes of fluid and solid scatterers. Ripoche and Maze [59] discussed the use of the method of isolation and identification of resonances (MIIR) for evaluation of elastic objects. Honarvar and Sinclair [45] used RAS for monitoring the health condition of clad rods. Ying et al. [60] implemented RAS in characterizing large-grained cylinders.

In the following sections, applications of RAS in the above-mentioned areas will be discussed in more detail. However, first we should introduce some experimental procedures usually used in RAS applications.

8.4 METHOD OF ISOLATION AND IDENTIFICATION OF RESONANCES (MIIR)

8.4.1 Introduction

RST theoretically shows the possibility of decomposing the backscattered amplitude spectrum of a penetrable scatterer into a smooth background component and a resonant component. Measurements of the backscattered pressure field from immersed elastic targets have been carried out since the early works of Faran [24] and Hickling [25]. Before 1981, the acoustic signatures of elastic targets usually were "time signature" and "form functions" (backscattered pressure versus frequency). The monostatic time signatures, obtained in the geometry described in Figure 8.1, contain a large number of backscattered echoes, which are due to the specular echo, Franz waves, Rayleigh waves, whispering gallery waves, etc. The computed far-field form functions have been compared to experimental form functions obtained by pulse methods by means of the fast Fourier transform (FFT) [61–64]. Although both short pulses [62, 65–67] and pulses of long duration [25, 68, 69] have been used for the measurement of the form function, it was not until 1981, when the *method of isolation and identification of resonances* (MIIR) was developed, that the real thrust in the experimental procedures started [36]. Using MIIR, the form function, the resonance spectrum, and the mode number of individual resonances of a penetrable scatterer can be experimentally measured using a delicate time-gating procedure. In the following sections we will discuss two approaches used in MIIR measurements which are 1) the quasi-harmonic MIIR method, and 2) the short-pulse MIIR method.

8.4.2 Quasi-Harmonic MIIR

The classical scattering theory gives the far-field form function of an elastic cylinder immersed in water and insonified by an infinite plane acoustic wave as,

$$f_\infty(\theta) = \frac{2}{\sqrt{\pi i x}} \sum_{n=0}^{\infty} \varepsilon_n A_n(x) \cos(n\theta). \tag{8.86}$$

The computed form function of an aluminum solid cylinder plotted versus $k_1 a$ is shown in Figure 8.6a [40].

The cylindrical target, with relatively high length-to-diameter ratio, is insonified by long incident tone-burst pulses with a carrier frequency of ω_o. In a monostatic situation, by slowly varying the frequency of the incident signal between one burst and the next, the quasi-steady state amplitude is recorded versus normalized frequency. This recording of the pressure amplitude gives the backscattered signal spectrum (Figure 8.6b) compared to the far-field form function. The theory agrees with the experiment. The spectrum consists of the superposition of a smooth background and peaks and valleys corresponding to resonances. The resonances are located at each minimum, but often, it is difficult to accurately locate them in this spectrum.

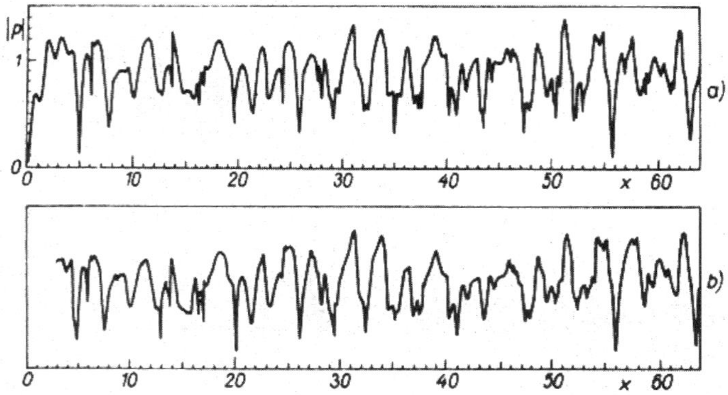

Figure 8.6 The form function of the acoustic pressure scattered by a solid elastic cylinder. a) computed result and b) measured result [40].

8.4.2.1 Resonance Isolation in Quasi-Harmonic MIIR

If the target is insonified by a rectangular wave pulse having a known frequency when this frequency does not coincide with one of the resonance frequencies of the target, the received backscattered echo will have a structure similar to that of the incident pulse with only a change in the amplitude of the signal (see Figure 8.7a). On the other hand, when the frequency of the incident wave coincides with one of the resonance frequencies of the cylinder, the backscattered echo will show a completely different structure compared to that of the incident pulse, Figure 8.7b.

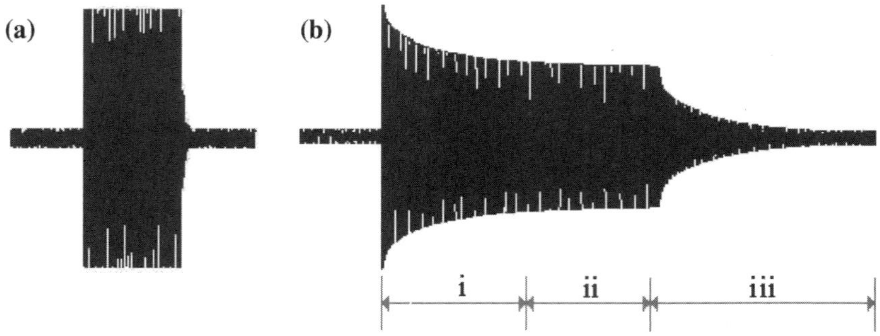

Figure 8.7 a) Incident tone-burst pulse used in quasi-harmonic MIIR. b) Echo scattered from the cylinder at a resonance frequency.

At a resonance frequency as shown in Figure 8.7b, three distinct parts can be distinguished in the backscattered echo: (i) a leading transient, (ii) a steady state vibration during which the amplitude of the signal remains almost constant (forced vibration part of the signal), and (iii) a second transient part after the end of the forced vibration during which the amplitude of the signal starts to decay exponentially (free vibration part of the signal) [70]. If one measures the amplitude of the signal in the steady state segment while sweeping the transmitter through a range of frequencies, then a plot of the measured amplitude versus frequency will produce the form function of the cylinder as shown in Figure 8.8a. If the same measurements are performed during the free vibration of the cylinder (second transient part of the signal), then the plot of the measured amplitude versus frequency is the resonance spectrum (Figure 8.8b).

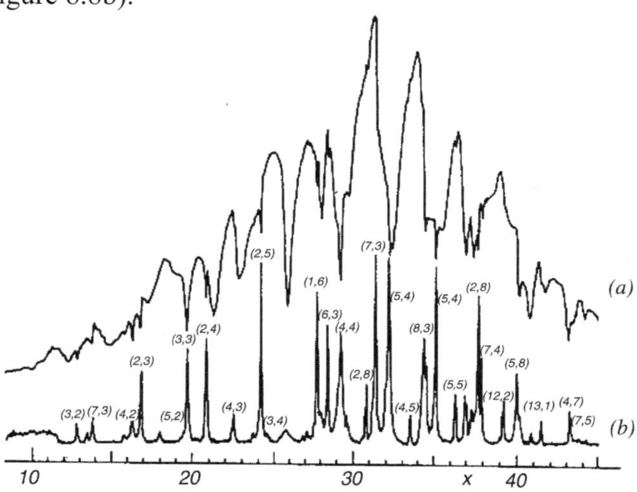

Figure 8.8 a) Measured form function, and b) resonance spectrum of an aluminum cylinder [40].

8.4.2.2 Resonance Identification in Quasi-Harmonic MIIR

The latter measurement (Figure 8.8b) can also be used to verify the mode shape and, accordingly, the mode number n of a specific resonance. According to RST [71], circumferential waves travel around the cylinder in two opposite directions and form standing waves. Therefore, it should be possible to observe nodal and antinodal points of displacement around the cylinder.

These nodal and antinodal points can be observed in a bistatic configuration where one transducer insonifies the cylinder with a long pulse modulated at a resonance frequency and another transducer (receiver) is rotated, at a constant radial distance, around the cylinder and measures the scattered signal. A typical arrangement is shown in Figure 8.9. The receiver measures the amplitude of the signal at a fixed point during the second transient part of the signal. A plot of amplitude versus angular position will then reveal the nodal and antinodal points. A typical plot of this kind for $n = 4$ is shown in Figure 8.10. The mode number n of the resonance is half the number of nodes or antinodes around the cylinder.

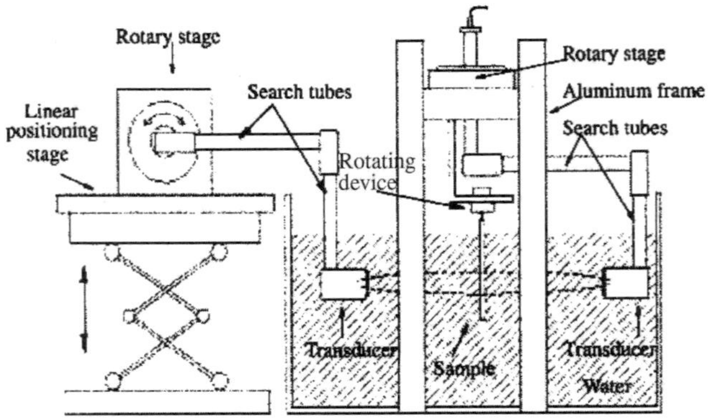

Figure 8.9 Test setup used for normal incidence MIIR measurements and mode shape measurements [60].

The experimental measurement of the mode number n gives an "identification of resonances" and allows a distinction of experimental resonances observed in resonance spectra. Besides, the knowledge of the mode number n for each resonance gives the experimental phase velocity of the circumferential waves [22]. Because the response of the target reaches the steady state, it is called "quasi-harmonic MIIR."

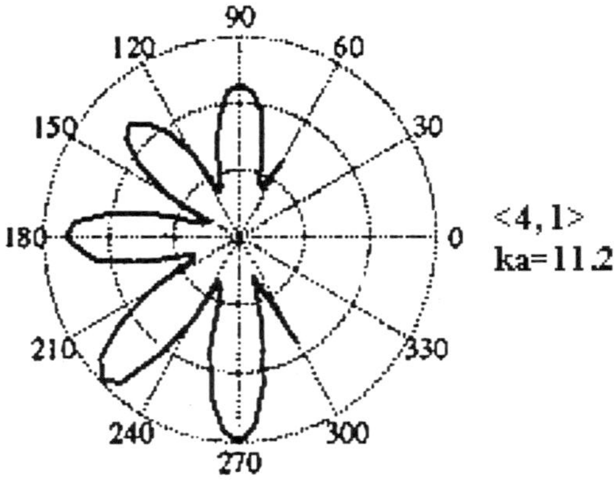

Figure 8.10 Measured mode shape for $n = 4$ [76].

Although tone-bursts of long duration were used in early MIIR studies, it was later shown that it is possible to obtain the same information with another, much less time-consuming, method. This technique is called short-pulse MIIR and will be discussed in the next section.

8.4.3 Short-Pulse MIIR

Several studies were developed independently by three laboratories (NRL, Washington; GPS. ENS, Paris; LEAH. US, Le Havre) to obtain resonance spectra by means of short-pulse methods. The pulse method of Dragonette, Numrich, and Frank [61, 62] offers some possibilities to obtain the total form function of targets immersed in water. This function is given after processing the time signatures by means of a fast Fourier transform (FFT) algorithm when the incident signal is a short pulse. In 1984, Numrich et al. [72] used this method, however only a selected portion of the time record of the echo was transformed by the FFT algorithm. The frequency signature, which was obtained for cylindrical bodies with hemispherical end caps was the elastic response of the target. A method was developed independently by de Billy [73] for the study of a cylinder ($2a$ = 0.7mm) where a portion of the time signature is isolated by a gate, then transformed into a frequency signature by a spectrum analyzer.

In Le Havre, Maze et al. [74] could obtain the form function and the resonance spectrum of the scatterer by calculating the FFT of the back-scattered echo of short-pulse MIIR with one single measurement.

8.4.3.1 Resonance Isolation in Short Pulse MIIR

In short-pulse MIIR, the scatterer is insonified by pulses having a duration that is short compared to the target's diameter divided by sonic velocity. The first arrival at the receiver of the backscattered echo of the short pulse corresponds to the specular reflection of the incident pulse from the vertex of the scatterer. The subsequent echoes following this first echo (see Figure 8.11) are due to the resonances generated by the surface waves encircling the body of the scatterer and producing standing waves.

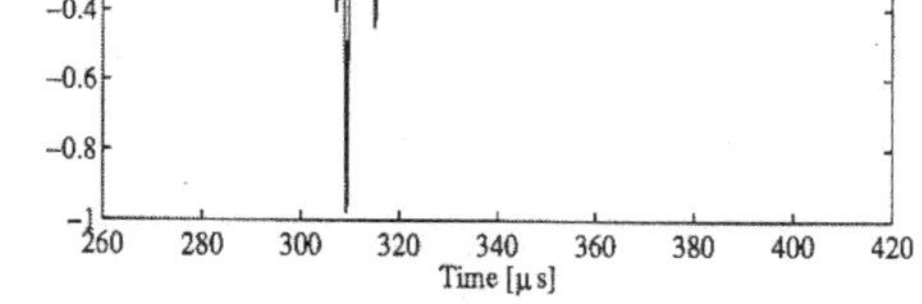

Figure 8.11 Backscattered echo from an aluminum cylinder [75].

By calculating the FFT of the backscattered echo, one can obtain the form function of the elastic scatterer for a broad range of frequencies with one single measurement. Moreover, if the specular echo is removed from the backscattered signal by proper time-gating, the FFT of the remaining part of the signal will be the resonance spectrum of the scatterer (see Figure 8.12 [75]).

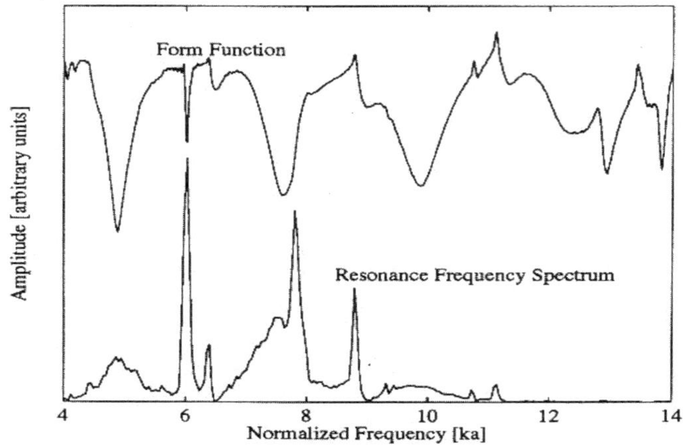

Figure 8.12 Measured form function and resonance spectrum of an aluminum cylinder at incident angle of $\alpha = 0^\circ$ [76].

8.4.3.2 Resonance Identification in Short-Pulse MIIR

The mode number of the resonances contained within the frequency range of the incident pulse can also be measured using short-pulse MIIR [77, 78]. The bistatic measurements are similar to those of the quasi-harmonic MIIR. For example, in the case of a long circular cylinder, returned echoes (with the specular echo removed) are measured around the cylinder (e.g., at 5° increments) and are stored in the computer memory. By finding the FFT of each echo, the mode shape of each resonance can be plotted in a polar diagram as in Figure 8.13 [76].

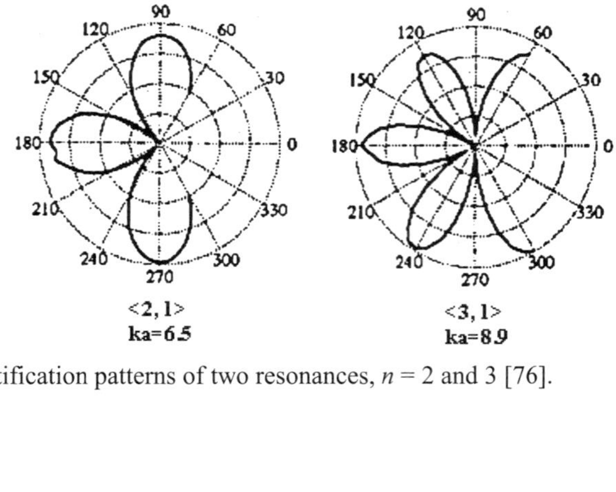

Figure 8.13 Identification patterns of two resonances, $n = 2$ and 3 [76].

Nowadays, short-pulse MIIR is more often used for the measurement of the scattered field, because it is much less time-consuming than quasi-harmonic MIIR.

8.5 EXPERIMENTAL AND NUMERICAL RESULTS

8.5.1 Introduction

Echoes of acoustic waves reflected from elastic targets carry within them certain resonance features caused by the excitation of the eigenvibrations of the target. By means of suitable background subtraction (either mathematically or experimentally), it is possible to obtain the resonance spectrum of the target. This resonance spectrum characterizes the target just as an optical spectrum characterizes the chemical element or compound that emits it. Extracting the resonance information from the echo provides the possibility of identifying the target as to its size, shape, composition, and integrity. In the following sections, first, the primary applications of RAS in characterizing the target shape will be presented. Then, various examples of engineering applications of RAS for nondestructive evaluation (NDE) and on-line monitoring of a number of cylindrical components including simple elastic rods and wires, elastic, viscoelastic and explosively welded clad rods, and fiber-reinforced composite (transversely isotropic) rods will be discussed.

8.5.2 Characterization of Target Shape by RAS

In this section, some early applications of RAS for identification of the shape of a target are discussed. The dependence of the resonance spectra on target shape for a fluid object in vacuo and elastic targets in a fluid were studied by Brill et al. [57, 58]. As these two cases have similar mathematical formulations, they will be considered together. Their eigenfrequency spectra have been illustrated in the form of "level schemes" similar to that given by atomic spectroscopy. The eigenfrequency spectrum of an object was tracked through various geometrical shape changes, starting with a spherical shape where the eigenfrequencies show azimuthal degeneracy. By removing the azimuthal degeneracy, as the sphere is deformed into a slightly eccentric, and finally, into a strongly eccentric prolate spheroid, it is observed that the eigenfrequencies begin to split. The spheroid spectrum is compared to that of a finite circular cylinder having the same ratio of length of axes. It is demonstrated that the two spectra approach each other as the spheroid and cylinder become longer, until they finally coalesce into the axially degenerate resonances of the infinite cylinder.

8.5.2.1 Mathematical Formulation

The mathematical equations are presented in three steps for 1) sphere, 2) prolate spheroid, and 3) finite cylinder.

8.5.2.1.1 Sphere: For the sphere, the solution of the wave equation for the acoustic field in the interior region in spherical coordinates (r,θ,ϕ) is [57]

$$\psi(r,\theta,\phi)=\sum_{lm} A_{lm} j_l(kr)e^{im\phi}P_l^m(\cos\theta), \tag{8.87}$$

where $k=\omega/c$, and c is the sound velocity inside the fluid object. The traction-free boundary condition at the sphere's surface, $r=a$, yields the characteristic equation,

$$j_l(ka)=0\,. \tag{8.88}$$

The solution to this equation leads to the eigenvalues of k (or frequency)

$$k_{lms}=x_{ls}/a \tag{8.89}$$

where x_{ls} are the zeros of the spherical Bessel functions $j_l(x)$ whose multiplicity is labeled by s = 1, 2, 3,.... These eigenvalues are degenerate in the azimuthal index m.

8.5.2.1.2 Prolate Spheroid: For the spheroid, the solution can be separated in prolate spheroidal coordinates (ξ,η,ϕ) defined by [49],

$$\begin{aligned} x&=(d/2)\{(1-\eta^2)(\xi^2-1)\}^{1/2}\cos\phi\\ y&=(d/2)\{(1-\eta^2)(\xi^2-1)\}^{1/2}\sin\phi\\ z&=(d/2)\eta\xi \end{aligned} \tag{8.90}$$

and prolate spheroidal surface is

$$\xi=\{1-(a/b)^2\}^{-1/2}=\text{const}, \tag{8.91}$$

where d is the distance between the focal points. Major and minor axes of the spheroid are $b=(d/2)\xi$ and $a=(d/2)(\xi^2-1)^{1/2}$, respectively. The solution of the wave equation in prolate spheroidal coordinates is of the form

$$\psi(\xi,\eta,\phi)=\sum_{lm} B_{ml}R_{ml}^{(1)}(h,\xi)S_{ml}(h,\eta)\Phi_m(\phi). \tag{8.92}$$

The standing-wave prolate spheroidal radial functions can be expanded as

$$R_{ml}^{(1)}(h,\xi)=\frac{(1-m)!}{(1+m)!}\left(\frac{\xi^2-1}{\xi^2}\right)^{m/2}\sum_{p=0,1}^{\infty}{}' i^{p+m-1}$$

$$\times d_p(h\,|\,ml)\frac{(p+2m)!}{p!}J_{p+m}(h\xi), \tag{8.93}$$

where the prime on the summation sign denotes p even (odd) if $l-m$ is even (odd), and $h=(kd/2)$. The numerical coefficients $d_p(h\,|\,m,l)$ are obtained from a recursion relation arising from solutions of the separated radial (ξ) and angular (η) differential equations [49, 79].

The pressure release boundary condition ($\psi = 0$) at the surface of a prolate spheroid ($\xi = \text{const}$) requires that the radial functions vanish at the value of ξ where

$$R_{ml}^{(1)}(h, \xi) = 0. \tag{8.94}$$

The roots of equation (8.94) are designated as h_{lms}, where s = 1, 2, 3…., labels their multiplicity). The eigenvalues of k (or of frequency) are,

$$k_{lms} = x_{lms} / a \tag{8.95}$$

where a is the semi-minor and b is the semi-major axes, and,

$$x_{lms} = h_{lms}(\xi^2 - 1)^{1/2}. \tag{8.96}$$

The eigenvalues are no longer degenerate in m.

8.5.2.1.3 Finite Cylinder: For the finite cylinder, the cylindrical coordinates (r, ϕ, z) are used and the wave equation is solved by

$$\psi(r, \phi, z) = \sum_{m\lambda} C_{m\lambda} J_m[(k^2 - \lambda^2)^{1/2} r] e^{im\phi} \sin \lambda z. \tag{8.97}$$

The traction-free boundary condition ($\psi = 0$) over the cylindrical surface, which includes the flat end surfaces, provides the expression for the eigenvalues [58],

$$k_{nms} \equiv \frac{x_{nms}}{a} = \left[\hat{x}_{ms}^2 + \left(\frac{n\pi a}{2b} \right)^2 \right]^{1/2} (a^{-1}) \tag{8.98}$$

$$\lambda_n = (\pi / 2)(n / b) \tag{8.99}$$

where $\hat{x}_{ms}$ are the roots of the characteristic equation

$$J_m(x) = 0, \tag{8.100}$$

and once again s = 1, 2, 3,…, labels the multiplicity of the zeros.

8.5.2.2 Identification of Fluid Targets in Vacuo

All eigenvalues for $x \equiv ka \leq 8$ were obtained for index values $l, m, n, s \leq 3$ for a sphere ($b / a = 1$), for a prolate spheroid ($b / a = 1.111...$, and $b / a = 4.0$), and for a cylinder ($b / a = 1.111...$, and $b / a = 4.0$).

The resulting eigenvalues $x \equiv ka$ are presented in the form of a spectroscopic level scheme in Figure 8.14. The dotted lines on the left side of Figure 8.14 show how the azimuthal (m) degeneracy of the spherical resonances x_{lms} with fixed values of l and s (first column) splits as the axes ratio b / a is increased from the value 1.0 for the sphere to the value 1.111…

for the prolate spheroid (column 2). The further passage to the prolate spheroid with $b/a = 4.0$ (column 4) is not indicated by any lines in the figure, but may be followed by the markings of the levels, and shows that, as expected, the splitting increases.

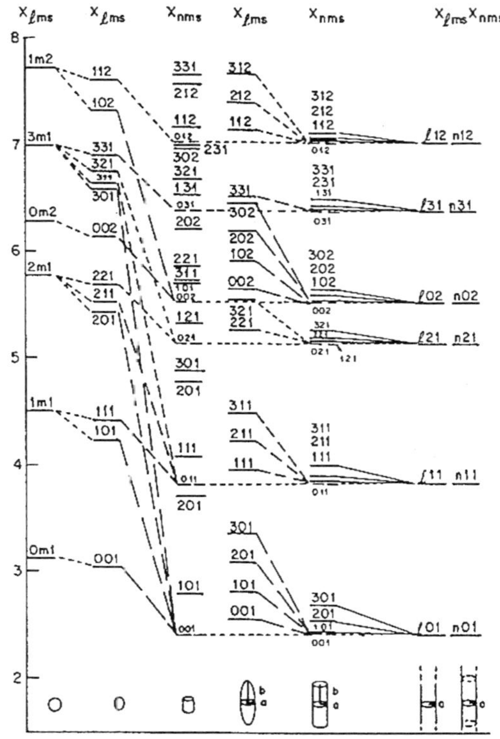

Figure 8.14 Spectroscopic level of the first few eigenfrequencies corresponding to internal resonances ($l, m, n, s \leq 3$ and $ka \leq 8$) for a fluid sphere, prolate spheroid, and cylinder in vacuo, with axes ratios b/a as indicated. (Nonphysical cylinder levels are shown as dashed lines [57]).

A comparison of the spheroid and cylinder levels for both $b/a = 1.111...$, and $b/a = 4.0$ shows that all x values for a given set of indices (*lms* for the spheroid and *nms* for the cylinder) decrease as b/a increases, except for the cylinder levels with $n = 0$. Equation (8.98) shows that x_{nms} is independent of b/a when $n=0$. These levels are connected by horizontal dotted lines in Figure 8.14, but they are not physically observable since they correspond to the trivial solution $\psi \equiv 0$ (see equations (8.97) and (8.98)).

Equation (8.98) indicates that the family of cylindrical levels *(n,m,s)* with fixed values of m and s but with varying values of n=1,2,3,..., all coalesce down to the $n = 0$ level *(0,m,s)* as $b/a \to \infty$. Consequently, the levels *(n,m,s)* shown in the last column of Figure 8.14 are obtained, which are degenerate in the axial index n for the infinite cylinder.

As b/a increases from 1.111... to 4.0, the prolate spheroid levels (l,m,s) of different l values $(l = m, m+1,...)$ but with the same value of m and s are seen to collect together. These families of different l values are illustrated in Figure 8.14 by lines of various dash lengths. As these levels collect together, they are also seen to be nesting down a cylindrical family (n,m,s). The nesting of the latter is indicated by solid lines. This can be justified by noting that as $b/a \to \infty$, the radial prolate spheroidal functions tend towards the Bessel functions of the cylinder. Similarly as $b/a \to 1$, the radial functions tend towards the spherical Bessel functions.

Figure 8.14 showed that the shifts and splitting of the eigenfrequencies under deformations of the target provide considerable information for shape determination of acoustic targets.

8.5.3 Material Characterization by Resonance Acoustic Spectroscopy (MCRAS)

In this section, an approach in resonance acoustic spectroscopy proposed by Honarvar and Sinclair [75] is presented. This method can be used for verifying the elastic constants of an isotropic elastic rod. The accuracy of this technique is verified by comparing it with other techniques. Experimental results on the identification of the mode number n of different resonances of an aluminum cylinder are also presented in this section.

It has been shown by previous researchers [27, 40, 80] that there exists a good agreement between numerically calculated and experimentally measured form functions for isotropic elastic cylinders. The elastic properties of the cylinder required for calculating its numerical form function are the density and two elastic constants of the material. These elastic properties can be obtained by matching a numerical form function to the experimentally measured form function. Although solving the *inverse scattering problem* (ISP) seems to be the desired approach to finding the elastic constants of the cylinder, application of ISP in practice is difficult. In this section, an alternative approach called *material characterization by RAS* (MCRAS) is used for extraction of the elastic properties of an isotropic elastic cylinder from its measured form function.

In MCRAS, the elastic properties of a solid elastic cylinder are found by matching the corresponding resonance frequencies on the measured and calculated form functions through an iterative computer algorithm.

The logic behind the iterative program used for matching the resonances of the measured form function to those of a numerically calculated form function is briefly described here. Figure 8.15 shows the form function of an aluminum rod calculated for a normally incident wave, i.e., $\alpha = 0^\circ$. Each resonance frequency on this curve is designated by a pair of integers according to the convention described in Section 8.3.2. Experience shows that all resonance frequencies are somewhat sensitive to perturbations in the shear wave velocity c_T (and accordingly μ) [59], but only some of them are significantly sensitive to variations of c_L (and accordingly λ) (see Figures 8.16 and 8.17). Therefore, one can first find c_T by matching the frequencies of those resonances which have relatively high sensitivity to variations of c_T through an iterative procedure. Knowing the value of c_T, the value of c_L can then be found by implementing a similar scheme.

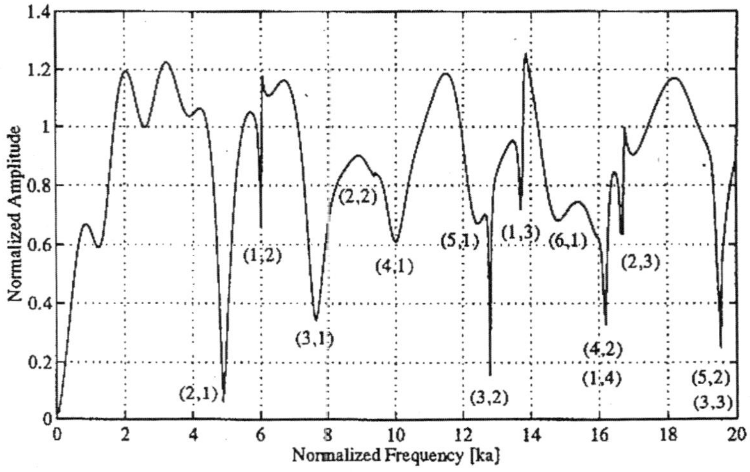

Figure 8.15 Calculated backscattered form function of the aluminum rod [75, 76].

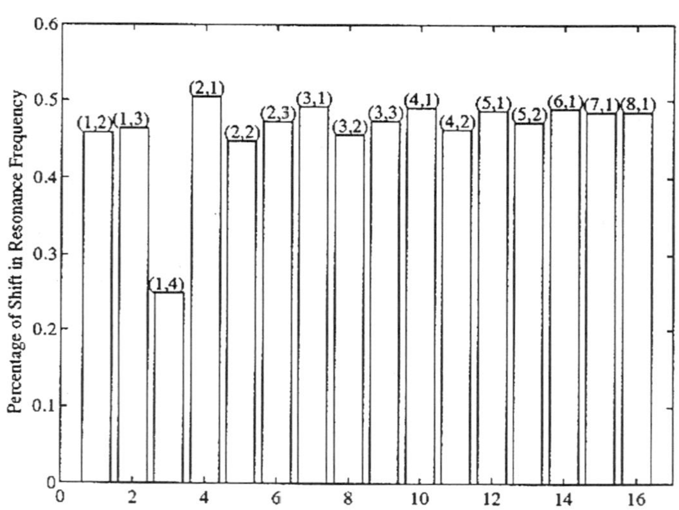

Figure 8.16 Shift of the resonance frequencies of the aluminum rod in the calculated form function due to 1% increase of c_T [75].

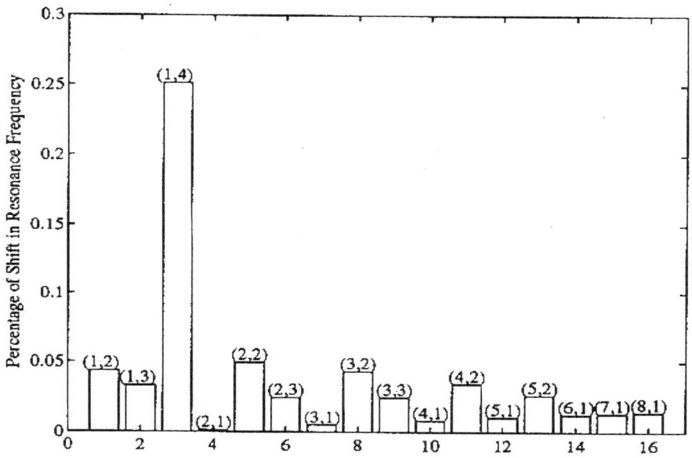

Figure 8.17 Shift of the resonance frequencies of the aluminum rod in the calculated form function due to 1% increase of c_L [75].

In the next section, the results obtained using MCRAS is compared to the results obtained from two other techniques, which are *ultrasonic time-of-flight* (UTOF) [81] and *resonance ultrasonic spectroscopy* (RUSpec) [82] measurements. The aluminum sample considered here serves as an example, and MCRAS has proven to be reliable for many other materials as well.

8.5.3.1 Accuracy of MCRAS Technique

The results from the MCRAS technique are given in the first row of table 8.1. An aluminum rod with a diameter of 6.67 mm and a length of 280 mm is tested in this case. The density of the material is measured by measuring the weight and volume of a coupon specimen made from the same material. The estimated uncertainties of the measured values are given in parentheses under each value in Table 8.1.

Table 8.1 Experimental results obtained for an aluminum sample using three different measurement techniques. Uncertainties are given in parentheses under each measured value [75, 76].

Technique	ρ (kg/m^3)	c_L (m/s)	c_T (m/s)
MCRAS	2694 (0.3%)	6427 (0.3%)	3112 (0.3%)
UTOF	2694 (0.3%)	6419 (0.2%)	3103 (0.3%)
RUSpec	2707 (n/av)	6536 (4%)	3102 (0.2%)

Another technique used for measuring the elastic properties of the aluminum sample is ultrasonic time-of-flight (UTOF) measurement where the compression and shear wave velocities in the coupon specimen are measured. The diameter of the coupon specimen is 17.04 mm and its length is 25.45 mm. The results of these measurements are given in the second row of Table 8.1 along with the estimated uncertainty of each measurement. The absolute inaccuracy of these measurements is calculated based on a *linear error propagation law* [83, (Vol. 1, page 257)]. The results from UTOF and MCRAS measurements agree very well and the wave velocities differ at most by 0.3%.

The third row of Table 8.1 lists the results of the measurements performed using RUSpec technique at Los Alamos National Laboratories on a parallelepiped sample. The difference between these values and those obtained by UTOF is less than 2%. Good agreement among the results obtained from the MCRAS, UTOF, and RUSpec techniques establishes the accuracy of the MCRAS technique. The better agreement of the UTOF and MCRAS results compared to RUSpec and MCRAS results is related to the high level of inaccuracy of RUSpec measurements in this case. The results obtained from UTOF measurements are considered more accurate for this case.

MCRAS can be a suitable technique for quick measurement of the elastic properties of isotropic elastic materials. Preparation of a cylindrical sample is usually easier than preparing samples with parallel surfaces like those required in UTOF and RUSpec measurements. Samples used in destructive tests are usually more complex, such as dumbbell-shaped samples. Moreover, in situations where the sample is only available in a cylindrical form, MCRAS might be the only choice for measuring the elastic properties of the material. Examples of such samples include the long whiskers used in fiber-reinforced composite materials where the diameter of the cylinder is much less than a millimeter. Figure 8.18 compares the measured and calculated form functions of the aluminum sample, and the agreement is seen to be excellent. It should be noted that the accuracy of MCRAS measurements is highly dependent on the accuracy of the measurement system.

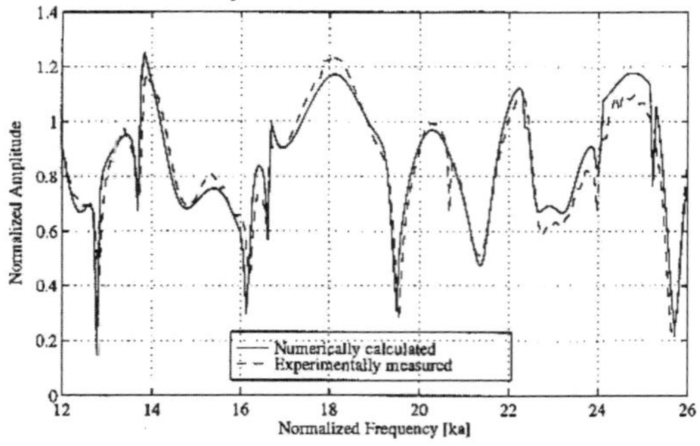

Figure 8.18 Comparison between measured and calculated form functions of the aluminum cylinder. The elastic constants used in calculating the numerical form function are obtained from UTOF measurements [75].

8.5.4 Nondestructive Evaluation (NDE) of Clad Rods by RAS

In this section, the monitoring of the perturbations of the cladding thickness of a cylindrical clad rod by RAS will be discussed [75]. In general, most of the resonances of a clad rod are sensitive to variations of the cladding thickness. This sensitivity is dependent on the material characteristics as well as the cladding thickness. For any specific material, those resonances which are most sensitive to variations of the cladding thickness can be identified and used for NDE and on-line monitoring purposes.

Physical properties of the cladding and the core materials are given in Table 8.2. The velocity of sound in water was calculated based on the water temperature [40]. The ultrasonic transducers had a diameter of 28.58 mm and a nominal center frequency of 500 kHz. The useful frequency range of each transducer spanned from approximately 200 to 800 kHz. In all the experiments, the sampling frequency was 50 MHz.

Frequency characteristics of the transmitting and receiving transducers were compensated by a deconvolution technique according to [28]. For this purpose a 0.25 mm diameter tungsten fiber was used to obtain a reference spectrum.

The dimensions of the clad rod are shown in Figure 8.19. The thickness of the cladding was reduced along a 50 mm length and the backscattered echo from the clad rod was measured at points 1, 2, and 3 along the axis. Figure 8.20a compares the measured form functions at points 1 and 3. The measured form functions at points 1 and 2 were identical, thus indicating that variations in test results were negligible. Comparison of the form functions measured at points 1 and 3, where a 30 μm difference in the cladding thickness exists, shows the sensitivity of resonance frequencies to the cladding thickness. The numerical model also predicts this shift in the resonance frequencies with the reduction of the cladding thickness as shown in Figure 8.20b. This indicates that resonances are sensitive to variations in the cladding thickness and can be used for on-line monitoring of variations of cladding thickness. Interpretation of the results of such on-line systems is aided by the mathematical model.

Table 8.2 Elastic properties of the core and the cladding materials [75].

Material	ρ (kg/m^3)	c_L (m/s)	c_T (m/s)
Aluminum	2694	6427	3112
Copper	8900	4600	2160

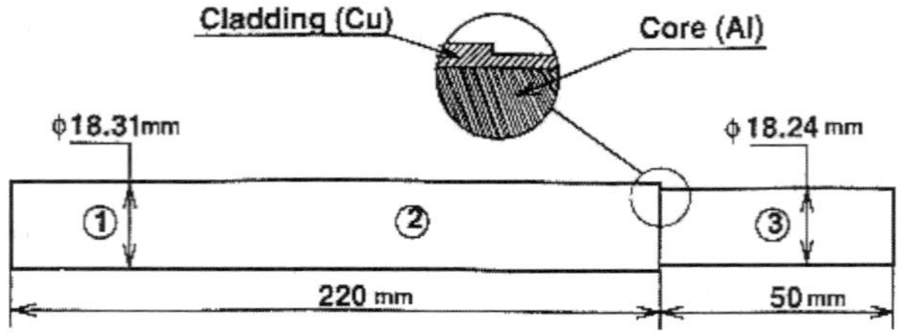

Figure 8.19 Schematic diagram of the clad rod sample [75].

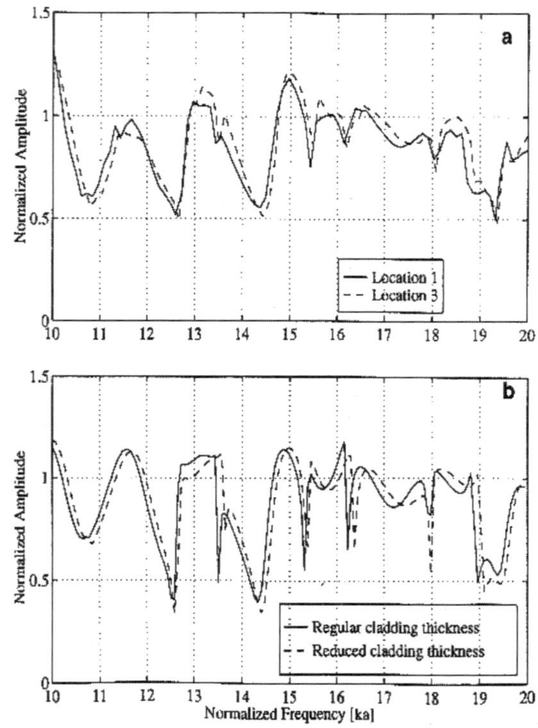

Figure 8.20 Form function of the clad rod sample at points 1 and 3. a) Experimentally measured. b) Numerically calculated [75].

8.5.5 Nondestructive Evaluation of Epon-815 Clad Steel Rod by RAS

The capability of RAS in detecting the delamination of the cladding from the core of a two-layered cylinder is investigated in this section [84]. Epon-815 is a transparent epoxy material which can easily be cast around a steel rod, thus producing a clad rod. Physical properties of Epon-815 and steel are given in Table 8.3. Epon-815 is a viscoelastic material and, therefore, its absorption coefficients should be taken into account in calculating the scattered field. This is accomplished by incorporating complex wave numbers as described in [85].

Table 8.3 Physical properties of Epon-815 and steel [75].

Material	Diameter (mm)	ρ (kg/m^3)	c_L (m/s)	c_T (m/s)	β_L (Neper)	β_T (Neper)
Epon-815	6	1177	2680	1200	–0.009	–0.015
Steel	1.592	9449	5890	3150	0	0

A spring model can also be incorporated at the interface of the core and cladding to represent the quality of the interfacial bond in the mathematical model. This is accomplished by writing the boundary conditions at the interface as follows [84]:

$$\begin{aligned} [\sigma_{rr}]_2 &= [\sigma_{rr}]_1 \quad ; \quad [\sigma_{rr}]_2 = K_r[(u_r)_2 - (u_r)_1] \\ [\sigma_{r\theta}]_2 &= [\sigma_{r\theta}]_1 \quad ; \quad [\sigma_{r\theta}]_2 = K_\theta[(u_\theta)_2 - (u_\theta)_1] \end{aligned} \tag{8.101}$$

where K_r and K_θ are the boundary stiffness constants. When $K_r = K_\theta = \infty$ the boundary conditions correspond to a perfect bond, while $K_r = K_\theta = 0$ corresponds to total debonding. Any values of K_r and K_θ taken between these two extremes will correspond to partial debonding of the two surfaces. Such a spring model can be physically interpreted as an interfacial zone with different physical properties from both the core and cladding materials. While, in theory $K_r = K_\theta = \infty$ corresponds to a perfect interfacial bond, in practice it was found that the lower limit for a "perfect" bond in the spring model of this sample is approximately $K_r = K_\theta = 10^{15}$ N/m^3. The measured and calculated form functions for Epon-815-clad steel rod at a nominal incident angle of $\alpha = 0^\circ$ are shown in Figures 8.21a and 8.21b, where the main resonances are marked by +."

Following the experimental measurements of the form function, the bonding between the core and the cladding was broken by sliding the steel core along the cladding sheath. The sample was then tested again and the form function of the system corresponding to its new condition was found (see Figures 8.22a and 8.22b). As expected, some changes in a number of resonance frequencies had taken place. The numerical form function for the new condition of the system was calculated by taking $K_r = K_\theta = 2.3 \times 10^{13}$ N/m (see Figure 8.22b). These values were found by an iterative process to give the best match between the resonances of the calculated and measured form functions.

This study shows that RAS is potentially capable of detecting the cladding delamination in a clad rod. It also demonstrates how the spring model can be used for evaluating the quality of the material bond.

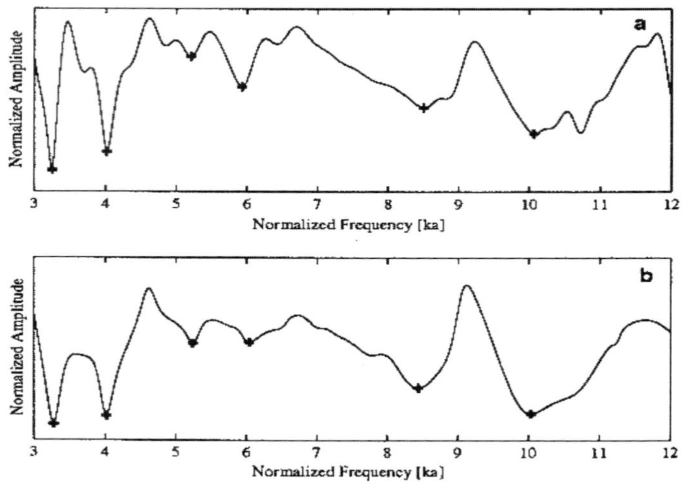

Figure 8.21 Form function of the Epon-815-clad steel rod. a) Experimentally measured. b) Numerically calculated [75].

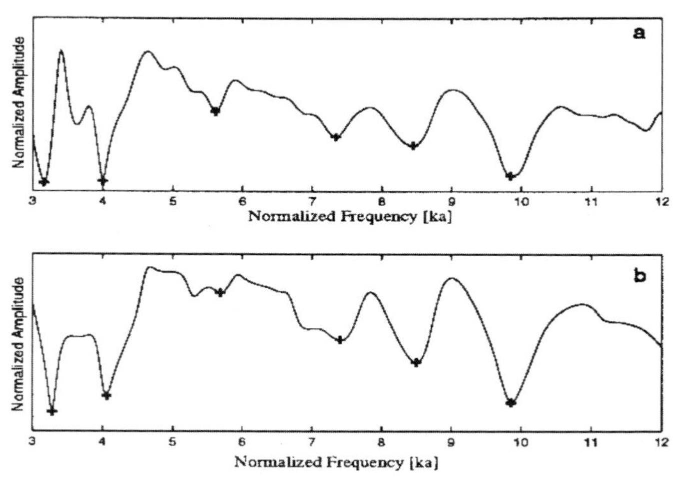

Figure 8.22 Form function of the Epon -815-clad steel rod after breaking the bond at the core-cladding interface. a) Experimentally measured, b) Numerically calculated with $K_r = K_\theta = 2.3 \times 10^{13}$ N/m^3 [75].

8.5.6 Characterization of Cladding Delamination

This section demonstrates how RAS can detect incipient cladding damage on a clad rod in an on-line system [75]. A piece of a copper-clad aluminum rod on which separation of the cladding from the core had initiated was scanned along its axis. At 1-cm intervals, the backscattered echoes from the rod were measured and their corresponding form functions were evaluated. Figure 8.23 shows the image obtained by plotting these measured form functions as a function of position of the probe along the rod axis. Grey color intensity on the image indicates the relative amplitude of the form function. Darker areas have lower amplitude compared to the lighter areas. The dark horizontal strips on the figure represent resonance dips on the form function.

For a good sample, these horizontal strips must be straight and uniform along the rod, i.e., the resonance frequencies should be the same at all axial positions. When a defect appears on the sample, some of these dark horizontal strips become distorted, hence indicating the existence of a problem.

Delamination of the cladding starts at approximately the 11-cm position on the horizontal axis where a noticeable shift in the dark strip at approximately 530 kHz is observed. This clearly indicates that damage in a clad rod can be detected by monitoring the variations of the form function along the rod during the manufacturing process.

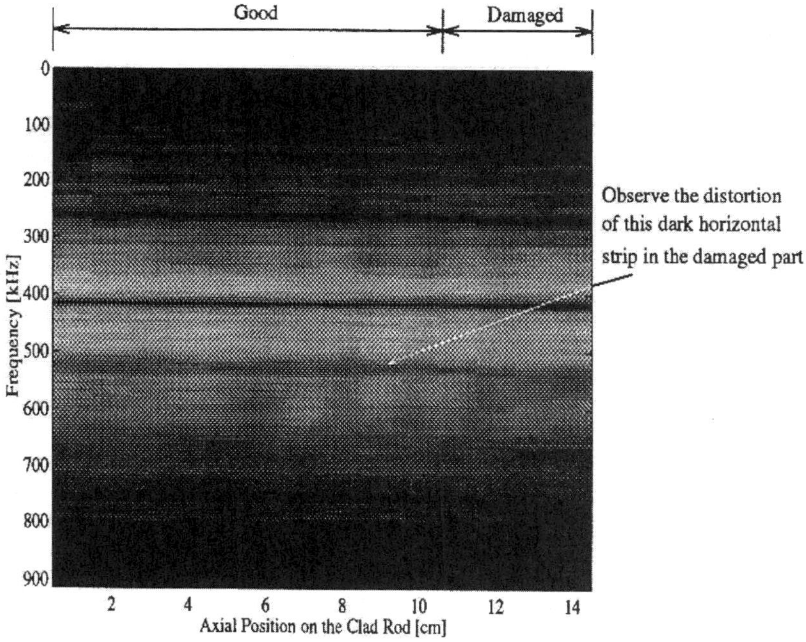

Figure 8.23 Scanned image of a copper-clad aluminum rod on which the delamination of the cladding has initiated [75].

8.5.7 Nondestructive Evaluation (NDE) of Explosively Welded Clad Rods by RAS

The applicability of RAS to the NDE of explosively welded clad rods is investigated in this section [86]. The focus is on copper-clad aluminum rods used as electrical conductors in high current applications. Any breakdown of the core/cladding interface may lead to hot spots, accelerated material degradation, or complete failure of the conductor. These clad rods are manufactured by explosive welding. Explosive welding is a solid-phase joining technique in which two or more metal components are driven together by a controlled explosion to form a very strong metallurgical bond [87, 88].

One of the limitations of traditional applications of resonance acoustic spectroscopy is the assumption of a perfect bond (continuity of all stress and displacement components) at material interfaces. This runs contrary to the experimental observation that a poor explosively welded specimen may feature bond discontinuities or the presence of interfacial contaminants consisting of bimetallic compounds.

8.5.7.1 Mathematical Model of the Interface

For the copper-clad aluminum rods, the interfacial transition zones are not always very sharp; they have finite mass and thickness, and their features must be included in the mathematical model to achieve an accurate result.

The thickness of the interlayer is still significantly smaller than one wavelength, such that the two surfaces of the interlayer move approximately in phase with each other when ultrasonic waves strike the cylinder. Under these circumstances, a spring-mass system is an appropriate first-order approximation for representing the thin interfacial layer. Thus, each rod is modeled as a two-layered cylinder with a spring-mass system to represent a thin interfacial layer containing the weld. At the core/cladding interface $r = b$, the boundary conditions are [76]

$$
\begin{aligned}
\frac{(\sigma_{1rr} + \sigma_{2rr})}{2} &= K_r (u_{2r} - u_{1r}), \\
\sigma_{2rr} + \sigma_{1rr} &= -m\omega^2 \frac{(u_{1r} + u_{2r})}{2}, \\
\frac{(\sigma_{1r\theta} + \sigma_{2r\theta})}{2} &= K_\theta (u_{2\theta} - u_{1\theta}), \\
\sigma_{2r\theta} + \sigma_{1r\theta} &= -m\omega^2 \frac{(u_{1\theta} + u_{2\theta})}{2},
\end{aligned}
\tag{8.102}
$$

where m is mass per unit area at the interface, and K_r and K_θ are the boundary stiffness constants. When $K_r = K_\theta = \infty$ and $m \to 0$ the boundary conditions correspond to a perfect bond with a vanishingly thin transition layer. The opposite extreme of $K_r = K_\theta = 0$ corresponds to total debonding. By applying the boundary conditions and taking the definition of form function defined in Section 3, the scattered pressure field can be evaluated. Details of the mathematical model can be found in [86].

8.5.7.2 Experimental Specimens

The explosive welding process can be described with the aid of the diagram shown in Figure 8.24. A copper tube is located concentrically around the aluminum core, the bore of the tube is so dimensioned that a small annular standoff gap exists between the two surfaces which are to be bonded. A plastic tube is located concentrically around the cylindrical components. Its diameter is carefully selected such that there remains a further annular gap into which explosive powder is poured. A detonator initiates the explosion at one end of the tube.

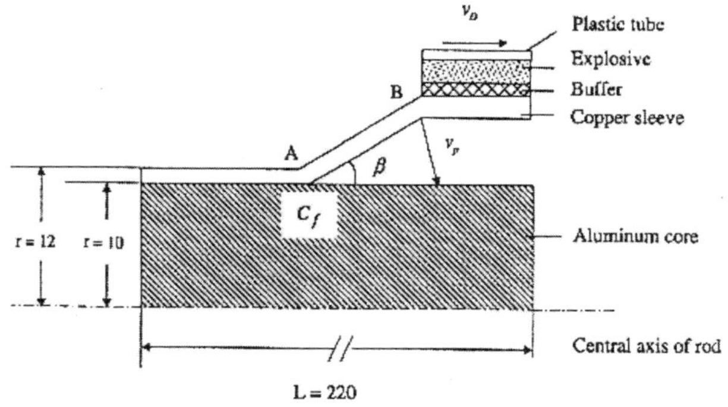

Figure 8.24 Arrangement for explosive welding of an aluminum core and a copper sleeve (all dimensions in mm) [86].

The detonator initiates the explosive powder and provides a detonation front. This travels down the length of the assembly and progressively swages the copper tube down upon the outer surface of the aluminum core. The explosion velocity, v_D, is controlled by the addition of a suitable inert diluting agent to the selected explosive powder. During the interval of time that any point ***A*** of the tube is traveling inward over the gap at a velocity v_p to collide at the collision front, the detonation front will have moved on to reach a point ***B*** (see Figure 8.24). Thus, the copper tube collides at an angle β with the aluminum core. As can be deduced from the diagram, the tangent of the angle β is v_D/v_p. It is this angle that controls the topography of the interface. From the relationship $\tan\beta = \nu_D/\nu_p$, it can be seen that a high detonation velocity will give a relatively large value of angle β, thereby creating a strongly turbulent system of solid waves at the aluminum-copper interface. Such waves have low pressure zones at the vortices that promote the generation of highly brittle copper-aluminum compounds. In general, it is desirable to dilute the explosive sufficiently to promote a slower explosion propagation velocity, yielding gently undulating waves or even a flat interface. However, excessive dilution of the explosive gives rise to frequent misfires; therefore, a practical limit to the dilution is imposed. Consequently, it is frequently found that the bond will be characterized by a wavy interface of some form. The pattern of these solid waves or interfacial oscillations provides the key to a nondestructive evaluation system for estimating interfacial integrity on the basis of the wave topography and the associated levels of intermetallics that can be expected. In this paper Ying et al. [86] manufactured four copper-clad aluminum rods by explosive welding. The sleeves were made of 99% pure copper and the core material was aluminum 6061. Initial dimensions of the Al core and Cu sleeves are shown in Figure 8.24. The detonation velocity ν_D was varied among the four specimens by appropriate dilution of the explosive mixture. For samples #1– #4, the concentration of the active explosive ingredient were 20%, 30%, 40%, and 50%, respectively. For sample #1, the value for ν_D was estimated at approximately 20 mm/μs.

Figure 8.25 shows *(r, z)*-planar cross-sections of the welds, which illustrate the intermetallic content associated with vorticity in the waves. Figure 8.25a features the specimen with the lowest detonation velocity, and correspondingly the highest value of β. Consequently, the interface is the most flat of the four samples and contains no intermetallics. By contrast, sample #2 (Figure 8.25b) contains moderate amounts of these compounds, while samples #3 and #4 (Figures 8.25c and 8.25d) show prominent waviness in the weld profile and significant amounts of intermetallics associated with the wave vortices. These intermetallics contain microcracks due to their brittleness and the effects of rapid cooling through the dissipation of heat from the weld to the surrounding metal mass in the immediate post-bonding period. A highly magnified image of this phenomenon is shown in Figure 8.26. Shear stresses may also arise from the welding process, due to a differential in the elastic recovery of the two metals; this effect also contributes to cracking. In extreme cases, the combined effects of these two factors in the immediate post-bonding period can entirely destroy the bond. The presence of intermetallics in the amounts seen in Figures 8.25c and 8.25d can also lead to problems should any fabrication process be imposed upon the welded specimen, such as drawing down of the rod to produce smaller diameters.

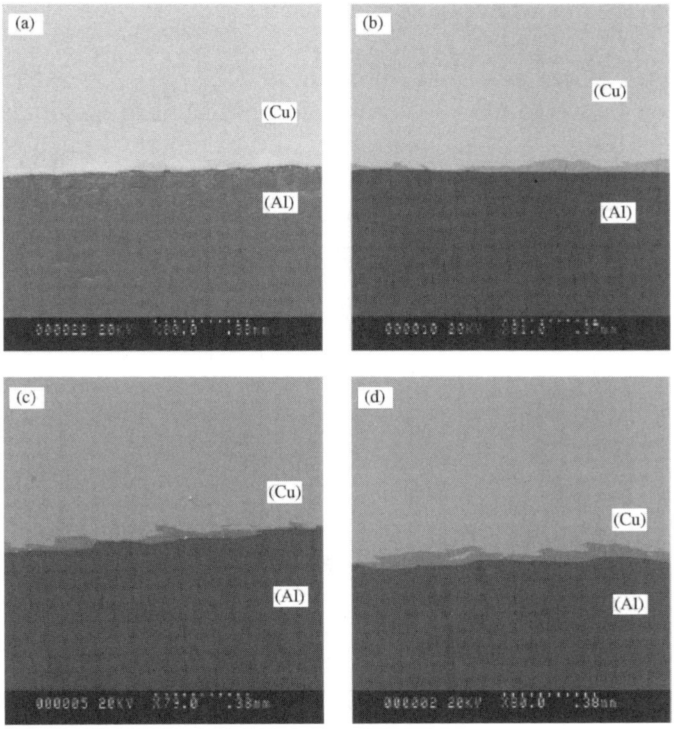

Figure 8.25 SEM results showing a cross-section of the welded area of explosively welded specimens: (a) sample #1 (20% explosive mixture); (b) sample #2 (30% explosive mixture); (c) sample #3 (40% explosive mixture); (d) sample #4 (50% explosive mixture). In (b), (c), and (d), intermetallic regions appear as semi-shaded gray areas between the bulk copper and aluminum [86].

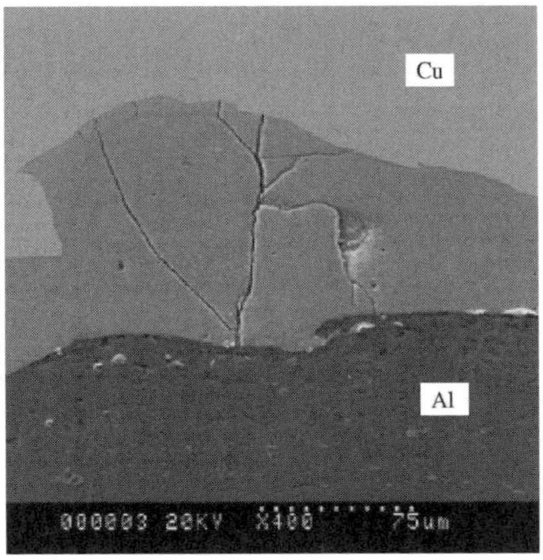

Figure 8.26 SEM result showing a magnified cross-section of the welded area of sample #4 (50% explosive mixture). Cracks can be seen clearly within the intermetallic region [86].

8.5.7.3 Resonance Acoustic Spectroscopy Results

The RAS form functions are shown in Figure 8.27 [86]. It was found that the form function showed minimal variation along the axis of each individual rod. Also shown in Figure 8.27 are the results of the numerical calculations of the form functions. The numerical solutions required parameters characterizing the interfacial region as inputs, namely, values of the interphase thickness t and interfacial spring constants K_r and K_θ . Their values are given in Table 8.4.

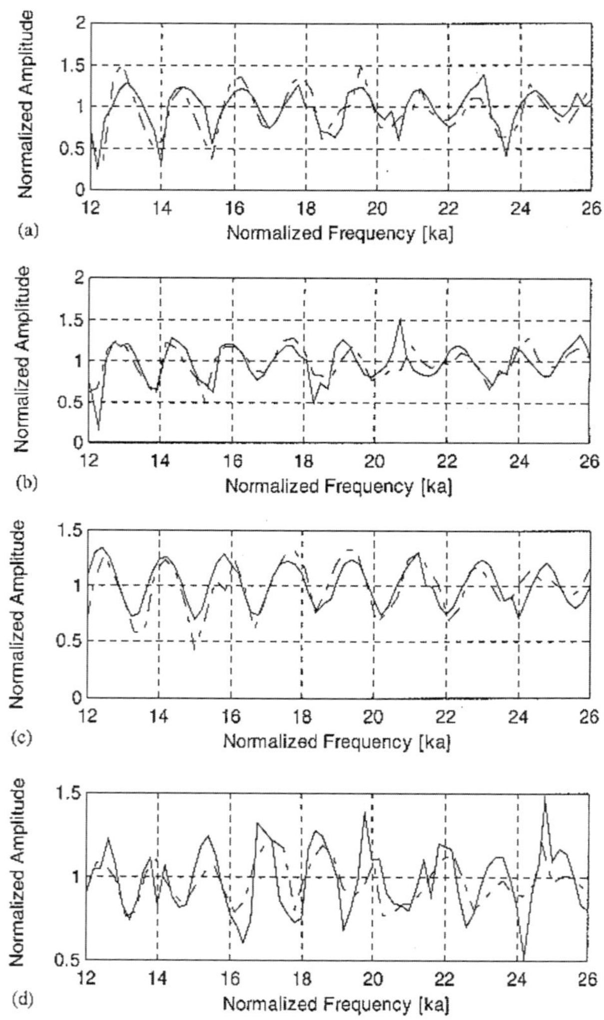

Figure 8.27 Calculated and measured form functions, (a), (b), (c), and (d) for samples #1, #2, #3, and #4, respectively, the solid lines for calculated and dotted lines for measured [86].

Table 8.4 Spring constants and average interlayer thickness for each specimen [86].

Sample #	Average thickness from SEM (μm)	Thickness used for numerical simulation (μm)	Longitudinal spring constant per unit area K_r (N/m^3)	Shear spring constant per unit area K_θ (N/m^3)
1	No intermetallics $\Rightarrow$ 0 thickness	0	≥6.5E16	≥2.5E16
2	32±7	20	6.5E15	2.5E15
3	60±15	60	1.0E15	4.0E14
4	80±18	90	5.0E14	2.0E14

They were selected for each rod by an iterative process to achieve the best possible match between resonance frequencies of numerical and experimental results in the four graphs of Figure 8.27. Examination of these four graphs revealed the following:

(1) By suitable selection of the input parameters to the numerical algorithm, excellent agreement was achieved between experimental and calculated values of the resonant frequencies.

(2) The form function shown in Figure 8.27a corresponds to the specimen manufactured at the lowest explosion propagation velocity. As seen in Figure 8.27a, no turbulence and intermetallic compounds are seen at this relatively smooth interface. For this specimen, extremely large spring constants and vanishingly small interphase thickness produced the best match between numerical and experimental results.

(3) For the three clad rods manufactured at higher explosion speeds, imperfections at the Cu–Al interface caused significant perturbations to the form functions (Figure 8.27b–d). This is readily apparent by noting the downshift in resonant frequencies, as compared to their values in Figure 8.27a. This downshift becomes increasingly apparent at higher values of ka, where the thickness of the interfacial region is of the same order as the ultrasonic wavelength.

Relatively low values of interfacial spring constants were required in order to achieve a good match between experimental and calculated resonant frequencies in specimens 2, 3, and 4. In addition, the calculated thickness t of the interlayer was larger than that for sample #1, as shown in Table 8.4.

8.5.8 Nondestructive Evaluation of Fiber-Reinforced Composite Rods

Application of RAS to the NDE of cylindrical components made from transversely isotropic materials is demonstrated in this section. Transverse isotropy is strongly evident in rods and wires made from fiber-reinforced composite materials. For such samples, the axial and transverse properties are significantly different and, therefore, characterization of both sets of properties is of interest.

RAS is able to characterize the sample in both axial and transverse directions. This is because some of the resonances are highly sensitive to axial properties of the cylinder while others are primarily sensitive to transverse properties. Rayleigh and whispering gallery waves are mostly dependent on the transverse properties, while guided waves show high sensitivity to the axial properties of the cylinder. Figure 8.28 shows the effect of a 5% increase in the Young's modulus E_{11} of a *continuous fiber aluminum matrix composite* (CF-AMC) rod with the following elastic properties (the G_{13} and G_{23} are shear modulus values):

$$E_{11} = E_{22} = 130 \quad \text{GPa}, \qquad E_{33} = 309 \quad \text{GPa}$$

$$G_{13} = G_{23} = 50 \quad \text{GPa}, \qquad \nu_{12} = 0.36, \quad \nu_{31} = 0.20,$$

where direction 3 is along the axis and directions 1 and 2 are in the transverse plane. The matrix of this composite material is aluminum and the reinforcing fibers are Nextel™ 610 Ceramic Fiber. Nextel™ 610 Ceramic Fiber is a high purity (above 99%) fine grain alumina fiber with high strength (2.7–3.4 GPa) and high stiffness (400 GPa) [75].

In Figure 8.28, the letters R, W, and G designate the resonance modes associated with Rayleigh, whispering gallery and guided waves, respectively. From Figure 8.28 it is clear that a 5% increase in E_{11} has primarily affected the R and W resonances, while resonances associated with guided waves G have remained almost unchanged. Figure 8.29 shows the effect of a 5% increase in the Young's modulus E_{33} and Figure 8.30 shows the effect of a 5% increase in the shear modulus G_{13}. These two elastic constants have a strong influence on waves propagating along the axis and, therefore, it can be observed that in these cases the G modes are most sensitive and R and W modes remain almost unchanged. It is also noticed that the sensitivity of G resonances to variations of E_{33} is less than their sensitivity to G_{13}. This is because the guided waves are mostly of a shear type and, therefore, more sensitive to variations of the shear modulus.

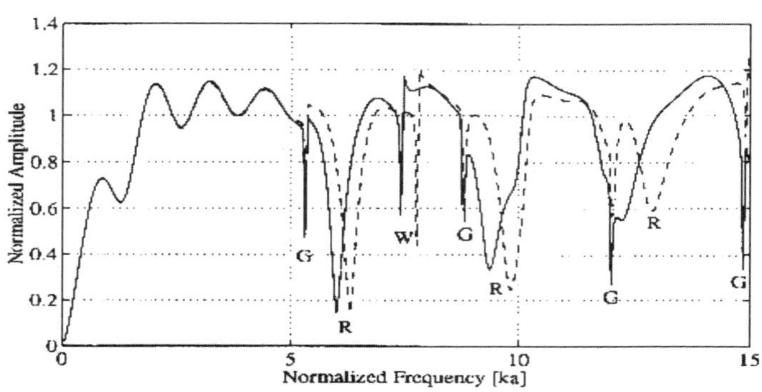

Figure 8.28 Solid line: Form function of the CF-AMC rod. Dashed line: 5% increase in Young's modulus in the transverse direction E_{11} [75].

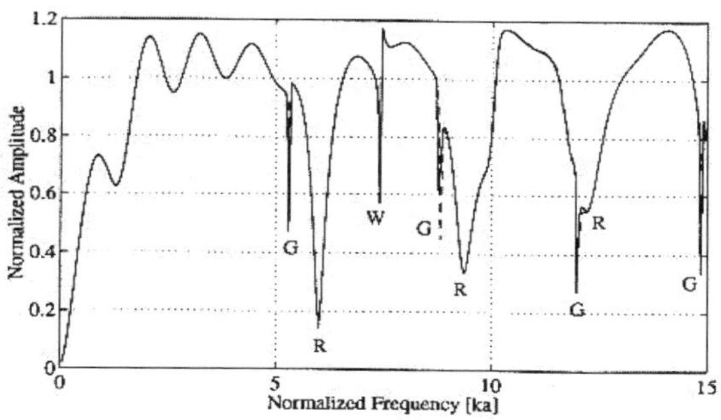

Figure 8.29 Solid line: Form function of the CF-AMC rod. Dashed line: 5% increase in Young's modulus in the transverse direction E_{33} [75].

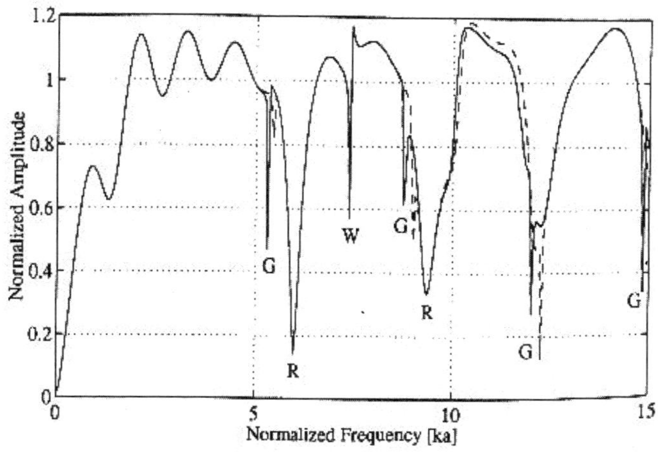

Figure 8.30 Solid line: Form function of the CF-AMC rod. Dashed line: 5% increase in shear modulus G_{13} [75].

To show the practical applicability of the mathematical model in the case of a transversely isotropic material, experimental results obtained from a CF-AMC rod are presented here. The length of the rod was 300 mm, and its surface was machined to a diameter of 1.52 mm. The density of the material was measured to be $\rho = 3220 \text{ kg/m}^3$. The elements of the stiffness matrix c_{ij} and the constants of equation (8.103) are related. The stiffness matrix for the material was calculated to be [76]

$$c_{ij} = \begin{bmatrix} 179 & 84 & 95 & 0 & 0 & 0 \\ 84 & 179 & 95 & 0 & 0 & 0 \\ 95 & 95 & 309 & 0 & 0 & 0 \\ 0 & 0 & 0 & 50 & 0 & 0 \\ 0 & 0 & 0 & 0 & 50 & 0 \\ 0 & 0 & 0 & 0 & 0 & 47.7 \end{bmatrix} \text{GPa.} \quad (8.103)$$

Figure 8.31 compares the measured and calculated form functions of the CF-AMC rod. Figure 8.31 serves as proof of the applicability of the RAS technique for the case of transversely isotropic cylinders. The resonances in the measured and calculated form functions (minima on the curves) agree very well and the maximum difference between the resonance frequencies is less than 1.5%. It is noted that the amplitude of the measured form function is affected by many different parameters of the measurement system while the resonance frequencies are primarily dependent on the properties of the rod. The sharpness of the measured resonances is limited by the resolution of the experimental system.

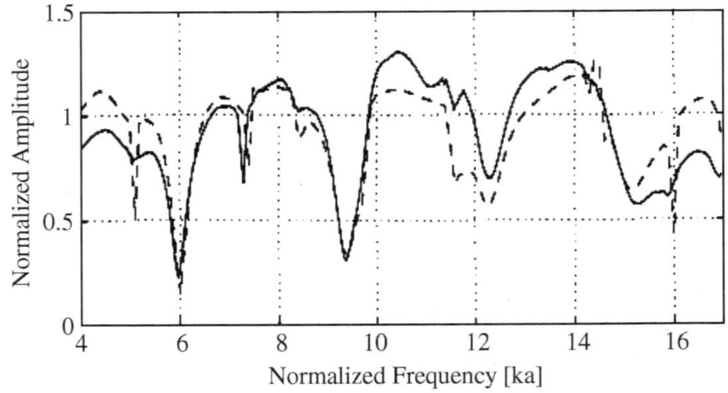

Figure 8.31 Form functions of CF-AMC rod. a) Dashed line: Measured, b) Solid line: Calculated [75].

The applicability of RAS for NDE of transversely isotropic cylinders has been demonstrated here. As was mentioned earlier, the resonances are fingerprints of the component and every component has its own set of resonances. For NDE purposes, it is important to identify the resonances that are most sensitive to the parameter that is to be evaluated or monitored. In the following section another application of RAS for NDE of transversely isotropic cylinders will be presented.

8.5.9 Nondestructive Evaluation of Continuously Cast Rods by RAS

In conventional casting processes, heat flows from the molten metal to the surrounding cooled mold. At the mold surface, nucleation of crystals can lead to appreciable friction and stress in products such as continuously cast rod. *Ohno continuous casting* (OCC) has successfully addressed this problem [89, 90]. In OCC, the mold is externally heated to a temperature just above the solidification point of the cast metal. This causes the heat flow to follow an axial direction as opposed to radial direction, and leads to the growth of long columnar grains, with excellent low-noise signal propagation characteristics. Current applications include copper rod and cable for video signal transmission. The texture of such rods is highly sensitive to the casting speed and mold temperature. It is, therefore, desirable to have an on-line system capable of measuring typical grain length, average grain cross-sectional area, and elastic stiffness constants.

In this section, RAS is used for characterization of long-grained cast copper rods produced by the OCC process. In particular, one desires to estimate the extent of anisotropy and the average cross-sectional dimensions of the copper grains corresponding to various casting speeds [91].

8.5.9.1 Experimental Specimens

The casting speed is the most significant variable affecting the grain structure [91]. Four OCC copper samples produced at casting speeds of 0.75, 1.0, 1.33 and 2.0 mm/s were used. The cast rods had a diameter of 8.0 mm and density $\rho = 8750\ \text{kg/m}^3$ and length of 20 cm. Decrease in average grain cross-section with increase in casting speed is illustrated in the micrographs of rod cross-sections in Figure 8.32. Figure 8.32 also shows that the average grain cross-section tends to be smaller near the perimeter as opposed to the central area of each rod. Over the entire cross-section, image analysis revealed the mean chord length of an individual grain cross-section to be 2.0, 1.3, 0.8, and 0.6 mm, corresponding to casting speeds of 0.75, 1.0, 1.33 and 2.0 mm/s, respectively. In addition to the four copper rods produced by the Ohno casting process, a fifth rod consisting of commercial grade copper was included as a reference for ultrasonic characterization.

To ensure that the measured form function and mode shape are independent of the specimens' radial direction, the measurements were repeated for eight values of radial angle β corresponding to 0, 30, 60, ..., 210°. In order to gain the axial material properties of transversely isotropic materials such as the elastic constant c_{33}, normal and oblique incidence short-pulse MIIR measurements were used [60].

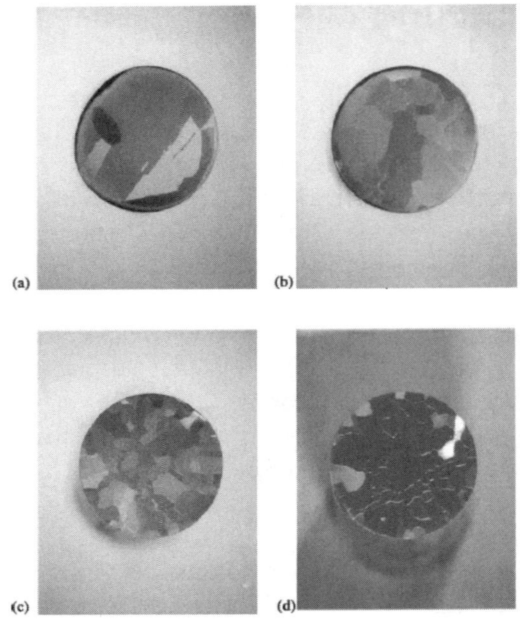

Figure 8.32 Micrographs showing changes in the cast microstructure of the rods with an increase in casting speed a) 0.75, b) 1.0, c) 1.33, d) 2.0 mm/s. The diameter of each rod is 8 mm [60].

8.5.9.1.1 Results corresponding to incidence angle $\alpha = 0^o$: The form function of a commercial copper rod, with a very fine grain structure, is shown in Figure 8.33. It was verified that the form function of this specimen is independent of the angle β of rotation around the rod. This is because the material can be considered homogeneous if spatial resolution is of the order of the ultrasonic wavelength. The calculated form function of this specimen is also shown in Figure 8.33 [60]. The MCRAS technique was used for adjusting the elastic constants of copper in the numerical model to produce good agreement between measured and calculated resonance frequencies [76]. Lame constants of λ = 105.7 GPa and μ = 45.6 GPa with uncertainties of ± 1% were obtained for this material by MCRAS. The material density of 8.90 g/cm^3 was obtained by dividing the mass of a rod segment by its volume. In Figure 8.33, the Rayleigh resonances are the most pronounced, particularly at lower mode numbers.

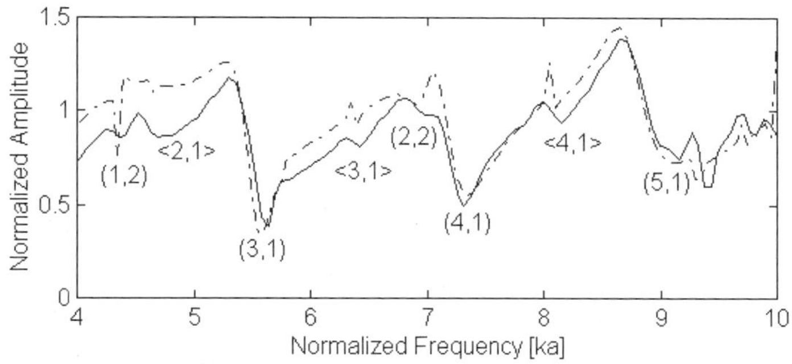

Figure 8.33 Measured and calculated form functions of a homogeneous copper sample for normal incidence $\alpha = 0^\circ$ [60].

Figure 8.34 shows the form functions obtained from single measurements on each of the four Ohno specimens. For the specimen produced with a casting speed of 0.75 mm/s, spectra are shown corresponding to two different values of azimuthal angle β around the cylinder.

The following observations are made:

1) Ohno rods produced at high casting speed of 2 mm/s (Figure 8.34a) have relatively simple spectra. A small number of prominent resonances can be easily identified and labeled on their form functions. Minor changes in the elastic constants of this specimen due to its oriented crystal structure has resulted in small shifts in its resonance frequencies compared to the commercial rod shown in Figure 8.33.
2) Pseudo-Rayleigh resonances labeled (3,1) and (4,1) appear in the form functions of all Ohno rods. On any single rod, the frequencies of these resonances vary by less than ±2% by changes in azimuthal angle. The corresponding circumferential wave velocities are approximately 2700 and 2600 m/s for the (3,1) and (4,1) resonances, respectively. Both wave modes have penetration depths of approximately 5–10 mm (corresponding to one wavelength), such that these modes penetrate to the center of the rod. Their propagation characteristics therefore represent a weighted averaged of elastic properties of all grains in the rod cross-section, and can be used to estimate average values of c_{11} and c_{12}, as shown in Table 8.5. In general, it is seen that values for c_{11} are lower for the slower casting speeds.
3) More resonances are observed in the form function of the large-grained Ohno rods compared to the homogeneous sample (Figure 8.34 versus Figure 8.33). These "rogue" resonances can be attributed to the reflection and/or refraction of waves by grain boundaries in cases where the grains are of the same order of magnitude or larger

than the resonance wavelength. By contrast, where the wavelength is much larger than the mean chord length D of the grains, there is minimal distortion of the form function compared to that of the homogeneous specimen.

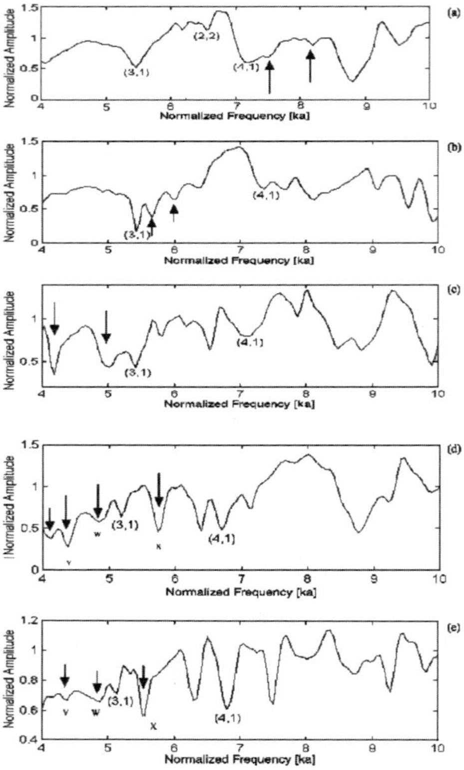

Figure 8.34 Form functions of Ohno specimens manufactured at a) 2.0, b) 1.33, c) 1.0, d) 0.75 mm/s at $\beta = 0^\circ$, e) 0.75 mm/s at $\beta = 30^\circ$. The arrows stand for the lowest rogue resonances in each form function [60].

Table 8. 5 Elastic constants for each Ohno specimen estimated by MCRAS [60].

Casting speed of sample (mm/s)	c_{11} (GPa)	c_{12} (GPa)
2.0	181.0 ± 2.0	97.0 ± 1.7
1.33	184.0 ± 3.0	97.3 ± 2.0
1.0	179.0 ± 2.8	95.9 ± 1.8
0.75	172 ± 2.6	94.5 ± 1.5

It can be concluded that for each OCC specimen, there exists a transition frequency below which there are no rogue resonances—only well-defined resonances that are predicted by the numerical model and that show little variation within any single OCC copper rod. Evaluation of the transition frequency is best done by comparing the location of the rogue resonance frequencies to those of the Rayleigh resonant modes, whose strong appearance makes them easy to identify in Figure 8.34. For the relatively fine-grained Ohno specimen produced at a casting speed of 2 mm/s shown in Figure 8.34a, the normalized transition frequency appears at $ka = 7$. This corresponds to a Rayleigh wavelength of approximately 6 mm in copper, or approximately 10 times the average grain dimension of 0.6 mm. This is consistent with scattering theory, which predicts that isotropic Rayleigh scattering will be the dominant scattering mode when the wavelength is greater than about 10 times the mean grain diameter.

For the larger-grained samples, the rogue resonances appear at lower frequencies, but in each case the transition frequency corresponds to a wavelength that is approximately 10 times the mean grain diameter. It becomes increasingly difficult to assign an accurate value to the transition frequency as the grains get larger and the specimens become more heterogeneous (Figure 8.34b–d). For the specimens produced at casting speeds of 1.33 and 1.0 mm/s, the transition frequencies appear at $ka = 5$ and $ka = 4$, respectively. For the specimen produced at 0.75 mm/s, the rogue resonances appear over the entire useful bandwidth of the transducer, $4 \leq ka \leq 10$ (see Figure 8.34d). From analysis of Figure 8.32a, the average grain size is estimated to be approximately 2 mm, and the transition frequency for Rayleigh mode resonances is predicted to correspond to a wavelength of 0.2 mm, or a normalized frequency of about $ka = 2$.

Mode shapes corresponding to several "rogue" resonances were measured for all four Ohno rods as functions of axial position and rotational angle β of the specimens. Results obtained from the 0.75 mm/s cast rod are presented in Figure 8.35, corresponding to the three rogue resonances designated by v, w and x in Figure 8.34d and V, W, and X in Figure 8.34e. The normalized magnitude and associated mode shapes of the rogue resonances were found to shift dramatically with a change in the rotational angle β. This strong azimuthal dependence, and the absence of these low frequency resonances in rods with fine grain structure, supports the theory that these rogue resonances originate from sound waves interacting with individual grain boundaries. The dependence of mode shape on rod rotation angle β for these three rogue resonances can be qualitatively correlated with the microstructure of this rod, shown in Figure 8.32a. These resonances are strongly dependent on the refraction angle of sound as it passes from the water into the cylinder; when the rod is rotated, the refraction angle will vary according to the orientations of the large grains at the rod's surface. By contrast, there is little variation of the resonances (3,1) or (4,1) with β. The mode shapes of these resonances are largely independent of rod rotation, as they are based on mode shapes that

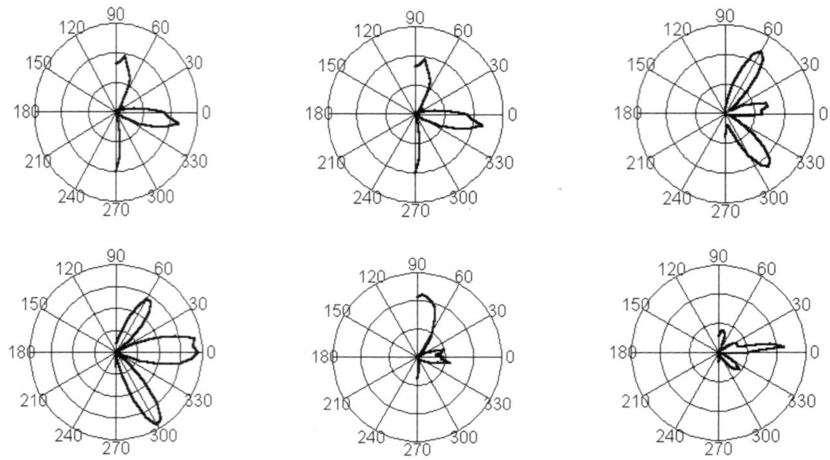

Figure 8.35 Mode shapes of the "rogue" resonances of the Ohno specimen produced at 0.75 mm/s, displayed as a function of the angle θ. Top— Resonances *v*, *w* and *x* in Fig. 34d were observed at $\beta = 0^{\circ}$. Bottom— Resonances *V*, *W* and *X* in Fig. 34e were observed at $\beta = 30^{\circ}$. [60].

penetrate deeply into the rod. Their associated frequencies are therefore suitable indicators for spatially averaged stiffness constants of the rod cross-section.

8.5.9.1.2 Results corresponding to non-zero angles of incidence: In general, RAS measurements performed at nominally non-zero angles of incidence of up to a few degrees are expected to yield all the resonances corresponding to $\alpha = 0^{\circ}$, plus additional resonances due to axially guided waves. Examples are shown in Figure 8.36 corresponding to a single cross-section for the isotropic copper rod and two cross-sections for the Ohno specimen produced at 2 mm/s. The angle of incidence was $\alpha = 2^{\circ}$. Resonances resulting from guided waves are assigned identification indices contained in angled brackets $< n,l >$.

The theoretical model and experiments are in good agreement as shown in Figure 8.36a. Values for the elastic constants used in the numerical model were the same as those used in Figure 8.33. For the 2-mm/s Ohno rod, the MCRAS technique was used to determine the constants c_{33} and c_{44} at two axial locations on the rod spaced approximately 25 mm apart. The fifth elastic constant, c_{13}, cannot be accurately determined from ultrasonic spectrum measurements since the resonant frequencies have very little sensitivity to the value of c_{13}.

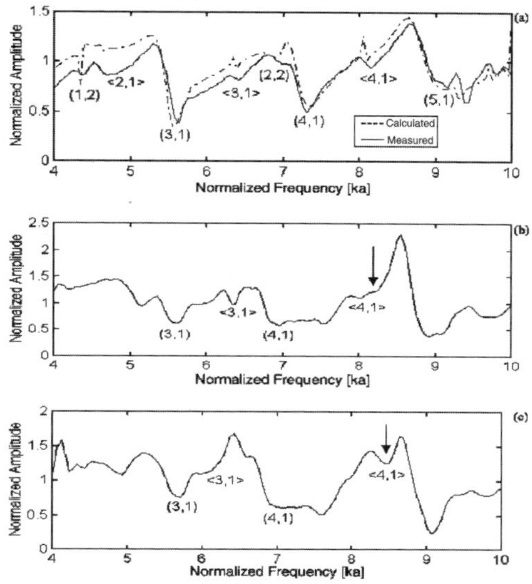

Figure 8.36 Form functions corresponding to non-zero angles of incidence (a) calculated and measured from a commercial isotropic rod; (b) and (c) measured from two cross-sections of the Ohno specimen produced at 2 mm/s. Some of the resonances identified by MIIR in (a) cannot be reliably located in (b) nor (c) [60].

Table 8.6 shows that there is only slight variation of the elastic constants between the two axial positions of this rod: both sets of measurements showed a measurable amount of anisotropy in the specimen (as gauged by the moderate differences between values of c_{11} and c_{33}). The lack of large variation of the elastic properties between the two axial positions was consistent with destructive analysis, which indicated highly elongated grains up to several centimeters in length.

Table 8.6 Elastic constants for the 2-mm/s Ohno specimen at two cross-sections estimated by MCRAS, as compared to measured values for the homogeneous, isotropic sample [60].

Cross section number	c_{11} (GPa)	c_{12} (GPa)	c_{33} (GPa)	c_{44} (GPa)
1	188.0 ± 2.2	96.0 ± 1.7	200.0 ± 3.0	44.5 ± 0.9
2	189.0 ± 2.2	96.8 ± 1.8	202.0 ± 3.0	45.2 ± 0.9
Homogeneous Cu sample	197.3 ± 1.5	106.0 ± 1.0	197.3 ± 1.5	46.0 ± 0.3

Measurements performed at $\alpha = 0^\circ$ on the three other Ohno rods produced very complicated spectra, with many resonances that were not predicted by the RAS theory for a homogeneous rod, e.g., Figure 8.37. Apparently, the large grains in these samples have introduced many additional resonances related to the local microstructure of the material. It was not possible to identify all of the expected resonances corresponding to guided waves on these samples, and there was evidence that even the resonances corresponding to $\alpha = 0^{\circ}$ may have had their apparent frequencies significantly shifted in the complex spectral pattern.

The measurable variation of resonance frequencies and mode shapes with azimuthal angle β of the Ohno rods produced at the lower casting speeds may appear inconsistent with the model of truly free vibration. In fact, the rod undergoes forced, damped oscillations for which mode shape and frequency clearly do depend on the point of entry of the incident plane wave relative to the arrangement of large grains; the numerically predicted pattern of standing waves around the periphery of the rod is also interrupted according to the relative positions of grain boundaries and oriented crystals with respect to the nodes and antinodes of the wave mode.

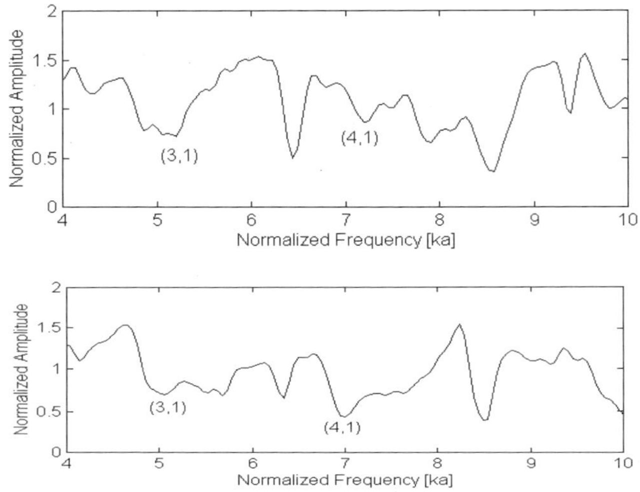

Figure 8.37 Measured form functions corresponding to non-zero angles of incidence from the Ohno specimen produced at (a) 1.33; and (b) 0.75 mm/s. There are several resonances that are not predicted by the numerical results shown in Figure 8.36a. In addition, some resonances predicted by the numerical model cannot be reliably located in this figure [60].

It can be concluded that certain resonance frequencies and associated mode shapes are primarily dependent on cross-sectional-averaged material properties and are relatively independent of grain size below a critical value. However, for cases where the average grain cross-sectional diameter is greater than 10% of the ultrasonic wavelength, it was found that the resonance spectrum is significantly distorted from its theoretical profile. This indicates that RAS could be applied in an on-line setting for nondestructive characterization of the grain structure.

REFERENCES

[1] Uberall, H. (1973) *Physical Acoustics, Vol.10* Surface waves in acoustics Ch. 1. New York, : Academic Press, Inc.

[2] Franz, W. & Naturforsch, Z. (1954) Uber die greenchen funktionen des zylinders und der kugel 9a 705-716.

[3] Frisk, G.V., Dickey, J.W. & Uberall, H. (1975) Surface wave modes on elastic cylinders, *J. Acoust. Soc. Am.* **58,** 996-1008.

[4] Strutt, J.W. (1945) (Third Baron Rayleigh), *The Theory of Sound, Vols. 1, 2.* New York: Dover Publications.

[5] Strutt, J.W. (1910) The problem of whispering gallery, *Phil. Mag.* **20,** 1001-1004.

[6] Uberall, H., Doolittle, R.D., McNicholas, J.V. (1966) *J. Acoust. Soc. Am.* **39**(3), 564-578.

[7] Grace, O.D. & Goodman, R.R. (1966) *J. Acoust. Soc. Am.* **39,** 173-174.

[8] Wait, J.R. (1967) *Canadian J. of Phys.* **45**, 1861-1869.

[9] Brekhovskikh, L.M. (1968) *Soviet Physics Acoustics* **13**, 462-472.

[10] Frisk, G.V. & Uberall, H. (1976) *J. Acoust. Soc. Am.* **59**, 46-54.

[11] Sitting, E.K. & Coquin, G.A. (1970) *J. Acoust. Soc. Am.* **48**(5), 1150-1158.

[12] Dardy, H.D., Flax, L. & Gaumond, C.F. (1987) *J. Acoust. Soc. Am.* **82**, 1378-1385.

[13] Maze, G., Izbicki, J.L. & Ripoche, J. (1985) *J. Acoust. Soc. Am.* **77,** 1352-1357.

[14] Izbicki, J.L., Maze, G. & Ripoche, J. (1986) *J. Acoust. Soc. Am.* **80,** 1215-1219.

[15] Leon, F., Lecroq, F., Decultot, D. & Maze, G. (1992) *J. Acoust. Soc. Am.* **91,** 1388-1397.

[16] Gaunaurd, G.C. & Uberall, H. (1979) *J. Acoust. Soc. Am.* **50**, 4642-4660.

[17] Vogt, R.H. & Neubauer, W.G. (1976) *J. Acoust. Soc. Am.* **60**, **15**-22.

[18] Flax, L., Dragonette, L.R. & Uberall, H. (1978) *J. Acoust. Soc. Am.* **63**, 723-731.

[19] Flax, L., Dragonette, L.R. & Uberall, H. (1981) *Physical Acoustics* (New York) **15** Ch. 3 p 723-731.

[20] Gaunaurd, G.C. (1989) *Appl. Mech. Rev.* **42**, 143-192.

[21] Doolittle, R.D., Uberall, H. & Ugincuis, P. (1968) *J. Acoust. Soc. Am.* **43**(1), 1-14.

[22] Uberall, H. (1985) *Traitement du Signal* **2,** 381-385.

[23] Hackman, R.H. (1993) Acoustic Scattering from Elastic Solids 22 Ch. 1 pp. 1-194, San Diego, CA: Academic Press.

[24] Faran, J.J. (1951) *J. Acoust. Soc. Am.* **23,** 405-418.

[25] Hickling, R. (1962) *J. Acoust. Soc. Am.* **34,** 1582-1592.

[26] White, R.M. (1958) *J. Acoust. Soc. Am.* **30,** 771-785.

[27] Flax, L., Varadan, V.K. & Varadan, V.V. (1980) *J. Acoust. Soc. Am.* **68,** 1832–1835.

[28] Li, T., Ueda, M. (1989) *J. Acoust. Soc. Am.* **86,** 2363-2367.

[29] Bao, X., Cao, H., Uberall, H. (1990) *J. Acoust. Soc. Am.* **87,** 106-110.

[30] Nagl, A., Uberall, H., Delsanto, P.P., Almar, J.D. & Rosario, E. (1983) *Wave Motion* **5,** 235-247.

[31] Veksler, N.D. (1990) *Acustica* **71,** 111-120.

[32] Conoir, J.M., Rembert, P., Lenoir, O. & Izbicki, J.L. (1993) *J. Acoust. Soc. Am.* **93,** 1300-1307.

[33] Conoir, J.M. (1988) In *Electromagnetic and Acoustic Scattering: Detection and Inverse Problem Conf. Proc.* 96–108, Marseille.

[34] Fan, Y., Honarvar, F., Sinclair, A.N. & Jafari, M.R. (2003) *J. Acoust. Soc. Am.* **113,** 1-12.

[35] Doolittle, R.D., Uberall, H. & Ugincuis. P. (1968) *J. Acoust. Soc. Am.* **43**(1), 1-14.

[36] Maze, G., Taconet, B. & Ripoche, J. (1981) *Phys. Lett.* **84 A**, 309-312.

[37] Maze, G. & Ripoche, J. (1983) *Rev. Phys. Appl.* **18,** 319-326.

[38] Maze, G. & Ripoche, J. (1983) *J. Acoust. Soc. Am.* **73,** 41-43.

[39] Derem, A., Rousselot. J.L., Maze, G. & Ripoche, J. (1982) *Acustica* **50,** 39-50.

[40] Maze, G., Ripoche, J., Derem, A. & Rousselot, J.L. (1984) *Acustica* **55,** 69-85.

[41] Maze, G., Izbicki, J.L. & Ripoche, J. (1985) *J. Acoust. Soc. Am.* **77,** 1352-1357.
[42] Ripoche, J., Maze, G. & Izbicki, J.L. (1985) In *Ultrasonics Int. Conf. Proc.* 364-369, Guildford, UK: Butterworth Scientific.
[43] Burving, R., Rousselot, J.L., Derem, A. & Maze, G. (1984) *Rev. CETHEDEC-Ondes Signal* **78**, 73–93.
[44] Izbicki, J.L., Maze, G. & Ripoche, J. (1986) *Phys. Lett.* A **115,** 393-397.
[45] Honarvar, F. & Sinclair, A.N. (1996) *J. Acoust. Soc. Am.* **100**(1), 57-63.
[46] Ahmad, F. & Rahman, A. (2000) *Int. J. Eng. Sci.* **38,** 325–335.
[47] Abramowitz, M. & Stegun, J.A. (1972) Handbook of Mathematical Functions New York: Dover Publications.
[48] Malvern, L.E. (1969) Introduction to the Mechanics of a Continuous Medium New Jersey: Prentice-Hall.
[49] Morse, P.M. & Feshbach, H. (1953) Methods of Theoretical Physics New York: McGraw-Hill.
[50] Honarvar, F., Enjilela, E., Mirnezami, S.A. & Sinclair, A.N. (2005) Submitted for publication.
[51] Veksler, N.D. (1993) *Resonance Acoustic Spectroscopy* New York: Springer-Verlag.
[52] Breit, G. & Wigner, E.P. (1936) *Phys. Rev.* **49,** 519-531.
[53] Uberall, H., Maze, G., & Ripoche, J. (1995) In *Experiments and Analysis of Sound-Induced Structural Vibrations* Eds. Guran A Inman D **1**. pp 35-91 Word Scientific.
[54] Uberall, H. (1993) History of resonance scattering theory in acoustics and its applications to target recognition in *SPIE- Automatic Object Recognition III* pp. 350-361.
[55] Junger, M.C. & Feit, D. (1972) Sound, Structures, and Their Interaction Cambridge, Massachusetts: MIT Press.
[56] Rhee, H. & Park, Y. (1997) *J. Acoust. Soc. Am.* **102**(6), 3401-3412.
[57] Brill, D., Gaunaurd, G.C. & Uberall, H. (1983) *Acustica* **53,** 11-18.
[58] Brill, D., Gaunaurd, G.C. & Uberall, H. (1982) *J. Acoust. Soc. Am.* **72,** 1067-1069.
[59] Ripoche, J. & Maze, G. (1992) A new acoustic spectroscopy: The resonance scattering pectroscopy by the method of isolation and identification of resonances (MIIR). in Acoustic Resonance Scattering Uberall H (Gordon and Breach Science Publishers). 69-103.
[60] Fan, Y., Sinclair, A.N. & Honarvar, F. (2003) *J. of Nondestructive Testing and Evaluation* **19,** 15-28.
[61] Dragonette, L.R. (1978) *Naval Research Laboratory Report NRL*, 8216.
[62] Dragonette, L.R. Numrich, S.K. & Frank, L. (1981) *J. Acoust. Soc. Am.* **69,** 1186-1189.
[63] Numrich, S.K., Frank, L. & Dragonette, L.R. (1982) *Prco. ICASSP* **82,** Paris 327-330.
[64] Veksler, N.D. (1980) *Acoustics Lett.* 15–18.
[65] Dragonette, L.R., Vogt, R.H., Flax, L. & Neubauer, W.G. (1974) *J. Acoust. Soc. Am.* **55,** 1130-1136.
[66] Dragonette, L.R. (1979) *J. Acoust. Soc. Am.* **65,** 1570-1572.
[67] Dragonette, L.R. & Flax, L. (1979) Resonance Scattering Predictions and Measurements, Office of Naval Research and the Ohio State U. p 401-411.
[68] Gaunaurd, G.C. & Uberall, H. (1980) Ultrasonics 261-269.
[69] Neubauer, W.G., Vogt, R.H. & Dragonette, L.R. (1974) *J. Acoust. Soc. Am.* **55,** 1123-1129.
[70] Ripoche, J., Maze, G. & Izbicki, J.L. (1985) *J. Nondestr. Eval.* **5,** (2) 69-79.
[71] Uberall, H., Dragonette, L.R. & Flax, L. (1977) *J. Acoust. Soc. Am.* **61,** 711-715.
[72] Numrich, S.K., Dale, N. & Dragonette, L.R. (1984) ASME PVP 78/AMD 64 59-74.
[73] De Billy, M. (1986) *J. Acoust. Soc. Am.* **79,** 219-221.
[74] Delestre, P., Izbicki, J.L., Maze, G. & Ripoche, J. (1986) *Acustica* **61,** 83–85.
[75] Honarvar, F. & Sinclair, A.N. (1998) *Ultrasonics* **36,** 845-854.
[76] Honarvar, F. (1997) PhD Thesis U. of Toronto, Dep. of Mech. and Indus. Eng.
[77] Ripoche, J., Maze, G. & Izbicki, J.L. (1989) *In Proc. Sym.* on Vibrations-Chocs 47-51.
[78] Quentin, G. & Cand, A. (1989) Electronics Lett. **25**, 353-354.

[79] King, B.J., Baier, R.V. & Hanish, S. (1970) NRL Report 7012.
[80] Gaunaurd, G.C. & Uberall, H. (1983) *J. Acoust. Soc. Am.* **73,** 1-12.
[81] Krautkramer, J. & Krautkramer, H. (1983) *Ultrasonic Testing of Materials* New York: Springer-Verlag.
[82] Migliori, A., Sarrao, J., Visscher, W., Bell, T., Lei, M., Fisk, Z. & Leisure, R. (1993) *Physica* **183,** 1-24.
[83] Sydenham, P.H. (1982) *Handbook of Measurement Science Vol. 1* New York: John Wiley and Sons.
[84] Honarvar, F., Sinclair, A.N., Piggot, M.R. & Ravichandran, M.V. (1995) In Proc. of the 1st Int. Conf. on Composite Materials and Energy, Canada.
[85] Schuetz, L.S. & Neubauer, W.G. (1977) *J. Acoust. Soc. Am.* **62**, 513-517.
[86] Fan, Y., Tysoe, B., Sim, J., Mirkhani, K., Sinclair, A.N., Honarvar, F., Sildva, H., Szecket, A. & Hardwick, R. (2003) *Ultrasonics* **41**, 369-375.
[87] Abrahamson, G.R. (1961) *J. Appl. Phys.* **28**, 519.
[88] Cowan, G.R. & Holtzman, A.H. (1963) *J. Appl. Phys.* **34,** 928.
[89] Ohno, A. (1986) *J. Metals* **4**, 14.
[90] Soda, H. & Wang, Z. (1995) *J. Material Sci.* **30**, 5438.
[91] Fan, Y. (1998) MASc Thesis U. of Toronto, Dep. of Mech. and Indus. Eng.

CHAPTER 9

FOURIER TRANSFORM INFRARED SPECTROSCOPY

Neena Jaggi and D.R. Vij
Kurukshetra University, Kurukshetra, India

9.1 INTRODUCTION

"Fourier spectroscopy" is a general term that describes the analysis of any varying signal into its constituent frequency components. The mathematical methods named after J.B.J. Fourier are extremely powerful in spectroscopy and have been discussed in detail [1–3]. Fourier transforms can be applied to a variety of spectroscopies including infrared spectroscopy known as *Fourier transform infrared* (FT-IR), *nuclear magnetic resonance* (NMR), and *electron spin resonance* (ESR) spectroscopy. FT-IR spectroscopy includes the absorption, reflection, emission, or photoacoustic spectrum obtained by Fourier transform of an optical interferogram. The power of the method derives from the simultaneous analysis of many frequency components in a single operation. When Fourier concepts are applied to various terms of spectroscopy, the resultant technology creates a spectrometer that gives the entire spectrum in the amount of time that a conventional spectrometer (using dispersive elements like prism and grating) would need to scan across just a single line in the spectrum. Fourier spectrometers utilizing interferometers are thus faster by a factor equal to the number of resolvable elements in the spectrum. Fourier-based methods are used over a wide spectral range [4–7]. FT spectroscopy can be employed for a long range of frequencies varying over ultraviolet, visible, near infrared, mid infrared and even far infrared regions by selecting different beam splitters and detectors for the required ranges. No other dispersive technique can be used for such a wide range of frequencies [8].

A variety of spectroscopic techniques have been used to study various samples, but FT-IR spectrometers are growing in popularity since they offer speed, accuracy and sensitivity previously impossible to achieve with wavelength dispersive spectrometers. This technique allows rapid analysis of micro-samples precise to the nanogram level in certain cases, making the

FT-IR an invaluable tool for problem solving in many studies. FT-IR spectroscopy requires some study because its principles are different from those of dispersive spectroscopy techniques and uses interferometers in the spectrometers.

In a conventional or continuous wave spectrometer, a sample is exposed to an electromagnetic radiation and usually the intensity of transmitted radiation is monitored. The energy of the incident radiation is varied over the desired range and the response is plotted as a function of frequency of the incident radiation. At certain resonant frequencies, which are characteristic of the sample, the radiation will be absorbed resulting in a series of peaks in the spectrum, which then can be used to identify the sample. In *Fourier transform spectroscopy* (FTS), instead of being subjected to varying energy of electromagnetic radiation, the sample is exposed to a single pulse of radiation consisting of frequencies in a particular range. The resulting signal contains a rapidly decaying composite of all possible frequencies. Due to resonance by the sample, resonant frequencies will be dominant in the signal; when Fourier transform on the signal is performed, the frequency response can be calculated. In this way the Fourier transform spectrometer can produce the same kind of spectrum as a conventional spectrometer, but in a much shorter time [9].

Different FT-IR spectrometers use different interferometers, such as the Michelson interferometer, lamellar grating interferometer, and Fabry-Perot interferometer. The spectrometers utilizing Fabry-Perot interferometers have low resolving power as compared to the two beam interferometers, namely the Michelson and lamellar grating interferometers. Both of the two beam interferometers have advantages and disadvantages one over the other with the basic difference being that in the Michelson interferometer division of wave amplitude takes place, whereas in the lamellar grating spectrometer division of wavefront takes place. The Michelson interferometer is often favored over the lamellar grating interferometer because of its easy construction and operation. Most of the commercially available FT-IR spectrometers use the Michelson interferometer. These have a number of advantages over the other techniques, such as high energy throughput, multiplexing, and high precision in frequency measurement. These will be discussed in detail in later sections. The principal disadvantage of FT-IR spectrometer is that from the interferometer, an interferogram is produced first rather than a spectrum, and little information is available from the interferogram without further processing. The conversion of the interferogram to a spectrum by Fourier analysis requires the use of a computer. Tremendous progress made in computer technology has allowed the method of Fourier spectroscopy to blossom. There is at present no serious technical reason that would prevent the numerical transformation of a large number of data points. Even for 10^6 sample transformations, the computation time is hardly a few seconds.

Nowadays, the main difficulty in performing Fourier spectroscopy depends upon the correct realization of the interferogram. FT-IR spectroscopy has been used as the dominant technique for measuring the infrared (IR) absorption and emission spectra of most materials [10]. The major advantage of the FT-IR technique over dispersive spectroscopy methods is that practically all compounds show characteristic absorption/emission in the IR spectral region, therefore, they can be analyzed both quantitatively and qualitatively. Although Fourier spectroscopy has been expanded in the last two decades, it still remains a new spectroscopic approach [7]. The interpretation of the effects observed in the spectrum due to features occurring in the interferogram often needs a lot of calculations in addition to Fourier transformation. The most important of these operations are apodization, phase correction, Fourier self deconvolution and curve fitting [11–15].

The far infrared region also plays an important role in the studies of low frequency molecular vibrations, derivation studies of molecular parameters via the observation of pure rotational spectra and research on molecular interactions in the solid state. The conventional IR spectrometers are not of much use for far IR region (20–400 cm^{-1}) as the sources are weak and detectors have lower sensitivity. FT-IR spectroscopy has made this energy-limited region more accessible [16, 17]. In this chapter first, the section on historical background is related to the development of Fourier transform spectroscopy. The important contributions of specific persons or groups are emphasized. In the next sections, theory, the basic integral equation used, and the various mathematical operations needed to construct the spectra from the interferogram are described. Advantages of the technique over the conventional spectrometers and some of its applications are discussed in detail in the later sections. The recent work done and the latest methodology developed are reviewed and included in the chapter.

9.2 HISTORICAL BACKGROUND

The science of Fourier transform interferometry was initiated in the late 1880s, when Albert A. Michelson invented the interferometer known as *Michelson interferometer*, on which he along with Morley, performed the well known experiment to determine the speed of light. The instrument and its uses were well received by the scientists of the day; Michelson received the Nobel Prize in Physics in 1907 for his efforts in measuring the exact speed of light and his inventions of precise scientific optical instruments [18]. As computers and electronic devices were not available back then, Michelson could not take advantage of the field of Fourier transform spectroscopy (FTS). However, using the techniques available to him, he resolved a number of doublet spectra [19]. Although Michelson knew the spectroscopic potential of his interferometer, yet the lack of sensitive detectors and nonexistence of Fourier transform algorithms were barriers for its practical application. Early investigators faced problems in computing Fourier transforms of interferograms and were not able to invert these directly. Rather they guessed a spectrum, computed the inverse Fourier transform, and then compared it to their measured interferogram. The guessed spectrum was then modified to bring it into better agreement with the data and the process was continued until sufficient agreement was obtained [9].

Practical FTS began to come into its existence only in the late 1940s. Interferometers were used to measure light from celestial bodies and scientists produced the first Fourier transform spectrum in 1949. By this time, it was possible to calculate the necessary Fourier transforms, however, it remained a laborious and time-consuming task. At this point, besides Michelson interferometer, researchers had developed different types of interferometers, namely *lamellar grating* and *Fabry-Perot interferometers* [9]. Figure 9.1 represents a basic Michelson interferometer.

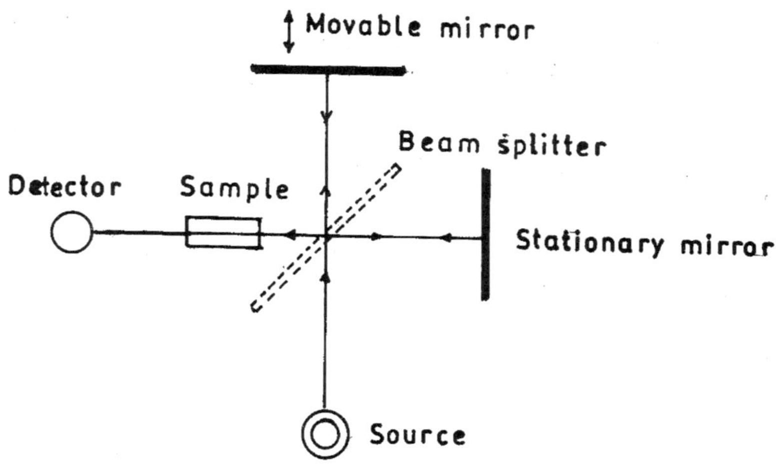

Figure 9.1 Schematic diagram of Michelson interferometer.

Lamellar grating spectrometers have much in common with the Michelson interferometer in that both are two-beam multiplex instruments of high optical throughput that produce interferograms, which are then Fourier-transformed to obtain the desired spectrum. The optical modulation device in the lamellar grating instrument is a pair of mirrors in tongue and groove arrangement as shown in Figure 9.2. When the surfaces of the two mirrors are in the same plane, they appear to be one large mirror divided into a dozen or more horizontal strips. The first mirror comprises one set of alternate strips, which are fixed, while the second mirror contains strips that are connected so they may be moved like a piston from a position forward of the fixed strips to a position behind the fixed strips [19]. In this interferometer the entire wavefront is used, whereas the Michelson interferometer loses one half of the total flux even if the beam splitter were perfectly efficient. In the far infrared or microwave region (below 100 cm^{-1}) lamellar grating spectrometers are preferred to the Michelson interferometer because of their high efficiency.

At the onset of 1960 interest in interferometric spectroscopy was growing and that decade saw many advances in the theory of interferometric measurements and its applications to physical systems. These developments were greatly aided by the works of Cooley and Tukey [20] on the fast Fourier Transform (FFT) algorithm. This algorithm allowed Fourier transforms to be computed efficiently and easily on electronic computers available at that time. This reduced the computation time by several orders of magnitude and made the transformation from interferogram to spectrum feasible. Application of these allowed Connes and Connes [21] to record in 1966 the first near-infrared planetary spectra. Later, in 1969, they along with others [22] obtained high resolution and high quality spectra of the planets. By this time, with the advancement of computers, scientists started comparing performances of Fourier transform interferometers with grating spectrometers. Jacquinot [23] openly stated that the planetary spectra of Connes and Connes [21] if recorded by good grating spectrometers could have taken a few thousand years. This improvement in speed was possible because of minicomputers, which could carry out the fast Fourier transforms in little time. Gibbie [24] commented that computers, like photographic plates, were wonderfully sophisticated and essential spectroscopic tools, which fortunately spectroscopists did not have to invent themselves. He compared the Fourier transform spectroscopy with grating spectroscopy in this paper in a very interesting manner, and he pointed out that Fourier spectroscopy gave experimentalists a new weapon of great effectiveness.

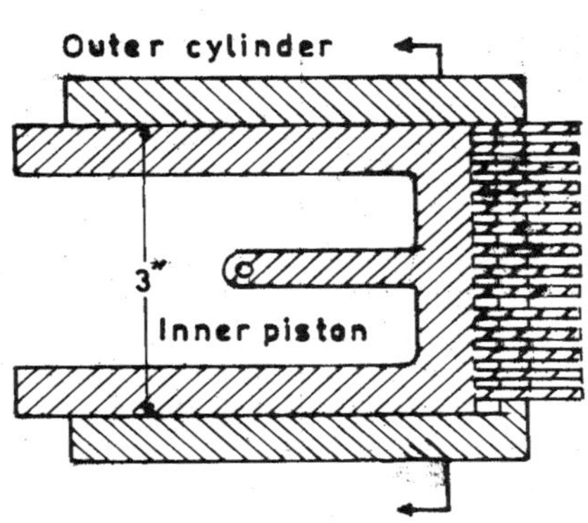

Figure 9.2 The lamellar grating modulator [19].

By 1970 commercial Fourier transform spectrometers had started to become widely accessible and the technique blossomed. However, the first FT-IR spectrometers were large and expensive and were found primarily in a few well-to-do research laboratories. Gradually, technology reduced the cost, increased the availability, and enhanced the capabilities of FT-IR spectrometers. Today Fourier transform spectrometers aided by fast computers, which perform Fourier transformation in a fraction of a second in the visible, infrared, and microwave regions, are common laboratory instruments. These are used for spectroscopic analysis in many diverse disciplines with reduced prices and increased performance.

9.3 FT–IR SPECTROSCOPY

The development and dominance of Fourier transform spectrometers over conventional spectrometers were discussed in the previous section. In the following, its basic principle, theory, and the various mathematical operations needed to generate spectrum will be discussed.

An optical system that utilizes an interferometer and a dedicated computer constitute the two basic parts of an FT-IR spectrometer. The direct acquisition of digitized infrared spectral information with the help of dedicated minicomputers has advantages. But the real advantage of an FT-IR spectrometer is obtained through the interferometer rather than the grating or prism used in the conventional spectrometers. The block diagram of an FT-IR spectrometer is represented in Figure 9.3. The three optical inputs, viz., a He-Ne laser, a white light, and the infrared source, are connected to the interferometer. All three optical signals share the same beam splitter and mirrors [25].

The computer controls optical components, collects and stores data, carries out calculations on data, and displays spectra. Because of the direct interfacing of the computer to the spectrometer, spectra can be arithmetically manipulated in an inventive way such that, for example, interfering absorbances can be eliminated by subtracting out from composite spectra the absorption bands due to interfering components [17, 26–28].

The principle of the FT-IR spectrometer is as follows: First a signal called an *interferogram* is generated by the interferometer. The interferogram obtained is a record of the signal (intensity) by the infrared detector as a function of the difference in the path for the two beams of the interferometer. The spectrum is obtained by carrying out the Fourier transform of the interferogram. In this, the intensity, which is a function of path difference x, is subjected to transformation as a whole to give the spectrum S, which depends only on frequency ν. Hence,

$$S(\nu) = \int_{-\infty}^{+\infty} I(x)e^{+i2\pi\nu x}\,dx = \mathbf{F}^{-1}[I(x)] \tag{9.1}$$

in which

$$I(x) = \int_{-\infty}^{+\infty} S(\nu)e^{-i2\pi\nu x}\, d\nu = \mathbf{F}[S(\nu)] \tag{9.2}$$

where the first integral is called the *inverse Fourier transform* and the second integral is called the *Fourier transform*. Thus, the integral given by equation (9.1) converts the interferogram $I(x)$, which is a function of path difference x to spectrum $S(\nu)$, which in turn is a function of frequency ν (from space domain to wave number domain). This calculation is carried out on a dedicated computer [29].

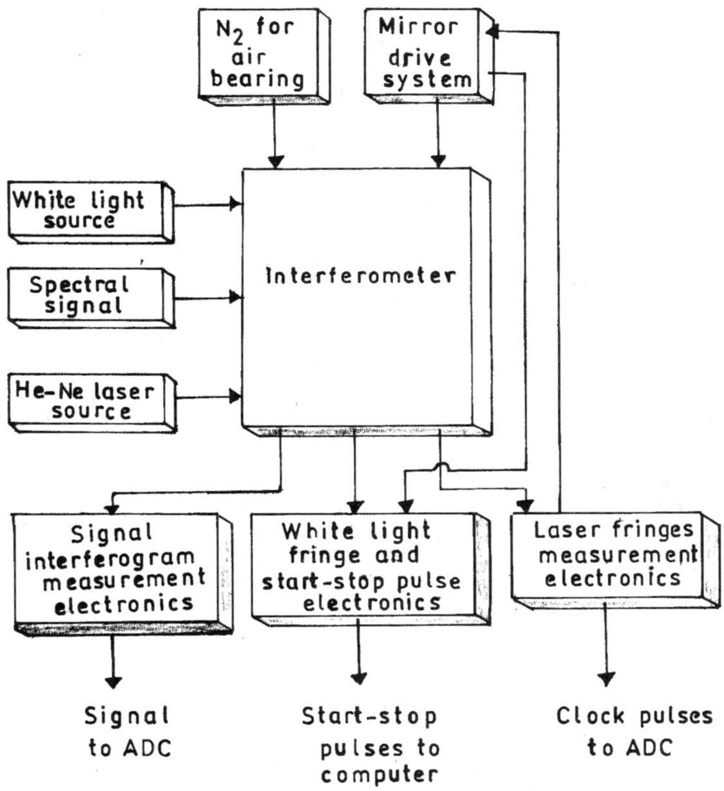

Figure 9.3 Block diagram of FT-IR spectrometer [25].

9.3.1 Basic Integral Equation

The basic integral equation used in the Fourier transform spectroscopy can be obtained from the definition of Fourier integral theorem and the principle of superposition of waves [19]. The basic equation used in the case of the Michelson interferometer can be derived mathematically as follows:

Let the amplitude of the wave (traveling in the z direction) incident on the beam splitter be given as

$$E(z,\nu)\, d\nu = E_0(\nu)\, e^{i(\omega t - 2\pi\nu z)}\, d\nu \tag{9.3}$$

where $E_0(\nu)$ is the maximum amplitude of the beam at $z = 0$. The amplitude of the beam is divided at the beam splitter and two beams are produced. Let z_1 and z_2 be the distances traveled by the beams when they recombine. Each beam undergoes one reflection from the beam splitter and one transmission through the beam splitter. If r and t are the reflection and transmission coefficients, respectively, of the beam splitter, then the amplitude of the recombined wave E_R is

$$E_R[z_1,\ z_2,\ \nu]\, d\nu = rt\ E_0(\nu)[e^{i(\omega t - 2\pi\nu z_1)} + e^{i(\omega t - 2\pi\nu z_2)}]d\nu \tag{9.4}$$

By definition, the intensity after recombination of the beams for the fixed spectral range $d\nu$ is given as

$$\begin{aligned} I(z_1, z_2, \nu)d\nu &= E_R(z_1, z_2, \nu)\, E_R^{*}(z_1, z_2, \nu)\, d\nu \\ &= 2E_0^2(\nu)|rt|^2\left[1 + \cos 2\pi(z_1 - z_2)\nu\right]d\nu \end{aligned} \tag{9.5}$$

and the total intensity at any path difference $x = (z_1 - z_2)$ for the whole spectral range is obtained by integrating equation (9.5) as

$$I_R(x) = 2|rt|^2 \int_0^{\infty} E_0^2(\nu)d\nu + 2|rt|^2 \int_0^{\infty} E_0^2(\nu)\cos(2\pi x\nu)d\nu\,. \tag{9.6}$$

Fourier cosine transform of equation (9.6) converts intensity into spectrum as

$$E_0^2(\nu) = (1/\pi|rt|^2)\int_0^{\infty}\left[I_R(x) - \tfrac{1}{2}I_R(0)\right]\cos(2\pi\nu x)dx. \tag{9.7}$$

In equation (9.7), $I_R\,(0)$ represents the flux associated with waves at zero arm displacement where the waves for all frequencies interact coherently. Thus, $I_R(0)$ is the flux associated with coherent interference and $I_R\,(x)$ is the flux associated at path difference x. $\left[I_R(x) - \frac{1}{2}I_R(0)\right]$ is called the *interferogram,* i.e., the oscillations of the signal about the value $\frac{1}{2}I_R\,(0)$. However, in various books $I_R\,(x)$ is sometimes referred to as interferogram. The spectrum $S(\nu)$, which is proportional to $E_0^2(\nu)$, can be given from equation (9.7) as

$$S(\nu) \propto E_0^2(\nu) = \text{constant} \int_0^{\infty} [I_R(x) - \tfrac{1}{2} I_R(0)] \cos(2\pi \nu x) dx \qquad (9.8)$$

The interferogram is Fourier transformed with the help of computer to convert the space domain into the wave number domain [29]. Before discussing the steps involved to compute the spectrum from the interferogram, a brief experimental setup will be mentioned in the next section.

9.3.2 Experimental Setup

Most of the available Fourier transform spectrometers make use of Michelson interferometer, therefore, discussion will be limited to the experimental setup of the Michelson interferometer. The main differences between various FT-IR spectrometers utilizing Michelson interferometer are based on the design and the manner in which it is operated. It can be operated by scanning in a discontinuous stepwise manner (*step-scan interferometer*), a slow continuous manner with chopping of the infrared beam (slow scanning) or rapidly without chopping of the infrared beam [17]. Most of the commercially available FT-IR spectrometers make use of a rapidly scanning interferometer, which is discussed here. Figure 9.4 illustrates the schematic diagram of a rapid scanning Fourier transform spectrometer.

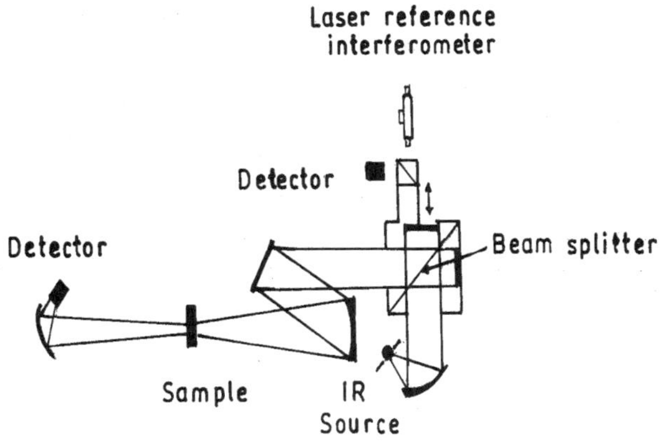

Figure 9.4 Schematic diagram of a rapid scanning Fourier transform infrared spectrometer [17].

The infrared source emits radiation of all wavelengths in the infrared region and is usually selected for desired spectral range, i.e., mercury lamps for the far infrared and glowers for the near infrared. The collimating mirror collimates the light from the infrared source. This parallel beam is then sent to the beam splitter (Fe_2O_3 in the near and far infrared regions and Ge or Si in

10,000–3000 cm^{-1}) of the Michelson interferometer. The thickness and coatings of the beam splitter must be considered carefully. The beam splitter divides the amplitude of the beam: one part of the radiation beam goes to the moving mirror and other part to the fixed mirror. The moving mirror must travel smoothly; a frictionless bearing is used with an electromagnetic drive for this purpose. The return beams from these mirrors recombine again at the beam splitter and undergo interference. The reconstructed beam is then passed through the sample (Figure 9.4) and focused on to the detector. The detector used in an FT-IR spectrometer must respond quickly because intensity changes are rapid as the moving mirror moves quickly. Pyroelectric detectors or liquid-cooled photon detectors must be used. Pyroelectric detecters have a fast response time and are used in most FT-IR spectrometers. Thermal detectors are very slow and are rarely used nowadays [30]. Pyroelectric detectors are made from a single crystalline wafer of a pyroelectric material, such as triglycerine sulphate. The heating effect of incident IR radiation causes a change in the capacitance of the material. Photoelectric detectors such as mercury-cadmium-telluride (MCT) detectors comprise a film of semiconductor material deposited on a glass surface, sealed in an evacuated envelope. Absorption of IR radiation by nonconducting valence electrons helps them to jump to a higher conducting state. This increases the conductivity of the semiconductor. These detectors have better response charactersistics than pyroelectric detectors and are used in most FT-IR instruments [31].

The motion of the mirror results in a path difference x between the two interfering beams. The path difference is twice the mirror displacement from the balanced position. The condition for maxima is $x = n\lambda$ and that for minima is $x = (n + 1/2)\lambda$, where n is the order of interference pattern, λ is the wavelength of incident radiation and ν is its frequency in cm^{-1}. Thus, each wavelength produces its own characteristic interference pattern as the mirror is displaced. For a monochromatic source there will be a cosine variation in the intensity of the combined beams at the detector. The angular frequency of the signal is $\omega = 2\pi V_m/\lambda = 2\pi V_m \nu$, where V_m is the velocity of mirror. The record of the detected signal versus optical path difference is the interferogram. Each wavelength will produce its own characteristic cosine intensity pattern of particular magnitude. Thus, for a source of many frequencies, the interferogram will be the sum of fluxes (variation of intensity) of each wavelength pattern at the detector.

To measure position and the direction of the moving mirror, the laser beam (usually an He-Ne laser) is used as shown in Figure 9.4. The laser beam undergoes the same change in the optical path as the infrared beam. Since the laser beam is monochromatic, the interferogram of the laser light will be a well-defined cosine wave. The mirror position is determined by counting the number of fringes in a highly defined laser pattern and the point of zero path length difference is obtained by finding the position of highest amplitude intensity using the white light in place of the laser beam. For the He-Ne laser the fringe intensity will go from a maximum to a minimum with a mirror motion of $\lambda/4$ or a path difference equal to $\lambda/2$. The wavelength of the He-Ne laser is 0.6328 μm, therefore, the mirror position can be easily located to within better than $\lambda/4 = 0.16$ μm [28]. Data from the additional scans can be co-added to the data stored in the computer memory to improve the signal-to-noise ratio (*S/N*) of the interferograms because by adding a large number of scans, the noise signal present in the interferogram can be averaged out. The averaged interferogram is prepared for computation by phase correction and apodization, and then the Fourier transform is performed on the interferogram to obtain the spectrum.

The experimental method of computing a spectrum from an interferometer is as follows:

1. Measure $I_R(x)$ by recording the signal versus the interferometer arm displacement. It must be noted that $x = 2$ (arm displacement).
2. Experimentally determine the intensity of the recombined beams at zero path difference, i.e., $I_R(0)$.
3. Substitute the value of $[I_R(x) - \frac{1}{2} I_R(0)]$ in equation (9.8), and evaluate the integral for a selected value of frequency ν. The integration is performed on the computer.
4. Compute the integration of Equation (9.8) for each selected frequency ν.
5. Compute $S(\nu)$ versus ν. This gives the required spectrum.

The rapid scan FT-IR spectrometer was discussed above. In recent years, interest in step-scan FT-IR spectrometers has greatly increased [32, 33]. Many reports have appeared in the literature indicating that the technique can be a useful tool for recording the vibration spectra of a variety of time-dependent systems. The main difference in operation between step-scan and rapid-scan FT-IR spectrometers is the average velocity of the interferometer mirror, which in turn determines the effective modulation frequencies of the infrared radiation and more importantly, the modulation frequency spacing between adjacent spectral elements. In rapid-scan FT-IR spectrometers the mirror velocity lies in range 0.16–3.16 cm/s, whereas in most step-scan FT-IR spectrometers the interferometer is stepped at a rate between 1–10 fringes/sec. For more details the reader can refer to [32].

9.3.3 Advantages

There are many advantages of FT-IR spectrometers over the conventional spectrometers, but the most important advantages of FT spectrometers over earlier instruments are:

(i) higher signal-to-noise ratios for spectra recorded for the same measurement time, and
(ii) higher accuracy in frequency measurement for the spectra.

The improved signal-to-noise ratio is a consequence of both the concurrent measurement of detector signal for all the resolution elements of the spectrum known as *multiplex* or *Fellgett advantage* and of the high optical throughput of the FT-IR spectrometer known as *throughput* or *Jacquinot advantage*. The improvement in frequency accuracy of the FT-IR spectrometer is a consequence of the use of a laser, which references the measurements made by the interferometer. This is known as *laser reference* or *Connes advantage*. These advantages of FT-IR spectrometers over the dispersive spectrometers are discussed in the following sections.

9.3.3.1 Multiplex (or Fellgett) Advantage

The multiplex advantage of the FT-IR spectrometer is a consequence of the simultaneous observation of all the resolution elements $\Delta\nu$ in the desired frequency range. Fellgett [34] was the first person to transform an interferogram numerically by using the multiplex advantage, and this advantage is the well-known Fellgett advantage. In the grating or prism spectrometers, as shown in Figure 9.5, the monochromator allows only one resolution element of the spectrum to be examined at a time. Frequencies above and below this frequency band $\Delta\nu$ are masked off from the detector.

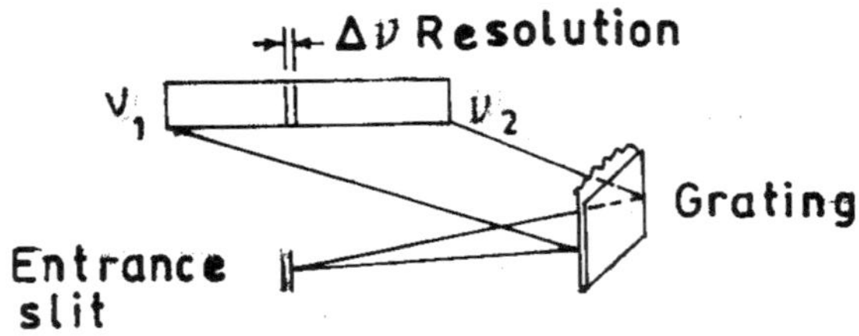

Figure 9.5 Illustration of frequency interval in a grating spectrometer [17].

The interferometer does not separate light into individual frequencies before measurement. This means each point in the interferogram contains information from each wavelength in the input signal. In other words, if 1000 data points are collected along the interferogram, each wavelength is sampled 1000 times, whereas, a dispersive spectrometer samples each wavelength only once [28]. The mathematical explanation of this is represented in the following.

Suppose one has to record a broad spectrum between the wave numbers ν_1 and ν_2 with resolution $\Delta\nu$. Let *M,* the number of spectral elements in the broad band, be represented as:

$$M = \frac{\nu_2 - \nu_1}{\Delta\nu}. \tag{9.9}$$

If grating or prism is used in the spectrometer then each small band of width $\Delta\nu$ can be observed for a time T/M, where T is the total time required for a scan from ν_1 to ν_2. Therefore, the signal received in a small band $\Delta\nu$ is proportional to T/M. If the noise is random and independent of signal level, the S/N should be proportional to $[T/M]^{1/2}$.

$$(S/N)_G \propto [T/M]^{1/2} \tag{9.10}$$

The situation for the interferometer is different because it detects in the broad band $(\nu_2 - \nu_1)$ all the small bands of resolution $\Delta\nu$ all the time. Therefore, the signal in the small band $\Delta\nu$ is proportional to T. If the noise is random and independent of the signal level, the noise is proportional to $T^{1/2}$. Thus, for the interferometer the signal-to-noise ratio is,

$$(S/N)_I \propto T^{1/2}. \tag{9.11}$$

Assuming the same proportionality constants in equations (9.10) and (9.11), the ratio of $(S/N)_I$ to $(S/N)_G$ is given as:

$$(S/N)_I/(S/N)_G = M^{1/2} \tag{9.12}$$

For a typical case $M \sim 10^4$. Thus, the signal-to-noise ratio for the interferometer is about hundred times the signal-to-noise ratio for the grating spectrometer. However, this improvement in the signal-to-noise ratio is achieved only if (i) the noise in the spectrum is only due to detector noise, and (ii) the detector noise is not proportional to the detector signal.

Studying the multiplex advantage, Voigtman and Winefordner [35] have noted that photon noise from the flux at each spectral interval contributes to the noise of each signal (or signals). Thus, a small signal may be buried in the noise from a large signal (or signals). They presented an analysis of the performance of a Fourier transform spectrometer with respect to source shot and source flicker noises. It was found that the source shot noise is uniformly distributed throughout the base whereas source flicker noise remains localized about the generating spectral region(s).

9.3.3.2 Throughput (or Jacquinot) Advantage

Like most spectroscopic techniques, the limiting aspect of infrared spectroscopy is the available energy per unit time. Infrared spectrometers should have a high energy throughput since they in particular suffer from energy-limited sources and noise-limited detectors. The increase in throughput obtainable by FT-IR spectrometers utilizing the Michelson interferometer was first observed by Jacquinot [36] and has been universally recognized as an advantage of FT-IR spectroscopy.

Monochromators, using either gratings or prisms as dispersive elements, allow the observation of a narrow, predetermined, nearly monochromatic frequency domain. Spectra are recorded by scanning the desired frequency range at successive resolution intervals. To achieve good resolution openings of the entrance and exit slits of monochromators are made narrow, which decreases the energy entering the monochromators. This allows the detector to view only a small portion of the energy emitted by the source. Measurable signals are achieved by increasing the time of measurement per frequency interval, thereby decreasing the scanning rate. The geometric dispersion in monochromators results in throwing away most of the valuable energy [37]. On the other hand, no slit is used in the interferometer. It utilizes a collimator to focus the radiations from the source onto the sample, which means more energy gets on to the sample than is possible with dispersive spectrometers. This results in more energy passed on to the detector, increasing the signal-to-noise ratio of the spectrum recorded.

The throughput τ is defined as the product of the cross-sectional area A and solid angle Ω of the beam at any focus in the optical system. The maximum throughput of the spectrometer also determines the maximum useful $A\Omega$ of the source. For a source of given brightness, $A\Omega$ determines the total radiant power accepted by the optical system. It is, therefore, desirable to maximize the throughput and hence the energy reaching the detector so that signal-to-noise ratio is maximized [19]. One can compare the values of throughput obtainable by an interferometer and a grating spectrometer as follows:

For the Michelson interferometer, if h is the diameter of the circular infrared source and f is the focal length of the collimating mirror, the solid angle subtended by the source at the collimating mirror can be expressed as

$$\Omega_{\mathrm{M}} = \frac{\pi h^2}{4 f^2} \tag{9.13}$$

and the resolving power for the interferometer is

$$R_{\mathrm{M}} = \frac{\nu}{\Delta\nu} = \frac{8 f^2}{h^2}. \tag{9.14}$$

On substituting equation (9.14) in equation (9.13) one gets:

$$\Omega_{\mathrm{M}} = \frac{2\pi}{R_{\mathrm{M}}}. \tag{9.15}$$

If A_{M} represents the area of the collimating mirror, the throughput obtained by the Michelson interferometer is given by

$$\tau_{\mathrm{M}} = A_{\mathrm{M}}\,\Omega_{\mathrm{M}} = \frac{2\pi A_{\mathrm{M}}}{R_{\mathrm{M}}}. \tag{9.16}$$

In the case of grating spectrometers, the slit area becomes the effective source area. If A_G is the area of grating, h is the height of the slit, f is the focal length of the collimating mirror, and R_G is the resolving power of the grating then the energy throughput obtained by the grating spectrometer can be expressed as

$$\tau_G = \frac{hA_G}{fR_G}. \tag{9.17}$$

Assuming the same area and focal length for collimating mirrors used in the Michelson interferometer and grating spectrometer and the same resolving powers, the ratio of throughput for interferometer and grating spectrometers becomes

$$\frac{\tau_M}{\tau_G} \approx 2\pi\left(\frac{f}{h}\right). \tag{9.18}$$

In the best available spectrometers, (f/h) is never less than 30 (typical), therefore, τ_M/τ_G ~190. This means about 200 times more power can be put through the interferometer than through the best grating spectrometer. However, the theoretical value of the throughput advantage depends upon certain other factors like the detector area and efficiency of the beam splitter. Detector area must be considered because the noise level of infrared detectors increases as the square root of the detector area. Thus, a small detector area will give better signal-to-noise ratio. It is also necessary to consider the efficiency of the beam splitter. It is often expedient to use a beam splitter to obtain a spectrum in a frequency region where its efficiency is low (< 20%). Although area is unaffected by the efficiency of the beam splitter, the signal is reduced and, thus the signal-to-noise ratio at the detector will vary directly as the efficiency of the beam splitter. Increase in throughput occurs when there is a large signal at the detector. In fact, for absorption spectroscopy, which now is the dominant field of application of FT-IR, the signal in which one is interested is the one that does not make it to the detector, i.e., the absorbed light [38]. This apparently trivial statement has profound consequences, since all that we can measure is the light that is left and what we really want is the difference between it and the original light level. For higher throughput, the instrument should be designed accordingly with larger sources, beam splitter, other optical components, and the detector. While the increase in the first three only harms the performance slightly, the detector noise level increases linearly with its diameter.

Although an FT-IR spectrometer has a theoretical throughput advantage over a grating spectrometer, the numerical value of the resultant signal-to-noise improvement is dependent upon several parameters of the essential components used in the Michelson interferometer. It has been verified that photon shot noise is the ultimate limit on S/N measurements of optical information. It represents the ultimate limit for an ideal FT-IR spectrometer, but the performances are a good order of magnitude short of those calculated

[39]. The causes for this shortfall are not well understood. Vague suspicions about detectors, pre-amplifiers, filters, and data processing are hard to investigate.

9.3.3.3 Laser Reference (or Connes) Advantage

An FT-IR spectrometer determines frequencies by direct comparison with a visible laser output, usually a He-Ne laser. Potentially, this offers an improvement in frequency accuracy and was determined by Connes; it is also known as *Connes advantage* [21]. It has been reported that the use of a laser reference presents a foremost advantage of FT-IR spectrometers over dispersive spectrometers.

With dispersive instruments, frequency precision and accuracy depend on (i) calibration with external standards and (ii) ability of electromechanical mechanisms to uniformly move gratings and slits. By contrast, in FT-IR spectrometers the direct laser reference has the advantage that no calibration is required. Here, both the mirror movement and detector sampling are clocked by the interferometer fringes from the monochromatic light of laser. All frequencies in the output spectrum are calculated from the known frequency of the laser light. The wavenumber scale of an interferometer is derived from the He-Ne laser. The wavenumber of this laser is known very accurately and is very stable. Thus, the wavenumber calibration of interferometers is much more accurate and has much better long term stability than the calibration of dispersive instruments. One of the most important strengths of FT-IR spectroscopy lies in the ease of measuring high accuracy difference spectra. This has been feasible only because of Connes advantage, which provides accurate alignment of the spectra to be subtracted.

9.3.3.4 Some Other Advantages

Another important advantage of FT-IR spectrometers is the stray light rejection efficiency. In FT-IR spectrometers the sample is usually positioned between the interferometer and detector to take advantage of the high stray light rejection efficiency. This arrangement allows only the infrared source radiation to be modulated by the interferometer; the emission from the sample would not be modulated, therefore, would not contribute to the spectrum. Depending upon the optical design of the spectrometer, this stray light rejection advantage in not valid. Emission from hot samples can also be modulated and can produce distortions in the spectrum. Reflections from the samples also lead to distortions. These effects can be eliminated by using a half-beam blocking aperture by tilting the sample and using corner-cubed FT-IR spectrometers [40]. This can be realized in practice easily and high stray light rejection efficiency can be achieved.

Another advantage of FT-IR spectrometers is that in these, solid state electronic components are used, which are more reliable and can be easily replaced if trouble arises. The use of a computer has the remarkable advantage of the data-handing. Computer-controlled spectral plots can be made and labeled on white paper ready for publication without the need for manual redrawing. This was not possible with dispersive spectrometers. The ability to add and subtract spectra in digital form is especially useful in eliminating interferences due to overlapping bands. This allows one to extract a maximum amount of information available in the spectrum. With the availability of increased speed and memory size of computers the use of programming in FT-IR spectrometers has also become possible.

FT-IR spectrometers are compact, simple and flexible for use in a large number of applications. The essential requirement of an FT-IR spectrometer is that the collimated beam from the source be passed through the interferometer and focused on the detector. The detector can be positioned within an experimental setup or at the end of a light pipe. To make use of the FT-IR spectrometer in various fields, some new sampling techniques such as attenuated total reflection (ATR) spectroscopy [41], photo-acoustic spectroscopy (PAS) [42], and diffused reflectance infrared fourier transform (DRIFT) spectroscopy [43] have been developed. The sample preparation is the most important feature in spectroscopic analysis. The above-referred techniques require very little sample preparation and can be used to study the samples without rigorous preparation by attaching suitable accessories with FT-IR spectrometers.

9.3.4 Other Aspects

The main objective of FT-IR spectrometers is to compute the spectrum from the interferogram. This involves a number of intermediate mathematical steps such as apodization, phase correction and Fourier self deconvolution (FSD). The excellence of the final spectrum makes these intermediate steps valuable and the use of a computer makes the operation practical. One has to devote a long time to understanding the theory of the FT-IR spectrometer, otherwise it will appear to be a black-box. Ideas about these mathematical operations are helpful for overcoming any difficulty in operating the spectrometers. A few of these are discussed here.

9.3.4.1 Apodization (or Truncation) of the Interferogram

The basic Fourier transform integral equation has infinite limits for the optical path difference (equation (9.8)), but the experimental limits are finite. The effect of this is that the truncation of interferogram occurs and is reflected in the spectrum when the interferogram is Fourier-transformed. Figure 9.6 shows the effects on spectral lineshape of various transactions of a pure sine wave.

The corrective procedure for modifying the basic Fourier transform integral is called *apodization*. In this the interferogram is usually multiplied by a function prior to Fourier transformation, which removes false sidelobes introduced into transformed spectra because of the finite limits of optical path displacement. The false sidelobes look like feet in the spectrum, so the word "apodization" is appropriate, since it means having "no feet". Codding and Horlick [11] discussed in detail the various apodization functions, which will be referred to in the further discussion. In their paper they reviewed the work done by various other researchers on apodization functions. Typical apodization functions include linear, triangular, exponential, and Gaussian truncation functions. The mathematical concept of *convolution theorem* is necessary for qualitative understanding of the truncation of the interferogram. The mathematical treatment of the convolution theorem is available in literature [1,19]. It states that the *product of the Fourier transforms of two functions is equal to the convolution of Fourier transforms of the functions and vice versa*.

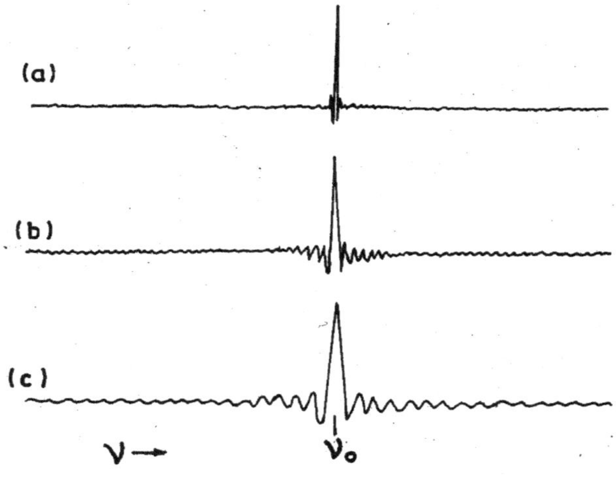

Figure 9.6 Truncated pure sine wave produced by truncating the interferogram a length equivalent to (a) 2, (b) 4, and (c) 8 cm^{-1} resolution and (a) 0.5, (b) 0.25, and (c) 0.125 cm retardation [17].

To see how apodization function is applied, first observe the effect of truncation on the spectrum of a monochromatic source. From equation (9.8) the spectrum is given by

$$S(\nu) = \int_{-\infty}^{+\infty} [I_R(x) - \tfrac{1}{2} I_R(0)] e^{-i2\pi\nu x}\, dx \tag{9.19}$$

and its Fourier transform is given by

$$\left[I_R(x) - \tfrac{1}{2} I_R(0)\right] = \int_{-\infty}^{+\infty} S(\nu) e^{+i2\pi\nu x} d\nu \qquad (9.20)$$

For the monochromatic source emitting radiation of frequency ν_1, the spectrum can be given as,

$$S(\nu_1) = \tfrac{1}{2}\left[\delta'(\nu - \nu_1) + \delta'(\nu + \nu_1)\right] \qquad (9.21)$$

where $\delta'(\nu - \nu_1)$ and $\delta'(\nu + \nu_1)$ are the Dirac delta functions. The two delta functions are used to assure that energy is conserved in going from the interferogram to the spectrum. On substituting $S(\nu)$ from equation (9.21) in equation (9.20), one gets:

$$\left[I_R(x) - \tfrac{1}{2} I_R(0)\right] = \tfrac{1}{2} \int_{-\infty}^{+\infty} \left[\delta'(\nu - \nu_1) + \delta'(\nu + \nu_1)\right] e^{+i2\pi\nu x} d\nu. \qquad (9.22)$$

Using the property of the Dirac delta function, i.e., if $f(x)$ is continuous, then $\int_{-\infty}^{+\infty} \delta'(x-a) f(x) dx = f(a)$, one gets

$$\left[I_R(x) - \tfrac{1}{2} I_R(0)\right] = 2\cos(2\pi\nu_1 x). \qquad (9.23)$$

It is clear from equation (9.23) that the interferogram of a monochromatic source is a cosine wave as shown in Figure 9.7 when the interferogram is scanned to infinity, i.e., $x = -\infty$ to $+\infty$.

The corresponding spectrum is obtained by Fourier transform of equation (9.23) as

$$S(\nu) = \int_{-\infty}^{+\infty} 2\cos(2\pi\nu_1 x) e^{-i2\pi\nu x} dx. \qquad (9.24)$$

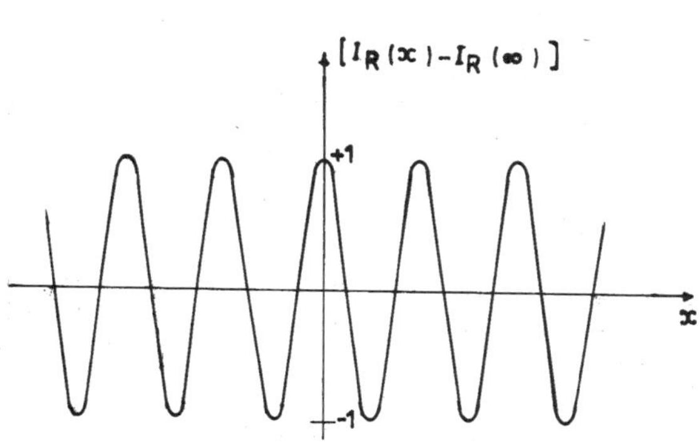

Figure 9.7 Interferogram of monochromatic source [19].

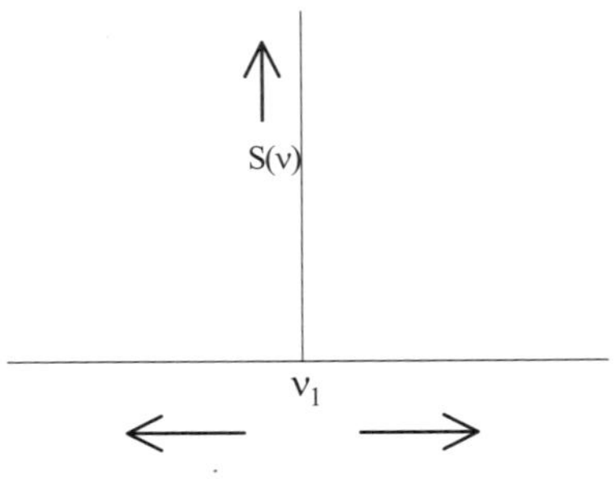

Figure 9.8 Spectrum of a monochromatic wave.

The integration of equation (9.24) results in a line spectrum at frequency ν_1 and is represented in Figure 9.8.

In actual practice, the interferogram cannot be scanned to infinity. Let the limits of x be $-L$ to $+L$. Then equation (9.24) can be written as

$$S(\nu) = \int_{-L}^{+L} 2\cos(2\pi\nu_1 x)\left[\cos 2\pi\nu \boldsymbol{x} - i\sin 2\pi\nu x\right] dx$$

$$= \int_{-L}^{+L} \left(\cos 2\pi x(\nu_1 + \nu) + \cos 2\pi x(\nu_1 - \nu)\right) dx. \tag{9.25}$$

Integration of equation (9.25) gives

$$S(\nu) = 2L\left(\frac{\sin 2\pi(\nu_1 + \nu)L}{2\pi(\nu_1 + \nu)L} + \frac{\sin 2\pi(\nu_1 - \nu)L}{2\pi(\nu_1 - \nu)L}\right). \tag{9.26}$$

The first term in equation (9.26) is small, ~ 0.01 or even less, and the second term in equation (9.26) has a peak value of unity at $\nu = \nu_1$. Therefore, ignoring the first term, the spectrum of a monochromatic source produced by truncating an interferogram within a finite maximum optical path displacement L can be given as

$$S(\nu) \equiv 2L\left(\frac{\sin 2\pi(\nu_1 - \nu)L}{2\pi(\nu_1 - \nu)L}\right)$$

$$\equiv 2L(\sin z)/z \equiv 2L \text{ sinc } z \tag{9.27}$$

where $z \equiv 2\pi(\nu_1 - \nu)L$. $2L$ sinc z is usually known as *instrument lineshape function* (ILS). The plot of the spectrum given by equation (9.27) is represented by a solid line in Figure 9.9.

Therefore, when an interferogram is scanned to finite limits, sinc z represents the spectrum of the monochromatic beam. The sidelobes or feet appear below the baseline. The central peak has a finite width as well. The feet appear as false sources of energy at nearby wavelengths. The process of apodization reduces the size of these feet. This can be carried out using different functions. Let the interferogram be corrected first by multiplying it with a triangular function $\Lambda = 1 - |x|/L$, then the spectrum will be given by

$$S(\nu) = 2\int_{-L}^{+L}\left(1-\frac{|x|}{L}\right)\cos(2\pi\nu_1 x)e^{-i2\pi\nu x}dx$$

$$= 2\int_{-L}^{+L}\left[\left(1-\frac{|x|}{L}\right)\cos\left[2\pi(\nu_1-\nu)x\right]\right]dx$$

$$= 2\left\{\frac{1-\cos[2\pi(\nu_1-\nu)L]}{[2\pi(\nu_1-\nu)]^2 L}\right\}$$

$$= 2\cdot\frac{2}{4}\frac{\sin^2\pi(\nu_1-\nu)L}{\left(\pi(\nu_1-\nu)L\right)^2}L$$

$$= L\frac{\sin^2\pi(\nu_1-\nu)L}{\left(\pi(\nu_1-\nu)L\right)^2} \tag{9.28}$$

Let $\pi(\nu_1-\nu)L = z/2$, then

$$S(\nu) = L\text{sinc}^2(z/2). \tag{9.29}$$

This function is plotted in Figure 9.9 by the dotted curve. It clearly indicates that the triangular apodization function on the interferogram reduces the feet or sidelobes. There is an increase in the width of the central peak but to a small extent. Here only the triangular apodization has been considered. Some other functions such as cosine, trapezoidal, Gaussian, etc., can also be used for apodization. From equations (9.27) and (9.29) it is clear that L appears in the spectrum, so one must take the same L for recording the background as well as the sample spectrum.

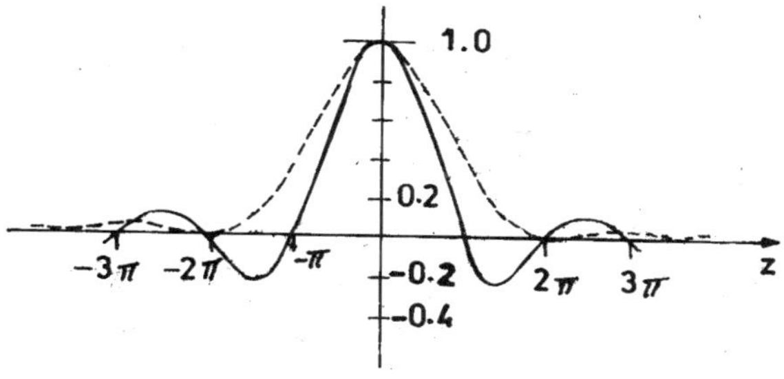

Figure 9.9 Plots of (1) sinc $z = (\sin z)/z$ versus z and (2) $\text{sinc}^2\, z/2$ versus z [17].

Now, with the help of a boxcar function it will be shown that the spectrum computed for a finite optical path difference is the convolution of the idealized spectrum produced by infinite optical path differences with the instrument lineshape function. Let $I_R(\infty)$ be the flux associated for infinite arm displacement, which corresponds to the incoherent interference. For the Michelson interferometer (the two-beam interferometer), the ratio of fluxes of incoherent interference and coherent interference is ½, i.e.,

$$\frac{I_R(\infty)}{I_R(0)} = \frac{1}{2}$$

or ½ $I_R(0) = I_R(\infty)$. Using this in equation (9.19), the idealized spectrum for infinite optical path difference and the instrumental spectrum for the finite optical path difference can be represented by equations (9.30) and (9.31), respectively:

$$S(\nu) = \int_{-\infty}^{+\infty} [I_R(x) - I_R(\infty)]\, e^{-i2\pi\nu x}\, dx \tag{9.30}$$

$$S_I(\nu) = \int_{-L}^{+L} [I_R(x) - I_R(\infty)]\, e^{-i2\pi\nu x}\, dx. \tag{9.31}$$

Let the rectangular function, or so-called boxcar function [44], be defined as

$$rect(x) = \begin{pmatrix} 1, & |x| \le L \\ 0 & |x| > L \end{pmatrix}. \tag{9.32}$$

Applying this boxcar function, the instrumental spectrum from equation (9.31) becomes

$$S_I(\nu) = \int_{-\infty}^{+\infty} [I_R(x) - I_R(\infty)]\, \mathrm{rect}(x)\, e^{-i2\pi\nu x}\, dx. \tag{9.33}$$

Taking the Fourier transforms of $S_I(\nu)$ and $S(\nu)$ given by equations (9.33) and (9.30), respectively:

$$\mathbf{F}[S_I(\nu)] = [I_R(x) - I_R(\infty)]\ \mathrm{rect}(x) \tag{9.34}$$

$$\mathbf{F}[S(\nu)] = [I_R(x) - I_R(\infty)] \tag{9.35}$$

Let function $G(\nu)$ be another function defined such that its Fourier transform is the rectangular function, i.e.,

$$\mathbf{F}[G(\nu)] \equiv \mathrm{rect}(x) \tag{9.36}$$

Using equation (9.36) in equation (9.34) one gets

$$\mathbf{F}[S_I(\nu)] = \mathbf{F}[S(\nu)]\ \mathbf{F}[G(\nu)] \tag{9.37}$$

Applying the convolution theorem here, which states that product of the Fourier transforms of two functions is equal to the Fourier transform of the convolution of two functions, then equation (9.37) can be written as:

$$\mathbf{F}\,[S_I(\nu)] = \mathbf{F}\,[S(\nu)*G(\nu)] \text{ or } S_I(\nu) = S(\nu)*G(\nu) \tag{9.38}$$

Using equations (9.32) and (9.36), the Fourier transform $G(\nu)$ of the rectangular function is given by:

$$G(\nu) = \int_{-L}^{+L} e^{-i2\pi\nu x}\,dx = 2L\sin(2\pi\nu L)/(2\pi\nu L)$$
$$= 2L\,\text{sinc}(2\pi\nu L) \tag{9.39}$$

which is the instrument lineshape function (ILS).
Therefore, $S_I(\nu) = S(\nu)*(\text{ILS})$.

Thus the instrumental spectrum is given by the convolution of the ideal spectrum with the instrumental lineshape function. This can be applied to find the Fourier transform of the truncated sine wave, which is the product of the sine wave of infinite extent and a box function of width $2x_{max}$ and is represented in Figure 9.10.

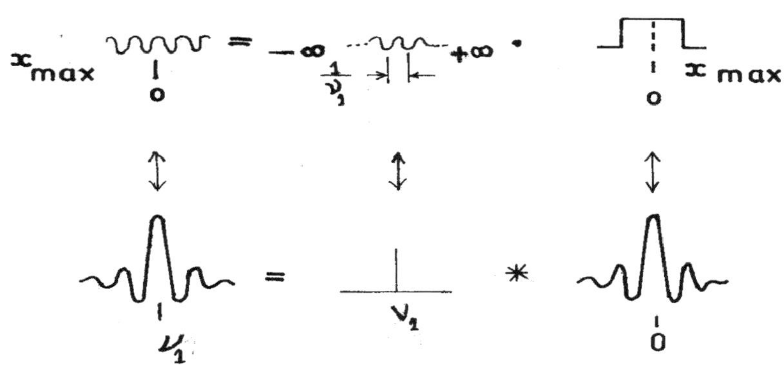

Figure 9.10 Illustration of the convolution theorem to show the effect of truncating an interferogram [17].

The Fourier transform of the infinite sine wave in distance space is just a delta function in frequency space at the value of the sine wave frequency. The Fourier transform of a boxcar function of width x_{max} is a function $\sin(2\pi x_{max}\nu)/(2\pi x_{max}\nu)$. This is by definition just $\text{sinc}(2\pi x_{max}\nu)$. The convolution of $\text{sinc}(2\pi x_{max}\nu)$ with the delta function is the mathematical equivalent of allowing the sinc function, to sweep over the line to generate a new function. This gives another sinc function, which is the Fourier transform of the truncated sine wave (Figure 9.10).

Figure 9.11 shows the instrument line function resulting from a triangular apodizing function using the convolution theorem.

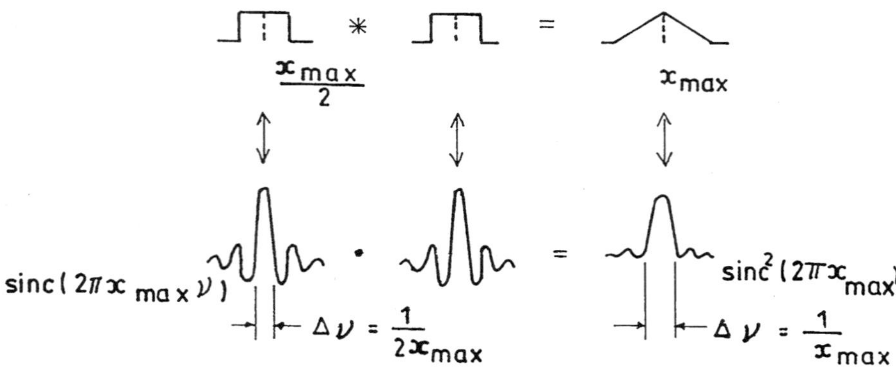

Figure 9.11 Instrumental line function resulting from a triangular apodization function using the convolution theorem [17].

9.3.4.2 Phase Information and Phase Correction

The phase information of interferograms is another important aspect of the conversion of interferograms to spectra [11]. The term "phase" refers to the phase of the individual frequencies that make up the interferogram. The phase of the frequencies may be calculated at any point on the interferogram, whereas the phase information is most easily interpreted if it is calculated at the point on the interferogram where the central maximum occurs. When calculated in this manner, the phase is said to be referenced to the central maximum. At this point the phase is stationary, i.e., all the frequency components of the interferogram have approximately the same phase and generate the central fringe maximum. The phase components of the Fourier transforms of the interferogram give information about the sample. Kawata et al. [13] developed a new idea of spectral library-searching in which only the phase components of Fourier transforms of both the sample and the reference spectra are used for spectral identification. This method has a high discrimination ability for distinguishing between similar spectra, and is very resistant to peak height variation and peak position shift due to experimental conditions. Figure 9.12 shows the spectra of toluene and benzene reconstructed from phase-only Fourier representation.

An ideal interferogram should be symmetric about the point of zero path length difference and on digitization should contain a data point at the peak of the interferogram. However, in general the interferogram is not symmetric around the strong peak at zero path difference because all the frequency components of the wave do not reach their maxima at exactly the same time. This may be due to improper construction of the beam splitter, which causes short wavelengths to be advanced or retarded with respect to long wavelengths and frequency-dependent delay in phase introduced by the detector. A correction for this asymmetry is obtained by a linear function that gives the relationship between the phase shift and the frequency [17].

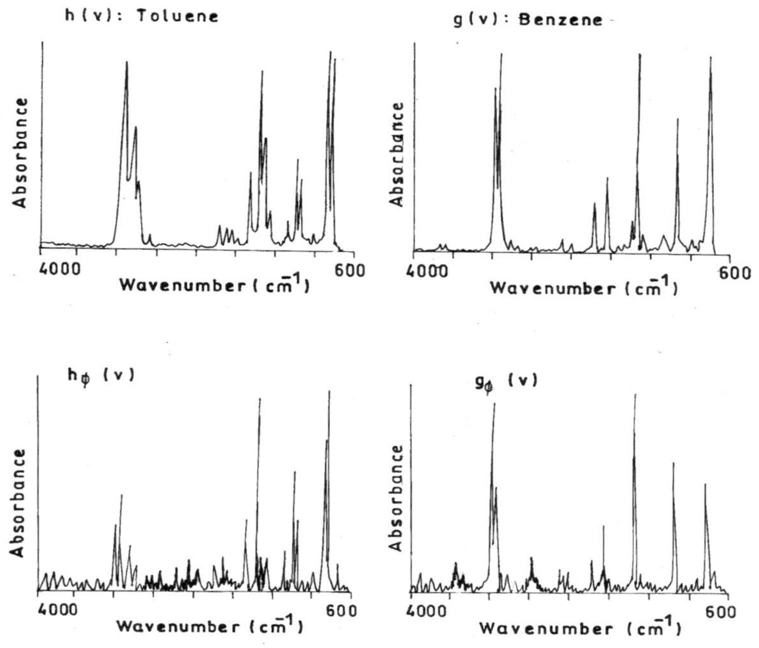

Figure 9.12 Spectra of toluene and benzene reconstructed from the phase-only Fourier representation [13].

9.3.4.3 Fourier Self-Deconvolution

The interpretation of spectral data is often impeded by the effect of overlapping bands. Earlier the mathematical technique known as *curve-fitting* was used for separation of overlapping bands. The scope and limitations of this technique were discussed in detail by Maddams [45]. Nowadays, *Fourier self-deconvolution* (FSD) is coming into increasing use as a method for band sharpening and a means for peak-finding in systems of overlapping infrared bands whose widths are significantly greater than the instrumental resolutions. This is the predominant case in a solution phase spectra, where bands are broadened because of intermolecular interactions and exhibit Lorentzian contours [46]. The method was put on a sound basis by Kauppinen and coworkers [15].

Application of FSD involves measurement of a spectrum followed by Fourier transformation of the region of interest back to the interferogram stage. If the original spectrum had bands of Lorentzian contour, the corresponding interferogram would be a sum of exponentially decaying cosine waves. The exponential decay is removed by multiplying the sum with an increasing exponential function, thus producing an interferogram of more cosinusoidal components. Fourier transform at this point is performed again, which regenerates the original spectrum but with significantly narrow bandwidths. This effectively resolves some overlapping features, although the signal-to-noise ratio is degraded by this process [44]. FSD offers some important advantages over other resolution enhancement techniques. The derivative technique, although effective, produces artifacts of opposite sign relative to the true features. The curve-fitting technique requires extensive user input and can produce highly subjective results. FSD, on the other hand, requires minimal user input. The two parameters that must be input, are bandwidth (full width at half maximum) and a parameter related to the factor by which bandwidths are to be narrowed. The optimal values for these parameters has to be determined since a bandwidth set too large will produce negative features, while a value of the factor by which bandwidths are to be reduced, if set too large, will produce excessive noise. Figure 9.13 shows the application of the technique to resolve the two overlapping bands.

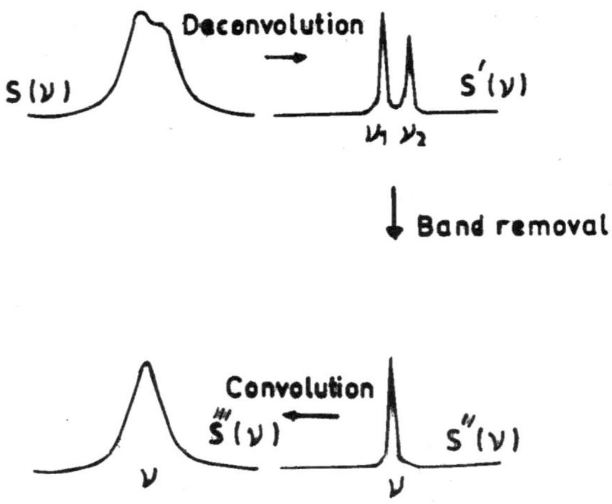

Figure 9.13 Illustration of band separation by deconvolution [15].

9.4 APPLICATIONS

Applications of FT-IR spectrometers are tremendous in almost every field, such as industry [47–51], research [52–55], and medical science [56–58]. Breakthroughs in analytical spectroscopic instrumentation in the last few decades have dramatically boosted its qualitative and quantitative analytical potential. The joint use of the FT-IR spectroscopic technique and suitable software for calculations allows one to fully exploit the vast amount of information contained in the IR spectra for organic substances with a view to the simultaneous resolution of complex mixtures. The widest use of FT-IR spectrometers is mainly in industry because here they are able to provide useful information in both time and frequency domains. Therefore, large amounts of data can be analyzed with high precision in a very small time. Nowadays microbeam FT-IR spectrometers allow spectra from a very small sample (a few nanograms) of material to be obtained quickly with little sample preparation. For many companies this is important economically since it results in more data at a lower cost for some materials; thin films of residue are identified with sensitivity that is highly competitive with electron- or ion-based surface analysis techniques. This is mainly achieved by measuring the attenuated total reflection from the material's surface. Areas as small as 10–15 microns can be analyzed. The technique has also been used to measure surface temperature, which is independent of material emissivity, surrounding radiation sources, and instrument calibration, as shown by Markhan et al. [49]. This use has applications in power generation, metal processing, chemical production, and many other areas. Many companies use FT-IR as an aid in the precise determination of the chemical identity of organic contamination in the disk drive, and also in biomedical and semiconductor industries. FT-IR spectroscopy can be used as a rapid and nondestructive technique for determination of constituents in many samples. The technique is a fast and cost effective tool for routine monitoring of multiple constituents such as sucrose, glucose, fructose, citric acid, etc. in the production of common juices, for example [59, 60]. Very recently, Van de Voort et al. [61] developed a practical FT-IR method to determine the moisture contents in lubricants, which has advantages in terms of sample size, speed, and the variety of oil types that can be handled. This technique has also been employed by Stewart [62] to study plant tissues. This allows the study of localized changes in cell wall composition and structure of individual cells and comparison of those with different and distant tissues. This was not possible with other spectroscopic techniques. Though there is a long list of areas where FT-IR spectrometers can be used (in identification of polymers, to confirm composition of raw materials, to verify compound composition, in surface analysis, in micro-contamination identification, in forensic departments, for quality control screening, in biomedical science, in semiconductor industry, etc.), yet the limiting applications of FT-IR spectrometers will be discussed which are of current interest. These will include studies made on atmospheric pollution, planetary atmospheres, optical fibers, as well as surface studies, and the study of polymers and bio-molecules.

9.4.1 Atmospheric Pollution

Atmospheric pollution is one of the greatest problems worldwide due to increase in the number of industries, oil refineries, vehicles, and so on. Atmospheric pollutants are very destructive. These damage crops and buildings and cause serious health problems to inhabitants. Therefore, it is essential to monitor these precisely. As the concentration of the air pollutants is very small—of ~ ppm (parts per million) level or even less such as of ~ ppb (parts per billion)—so a very sensitive method is required to record their presence in air. The advantages of FT-IR spectroscopy can be utilized to take the spectrum of polluted air (near factories or ponds, etc.), which can show the presence of pollutants even if their concentration is of the order of ppm or less. The fast and the intensive development of FT-IR spectroscopic technique has propelled the progress of trace gas analysis of the atmosphere. Basick et al. [10] reviewed the results of the most significant contributions in this field by FT-IR spectroscopy. The techniques reviewed include *in situ* IR absorption measurements over open paths in the field, such as remote sensing using the sun, the sky or natural hot objects as IR source of radiation and also IR emission measurements of hot trace gas sources, e.g., stack emissions, exhaust gases of combustion sources, and other industrial effluents.

The measurements of gaseous air pollutants by infrared absorption has been possible only through the use of multiple reflection long path cells or by some remote sensing instruments. The detectors used should be highly sensitive and nitrogen-cooled. The fundamental problem in investigating gaseous atmospheric pollutants is the detection of small concentrations at partial pressures as low as 10^{-8} atm (or lower). Therefore, it is necessary that these have strong infrared bands that do not overlap with water or CO_2 absorption bands. It becomes essential to concentrate the pollutant first by a separation technique that separates it from nitrogen, oxygen, water vapor, and CO_2. Hanst and coworkers [63] developed a separation technique operating on the principle of vapor pressure differences. At low temperatures, the vapor pressure of pollutants are lower than the vapor pressure of nitrogen and oxygen. Cryogenic condensations followed by distillation can be used for concentrating the pollutant molecules. CO_2 can be absorbed chemically. Trace gases in the atmosphere have been measured at 10^{-11} atm using this concentration technique along with long-path infrared spectroscopy. The cell used by Hanst and coworkers and the spectrum of the polluted air [64] are shown in Figures 9.14 and 9.15 respectively.

The spectrum between 700–1200 cm^{-1} shows identifiable bands of eleven species of pollutants in addition to the naturally occurring nitrous oxide. Here the spectrum subtraction routines are applied to obtain the presence of weaker bands, which is a characteristic of Fourier transform spectrometers. In the upper spectrum, the dominant bands of acetylene, trichlorofluoromethane, dichlorodifluoromethane, ethylene, and nitrous oxide have been subtracted so

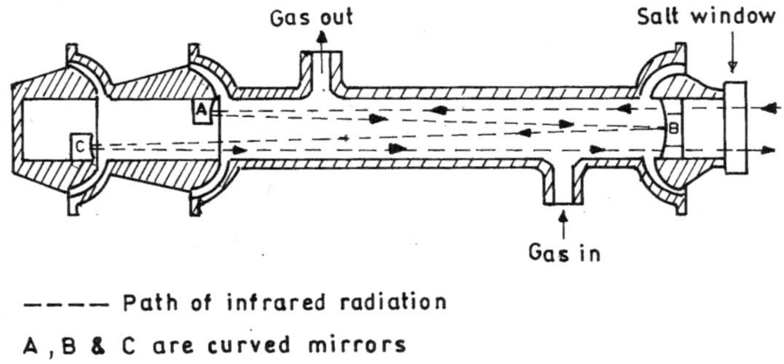

Figure 9.14 Multiple-pass cell with ball joint mounting of mirrors [64].

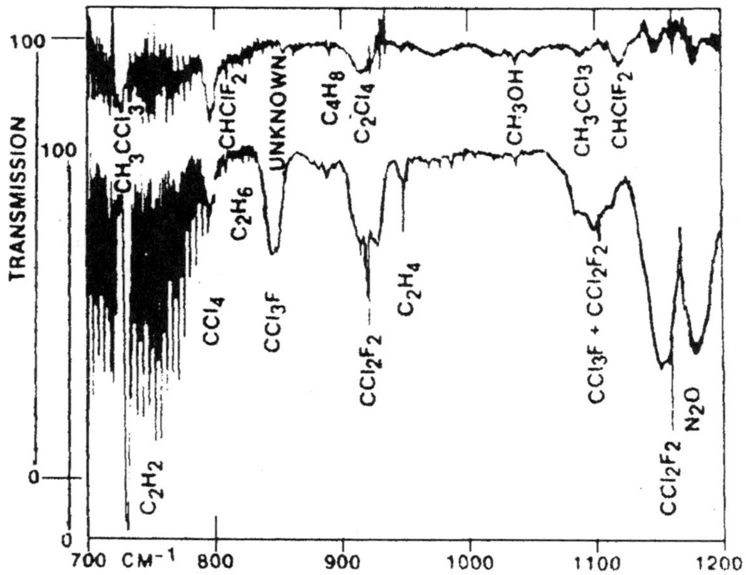

Figure 9.15 Spectra of pollutants recorded utilizing the cryogenic concentration technique [64].

that weaker bands of other pollutants could be measured which otherwise were hidden in the strong bands of other pollutants.

Nowadays, with the advancement in instrumentation, the scenario is quite different. Hong and Cho [65] used *open-path FT-IR* (OP-FT-IR) spectrometry, which improves the temporal and spatial resolution in air pollutant measurement over the conventional methods. However, a successful OP-FT-IR operation requires an experienced analyst to resolve chemical interference as well as derive a suitable background spectrum. Recently Kraft et al. [66] developed an FT-IR sensor system for the continuous determination of a range of environmentally relevant volatile organic compounds in sea water. They developed a prototype of a robust, miniaturized FT-IR spectrometer for *in situ* underwater pollution monitoring. The assembled instrument is enclosed in a sealed pressure vessel and capable of operation in an oceanic environment down to depths of at least 300 m. They detected a range of chlorinated hydrocarbons and monocyclic aromatic hydrocarbons in sea water down to a concentration level of ppb.

9.4.2 Study of Planetary Atmosphere

Planetary atmospheres consist of chemical systems that can be observed remotely or directly sampled with entry probes. Earlier it was possible only to study the atmosphere of our planet earth with the spectroscopic techniques. With the advent of FT-IR spectroscopy, it is possible to explore the atmosphere of every planet, satellite, or asteroid. FT-IR spectrometers have been shown to offer significant improvements over classical dispersive methods in the study of infrared spectral features in astronomical measurements [67]. These improvements result from the three well known advantages of the technique, i.e., spectral multiplexing, high throughput, and significantly improved wavenumber scale.

Fourier transform infrared spectrometers used for astronomical observations differ in some important ways from their laboratory counterparts. The most evident difference is seen in the need to interface the spectrometer to the telescope optics and to minimize its response to the thermal radiation background and to the fluctuations in atmospheric transmission. The spectroscopist in the laboratory can change the brightness of the source or repeat the experiment a number of times. But an astronomer has to complete the experiment in limited time and there is no opportunity to repeat the experiment. Thus the astronomical Fourier spectroscopist pays stress on high optical efficiency, optimal conversion of observing time into integration time, and high levels of stability and reliability of all components of the spectrometer. Although the spectra of Venus were recorded by Connes [21] in 1966 and still earlier in 1962 by grating spectrometers, the real analysis could only be made in 1965 with the advancement of the technique and availability of computers [68]. A section of spectra of Venus observed with various instruments by different scientists is shown for comparison (Figure 9.16).

It is clear from Figure 9.16 that many weak lines between the major CO_2 lines are also visible. A comparison of Figure 9.16a with Figure 9.16d shows the advantages (multiplex) of FT-IR spectrometers as both spectra had been recorded using the PbS detector with approximately the same detector sensitivity. Further studies of FT-IR spectra of Venus showed the presence of water. Figure 9.17 shows a section of the Venus spectra obtained by an improved interferometer having resolution 8 cm^{-1}, carried by flights of CV999 by National Aeronautics and Space Administration (NASA). These spectra were analyzed using the technique of curve-fitting [68]. The spectrum shows several CO_2 bands marked on the Figure 9.17 and the 1.9-μm water vapor band in Venus's atmosphere.

The atmospheric spectra of all the planets have been recorded. The studies can be broadly divided into two categories, viz., that of terrestrial planets which comprise of Mercury, Venus, Mars, and the moon (which defines earth's atmosphere) and major planets, which include Jupiter, Saturn, Uranus, and Neptune. The difference between the terrestrial planets and the major planets is that of densities of the materials present on the two groups of planets. The density of the terrestrial planets is due to a mixture of rocky materials (silicates) and the density of the major planets is due to the presence of light elements like hydrogen and helium. Comparative spectra of the four terrestrial planets and the major planets are shown in Figures 9.18 and 9.19, respectively. Strong methane absorptions are observed in the spectrum of the major planets. A laboratory spectrum of CH_4 is also shown in Figure 9.19e. The spectrum of Jupiter has additional absorptions due to presence of NH_3 gas at 1.5 μm.

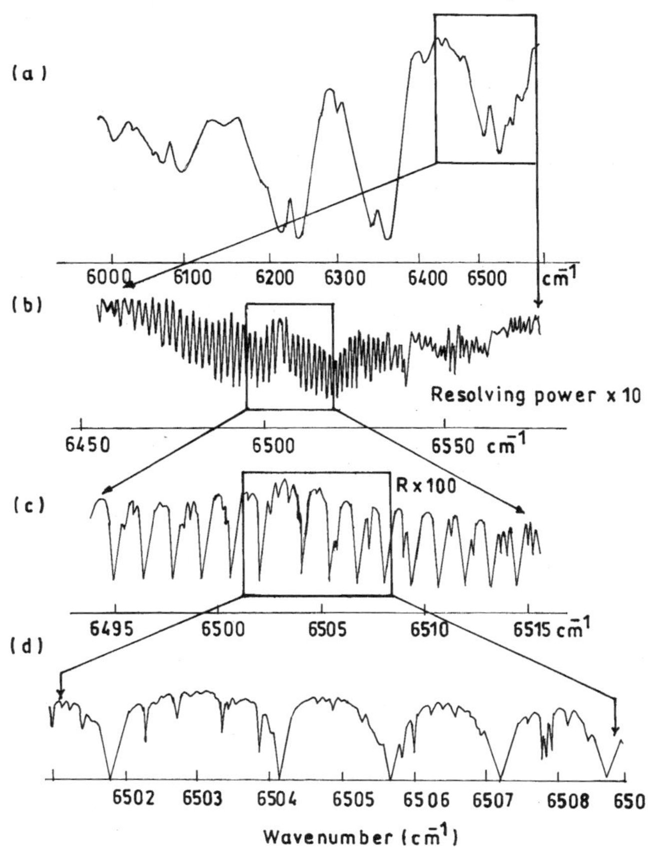

Figure 9.16 Sections of Venus spectra (a) by grating, (b) the first FT-IR spectra by Connes and Connes (1966), (c) an excerpt of the Connes et al. (1969) and (d) spectra recorded by Connes and Michal (1975) [68].

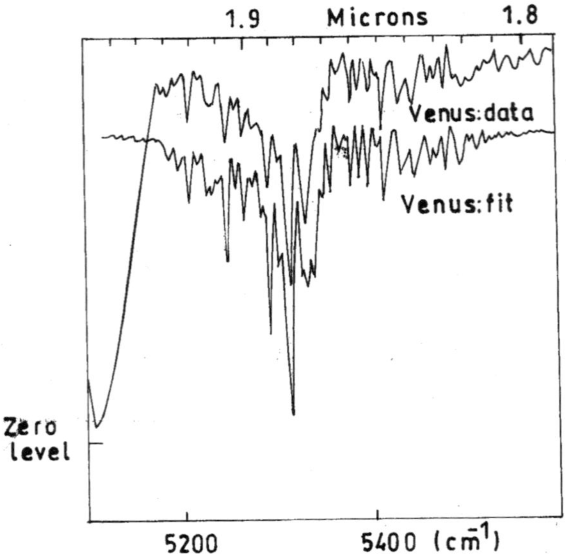

Figure 9.17 A section of the Venus spectra at resolution 8 cm^{-1} [68].

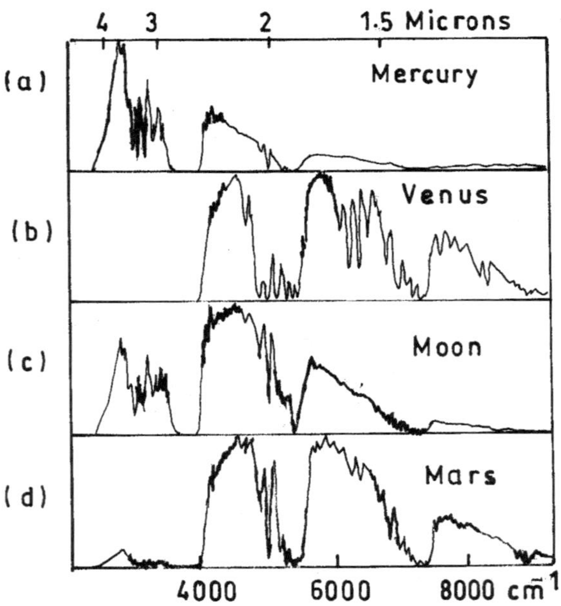

Figure 9.18 Comparative spectra of the terrestrial planets [68].

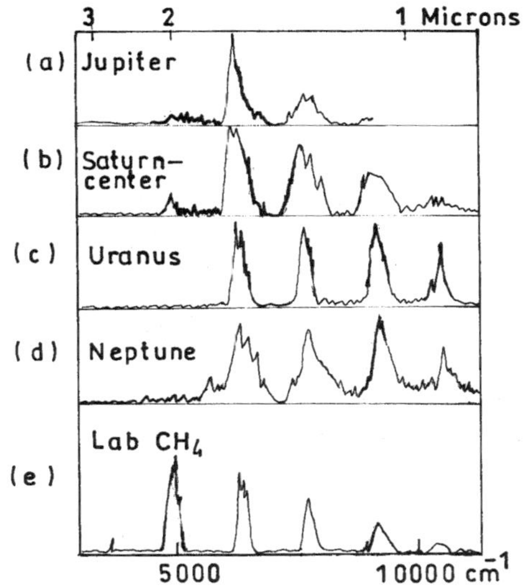

Figure 9.19 Comparative spectra of the major planets and CH_4 [68].

Recently, the planetary Fourier spectrometer on the European Space Agency's Mars express spacecraft was used to investigate the chemical composition of the Martian atmosphere. It is capable of measuring the distribution of the major gaseous components of the atmosphere, the vertical distribution of temperature, and the pressure and determining their variation and global circulation during the different Martian seasons.

9.4.3 Surface Studies

These days, much attention has been paid to studies of thin films deposited on solid substrates for microelectronics applications [69]. This interest largely lies in the potential applications of ordered nanostructures in a number of fields such as molecular electronics, nonlinear optics, and sensors. The FT-IR spectrometer allows us to handle data in a variety of ways. The study of thin films becomes more reliable and efficient by FT-IR spectrometers than by any other technique. Vibrational spectroscopy by FT-IR instruments in particular has been applied successfully to such films deposited on a flat metal substrate using reflection-absorption spectroscopy (RAS) and attenuated total reflectance (ATR). Sometimes, a transmission spectrum is needed if detailed information about a molecular spectrum is required [70]. ATR and RAS give a quasi-absorption spectrum which is not identical but is similar to a transmission spectrum, the interpretation of results may be much easier than in the case for other reflection techniques, which give dispersive spectra [69]. In addition, these techniques facilitate sample preparation and an enhancement is often observed with respect to transmission under proper conditions.

As the quality of spectra has improved with the progress of instrumentation, it has become possible for one to observe the reflection spectra for even a single monolayer on dielectric materials; however, interpretation of the spectra becomes difficult due to optical effect. It has been shown that it is possible to determine the thickness of the thin surface layers by FT-IR/ATR spectroscopy. A method was proposed by Ohta and Iwamoto [71] to give the relation between thickness of the probed surface layer and absorbance in FT-IR/ATR spectroscopy.

9.4.4 Characterization of Optical Fibers

Optical fibers are becoming increasingly important in the areas of data communication and sensing as advantage is taken of their low attenuations, which result in little signal loss over long distances of fibers. However, the fiber attenuations are wavelength-dependent and there are spectral regions where the fibers have lower attenuations as well as areas where the light losses are very severe. In order for one to select the best regions for transmission, the fiber must be well characterized spectroscopically [72]. The FT-IR spectrometer offers considerable attraction as an instrument for the characterization of optical fibers.

Lowry et al. [73] have developed an accessory for FT-IR spectroscopy to measure the spectral response from various types of optical fiber cables. The study has proved to be important for use in fiber-optic sensor-based applications. This work has made extensive use of the flexibility of the FT-IR spectrometer, and easily changed optical components to allow for rapid measurements without changing the position of the optical fiber under investigation.

In recent years optical fiber-based biosensors have been developed and are used in biomedicine and biotechnology [74]. The measurement of amounts of biomolecules such as proteins in solutions is one of the numerous applications of these biosensors. Most recently Kraft et al. [66] and Steiner et al. [75] used fiber-optic cables to develop a fiber-optic sensor system, which can be used to measure sea water pollution with concentration of the order of ppb by FT-IR spectrometers. Sol-gel coated fiber-optic sensors have also been developed and the spectroscopic properties of this important class of fibers have been monitored by FT-IR technology [76].

9.4.5 Vibrational Analysis of Molecules

The absorption of IR radiation is associated with the vibration of chemical bonds, therefore, the absorption spectrum contains a wealth of information about the vibrational levels of the molecule. Infrared spectra of n-alkanes and branched alkanes have interested a number of workers during the last few

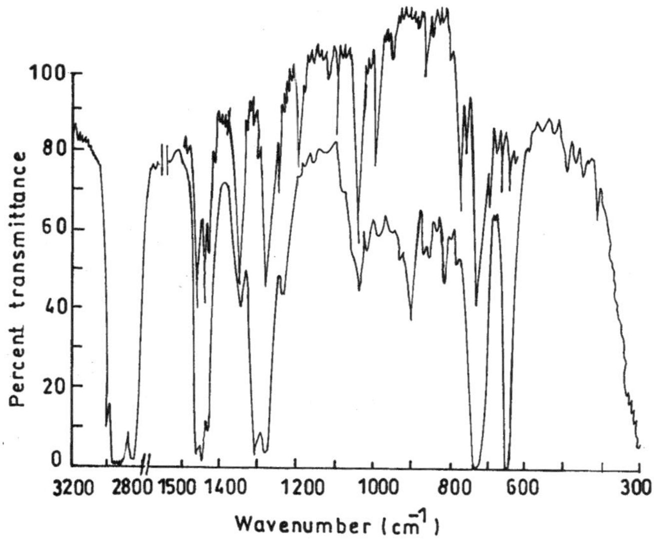

Figure 9.20 Infrared spectra of 1,6-dichlorohexane. Lower curve, liquid; upper curve, annealed solid at about 80K.

decades [77–79]. As n-alkanes and branched alkanes are generally liquids at ambient temperatures, they are expected to exist in many conformations; therefore, their vibrational (infrared and Raman) spectra in the liquid phase are a superposition of the spectra due to all the conformations. However, when the spectrum is recorded at subambient temperatures (at which these substances crystallize) it contains information for only one conformation—the most stable. The FT-IR spectrometer is a powerful tool for recording the infrared spectra of liquids at ambient temperatures and crystalline solids at subambient temperatures. This method has been used to make a complete assignment of observed bands to definite modes of vibrations in different conformations of some alkanes by normal coordinate calculations [80–82]. Jaggi and Jaiswal [83–87] have also carried out the vibrational analysis of some chloro-substituted alkanes. Figure 9.20 shows the infrared spectra recorded by the FT-IR spectrometer of 1,6-dichlorohexane in liquid as well as in crystalline phase.

9.4.6 Study of Bio-Molecules

In the analysis of bio-molecules such as proteins, the region of the spectrum that gives the greatest amount of information on structure and composition is usually the biological fingerprint region from 2000 cm^{-1} to 900 cm^{-1}. This region can be very well studied with FT-IR spectrometers. The fact that most proteins in their native state are in aqueous solution raises the problem of resolving the weaker protein absorption bands from the stronger water absorption bands. One method of avoiding this problem is to use D_2O as the solvent instead of water. However, this defeats the original objective of observing the protein in native environment. Powell et al. [88] developed an important algorithm for the spectral subtraction of water from the FT-IR spectra of proteins in dilute solutions and observed monolayers, which is a very useful way to study biological molecules.

During recent years, FT-IR spectroscopy has been applied to a large number of problems in clinical chemistry. Due to improvements in spectrometric equipment and efficient data evaluation and software, more accurate and reliable analytical results are now obtained, compared to earlier studies. Heise et al. [89] gave an analytical multicomponent method for the blood substrates total protein, glucose, total cholesterol, triglycerides, and urea in human plasma by FT-IR spectroscopy. The obvious advantages are that, for this multicomponent method, no reagents apart from rinsing and cleaning agents are necessary. In addition, this method could be used for monitoring purposes.

Breast cancer is a worldwide and common disease whose fatal outcome can be prevented only by early diagnosis. Infrared techniques such as thermography and transmission imaging, which are based on infrared radiation emission and absorption between normal and diseased breast tissue, have already been used as clinical diagnostic tools for breast cancer. Wallon et al. [90] have applied near infrared spectroscopy to detect the presence of cancer cells in breast tissue in correlation with the pathological diagnosis. Zhang et al. [91] have recently applied cluster analysis and artificial natural networks (ANN) to the automated assessment of a disease state by Fourier transform infrared microscopic imaging measurements of normal and carcinomatous human breast cells.

Sowa and Mantsch [92] have utilized rapid scan and step-scan-based FT-IR photoacoustic techniques for depth profiling of an extracted but intact human tooth. As photoacoustic detection is not directly affected by light scattering or high optical sample density, the mineral phase of intact calcified tissues can be nondestructively characterized. In addition, photoacoustic depth profiling can potentially provide information on the interaction between the mineral phase and the underlying protein matrix during tissue mineralization.

9.4.7 Study of Polymers

Polymers are present in our every day life, having different morphologies such as films, powders, or fibers [93]. Koeing described numerous applications of FT-IR spectroscopy to study a variety of polymer systems. O'Neil and Fateley [94] reviewed the achievements of Koeing in the field of polymer science. Koeing has been a remarkable leader in the field of polymer systems. His innovative use of various modern spectroscopic techniques including FT-IR has revolutionized the understanding of the characterization of polymer systems. The research has greatly influenced work in basic structural measurements of newly discovered polymers to quality control measurements of polymers. He had also carried out the vibrational analysis of polymers by recording the vibrational spectra of polymers. In the determination of structure of various polymers, FT-IR technique has been a very useful analytical method. But with the development of FT-IR microspectroscopy, detailed analysis of the spatial distribution of chemical species at the fiber-matrix interphase can be performed. Recently Arvanitopoulos and Koeing [95] applied this technique to study the interphase of glass fiber-epoxy composites.

Polyethylene terphthalate (PET) is one of the most widely utilized industrial polymer materials. Detailed knowledge about structures and physical properties of the polymer is indispensable for its application to industrial products such as base films of magnetic tapes, plastic bottles, synthetic fibers, and so on. Investigation of the surface structure of PET film has been conducted with an attenuated total reflection method [96] and by Sonoyama et al. [33] by dynamic step-scan two-dimensional Fourier transform infrared studies. Very recently Kandilioti et al. [97] made the conformational analysis of this polymer.

REFERENCES

[1] Ghatak, A.K., Goyal, I.C. & Chua, S.J. (1995) *Mathematical Physics* New Delhi: MacMillan Academic Press.

[2] Harper, C. (1993) *Introduction to Mathematical Physics* New Delhi: Prentice-Hall of India Private Ltd.

[3] Arfken, G. (1985) *Mathematical Methods for Physicists* Orlando: Academic Press, Inc.

[4] Lee, J.P. & Comisarow, M.B. (1987) *Appl. Spectrosc.* **41**, 93-98.

[5] Steward, E.G. (1983) *Fourier Optics: An Introduction* New York: John Wiley & Sons.

[6] Banwell, C.N. & McCash, E.M. (1999) *Fundamentals of Molecular Specroscopy* New Delhi: Tata McGraw-Hill Publishing Co. Ltd.

[7] Guelachvili, G. (1981) *Spectroscopic Techniques*; G. A. Vanasse; Volume **II** New York: Academic Press.

[8] Berry, A., www.FT-IR.htm

[9] www. ericweisstein.com

[10] Bacsik, Z., Mink, J. & Keresztury, G. (2004) *Appl. Spectrosc. Reviews* **39**, 295-363.

[11] Codding, E.G. & Horlick, G. (1973) *Appl. Spectrosc.* **27**, 85-92.

[12] Lipp, E.D. (1986) *Appl. Spectrosc.* **40**, 1009-1011.

[13] Kawata, S., Noda, T. & Minami, S. (1987) *Appl. Spectrosc.* **41**, 1176-1182.

[14] James, D.I., Maddams, W.F. & Tooke, P.B. (1987) *Appl. Spectrosc.* **41**, 1362-1370.

[15] Kauppinen, J.K., Moffatt, D.J., Mantsch, H.H. & Cameron, D.G. (1981) *Appl. Spectrosc.* **35**, 271-276.

[16] Aruldhas, G. (2004) *Molecular Structure and Spectroscopy* New Delhi, Prentice- Hall of India Private Ltd.

[17] Green, D.W. & Reedy, G.T. (1978) *Fourier Transform Infrared Spectroscopy Applications to Chemical Systems Vol. 1*, Ferraro, J.R. & Basile; L.J., p18–38 New York, Academic Press.
[18] Grasselli, J. (1987) *Appl. Spectrosc.* **41**, 933-935.
[19] Bell, R.J. (1972) *Introductory Fourier Transform Spectroscopy* New York, Academic Press.
[20] Cooley, J.W. & Tukey, J.W. (1965) *Math. Comput.* **19**, 291.
[21] Connes, J. & Connes, P. (1966) *J. Opt. Soc. Am.* **56**, 896-910.
[22] Connes, P., Connes, J. & Maillard, J.P. (1969) *Atlas des Spectres dans le Proche Infrarouge de Venus, Mars, Jupiter, et Saturne* Paris: Editions des Centre National de Recherche Scientifique.
[23] Jacquinot, P. (1969) *Appl. Optics* **8**, 497-499.
[24] Gibbie, H.A. (1969) *Appl. Optics* **8**, 501-504.
[25] Horlick, G. & Yuen, W.K. (1978) *Appl. Spectrosc.* **32**, 38-46.
[26] Koeing, J.L. (1975) *Appl. Spectrosc.* **29**, 293-308.
[27] Hirschfeld, T. (1976) *Appl. Spectrosc.* **30**, 68-69.
[28] www.photometrices.net
[29] Hoffmann, P. & Knözinger, E. (1987) *Appl. Spectrosc.* **41**, 1303-1306.
[30] Thackeray, P.P.C. (1972) *Laboratory Methods in Infrared Spectroscopy* Miller, R.G.J., Stace, B.C., p 8 London; Heyden and Son Ltd.
[31] Kember, D., Chenery, D.H., Sheppard, N. & Fell, J. (1979) *Spectrochimica Acta.* **35A** 455-459.
[32] Manning, C.J. & Griffiths, P.R. (1997) *Appl. Spectrosc.* **51**, 1092-1099.
[33] Sonoyama, M., Shoda, K., Katagiri, G. & Ishida, H. (1996) *Appl. Spectrosc.* **50**, 377-381.
[34] Fellgett, P. (1958) *J. Phys. Radium* **19***,* 187.
[35] Voigtman, E. & Winefordner, J.D. (1987) *Appl. Spectrosc.* **41**, 1182-1184.
[36] Jacquinot, P. (1954) *Proceedings of the 17th Congress du Gaurs CNRS* (Paris)
[37] Koeing, J.L. (1975) *Appl. Spectrosc.* **29**, 293-308.
[38] Hirschfeld, T. (1985) *Appl. Spectrosc.* **39**, 1086-1087.
[39] Hirschfeld, T. (1986) *Appl. Spectrosc.* **40**, 1239-1240.
[40] Tripp, C.P. & McFarlane, R.A. (1994) *Appl. Spectrosc.* **48**, 1138-1142.
[41] Parry, D.B. & Harris, J.M. (1988) *Appl. Spectrosc.* **42**, 997–1004.
[42] Johnson, S.A., Rinkus, R.M., Diebold, T.C. & Maroni, V.A. (1988) *Appl. Spectrosc.* **42**, 1369-1375.
[43] Sergides, C.A., Chughtai, A.R. & Smith, D.M. (1987) *Appl. Spectrosc.* **41**, 157-160.
[44] Rahmelow, K. & Hübner, W. (1996) *Appl. Spectrosc.* **50**, 795-804.
[45] Maddams, W.F. (1980) *Appl. Spectrosc.* **34**, 245–267.
[46] Bowley, H.J., Collin, S.M.H., Gerrard, D.L., James, D.I., Maddams, W.F., Tooke, P.B. & Wyatt, I.D. (1985) *Appl. Spectrosc.* **39**, 1004-1009.
[47] Notingher, I., Imhof, R.E., Xiao, P. & Pascut, F.C. (2003) *Appl. Spectrosc.* **57**, 1494-1501.
[48] Cran, M.J. & Bigger, S.W. (2003) *Appl. Spectrosc.* **57**, 928-932.
[49] Markham, J.R., Best, P.E. & Solomon, P.R. (1994) *Appl. Spectrosc.* **48**, 265-270.
[50] Iwaoka, T., Tabata, F. & Tsutsumi, S. (1994) *Appl. Spectrosc.* **48**, 818-826.
[51] Lutz, E.T.G., Luinge, H.J., Maas, J.H. van der & Agen, R. van (1994) *Appl. Spectrosc.* **48**, 1021-1025.
[52] Ouyang, H., Sherman, P.J., Paschalis, E.P., Boskey, A.L. & Mendelsihn, R. (2004) *Appl. Spectrosc.* **58**, 1-9.
[53] Thompson, S.E., Foster, N.S., Johnson, T.J., Valentine, N.B. & Amonette, J.E. (2003) *Appl. Spectrosc.* **57**, 893-899.

[54] Horn, B.A., Qiu, J., Owen, N.L. & Feist, W.C. (1994) *Appl. Spectrosc.* **48**, 662-668.

[55] Chirsty, A.A., Nodland, E., Burnham, A.K., Kvalheim, O.M. & Dahl, B. (1994) *Appl. Spectrosc.* **48**, 561-568.

[56] Heberle, J. & Zscherp, C. (1996) *Appl. Spectrosc.* **50**, 588-596.

[57] Lewis, E.N., Gorbach, A.M., Marcott, C. & Levin, I.W. (1996) *Appl. Spectrosc.* **50**, 263-269.

[58] Sowa, M.G. & Mantsch, H.H. (1994) *Appl. Spectrosc.* **48**, 316-319.

[59] Conde-Gallardo, A., Cruz-Orea, A. & Tomas, S.A. (2004) *Appl. Spectrosc.* **58**, 917–921.

[60] Irudayaraj, J. & Tewari, J. (2003) *Appl. Spectrosc.* **57**, 1599-1604.

[61] Van de Voort, F.R., Sedman, J., Yaylayan, V., Laurcnt, C.S. & Mucciardi, C. (2004) *Appl. Spectrosc.* **58**, 193-198.

[62] Stewart, D. (1996) *Appl. Spectrosc.* **50**, 357-381.

[63] Hanst, P.L., Spiller, L.L., Watts, D.M., Spence, J.W. & Miller, M.F. (1975) *J. Air Pollut. Contr. Assoc.* **25**, 1220.

[64] Hanst, P.L. (1978) *Fourier Transform Infrared Spectroscopy Applications to Chemical Systems Vol. 2,* Ferraro, J.R., Basile, L.J. (Eds.), p 88–89 New York, Academic Press.

[65] Hong, D.W. & Cho, S.Y. (2003) *Appl. Spectrosc.* **57**, 299-308.

[66] Kraft, M., Jakusch, M., Karlowatz, M., Katzir, A. & Mizaikoff, B. (2003) *Appl. Spectrosc.* **57**, 591-599.

[67] Mantz, A.W. (1976) *Appl. Spectrosc.* **30**, 459-461.

[68] Fink, U. & Larson, H.P. (1978) *Fourier Transform Infrared Spectroscopy Applications to Chemical Systems Vol. 2,* Ferraro, J. R. Basile; L.J., p 247–253, 259 New York, Academic Press.

[69] Ochiai, S., McClelland, J.F., Kobayashi, K. & Takaoka, K. (1994) *Appl. Spectrosc.* **48**, 1287-1289.

[70] Yamamoto, K. & Ishida, H. (1994) *Appl. Spectrosc.* **48**, 775-787.

[71] Ohta, K. & Iwamoto, R. (1985) *Appl. Spectrosc.* **39**, 418-425.

[72] Fredericks, P.M., Samson, P.J. & Stuart, A.D. (1987) *Appl. Spectrosc.* **39**, 327-329.

[73] Lowry, S., May, T., Bornstein, A., Weissman, Y., Harman, R. & Tugenthaft, I. (1994) *Appl. Spectrosc.* **48**, 852-856.

[74] Poirier, M.A., Lopes, T. & Singh, B.R. (1994) *Appl. Spectrosc.* **48**, 867-870.

[75] Steiner, H., Jakusch, M. & Kraft, M. et al. (2003) *Appl. Spectrosc.* **57**, 607-613.

[76] Katzir, A. & Mizaikoff, B. (2003) *Appl. Spectrosc.* **57**, 823-828.

[77] Schachtschneider, J.H. & Snyder, R.G. (1963) *Spectrochimica Acta.* **19**, 117-168.

[78] Snyder, R.G. & Schachtschneider, J.H. (1965) *Spectrochimica Acta.* **19**, 169-195.

[79] Schachtschneider, J.H. (1964) *Shell Development Co. Tech. Repts*. 231-264.

[80] Jaiswal, R.M.P. & Crowder, G.A. (1983) *Candian J. Spectrosc.* **28**, 160-164.

[81] Crowder, G.A. & Jaiswal, R.M.P. (1983) *J. Mol. Struct.* **102**, 145-164.

[82] Crowder, G.A. & Jaiswal, R.M.P. (1983) *J. Mol. Struct.* **99**, 93-100.

[83] Jaggi, N. & Jaiswal, R.M.P. (2002) *Ind. J.Pure & Appl. Phys.* **40**, 385-392.

[84] Jaggi, N. & Jaiswal, R.M.P. (2002) *Ind. J. Phys.* **76B**, 297-306.

[85] Jaggi, N. & Jaiswal, R.M.P. (2001) *Ind. J. Pure & Appl Phys.* **39**, 123-129.

[86] Jaggi, N. & Jaiswal, R.M.P. (2000) *Ind. J. Phys.* **74B**, 493-496.

[87] Jaggi, N. & Jaiswal, R.M.P. (2000) *Ind. J. Pure & Appl Phys.* **38**, 69-80.

[88] Powell, J.R., Wasacz, F.M. & Jakobsen, R.J. (1986) *Appl. Spectrosc.* **40**, 339-344.

[89] Heise, H.M., Marbach, R., Koschinsky, T.H. & Gries, F.A. (1994) *Appl. Spectrosc.* **48**, 85-95.

[90] Wallon, J.W., Yan, S.H., Tong, J., Meurens, M. & Haot, J. (1994) *Appl. Spectrosc.* **48**, 190-193.

[91] Zhang, L., Small, G.W., Haka, A.S., Kidder, L.H. & Lewis, E.N. (2003) *Appl. Spectrosc.* **57**, 14-22.

[92] Sowa, M.G. & Mantsch, H.H. (1994) *Appl. Spectrosc.* **48**, 316-319.
[93] Papini, M. (1994) *Appl. Spectrosc.* **48**, 472-476.
[94] O'Neil, S.E. & Fateley, W. (1988) *Appl. Spectrosc.* **42**, 1177-1180.
[95] Arvanitopoulos, C.D. & Koeing, J.L. (1996) *Appl. Spectrosc.* **50**, 1-10.
[96] Walls, D. (1991) *Appl. Spectrosc.* **45**, 1113.
[97] Kandilioti, G., Govaris, G.K. & Gregoriou,V.G. (2004) *Appl. Spectrosc.* **58**, 1082-1092.

CHAPTER 10

AUGER ELECTRON SPECTROSCOPY

Richard P. Gunawardane and Christopher R. Arumainayagam
Department of Chemistry, Wellesley College, Wellesley, MA 02481, U.S.A.

10.1 INTRODUCTION

Auger electron spectroscopy (AES) is a nondestructive core-level electron spectroscopy for semi-quantitative determination of the elemental composition of surfaces, thin films, and interfaces. The popularity of this ultrahigh vacuum technique may be attributed to high surface sensitivity (an analysis depth of less than 100 Å) and a relatively low detection limit (~0.1 atomic percent). In addition to having an elemental coverage from lithium to uranium and beyond, AES has the ability to distinguish between two elements that are close to each other in the periodic table. In addition, AES has an atomic number-dependent sensitivity that varies at most by one order of magnitude. AES chemical shifts and line shapes can also yield bonding (chemical state) information, albeit with less precision than is possible with X-ray photoelectron spectroscopy (XPS) (Chapter 11), another core-level electron spectroscopy. Auger electron spectroscopy has a depth resolution of 5–25 Å, and can be used, with simultaneous ion sputtering, for depth profiling. With a lateral resolution (<100 Å) that is significantly better than that of XPS, *scanning Auger microscopy* (SAM) can be used effectively for imaging nanoscale structures and to produce two-dimensional maps of surface elemental composition. Survey Auger spectra typically take less than five minutes, providing for rapid data acquisition. Although somewhat sophisticated and expensive, Auger instrumentation is relatively simple to use and is readily available from many different commercial sources. The reasons enumerated above explain why Auger electron spectroscopy has become perhaps the most widely used surface analytical technique.

Recent developments in AES have expanded the scope of this technique beyond the probing of surface elemental composition. For example, spin polari-zation of Auger electrons can be used to study magnetized solid surfaces [1]. Moreover, results of resonant Auger electron spectroscopy experiments provide information relevant to femtosecond charge transfer dynamics [2]. Auger electron diffraction can also be used to determine surface structure [3]. Finally, results of recent experiments have demonstrated that angle-resolved Auger electron spectroscopy can provide a means to study excitation processes in solids [4].

The Auger process is a three-electron process. When a beam of electrons, typically with an energy range of 3–20 keV, strikes a solid atom, a core-level (inner) electron is ejected producing a singly ionized excited atom. An outer level electron can fill the resulting vacancy in the core level. Following this radiationless transition, the excess energy of the resulting excited state ion may be removed by emitting either (i) an X-ray (the basis for X-ray fluorescence (XRF)/electron microprobe (EMP) analysis) or (ii) another electron from the atom. The emitted electrons in process (ii) are called *Auger electrons*, after Pierre Auger, who discovered this process in the 1920s [5]. Although Lise Meitner independently discovered the effect around the same time [6], she is given very little recognition in the literature. While the emission of X-rays (process (i)) produces singly ionized atoms, the emission of Auger electrons (process (ii)) results in doubly ionized atoms.

Because Auger is a three-electron process, hydrogen and helium cannot be detected by this technique. Although Li has three electrons, an *isolated ground state* Li atom does not yield Auger peaks because the atom has only two energy levels that contain electrons. Auger peaks, however, have been detected from multiply excited Li atoms [7]. The presence of electrons in the valence band of *solid* Li also allows for Auger transitions of the type KVV.

The surface sensitivity of AES is due to the short mean free path of the relatively low energy Auger electrons. Although atomic excitations can take place to a depth of ~10,000 Å below the surface, the Auger electrons from only the uppermost atomic layers, down to a depth of ~100 Å, are ejected from the specimen without undergoing any energy loss. In contrast, electron microprobe analysis, involving the detection of X-ray photons, is more of a *bulk*, rather than a *surface*, analysis tool (Figure 10.1).

Because of the very short lifetime of the electronic states associated with the Auger process, the Auger peaks are relatively wide (typically 1–2 eV), consistent with Heisenberg's uncertainty principle.

The kinetic energies of the Auger electrons are characteristic of each emitting atom. Thus, the measurement of the kinetic energies of Auger electrons can be used to identify the elements present on the surface of the sample. Because the kinetic energies of Auger electrons depend on the binding energies of the electron levels involved in the Auger process, the shifts in these kinetic energies can, in principle, provide useful information on the oxidation states and bonding environment of the surface atoms. In addition to the above qualitative analysis, quantitative information may also be determined from the intensities of the Auger peaks.

A *direct* Auger spectrum is represented as a plot of the number of electrons detected as a function of electron kinetic energy. However, to make the small Auger peaks more prominent, often AES spectra are displayed as the first derivative of the number of electrons emitted as a function of electron

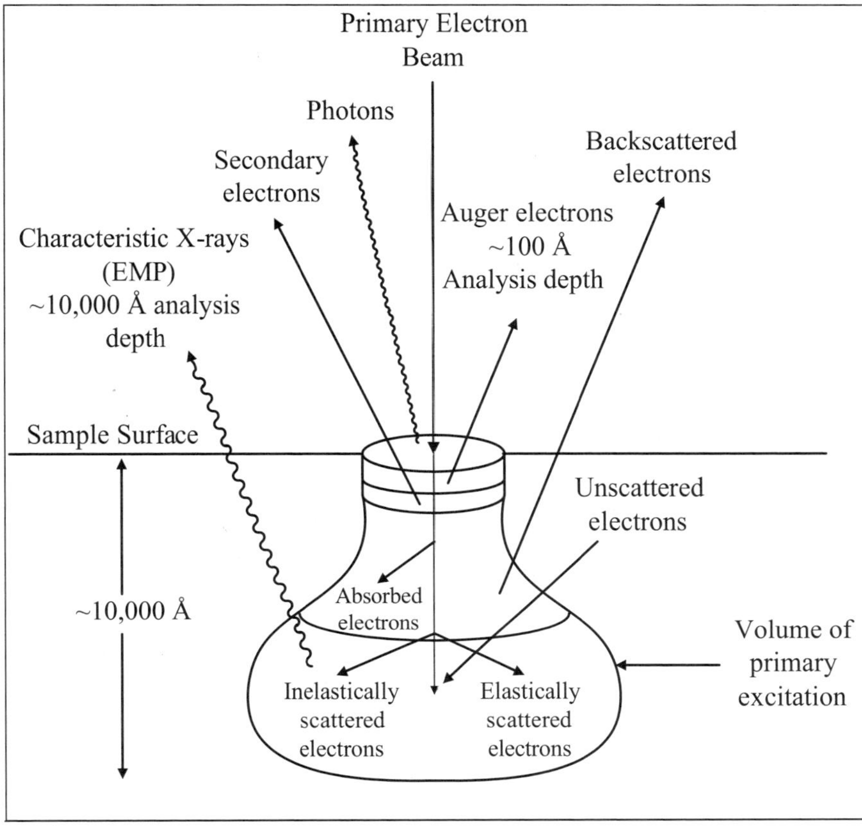

Figure 10.1 Schematic diagram showing the various locations and outcomes of electron-solid interactions (a composite diagram based on various sources) [8, 9].

kinetic energy. The derivative form of the Auger spectrum enhances the Auger peaks and suppresses the background arising from the secondary and backscattered electrons.

The AES analysis is carried out in an *ultrahigh vacuum* (UHV) chamber in which a pressure of $\sim 10^{-9}$ torr or below is continuously maintained. High vacuum is necessary to allow uninterrupted passage of the electron beam and ultrahigh vacuum is necessary to avoid contamination of the surface by atmospheric gases during analysis.

The goal of this chapter is to provide a detailed introduction to Auger electron spectroscopy. Topics covered emphasize physical principles, experimental techniques and procedures, research and industrial applications, and new developments in Auger electron spectroscopy. For a more detailed treatment of the technique and its applications, the reader is referred to a 900-page comprehensive treatise on Auger and X-ray photoelectron spectroscopy published recently [10]. A less extensive text devoted to the above two techniques was published in 2003 [11].

10.2 HISTORICAL PERSPECTIVE

The historical developments of Auger electron spectroscopy have been comprehensively reviewed [12]. Here we provide a brief synopsis. The name "Auger electron spectroscopy" is derived from the effect reported in 1923 by Pierre Auger, a French physicist. Three decades passed before J.J. Lander first applied this phenomenon to solids in 1953 [13]. L.A. Harris demonstrated in 1968 the utility of taking derivatives for plotting AES spectra [14]. Initially, the retarding field analyzer (RFA) used in low energy electron diffraction (LEED) experiments was modified by many research groups to obtain Auger spectra [15]. In 1969 Palmberg and coworkers developed the cylindrical mirror analyzer (CMA) for detecting Auger electrons [16]. The first commercial Auger electron spectrometers also became available in 1969. Scanning Auger microscopy (SAM) was first demonstrated in 1971 by MacDonald and Waldrop [17] and later developed extensively by Prutton and coworkers [18, 19]. Auger depth profiling with noble gas ion sputtering was first demonstrated by Palmberg in 1972 [20]. AES instrumentation has undergone considerable improvement over the years leading to automation with computer control and use of modern software for sophisticated data analysis.

10.3 BASIC PRINCIPLES OF AES

10.3.1 X-Ray Notation

In Auger electron spectroscopy, electron energy states are denoted by using X-ray notation. Because removing an electron from a complete shell is equivalent to placing a single electron in an empty shell, X-ray spectra are similar to one-electron alkali atom spectra. Hence, we first examine the *fine structure* in the optical spectra of alkali atoms. The fine structure, the splitting of lines (with the exception of those due to s-state electrons) in the spectra of alkali atoms into doublets, is due to *spin-orbit coupling*, the interaction of the spin magnetic moment with the magnetic field arising from the orbital angular momentum. Spin-orbit coupling splits non-s energy *terms* in alkali atoms into two *levels*.

We now discuss the terminology used for electronic energy *levels* for light atoms, for which Russell-Saunders coupling (*also called L-S coupling*) is a valid approximation. In the general case, each level is specified by the principle quantum number (n) and a level symbol ($^{2S+1}L_J$). In this symbol, S is the total electronic *spin angular* momentum quantum number, L is the code for the total electronic *orbital angular* momentum quantum number, and J is the total electronic *angular* momentum quantum number. Because X-ray spectra are similar to one-electron alkali atom spectra, the following simplification is made to yield the *XPS notation.* In addition to the principal quantum number (n), the energy level can be specified with the orbital angular momentum (ℓ) and the total angular momentum quantum number (j) of a *single* electron. The spin multiplicity $2S+1$ can be ignored because it is always 2 for a one-electron (one-hole) atom since $S = ½$. The total angular momentum of a single electron is obtained by using the Clebsch-Gordon series:

$$j = \ell + s,\ \ \ell + s - 1, \ldots, \left|\ell - s\right|. \tag{10.1}$$

In the above expression, ℓ is the orbital angular momentum quantum number and s is the spin angular momentum quantum number, which is ½ for all electrons. The general form of the XPS notation is $n\ell_j$.

We illustrate the above discussion with a specific example involving the fine structure of the sodium D line. The excited Na atom electron configuration ($1s^2 2s^2 2p^6 3p$) yields the ^2P term because $L = \ell = 1$ and $S = s = ½$. Using the Clebsch-Gordon series we obtain $J = 1 - ½$ and $1 + ½$, yielding two levels ($^2P_{1/2}$ and $^2P_{3/2}$) for the ^2P term.

The principal quantum number 3 is often omitted. As is often done in X-ray photoelectron spectroscopy, the above two energy levels may be written alternatively as $3p_{1/2}$ and $3p_{3/2}$.

Specifying the energy levels with *Auger notation* involves using the letters K, L, M,… for the principal quantum number 1, 2, 3,… and a subscript that depends on the orbital quantum number (ℓ) and the total angular momentum quantum number (j). For example, the two levels $^2P_{1/2}$ and $^2P_{3/2}$—the energy levels of the excited Na atom—may be written as M_2 and M_3. Note that M_1 is the Auger notation for the energy level of the ground state electron configuration ($1s^2 2s^2 2p^6 3s$) of the Na atom. The level symbol for the ground state electron configuration is $^2S_{½}$ corresponding to the XPS notation of $3s_{½}$. When the energy levels are very close to each other they are not usually resolvable experimentally. These unresolvable energy levels are normally designated with a comma between the subscripts (e.g., $L_{2,3}$ and $M_{4,5}$). We summarize the above discussion in Table 10.1

10.3.2 Auger Transitions

The Auger process for a solid is schematically illustrated in Figure 10.2. The KL_2L_3 Auger transition, illustrated in this diagram, involving ionization, relaxation, and emission, may be visualized as follows:

(1) A core-electron in the atom is removed by the high-energy incident electron creating a vacancy in the K shell and yielding an electronically excited ion (ionization).

(2) An electron from the L_2 level falls down almost immediately in a radiationless transition to fill the vacancy in the K shell (relaxation).

Table 10.1 X-ray notation of electron energy states.

Quantum Numbers			Level Symbol	Auger Notation	XPS Notation
n	ℓ	j			
1	0	1/2	$^2S_{1/2}$	K	$1s_{1/2}$
2	0	1/2	$^2S_{1/2}$	L_1	$2s_{1/2}$
2	1	1/2	$^2P_{1/2}$	L_2	$2p_{1/2}$
2	1	3/2	$^2P_{3/2}$	L_3	$2p_{3/2}$
3	0	1/2	$^2S_{1/2}$	M_1	$3s_{1/2}$
3	1	1/2	$^2P_{1/2}$	M_2	$3p_{1/2}$
3	1	3/2	$^2P_{3/2}$	M_3	$3p_{3/2}$
3	2	3/2	$^2D_{3/2}$	M_4	$3d_{3/2}$
3	2	5/2	$^2D_{5/2}$	M_5	$3d_{5/2}$

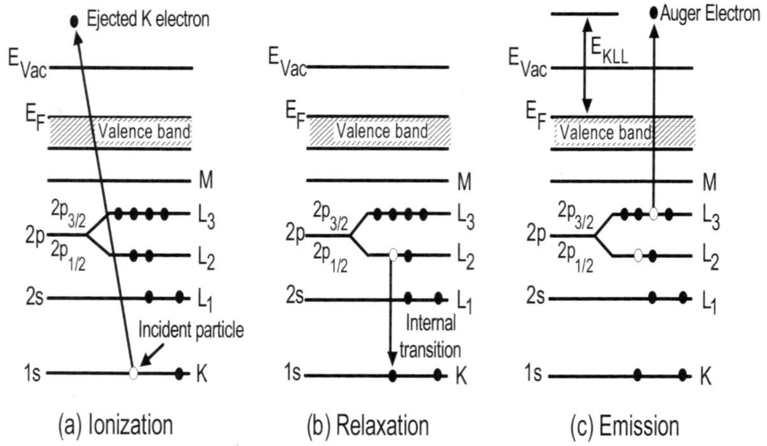

Figure 10.2 Schematic showing the three steps involved in the Auger process. The KL_2L_3 Auger transition is illustrated. The open circles symbolize holes (absence of electrons).

(3) Excess energy of the excited state ion is removed by the ejection of an Auger electron from the L_3 level (emission).

The nomenclature of the Auger transition indicates the energy levels in the order in which they are involved in the whole process. Thus, the transition described above may be designated as KL_2L_3. In the context of Russell-Saunders coupling, there are six KLL transitions corresponding to the three final electron configurations, as shown in Table 10.2.

Table 10.2 KLL Auger transitions corresponding to the different final electron configurations.

Final Electron Configuration	Auger Transition
$2s^0 2p^6$	KL_1L_1
$2s^1 2p^5$	KL_1L_2 KL_1L_3
$2s^2 2p^4$	KL_2L_2 KL_2L_3 KL_3L_3

Although many Auger transitions are available, especially for atoms with high atomic number, most have low probabilities. Some transitions, although energetically allowed, are forbidden due to selection rules. When a valence electron is involved, the letter V is often used (e.g., KLV, KVV, and LMV). The letter C is sometimes used to denote a core level (e.g., CVV). The strongest Auger transitions are of the type ABB (e.g., KLL and LMM). Special transition of the type AAB, commonly termed *Coster-Kronig transitions*, are also very strong.

X-ray emission is a competing process for Auger emission because the energy difference between the core and outer levels can also be released in the form of a characteristic X-ray. The sum of the Auger yield and X-ray emission yield is unity. It has been observed that the probability for X-ray emission is much lower than the Auger emission for the range of energies normally measured in AES [34]. Moreover, for elements with low atomic numbers, the cross-section for the Auger process is much higher than that for emission of X-ray photons.

10.3.3 Kinetic Energies of Auger Electrons

Auger electron spectroscopy involves measurement of the Auger electron's kinetic energies that are characteristic of the elements present in the sample. An uncertainty of 1 to 2 eV in the measurement of the kinetic energy is acceptable for elemental identification. For the purpose of Auger chemical shift analysis, however, the uncertainty in kinetic energy measurements must be reduced to 0.1 to 0.2 eV [21]. Extensive work has been performed to calibrate and standardize Auger electron energies [22–25]. Before the advent of Auger databases, the identification of elements via AES required the computation of Auger electron energies, as summarized below.

The kinetic energy of an Auger electron is the energy difference between the doubly ionized final state and the singly ionized initial state. Because calculating Auger electron energies based on first principles is highly complex, and because errors on the order of ~10 eV are acceptable for most practical purposes, we make several approximations. Let us consider an ABC

Auger transition in which the first electron is ejected creating a hole in the A level, the second electron falls from the B level to the A level, and the third electron (the Auger electron) is ejected from the C level. Let E_A, E_B, and E_C be the binding energies of electrons in A, B, and C levels, respectively, of the neutral atom. The energy released ($E_A - E_B$) when the second electron falls from level B to level A is transferred to the third electron. Hence, the kinetic energy (E_{ABC}) of this Auger electron may be approximated as follows:

$$E_{\mathrm{ABC}} \approx E_{\mathrm{A}} - E_{\mathrm{B}} - E_{\mathrm{C}} \tag{10.2}$$

Equation (10.2) demonstrates that the Auger electron energy is independent of the primary beam energy and is dependent only on the atomic energy levels. Therefore, the measured Auger electron energies are representative of the elemental composition of the sample surface. Because each element has a unique set of energy levels, each element has a unique set of Auger peaks. The KL_1L_3 Auger transition energy for Al, for example, may be calculated as follows:

$$\begin{aligned} E_{\mathrm{KL_1L_3}} &\approx E_{\mathrm{K}} - E_{\mathrm{L_1}} - E_{\mathrm{L_3}} \\ &\approx 1560 - 118 - 73 \\ &\approx 1369\,\mathrm{eV}. \end{aligned}$$

The corresponding measured value is 1354 eV. We discuss two corrections that may account for this discrepancy.

Equation (10.2) must be modified by taking into account the spectrometer work function because Auger energies are typically referenced to the Fermi level. If the sample is in good electrical contact with the sample holder, the Fermi levels of the sample and instrument are identical. Hence, the kinetic energy of the ABC Auger electron may be approximated as follows:

$$E_{\mathrm{ABC}} \approx E_{\mathrm{A}} - E_{\mathrm{B}} - E_{\mathrm{C}} - \phi_A\,. \tag{10.3}$$

In the above expression, φ_A is the work function of the analyzer. Because the typical electron analyzer's work function is approximately 4 eV, equation (10.3) yields a value of 1365 eV for the KL_1L_3 Auger transition energy for Al. Now the discrepancy between the measured and calculated value has been reduced from 15 eV to 11 eV.

Equation (10.2), based on the binding energies of a neutral atom, can be further refined by taking into account the change in binding energy of a level that accompanies the formation of an ion.

$$E_{\mathrm{ABC}} \approx E_{\mathrm{A}} - E_{\mathrm{B}} - E_{\mathrm{C}}^{*} - \phi_{\mathrm{A}}\,. \tag{10.4}$$

In the above expression, E_C^* is the binding energy of a level in the presence of a core hole. Equation (10.4) can be used to estimate Auger electron energies based on various empirical approximations [26]. More sophisticated semi-empirical methods have also been used to perform Auger electron kinetic energy calculations. The highest level of such Auger energy calculations involve relativistic *ab initio* computations, which take into account many body effects [27].

10.4 INSTRUMENTATION

Because an atomically clean surface needs to be maintained throughout the AES analysis process and because the electron beam needs to reach the sample without collisions with the intervening gas molecules, *turbo molecular pumps* (TMP) and/or ion pumps are usually employed to achieve ultrahigh vacuum conditions (10^{-9} torr and below) inside a stainless steel chamber equipped with metal seals. Vacuum locks in modern ultrahigh vacuum equipment allow for the introduction of new samples in less than 30 minutes without breaking the vacuum. Mu metal shielding inside the ultrahigh vacuum chamber may be necessary to minimize the effect of stray magnetic fields on the trajectory of electrons. In order to perform Auger analysis of grain boundaries, samples can be fractured *in situ* in ultrahigh vacuum [28]. Clean room compatible AES instrumentation for semiconductor analysis is now available from several manufacturers. Because the ability to finely focus electron beams allows for excellent spatial resolution, electron beams are the most popular choice in AES, although ionizing sources such as X-rays, ions, and positrons can also be used for Auger analysis.

The basic components of a typical Auger electron spectrometer (Figure 10.3) consist of the following: (1) electron source and electron optical column to form an electron probe onto the specimen surface; (2) an ion optical column for cleaning the sample surface and/or sputtering for depth profiling; (3) an electron energy analyzer; (4) a secondary electron detector and a pulse counter; (5) computer control and data display systems. The first four modules are located within the UHV chamber.

10.4.1 Electron Optical Column

In AES instrumentation, the electron beam from an electron source is focused onto the specimen surface by a suitable optical column. In addition to being monoenergetic, the electron beam used in AES instrumentation should be small in size with high brightness. The sample to be analyzed is irradiated with electrons with energy of 2–10 keV and beam current of 10^{-8} to 10^{-5} A. In the case of scanning Auger microscopy, energies as high as 35 keV and currents as low as 10^{-9} A are used to produce beam diameters as small as 100 Å. The electron source also should have a long life and high temporal stability. There are four main types of electron sources:

1. The tungsten thermionic emitter operates at a temperature of ~2700 K producing a low current density. Because of the low brightness and concurrent large beam size, the tungsten thermionic emitter is not optimal for Auger experiments that require good lateral resolution.
2. The lanthanum hexaboride (LaB_6) thermionic emitter, with a lower work function than tungsten, operates at a lower temperature (~1850 K)

and provides for good lateral resolution because it produces relatively high current density.

3. The cold field emitter, made of a tungsten single crystal, operates at room temperature in the presence of a high electrostatic field and produces high brightness. The cold field emitter is unstable in the presence of residual gases and hence requires pressures on the order of 10^{-10} torr.

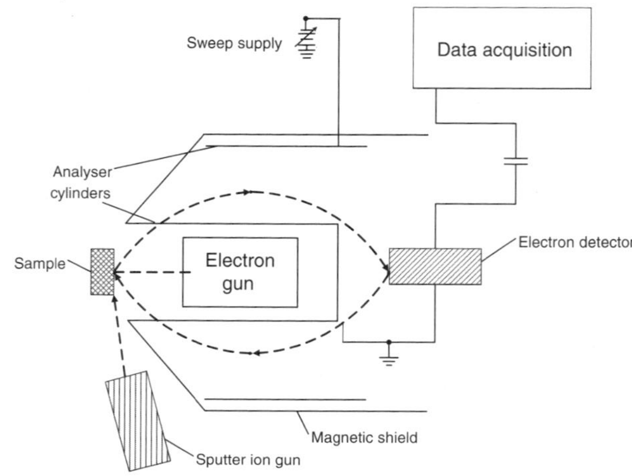

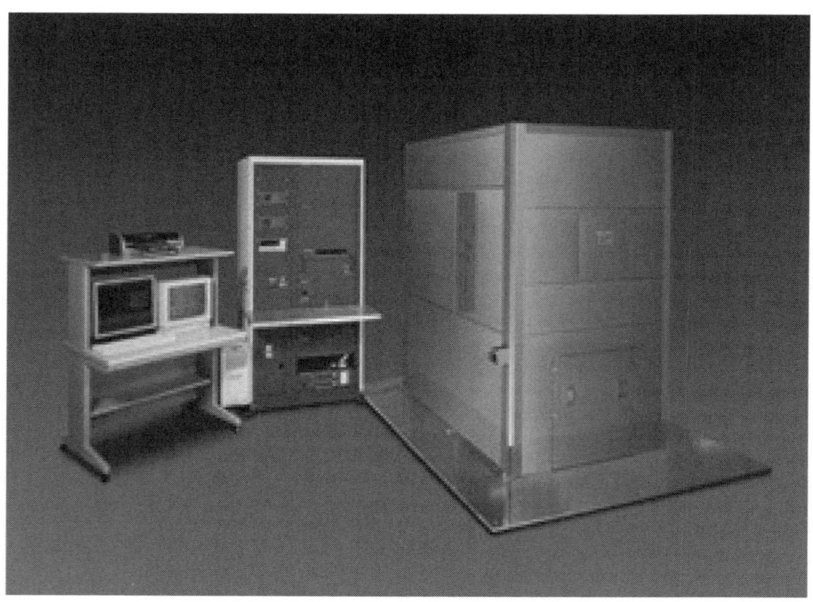

Figure 10.3 Top—A schematic representation of an Auger electron spectroscopy instrument. Bottom—A photograph of a commercial AES instrument (Physical Electronics).

4. The hot field emitter, also known as ZrO_2/W Schottky-type field emitter, operates at ~1800 K and produces high current densities.

Most modern AES instruments use either a LaB_6 or a Schottky type emitter as an electron source. The Schottky-type emitter is the preferred electron source for the highest resolution Auger instruments. A typical electron optical column with an electron source and other essential components is shown in Figure 10.4. It should be noted that some materials are susceptible to electron beam damage and therefore care must be exercised in examining such materials by AES.

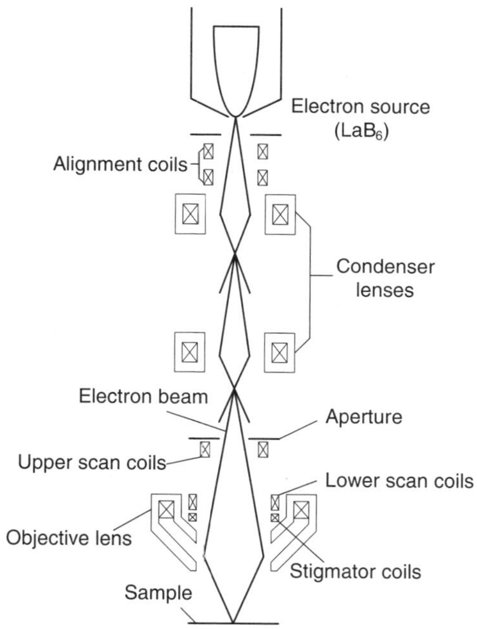

Figure 10.4 Schematic of electron gun used for Auger electron spectroscopy. Adapted from [10].

10.4.2 Ion Optical Column

For depth profiling and/or sputter cleaning, an electron impact-type ion source [16] is usually employed in conjunction with AES instruments. Electrons from a heated filament are accelerated by a cylindrical grid to an energy sufficient to ionize gas atoms by collisions. The resulting ions are accelerated into a focusing lens column. Inert gases such as argon (Ar) or xenon (Xe) are used in a typical electron impact-type ion source with a hot tungsten filament. Figure 10.5 shows a typical ion optical column with an electron impact ion source.

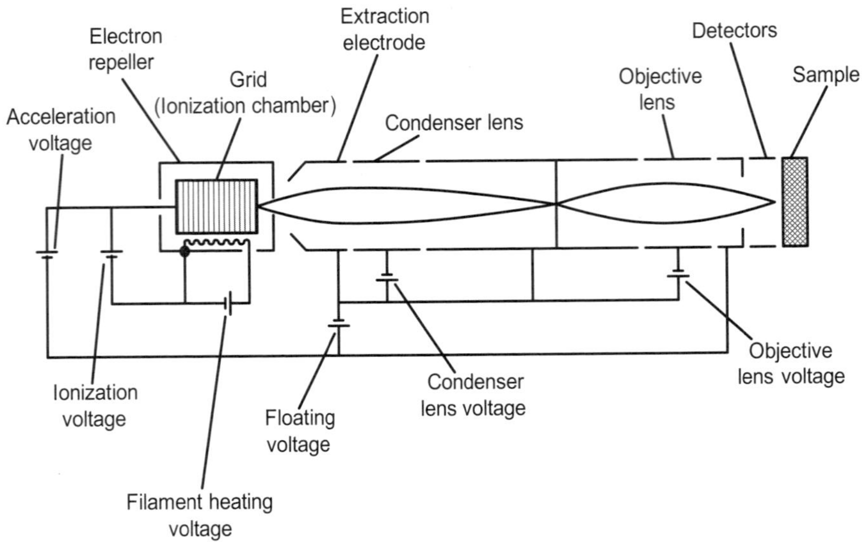

Figure 10.5 Schematic diagram of an ion optical column. Adapted from [10].

10.4.3 Electron Energy Analyzers

Electron energy analyzers are used to measure the number of ejected electrons (*N*) as a function of electron energy (*E*). The most commonly used energy analyzers in AES are: (i) the retarding field analyzer (RFA), (ii) the cylindrical mirror analyzer (CMA), and (iii) the concentric hemispherical analyzer (CHA).

Because of the limited energy resolution and poor signal-to-noise ratio resulting from mediocre transmission efficiency, the retarding field analyzer (Figure 10.6), commonly used for low energy electron diffraction studies, is not an optimal choice for Auger electron spectroscopy.

The high transmission efficiency, compact size, and ease of use of the cylindrical mirror analyzer (Figure 10.7) combine to make it the analyzer of choice for Auger electron spectroscopy.

Because of its higher resolution, the concentric hemispherical analyzer (Figure 10.8) is used in Auger electron spectroscopy when chemical state information is desired. The CHA consists of an input lens and the hemispherical analyzer. All XPS energy analyzers are concentric hemispherical analyzers.

The cylindrical mirror and concentric hemispherical analyzers are *band pass filters* in contrast to the retarding field analyzer (RFA), which is a *high pass filter*. The band pass filter allows passage of electrons within a band of energy (ΔE) at a pass energy (E), resulting in an energy resolution of $\Delta E/E$. Because the retarding field analyzer collects electrons with an energy greater than the specified energy E, the spectrum must be differentiated once to obtain the $N(E)$ spectrum.

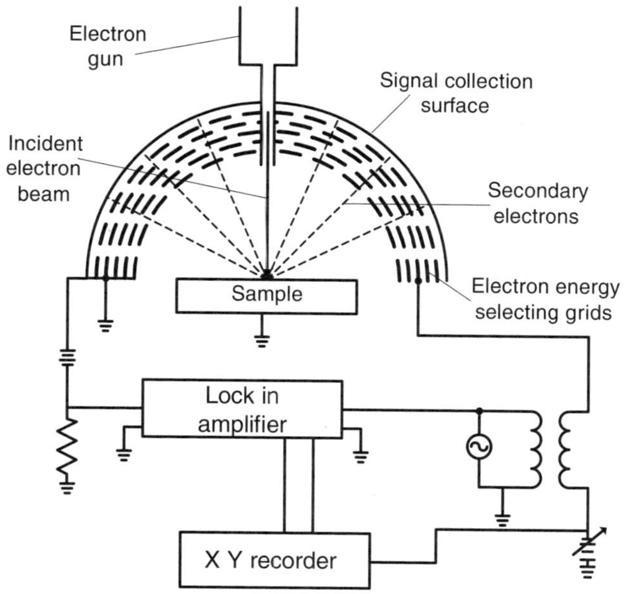

Figure 10.6 A schematic diagram of the retarding field analyzer. Adapted from [10].

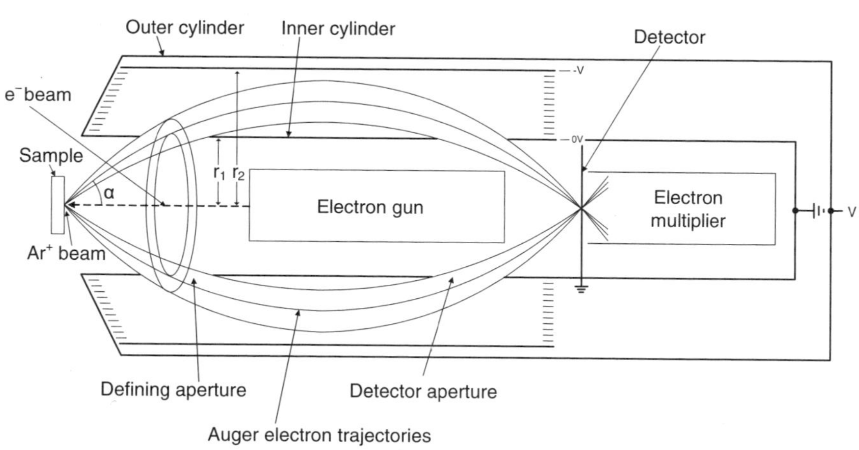

Figure 10.7 A schematic diagram of the cylindrical mirror analyzer. Adapted from various sources [10, 11].

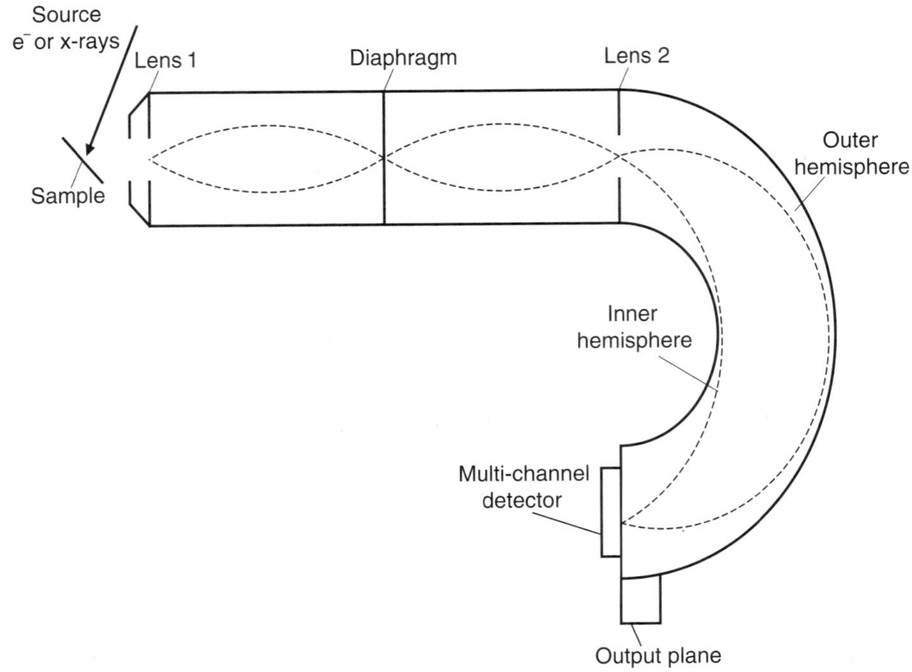

Figure 10.8 A schematic diagram of the concentric hemispherical analyzer. Adapted from [10].

10.4.4 Electron Detector

Electrons exiting the analyzer and arriving at the detector are amplified and counted by an electron multiplier, either a channeltron or a microchannel plate (MCP). Electrons striking the specially coated inside of a channeltron, a cone-shaped dynode, produce many secondary electrons, which are accelerated toward the anode. The many intervening collisions produce an avalanche effect, which results in a gain of ~10^8 for a channeltron. A microchannel plate consists of many tiny channeltrons fused together to form a disc. Electron intensity measurements are usually done by pulse counting.

10.4.5 Computer Control and Data Display Systems

Modern AES instrumentation has sophisticated computer control, data processing, and display systems. The computer control system has four major functions: (i) setting up conditions for analysis; (ii) acquiring and storing data efficiently; (iii) processing data; and (iv) displaying results in the form of spectra.

10.5 EXPERIMENTAL PROCEDURES INCLUDING SAMPLE PREPARATION

10.5.1 Sample

Both single crystal and polycrystalline solid samples can be analyzed with AES. Because a flat smooth sample enhances the quality of the Auger spectra, powders are pressed into the shape of a wafer before AES analysis. The specimen sample to be analyzed by Auger must be compatible with ultrahigh vacuum conditions. For example, a sample containing a significant amount of Zn is not suitable because of the high vapor pressure of Zn. Even if a suitable sample is placed in an ultrahigh vacuum chamber, additional treatment such as inert gas sputtering or ion etching to remove surface contaminants may be required before analysis. State-of-the-art AES instruments allow for samples as large as the 300 mm wafers used in the semiconductor industry.

While most metals and semiconductors are amenable to Auger analysis, insulators present a special problem. Charging of insulator samples during electron spectroscopy is problematic because: (1) kinetic energy measurements may be in error by as much as tens of eV, and (2) spectral peaks may be distorted due to inhomogeneous surface charge distribution [29]. Because an electron beam is used in Auger electron spectroscopy, charge compensation of insulator samples must be achieved by one or more methods: (1) lowering the incident electron beam energy to increase the emission of secondary electrons; (2) tilting the sample to decrease the angle between the sample surface and the beam and hence increase the number of electrons leaving the sample; (3) neutralizing the charge with low energy (~50 eV) positive ions such as Ar^+; (4) placing thin films of the insulating sample on a conductive surface such as graphite; and (5) decreasing the incident current density [29, 30]. Such techniques allow the analysis of insulators such as ceramics [31].

10.5.2 Beam Effects and Surface Damage

Electron beam damage to specimens is a concern with Auger electron spectroscopy. Examples of electron beam-induced surface damage include: (1) creation of defects, (2) change of crystal structure, (3) change of surface topography, (4) change of oxidation state, (5) bond cleavage, (6) adsorption, (7) desorption, and (8) segregation [32].

In order to minimize damage, checks for the presence of specimen damage must be performed routinely during analysis. If damage is detected, the experimental conditions can be adjusted to minimize surface damage. Electron beam effects may be reduced by: (1) decreasing the electron beam energy, (2) decreasing the current density at the surface by defocusing the beam, and/or (3) reducing the time of exposure by rastering the beam over a selected area. Reducing sample temperature during electron irradiation was found to minimize radiation damage to polymers during Auger analysis [33].

10.5.3 AES Modifications and Combinations with Other Techniques

The basic technique of AES as described earlier has also been adapted for use in Auger depth profiling, which provides quantitative compositional information as a function of depth below the surface, and scanning Auger microscopy, which provides spatially resolved information on the composition of surfaces. Auger depth profiling and scanning Auger microscopy will be discussed later in this chapter.

Auger electron spectroscopy can be combined with other analytical techniques for simultaneous analysis. For example, AES is combined with energy dispersive X-ray spectroscopy (EDS or EDX) for elemental analysis and secondary ion mass spectroscopy (SIMS) for trace impurity analysis [10]. Auger instruments can also be equipped with an X-ray source for X-ray photoelectron spectroscopy (XPS). Similarly, XPS instruments can also be equipped with an electron gun for simultaneous AES analysis.

10.6 AUGER SPECTRA: DIRECT AND DERIVATIVE FORMS

Auger spectra are usually acquired and displayed in one of two ways: (a) *direct form,* where the total electron signal is measured as a function of the kinetic energy of the electrons leaving the sample; (b) *derivative form,* where the derivative of the total electron signal is measured as a function of kinetic energy. The derivative spectrum helps to accentuate the Auger signal by suppressing the background due to secondary and backscattered electrons, as described below.

The interaction of an electron beam with a solid sample results in the emission of secondary and backscattered electrons whose distribution plotted as a function of kinetic energy is shown in Figure 10.9.

The schematic secondary and backscattered electron distribution shown in this figure demonstrates four features (right to left):

(1) The strong peak at the incident electron energy is due to elastically back- scattered primary electrons.
(2) The loss peaks at discrete energies on the low energy side of the elastic peak are due to electrons that have lost energy to surface and bulk plasmons.
(3) The sharp peaks that are independent of primary beam energy are due to Auger electrons.

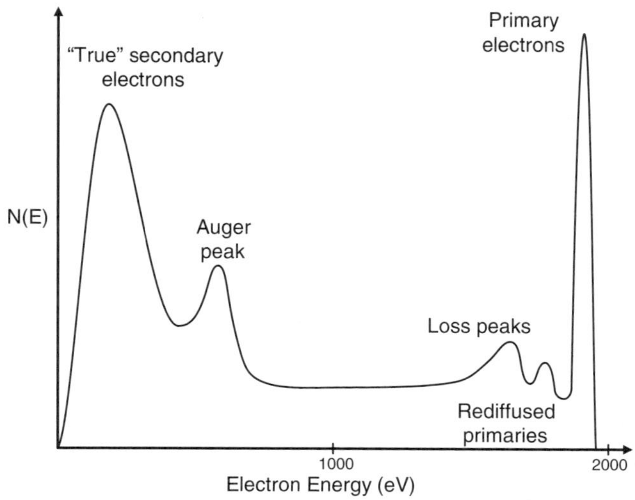

Figure 10.9 Schematic diagram of the energy distribution of secondary and backscattered electrons produced by the interaction of a nearly monochromatic electron beam with a solid. Adapted from [9].

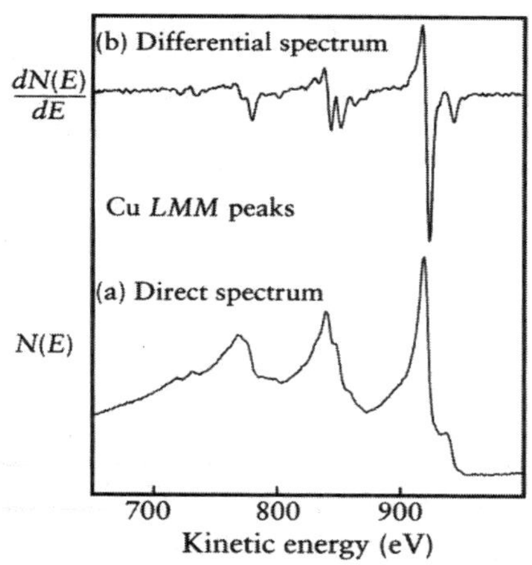

Figure 10.10 The direct and differentiated spectrum of copper [11]. (Copyright John Wiley & Sons, Ltd. Reproduced with permission.)

(4) The broad peak at low energies (< 50 eV) is due to secondary electrons that have undergone multiple inelastic collisions within the solid.

Because the low intensity Auger peaks are superimposed on a large slowly varying background signal, the $N(E)$ vs. E spectrum is differentiated to yield the differentiated spectrum, $\mathrm{d}N(E)/\mathrm{d}E$ vs. E, in which the Auger peaks are accentuated with respect to the background signal, as exemplified in Figure 10.10. The main peaks, which are characteristic of Cu, appear between 700 and 1000 eV. The background, due to the secondary electrons generated by inelastic scattering of electrons, is fairly extensive in the direct spectrum. The AES spectrum for Cu demonstrates that because the background is considerably reduced in the differentiated form the sensitivity of detection is enhanced. Direct spectra, however, are now being used more frequently because of easier quantification. The absence of secondary electrons in the gas phase makes direct spectra the ideal choice.

10.7 APPLICATIONS

10.7.1 Qualitative Analysis

Elemental analysis of surfaces is based on the kinetic energies of the observed Auger transitions. For positive identification it is necessary to match not only energies but also the shapes and relative strengths of the observed Auger peaks [34]. Modern AES instrumentation makes use of computer programs for rapid Auger peak identifications. However, such identification must be done with care, especially when two or more elements have overlapping or nearly overlapping Auger peaks. Given the new international standards [25], incorrect calibration of the kinetic energy scale should in the future not contribute to additional uncertainty in element identification. As in most other spectroscopic methods, conclusive identification of elements is facilitated if the reference and sample Auger spectra are both obtained on the same instrument [34].

The first step in qualitative Auger analysis involves obtaining a survey spectrum with relatively modest resolution in a fairly short time. Survey Auger spectra are typically recorded at energies between 0 eV and 1000 eV because most elements have significant Auger transitions in this range. Moreover, Auger electrons with energies greater than 1000 eV are less surface-sensitive because these electrons have longer inelastic mean free paths. Because Auger features are much more pronounced in derivative spectra, such spectra are commonly used in qualitative AES investigations.

Each element has a set of characteristic Auger peaks, as demonstrated by a plot of the Auger electron energy as a function of atomic number (Z) (Figure 10.11). The figure demonstrates that the Auger transitions of choice for the different elements can be summarized as follows: $3 < Z < 14$ (KLL transitions), $14 < Z < 40$ (LMM transitions), $40 < Z < 82$ (MNN transitions), and $82 < Z$ (NOO transitions).

10.7.2 Quantitative Analysis

The goal of quantitative analysis by AES is to determine the chemical composition of solid surfaces by calculating atomic concentrations from Auger peak intensity measurements. In the case of a direct Auger spectrum, peak intensities are obtained from peak area measurements following a suitable baseline subtraction. In the case of a derivative spectrum, peak intensities are characterized by the peak-to-peak heights. The direct spectrum is preferred for most AES quantification studies. Estimating the in-depth distribution of elements is important in the AES quantification of surface elemental concentration. Even if the solid composition is uniform down to a depth of a few nanometers, the surface concentration may not be proportional to the measured peak intensity. More accurate quantification can be achieved if the peak shape is also taken into consideration when performing Auger surface analysis.

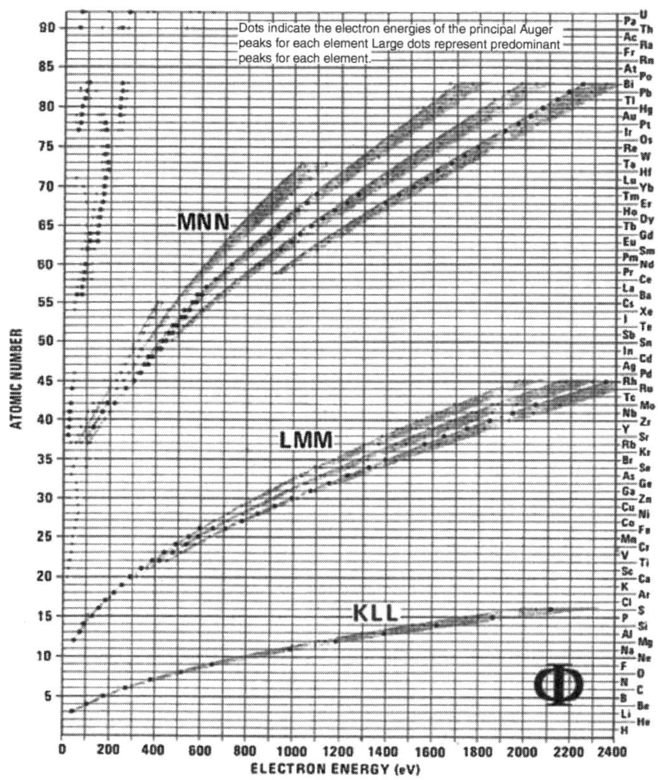

Figure 10.11 Principal Auger electron energies for each element. The larger dots correspond to prominent Auger peaks [35]. (Copyright Physical Electronics, Eden Praire, MN. Reproduced with permission.)

Reproducibility of Auger intensity ratios may be compromised by: (1) problems with sample alignment, (2) deflections of electrons by stray magnetic fields, and (3) energy-dependent detector efficiencies [36]. A comprehensive review of Auger and XPS quantification methods was published in 1996 [37].

We describe in detail below the three methods used for Auger quantification: (1) first principles Auger intensity calculations, (2) standards of known composition, and (3) elemental relative sensitivity factors.

10.7.2.1 First Principles Auger Intensity

The measured Auger current intensity (I_A) can be related to the depth (z) dependent atomic density $N_A(z)$ by considering the physical principles underlying (1) Auger electron generation, (2) transport of the Auger electron through the solid matrix, and (3) detection of the Auger electron. The resulting relationship is as follows:

$$I_A = I_o G\sigma_A G_{CK}\gamma_A K(1+r_M)\int_{z=0}^{\infty} N_A(z)\exp\left[\frac{-z}{\lambda_A\cos\theta}\right]dz\, fTD. \quad (10.5)$$

In the above are the following variables: I_A is the measured Auger current from element A, I_o is the incident electron current (primary beam current), G is a geometrical factor accounting for the area under irradiation (depends on angle of incidence of primary beam), σ_A is the ionization cross-section (a function of incident electron energy and the atomic level being ionized), G_{CK} is the Coster-Kronig yield correction, γ_A is the Auger yield factor (corrects for X-ray fluorescence), K is the sample condition factor (surface roughness, contamination, etc.), r_M is the matrix-dependent backscattering factor that accounts for additional ionizations due to backscattered primary electrons, z is the depth below the surface, N_A is the number density of atoms of type A, λ_A is the matrix- and Auger electron energy-dependent inelastic mean free path (related to attenuation length and the effective escape depth), θ is the Auger electron's take-off angle, measured from the surface normal, f is the analyzer retardation factor (for RFA analyzer; not needed for CMA), T is the analyzer transmission factor (a function of energy), and D is the detector efficiency factor (a function of energy and time).

Most of the above terms are constants for a given set of experimental conditions and for a particular Auger transition. Moreover, the integral can be simplified by assuming that the concentration is uniform over a depth of ~$5\lambda_A$, as shown below:

$$\begin{aligned} I_A &= I_o G\sigma_A G_{CK}\gamma_A K(1+r_M)N_A fTD\int_{z=0}^{\infty}\exp\left[\frac{-z}{\lambda_A\cos\theta}\right]dz \\ &= I_o G\sigma_A G_{CK}\gamma_A K(1+r_M)N_A fTD\lambda_A\cos\theta. \end{aligned} \quad (10.6)$$

Although absolute Auger quantification using the above equation is not practical for most cases, this equation provides a theoretical basis for understanding other quantification methods described below.

10.7.2.2 Standards of Known Composition

Relative Auger quantification is performed by keeping many of the experimental parameters in equation (10.6) fixed and by using locally produced standards of known composition. Instrumental variables are eliminated in this ratio technique because the same instrument is used to analyze both the *test* and *standard* samples. The ratio of the measured Auger current from the test (I_A^T) and standard samples (I_A^S) is given by

$$\frac{I_A^T}{I_A^S} = \frac{(1+r_{M,A})_T}{(1+r_{M,A})_S} \frac{\lambda_A^T}{\lambda_A^S} \frac{N_A^T}{N_A^S}. \quad (10.7)$$

In the above expression, N_A^T and N_A^S are the number densities of A in the test and the standard samples. Equation (10.7) may be rewritten as follows:

$$\frac{N_A^T}{N_A^S} = \frac{(1+r_{M,A})_S}{(1+r_{M,A})_T} \frac{\lambda_A^S}{\lambda_A^T} \frac{I_A^T}{I_A^S}. \quad (10.8)$$

If the composition of the test sample is close to that of the standard, the influence of the matrix on electron backscattering and inelastic mean free path may be ignored and the above equation may be used to obtain the number density of the test sample directly from the ratio of the Auger yields:

$$\frac{N_A^T}{N_A^S} = \frac{I_A^T}{I_A^S}. \quad (10.9)$$

More generally, it is necessary to evaluate the ratios of the backscattering factors and the inelastic mean free paths to determine the composition of the test sample. Because evaluating these matrix effects can be rather complicated, a new method to quantify AES and XPS data has been proposed based on angle-averaged reflected electron energy loss spectroscopy (REELS) spectra and involving "average matrix sensitive factors" [40].

10.7.2.3 Elemental Sensitivity Factors

This quantitative Auger method is based on the assumption that matrix effects can be ignored. With this assumption, equation (10.6) can be simplified as follows:

$$I_A = s_A N_A. \quad (10.10)$$

In the above expression, s_A is the Auger sensitivity factor of element A. Relative Auger sensitive factors are defined relative to A and tabulated. If the sample contains two elements A and B, the number density N_A of A is given by the following expression:

$$N_A = \frac{\left(\frac{I_A}{s_A}\right)}{\left(\frac{I_A}{s_A} + \frac{I_B}{s_B}\right)}. \quad (10.11)$$

10.7.3 Chemical Information

While elements present on the surface can be positively identified by the Auger peak energies, changes in chemical state (e.g., oxidation state) or chemical environment can be deduced from changes in Auger peak positions, intensities, and shapes. As such, AES spectra, particularly derivative spectra, are used as "chemical fingerprints" for identification and characterization. The classic example of such *qualitative* analysis involves the carbon KLL Auger spectra for molybdenum carbide, silicon carbide, graphite, and diamond (Figure 10.12) [39]. The spectra show variations in lineshapes due to differences in the chemical environments of the carbon atoms in the four samples. Although the data clearly show that bonding information can be obtained from Auger spectroscopy, obtaining chemical state and chemical environment information from Auger spectra is challenging because the Auger process involves three energy levels. More detailed electronic structure information (e.g., hybridization, electron delocalization, screening effects) may be obtained from *quantitative* spectral lineshape analysis, a daunting task that is currently beyond the scope of most Auger practitioners [40].

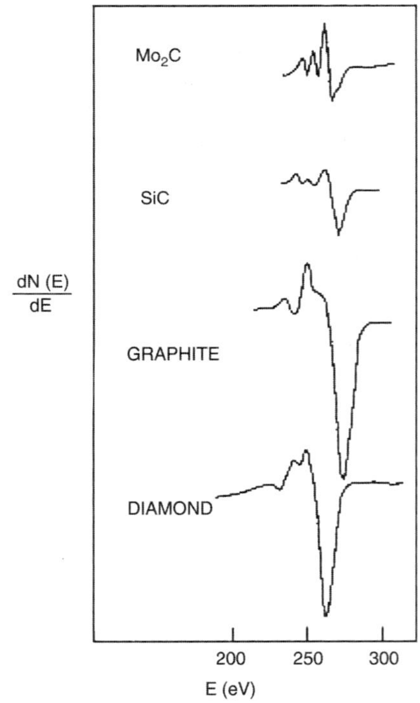

Figure 10.12 Effect of chemical environment on the KLL Auger spectra of carbon [39]. The shift in the Auger peak energy for diamond is due to charging of the insulator. (Reproduced with permission from the American Institute of Physics.)

Chemical shifts, or, equivalently, energy shifts, occur when there is a charge transfer from one atom to another. For example, atomic oxygen adsorbed on clean metal surfaces can produce measurable shifts in energies of Auger peaks from metals. Auger spectra of aluminum in AlN, AlF_3, AlB_2 and Al_2O_3 (Figure 10.13) clearly show the phenomena of chemical shifts in Auger spectroscopy.

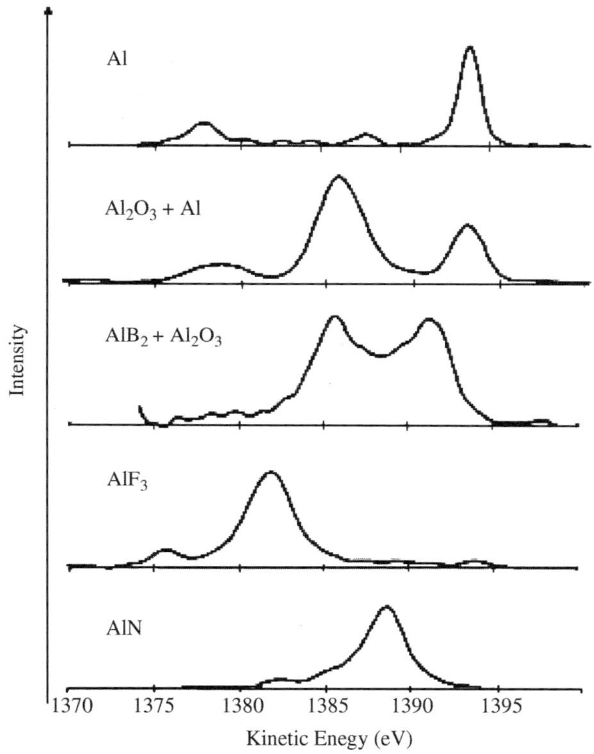

Figure 10.13 Chemical shifts demonstrated in the Auger KLL spectra of several Al compounds [41]. (Copyright John Wiley & Sons Ltd. Reproduced with permission.)

Several metal LVV and MVV lineshapes have also been used extensively for Auger chemical fingerprinting to study the effects of hydrogen or oxygen exposure, intercalation, or changes in crystal structure. Auger chemical fingerprinting has also been used to analyze the $L_3M_{23}V$ Auger lineshape of high-temperature superconductors [42].

10.7.4 Auger Depth Profiling

Obtaining the chemical composition as a function of depth below the surface using Auger electron spectroscopy is referred to as "Auger depth profiling," one of four modes of Auger operation (point analysis, line scan, and surface imaging the other three modes of Auger operation). Industrial applications of Auger depth profiling include analyzing microelectronics devices, investigating corrosion-resistant surfaces, and characterizing plasma-modified surfaces [43]. Auger depth profiling methods may be broadly categorized as follows: (1) nondestructive, (2) sputtering by noble gas ions, and (3) mechanical sectioning. Brief summaries of methods 1 and 2 are given below. Interestingly, Auger sputter depth profiling (method 2) has become the most popular choice for chemical analysis of thin films [44].

10.7.4.1 Nondestructive Methods

Angle-resolved Auger electron spectroscopy (ARAES) allows for non-destructive depth profiling but works for only very thin layers up to a thickness of approximately 100 Å. Auger depth profiling may be accomplished by changing the geometry of the experiment because the depth of analysis depends on the emission angle of the Auger electron. While angle-resolved X-ray photoelectron spectroscopy has been extensively used for non-destructive depth profiling, the use of angle-resolved Auger for such analysis has been somewhat limited [45, 46].

10.7.4.2 Sputter Depth Profiling

Auger sputter depth profiling usually involves simultaneous Auger elemental analysis and inert gas (argon or xenon) ion bombardment sputtering (etching) with an ion gun to remove material from the surface in a controlled manner to expose underlying atomic layers. Although Auger *survey* scans can be performed during sputter depth profiling, Auger analysis within pre-selected energy windows allows for more rapid data acquisition. Depth profiling with Auger may also be performed sequentially with alternating cycles of sputtering and analysis by AES. An example of Auger sputter depth profiling is shown in Figure 10.14.

Despite the popularity of sputter profiling with Auger, many studies have documented the complexities inherent in this technique [47]. Auger sputter depth profiling is challenging due to ion beam-induced changes in surface roughness and composition, changes that are associated with effects such as preferential sputtering (one element is sputtered faster than another element in the matrix) [48], collisional mixing, and ion-induced reactions (e.g., metal surface oxide is reduced to metal [8]). Such complexities associated with AES sputter depth profiling experiments make difficult the task of converting Auger intensities measured as a function of sputtering time into elemental concentration as a function of depth. Despite formidable challenges, recent developments in the modeling of sputtering, such as the so-called *mixing roughness information* (MRI) depth model, allow for depth analysis of nanostructures [49].

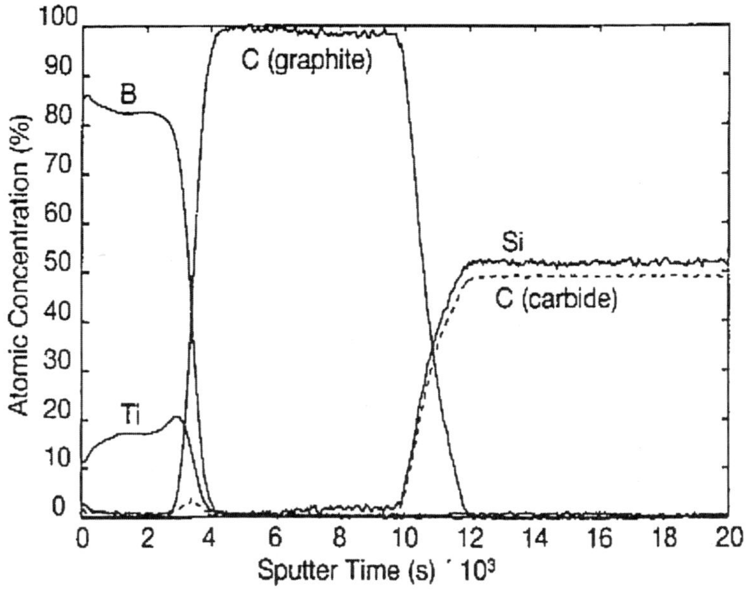

Figure 10.14 Auger sputter depth profile of a TiB_2/C-coated SiC filament [9]. (Reproduced by permission of Taylor & Francis Group, LLC, http:/www.taylorandfrancis.com)

In addition to advances in theory, several experimental innovations have made sputter depth profiling with Auger more precise. For example, collisional atomic mixing, which degrades depth resolution, may be minimized by using ultralow ion energies (<500 eV) instead of the typical high ion energies (several keV). The low ion energies, however, result in low beam current densities, which yield etching rates below 0.2 nm min^{-1} [50]. A more acceptable etching rate of 1 nm min^{-1} and a high depth resolution of 2 nm has been achieved with a specially designed ion gun capable of producing ion currents of $\sim 10^{-6}$ A at ion energies as low as 100 eV [50]. The use of such ultralow ion energies, however, is complicated by enhanced preferential sputtering, as demonstrated recently for the GaAs/AlAs superlattice, an ISO reference material [51].

A second example of experimental innovation involves sample rotation during Auger sputter depth profiling [52]. Because this so-called Zalar rotation technique time-averages the angles between the ion beam and the various crystalline planes, sample roughening is minimized and, consequently, depth resolution is enhanced. This feature in now available in commercial Auger instruments such as the PHI 700 manufactured by Physical Electronics.

Auger sputter depth profiling requires stringent ultrahigh vacuum conditions because the surface exposed by sputtering may act as a getter for residual gases. Moreover, for the same reason, the noble gas used for sputtering must be of the highest purity [11].

10.7.5 Auger Images and Linescans

Scanning Auger microscopy involves the acquisition of Auger data as a function of two-dimensional position within a defined area on a specimen. With an excellent lateral resolution of less than 10 nm, scanning Auger microscopy can be used effectively to image nanoscale structures and to produce two-dimensional maps of surface elemental composition. In combination with XPS, scanning Auger microscopy has been used recently to successfully analyze archeological artifacts such as ancient metal surfaces [53]. The large amount of data collected during each scan poses a significant challenge for data storage [54]. Adequate image quality also requires times on the order of hours to acquire Auger maps [54]. Progress in scanning Auger microscopy has been recently reviewed [55].

Sophisticated scanning Auger microscopy instrumentation is commercially available from several vendors, including Physical Electronics, JEOL, SPECS and STAIB Instruments. In addition to the standard components found in a typical Auger spectrometer, a scanning Auger microscope contains a secondary electron detector (SED) for performing secondary electron microscopy (SEM) imaging (Figure 10.15). Such imaging provides surface topographic information that may be correlated to information obtained from Auger electron spectroscopy. Moreover, by taking account of the background signal, secondary electron microscopy may be used to suppress topographic information and enhance chemical information when performing scanning Auger microscopy, as demonstrated in Figure 10.16.

While an Auger map shows the relative elemental concentration as a function of x and y, an Auger linescan measures the relative elemental concentration as a function of x along a straight line on a sample. As with scanning Auger microscopy, a scanning electron image provides the means to choose the analysis line.

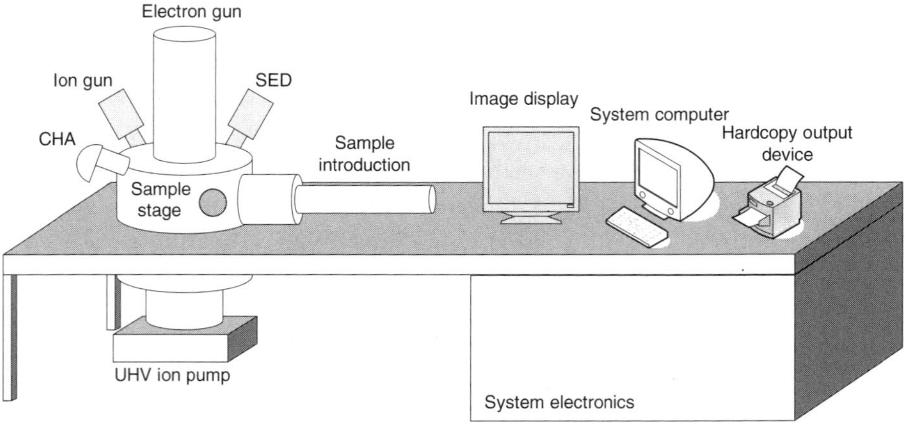

Figure 10.15 Schematic diagram of a scanning Auger microscope. Adapted from [9].

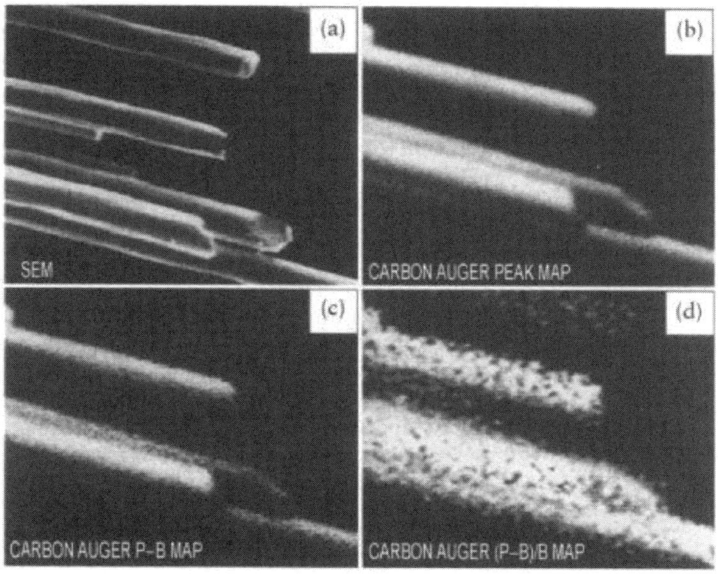

Figure 10.16 Scanning Auger microscopy of carbon fibers: (a) SEM image, (b) peak map of carbon Auger electrons, (c) peak background map, and (d) correction for topographic effects [11]. (Copyright John Wiley & Sons Ltd. Reproduced with permission.)

10.7.6 Research and Industry

In addition to being used as a tool in surface science research, both basic and applied, Auger electron spectroscopy is widely used in industry as an analytical tool to investigate surfaces, interfaces, thin films, and submicron features. In the thin film industry, AES depth profiling is routinely carried out for monitoring chemical composition. Auger is especially useful for detecting small defects (< 500 Å) that may play a critical role in the fabrication of small semiconductor devices. Recent experiments have demonstrated the utility of AES for investigating biomaterials such as hydroxyapatite [56]. The other areas of Auger applications are found in catalysis, solid state reactions, metallurgy, corrosion, advanced ceramics, and structural materials. A few of these applications are discussed in more detail below.

10.7.6.1 Corrosion

While XPS is ideally suited to characterize the growth of passivating thin films that inhibit the corrosion of stainless steel, submicron scanning Auger spectroscopy, with its high spatial resolution, is very useful for studying microscopic inclusions (chemical inhomogeneities) present in these protective thin films [11]. These inclusions, acting as pit initiation sites, are known to compromise the corrosion resistance of stainless steel in the presence of chloride ions [57].

10.7.6.2 Ceramics

Despite problems with charging of insulator samples, Auger electron spectroscopy has been used to study ceramic thin film materials. For example, Auger depth profiling was recently employed to characterize sol-gel-derived $Pb(Mg_{1/3}Nb_{2/3})_{0.7}Ti_{0.3}O_3$ (PMNT) thin films, a type of ferroelectric oxide thin film that has potential applications in electro-optic, pyroelectric, and micro electromechanical devices [58]. The Auger depth profile (intensity as a function of sputter time) shown in Figure 10.17 demonstrates the uniform deposition of a PMNT thin film on a platinum electrode.

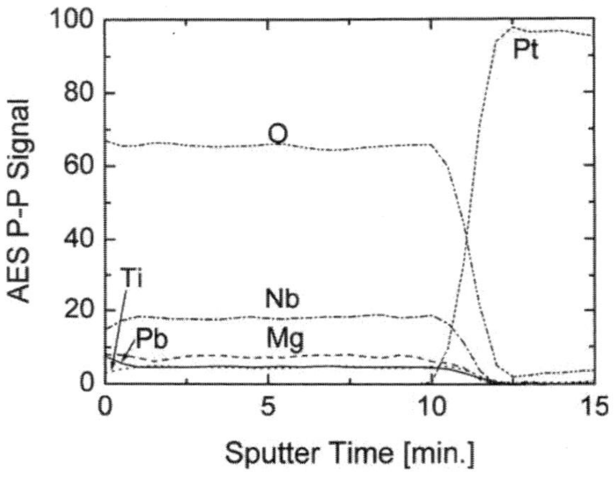

Figure 10.17 An Auger depth profile of a lead magnesium niobium titanate (PMNT) thin film [58]. (Reproduced with permission from the Institute of Applied Physics, Japan.)

10.7.6.3 Semiconductor Industry

The surface sensitivity of Auger electron spectroscopy has been extensively exploited in many laboratories for semiconductor research. In addition to surface sensitivity, Auger electron spectroscopy is also very useful for performing high resolution depth analysis of the layered material in semiconductor devices. The relatively flat interfaces present in such devices are especially amenable to analysis by Auger electron spectroscopy sputter depth profiling [11]. Such Auger depth profile studies have yielded information important for understanding interface purity, thickness, and diffusion [11]. Moreover, Auger electron spectroscopy can be used to combine depth and chemical state information in semiconductor devices

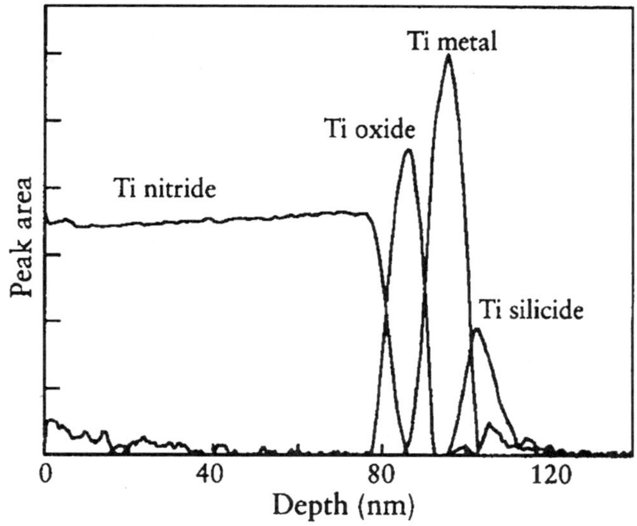

Figure 10.18 Depth profile of a semiconductor material consisting of silicon and various compounds of titanium [11]. (Copyright John Wiley & Sons Ltd. Reproduced with permission.)

(Figure 10.18). It is interesting to note that AES has also been used to characterize room temperature single electron transistors (SET), which represent a potential breakthrough in the quest for even smaller devices [59].

Because of shrinking device dimensions, AES has recently become a vital tool in the semiconductor industry for process control and failure analysis [60]. For example, because of its excellent spatial resolution and non-destructive nature, AES is ideally suited to identify and characterize submicron semiconductor defects (as small as 500 Å) which are beyond the capabilities of conventional methods based on X-ray dispersive analysis tools [61]. An Auger defect review tool (DRT) capable of analyzing 200-mm thick wafers has been developed and evaluated [62–64]. Despite the demanding requirements for ultrahigh vacuum, a recent publication has recommended that AES and XPS analysis be moved from laboratory settings to the production line in integrated circuit manufacture [65].

10.8 RECENT ADVANCES

Recent developments in the theory of Auger spectroscopy, especially those dealing with electron correlation, have been comprehensively reviewed [66]. We briefly review two of the Auger experimental techniques that have undergone rapid growth in the last decade or two.

10.8.1 Positron-Annihilation-Induced AES

First demonstrated in 1988 [67], positron-annihilation-induced Auger electron spectroscopy (PAES) involves the detection of electron emission due to Auger transitions initiated by positrons (the antiparticles of electrons) annihilating core electrons. The basic principles of AES and PAES are compared in Figure 10.19.

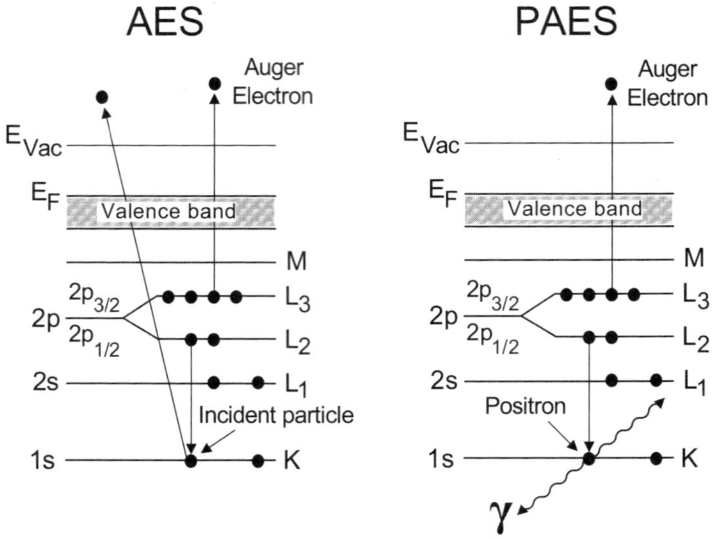

Figure 10.19 Comparison of basic principles of AES and PAES.

Compared to the traditional high energy electron-induced AES, PAES offers several advantages: (1) surface sensitivity is enhanced because positrons are trapped in an image potential on the surface prior to annihilation; (2) beam-induced damage is minimized because very low energy (~15 eV) positrons are used; and (3) background of secondary electrons is drastically reduced because the energy of the positrons is too low to excite secondary electrons. Recent applications of PAES include distinguishing the different crystallographic planes in semiconductors [68], determining Auger lineshapes [69], detecting surface impurities [70], and studying the growth dynamics of thin films [71].

10.8.2 Auger Photoelectron Coincidence Spectroscopy

Auger photo electron coincidence spectroscopy (APECS) is a powerful new experimental technique to characterize a number of important properties such as local electronic structure in novel solids including high temperature superconductors [72]. In addition, despite the many challenges associated with these measurements, APECS is considered a valuable tool for simplifying complex Auger spectra, thereby enabling a fundamental understanding of the physical processes, including electron-electron correlation, governing such spectra [73]. In a typical APECS experiment the sample is irradiated with X-rays that excite both a core photoelectron and an Auger electron, both of which are detected in coincidence (at the same time) using two electron energy analyzers. By allowing the separation of overlapping Auger peaks, APECS is particularly useful for analyzing samples containing more than one element [74]. Very recent results of APECS experiments have demonstrated the technique's extreme surface sensitivity [75] and ability to discern the emission depth of Auger electrons [76].

10.9 CONCLUSIONS

Auger electron spectroscopy has been used to study the surface and interfacial properties of semiconductor devices, catalysts, metals, advanced ceramics, biomaterials, and other novel materials. Advances during the past four decades in Auger instrumentation have made possible the transition from research and development to more practical applications such as quality control and failure analysis [21]. Auger electron spectroscopy will continue to make pivotal contributions to the continuing revolution in material science.

ACKNOWLEDGMENTS

The authors thank Wesley R. Nieveen and Ronald E. Negri for their assistance in review of the manuscript and Lin Zhu for assistance with preparation of figures. R.P.G was supported by a grant from the National Science Foundation (NSF grant number CHE–0455278).

REFERENCES

[1] Allenspach, R., Mauri, D., Taborelli, M. & Landolt, M. (1987) *Phys. Rev. B* **35** (10), 4801.

[2] Bruhwiler, P.A., Karis, O. & Martensson, N. (2002) *Reviews of Modern Physics* **74** (3), 703.

[3] Yue, F., Swineford, R.S. & Pappas, D.P. (1996) *Phys. Rev. B* **53** (12), 8036.

[4] Ramaker, D.E., Fry, R.A. & Idzerda, Y.U. (1995) *J. Electron Spectrosc. Relat. Phenom.* **72**, 169.

[5] Auger, P. (1925) *J. Phys. Radium* **6**, 205.

[6] Meitner, L. (1923) *Z. Phys.* **17**, 54.

[7] Diehl, S., Cubaynes, D., Kennedy, E.T., Wuilleumier, F.J., Bizau, J.M., Journel, L., VoKy, L., Faucher, P., Hibbert, A., Blancard, C., Berrah, N., Morgan, T.J., Bozek, J. & Schlachter, A.S. (1997) *J. Phys. B-At. Mol. Opt. Phys.* **30** (18), L595.

[8] Dastoor, P.C. (2003) *Springer Series in Surface Sciences* **23**, Surface Analysis Methods in Materials Science (2nd Ed.), 155.

[9] Narumand, D.H. & Childs, K.D. (1999) *Applied Spectroscopy Reviews* **34** (3), 139.
[10] Briggs, D. & Grant, J.T. (Eds.) (2003) *Surface Analysis by Auger and X-Ray Photoelectron Spectroscopy* IMPublications and SurfaceSpectra, Chichester, UK and Manchester, UK.
[11] Watts, J.F. & Wolstenholme, J. (2003) *An Introduction to Surface Analysis by XPS and AES*. John Wiley & Sons Ltd., Chichester.
[12] Gergely, G. (1994) *Vacuum* **45** (2-3), 311.
[13] Lander, J.J. (1953) *Phys. Rev.* **91** (6), 1382.
[14] Harris, L.A. (1968) *Appl. Phys.* **39** (3), 1419.
[15] Taylor, N.J. (1969) *Rev. Sci. Instrum.* **40** (6), 792.
[16] Palmberg, P.W. & Bohn, G.K. (1969) Tracy, J.C. *Appl. Phys. Lett.* **15** (8), 254.
[17] Macdonal. Nc, Waldrop, J.R. (1971) *Appl. Phys. Lett.* **19** (9), 315.
[18] Barkshire, I.R., Roberts, R.H. & Prutton, M. (1997) *Appl. Surf. Sci.* **120** (1-2), 129.
[19] Prutton, M., Barkshire, I. R. & Crone, M. (1995) *Ultramicroscopy* **59** (1-4), 47.
[20] Palmberg, P. W. (1972) Vac. J. *Sci. Technol.* **9** (1), 160.
[21] Powell, C. J. (2003) Vac. J. *Sci. Technol. A* **21** (5), S42.
[22] Seah, M. P. (2003) *Surface Analysis by Auger and X-Ray Photoelectron Spectroscopy*, 167.
[23] Seah, M.P. (2004) *Surf. Interface Anal.* **36** (13), 1645.
[24] Seah, M.P. (2003) *Surf. Interface Anal.* **35** (3), 327.
[25] Seah, M.P. (2003) *Surf. Interface Anal.* **35** (3), 329.
[26] Chung M.F. & Jenkins, L.H. (1970) *Surf. Sci.* **22** (2), 479.
[27] Ishii, T., Kover, L., Berenyi, Z., Cserny, I., Ikeno, H., Adachi, H. & Drube, W. (2004) *J. Electron Spectrosc. Relat. Phenom.* **137-40**, 451.
[28] Smith J.F. & Southworth, H.N. (1981) *Journal of Physics E- Scientific Instruments* **14** (7), 815.
[29] Kelly, M.A. (2003) in *Surface analysis by Auger and X-Ray Photoelectron Spectroscopy*, edited by D. Briggs and J. T. Grant IMPublications and SurfaceSpectra, Chichester, UK and Manchester, UK.
[30] Hofmann, S. (1992) *J. Electron Spectrosc. Relat. Phenom.* **59** (1), 15.
[31] Watts, J.F. (1993) *British Ceramic Proceedings* **51**, (Nanoceramics), 195.
[32] Baer, D.R., Engelhard, M.H., Gaspar, D.J. & Lea, A.S. (2003) in *Surface Analysis by Auger and X-Ray Photoelectron Spectroscopy*, edited by D. Briggs and J.T. Grant (IMPublications and SurfaceSpectra, Chichester, UK and Manchester, UK,).
[33] Sekine, T., Ikeo, N., Nagasawa, Y. & Kikuma, J. (1995) *Surf. Interface Anal.* **23** (6), 386.
[34] Grant, J.T. (2003) in *Surface analysis by Auger and X-Ray Photoelectron Spectroscopy*, edited by Briggs D. & Grant J.T. (IMPublications and SurfaceSpectra, Chichester, UK and Manchester, UK,).
[35] Childs, K.D., Vanier, L.A., Moulder, J.F., Paul, D.F., Stickle, W.F. & Watson, D.G. (1995) *Handbook of Auger Electron Spectroscopy*. (Physical Electronics, Eden Prairie, MN,).
[36] Powell, C.J., Erickson, N.E. & Madey, T.E. (1982) *Prog. Surf. Sci.* **25** (2-3), 87.
[37] Tilinin, I.S., Jablonski, A. & Werner, W.S.M. (1996) *Prog. Surf. Sci.* **52** (4), 193.
[38] Seah, M.P., Gilmore, I.S. & Spencer, S.J. (2001) *Surf. Interface Anal.* **31** (8), 778.
[39] Haas, T.W., Grant, J.T. & Dooley, G.J. (1972) *J. Appl. Phys.* **43** (4), 1853.
[40] Remaker, D.E. (2003) In *Surface Analysis by Auger and X-Ray Photoelectron Spectroscopy,* edited by D. Briggs and J. T. Grant (IMPublications and SurfaceSpectra, Chichester, UK and Manchester, UK,).
[41] Timmermans, B., Vaeck, N., Hubin, A. & Reniers, F. (2002) *Surf. Interface Anal.* **34** (1), 356.
[42] Ramaker D E. (1994) *J. Electron Spectrosc. Relat. Phenom.* **66** (3-4), 269.
[43] Zalar, A. (2000) *Inf. Midem-J. Microelectron. Electron. Compon. Mater.* **30** (4), 203.
[44] Hofmann, S. (2003) *Surf. Interface Anal.* **35** (7), 556.

[45] Derry, G.N. & Vanderlinde, W.E. (1992) *J. Vac. Sci. Technol. A-Vac. Surf. Films* **10** (4), 2826.
[46] Davidson, M.R., Hoflund, G.B. & Outlaw, R.A. (1991) *J. Vac. Sci. Technol. A-Vac. Surf. Films* **9** (3), 1344.
[47] Cao, Z.X. (2000) *Surf. Sci.* **452** (1-3), 220.
[48] Holloway, P.H. (1977) *Surf. Sci.* **66** (2), 479.
[49] Hofmann, S. (2005) *Appl. Surf. Sci.* **241** (1-2), 113.
[50] Inoue, M., Shimizu, R., Lee, H.I. & Kang, H.J. (2005) *Surf. Interface Anal.* **37** (2), 167.
[51] Mizuhara, Y., Bungo, T., Nagatomi, T. & Takai, Y. (2005) *Surf. Interface Anal.* **37** (3), 343.
[52] Zalar, A. (1985) *Thin Solid Films* **124** (3-4), 223.
[53] Paparazzo, E. (2003) *Archaeometry* **45** 615.
[54] Prutton, M. (2003) in *Surface Analysis by Auger and X-Ray Photoelectron Spectroscopy*, edited by D. Briggs & J.T. Grant (IMPublications and SurfaceSpectra, Chichester, UK and Manchester, UK).
[55] Jacka, M. (2001) *J. Electron Spectrosc. Relat. Phenom.* **114** 277.
[56] Ong J.L. & Lucas, L.C. (1998) *Biomaterials* **19** (4-5), 455.
[57] Rossi, A., Elsener, B., Hahner, G., Textor, M. & Spencer, N.D. (2000) *Surf. Interface Anal.* **29** (7), 460.
[58] Fan, H.Q. & Kim, H.E. (2002) *Japanese J. Appl. Phys.* **41** (11B), 6768.
[59] Matsumoto, K. (1998) *Scanning Microscopy* **12** (1), 61.
[60] Dittmar, K. (2004) *Surf. Interface Anal.* **36** (8), 837.
[61] Childs, K.D., Paul, D.F. & Schauer, S.N. (2001) *AIP Conference Proceedings* **550**, Characterization and metrology for ULSI technology, 312.
[62] Childs, K.D., Watson, D.G., Paul, D.F. & Clough, S.P. (1998) *AIP Conference Proceedings* **449**, Characterization and metrology for ULSI technology, 810.
[63] Childs, K.D., Paul, D.F. & Clough, S.P. (1996) *Proceedings Institute of Environmental Sciences* **42**, Contamination control; Symposium on minienvironments, 147.
[64] Fillmore, D.K. & Krasinski, H.A. (1998) *Surf. Interface Anal.* **26** (2), 109.
[65] Savage, R.N., (2003) *Solid State Technology* **46** (8), 57.
[66] Weightman, P. (2003) in *Surface analysis by Auger and photoelectron spectroscopy*, edited by D. Briggs & J. T. Grant (IMPublications and SurfaceSpectra, Chichester, UK and Manchester, UK,).
[67] Weiss, A., Mayer, R., Jibaly, M., Lei, C., Mehl, D. & Lynn, K.G. (1988) *Phys. Rev. Lett.* **61** (19), 2245.
[68] Fazleev, N.G., Kim, J., Fry, J.L. & Weiss, A.H. (2003) *Phys. Rev. B* **68** (24).
[69] Weiss, A.H., Yang, S., Zhou, H.Q., Jung, E. & Wheeler, S. (1995) *J. Electron Spectrosc. Relat. Phenom.* **72**, 305.
[70] Ohdaira, T., Suzuki, R., Mikado, T., Ohgaki, H., Chiwaki, M., Yamazaki, T. & Hasegawa, M. (1996) *Appl. Surf. Sci.* **101**, 73.
[71] Kim J.H. & Weiss, A.H. (2000) *Surf. Sci.* **460** (1-3), 129.
[72] Stefani, G., Gotter, R., Ruocca, A., Offi, F., Da Pieve, F., Iacobucci, S., Morgante, A., Verdini, A., Liscio, A., Yao, H. & Bartynski, R.A. (2004) *J. Electron Spectrosc. Relat. Phenom.* **141** (2-3), 149.
[73] Thurgate, S.M., Lund, C.P. & Wedding, A.B. (1994) *Nucl. Instrum. Methods Phys. Res. Sect. B–Beam interact. Mater. Atoms* **87** (1-4), 259.
[74] Robins, J.L. (1995) *Prog. Surf. Sci.* **48** (1-4), 167.
[75] Liscio, A., Gotter, R., Ruocco, A., Iacobucci, S., Danese, A.G., Bartynski, R.A. & Stefani, G. (2004) *J. Electron Spectrosc. Relat. Phenom.* **137-40**, 505.
[76] Werner, W.S.M., Smekal, W., Stori, H., Winter, H., Stefani, G., Ruocco, A., Offi, F., Gotter, R., Morgante, A. & Tommasini, F. (2005) *Phy. Rev. Lett.* **94** (3).

CHAPTER 11

X-RAY PHOTOELECTRON SPECTROSCOPY

[1]Hsiao-Lu Lee and [2]Nolan T. Flynn
[1]Department of Chemistry, Stanford University, Stanford, CA 94305
[2]Department of Chemistry, Wellesley College, Wellesley, MA 02481

11.1 INTRODUCTION AND BASIC THEORY

Hailed as the "NMR of the periodic table," X-ray photoelectron spectroscopy (XPS), also known as *electron spectroscopy for chemical analysis* (ESCA), is one of the most powerful and common chemical analysis techniques. XPS is based on the photoelectric effect in which the binding energy (E_B) of a core-level electron is overcome by the energy ($h\nu$) of an impinging soft X-ray photon, and the core-level electron is excited and ejected from the analyte. The kinetic energies of the ejected photoelectrons, E_K, are measured by an electron spectrometer whose work function is ϕ. Invoking conservation of energy, the following relationship is obtained:

$$E_B = h\nu - E_K - \phi. \tag{11.1}$$

The binding energy of the photoelectron is characteristic of the orbital from which the photoelectron originates. The binding energy of the ejected photoelectron depends on the final state configuration after photoemission. The initial and final state configurations are depicted in Figure 11.1. The final state is characterized by full relaxation of all atomic orbitals towards the hole in the core level.

The study of the ejected photoelectrons provides a wealth of information about the sample. XPS enables qualitative elemental identification for the entire periodic table save H and He. Simple identification can be achieved by recording low resolution spectra, often called *survey scans*, over a broad binding energy range. Beyond simple identification, acquisition of high resolution spectra in binding energy regions of interest followed by peak-fitting can provide information concerning the chemical (oxidation) state or environment for these atoms. Additionally, quantitation for the near-surface region of the sample is commonly achieved. Angle-resolved XPS and XPS imaging provide the potential to learn about spatial atomic distribution within the sample and to understand the surface and bulk chemistry of the sample. If the sample is presented as a thin film, XPS may help answer questions regarding the thickness of the film, the mode of adsorption (*chemisorption* or

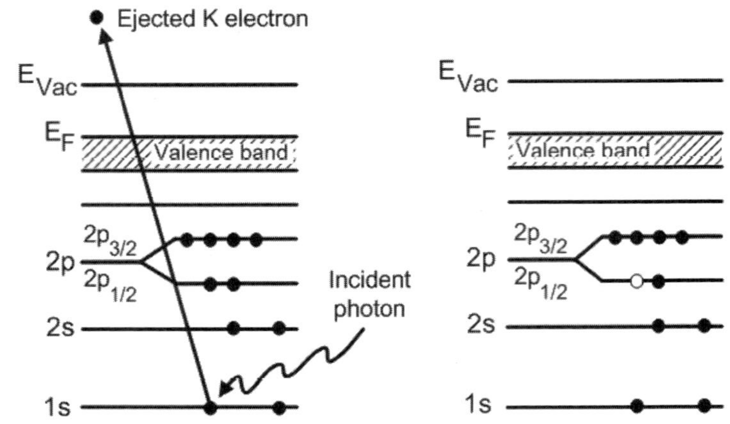

Figure 11.1 Initial and final states in the photoelectric effect in XPS.

physisorption), and adsorbate morphology (i.e., clusters/islands or uniform and conformal layer).

11.2 HISTORICAL PERSPECTIVE

Heinrich Hertz first observed the photoelectric effect in 1887 while working with metal contacts in electrical systems [1]. His observations, along with those of others working on the photoelectric effect in the later part of the 19th century, were not described theoretically until 1905 when Albert Einstein published his paper "Heuristic viewpoint on the production and conversion of light" [2]. This work proved to be one of the seminal events in the evolution of quantum mechanics and earned Einstein the 1921 Nobel Prize in Physics.

The first uses of XPS as an analytical tool occurred in the early 1950s [1]. Over the course of the next two decades, researchers developed the theory and instrumentation of XPS to produce the powerful technique currently in use. Kai Siegbahn, who coined and advocated the "electron spectroscopy for chemical analysis" (ESCA) name, is responsible for much of the pioneering work [3–7]. Siegbahn was awarded the 1981 Nobel Prize in Physics for his "contribution to the development of high-resolution electron spectroscopy."

11.3 INSTRUMENTATION

A simplified schematic diagram of an X-ray photoelectron spectrometer is shown in Figure 11.2. The X-rays generated by the source impinge upon the sample, resulting in the annihilation of the photons and ejection of photoelectrons. The photoelectrons are collected by electron optics and directed into an electron energy analyzer, where they are sorted by energy. The number of electrons per energy interval is then transduced to a current by

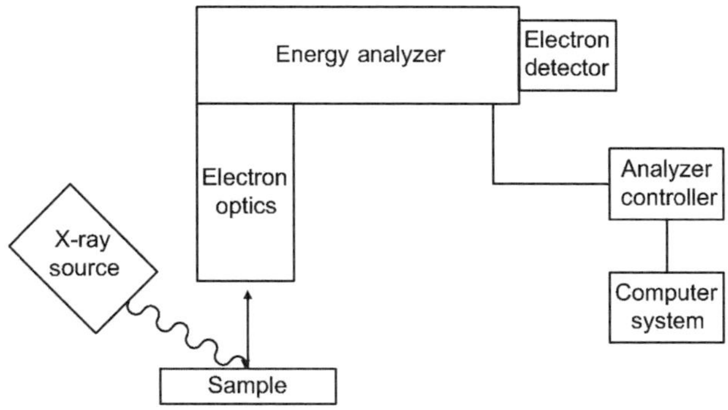

Figure 11.2 Schematic design of an X-ray photoelectron spectrometer.

an electron detector. The photocurrent is subsequently converted and processed into a spectrum by suitable electronics. The experiment is typically conducted under ultrahigh vacuum (UHV) conditions.

At the sample, the X-ray beam can have a diameter ranging from 5 mm down to 1–5 μm. Figure 11.3 shows a schematic representation of the X-ray excitation and photoelectron collection volumes. The photoelectrons are produced in the sample through the entire X-ray penetration volume. Because of the relatively short inelastic mean free path in the irradiated material and the typical kinetic energies possessed by the photoelectrons, only the photoelectrons produced in the top several nanometers of the sample are observed at their characteristic energies, while those from below this depth contribute to the background. This depth corresponds to approximately 10 atomic layers for heavier elements.

11.3.1 Vacuum System

X-ray photoelectron spectroscopy is performed in UHV, typically 10^{-9}–10^{-11} torr, for several reasons. The primary reason is to maintain sample surface integrity. Although the time it takes for the surface to be covered with containments varies depending on the gas molecules present in the chamber background and the reactivity of sample surface, the low pressure aids in surface cleanliness. Invoking the kinetic theory of gases and assuming a sticking coefficient of 1 for a monatomic gas, we can calculate that at medium vacuum (10^{-3} torr) only 1 ms is needed to cover a clean sample surface with a monolayer of adsorbate, at high vacuum (10^{-6} torr) only 1 s is needed, and at UHV (10^{-9} torr) 1000 s is needed. In addition, the vacuum condition minimizes the scattering of photoelectrons by other gas molecules that are present and maximizes the mean free path of the ejected photoelectrons.

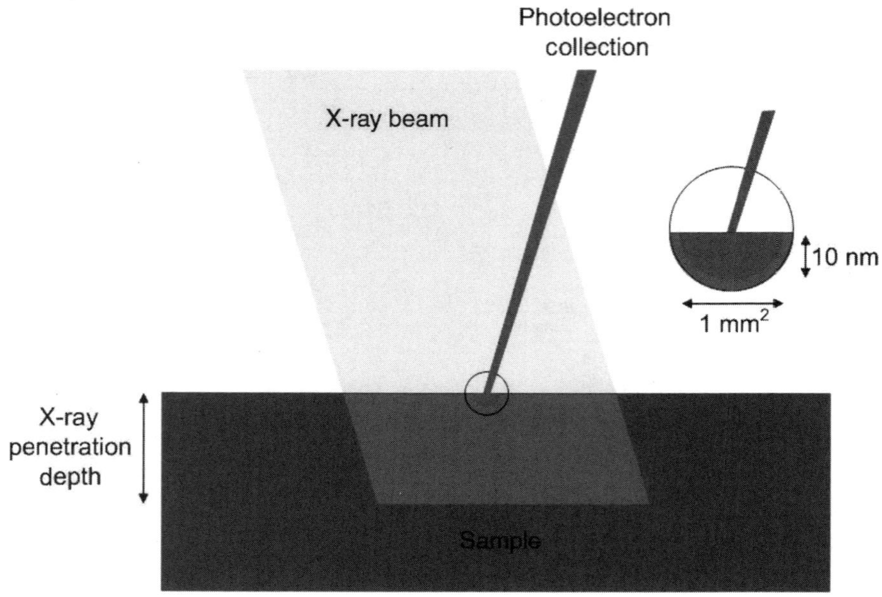

Figure 11.3 X-ray photoelectron collection scheme for an analyzer-lens system that defines analysis spot area.

Moreover, the vacuum condition facilitates the use of tungsten/coated-tungsten filament, LaB_6, or other electron sources used in the X-ray source cathode.

Ultrahigh vacuum conditions are commonly achieved and maintained with all-metal, leak-free components, minimizing or eliminating polymer and/or high vapor pressure materials, combinations of sorption and capture pumps, and baking of the chamber to remove adsorbed water vapor and other gases from the chamber and component walls. To enable high sample throughput, load lock systems are used to pre-pump a sample to high vacuum levels in sample insertion chambers separated from the main analysis chamber by gate valves, and then the sample is introduced into the main analysis chamber. The sample insertion chambers are often equipped with direct-drive rotary vane mechanical oil pumps and turbo molecular pumps, whereas the main analysis chamber vacuum is commonly maintained by ion pumps, titanium sublimation pumps, nonevaporative getter pumps, cryopumps, or sputter ion pumps. The pressure inside the chamber is usually measured with gauges such as cold cathode and nude ion gauges [8–11]. Pressure gauges should be placed in proximity to the sample stage to most accurately measure the sample environment.

11.3.2 X-ray Source

The X-ray source used in XPS must produce beams energetic enough to overcome the binding energies of core electrons. For best spectroscopic and analytical results, the X-ray emission from the source should be high intensity with a narrow linewidth.

In the X-ray source, a large electric potential is applied between a cathode, usually a heated tungsten filament, and the anode, producing a high current flow of electrons onto the target anode surface [8, 11]. As electrons from the cathode strike the anode, X-rays are generated through processes known as *Bremsstrahlung radiation* and *X-ray fluorescence.* The Bremsstrahlung process is attributed to the continuous electron deceleration towards the atomic nuclei of atoms in the anode; therefore it contributes a continuous spectrum. The X-ray fluorescence process involves the ejection of a core-level electron in the anode material, and as a higher level electron relaxes to fill the core-level hole X-ray photons are emitted. Therefore, the emission spectrum is characteristic of the anode material. In sum, the X-ray produced directly from the source may exhibit a spectrum of characteristic peaks superimposed on a continuous background.

A partial list of some common anode materials for X-ray sources is given in Table 11.1.

Table 11.1 Common X-ray sources used in XPS.

X-Ray Source	Energy (eV)	Linewidth (eV)
Y Mζ	132.3	0.47
Zr Mζ	151.4	0.77
Na Kα	1041.0	0.70
Mg Kα	1253.6	0.70
Al Kα	1486.6	0.85
Zr Lα	2042.4	1.7
Ti Kα	4510.0	2.0
Cu Kα	8048.0	2.6

Mg Kα (1253.6 eV) and Al Kα (1486.6 eV) are the two most commonly used X-ray photon sources in commercial XPS systems. Frequently, dual-anode X-ray sources such as Al/Mg, Mg/Zr, and Al/Zr with special properties such as enhanced sensitivity towards specific elements are also used for XPS.

Furthermore, the dual-anode X-ray source can also be used in furthering spectral interpretation, especially in identifying Auger electron peaks. For example, a comparison between the Mg Kα and Al Kα sources for Cu spectra are shown in Figure 11.4. Note that the Cu 2p and Cu 3p peak positions remain unchanged in both spectra, whereas the Auger peaks are shifted on the binding energy scale due to the energy difference between the Mg Kα and Al Kα lines. Cathode filament alignment is crucial in dual-anode sources because a misaligned cathode filament may result in "cross-talk," a spectral defect characterized by additional ghost peaks (i.e., the appearance of Zr Lα excited peaks when using the Mg Kα anode) [8, 11, 12].

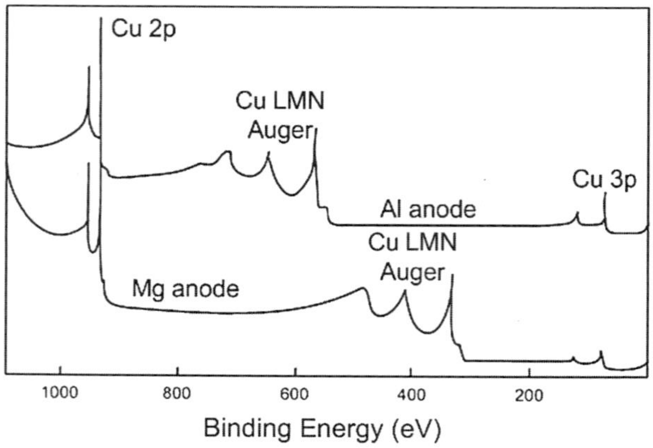

Figure 11.4 Cu XP spectra illustrating the shift in Auger peak positions with the change from Mg to Al anodes in the X-ray source. Adapted from [11].

These direct X-ray sources produce a high intensity photon flux reaching the sample surface, resulting in high photoelectron throughput to the detector. However, several disadvantages are associated with these sources. Bremsstrahlung radiation can induce sample damage. Emission from multiple X-ray fluorescence lines leads to satellite peaks in XP spectra. Finally, the XPS system's resolution is limited by the natural linewidth of the source, typically 1–2 eV. To alleviate the problems associated with direct application from X-ray sources, monochromators are often employed in XPS systems. Figure 11.5 shows one configuration for a monochromator used in XPS. Monochromators are made with crystals such as quartz, silicon, and germanium, which are used to select desired emission lines and/or remove the Bremsstrahlung continuum [8, 11]. The lattice spacing within these crystals is such that the characteristic X-rays emitted from the source will satisfy the Bragg relationship given below, equation (11.2), where n is the diffraction order, λ the wavelength of light, d the lattice plane spacing, and θ the angle of incidence:

$$n\lambda = 2d\sin\theta \tag{11.2}$$

Specific crystal materials are chosen for mechanical integrity to withstand the X-ray intensity. These materials must possess small mosaicity and high intensity output at the desired wavelength. Advances in manufacturing X-ray optics, such as silicon and germanium crystal monochromators, facilitates the use of XPS in providing information regarding lateral and spatial variation in surface composition, chemical state information such as oxidation states, and chemical bonding configurations of the atoms based on chemical shifts [8, 11]. Use of a focused X-ray gun and a well-engineered monochromator crystal can also reduce the size of the spot size on the sample from the mm to the μm range. Perhaps the most common configuration for X-ray generation is an Al X-ray source and a quartz crystal monochromator.

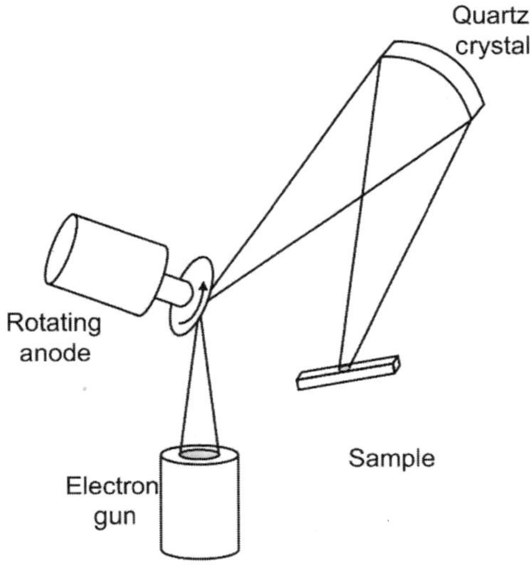

Figure 11.5 Schematic of components in a monochromatic XPS source with a rotating anode.

The development of XPS has been aided by the advent of third generation synchrotron radiation [9]. Synchrotron radiation is generated through magnetic field pulses that induce electrons to move in circular orbits at speeds near that of light. Continuous energy radiation is emitted as these electrons move in circular orbits to provide photons ranging from 10 eV to 10 keV. X-ray beams of specific energies may be selected using monochromators. However, the high monetary and time costs involved in using synchrotrons greatly limit accessibility for routine analysis. Additionally, the high intensity of synchrotron radiation may increase X-ray beam-induced sample damage. The use of synchrotron radiation for XPS analysis has recently been reviewed elsewhere [13, 14].

In some applications, particularly those involving investigation of coordination environment, an ultraviolet light source is used in the place of the X-ray source. This complementary technique to XPS is called *ultraviolet photoelectron spectroscopy* (UPS) [15].

11.3.3 Electron Energy Analyzer

Photoelectrons are produced as the incident X-rays strike the sample surface. The photoelectrons are then directed into the electron energy analyzer usually through a collecting lens (electrostatic or electromagnetic) to be sorted according to their respective energies. Although the linewidths of the measured photoelectron peaks depend mostly on the X-ray source's natural linewidth for non-monochromatic systems, the quality of the analyzer dictates the sensitivity and resolution of the X-ray photoelectron spectrometer. The most common types of electron energy analyzers for XPS are the *cylindrical mirror analyzer* (CMA) (see Figure 10.7 in Chapter 10, Auger Electron Spectroscopy) and the *concentric hemispherical analyzer* (CHA), which is also called the hemispherical sector analyzer or spherical capacitor analyzer (see Figure 10.8, Chapter 10). The CHA is the most frequently employed analyzer in XPS due to its high resolution, which is important in obtaining chemical state information [8–11].

Both types of analyzers are also commonly used in Auger electron spectroscopy. As described in Chapter 10, the cylindrical mirror analyzer has high transmission efficiency, and it is also compact in size and easy to use. However, the CMA is very sensitive to sample-to-analyzer distance in energy scale linearity and complicates the precise determination of Auger or photoelectron energy. In the hemispherical sector analyzer, the photoelectrons ejected by the sample are focused using electron optics into the entrance slits, and the photoelectrons are separated as they travel through the electric field between the inner and outer spheres. As the photoelectrons exit the analyzer they enter electron detectors such as electron multiplier tubes, multichannel plates, or anode electrodes. The electron detectors act as signal transducers to yield a current, which is then processed to produce XP spectra.

11.4 SAMPLE SELECTION AND PREPARATION

To enable analysis of the true sample surface, cleaning and etching are typically avoided prior to XPS analysis because these processes will alter the near surface chemistry and topology. However, in some cases the sample for XPS analysis needs to be cleaned, outgassed, and reduced in size prior to introduction into the vacuum chamber. Additional care is generally given to sampling. The goal of sampling is to select a portion of the sample that is representative of its entirety and to avoid undesired sample modification and contamination. When necessary, samples may be cleaned using combinations of filtered solvents, chemical etching, CO_2 "snow," plasma cleaning, and ion sputtering cleaning [8].

Sample mounting is commonly accomplished by affixing the specimen to the sample holder using one of a variety of commercially available electrically conductive adhesive tapes. The contamination introduced by the adhesive tape may be characterized prior to sample analysis by the collection of a contamination depth profile [16]. Additionally, although small particles and powderlike samples may sometimes be analyzed without affixing them to the sample holder, at other times it may be necessary to secure the small specimens on the sample stage. This type of sample mounting may be achieved by evacuating between the sample and sample stage to create airlocks, and by using silver paint, silver epoxy, indium foil, or conductive tape [8].

11.4.1 Sample Charging

As photoelectrons are emitted by the sample after X-ray bombardment, the sample becomes positively charged. The charged surface possesses a potential that may alter the velocity of the ejected photoelectrons as they are directed into the electron energy analyzer and distort the energy measurement of the electrons. With conductive samples, this charging effect may be compensated by electrically connecting the sample to the spectrometer. The electrical contact equilizes the Fermi level of the sample and spectrometer. Therefore, the kinetic energy of the photoelectrons can be accurately measured if the spectrometer is set to measure the kinetic energy of the entering electrons with respect to the spectrometer's own Fermi level [8–10, 17–19]. Figure 11.6a depicts the energy levels for conducting samples.

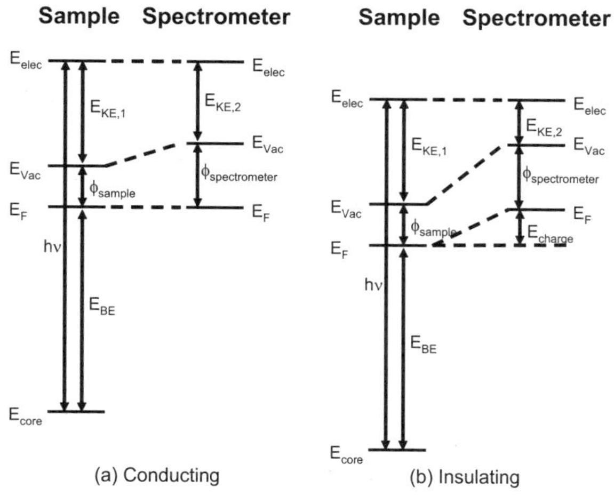

Figure 11.6 Schematic for energy levels between the instrument and (a) conducting or (b) insulating samples. Here E_{elec}, E_{Vac}, E_F, E_{core} are the energy levels of the free electron, the vacuum, the Fermi level, and the core level, respectively, from which the photoelectron was generated, $h\nu$ is the energy of the incident X-ray photon, E_{BE} the binding energy of the electron, $E_{KE,1}$ the kinetic energy of the electron, $E_{KE,2}$ the measured kinetic energy of the electron, ϕ_{sample} the sample's work function, $\phi_{spectrometer}$ the spectrometer's work function, and E_{charge} the magnitude of the sample charging. Adapted from [20].

Additional procedures are needed to treat insulating samples to account for the charging effect. Electrically connecting the sample and spectrometer is obviously not possible. As a result the Fermi level of the sample and spectrometer may be different with insulting samples, leading to error in the binding energies calculated from the photoelectrons' kinetic energies. The relative energy levels involved are illustrated in Figure 11.6b. The two main problems in analyzing insulating materials through XPS stem largely from variable surface potential of insulating materials caused by differential charging. The first problem results from the lack of knowledge about the surface potential of the insulating material as charge accumulates, which leads to difficulties in matching measured photoelectron binding energies to tabulated values. The second problem is attributed to the heterogeneity, both with respect to location and time, in the surface potential of the insulating material, which may greatly distort the spectral lineshapes. Because the aforementioned problems contribute to lower spectral resolution despite the use of monochromatic X-ray beams, the surface potential of insulating materials must be controlled.

To control the surface potential of insulating samples, the accumulated sample charges need to be compensated. The techniques involved in compensating the sample charges are known as "charge neutralization" or "charge compensation." The goal of charge neutralization is to provide uniform and stable surface potential. Traditionally, flood guns are used to provide a source of low energy electrons to take the place of the ejected photoelectrons in XPS, as shown in Figure 11.7.

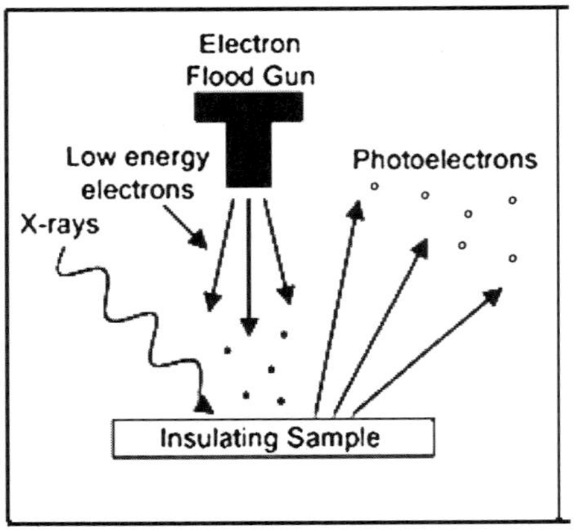

Figure 11.7 The use of a flood gun in XPS. Adapted from [8].

The use of a non-monochromatic X-ray source may help in minimizing problems associated with insulating samples. The non-monochromatic X-ray beams are divergent and are therefore broader than monochromatic X-ray beams [8, 10, 11]. As a result, the non-monochromatic X-ray beams may bombard other surfaces in the sample chamber to create copious numbers of low energy secondary electrons that may be used to compensate the emitted photoelectrons. Additionally, the differential charging may also be reduced through placing additional sources of low energy electrons on the sample holder in proximity to the insulating sample. For example, the placement of a metallic wall surrounding the sample was shown to reduce differential charging [19]. Other techniques to reduce differential charging may include placing the sample on a conductive platform and mixing the sample with another substance, such as carbon black powder in a 50/50 ratio [8, 21].

There are several considerations one makes in achieving ideal charge neutralization for insulating samples [8]:

- To achieve uniform surface potential, the low energy electrons emitted by flood guns should exhibit narrow energy distribution.
- The flood gun should be positioned such that the low energy electron beam emitted by the flood gun is normal to the plane of the sample surface to avoid deflection.
- The flood gun electron energy should be set such that it is high enough to eliminate space charge effect, but low enough to avoid overcharging sample.

Charge neutralization by the application of an external bias has recently been investigated. The external bias technique, involving applying a 0–10-V bias to the sample during data collection, may be used to increase the separation of overlapping peaks to improve spectral resolution [18]. The external bias alleviated differential sample charging through controlling the flow of low energy electrons to the sample. The researchers successfully utilized the external bias to resolve Au 4f peaks from gold nanoparticles in a SiO_2 matrix and Au 4f peaks from gold metal strips, resulting in a facile method for analyzing heterostructures [18].

11.4.2 X-Ray Beam Effects

Because of the penetrative nature of X-ray, sample damage may be of concern in conducting XPS experiments. Materials such as metals and ceramics are usually stable under X-ray irradiation. However, the radiation can induce alterations in certain polymers, such as poly vinyl chloride (PVC) and soft biological specimens, especially when the samples are in thin films [8]. Alterations of the sample induced by X-ray may be physical and/or chemical. Some physical structural changes may include creation of point defects, changes in near surface crystal structure, and changes in the surface topography. Point defects are of concern when analyzing metallic oxides and multielement samples [8]. The possible X-ray-induced point defects include lattice and surface vacancies and dislocations that may complicate spectral interpretation [8].

X-ray-induced chemical changes are of particular concern when using XPS to investigate the effects of chemical treatment on a sample. Decoupling of X-ray-induced chemical changes from those resulting from the chemical treatment is challenging. In addition to chemical changes brought about by the X-ray irradiation itself, the secondary electrons that may be generated as a by-product of the irradiation could also cause additional chemical changes. Specifically, both secondary electron and X-ray irradiation may cause chemical changes, including changes in oxidation states and the cleavage of chemical bonds. Additionally, the surface chemical composition may be altered after X-ray irradiation. The radiation may enhance the surface reactivity and increase adsorption of background gas molecules, break chemisorption bonds to increase desorption of surface species, or induce migration of surface species on the surface. Generally, the use of monochromatic X-ray sources may reduce the amount of sample damage because much of the energy in the original X-ray beam does not contribute to the useful spectral information. The monochromator attenuates the beam, making it a more effective X-ray source from a damage standpoint [8, 11].

11.5 SPECTRAL ANALYSIS

As Figure 11.4 demonstrates, XP spectra are generally plotted with respect to binding energy, increasing from right to left. The binding energy scale may be thought of as the reverse of kinetic energy, which would increase from left to right (see equation (11.1)). Typically, the peaks in the range from 0 eV to ~15 eV in binding energy are associated with valence electrons. These valence electron peaks are followed by characteristic peaks of the core-level electron ejection at higher binding energies. The valence electron orbitals are often delocalized or hybridized such that the ejection of the valence electrons may result in peaks so closely spaced as to present a valence band structure.

The valence band spectra may provide additional information to the core spectra to yield valuable information in differentiating similar compounds. For example, valence spectra were used to differentiate industrial lubricants on an aluminum surface that exhibit extremely similar core spectra [22]. These valence band peaks are often useful for investigating not simply the chemical environment of an atom but also changes in coordination around the atom. An example of this type of analysis is the differentiation of molybdenum in the 6+ oxidation state in a tetrahedral versus octahedral coordination environment [23]. Researchers have successfully identified sodium molybdate, which contains tetrahedral Mo^{6+}, and ammonium heptamolybdate, which contains octahedral Mo^{6+}, using valence band XPS [23]. These valence band peaks are often investigated using an ultraviolet light source in UPS [15].

In the low binding energy region, the spectral background may be attributed to the Bremsstrahlung radiation and secondary electron cascade background; at higher binding energy the background may be attributed mainly to inelastic electron scattering, and the steplike rise character may be attributed to photoelectrons losing energy as they reach the electron detector

[8, 10]. The increasing background from low to high binding energy that is seen in Figure 11.4 is significantly reduced with the use of monochromatic sources. Prior to analysis such as peak-fitting and peak area quantification, XP spectral backgrounds are commonly removed with either linear or Shirley background removal schemes found in commercially available XPS systems [16, 21, 24–26]. Other background removal schemes may involve principle component analysis, polynomial approximation, and the use of reflected electron energy loss spectroscopy (REELS) data [26].

Auger electron peaks may also be observed, often in series in the XP spectra. The major Auger series that are commonly observed in XPS include the KLL, the LMM, and two types of the MNN series [8, 10, 11]. Because the Auger electron's peak position on the binding energy scale is dependent upon the energy of the excitation source, while the photoelectron's peak position on the binding energy scale is independent of the source energy, the identification of Auger electron peaks may be accomplished by switching the X-ray source anode as is illustrated in Figure 11.4.

Although the binding energies of the photoelectrons are usually adequate for elemental identification, XPS offers information that enables identification of atoms of the same element in different chemical states. Surrounding species can cause changes in the binding energies of the core electrons; these changes are called "chemical shifts." Since the chemical shifts caused by surrounding species are generally less than 10 eV and the linewidth of the peaks are usually ~1 eV, complications such as spectral overlaps of the atoms in different chemical states, difficulties in eliminating spectral background, and further difficulties in analyzing insulating samples may arise [8]. Possible solutions include curve-fitting, careful choice in background correction, and use of an internal reference.

Figure 11.8 contains an XP spectrum of the C 1s region for polymethylmethacrylate (PMMA), which illustrates the differentiation of the carbon atoms in various chemical states. Curve-fitting identified four peaks, labeled 1–4 in the spectrum, in the C 1s binding energy region near 285 eV. The four peaks are assigned to specific carbon atoms within PMMA. For carbon, as with most atoms, the atom in the most electron-withdrawing environment will have the highest binding energy. For PMMA, this atom corresponds to the carbon in the ester group (number 4 carbon in the inset to Figure 11.8). The next highest binding energy, ~287 eV, corresponds to the carbon bound to the oxygen of the ester, number 3 in the inset to Figure 11.8. The carbon adjacent to the carbonyl group is in the next most electron-withdrawing environment and corresponds to the peak labeled 2 in Figure 11.8. Although the two carbons assigned as number 1 carbon are in slightly different chemical environments, they are similar enough that the corresponding peaks are unresolved and fit as a single peak. In addition, the aforementioned rationale is corroborated by the relative intensities of the peaks—the areas of peaks 2, 3, and 4 are relatively similar, implying similar

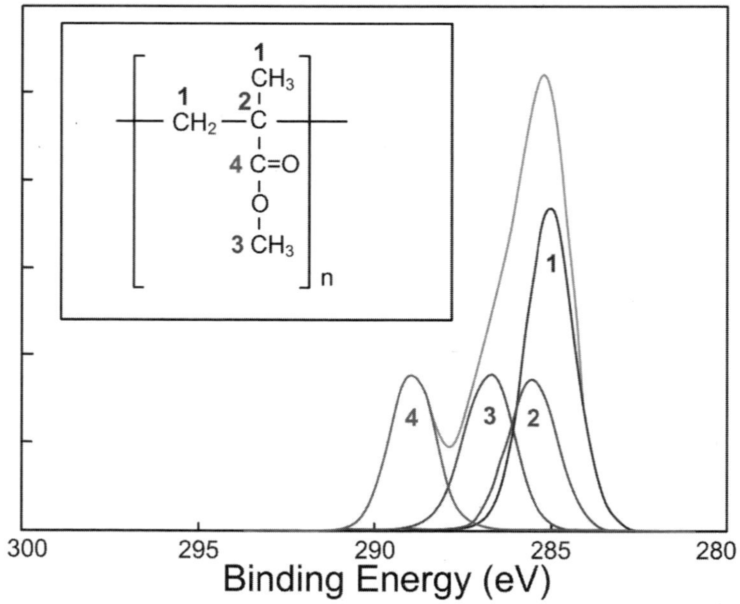

Figure 11.8 C 1s region XP spectrum for polymethylmethacrylate (PMMA). The inset shows the repeat unit for PMMA along with identification of the peaks in the spectrum.

percentage of carbon atoms in the sample volume analyzed, and peak 1 has approximately twice the total area of the other peaks.

Other factors affecting the binding energies of the ejected photoelectrons may include sample size, morphology, and thermal history, as well as the surrounding matrix, means of charge compensation, and application of external bias [8, 9, 18].

11.5.1 Core Level Splitting

When the photoelectrons are emitted from levels with the orbital angular momentum quantum number (l) greater than 0, a doublet in the spectrum may be observed because of *spin-orbit coupling.* Spin-orbit coupling arises when an orbital possesses more than one possible state with different total angular momentum quantum number ($j = l + s, j =| l - s|$) values. These two states are energetically different depending on whether the spin vector and orbital angular momentum vector of the unpaired electron resulting from the ejection of the photoelectron are aligned or not. Examples of the spin-orbit coupling phenomenon are shown in Figure 11.9, which is discussed in more detail below.

The energy difference between the doublet, ΔE_j, is dependent on the spin-orbit constant, ξ_{nl}, which is related to the expectation value of the inverse of radius cubed, $\langle 1/r^3 \rangle$, of the particular orbital involved [8]. The relative peak areas of the two doublet peaks are ultimately dependent on the atomic photoemission cross-section, σ. The atomic photoemission cross-section is related to the atomic number Z, principal quantum number n, orbital angular momentum quantum number l, the total angular momentum quantum number j, and how close the X-ray energy is to the photoemission threshold [8, 11].

Table 11.2 lists common area ratios of doublet splitting. The core level splitting resulting in doublet formation is also illustrated in Figure 11.9. The C 1s shell photoelectron, with only one possible j value, exhibits a single spectral peak, whereas the Cu 2p, Ag 3d, and Au 4f photoelectrons exhibit doublet peaks at approximately the ratio indicated in Table 11.2. Since the binding energy difference between the doublets is specific to the metal and orbital involved, ΔE_j can be a helpful tool in element identification.

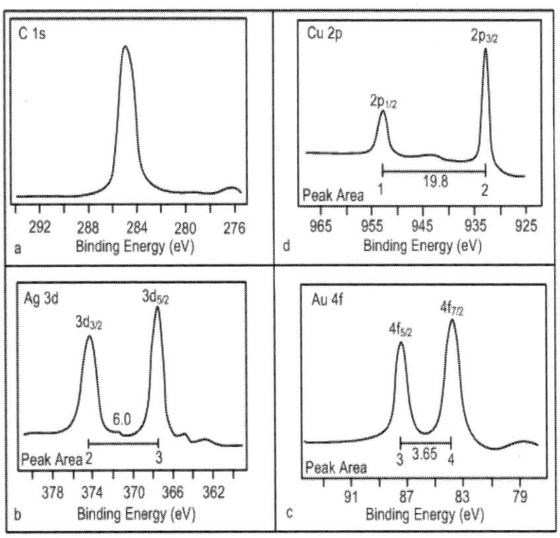

Figure 11.9 Core level splitting patterns for (a) C 1s, (b) Ag 3d, (c) Au 4f, and (d) Cu 2p.

Table 11.2 Summary for core level splitting relative intensities.

Subshell	j Values	Area Ratio; $(2j_1 + 1):(2j_2 + 1)$
s	1/2	
p	1/2, 3/2	1:2
d	3/2, 5/2	2:3
f	5/2, 7/2	3:4

11.5.2 Linewidths

The linewidths of the XP spectral peaks, ΔE, is generally on the order of 1 eV and is defined at half maximum and may be roughly approximated by equation (11.3) [8].

$$\Delta E = \sqrt{\Delta E_n^2 + \Delta E_p^2 + \Delta E_a^2}\,. \tag{11.3}$$

Here, ΔE_n is the natural linewidth, ΔE_P is the linewidth of the X-ray source, and ΔE_a is the energy resolution of the electron analyzer. Invoking the uncertainty principle, the ΔE_n may be defined as shown in equation (11.4) [8].

$$\Delta E_\mathrm{n} = \frac{h}{\tau} = \frac{4.1 \times 10^{-15}}{\tau}. \text{ (in eV)}. \tag{11.4}$$

Here, h is Planck's constant and τ is the lifetime of the hole state left behind by the ejected photoelectron. The lifetime of the hole state is determined by the rate for core-hole elimination as illustrated in Figure 11.10.

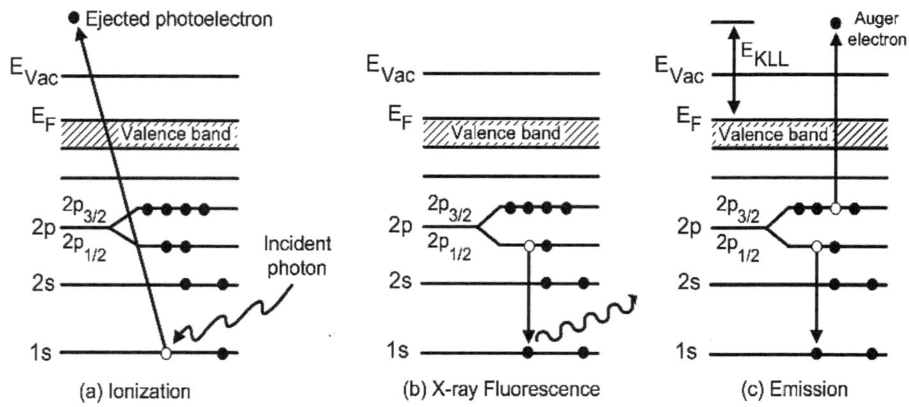

Figure 11.10 (a) Schematic representations of the photoelectric process, (b) elimination of core-hole through X-ray fluorescence, and (c) elimination of core-hole through ejection of an Auger electron.

The emission of the Auger electron through the Coster-Konig process as depicted in Figure 11.10c occurs within approximately 10^{-15} s [8].

11.5.3 Elemental Analysis: Qualitative and Quantitative

Generally, qualitative elemental identification is readily achieved by comparing the peak binding energies to tabulated values. Numerous sources of tabulated XPS binding energies are available both in text and on-line [11, 27]. In many cases elemental identification can be accomplished with the acquisition of survey spectra. For chemical state analysis, high resolution spectra are acquired and fit.

The relative intensities of the spectral peaks, typically obtained by integrating peak areas from curve-fitting, are directly related to the relative amount of the corresponding atom in the sample. This direct relationship allows for facile quantification of different types of atoms after background elimination [8]. In the simplest case of a homogenous solid, the concentration of any chosen atom A may be found using equation (11.5) [8].

$$C_{\mathrm{A}} = \frac{\dfrac{I_{\mathrm{A}}}{S_{\mathrm{A}}}}{\sum_{n} \left(\dfrac{I_n}{S_n} \right)}. \tag{11.5}$$

Here C_{A} is the concentration of A in the volume sampled, I_{A} is the intensity of the spectral line attributed to atom *A*, and S_{A} is the sensitivity of the spectrometer to *A*. The sensitivity of a spectrometer varies for different atoms as illustrated in Figure 11.11. The I_n and S_n notations are used to represent all atoms found in the XPS survey.

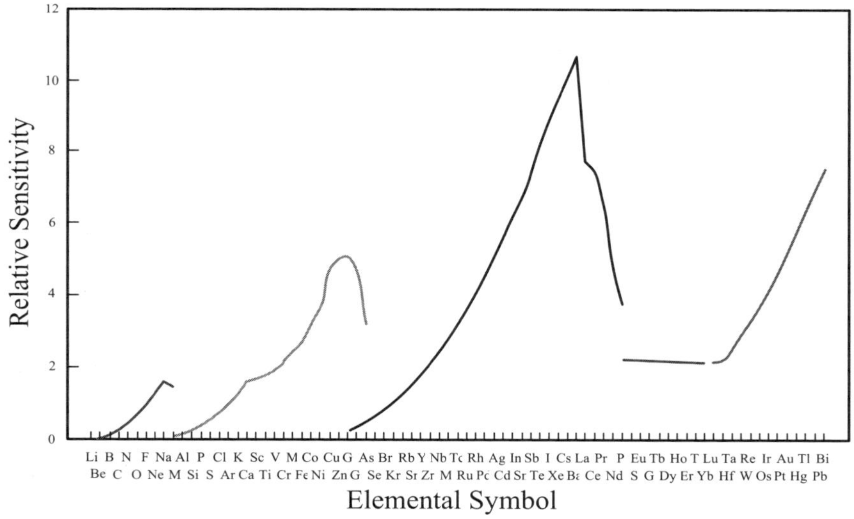

Figure 11.11 Relative sensitivity for different elements in XPS. Trends may vary depending on the spectrometer. Adapted from data in [11].

11.5.4 Secondary Structure

Exchange splitting, also referred to as *electrostatic splitting*, occurs when electrons in the valence levels are unpaired [8]. In the simplest case of an s orbital, the unpaired electron in the valence level interacts with the unpaired electron left in the core level following photoemission. Thus, the final state of photoemission, which determines the kinetic energy that will be measured, depends on whether the spins of the aforementioned unpaired electrons are parallel or antiparallel. This exchange interaction results in splitting of the photoelectron peak into a doublet separated by the exchange energy. The exchange energy is defined as the difference between the binding energy of the photoelectron and the ground state orbital energy in the final state configuration. The exchange energy may be written as shown in Equation 11.6 [8].

$$E = (2S + 1)K_{\mathrm{CO,VO}}. \tag{11.6}$$

Here S represents the sum of the unpaired electron spin and $K_{\mathrm{CO,VO}}$ denotes the exchange integral of the core orbital and the valence orbital. When dealing with non s-orbitals, angular momentum coupling in addition to the exchange splitting causes the photoelectron peak to be split into multiplets. Furthermore, the exchange splitting is most dominant when the valence level and the core level from which the photoelectron originates are of the same principal quantum number n [8].

"Shake-up lines" are important secondary spectral structures. Following photoemission, the valence level electrons experience a greater effective nuclear charge as compared to prior photoemission [28]. The change in the nuclear attraction causes valence electron relaxation and reorganization, which may cause valence electrons to be excited to a higher unfilled orbital, resulting in an overall electronically excited system as its final state configuration. In this case the ejected photoelectron experiences a reduction of its kinetic energy to compensate for the excited final state. The photoelectrons that undergo the kinetic energy reduction as mentioned above are recorded as shake-up lines in XPS, appearing a few eV above its characteristic peak on the binding energy scale. The shake-up lines may be used in differentiating atoms of the same element in various oxidation states. These lines are particularly strong and therefore valuable with highly conjugated organic species and certain transition and rare earth metals [8]. An example of shake-up satellite lines is shown in Figure 11.12.

11.6 XPS IMAGING

Imaging or mapping surface composition in two dimensions is common to many surface analysis methods such as Auger electron spectroscopy and secondary ion mass spectrometry. On the other hand, XPS has long suffered from poor lateral resolution relative to these other techniques because of difficulties in constructing and utilizing X-ray optics. However, significant advances in spatially-resolved XPS have occurred over the last several decades [14, 30–33]. Two different methods have been employed to produce spatially resolved images [1]. The first involves focusing the X-ray source to the smallest possible spot size. With this method, images are collected in one of two ways, either by scanning the X-ray beam across the sample surface or moving the sample under a fixed position X-ray beam. The second method to achieve spatially resolved images utilizes position-sensitive electron detection. Generally, the best spatial resolution, which is on the order of a few μm, is achieved with this latter method. The XPS imaging technique provides information for facile understanding of surface species and morphology. Figure 11.13 is an XPS image of a gold substrate upon which copper has been deposited in a spatially controlled fashion [34].

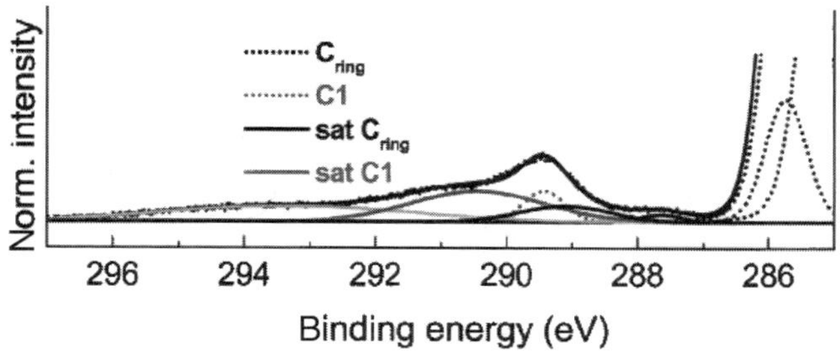

Figure 11.12 XP spectrum of C 1s region for benzoperylene-(1,2)-dicarboxylic acid anhydride, an aromatic organic molecule [29]. Carbon peaks are fitted with dotted lines and the shake-up satellite peaks are fitted with solid lines. Reused with permission from A. Schöll, Y. Zou, M. Jung, Th. Schmidt, R. Fink, and E. Umbach, (2004). *Journal of Chemical Physics*, **121**, 10, 260. Copyright 2004, American Institute of Physics.

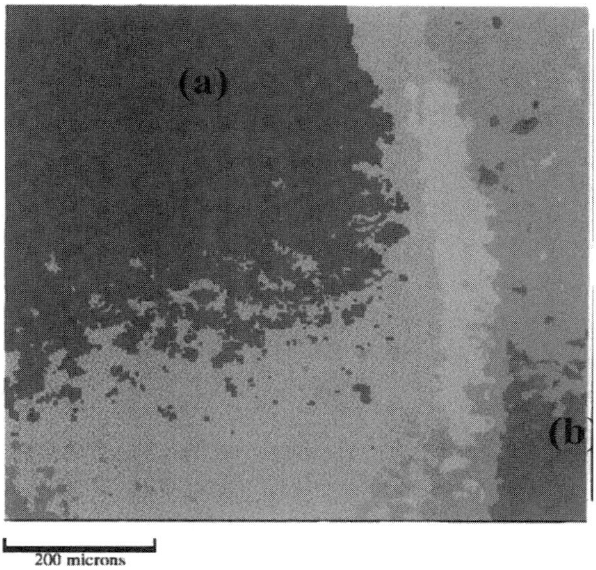

Figure 11.13 XPS image showing a Au substrate on which Cu has been deposited with a potential gradient method. The region labeled (a) is high in Au content and that labeled (b) is high in Cu [34]. Reprinted with permission from K.M. Balss, B.D. Coleman, C.H. Lansford, R.T. Haasch, and P.W. Bohn. (2001) *J. Phys. Chem. B* **105** 8970–8978. Copyright 2001 American Chemical Society.

11.7 ANGLE-RESOLVED XPS

Angle-resolved XPS offers the ability to tune the surface specificity of the XPS signal over the top 0–10 nm of the sample by varying the emission angle of the ejected photoelectrons. Note that the emission angle is measured with respect to the surface normal—differentiating it from the take-off angle, which is measured with respect to the surface plane. As the emission angle θ is altered the escape depth of photoelectrons changes as described in equation (11.7), where λ_e is the inelastic mean free path for electrons in the material [35].

$$\text{depth} = \lambda_e \cos\theta \tag{11.7}$$

Therefore, the least surface-specific information is obtained when the emission angle for the photoelectron escape is along the surface normal, $\theta = 0^\circ$. This orientation maximizes the depth from which photoelectrons are able to escape from the sample and be collected by the electron optics while retaining their characteristic energies. Changing the emission angle toward the surface as illustrated in Figure 11.14 reduces the absolute depth along the surface normal from which electrons can escape with their characteristic energies. A change from 0° to 75° results in an approximately four-fold reduction in the probe depth along the surface normal. Recording and analyzing spectra as a function of depth into the sample provides a glimpse of the near-surface compositional variation *without creating the need to destructively remove material from the surface* [16]. However, care must be taken with data interpretation for several reasons [35]. First, particularly in layered materials, λ_e is dependent on the identity of the material through which the photoelectrons pass. This phenomenon results in deviations from the simple cos θ dependence predicted by equation (11.7). Additionally, the use of *grazing angles*—high emission angles—increases photoelectron energy lost to surface plasmon bands. Finally, the high surface specificity at low collection angle can reduce the expected width for peaks due to a decrease in the coordination number for the surface atoms [35].

Angle-resolved XPS allows for depth-profiling of many samples and has been used extensively for the analysis of self-assembled monolayers on various substrates [36–38]. Additionally, angle-resolved XPS combined with XPS imaging enables the acquisition of three-dimensional information from the sample. This method is referred to as "3D XPS".

11.8 RECENT ADVANCES AND APPLICATIONS

A wide variety of advancements in XPS instrumentation and application has taken place recently. Many of the instrumental advances have focused on the aforementioned improvements afforded by the use of synchrotron radiation. The spatial resolution attainable with XPS imaging continues to improve as described previously. Continued efforts on engineering X-ray and electron optics, analyzers, and detectors will likely serve to broaden the influence of XPS. Two additional areas in which advancement and novel application have been made are fast or time-resolved XPS and high pressure XPS, as described below.

The ability to monitor chemical reactions is incredibly important to the understanding and designing of catalysts; historically, XPS has played a role in studying catalysts, but recent efforts toward real-time monitoring of surface reactions have changed that role [39, 40]. The advent of third generation synchrotron X-ray sources and advances in high speed electronics have helped reduce the time needed for acquisition of high resolution spectra from the minute to second time range. Now reactions such as molecular adsorption and desorption [40], the trimerization of ethylene to form benzene [41], and the epoxidation of alkenes can be monitored in real time [42]. Although the temporal resolution for fast XPS does not rival that of other analytical methods, the type of information XPS provides—namely, chemical state information—is very useful in these surface reaction studies. A new method using the application of a pulsed voltage to the sample may enable the study of millisecond timescales for specific applications [43].

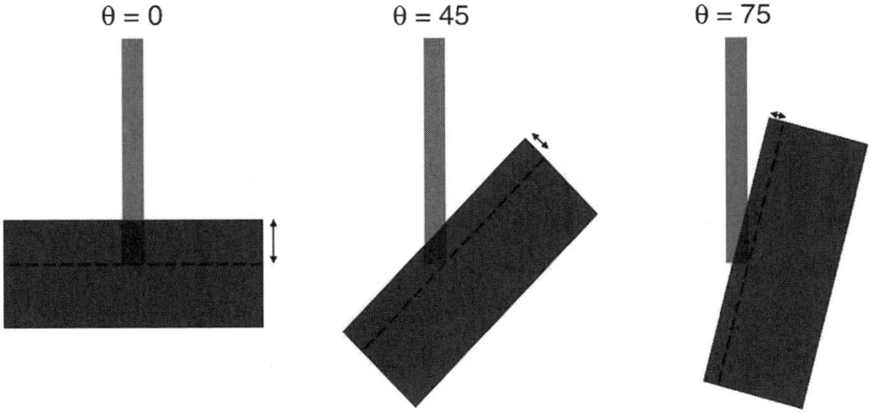

Figure 11.14 Illustration of the variation in photoelectron escape depth as a function of emission angle.

High pressure systems for XPS studies have an extensive history, dating back decades [44–46]. Recently, however, researchers presented a new system for XPS analysis in environments up to the mbar pressure range [47]. The system relies on a combination of differential pumping and electrostatic focusing of the photoelectrons. The authors speculate that this method may find extension up to a pressure near 100 mbar. This high pressure XPS system was recently used to investigate the enhancement of halide concentrations at the liquid/vapor interface [48]. Use of this technique may advance the understanding of the interaction of water with solid interfaces.

11.9 CONCLUSIONS

Since its proliferation as an analysis method, XPS has been applied to study many material types used in a wide range of fields from metallurgy to bio-medical engineering and semiconductor processing to archeology. The main strength of XPS lies in its ability not simply to identify elements but also to provide information regarding chemical environments, oxidation states, and morphology. Combine this wealth of materials information with a rapidly progressing ability to spatially map composition across the surface plane and normal to the plane and XPS promises to retain its place as a vital spectroscopic tool for those working in a wide range of fields in both industry and academia.

ACKNOWLEDGMENTS

The authors thank Wesley R. Nieveen, William F. Coleman, John F. Moulder, and Craig M. Teague for their assistance in review of the manuscript. The authors also thank Lin Zhu for assistance with preparation of figures and Richard T. Haasch for the use of Figures 11.3, 11.8, and 11.11.

REFERENCES

[1] Ratner, B.D. & Castner, D.G. (1997) Surface analysis: *The principal techniques JC Vickerman* (Chichester [England]; New York: John Wiley) p 43-98.
[2] Einstein, A. (1905) *Annalen der Physik* **17,** 132-148.
[3] Sokolowski, E., Nordling, C. & Siegbahn, K. (1958) *Physical Review* **110**, 776.
[4] Hagstrom, S., Nordling, C. & Siegbahn, K. (1964) *Physics Letters* **9,** 235-236.
[5] Nordling, C., Hagstrom, S. & Siegbahn, K. (1964) *Zeitschrift fuer Physik* **178,** 433-438.
[6] Siegbahn, K., Hammond, D., Fellner-Feldegg, H. & Barnett, E.F. (1972) *Science* **176,** 245-252.
[7] Siegbahn, K. (1982) *Science* **217,** 111-121.
[8] Briggs, D. & Grant, J.T. (2003) In *Surface Analysis by Auger and X-Ray Photoelectron Spectroscopy* (pp. 1-30).
[9] Barr, T.L. (1994) Modern ESCA *The Principles and Practice of X-Ray Photoelectron Spectroscopy* Boca Raton: CRC Press.
[10] Watts, J.F. & Wolstenholme, J. (2003) *An Introduction to Surface Analysis by XPS and AES* (Chichester, West Sussex; New York, NY: J. Wiley) Vol. X.
[11] Moulder, J.F., Stickle, W.F. & Sobol, P.E. (1993) *Handbook of X-Ray Photoelectron Spectroscopy* (Eden Prairie, Minn.: Perkin-Elmer, Physical Electronics Division).
[12] http://www.thermo.com/com/cda/product/detail/1, 15955,00.html.
[13] Margaritondo, G. (2003) In *Surface Analysis by Auger and X-Ray Photoelectron Spectroscopy* pp. 733-748.
[14] Kinoshita, T. (2002) *J. Electron Spectrosc. Relat. Phenom.* **124,** 175-194.
[15] Leckey, R. (2003) *Springer Series in Surface Sciences* **23,** 337-345.
[16] Oswald, S. & Baunack, S. (2003) *Thin Solid Films* **425,** 9-19.

[17] Fulghum, J.E. (1999) *Journal of Surface Analysis* **6,** 13-21.
[18] Ertas, G., Korcan Demirok, U. & Suzer, S. (2005) *Appl. Surf. Sci.* **249,** 12-15.
[19] Oyama, T., Nishizawa, S. & Yamamoto, H. (1997) *Journal of Surface Analysis* **3,** 558-564.
[20] Kelly, M.A. (2003) In *Surface Analysis by Auger and X-Ray Photoelectron Spectroscopy.* (pp. 191-210).
[21] Turner, N.H. & Schreifels, J.A. (2000) *Anal. Chem.* **72,** 99R-110R.
[22] Havercroft, N.J. & Sherwood, P.M.A. (1998) *J. Vac. Sci. Technol., A* **16,** 1112-1116.
[23] Fiedor, J.N., Proctor, A., Houalla, M., Sherwood, P.M.A., Mulcahy, F.M. & Hercules, D.M. (1992) *J. Phys. Chem.* **96,** 10967-10970.
[24] Flamia, R., Lanza, G., Salvi, A.M., Castle, J.E. & Tamburro, A.M. (2005) *Biomacromolecules* **6,** 1299-1309.
[25] Scrocco, M. (1979) *Chem. Phys. Lett.* **63,** 52-56.
[26] Seah, M.P. (2001) *Surf. Sci.* **471,** 185-202.
[27] http://srdata.nist.gov/xps/.
[28] Kover, L. (2003) In *Surface Analysis by Auger and X-Ray Photoelectron Spectroscopy* (pp. 421-464).
[29] Scholl, A., Zou, Y., Jung, M., Schmidt, T., Fink, R. & Umbach, E. (2004) *J. Chem. Phys.* **121,** 10260-10267.
[30] Vohrer, U., Blomfield, C., Page, S. & Roberts, A. (2005) *Appl. Surf. Sci.* **252,** 61-65.
[31] Reniers, F. & Tewell, C. (2005) *J. Electron Spectrosc. Relat. Phenom.* **142,** 1-25.
[32] Escher, M. et al. (2005) *J. Electron Spectrosc. Relat. Phenom.* **144-147,** 1179-1182.
[33] Blomfield, C.J. (2005) *J. Electron Spectrosc. Relat. Phenom.* **143,** 241-249.
[34] Balss, K.M., Coleman, B.D., Lansford, C.H., Haasch, R.T. & Bohn, P.W. (2001) *J. Phys. Chem. B* **105,** 8970-8978.
[35] Ghosh, P.K. (1983) *Introduction to Photoelectron Spectroscopy* (New York: Wiley) Vol. X.
[36] Baunach, T., Ivanova, V., Kolb, D.M., Boyen, H.-G., Ziemann, P., Buettner, M. & Oelhafen, P. (2004) *Adv. Mater.* **16,** 2024-2028.
[37] Ye, S., Li, G., Noda, H., Uosaki, K. & Osawa, M. (2003) *Surf. Sci.* **529,** 163-170.
[38] Doron-Mor, I., Hatzor, A., Vaskevich, A., Van Der Boom-Moavt, T., Shanzer, A., Rubinstein, I. & Cohen, H. (2000) *Nature* **406,** 382-385.
[39] Paolucci, G. (2001) *J. Phys.: Condens. Matter* **13,** 11293-11303.
[40] Baraldi, A., Comelli, G., Lizzit, S., Kiskinova, M. & Paolucci, G. (2003) *Surf. Sci. Rep.* **49,** 169-224.
[41] Lee, A.F., Wilson, K., Middleton, R.L., Baraldi, A., Goldoni, A., Paolucci, G. & Lambert, R.M. (1999) *J. Am. Chem. Soc.* **121,** 7969-7970.
[42] Cropley, R.L., Williams, F.J., Urquharta, A.J., Vaughan, O.P.H., Tikhov, M.S. & Lambert, R.M. (2005) *J. Am. Chem. Soc.* **127,** 6069-6076.
[43] Demirok, U.K., Ertas, G. & Suzer, S. (2004) *J. Phys. Chem. B* **108,** 5179-5181.
[44] Siegbahn, H., Asplund, L., Kelfve, P. & Siegbahn, K. (1975) *J. Electron Spectrosc. Relat. Phenom.* **7,** 411-419.
[45] Fellner-Feldegg, H., Siegbahn, H., Asplund, L., Kelfve, P. & Siegbahn, K. (1975) *J. Electron Spectrosc. Relat. Phenom. 7* 421-428.
[46] Siegbahn, H. & Siegbahn, K. (1973) *J. Electron Spectrosc. Relat. Phenom.* **2,** 319-325.
[47] Ogletree, D.F., Bluhm, H., Lebedev, G., Fadley, C.S., Hussain, Z. & Salmeron, M. (2002) *Rev. Sci. Instrum.* **73,** 3872-3877.
[48] Ghosal, S., Hemminger, J.C., Bluhm, H., Mun, B.S., Hebenstreit, E.L.D, Ketteler, G., Ogletree, D.F., Requejo, F.G. & Salmeron, M. (2005) *Science* **307,** 563-566.

CHAPTER 12

LUMINESCENCE SPECTROSCOPY

[1]Baldassare Di Bartolo and [2]John Collins
[1]Department of Physics, Boston College, Chestnut Hill, MA 02467, USA
[2]Department of Physics, Wheaton College, Norton, MA 02766, USA

12.1 INTRODUCTION

12.1.1 Basic Concepts

A material, after absorbing some energy, can become a source of light by two processes:

1. The absorbed energy is converted into *low* quantum energy heat that diffuses through the material, which then emits *thermal radiation*.
2. The absorbed energy is in part *localized* as high quantum energy of atoms, which then emits radiation called *luminescence radiation*.

We note the following:

1. The quantity and quality of thermal radiation depend primarily on the *temperature* rather than on the nature of the *body*.
2. The quality and quantity of luminescence radiation are strongly dependent on the *nature* of the emitting material.

Therefore, the results of investigations in the field of luminescence are relevant to the study of material bodies. The process of luminescence is started by exciting the material with UV radiation, X-rays, electrons, alpha particles, electric fields, or energy that is liberated in chemical reactions. Accordingly, luminescence is qualified as photoluminescence, Roentgen luminescence, electroluminescence, and so on. We shall limit ourselves to examining the process of photoluminescence in solids, i.e., the luminescence emission that follows the absorption of UV or optical radiation.

We have found it convenient from a pedagogical point of view to subdivide the luminescent systems into two categories, *localized* and *delocalized*. For the first category, the absorption and emission processes are associated with quantum states of optically active centers that are spatially localized to particular sites in the solid. For the second category, these processes are associated with quantum states of the entire solid.

Regardless of the type of excitation or the nature of the system, the common character of the phenomenon of luminescence resides in the fact that, once excited, the system is left alone, without any external influence and, in this condition, it emits radiation that we may appropriately call *spontaneous*.

The purpose of this chapter is to describe the basic themes that motivate the investigations in luminescence spectroscopy, the techniques that are used to conduct these investigations and the challenges that are presented by this field of research.

In the introduction we consider the nature of luminescence research and emphasize its material-related character. We also recount briefly the history of the research in this field. In Section 12.2, we set up the theoretical bases for the treatment of the interaction of radiation with material systems. In Section 12.3, we consider the measurements that are conducted in lumine-scence spectroscopy and examine their simplest and fundamental aspects, apart from their more advanced technical features. We dedicate the next two Sections, 12.4 and 12.5, to the treatment of localized luminescent systems in which the radiative processes are associated with optically active centers. In the following two Sections, 12.6 and 12.7, we treat the delocalized systems in which the radiative processes are associated with quantum states of the entire solid. Finally, in Section 12.8 we consider the challenges presented by this field of research, suggesting directions of future efforts and recognizing the fact that luminescence spectroscopy is an inexhaustible source of problems for future investigations and educational applications.

12.1.2 History

The first reported observation of luminescent light from glowworms and fireflies is in the Chinese book *Shih-Ching* or “Book of Poems” (1200–1100 B.C.). Aristotle (384–322 B.C.) reported the observation of light from decaying fish. The first inquiry into luminescence dates circa. 1603 and was made by Vincenzo Cascariolo. Cascariolo’s interests were other than scientific: He wanted to find the so-called “philosopher’s stone” that would convert any metal into gold. He found some silvery white stones (barite) on Mount Paderno, near Bologna, which, when pulverized, heated with coal, and cooled, showed a purple-blue glow at night. This process amounted to reduction of barium sulfate to give a weakly luminescent barium sulfide:

$$BaSO_4 + 2C \rightarrow BaS + 2CO_2$$

News of this material, the “Bologna stone,” and some samples of it reached Galileo, who passed them to Giulio Cesare Lagalla. Lagalla called it “lapis solaris” and wrote about it in his book *De phenomenis in orbe lunae* (1612).

It is rare the case when one can insert in a scientific article a literary citation. This happens to be the case here. In *The Sorrows of Young Werther* of Goethe, the protagonist is unable to see the woman he loves because of an engagement he cannot refuse, and sends a servant to her, "only so that I might have someone near me who had been in her presence…" This is then his reaction when the servant comes back:

> It is said that the *Bologna stone*, when placed in the sun, absorbs the sun's rays and is luminous for a while in the dark. I felt the same with the boy. The consciousness that her eyes had rested on his face, his cheeks, the buttons of his jacket and the collar of his overcoat, made all these sacred and precious to me. At that moment I would not have parted with him for a thousand taler. I felt so happy in his presence [1].

The second important investigation on luminescence is due to Stokes and dates to the year 1852. Stokes observed that the mineral fluorspar (or fluorite) when illuminated by blue light gave out yellow light. Fluorite (CaF_2) is colorless in its purest form, but it absorbs and emits light when it contains such impurities as Mn, Ce, Er, etc. The term "fluorescence" was coined by Stokes and has continued to be used to indicate short-lived luminescence. A *Stokes' law* has been formulated according to which the wavelengths of the emitted light are always longer than the wavelength of the absorbed light.

12.2 SPONTANEOUS EMISSION, ABSORPTION, AND INDUCED EMISSION

12.2.1 Classical Bound, Radiating Electron

Let us consider first the equation of motion of a classical *bound* electron that we assume to be *nonradiating*:

$$F = -kx = m\frac{d^2x}{dt^2} \tag{12.1}$$

or

$$\frac{d^2x}{dt^2} + \omega_0{}^2 x = 0 \tag{12.2}$$

where

$$\omega_0 = \sqrt{\frac{k}{m}}\,. \tag{12.3}$$

The energy is given by

$$E = KE + PE = \frac{1}{2}mv^2 + \frac{1}{2}kx^2 = \text{const}. \tag{12.4}$$

We take as solution to the equation of motion (12.1),

$$x = x_0 \cos(\omega_0 t). \tag{12.5}$$

This means that at time $t = 0$, $x = x_0$, and $x = \mathrm{v} = 0$; the energy is then given by

$$E = \frac{1}{2}kx_0^2 = \frac{1}{2}m\omega_0^2 x_0^2 \tag{12.6}$$

A *radiating* electron will produce radiation due to the vibrating dipole

$$d = ex = ex_0 \cos(\omega_0 t) = \mathrm{Re}\ ex_0 e^{i\omega_0 t} \tag{12.7}$$

The average energy radiated in the unit time is given by

$$\overline{S} = \frac{2\overline{\left(\ddot{d}\right)^2}}{3c^3} = \frac{\omega_0^4 e^2 x_0^2}{3c^3}. \tag{12.8}$$

The radiated energy is given out at the expense of the internal energy of the electron:

$$\overline{S} = -\frac{\partial E}{\partial t} \qquad \left(E = \frac{1}{2}m\omega_0^2 x_0^2\right) \tag{12.9}$$

or

$$\frac{\omega_0^4 e^2 x_0^2}{3c^3} = -m\omega_0^2 x_0 \frac{\partial x_0}{\partial t}$$

$$\dot{x}_0 = -\frac{\omega_0^2 e^2}{3c^3} x_0 \tag{12.10}$$

We can write

$$\dot{x}_0 = -\frac{1}{2}\gamma x_0 \tag{12.11}$$

where

$$\gamma = \frac{2\omega_0^2 e^2}{3mc^3} = \frac{2\omega_0^2}{3c}\frac{e^2}{mc^2} = \frac{2\omega_0^2}{3c} r_0 = \frac{2\omega_0^2}{3}\tau \tag{12.12}$$

and

$$r_0 = \frac{e^2}{mc^2} = 2.8 \times 10^{-13}\,\text{cm},$$

$$\tau = \frac{r_0}{c} = 9.38 \times 10^{-24}\,\text{sec}.$$

If $\lambda = 6000$ Å, $\omega_o = 3 \times 10^{15}$, and $\gamma = 0.6 \times 10^{8}\text{sec}^{-1}$, the solution of (12.11) is

$$x_0 = X_0 e^{-\frac{1}{2}\gamma t}, \qquad X_0 = \text{const.} \tag{12.13}$$

Therefore,

$$\begin{aligned} E &= \frac{1}{2} m\omega_0^{\;2} x_0^{\;2} = \frac{1}{2} m\omega_0^{\;2} \left(X_0 e^{-\frac{1}{2}\gamma t} \right)^2 \\ &= \frac{1}{2} m\omega_0^{\;2} X_0^{\;2} e^{-\gamma t} = E_0 e^{-\gamma t} \end{aligned} \tag{12.14}$$

where

$$E_0 = \frac{1}{2} m\omega_0^{\;2} X_0^{\;2}. \tag{12.15}$$

γ is the rate of radiative decay for the classical electron.

The equation of motion, equation (12.1), needs a revision on account of the presence of radiation. We proceed as follows:

$$x = X_0 e^{-\frac{1}{2}\gamma t} \cos(\omega_0 t) \tag{12.16}$$

$$\dot{x} = -\frac{\gamma}{2} x - \omega_0 X_0 e^{-\frac{1}{2}\gamma t} \sin(\omega_0 t) \tag{12.17}$$

and

$$\frac{\gamma}{2} \dot{x} = -\left(\frac{\gamma}{2}\right)^2 x - \omega_0 \frac{\gamma}{2} X_0 e^{-\frac{1}{2}\gamma t} \sin(\omega_0 t) \tag{12.18}$$

$$\ddot{x} = -\frac{\gamma}{2} \dot{x} + \omega_0 \frac{\gamma}{2} X_0 e^{-\frac{1}{2}\gamma t} \sin(\omega_0 t) - \omega_0^{\;2} x. \tag{12.19}$$

Summing the above two relations we obtain

$$\ddot{x} + \gamma \dot{x} + \left[\omega_0^{\;2} + \left(\frac{\gamma}{2}\right)^2 \right] x = 0, \tag{12.20}$$

which represents the equation of motion of the radiating, bound electron.

12.2.2 Quantum Mechanical Radiative Decay

Consider a two-level quantum system, where the energy and the eigenfunction of the lower (upper) level are E_a (E_b) and Ψ_a (Ψ_b), respectively. If ψ is the wavefunction of the system, then

$$i\hbar\frac{\partial\psi}{\partial t} = H\psi \tag{12.21}$$

where

$$\begin{cases} H\psi_a = E_a\psi_a \\ H\psi_b = E_b\psi_b\,. \end{cases} \tag{12.22}$$

In general, a quantum state of the system is given by

$$\psi = ae^{-\frac{i}{\hbar}E_a t}\psi_a + be^{-\frac{i}{\hbar}E_b t}\psi_b\,. \tag{12.23}$$

For a non-radiating system, H is independent of time:

$$\begin{cases} \dot{a} = 0 \rightarrow a = \text{const} \\ \dot{b} = 0 \rightarrow b = \text{const} \end{cases} \tag{12.24}$$

and

$$\langle H\rangle = |a|^2 E_a + |b|^2 E_b = \text{const.} \tag{12.25}$$

Let us now allow for dipolar radiation

$$\begin{aligned} \langle ex\rangle &= \langle\psi|ex|\psi\rangle \\ &= \int\left[ae^{-i\frac{E_a}{\hbar}t}\psi_a + be^{-i\frac{E_b}{\hbar}t}\psi_b\right]^* ex\left[ae^{-i\frac{E_a}{\hbar}t}\psi_a + be^{-i\frac{E_b}{\hbar}t}\psi_b\right]d\tau \\ &= a^*be^{-i\omega_0 t}\mu_{ab} + ab^*e^{i\omega_0 t}\mu_{ba} \\ &= \text{Re}\left[d_0 e^{-i\omega_0 t}\right] \end{aligned} \tag{12.26}$$

where

$$\omega_0 = \frac{(E_b - E_a)}{\hbar} \tag{12.27}$$

$$\mu_{ab} = \langle\psi_a|ex|\psi_b\rangle \tag{12.28}$$

$$\mu_{ba} = \langle\psi_b|ex|\psi_a\rangle \tag{12.29}$$

$$d_0 = 2a^*b\mu_{ab} \tag{12.30}$$

and where we have assumed

$$\mu_{aa} = \mu_{bb} = 0. \tag{12.31}$$

We have now from equation (12.8)

$$\begin{aligned} \overline{S} &= \frac{\omega_0{}^4 d_0 d_0{}^*}{3c^3} = \frac{4\omega_0{}^4|\mu_{ab}|^2}{3c^3}|a|^2|b|^2 \\ &= \frac{4\omega_0{}^4|\mu_{ab}|^2}{3c^3}|b|^2\left(1-|b|^2\right) = \frac{4\omega_0{}^4|\mu_{ab}|^2}{3c^3}|b|^2 \end{aligned} \tag{12.32}$$

assuming $|b|^2 <<1$. We can choose the phase of Ψ_b that makes b real:

$$\overline{S} = \frac{4\omega_0{}^4 |\mu_{ab}|^2}{3c^3} b^2. \tag{12.33}$$

The energy of the system is given by

$$\begin{aligned} E = \langle H \rangle = b^2 E_b + a^2 E_a = b^2 E_b + (1-b)^2 E_a \\ = b^2 (E_b - E_a) + E_a \end{aligned} \tag{12.34}$$

If we set $E_a = 0$, we obtain

$$E = b^2 \hbar\omega_0, \tag{12.35}$$

which can be contrasted with the classical formula (equation (12.6)). For non-radiating systems, E remains constant. For radiating systems

$$\overline{S} = -\frac{\partial E}{\partial t} = -2b\frac{\partial b}{\partial t}\hbar\omega_0 = \frac{4\omega_0{}^4 |\mu|^2}{3c^3} b^2 \tag{12.36}$$

where we have dropped the subscript ab from μ. Then

$$\frac{\partial b}{\partial t} = -\frac{1}{2}\gamma b \tag{12.37}$$

where

$$\gamma = \frac{4\omega_0{}^3 |\mu|^2}{3c^3\hbar}. \tag{12.38}$$

The solution of equation (12.37) is

$$b = b_0 e^{-\frac{\gamma}{2}t} \tag{12.39}$$

and

$$b^2 = b_0{}^2 e^{-\gamma t}. \tag{12.40}$$

b^2 represents the probability of occupancy of the excited state. γ is the quantum mechanical rate of *radiative decay* or *spontaneous emission*. We may equate it to the Einstein coefficient A:

$$\gamma = A = \frac{4\omega_0{}^3 |\mu|^2}{3\hbar c^3} = \frac{64\pi^4 \nu^3 |\mu|^2}{3hc^3} = \frac{64\pi^4 |\mu|^2}{3h\lambda^3} \tag{12.41}$$

In a nonmagnetic medium of index of refraction n an oscillating dipole emits the power

$$S = \frac{2(\ddot{\mathrm{d}})}{3c^3} n \tag{12.42}$$

whose time average is

$$\overline{S} = \frac{\omega_0{}^4 d_0 d_0{}^*}{3c^3} n. \tag{12.43}$$

Therefore, in this case,

$$A = \frac{4\omega_0^{\,3}\left|\mu\right|^2}{3\hbar c^3}n. \tag{12.44}$$

12.2.3 Absorption and Emission

Consider a cavity whose walls are at temperature T, and containing radiation and an ensemble of atoms, and let each atom be represented by a two-level quantum mechanical system with an energy level separation of $\hbar\omega_0$. In thermal equilibrium, then, energy density per unit angular frequency range at ω_0 is given by [2]:

$$\rho_{\omega_0} = \frac{\hbar\omega_0^{\,3}}{\pi^2 c^3}\frac{n^3}{e^{\hbar\omega_0/kT}-1} = \frac{\omega_0^{\,2}}{\pi^2 c^3}\frac{\hbar\omega_0 n^3}{e^{\hbar\omega_0/kT}-1} \tag{12.45}$$

where $n =$ index of refraction of the medium inside the cavity. We can write

$$\rho_{\omega_0}\left(e^{\hbar\omega_0/kT}-1\right) = \frac{\hbar\omega_0^{\,3}n^3}{\pi^2 c^3}. \tag{12.46}$$

In addition, because of detailed balance,

$$AN_2^{\mathrm{e}} + B_{21}\rho_{\omega_0}N_2^{\mathrm{e}} = B_{12}\rho_{\omega_0}N_1^{\mathrm{e}}, \tag{12.47}$$

where N_1^{e} and N_2^{e} are the equilibrium populations of atoms in the lower and upper levels, respectively, and $B_{21}\rho_{\omega_0}$ and $B_{12}\rho_{\omega_0}$ are the probabilities per unit time of induced downward and upward transitions, respectively. We can write

$$A\frac{N_2^{\mathrm{e}}}{N_1^{\mathrm{e}}} + B_{21}\rho_{\omega_0}\frac{N_2^{\mathrm{e}}}{N_1^{\mathrm{e}}} = B_{12}\rho_{\omega_0} \tag{12.48}$$

and

$$Ae^{-\hbar\omega_0/kT} + B_{21}\rho_{\omega_0}e^{-\hbar\omega_0/kT} = B_{12}\rho_{\omega_0} \tag{12.49}$$

or

$$\rho_{\omega_0}\left(B_{12}e^{\hbar\omega_0/kT} - B_{21}\right) = A. \tag{12.50}$$

We set

$$B_{12} = B_{21} = B. \tag{12.51}$$

Then

$$\rho_{\omega_0}\left(e^{\hbar\omega_0/kT}-1\right)B = A \tag{12.52}$$

$$\frac{A}{B} = \rho_{\omega_0}\left(e^{\hbar\omega_0/kT}-1\right) = \frac{\hbar\omega_0^{\,3}n^3}{\pi^2 c^3} \tag{12.53}$$

and

$$B = \frac{\pi^2 c^3}{\hbar\omega_0^3 n^3} A = \frac{\pi^2 c^3}{n^3 \hbar\omega_0^3} \frac{4n\omega_0^3 |\mu|^2}{3\hbar c^3} = \frac{4\pi^2}{3\hbar^2 n^2} |\mu|^2 \tag{12.54}$$

or

$$\begin{cases} A = \dfrac{4n\omega_0^3}{3\hbar c^3} |\mu|^2 & \left(\sec^{-1}\right) \\ B = \dfrac{4\pi^2}{3\hbar^2 n^2} |\mu|^2 & \left(\dfrac{\text{cm}^3}{\text{erg-sec}^2}\right) \end{cases} \tag{12.55}$$

$$\frac{A}{B} = \frac{\hbar\omega_0^3 n^3}{\pi^2 c^3} \quad \left(\frac{\text{erg-sec}}{\text{cm}^3}\right). \tag{12.56}$$

Let us now consider the more realistic situation in which the energies of the two atomic levels are not sharply defined, but have a certain width $\Delta\omega$ such that $\Delta\omega << \omega_0$. We can set the following:

$N_2 A_\omega d\omega$ = number of atoms that per unit time that decay by *spontaneous* emission, giving out a photon with angular frequency in (ω, $\omega + d\omega$);

$N_2 B_\omega d\omega$ = number of atoms that in the unit time undergo a downward transition by *induced emission*, giving out a photon with angular frequency in (ω, $\omega + d\omega$); and

$N_1 B\rho_\omega d\omega$ = number of atoms that in the unit time undergo an upward transition by the *absorption* of a photon with angular frequency in (ω, $\omega + d\omega$).

If we put a filter between the atoms and the walls that allows only the radiation in the narrow band $d\omega$ to interact with the atoms we have

$$N_2^e A_\omega d\omega + N_2^e B_\omega \rho_\omega d\omega = N_1^e B_\omega \rho_\omega d\omega \tag{12.57}$$

or

$$A_\omega \frac{N_2^e}{N_1^e} + B_\omega \rho_\omega \frac{N_2^e}{N_1^e} = B_\omega \rho_\omega . \tag{12.58}$$

Then

$$A_\omega e^{-\frac{\hbar\omega_0}{kT}} + B_\omega \rho_\omega e^{-\frac{\hbar\omega_0}{kT}} = B_\omega \rho_\omega \tag{12.59}$$

$$\rho_\omega \left(e^{\hbar\omega_0/kT} - 1\right) B_\omega = A_\omega \tag{12.60}$$

and

$$\frac{A_\omega}{B_\omega} = \rho_\omega\left(e^{\hbar\omega_0/kT} - 1\right) \approx \rho_{\omega_0}\left(e^{\hbar\omega_0/kT} - 1\right) = \frac{\hbar\omega_0^3 n^3}{\pi^2 c^3} = \frac{A}{B}. \qquad (12.61)$$

If $g(\omega)$ indicates the spectral lineshape:

$$\int g(\omega)\,d\omega = 1 \qquad (12.62)$$

we can write

$$\begin{cases} A_\omega = Ag(\omega)\,; & \int A_\omega d\omega = A \\ B_\omega = Bg(\omega)\,; & \int B_\omega d\omega = B \end{cases} \qquad (12.63)$$

If we call $w(\omega)d\omega$ the probability per unit time that an atom undergoes an *induced* transition by absorbing or emitting a photon with angular frequency in $(\omega, \omega + d)$, we find

$$\begin{aligned} w(\omega)\,d\omega &= Bg(\omega)\rho_\omega d\omega = \frac{4\pi^2}{3\hbar^2 n^2}|\mu|^2 \rho_\omega g(\omega)\,d\omega \\ &= \frac{4\pi^2}{3nc\hbar^2} I(\omega)\,g(\omega)\,d\omega \end{aligned} \qquad (12.64)$$

where

$$I(\omega)\,d\omega = \rho_\omega \frac{c}{n} d\omega = \text{ intensity of radiation with angular frequency in } (\omega, \omega + d\omega). \qquad (12.65)$$

We can write

$$\int w(\omega)\,d\omega = \frac{4\pi^2}{3\hbar^2 n^2}|\mu|^2 \rho_{\omega_0} = \frac{4\pi^2}{3\hbar^2 nc}|\mu|^2 I(\omega_0) \qquad (12.66)$$

12.2.4 Absorption Coefficient and Absorption Cross-Section

Let us assume that a plane wave goes through a certain medium in the x direction. Let the medium consist of atoms that have two possible energy levels and let N_1 (N_2) be the concentration of atoms in the lower (higher) energy level.

The energy intensity per unit angular frequency range $I(\omega)$, when the wave travels through the medium distance dx, undergoes a change given by

$$dI(\omega) = -w(\omega)(N_1 - N_2)\hbar\omega dx \qquad \left(\frac{\text{erg}}{\text{cm}^2}\right) \qquad (12.67)$$

But, from equation (12.64):

$$w(\omega) = \frac{4\pi^2}{3nc\hbar^2}|\mu|^2 I(\omega)\,g(\omega). \qquad (12.68)$$

Then

$$dI(\omega) = -\left[\frac{4\pi^2}{3nc\hbar^2}|\mu|^2 I(\omega) g(\omega)\right](N_1 - N_2)\hbar\omega dx$$
$$= -\left[\frac{4\pi^2 (N_1 - N_2)}{3nc\hbar}|\mu|^2 \omega g(\omega)\right] I(\omega) dx = -\alpha(\omega) I(\omega) dx \tag{12.69}$$

where the *absorption coefficient* $\alpha(\omega)$ is given by

$$\alpha(\omega) = \frac{4\pi^2 (N_1 - N_2)}{3nc\hbar}|\mu|^2 \omega g(\omega) = \frac{n\hbar\omega}{c} B(\omega)(N_1 - N_2) \quad \left(\text{cm}^{-1}\right) \tag{12.70}$$

The solution of equation (12.69) is

$$I(\omega; x) = I(\omega; x = 0)\ e^{-\alpha x} \tag{12.71}$$

We define the *absorption cros-section* of a radiative transition as follows

$$\sigma(\omega) = \frac{\alpha(\omega)}{N_1 - N_2} \frac{4\pi^2}{3nc\hbar}|\mu|^2 \omega g(\omega) \quad \left(\text{cm}^2\right) \tag{12.72}$$

Note the following

$$\int \alpha(\omega) d\omega = \frac{4\pi^2 (N_1 - N_2)}{3nc\hbar}|\mu|^2 \omega_0 = \frac{n\hbar\omega_0}{c} B(N_1 - N_2) \quad \left(\text{cm}^{-1}\,\text{sec}^{-1}\right) \tag{12.73}$$

In Table 12.1, we have summarized the results obtained in the previous sections.

12.3 MEASUREMENTS AND TECHNIQUES

Luminescence spectroscopy relies on four fundamental measurements: (1) absorption spectra, (2) luminescence spectra, (3) excitation spectra, and (4) response to pulsed excitation. Technical improvements or breakthroughs have not and probably will not produce any conceptually new additions to these four basic measurements. The data that they make available can be considered "scarce" when one contrasts them with the complexity of the processes. However, an attentive and appropriate usage of the different techniques generally yields a satisfactory model for the system under investigation.

12.3.1 Absorption Spectra

The main purpose of absorption measurements is to set the energy level scheme and to identify particular levels that may be convenient to "pump" the system for the purpose of obtaining luminescence or laser emission. If the concentration of the optically active centers is known, the ratio of the absorption coefficient and of the concentration gives the *absorption cross-section*.

Table 12.1 Summary of Section 12.2 formulas.

$$w(\omega) = \frac{4\pi^2}{3n^2\hbar^2}|\mu|^2 \rho_\omega g(\omega) = \frac{4\pi^2}{3nc\hbar^2}|\mu|^2 I(\omega) g(\omega) = \frac{\sigma(\omega) I(\omega)}{\hbar\omega}$$

$$w = \int w(\omega)\, d\omega = \frac{4\pi^2}{3n^2\hbar^2}|\mu|^2 \rho_{\omega_0} = \frac{4\pi^2}{3nc\hbar^2}|\mu|^2 I(\omega_0) = B\rho_{\omega_0} \qquad \left(\text{sec}^{-1}\right)$$

$$\alpha(\omega) = \frac{4\pi^2 (N_1 - N_2)}{3nc\hbar}|\mu|^2 \omega g(\omega) = \frac{n\hbar\omega}{c} B_\omega (N_1 - N_2) \qquad \left(\text{cm}^{-1}\right)$$

$$\sigma(\omega) = \frac{\alpha(\omega)}{N_1 - N_2} = \frac{4\pi^2}{3nc\hbar}|\mu|^2 \omega g(\omega) \qquad \left(\text{cm}^2\right)$$

$$\int \alpha(\omega)\, d\omega = \frac{4\pi^2 (N_1 - N_2)}{3nc\hbar}|\mu|^2 \omega_0 = \frac{n\hbar\omega_0}{c} B (N_1 - N_2) \qquad \left(\text{cm}^{-1}\,\text{sec}^{-1}\right)$$

$$\int \sigma(\omega)\, d\omega = \frac{4\pi^2}{3nc\hbar}|\mu|^2 \omega_0 = \frac{n\hbar\omega_0}{c} B = \frac{2\pi^2 e^2}{mcn} f \qquad \left(\text{cm}^2\text{sec}^{-1}\right)$$

where the f-number is

$$f = \frac{2m\omega_0}{3\hbar e^2}|\mu|^2 \qquad \text{(pure number)}$$

$$A = \frac{1}{\tau_{sp}} = \frac{4n\omega_0^3}{3\hbar c^3}|\mu|^2 = \frac{2e^2\omega_0^2 n}{mc^3} f \qquad \left(\text{sec}^{-1}\right)$$

$$B = \frac{4\pi^2}{3n^2\hbar^2}|\mu|^2 = \frac{2\pi^2 e^2}{m\hbar\omega_0 n^2} f \qquad \left(\frac{\text{cm}^3}{\text{erg-sec}^2}\right)$$

$$\frac{A}{B} = \frac{\hbar\omega_0^3 n^3}{\pi^2 c^3} \qquad \left(\frac{\text{erg-sec}}{\text{cm}^3}\right)$$

$$f\tau_{sp} = 1.5\frac{\lambda^2}{n}$$

$$A_\omega = A g(\omega) \qquad \text{(pure number)}$$

$$B_\omega = B g(\omega) \qquad \left(\frac{\text{cm}^2}{\text{erg-sec}}\right)$$

$$\frac{A_\omega}{B_\omega} = \frac{A}{B}$$

The conventional absorption measurements are related to transitions originating in the ground state of the material. Transitions originating in an excited state and ending up in still higher states may be observed if one is able to populate the initial state. This procedure is called *excited state absorption* (ESA) and has been found useful towards elucidating the process by which certain systems are excited prior to the emission of luminescence. It is worth noticing that the ESA signal is dependent on the strength of the excitation of the initial level.

The absorption spectra are generally obtained by using an instrument called an *absorption spectrophotometer*, which compares the intensity of a beam of light impinging on a sample with the intensity of the same beam after traversing the sample:

$$I_0(\lambda) = I_i(\lambda) e^{-\alpha(\lambda)L} \tag{12.74}$$

Here

$I_i(\lambda)$ = intensity of the input beam at wavelength λ
$I_o(\lambda)$ = intensity of the output beam at wavelength λ
$\alpha(\lambda)$ = absorption coefficient of the material at wavelength λ
L = thickness of the sample.

Information about the absorption coefficient is often given as *optical density* (*OD*):

$$OD = \log \frac{I_i(\lambda)}{I_o(\lambda)}. \tag{12.75}$$

An $OD = 1$ corresponds to an absorption of 90%, an $OD = 2$ to an absorption of 99%. Given *OD*, the coefficient of absorption can be obtained as follows

$$\alpha(\lambda) = \frac{1}{L} OD \ln 10. \tag{12.76}$$

We note that $\alpha(\lambda)$ is independent of the intensity of the incoming beam I_i. The absorption measurement can then be considered an *absolute* measurement.

A common feature of the absorption spectrophotometer is the presence of two chambers, one for the sample and one for the reference material. This arrangement allows the apparatus to sort out all the features of the material that may reduce the intensity of the output beam, I_o, such as reflection from the walls of the material, not due to the absorption.

12.3.2 Luminescence Spectra

We measure the luminescence spectra to identify the luminescent states. They may also help to establish energy levels, especially the position of the ground state multiplets. The changes of spectra with temperature provide information regarding such temperature time-dependent processes as thermal line broadening and thermal change of line positions. The dependence of the spectral intensity on the concentration of luminescent centers may provide information on the threshold of quenching and on up-conversion.

The step preceding the emission of the luminescence is an excitation process that has the purpose of populating the luminescent level(s). This excitation may be achieved by lamps or lasers. Accordingly, we may use *wide band* or *selective excitation.* Wide band excitation is useful for pumping systems with wide absorption bands, such as those generally presented by transition metal ions in solids.

Selective excitation may be achieved by filtering the light of a lamp or by using a laser. Lasers have become common in spectroscopy laboratories; they provide high light intensity in a very narrow spectral band. The exciting light sources can be continuous or pulsed.

For continuous sources generally the beam of light is chopped and the signal from the photomultiplier is sent to a lock-in amplifier before being recorded. The lock-in amplifier is tuned at the chopping frequency and selects out signals only at this frequency coming from the photomultiplier, reducing considerably the presence of noise.

For pulsed sources the signal from the photomultiplier is sent to a boxcar integrator that integrates the signal over time. The boxcar is triggered at the same rate as the pulsed source. It is provided with a "time window" whose length and position with respect to thė trigger pulse can be varied, making this device very useful for obtaining time-resolved spectra. The time window of the boxcar can also be moved continuously making the boxcar adapt to perform lifetime measurements.

12.3.3 Excitation Spectra

The excitation spectra tell us in what spectral region we have to pump the system in order to obtain the emission that is being monitored. Excitation spectra are essential for recognizing the presence of energy transfer among different centers; say those of type *A* and those of type *B*. Energy transfer from *A* to *B* is present if the excitation spectrum of *B* contains bands typical of *A*. Excitation spectra can also help in the assignment of luminescence transitions. A study of the temperature dependence of the monitored luminescence of *A* when pumped via the absorption bands of *A* and when pumped via the absorption bands of *B* gives information about the role played by phonons in the energy transfer process.

In order to acquire an excitation spectrum we monitor the intensity of a specific luminescence line while continuously changing the wavelength of the exciting light. This is generally achieved by putting a monochromator between a wide-spectrum source and the sample and a filter between the sample and the detector. The detection system may use a combination chopper/lock-in if the source is continuous or a boxcar if the source is pulsed.

12.3.4 Responses to Pulsed Excitation

The decay constant τ of an exponentially decaying luminescence signal is the lifetime of the level from which the transition originates. It includes a *radiative* part and a *non-radiative* part as follows:

$$\frac{1}{\tau}=\frac{1}{\tau_{\text{rad}}}+\frac{1}{\tau_{\text{non-rad}}(T)} \tag{12.77}$$

where

$$\frac{1}{\tau_{\text{rad}}}=\text{probability of radiative decay,}$$

$$\frac{1}{\tau_{\text{non-rad}}}=\text{probability of non-radiative decay.}$$

The probability of non-radiative decay is affected by the temperature of the sample. If τ is found to be independent of temperature, then

$$\left(\tau_{\text{non-rad}}\right)^{-1}=0. \tag{12.78}$$

The decay pattern of a luminescence signal may not be exponential when the excitation energy reaches the luminescence level after undergoing a number of downward steps. If the number of this step is n, then the decay signal will contain $(n + 1)$ exponentials. It may be difficult to disentangle all the exponentials because the components faster than the response time of the detecting apparatus leave no trace.

An additional mechanism that may contribute to the decay of a luminescent level is due to the emission of vibronic lines, corresponding to transitions that involve the emission of one photon and the emission or absorption of a photon. These transitions have been found to contribute to the probability of decay of such systems as Al_2O_3:Cr (ruby) and YAG:Cr [3].

Pulsed excitation is generally achieved by sending a pulse of light at a selected wavelength in order to excite a particular level of the emitting center. The response of a particular emission line is monitored by filtering the luminescence emission. The detection of the response signal can be done by looking, on a scope, directly at the signal following the exciting pulse or by using a boxcar integrator and moving the time window.

12.4 LOCALIZED SYSTEMS

12.4.1 Introduction

The most important classes of localized luminescent centers are transition metal ions (TMI) and rare earth ions (REI) that have been intentionally doped into ionic insulating host materials. The luminescence properties of these systems depend on both the dopant ion and the host. Another class of localized centers is defects in solids. One such center is an electron trapped at a vacant lattice site. These defects often absorb in the optical region giving the crystal color, and so are called *color centers*.

The host materials for localized, optically active centers are large band gap (usually > 6 eV) ionic solids. The large gap renders the host transparent to visible radiation, and also prevents any electrons from bridging the gap thermally, so that an undoped host is optically and electrically inert. Common host materials include alkaline halides, oxides, fluorides, chlorides, tungstates, phosphates, and garnets, to name but a few.

12.4.2 The Hamiltonian of an Ion in a Solid

For an optically active ion in a solid, the Hamiltonian can be written as

$$H = H_{FI} + H_L + H_{CF} \ , \tag{12.79}$$

where H_{FI}, H_L and H_{CF} are the Hamiltonians for the free ion, lattice, and crystal field interaction, respectively.

The free ion Hamiltonian, H_{FI}, includes all electric and magnetic interactions in the ion, and for many-electron atoms is quite cumbersome. Extensive treatments on the various interactions contained in H_{FI} are given in [4–6]. Considering only the most important interactions, H_{FI} can be written in the following form:

$$H_{FI} = H_0 + H_C + H_{SO} \, . \tag{12.80}$$

H_0 includes the kinetic energy of the electrons, and the electrostatic interaction of each electron with an average (spherically symmetric) potential due to the nucleus and the other electrons. Since all valance electrons are subject to the same potential, the eigenstates of H_0 corresponding to a particular configuration are degenerate.

H_C accounts for the electrostatic interaction among the electrons in the unfilled shell. This interaction splits the ground configuration into different spectral terms, that is, energy levels with common values of S and L. Such terms are identified using the ^{2S+1}L notation. L^2, S^2, L_z, and S_z all commute with the H_C and so the corresponding quantum numbers (L, S, M_L, and M_S, respectively) are valid. The energies of the states are independent of M_L and M_S, and so have a degeneracy of (2L + 1) (2S + 1).

For rare earth ions, the spin-orbit term (H_{SO}) is the next most important interaction, followed by the crystal field interaction, H_{CF}. This is the so-called *weak crystal field scheme*. For transition metal ions H_{CF} is larger that H_{SO}, so that either the medium or strong field scheme is relevant. We first consider the rare earth ions.

12.4.3 Rare Earth Ions in Solids

12.4.3.1 Energy Levels of Rare Earth Ions in Solids

The rare earth elements include the lanthanides and the actinides. By far the most important of these as luminescent centers are the lanthanides, particularly when incorporated into a solid in the trivalent state. Each trivalent REI has a Xe core, filled 5s- and 5p-shells, and an unfilled 4f-shell, with the number of 4f-electrons running from 1 (Ce^{3+}) to 13 (Yb^{3+}). The 4f-electrons are responsible for the optical activity of the center. The 5s- and 5p-shells are located farther from the nucleus than the 4f-shell, and partially shield the 4f-electrons from the nearby ligands. This shielding renders the crystal field interaction (H_{CF}) much smaller than the spin orbit interaction, H_{SO}.

For REI, then, the next most important term in the Hamiltonian is H_{SO}, which is given by

$$H_{SO} = \Sigma_i\, \xi_i\, \mathbf{l}_i \cdot \mathbf{s}_i\ . \tag{12.81}$$

This interaction splits each spectral term into different levels, called *J-multiplets* (or *J-manifolds*). The operators $J^2 = (L + S)^2$ and $J_z = (L_z + S_z)$ commute with the H_{SO}, so that J and M_J are good quantum numbers. Each J-multiplet has a degeneracy of 2J + 1. For rare earth ions, these splittings are large enough to cause some mixing of states of different L and S and of equal J-values. Nevertheless, the J-multiplets are commonly labeled using S, L, and J quantum numbers according to the usual spectral designation $^{2S+1}L_J$, where J runs from L + S to L – S.

The remaining interaction is that of the crystal field with the rare earth ion. H_{CF} consists of a static term ($H_{CF\text{-static}}$) and a dynamic term ($H_{CF\text{-dymanic}}$). We consider only the static term here since it makes a more important contribution to the energy.

Due to the shielding of the 4f-electrons, there is little or no overlap of their wavefunctions with those of the ligands. Thus, we consider the ion to be under the influence of an external field—this is the crystalline field approximation. In this approximation, $H_{CF\text{-static}}$ is the interaction of the electrons with the potential due the ligands, V(**r**).

$$H_{CF\text{-static}} = \Sigma_i\, e\, V(\mathbf{r}_i)\ . \tag{12.82}$$

The sum is over all 4f electrons, and $V(\mathbf{r}_i)$ reflects the symmetry at the site of the REI. This interaction splits each J-multiplet into no more than (2J + 1) levels for ions with an even number of f-electrons and no more than (J + 1/2) levels for ions with an odd number of f-electrons. The number of levels into which each multiplet splits depends on the symmetry of the crystal field—for higher ion site symmetries there are fewer levels. The splittings due to H_{CF} typically range from ten to a few hundred cm^{-1}.

The observed energy levels of the rare earth ions in $LaCl_3$ are shown in Figure 12.1. Because H_{CF} is small, the energy levels are similar in other ionic solids [5, 6]. Most of the energy levels of the REI are known up to about 40,000 cm^{-1}. Research continues to locate and identify the higher lying 4f-levels [7, 8].

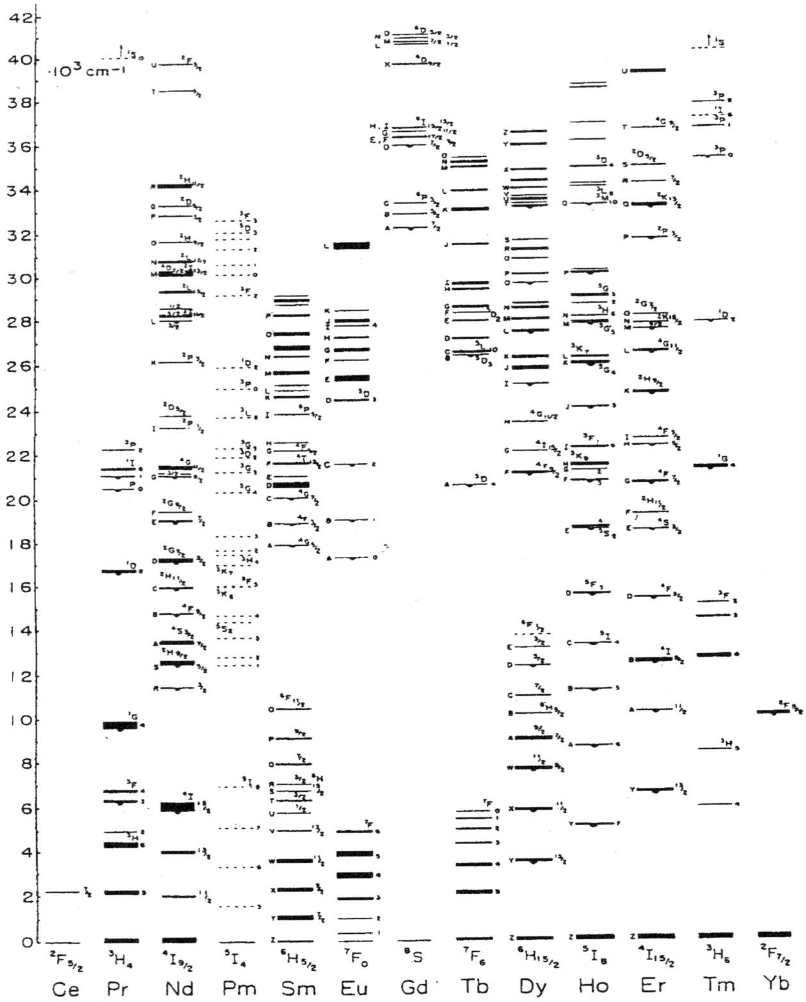

Figure 12.1 Energy levels of trivalent rare earth ions. The width of the levels indicates the total splitting due to the crystalline field in anhydrous $LaCl_3$ [4].

Although the crystal field interaction makes only a minor contribution to the static energy of the system, its affect on the dynamical processes is profound. The crystal field is responsible for the radiative transitions between 4f states, and also drives the non-radiative processes that allow for energy to be exchanged between the ion and the lattice. These processes are discussed in Section 12.5.

12.4.3.2 Spectral Features of REI in Solids

The small crystal field interaction implies that the energy levels of REI are not very sensitive to the motion of the lattice. Thus, f→f transitions are characterized by sharp lines, with linewidths on the order of 1 cm^{-1} at low temperatures. A sample REI emission spectrum is shown in Figure 12.2. Typical radiative lifetimes range from 10 μsec to a few msec.

Although radiative transitions do occur between f-levels, luminescence spectra show certain lines very weak or missing entirely. This is due to one or more of the following factors:

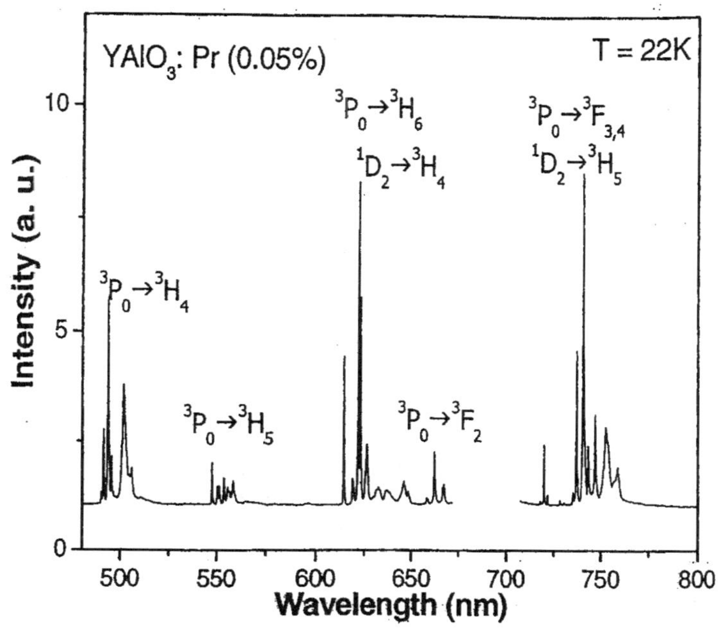

Figure 12.2 Luminescence spectrum of YALO : Pr (.05%) at 22 K. Excitation was into the $^3P_0$ level at 460 nm.

1. Certain transitions may be only weakly allowed. This depends, in part, on the symmetry of the crystal field.
2. An excited REI can decay via the emission of phonons instead of a photon. Such transitions are called *multiphonon transitions*, and are discussed in Section 12.5.3.
3. The REIs are in thermal equilibrium; each ion reaches thermal equilibrium with the lattice, and so they are in equilibrium with one another. Thus, the levels of the ground multiplet (or of an excited metastable multiplet) are populated according to the Boltzmann distribution. (The idea of the lattice as a heat reservoir is found to be widely true, but may not hold on very short timescales during which local nonequilibrium phonon populations can be created [9, 10].)

REI spectra also exhibit sidebands to the zero-phonon line. These sidebands represent vibronic transitions (Section 12.5.4), which involve the emission of a photon and the simultaneous emission or absorption of one or more phonons. For

REI, such sidebands usually involve a single phonon, and are much weaker than the zero-phonon line. The structure of the sidebands reflects somewhat the phonon spectrum of the host lattice.

In many systems, the REIs are situated at two or more types of lattice sites. Due to differences in the local crystal field, the same transition at the different sites will generally appear at different energies, complicating the spectra considerably. In glasses, the random positioning of the ions leads to a continuum of crystal field interactions and to bands on the order of 100 cm^{-1} wide.

We conclude this section with the observation that the spectral features of REI can be traced back to their peculiar charge distribution, namely, that the optically active 4f electrons are shielded from the crystalline field by the outer 5s- and 5p-shells. The next group of ions to be considered, the transition metal ions, has valence electrons that are exposed to the crystalline field, and exhibit a vastly different behavior.

12.4.4 Transition Metal Ions in Solids

12.4.4.1 Introduction

The class of transition metals consists of those elements with unfilled d-shells. In pure form they are metals, with electrical and thermal properties usually associated with metals. When doped into insulators, however, such similarities are less evident. Their spectroscopic properties vary significantly ion to ion and host to host. We restrict our discussion to those elements most commonly used as dopants in optical materials, the 3d elements.

When doped into a solid, these elements form positive ions, the valency of which is host-dependent. The outer 4s-electrons are stripped from the ion and reside closer to the anions of the solid. Common valencies and shell configurations of each ion are shown in Table 12.2.

Table 12.2 Valencies and electronic configurations of 3d-ions in solids.

Transition Metal Ion	Electronic Configuration	
Ti^{3+}, V^{4+}	$1s^2\ 2s^2\ 2p^6\ 3s^2\ 3p^6$	$3d^1$
V^{3+}, Cr^{4+}	"	$3d^2$
V^{2+}, Cr^{3+}, Mn^{4+}	"	$3d^3$
Cr^{2+}, Mn^{3+}	"	$3d^4$
Mn^{2+}, Fe^{3+}	"	$3d^5$
Fe^{2+}, Co^{3+}	"	$3d^6$
Co^{2+}	"	$3d^7$
Ni^{2+}	"	$3d^8$
Cu^{2+}	"	$3d^9$

12.4.4.2 Energy Levels of Transition Metal Ions in Solids

Optical transitions in these centers involve the 3d-electrons, which are in the outermost shell and so are exposed to the crystal field. Consequently, H_{CF} is larger than H_{SO}, and is often on that same order of magnitude as the interaction H_C among the 3d-electrons.

The simplest transition metal ion (TMI) has only one 3d-electron, so that H_C is zero. Following the interaction with the central field, the next strongest term in the Hamiltonian is the H_{CF}. Though several crystal field symmetries are possible, in many oxide hosts the TMI is situated at a site with octahedral (or near octahedral) symmetry (Figure 12.3). At an octahedral site the angular parts of the wavefunctions are linear combinations of spherical harmonics [11], the plots of which are shown in the Figure 12.4. The following observations can be made regarding these wavefunctions.

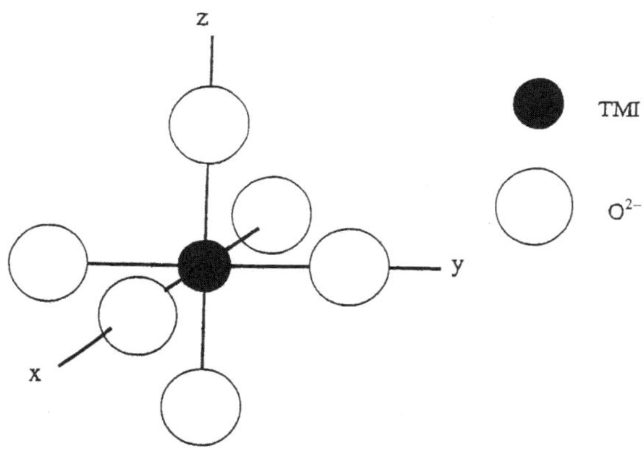

Figure 12.3 Transition metal ion in an oxide at a site of octahedral symmetry.

1. The wavefunctions are even. In fact, each wavefunction is invariant under all symmetry operations of the octahedral field.
2. The wavefunctions d_{xy}, d_{xz}, and d_{yz} are identical both in shape and in their relative orientation with respect to the surrounding ions, and so have the same energy. These states are labeled as t_{2g}.
3. The wavefunctions $d_{r^2-z^2}$ and $d_{x^2-y^2}$ are directed toward the O^{2-}ions and so have higher interaction energies than the t_{2g} states. The energy of these two states is identical. These states are labeled e_{2g}.

Given observations 2. and 3. above, this system exhibits two energy levels as shown in Figure 12.5. The difference in energy between the two levels is commonly designated as 10 Dq, where Dq is a measure of the strength of the crystal field. We note that for a TMI in a site having cubic symmetry, the system also splits into two energy levels. For example, in Ti^{3+}-doped Al_2O_3, the crystal field is mainly cubic, with a slight trigonal distortion. The splitting between the ground state ($^2T_2$) and the excited state (^2E) is around 19,000 cm^{-1} [12]. The absorption band occurs in the blue-green region, accounting for the crystal's pink color.

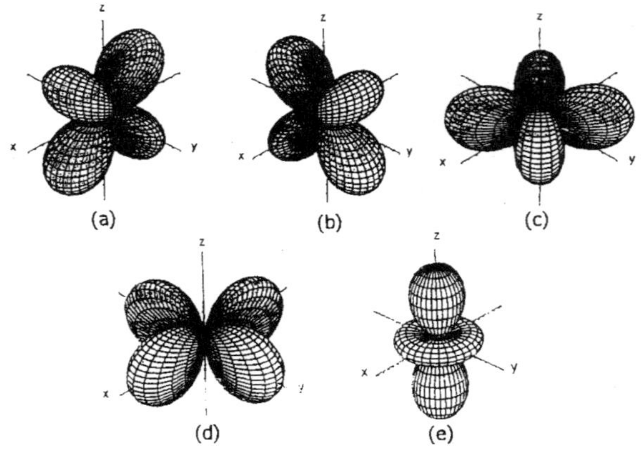

Figure 12.4 The wavefunctions of a $3d^1$ system in an octahedral environment. The ground state wavefunctions are labeled (a) d_{xz}, (b) d_{yz}, and (c) d_{xy}, and the excited states are (d) $d_{r^2-z^2}$ and (e) $d_{x^2-y^2}$.

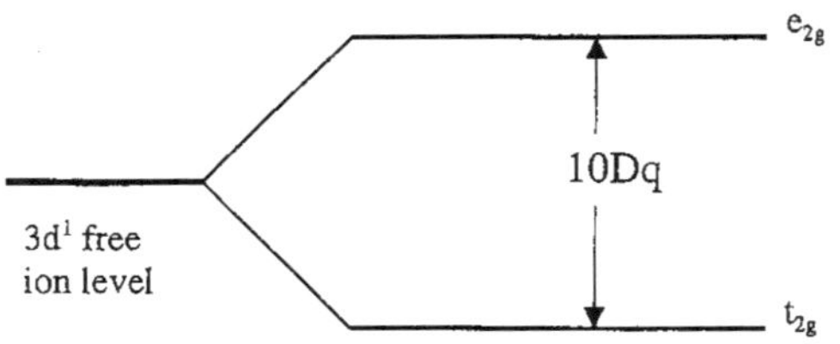

Figure 12.5 Splitting of the ground configuration due to an octahedral field for a $3d^1$ system.

When there is more than one 3d electron, the configuration interaction H_C must be included. Taking into account the crystal field, but before including the H_C, each 3d-electron will be in either the t_{2g} or e_{2g} state. The state of the system, then, is a product of one-electron states. For the case of a $3d^2$ system (Cr^{4+} or V^{3+}) the states are $(t_{2g})^2$, $t_{2g}e_{2g}$, and $(e_{2g})^2$.

When H_C is included, the one-electron product states are split. For a $3d^2$ system, the new states are assigned the group theoretical labels: A_1, A_2, E, T_1, and T_2. Since the spins of this system are parallel or antiparallel, the total spin quantum number, S, is either 1 or 0, and 2S + 1 is 3 either or 1, respectively.

Including spin, the following labels describe the states: $^1A_1$, $^3A_2$, ^1E, $^3T_1$, $^1T_1$, $^3T_2$, and $^1T_2$.

The energy of each state depends on the strength of the crystal field. This dependence is usually depicted in a Tanabe-Sugano diagram [13] of the type shown in Figure 12.6, which shows the $3d^2$ splittings in an octahedral field. At zero crystal field, the levels are designated according to the free ion spectral terms, shown to the left of the diagram. In the strong field limit, the slopes of the lines approach one of three possible values, which are associated with the $(t_{2g})^2$, $t_{2g}e_{2g}$, and $(e_{2g})^2$ orbitals.

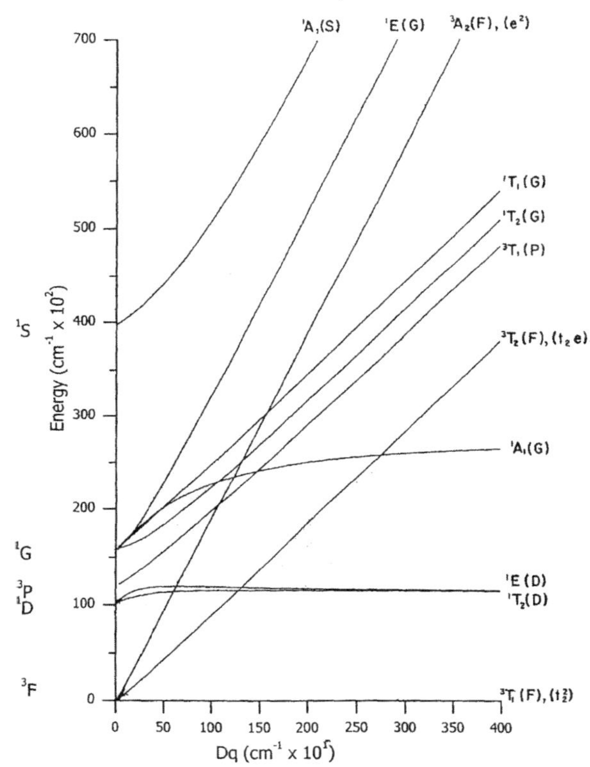

Figure 12.6 Tanabe-Sugano diagram for a $3d^2$ ion in an octahedral field.

12.4.4.3 Spectral Features of TMI in Solids

Absorption and emission spectra of TMI in solids generally show broad bands, though sharp lines are occasionally observed. The absorption and emission spectra of $Ti^{3+}:Al_2O_3$ are shown in Figure 12.7. The broad bands result from the 3d-electrons' exposure to the neighboring ions, making the wavefunctions sensitive to the position of the nearby ions.

Note the large Stokes shift between the peaks of the absorption and emission bands. This can be understood in the following way. Absorption of a photon results in a redistribution of the charge of the 3d-electrons. The neighboring ions then "see" the new potential and relax to a new equilibrium position. During relaxation, the center gives up energy to the lattice, usually by emitting phonons. A similar relaxation process occurs following the emission of a photon. Since part of the absorbed energy is given up to the lattice, less is available for the emitted photon.

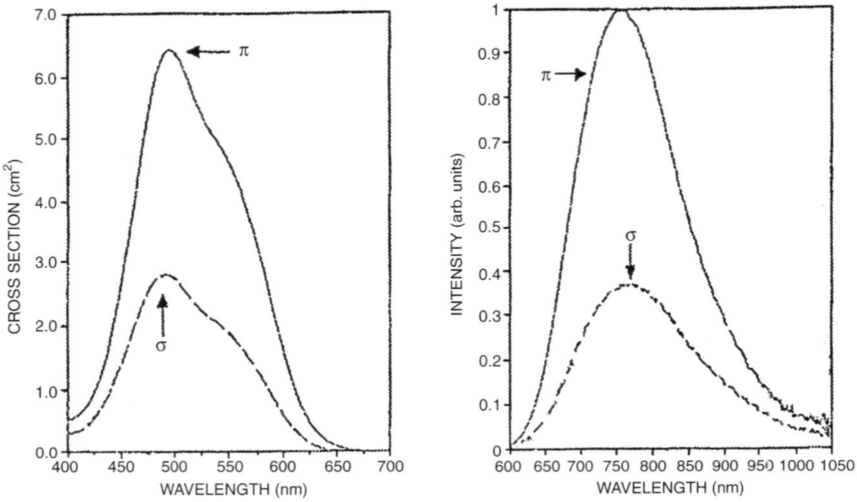

Figure 12.7 Polarized (a) absorption spectra and (b) emission spectra of Ti^{3+} in Al_2O_3 [12].

12.4.4.4 The Configurational Coordinate Model

The broad bands and large Stokes shift are related to one another, and can be understood using a *configurational coordinate* (CC) model [14, 15]. In this model, each electronic energy level is represented by a parabola on an energy vs. position diagram (Figure 12.8). The position coordinate can be considered one of the normal coordinates of the crystal, though it is often associated with the "breathing mode" of the ligands about the central TMI. The equilibrium position depends on the electronic state of the system. Horizontal lines within each parabola represent vibrational levels.

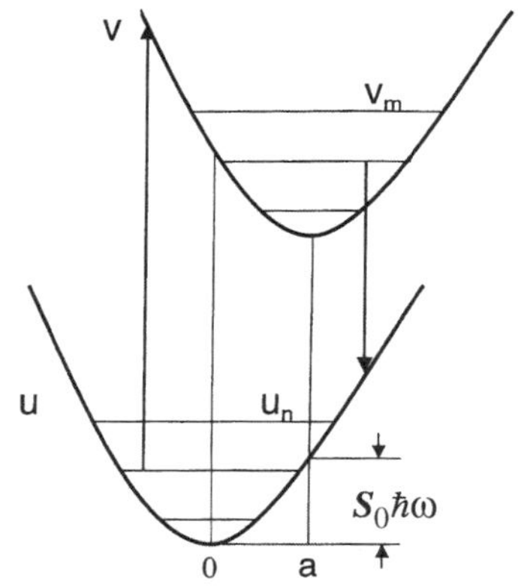

Figure 12.8 Configuration coordinate (CC) diagram of a two-level system.

The assumptions included in the model are the following:

1. The adiabatic approximation: The electrons and nuclei are separated into fast and slow subsystems, respectively. The electronic wavefunctions depend parametrically on the position of the nearby nuclei. The motion of the nuclei is determined by a potential defined by an average position of the electrons. Physically, this is justified by the fact the mass of the electron is much less than the nuclear mass.

2. The harmonic approximation: The electronic states that result from the adiabatic approximation have potential curves that are parabolic. Physically, this assumes that the amplitude of the TMI vibrations is small.

3. The Franck-Condon principle: This asserts that during a radiative transition the electronic charge is redistributed before the nuclei can react. Thus, transitions are represented by vertical arrows in the CC diagram. This is justified on the same basis as the adiabatic approximation.

Further discussions of these assumptions can be found in [14] and [16].

In Figure 12.8 the ground and excited electronic levels are labeled u and v, respectively, and their parabolas are offset from one another by a distance a. We assume identical force constants for each parabola. The vibrational levels of u and v are labeled as n and m, respectively, and the associated vibrational wavefunctions are ϕ_n and ϕ_m. Transitions in absorption and emission are represented by upward and downward arrows, respectively. In absorption, the electron moves from the state $|u_n\rangle$ to $|v_m\rangle$, creating (m – n) phonons in the process. After a transition (absorption or emission), the ion relaxes to a new equilibrium position by releasing phonons to the surrounding lattice.

The Stokes shift between absorption and emission and the shape of the spectral bands are determined by both the force constant of the vibration, k, and the offset, a. The most important parameter in the model is the Huang-Rhys parameter S_0 which is defined in

$$S_0 \hbar\omega = \frac{1}{2} k a^2 \ , \tag{12.83}$$

where ω is the angular vibrational frequency of the system. S_0 is also the average number of phonons created or destroyed in the transition [17]. The physical meaning of S_0 can be understood in the following manner. The lattice motion induces a local strain, which depends on the electronic state of the ion. If the strain is similarly affected for two electronic states, the value of S_0 between those two states is small. If two states affect the strain very differently, then S_0 will be large. The constant S_0, then, describes the *relative* coupling of the electronic wavefunctions to the vibrationally induced strains in the lattice.

Enforcement of the Franck-Condon principle leads to the result that the strength of a transition from $|v_m\rangle$ to $|u_n\rangle$ is proportional to the square of the overlap between the vibrational wavefunctions of the two states:

$$W_{m,n} \propto \left|\langle \phi_m | \phi_n \rangle\right|^2 . \tag{12.84}$$

These overlap integrals are called the *Franck-Condon factors*. In thermal equilibrium, the vibrational levels of the initial electronic state are populated according to the Boltzmann distribution, so that $W_{m,n}$ has an inherent temperature dependence.

A CC model for ruby (Cr^{3+}:Al_2O_3) is shown in Figure 12.9, along with the absorption and emission spectra. Note that the transitions from the ground $^4A_2$ state to the upper $^4T_1$ and $^4T_2$ states, are spin-allowed. The $^4T_1$ and $^4T_2$ states are offset from the $^4A_2$ state, and so exhibit broad absorption bands. In emission, the situation is different. Following absorption into the $^4T_2$ state, the ion relaxes to the base of the $^4T_2$ parabola, emitting phonons in the process. From that point, there is a feeding into the ^2E state, which is accompanied by the further emission of phonons. Because there is a small offset between the ^4A and ^2E states, the $^2E \rightarrow {}^4A$ emission is a sharp line. (This is the 694-nm ruby laser emission line.) This transition is between states of different spin, and so it is weak. The lifetime of the ^2E state is correspondingly long at 3 ms.

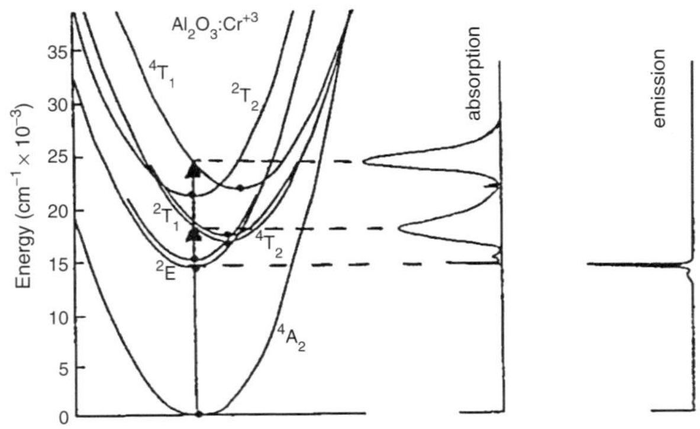

Figure 12.9 The configurational coordinate diagram and absorption and emission spectra of Cr^{+3}: Al_2O_3 [18].

12.4.5 Color Centers in Solids

12.4.5.1 Introduction

Even the purest crystals are found to have defects in the crystalline structure. The defect centers of concern in this section are color centers, some of which are given in Table 12.2. These are most widely found in alkaline halides, though they also occur in oxides, fluorides, and other miscellaneous hosts. Unless otherwise stated, our discussion assumes alkaline halide host materials.

Table 12.2 Color centers in alkaline halides.

Color Center	Description in Alkaline Halides
F	One electron trapped at an anion site
F^-	Two electrons trapped at an anion site
F^+	Vacant anion site
F_2 (M)	Two neighboring F-centers along the 110 axis
F_2^+	Two neighboring vacant anion sites (along 110 axis) containing a single electron
F^3 (R)	Three neighboring F-centers forming an equilateral triangle
F_A	F-center adjacent to a smaller alkali impurity ion
V_K	Singly ionized halogen-ion molecule formed from two adjacent halides

12.4.5.2 Energy Levels of F Centers

A single crystal often contains more than one type of color center. Consequently, the spectra often show several broad bands in absorption. The simplest of these centers is the *F center*, which consists of a single electron trapped at an anion vacancy (Figure 12.10). F centers can be formed either chemically or by bombardment with particles (e^-, p^+, neutrons) or high-energy radiation (ultraviolet, X-ray, γ-ray). A more complete discussion of F centers can be found in [19].

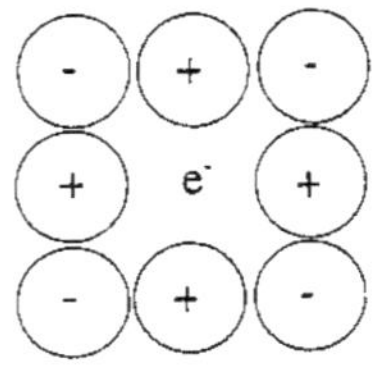

Figure 12.10 The F center, a trapped electron at an anion vacancy.

In cubic lattices, such as alkaline halides, the obvious physical model for the F center is that of a particle in a three-dimensional box of length $2l$, where l is the distance from the center of the vacancy to the nearest alkaline cation. The energy difference between the two lowest levels of such a particle is

$$\Delta E = \frac{3\pi^2\hbar^2}{8ml^2}\,, \tag{12.85}$$

where we note the l^{-2} dependence. Data on the low energy absorption band of F centers in several alkaline halides are found to obey the following (Mollwo-Ivey) law: $\Delta E = 17.7(l)^{-1.84}$. The nearly l^{-2} dependence lends some credibility to the particle-in-a-box model. Generally, absorption spectra show only one excited level associated with an isolated F center before the onset of the conduction band.

The F center has also been described as a hydrogen-like atom, where the electron is held at a central location by the surrounding charge rather than pulled in by the nucleus. In the so-called "point-ion" approximation, the electron moves in a spherically symmetric potential. The energy level structure is similar in some ways to that of a hydrogen atom; the first energy level is well separated from the ground state, and then numerous closely spaced levels entering the conduction band of the solid.

The states of the color center are labeled according to the irreducible representations of the site symmetry. For alkaline halides the ground and first excited states are Γ_1^+ and Γ_4^-, respectively. It is found that the lowest electronic states exhibit character consistent with 1s and 2p states of the hydrogen atom. The s- and p-type states emerge naturally from the point-ion model. Even the particle-in-a-box ground and first excited state wavefunctions show s and p character, respectively (Figure 12.11).

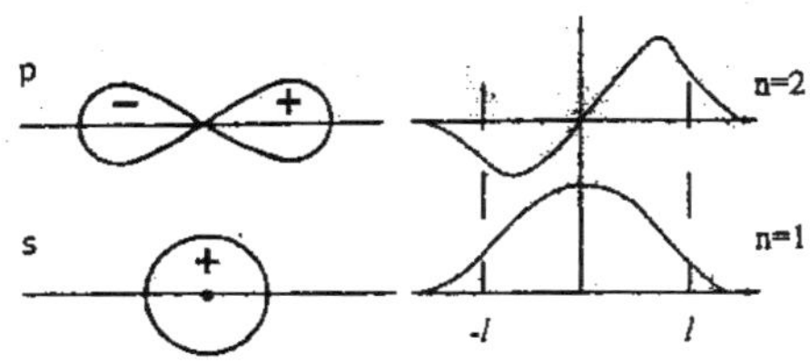

Figure 12.11 Diagram comparing the s- and p-type hydrogenic wavefunctions to the n = 1 and n = 2 wavefunctions of a particle in a finite well.

12.4.5.3 Spectral Features of F Centers

The potential in which the electron is situated is determined completely by the charge distribution of the nearby ions. As a result, the coupling between the F center and the lattice is strong, and absorption and emission spectra exhibit broad bands and large Stokes shifts, on the order of 1 eV. Figure 12.12 shows the absorption and emission of an F center in KBr. No zero phonon line is observed, and absorption and emission bands show no structure, even at low temperature. A configurational coordinate diagram of an F center has a large offset between the ground and excited state parabolas, and a Huang-Rhys parameter S_0 on the order of 30.

Absorption from the 1s state to the 2p state is an allowed transition with an f-number on the order of unity. Because of the relaxation of the nearby ions following excitation, the matrix element driving the emission process can be very different from that responsible for absorption. If the transition is from a p-state to an s-state, the f-number will again be close to 1.

The quantum efficiency of emission from the 2p state is close to unity at low temperatures. As the temperature is raised, the quantum efficiency decreases due to the presence of one or more quenching processes. One such process is nonradiative decay from the 2p to the 1s state. Another process is thermal activation from the occupied 2p state into the conduction band. Following absorption, the 2p state moves to within a few tenths of an eV of the conduction band. Photoconductivity experiments show thermal activation into the conduction band at temperatures around 100 K, depending on the host. Other quenching processes, including the interaction with other centers, have also been proposed [21]. At room temperature, the luminescence from F centers in alkaline halides is often severely quenched. Most color center lasers operate at liquid nitrogen temperature for this reason.

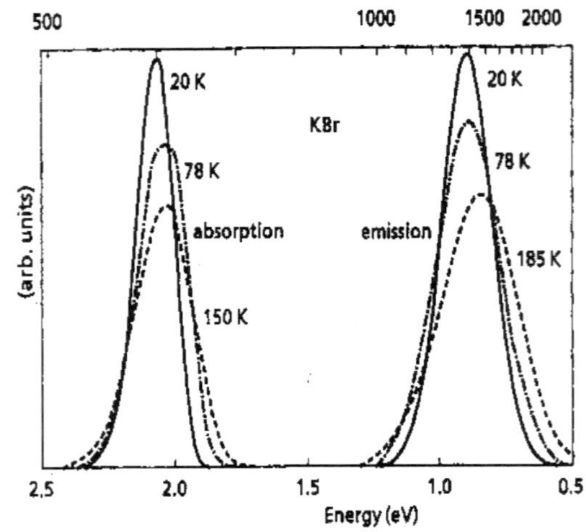

Figure 12.12 Absorption and emission bands of an F center in KBr at different temperatures [20].

12.4.5.4 Other Color Centers

We briefly describe the optical properties of two other color centers, the F_A center in alkaline halides and the F^+ center in oxides. The F_A center (Figure 12.13) is an F center in which a neighboring halide ion is replaced by a smaller halide ion, e.g., Na^+ replaces K^+ in KCl. This reduces the symmetry of the F center from cubic to tetragonal, and splits the excited 2p triplet into two levels. The splitting is on the order of 0.2 eV. An energy level diagram is shown in Figure 12.13b. Optical absorption measurements show one transition polarized along the vacancy-impurity ion axis and a twofold degenerate transition polarized perpendicular to that axis. The emission spectrum shows a singe band.

Finally, we consider an F^+ center in oxides (Figure 12.14). By definition, an F^+ center is one in which the net charge of the center is +1 with respect to the ion that would normally occupy that site. Since the oxygen has a charge of –2 in solids, the F^+ center has one electron at an oxygen vacancy, analogous to the F center in alkaline halides. Because of the net positive charge at the vacancy site, the electron is bound more tightly to the vacancy than in alkaline halides. Both the ground (1s) and excited state (2p) wavefunctions are more localized, and so the coupling to the lattice is less than for an F center in alkaline halides. Consequently the absorption and emission bands show vibronic structure and a relatively small Stokes shift, a few tenths of an eV (e.g., Figure 12.14b).

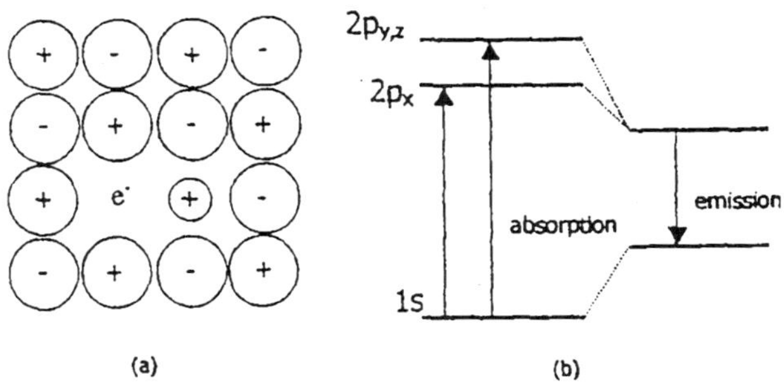

Figure 12.13 (a) F_A center and (b) its energy level diagram.

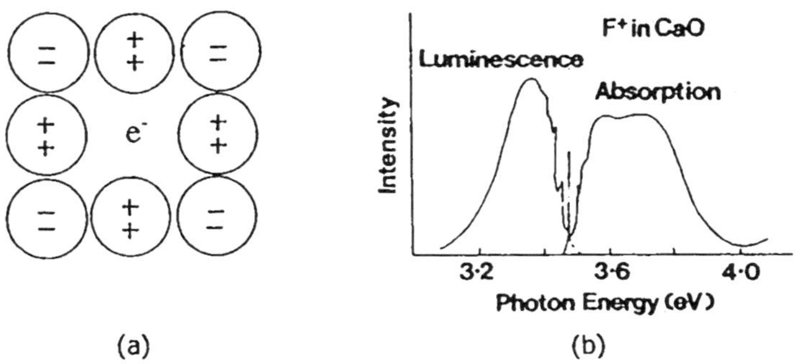

Figure 12.14 (a) The F^+ center and (b) absorption and emission spectra of an F^+ center in CaO at 4K [22].

12.5 PROCESSES IN LOCALIZED SYSTEMS

12.5.1 Introduction

The processes affecting the spectral characteristics of optically active centers include both radiative and non-radiative types. In this section we discuss the origin of these processes and their effects on the luminescence properties of the system.

Of the non-radiative processes, many of them include the emission and/or absorption of phonons. In fact, phonon-related processes are important to the luminescence properties of virtually all optically active centers, even those where the ion is only weakly coupled to the lattice. To understand why this is so, recall that the density of phonons for a monatomic solid in the frequency range between ω and $\omega + d\omega$ is:

$$n(\omega)d\omega = \left[\frac{1}{\exp(\hbar\omega / kT) - 1}\right]\frac{3\omega^2}{2\pi^2 c_s^3} d\omega. \tag{12.86}$$

With the velocity of sound in a solid (c_s) on the order of 5×10^5 cm/sec, assuming T=300 K and a Debye temperature of 1000 K, the total number of phonons is on the order of 10^{23}/cm^3. This high phonon density makes phonon-related processes ubiquitous, affecting all the spectral characteristics (e.g., intensity, lineshape, linewidth, and lifetime) of optical centers in solids.

12.5.2 Radiative Decay

12.5.2.1 Radiative Transitions Between f-Levels of REI in Solids

The probability of absorption or emission of radiation is determined by matrix elements of the form $\langle\psi_a|\mathbf{D}|\psi_b\rangle$, where **D** is the operator responsible for the transition. For rare earth ions in gaseous form, the 4f-states are of odd parity. If **D** is the electric dipole operator, which is odd, then the matrix element vanishes. Thus, electric dipole transitions between f-states are forbidden. In a crystal, however, such transitions are frequently observed. In this section, we show that this is because the crystal field interaction mixes some of the character of the upper nearby configurations of opposite parity (e.g., 5d and 6s) into the 4f states.

Because $H_{CF\text{-static}}$ is small, we treat it as a perturbation. We define $|u\rangle$ and $|u'\rangle$ to be the free ion states of the 4f and 5d configurations, respectively. To the first order in the perturbation, the 4f state of the ion in a solid, $|\psi\rangle$, is

$$|\psi_k\rangle = |u_k\rangle + \sum_j \frac{\langle u'_j|V_{\text{odd}}|u_k\rangle}{E_k - E'_j}|u'_j\rangle . \tag{12.87}$$

The sum is over all free ion states belonging to the 5d configuration, V_{odd} represents the odd components of $H_{CF\text{-static}}$, and E_k and E'_j are the energies of $|u_k\rangle$ and $|u'_j\rangle$, respectively. When there is inversion symmetry at the rare earth ion site, $V_{\text{odd}} = 0$ and no mixing occurs. If the ion site lacks inversion symmetry, mixing is always present.

The matrix element of the electric dipole operator between two states of the type given by equation (12.87) is:

$$\langle\psi_l|e\mathbf{r}|\psi_k\rangle = \sum_m \frac{\langle u_l|V_{\text{odd}}|u'_m\rangle}{E_l - E'_m}\langle u'_m|e\mathbf{r}|u_k\rangle + \sum_j \frac{\langle u'_j|V_{\text{odd}}|u_k\rangle}{E_k - E'_j}\langle u_l|e\mathbf{r}|u'_j\rangle . \tag{12.88}$$

Transitions are driven by the square of equation (12.88). Since the terms of the form $|\langle u_l|e\mathbf{r}|u'_j\rangle|^2$ are nonzero, electric dipole transitions between 4f states can occur whenever $V_{\text{odd}} \neq 0$. Transitions of this type have f-numbers on the order of 10^{-6}.

We note that magnetic dipole and electric quadrupole transitions are allowed even among pure f-states, and obey the selection rules, $\Delta L, \Delta J \leq 1$ and $\Delta L, \Delta J \leq 2$, respectively. Transitions are frequently observed, however, with ΔL and ΔJ as large as 6. Based on both theoretical and experimental evidence, it is generally agreed that the vast majority of f→f transitions are electric dipole in nature.

12.5.2.2 Judd-Ofelt Theory of Radiative Transitions for REI in Solids

While the strength of a radiative transition from the ground state to an excited state is readily measured in absorption, the strength of a radiative transition between excited levels is not easily determined experimentally. In this section, we summarize a method (called the *Judd-Ofelt method*) of predicting the strength of such transitions for REI in solids [23, 24].

The task is to simplify equation (12.88) into an expression whereby radiative transition rates can be estimated. We make the first simplification by invoking the property of closure over the terms of the form $|u'_j\rangle\langle u'_j|$. To do this, however, the term E'_j must be held constant for all values of j. Thus, the first assumption in the Judd-Ofelt theory is to set $E'_j = E'$, making all states of the $4f^{n-1}5d$ configuration degenerate. We are left with

$$\langle \psi_k | e\mathbf{r} | \psi_l \rangle = \frac{\langle u_l | V_{\text{odd}} e\mathbf{r} | u_k \rangle}{E_l - E'} + \frac{\langle u_k | V_{\text{odd}} e\mathbf{r} | u_l \rangle}{E_k - E'} \quad . \tag{12.89}$$

This assumption greatly simplifies the calculations. The crystal field operator and electric dipole operator have been combined, and only the knowledge of the free-ion 4f wavefunctions is required.

The next assumption is to set $E_l = E_k$, which essentially assumes that the energy difference between the 4f-levels is much smaller than the interconfigurational energy. This allows the terms in equation (12.89) to be combined.

By squaring the resulting matrix element, separating the radial and angular parts of the integral, and carrying out the necessary tensor algebra, the equation for the radiative transition probability becomes

$$P = \sum_{\lambda=2,4,6} \Omega_\lambda \langle \psi_k \| U^{(\lambda)} \| \psi_l \rangle^2 \quad . \tag{12.90}$$

The $\langle \psi_k \| U^{(\lambda)} \| \psi_l \rangle$ are reduced matrix elements and have been tabulated [25]. They do not vary significantly from host to host. The Ω_2, Ω_4, and Ω_6, are known as the Judd-Ofelt parameters, and are experimentally determined using a best fit to absorption data. In spite of the assumptions, the Judd-Ofelt method has been generally successful in predicting the radiative rates between REI 4f-levels [26].

12.5.2.3 Radiative Transitions Between d-Levels of TMI in Solids

In Section 12.4.4.2 it was noted that for a TMI at a location of inversion symmetry (e.g., octahedral), the 3d-orbitals are even. Thus, electric dipole matrix elements between states belonging to the 3d configuration are zero. For electric dipole transitions to occur, there must be a distortion to the inversion symmetry. This distortion introduces odd components into H_{CF} and mixes the character of crystal field states of higher configurations (e.g., $3d^{n-1}4p$) into those of the $3d^n$ configuration. This can occur through vibrational motion of the nearby ligands, but here we assume that the crystal contains a static distortion, and that it can be treated using perturbation techniques.

Considering the $^3T_2$ state of V^{3+} in an octahedral field, for example, the corrected wavefunction, $|\psi(^3T_2)\rangle$ to the first order, is

$$\left|\psi\left(^3\mathrm{T}_2\right)\right\rangle = \left|u\left(^3\mathrm{T}_2\right)\right\rangle + \sum_j \frac{\left\langle u'_j\right|V_{\mathrm{odd}}\left|u\left(^3\mathrm{T}_2\right)\right\rangle}{E\left(^3\mathrm{T}_2\right)-E'_j}\left|u'_j\right\rangle \tag{12.91}$$

where $|u(^3T_2)\rangle$ is the state of the system in the octahedral field, $|u'_j\rangle$ represents a state of an upper (odd) configuration in the same field, and $E(^3T_2)$ and E'_j are their respective energies. The odd components of $H_{CF\text{-static}}$ are labeled V_{odd}. A similar expression can be obtained for $|\psi(^3T_1)\rangle$. The matrix element of the electric dipole operator between the corrected $^3T_2$ and $^3T_1$ states is

$$\left\langle\psi\left(^3\mathrm{T}_1\right)\right|e\mathbf{r}\left|\psi\left(^3\mathrm{T}_2\right)\right\rangle = \sum_j \frac{\left\langle u'_j\right|V_{\mathrm{odd}}\left|u\left(^3\mathrm{T}_2\right)\right\rangle}{E\left(^3\mathrm{T}_2\right)-E'_j}\left\langle u\left(^3\mathrm{T}_1\right)\right|e\mathbf{r}\left|u'_j\right\rangle + \sum_m \frac{\left\langle u\left(^3\mathrm{T}_1\right)\right|V_{\mathrm{odd}}\left|u'_m\right\rangle}{E\left(^3\mathrm{T}_1\right)-E'_m}\left\langle u'_m\right|e\mathbf{r}\left|u\left(^3\mathrm{T}_2\right)\right\rangle. \tag{12.92}$$

Typically, the energy denominators in equation (12.92) are large (several eV) and V_{odd} is small, so that the transitions are weak: f-numbers are on the order of 10^{-4}.

We note that since the spin-orbit interaction is small in TMI, to a large degree S remains a good quantum number. Thus, transitions between states of different S values are very weak.

Magnetic dipole transitions are allowed between 3d states, even in a perfect octahedral field. In spite of this, they are not as important as electric dipole transitions unless the TMI is at a site having perfect octahedral symmetry, such as in the host MgO, where electric dipole transitions are strictly forbidden. Magnetic dipole transitions in TMI have f-numbers on the order of 10^{-6}.

12.5.3 Multiphonon Decay

The process considered here is the decay of an excited ion by the emission of phonons, commonly called *multiphonon decay* (Figure 12.15). In this section, we find an expression for the multiphonon decay rate of ions in solids, considering first the case of rare earth ions.

The first step is to express the crystal field as a Taylor expansion about the equilibrium position of the ion:

$$\mathrm{H}_{\mathrm{CF}} = \mathrm{H}_{\mathrm{CF\text{-}static}} + \sum_i \frac{\partial V}{\partial Q_i} Q_i + \text{ higher order terms}\,. \qquad (12.93)$$

In equation (12.93) V is the potential seen by the electrons and the Q_i are the normal modes of vibration of the lattice. At a lattice site with no inversion symmetry, the zeroth order term, $\mathrm{H}_{\mathrm{CF\text{-}static}}$, is responsible for breaking the f→f parity selection rule, allowing radiative electric dipole transitions within the 4f configuration. The first order and higher order terms describe modulations of the crystal field due to vibrations of the ligands, and is what we have termed $\mathrm{H}_{\mathrm{CF\text{-}dynamic}}$. It is also called the *ion-lattice* or *electron-phonon interaction*. This couples the motion of the electron to that of the lattice, allowing for phonon transitions between electronic states.

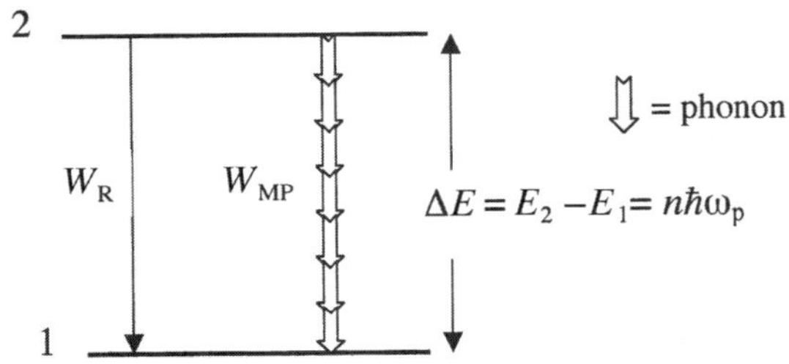

Figure 12.15 Two-level system showing the radiative (W_R) and multiphonon (W_{MP}) decay processes.

To obtain an expression for the multiphonon decay rate, one may take the first order term in equation (12.93) to the *n*th order of the perturbation. Alternatively, one can use the higher ordered term in equation (12.93) to a lower order of perturbation. No matter the approach, the result takes the following form:

$$W_{\mathrm{MP}}(n) = W_{\mathrm{MP}}(n-1)C\ , \qquad (12.94)$$

where $W_{\mathrm{MP}}(n)$ and $W_{\mathrm{MP}}(n\text{-}1)$ are the multiphonon decay rates of an n and n–1 phonon process, respectively, and C is a constant much less than unity. This leads to the so-called energy gap law [27]:

$$W_{\mathrm{MP}}(n) = W_0 e^{-\alpha\Delta E}\,. \qquad (12.95)$$

W_0 is a constant, and ΔE is the energy gap to the next lowest level. If $\hbar\omega$ is the energy of the phonons involved, then $\Delta E = n\hbar\omega$. The high energy phonons play the dominant role since fewer phonons are required, thereby lowering the order of the process. The constant α in equation (12.95) describes the ion-lattice coupling strength, and is determined experimentally. The energy gap law has been found to be valid for REI in solids [28].

An alternative starting point for describing the multiphonon decay rate is to express a non-adiabatic interaction Hamiltonian in the adiabatic approxi-mation. This leads to a similar energy gap law for the case of REI [29].

The temperature dependence of this process is related to the number of available phonons in the host, particularly the high energy phonons. The average occupation number of phonons in the ith mode is $N(\omega_i) = [\exp(\hbar\omega_i/kT) - 1]^{-1}$, and the emission of a phonon is proportional to $(N(\omega_i) + 1)$. The nth order multiphonon decay rate goes as $(N(\omega_i) + 1)^n$. This temperature dependence has been verified [30].

Observations have lead to the following rule of thumb for REI: If the energy difference to the next lowest level requires seven phonons or more, the dominant decay mechanism is radiative; otherwise it prefers to emit non-radiatively.

For TMI, a perturbation approach to multiphonon decay is not feasible. The multiphonon decay process and its temperature dependence are well described, however, using the configurational coordinate model. In this model, the ion first undergoes a non-radiative transition from state $|v_m\rangle$ to state $|u_n\rangle$, as shown in Figure 12.16. This transition is driven by the ion-lattice interaction, and the rate at which it occurs can be accurately calculated by summing by the Franck-Condon factors, equation (12.84), weighted of course by the population of the initial state $|v_m\rangle$. The transition is favored at the point where two parabolas cross, since that is where the Franck-Condon factors are greatest. Once in the electronic state $|u\rangle$, the system decays rapidly by giving up phonons to the lattice, resulting in the thermalization of $|u\rangle$. The work of Struck and Fonger [31] provides an in-depth discussion of this method.

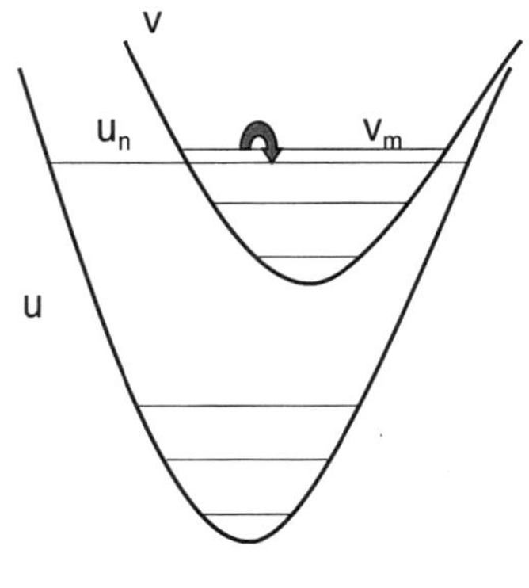

Figure 12.16 Configurational coordinate diagram showing a non-radiative transition from $|v_m\rangle$ to $|u_n\rangle$.

12.5.4 Vibronic Transitions

A vibronic transition in emission involves the creation of a photon and a simultaneous creation or absorption of one or more phonons. Schematic diagrams of the vibronic transitions are shown in Figure 12.17. At low temperatures, the vibronic sidebands appear on the high (low) energy side of the zero-phonon line in absorption (emission). At high temperatures, vibronic lines appear on both the high and low energy sides of the zero-phonon line. The vibronic spectrum of Pr-doped YAG is shown in Figure 12.18.

As with multiphonon decay, vibronic transitions are driven by the ion-lattice interaction. These transitions occur via two different types of processes, commonly labeled the M and the Δ processes [32]. Whereas both processes apply to the rare earth ions, the Δ process is most relevant to transition metal ions and color centers.

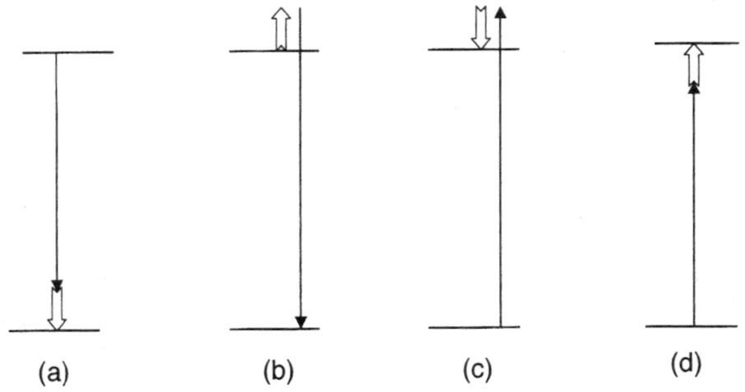

Figure 12.17 Vibronic transitions in (a), (b) emission and (c), (d) absorption. (b) and (d) are anti-Stokes processes and occur only at high temperatures. (a) and (c) occur at all temperatures.

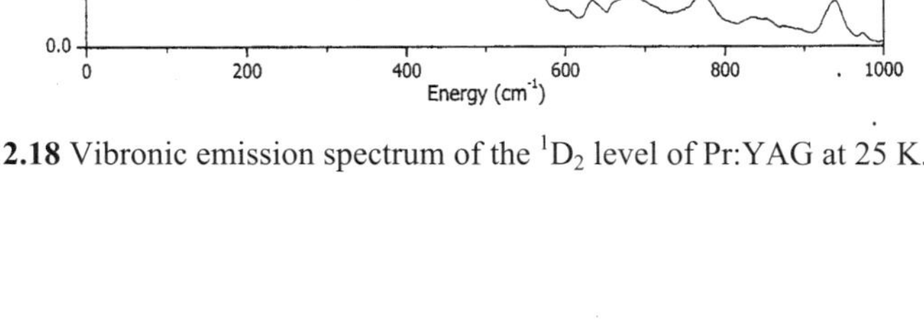

Figure 12.18 Vibronic emission spectrum of the $^1D_2$ level of Pr:YAG at 25 K.

In equation (12.87) we saw that the odd components of $H_{CF\text{-}static}$ causes the mixing of opposite parity states into the 4f-states of REI, allowing radiative transitions to occur. The M process occurs when the vibrations of the nearby ligands introduce odd components into $H_{CF\text{-}dynamic}$, thus leading to radiative transitions. Because they are vibrationally induced, these transitions are accompanied by a phonon. The M process is especially important in systems where the ion is at a site of inversion symmetry.

The Δ process occurs in systems where the equilibrium position of the optical center with respect to the ligands changes following an electronic transition. In the configurational coordinate model, this means that S_0 is greater than zero. For Δ processes, the intensity I of the vibronic transitions is given by

$$I = I_0 \frac{e^{-S_0} S_0^N}{N!} \tag{12.96}$$

where I_0 is the intensity of the zero-phonon line and N is the number of phonons involved in the process. This function is called a *Pekarian distribution*, and is plotted in Figure 12.19 for different values of S_0. For REI, $S_0 \approx 0.1$, so that that most of the energy is in the zero-phonon line.

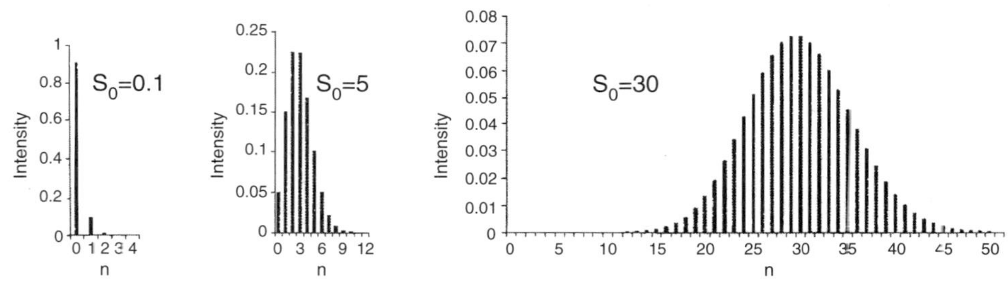

Figure 12.19 Plots of vibronic emission intensity versus number of phonons for various values of S_0 according to the Pekarian distribution in equation (12.96).

Vibronic transitions of REI can exhibit a strong host-dependence, with the ratio of the area under the zero-phonon line to the area under the vibronic sidebands changing by as much as two orders of magnitude depending on the host [33]. Equation (12.96) is also applicable to transition metal ions and color centers. For TMI S_0 is between 2 and 10, so they often exhibit a zero-phonon line and strong vibronic sidebands. In color centers S_0 is often on the order of 30, and the transitions are almost always purely vibronic.

12.5.5 Energy Transfer

When more than one optically active center is present, an excited center may impart all or part of its energy to a nearby center, a process called *energy transfer*. This can happen radiatively or non-radiatively.

In the radiative process (Figure 12.20a), the excited ion emits a photon and another ion absorbs that photon before it leaves the crystal. This process obeys a $1/R^2$ law and has little effect on the lifetime of the radiating level.

Some non-radiative energy transfer processes are shown in Figures 12.20b–d. The ion from which the energy is being transferred is called the *sensitizer* S. The ion to which the energy is transferred is called the *activator* A. We first consider the case of resonant energy transfer, shown is Fig. 12.20b.

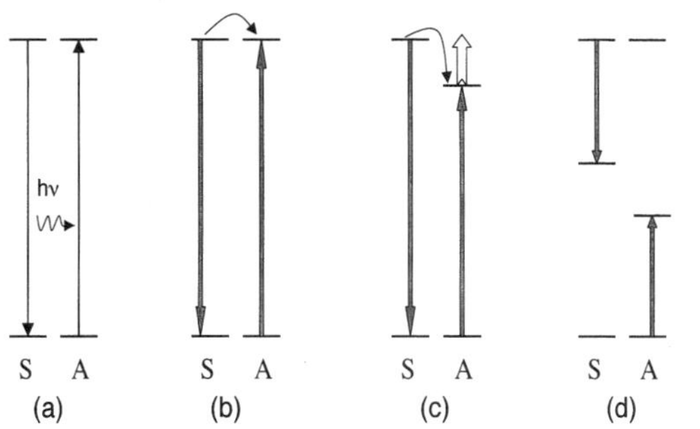

Figure 12.20 Energy transfer processes: (a) radiative, (b) resonant, non-radiative, (c) phonon-assisted, non-radiative and (d) non-radiative cross-relaxation.

The initial state of the system is an excited sensitizer and an activator in the ground state; the final state of the system is a sensitizer in ground state and an activator in excited state. The relevant Hamiltonian is the coulombic interaction between the electrons of the sensitizer and those of the activator. Forster [34] and Dexter [35] developed the following equation for the resonant energy transfer rate W_{SA} between S and A:

$$W_{SA} = K \frac{Q_A}{R^n \tau_S} \int \frac{g_S(E) g_A(E)}{E^4} dE \ . \qquad (12.97)$$

In equation (12.97), K is a constant, Q_A is the area under the absorption band of the activator, τ_S is the lifetime of the sensitizer, and R is the distance between the sensitizer and the activator. g_S and g_A are the normalized shape functions of the emission bands of the sensitizer and the absorption bands of the activator, respectively. The integral includes the overlap of these two functions, and is essentially a conservation of energy statement.

The R^n term is determined by the multipolar interaction between S and A. In most cases, the dipole-dipole ($n = 6$) mechanism is dominant. In cases where the dipole-dipole term vanishes due to symmetry, then the dipole-quadrupole ($n = 8$) or quadrupole-quadrupole ($n = 10$) terms may dominate. The exchange interaction may also drive the transfer, but is important only when S and A are in close proximity to one another [36].

Experimentally, one can gain insight into the nature of the interaction by examining the response of the system to pulsed excitation. Based on the work of Förster [37], assuming an even distribution of sensitizers and activators, the evolution in time of the population density (n_S) of excited sensitizers goes as

$$n_S(t) = n(0)\exp\left(-\frac{t}{\tau_S} - \Gamma\left(1-\frac{3}{q}\right)\frac{C_A}{C_0}\left(\frac{t}{\tau_S}\right)^{3/q}\right). \quad (12.98)$$

In equation (12.98), C_A is the acceptor concentration, C_0 is the sensitizer concentration at which energy transfer is equally likely as spontaneous deactivation, τ_S is the lifetime of the sensitizer in the absence of the acceptor, and $q = 6$, 8, or 10 depending on whether the multipolar interaction is dipole-dipole, dipole-quadrupole, or quadrupole-quadrupole, respectively. A best fit of the decay curve to this equation can be used to determine the type of interaction.

In *phonon-assisted energy transfer* (Figure 12.21c) the overlap integral in equation (12.97) is close to zero, and energy difference between the emission of the sensitizer and absorption of the activator must be made up for by the absorption or creation of one or more phonons. The following observations can be made regarding phonon-assisted transfer.

1. The transfer rate obeys an exponential energy gap law if the gap is much larger than the phonon energy [38], that is, if more than one phonon is required. This situation is frequently encountered in REI.
2. If W_{SA} is much faster than the decay rate of the upper levels of both S and A, the two levels become thermalized according to the Boltzmann distribution.
3. For TMI, the phonon-assisted energy transfer rate generally increases with temperature. For REI, the rate can increase or decrease with temperature depending on the energy levels involved [39].

12.5.6 Upconversion

The term "upconversion" applies to processes whereby a system excited with photons of energy $\hbar\omega_1$ emits photons at a frequency $\hbar\omega_2$, where $\hbar\omega_2 \gg \hbar\omega_1$ The importance of this phenomenon is due, in part, to the fact that it can generate short wavelength emission without having to rely on UV excitation, which can have deleterious secondary effects (such as creation of color centers and excitation of deep traps).

The scheme in Figure 12.21a is called *excited state absorption* (ESA) and involves a single ion. In this process two photons are absorbed sequentially, and the intermediate state is real. Non-radiative relaxation to a lower level may or may not occur following the absorption of the first photon.

Figure 12.21b shows a process called *two-photon absorption* (TPA). In this process, the two photons are absorbed simultaneously, and the intermediate state is a virtual one.

Another type of upconversion is a two-ion energy transfer process in which the activator is initially in an excited state (Figure 12.21c). This is referred to as *upconversion by energy transfer* (ETU). Three-ion up-conversion, an example of which is shown in Figure 12.21d, has been observed, but such processes are very weak. All upconversion processes are nonlinear, since they require the absorption of at least two photons. ESA and ETU are the most common. Experimentally, the two mechanisms can be distinguished from one another by examining the system's response to pulsed (laser) excitation. Both systems show a rise followed by a decay. For ESA-type upconversion, the rise time is within the lifetime of the pump (laser) pulse. For ETU, the luminescence peak occurs after the laser pulse is over, with the rise time depending on the decay time of the luminescent state and the ETU rate. The reader is referred to a recent review of upconversion processes by Auzel [40].

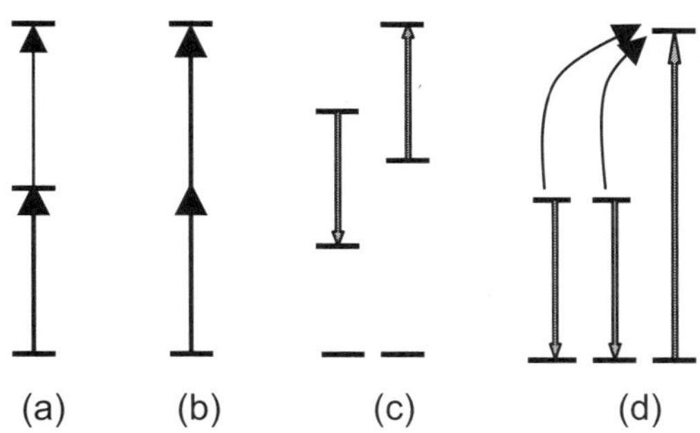

Figure 12.21 Upconversion processes: (a) excited state absorption (ESA), (b) two-photon absorption (TPA), (c) upconversion by energy transfer (ETU), and (d) three-ion cooperative energy transfer.

12.5.7 Line Broadening and Shifting with Temperature

In this section we discuss the affect of temperature on the width and position of the zero-phonon lines, particularly in REI-doped solids.

12.5.7.1 Spectral Line Broadening with Temperature

The width of a spectral line is affected by several mechanisms, including inhomogeneous broadening, direct one-phonon processes, multiphonon decay, radiative decay, and Raman scattering. Inhomogeneous broadening is caused by variations in the crystal field local to the REI. This effect is not temperature-dependent, and so will not be discussed here. Direct one-phonon processes, multiphonon decay, and radiative decay all affect width of the level by reducing its lifetime. With lifetimes of luminescent REI levels generally longer than a 10^{-7} sec, the sum of the contributions of these processes as given by the uncertainty principle is less than 0.01 cm^{-1}. Raman scattering involves the simultaneous absorption and emission of phonons, usually of two different frequencies. The electronic state of the ion is unchanged in the process, so that the lifetime is not affected. Experience has shown that the main contribution to the linewidth is from Raman processes and is on the order of a few cm^{-1}.

Raman scattering is a second order process. Using a perturbation approach, the probability of the ion absorbing an ion at frequency ω_i and emitting a phonon of ω_j is [41]

$$P_{ij} = A\omega_i\omega_j N(\omega_i)(N(\omega_j)+1) \ . \tag{12.99}$$

$N(\omega_i)$ and $(N(\omega_j) + 1)$ refer to the absorption and emission of a phonon, respectively, where $N(\omega_i) = [\exp(\hbar\omega_i/kT) - 1]^{-1}$ is the number of phonons in the ith mode. A is a constant describing the ion-lattice interaction strength.

Assuming a phonon density of states given by the Debye distribution and integrating over all allowed phonon frequencies, the contribution to the linewidth becomes

$$\Delta E = A'\left(\frac{T}{T_D}\right)^7 \int_0^{T_D/T} \frac{x^6 e^x}{(e^x-1)^2}\, dx \ . \tag{12.100}$$

T_D is the Debye temperature of the solid. Experimental data of linewidth versus temperature are usually fitted to equation (12.100) treating A' and T_D as adjustable parameters [42].

12.5.7.2 Shift of Spectral Lines with Temperature

In Section 12.4 we discussed the effect of $H_{\text{CF-static}}$ on the energy of the system. The contribution of $H_{\text{CF-dynamic}}$ to the energy of an ion in a solid is revealed in the shift of the spectral lines with temperature. As in the case of line broadening, Raman scattering is found to be the dominant process.

The contribution of Raman processes to the energy of the ith level is

$$\delta E_i = \sum_{j,k} \frac{\langle i,k|\mathrm{H}_{\text{CF-dyn}}|j,k\pm 1\rangle\langle j,k\pm 1|\mathrm{H}_{\text{CF-dyn}}|i,k\rangle}{E_i - E_j + \hbar\omega_k} + \langle i,k\,|\,\mathrm{H}_{\text{CF-dyn}}|i,k\rangle. \tag{12.101}$$

The sum over j includes all the electronic states of the system except the ith state. The sum over k includes all phonon modes.

Assuming a Debye phonon distribution, and also that the energy difference between electronic levels is greater than the phonon energies, equation (12.101) reduces to

$$\delta E_i = A''\left(\frac{T}{T_\mathrm{D}}\right)^4 \int_0^{T_\mathrm{D}/T} \frac{x^3}{e^x - 1}\, dx \ , \tag{12.102}$$

where A'' is a constant. A fit of this equation to lineshift data allows one to determine the Debye temperature of the solid. We note the following:

1. Raman scattering involves the participation of real intermediate states. For REI, these states include all unoccupied 4f-levels as well as those of the upper configurations.
2. The experimentally observed shift is the difference in the energy shifts of the two levels involved in the transition.

Experimental data show that lines shift generally toward the red, though blue shifts are sometimes observed. Typically, the observed shift is on the order of 10 cm^{-1} as the temperature is varied from 4 K to 500 K [42].

12.6 DELOCALIZED SYSTEMS

12.6.1 Density of One-Electron States and Fermi Probability Distribution

Given a volume $V = L^3$ the number of one-particle states in the range $dp_x dp_y dp_z$ is

$$\frac{V}{8\pi^3} dk_x dk_y dk_z = \frac{V}{h^3} dp_x dp_y dp_z . \tag{12.103}$$

The number of one-particle states in the range $(p, p + dp)$ is

$$g(p)dp = \frac{V}{h^3}\int_0^{\pi}\int_0^{2\pi} p^2 \sin\theta d\theta d\varphi dp = \frac{4\pi V}{h^3} p^2 dp \tag{12.104}$$

and, if the particles are electrons, taking the spin into account,

$$\begin{aligned} 2g(E)dE &= 2g(p)\frac{dp}{dE}dE \\ &= 2\frac{4\pi V}{h^3} p^2 \frac{1}{2}\sqrt{\frac{2m}{E}} dE = \frac{4\pi V}{h^3} 2mE \frac{1}{2}\sqrt{\frac{2m}{E}} dE \\ &= \frac{4\pi V}{h^3}(2m)^{3/2} E^{1/2} dE. \end{aligned} \tag{12.105}$$

Given a system of Fermions at temperature T, the probability distribution that specifies the occupancy probability is

$$F(E) = \frac{1}{e^{(E - E_\mathrm{F})/kT} + 1}. \tag{12.106}$$

In metals E_F, the *Fermi energy*, is the energy of the most energetic quantum state occupied at $T = 0$. At $T \neq 0$, E_F is the energy of a quantum state that has the probability 0.5 of being occupied. The number of available states in $(E, E + dE)$ for a system of electrons is

$$2g(E)dE = \frac{4\pi V}{h^3}(2m)^{3/2} E^{1/2} dE = \frac{8\sqrt{2}\pi m^{3/2}}{h^3} V E^{1/2} dE \quad (12.107)$$

The Fermi energy at $T = 0$ is determined by

$$\begin{aligned} N = \int_0^{\infty} 2g(E)dE &= \frac{8\sqrt{2}\pi m^{3/2}}{h^3} V \int_0^{E_F} E^{1/2} dE \\ &= \frac{16\sqrt{2}\pi m^{3/2}}{3h^3} V E_F^{3/2}. \end{aligned} \quad (12.108)$$

Then

$$E_F = \frac{\hbar^2}{2m}\left(3\pi^2 \frac{N}{V}\right)^{2/3} = \frac{0.121 h^2}{m} n^{2/3} \quad (12.109)$$

where $n = N/V$.

12.6.2 Classification of Crystalline Solids

Crystalline solids are arranged in a repetitive 3D structure called a *lattice*. The basic repetitive unit is the *unit cell*. Prototypes of crystalline solids are: (i) copper—metal, (ii) diamond—insulator, and (iii) silicon—semiconductor. We can classify the solid according to three basic properties:

1. Resistivity ρ at room temperature:

$$\rho = \frac{E}{J} \quad (\Omega\text{m})$$

where E = electric field, J = current density.

2. Temperature coefficient of resistivity

$$\alpha = \frac{1}{\rho}\frac{d\rho}{dt} \quad (\text{K}^{-1})$$

3. Number density of charge carriers, n (m^{-3}).

The resistivity of diamond is greater than 10^{24} times the resistivity of copper. Some typical parameters for metals and undoped semiconductors are reported in Table 12.3.

Table 12.3 A comparison of the properties of metals and semiconductors.

Property	**Unit**	**Copper (metal)**	**Silicon (semiconductor)**
n	m^{-3}	9×10^{28}	1×10^{16}
ρ	Ωm	2×10^{-8}	3×10^{3}
α	K^{-1}	4×10^{-3}	-70×10^{-3}

If we assemble N atoms, each level of an isolated atom splits into N levels in the solid. Individual energy levels of the solid form *bands*, with adjacent bands being separated by *gaps*. A typical band is only a few eV wide. Since the number of levels in one band may be on the order ~10^{24}, the energy levels within a band are very close.

12.6.2.1 Insulators

The electrons in the filled upper band have no place to go: The vacant levels of the band can be reached only by giving an electron enough energy to bridge the gap. For diamond the gap is 5.5 eV. For a state at the bottom of the conduction band, the energy difference $E - E_F$ is $0.5E_g$, since, as we shall see later, the Fermi energy for undoped semiconductors is approximately at the middle of the gap. Therefore we have $(E - E_F) \gg kT$, and we can write for the probability that one electron occupies a quantum level at the bottom of the *conduction band* as

$$P(E) = \frac{1}{e^{(E-E_F)/kT} + 1} \approx e^{(E-E_F)/kT} = e^{-\frac{E_g}{2kT}}. \qquad (12.110)$$

At room temperature such a probability has the value

$$P(E) = e^{-5.5/2(0.026)} \approx 1.2 \times 10^{-46}$$

and is negligible.

12.6.2.2 Metals

The feature that defines a metal is that the highest occupied energy level falls near the middle of an energy band. Electrons have empty levels to which they can go!

A classical free electron model can be used to deal with the physical properties of metals. This model predicts the functional form of Ohm's law and the connection between the electrical and thermal conductivity of metals, but does not give correct values for the electrical and thermal conductivities. This deficiency can be remedied by taking into account the wave nature of the electron.

12.6.2.3 Semiconductors

In this section we shall treat semiconductors that do not contain any impurity, generally called *intrinsic semiconductors*. We shall see later how the presence of impurities affects greatly the properties of semiconductors. The band structure of a semiconductor is similar to that of an insulator: The main difference is that a semiconductor has a much smaller energy gap E_g between the top of the highest filled band (*valence band*) and the bottom of the lowest empty band (*conduction band*) above it. For diamond, $E_g = 5.5\text{eV}$ whereas for Si, $E_g = 1.1\text{eV}$.

The charge carriers in Si arise only because, at thermal equilibrium, thermal agitation causes a certain (small) number of valence band electrons to jump over the gap into the conduction band. They leave an equal number of vacant energy states called *holes*. Both electrons in the conduction band and holes in the valence band serve as charge carriers and contribute to the conduction. The resistivity of a material is given by:

$$\rho = \frac{m}{e^2 n\tau} \tag{12.111}$$

where m is the mass of the charge carrier, n is the number of charge carriers/V, and τ is the mean time between collisions of charge carriers. Now, $\rho_C = 2 \times 10^{-8}$ Ω, $\rho_{Si} = 3 \times 10^3\ \Omega\text{m}$, and $n_C = 9 \times 10^{28}\ \text{m}^{-3}$, $n_S = 1 \times 10^{16}\ \text{m}^{-3}$, so that

$$\frac{\rho_{Si}}{\rho_{Cu}} = 10^{11} \qquad \text{and} \qquad \frac{n_{Cu}}{n_{Si}} = 10^{13}.$$

The vast difference in the density of charge carriers is the main reason for the great difference in ρ.

We note than the temperature coefficient of resistivity is positive for Cu and negative for Si. The atom Si has the following electronic configuration

$$\text{Si}: \underbrace{1s^2 2s^2 2p^6}_{core} 3s^2 3p^2.$$

Each Si atom has a core containing 10 electrons and contributes its $3s^2 3p^2$ electrons to form a rigid two-electron covalent bond with its neighbors. The electrons that form the Si-Si bonds constitute the valence band of the Si sample. If an electron is torn from one of the four bonds so that it becomes free to wander through the lattice, we say that the electron has been raised from the valence to the conduction band.

12.6.3 Intrinsic Semiconductors

We shall now present a model for an intrinsic semiconductor. In general the number of electrons per unit volume in the conduction band is given by

$$n_c = \int_{E_c}^{top} N(E) F(E)\, dE \tag{12.112}$$

where $N(E)$ = density of states and E_c = energy at the bottom of the conduction band.

We expect E_F to lie roughly halfway between E_v and E_c: The Fermi function decreases strongly as E moves up in the conduction band. To evaluate the integral in equation (12.112), it is sufficient to know $N(E)$ near the bottom of the conduction band and integrate from $E = E_c$ to $E = \infty$. Near the bottom of the conduction band, according to equation (12.105), the density of states is given by:

$$N(E) = \frac{4\pi}{h^3}\left(2m_e^*\right)^{3/2}\left(E - E_c\right)^{1/2} \tag{12.113}$$

where m_e^* = effective mass of the electron near E_c. Then

$$\begin{aligned} n_c &= \frac{4\pi}{h^3}\left(2m_e^*\right)^{3/2} \int_{E_c}^{\infty} \frac{\left(E - E_c\right)^{1/2}}{e^{(E-E_F)/kT} + 1} dE \\ &\xrightarrow[(E_c - E_F) \gg kT]{} \frac{4\pi}{h^3}\left(2m_e^*\right)^{3/2} \int_{E_c}^{\infty} \frac{\left(E - E_c\right)^{1/2}}{e^{(E-E_F)/kT}} dE. \end{aligned} \tag{12.114}$$

The integral may then be reduced to one of type

$$\int_0^{\infty} x^{1/2} e^{-x} dx = \frac{\pi^{1/2}}{2} \tag{12.115}$$

and we obtain the number of electrons per unit volume in the conduction band:

$$n_c = 2\left(\frac{2\pi m_e^* kT}{h^3}\right)^{3/2} e^{(E_F - E_c)/kT}. \tag{12.116}$$

Let us now consider the number of holes per unit volume in the valence band:

$$n_h = \int_{\text{bottom}}^{E_v} N(E)\left[1 - F(E)\right] dE \tag{12.117}$$

where E_v = energy at the top of the valence band. $1 - F(E)$ decreases rapidly as we go down below the top of the valence band (i.e., holes reside near the top of the valence band). Therefore, in order to evaluate n_h, we are interested in $N(E)$ near E_v:

$$N(E) = \frac{4\pi}{h^3}\left(2m_h^*\right)^{3/2}\left(E_v - E\right)^{1/2} \tag{12.118}$$

where m_h^* = effective mass of a hole near the top of the valence band. For $E_F - E_v \gg kT$,

$$1 - F(E) = 1 - \frac{1}{e^{(E-E_F)/kT} + 1} \approx e(E - E_F)/kT \tag{12.119}$$

Substituting (12.118) and (12.119) into (12.117), we obtain

$$\begin{aligned} n_h &= \int_{\text{bottom}}^{E_v} N(E)\left[1 - F(E)\right] dE \\ &= \frac{4\pi}{h^3}\left(2m_h^*\right)^{3/2} \int_{-\infty}^{E_v} \left(E_v - E\right)^{1/2} e^{(E-E_F)/kT} dE \\ &= 2\left(\frac{2\pi m_h^* kT}{h^2}\right)^{3/2} e^{(E_v - E_F)/kT}. \end{aligned} \tag{12.120}$$

We now use the fact that

$$n_c = n_h \tag{12.121}$$

and equate the two expressions for n_c and n_h given by equations (12.116) and (12.120), respectively. We find

$$E_F = \frac{E_c + E_v}{2} + \frac{3}{4} kT \ln \frac{m_h^*}{m_e^*}. \tag{12.122}$$

If $m_e^* = m_h^*$, E_F lies exactly halfway between E_c and E_v. Replacing equation 12.122 in the expression for $n_c = n_h$ we find

$$n_c = n_h = 2\left(\frac{2\pi kT}{h^2}\right)^{3/2} \left(m_e^* m_h^*\right)^{3/4} e^{-\frac{E_g}{2kT}} \tag{12.123}$$

At room temperature,

$$2\left(\frac{2\pi kT}{h^2}\right)^{3/2} m^{3/2} \approx 10^{19}\, cm^{-3},$$

where m = mass of the electron.

12.6.4 Doped Semiconductors

12.6.4.1 *n*-Type Semiconductors

Consider the phosphorus atom's electronic configuration:

$$\text{P} : 1s^2 2s^2 2p^6 3s^2 3p^3 \qquad (Z = 15).$$

If a P atom replaces a Si atom it becomes a *donor*. The fifth extra electron is only loosely bound to the P ion core: It occupies a localized level with energy $E_d \ll E_g$ below the conduction band. By adding donor atoms, it is possible to increase greatly the number of electrons in the conduction band. Electrons in the conduction band are *majority carriers*. Holes in the valence band are *minority carriers*.

Example: In a sample of pure Si the number of conduction electrons is $\approx 10^{16}\text{m}^{-3}$. We want to increase this number by a factor 10^6. We shall dope the system with P atoms creating an *n*-type semiconductor. At room temperature the thermal agitation is so effective that practically every P atom donates its extra electron to the conduction band.

The number of P atoms that we want to introduce in the system is given by

$$10^6 n_0 = n_0 + n_P,$$

or

$$n_P = 10^6 n_0 - n_0 \approx 10^6 n_0 = 10^6 \times 10^{16} = 10^{22}\,\text{m}^{-3}.$$

The number density of Si atoms in a pure Si lattice is

$$n_{\text{Si}} = \frac{N_a \rho}{A} = 5 \times 10^{28} m^{-3}$$

because N_a = Avogadro number, ρ = density of Si = 2,330 kg/m^3 and A = molar mass = 28.1g/mol = 0.028 *kg/mol.* The fraction of P atoms we seek is approximately

$$\frac{n_{\text{P}}}{n_{\text{Si}}} = \frac{10^{22}}{5 \times 10^{28}} = \frac{1}{5 \times 10^{6}}.$$

Therefore, if we replace only one Si atom in five million with a phosphorous atom, the number of electrons in the conduction band will be increased by a factor of 10^6.

12.6.4.2 *p*-Type Semiconductors

Consider the electronic configuration of an aluminum atom:

$$\text{Al} : 1\text{s}^2 2\text{s}^2 2\text{p}^6 3\text{s}^2 3\text{p} \qquad (Z = 13).$$

If an Al atom replaces a Si atom it becomes an *acceptor*. The Al atom can bond covalently with only three Si atoms; there is now a missing electron (a hole) in one Al-Si bond. With a little energy an electron can be torn from a neighboring Si-Si bond to fill this hole, thereby creating a hole in that bond. Similarly, an electron from some other bond can be moved to fill the second hole; in this way the hole can migrate through the lattice. It has to be understood that this simple picture should not be taken as indicative of a hopping process, since a hole represents a state of the whole system. Holes in the valence band are now *majority carriers*. Electrons in the conduction band are *minority carriers*. We compare the properties of an *n*-type semiconductor and of a *p*-type semiconductor in Table 12.4.

12.6.5 Model for a Doped Semiconductor

Most semiconductors owe their conductivity to impurities, i.e., either to foreign atoms put in the lattice or to a stoichiometric excess of one of its constituents. Energy level scheme for an *n*-type semiconductor and a *p*-type semiconductor is shown schematically in Figure 12.22.

Table 12.4 A comparison of the properties of an n-type and a p-type semiconductor.

Property	Type of Semiconductor	
	n	*P*
Matrix material	Silicon	Silicon
Matrix nuclear charge	14*e*	14*e*
Matrix energy gap	1.2 eV	1.2 eV
Dopant	Phosphorous	Aluminum
Type of dopant	Donor	Acceptor
Majority carriers	Electrons	Holes
Minority carriers	Holes	Electrons
Dopant energy gap	0.045 eV	0.067 eV
Dopant valence	5	3
Dopant nuclear charge	+15*e*	+13*e*

12.6.5.1 *n*-Type Semiconductors

At $T = 0$ all the donor levels are filled. At low temperatures only a few donors are ionized; the Fermi level is halfway between donor levels and the bottom of the conduction band. If we assume that E_F is below the bottom of the conduction band by more than a few kT, then we can use in this case a formula similar to equation (12.116) and find

$$n_c = 2\left(\frac{2\pi m_e^* kT}{h^2}\right)^{3/2} e^{(E_F - E_c)/kT} \qquad (12.124)$$

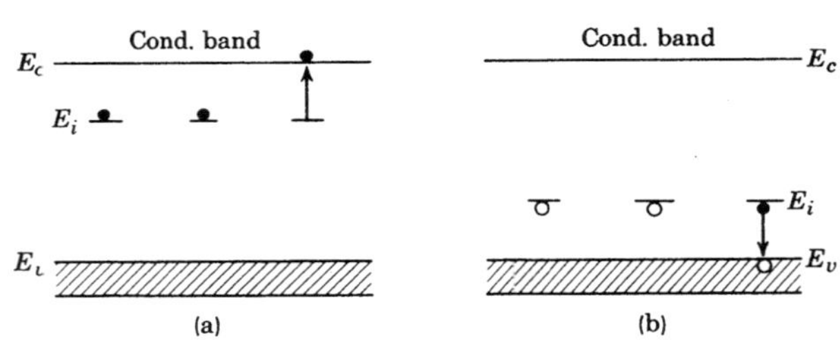

Figure 12.22 Energy level scheme for (a) an *n*-type semiconductor, and (b) a *p*-type semiconductor. E_i is the energy of the donor level (a) or the acceptor level (b).

If E_F lies more than a few kT above the donor level at E_d, the density of empty donors is equal to

$$n_d\left[1 - F(E_d)\right] \approx n_d e^{(E_d - E_F)/kT}. \qquad (12.125)$$

Equating (12.124) and (12.125) we obtain

$$E_F = \frac{1}{2}\left(E_d + E_c\right) + \frac{kT}{2}\ln\left[\frac{n_d}{2}\left(\frac{2\pi m_e^* kT}{h^2}\right)^{-3/2}\right]. \tag{12.126}$$

At $T = 0$, E_F lies halfway between the donor level and the bottom of the conduction band. As T increases, E_F drops (see Figure 12.23). Putting the expression for E_F from equation (12.126) in n_c given by equation (12.124), we find

$$n_c = \left(2n_d\right)^{1/2}\left(\frac{2\pi m_e^* kT}{h^2}\right)^{3/4} e^{-\frac{E_c - E_d}{2kT}}. \tag{12.127}$$

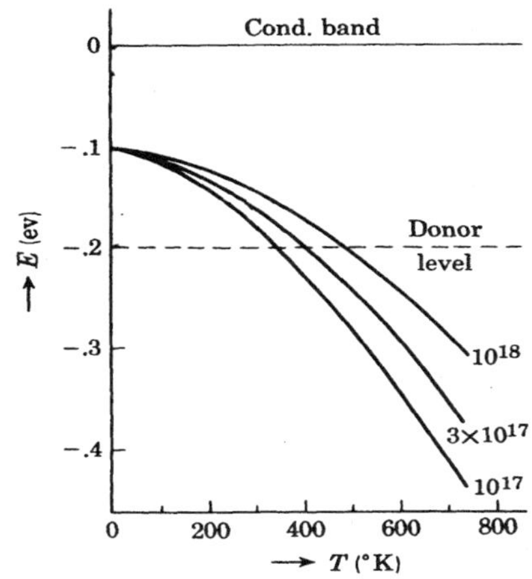

Figure 12.23 The variation of the position of the Fermi level with temperature with a donor level 0.2 eV below the bottom of the conduction band for three different values of n_d [43]

12.6.5.2 *p*-Type Semiconductors

The case of *p*-type semiconductors can be treated in a similar way. n_h has an expression similar to that for n_c. The Fermi level lies halfway between the acceptor level and the top of the valence band at $T = 0$. As T increases, E_F rises.

Figure 12.24 represents schematically the behavior of the Fermi level for an *n*-type and for a *p*-type semiconductor. The figure illustrates the fact that as the temperature increases the Fermi level for an *n*-type semiconductor does not drop indefinitely as indicated by (12.127). As the temperature increases the intrinsic excitations of the semiconductor become more important and the Fermi level tends to set in the middle of the gap. Similar effects take place for the *p*-type semiconductor. For additional considerations, the reader is referred to the book by Dekker (see bibliography at the end of this chapter).

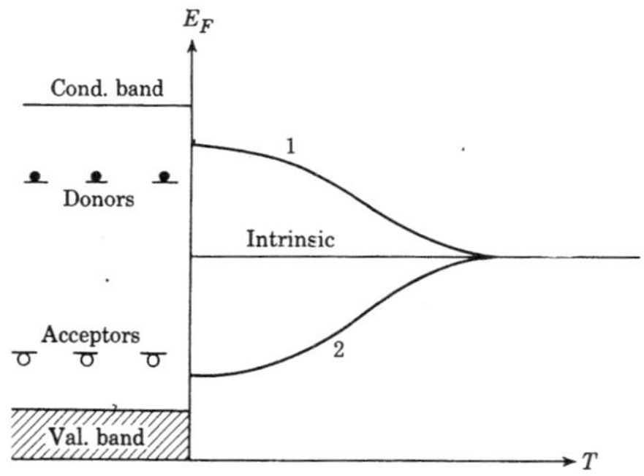

Figure 12.24 The variation of the position of the Fermi level with temperature. Curve 1 relates to insulators with donors and curve 2 relates to insulators with acceptors.

12.7 PROCESSES IN DELOCALIZED SYSTEMS

12.7.1 Direct Gap and Indirect Gap Semiconductors

The energy of the band gap of a semiconductor determines the spectral region in which the electronic transitions, both in absorption and emissions, take place. For visible or near infrared transitions we need materials with gaps ~1-1.7eV. A list of such materials is provided in Table 12.5.

Table 12.5 A list of typical semiconductors.

Material	Type	Band Gap (eV)
Si	Indirect	1.16
InP	Direct	1.42
GaAs	Direct	1.52
GaP	Indirect	2.3
AlP	Indirect	2.5
SiC	Indirect	3.0

Direct gap transitions take place when the maximum energy of the valence band and the minimum energy of the conduction band both occur in correspondence to a value of the linear momentum equal to zero or at the same $\vec{k} \neq 0$. Such semiconductors are called *direct gap semiconductors*.

In other materials, the maximum of the valence band and the minimum of the conduction band occur at different values of $\vec{k}$. Such materials are called *indirect gap semiconductors*.

It is interesting to consider the case of the semiconductor GaAs. By changing the chemical composition of this material according to the formula $GaAs_{1-x}P_x$ it is possible to change the band gap from 1.52 eV with $x = 0$ to 2.3 eV with $x = 1$. In addition, for $x > 0.4$ the material changes its character from that of a direct gap to an indirect gap semiconductor. Mixtures of InP and AlP can also yield gaps from 1.42 to 2.5 eV.

12.7.2 Excitations in Insulators and Large Band Gap Semiconductors

If a beam of light with photons exceeding in energy the energy gap goes through an insulator or a semiconductor, it raises an electron from the valence band into the conduction band for each photon absorbed, leaving behind a hole. The electron and the hole may move away from each other contributing to the *photoconductivity* of the material. On the other hand, they may combine producing an *exciton*, a hydrogen-like or a positron-electron pair-like structure. Excitons are free to move through the material. Since the electron and the hole have opposite charge, excitons are neutral and, as such, are difficult to detect. When an electron and a hole recombine, the exciton disappears and its energy may be converted into light or it may be transferred to an electron in a close-by atom, removing an electron from this atom and producing a new exciton.

Excitons are generally more important in insulators and in semiconductors with large gaps, even if some excitonic effects in small gap materials have been observed. Excitons do not obey the Fermi-Dirac statistics and therefore it is not possible to obtain a filled band of excitons. Excitons may also be created in doped semiconductors. In these, however, the free charges provided by the impurities tend to screen the attraction between electrons and holes, and excitonic levels are difficult to detect.

Two models are generally used to deal with excitons in solids. These models are more than two different ways of looking at the same problem; rather, they reflect two extreme physical situations:

a) a model in which the electron, after its excitation, continues to be bound to its parent atom; and
b) a model where the electron loses the memory of its parent atom and binds together with a hole.

The first case corresponds to the so-called *Frenkel exciton* and the second case to the *Wannier exciton*. Experimentally, the Frenkel exciton is in principle recognizable because the optical transitions responsible for the production of the exciton occur in the same spectral region of the atomic transitions. Experimentally, the transitions responsible for the production of a Wannier exciton fit a hydrogen-like type of behavior.

12.7.3 Radiative Transitions in Pure Semiconductors

12.7.3.1 Absorption

The absorption optical spectra of pure semiconductors generally present the following features (see Figure 12.25):

a) A region of strong absorption is present in the ultraviolet with a possible extension to the visible and infrared, due to electronic transitions from the valence to the conduction band. These *interband transitions* produce mobile electrons and holes that contribute to the photoconductivity. The value of the absorption coefficient is typically 10^5–10^6cm^{-1}. On the high energy side, the absorption band (~20 eV) decreases in value smoothly in a range of several eV. On the low energy side, the absorption decreases abruptly and may decrease by several orders over a range of a few tenths of an eV. In semiconductors, this region of the absorption spectrum is referred to as the *absorption edge*.

b) The low energy limit of the absorption edge corresponds to the photon energy necessary to move an electron across the minimum energy gap E_g. The exciton structure appears in the absorption edge region. It is more evident in insulators such as ionic crystals than in semiconductors.

c) At longer wavelengths the absorption rises again due to *free-carrier absorption*, i.e., electronic transitions within the conduction or valence bands. This absorption extends to the infrared and microwave regions of the spectrum.

d) A set of peaks appear at energies 0.02 to 0.05 eV (λ = 50–20 μm), due to the interaction between the photons and the vibrational modes of the lattice. In ionic crystals, the absorption coefficient may reach 10^5cm^{-1}; in homopolar crystals the absorption coefficient is generally much lower, such as 10–10^2cm^{-1}.

e) Impurities, if present in the semiconductor, may be responsible for absorption in the region of, say, 10^{-2}eV or so. This absorption is observable for kT lower than the ionization energy.

f) If the semiconductor contains paramagnetic impurities the absorption spectrum will present absorption lines in the presence of a magnetic field that splits the Zeeman levels.

g) An absorption peak in the long wavelength region may be present in the presence of a magnetic field, due to *cyclotron resonance* of the mobile carriers.

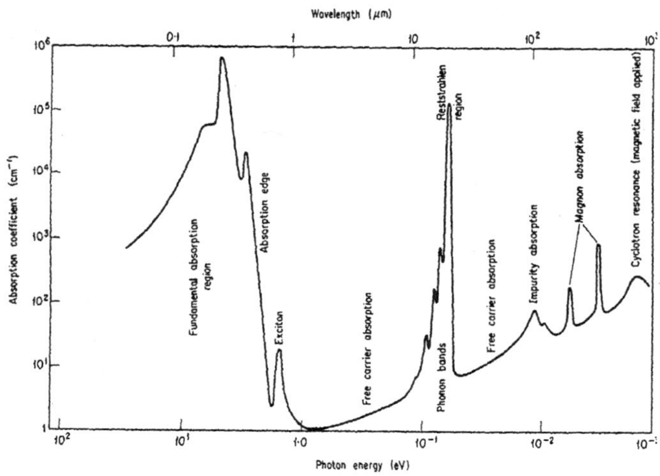

Figure 12.25 Absorption spectrum of a hypothetical semiconductor [44].

We want to make some additional considerations regarding the features a) and b) of the absorption spectrum. Interband transitions can take place subject to the two conditions of energy and conservation of wavevector:

$$\begin{cases} E_{\mathrm{f}} - E_{\mathrm{i}} = \hbar\omega\,(\vec{k}') \\ \vec{k}_{\mathrm{f}} - \vec{k}_{\mathrm{i}} = \vec{k}\,' \end{cases} \tag{12.128}$$

where the subscripts f and i refer to the final and initial one electron states and $\vec{k}'$ is the wave vector of the absorbed photon of energy $\hbar\omega(\vec{k}')$. Since the wavelength of the radiation is much longer than the lattice constant, $\vec{k}'$ is much smaller than the size of the reciprocal lattice constant and $\vec{k}'$ can then be neglected in the second equation (12.128). This means that in an $(E,\vec{k})$ diagram we should rely on *vertical* transitions.

The interband absorption is restricted by the conditions (12.128) and shows a structure that depends on the density of final states. The peaks can be presumably associated with values of $\vec{k}$ about which the empty and the filled bands run parallel:

$$\vec{\nabla}_k E_{\mathrm{v}}\left(\vec{k}\right) = \vec{\nabla}_k E_{\mathrm{c}}\left(\vec{k}\right). \tag{12.129}$$

In such a case there is a large density of initial and final states available for the transitions in a small range of energies.

In Section 12.7.1 we have made a distinction between direct gap and indirect gap semiconductors. For the former, the maximum of the valence band energy and the minimum of the conduction band energy occur frequently at $\vec{k} = 0$, (but not always, e.g., this is not true for Ta-doped halides of lead salts), whereas for the latter they occur at different values of $\vec{k}$. In indirect gap semiconductors, the absorption transitions at the band edge are phonon-assisted and have probability smaller than that of the direct gap absorption transitions. The absorption edge of the indirect gap absorption may show features related to the available phonon energies.

We now turn our attention to the excitonic structure of the absorption band. We have two models at our disposal: the Frenkel model and the Wannier model. In the Frenkel model, an excited electron describes an orbit of atomic size around an atom with a vacant valence state; this model is more appropriate for ionic insulators. The Wannier model represents an exciton as an electron and a hole bound by the Coulomb attraction, but separated by several lattice sites. This model is more appropriate for semiconductors.

An example of the Frenkel exciton is given by the crystal MnF_2 [45] in which the excited state may be considered to consist of an electron and a hole residing in the same ion. The excitation can travel throughout the system via the energy transfer mechanism. A good example of the Wannier exciton is given by Cu_2O, which presents absorption lines up to $n = 11$. [46]

12.7.3.2 Emission

Following the absorption process, an excited electron can decay radiatively by emitting a photon (possibly accompanied by a phonon) or non-radiatively by transforming its excitation energy entirely into heat (phonons). The following reasons make the emission data relevant:

1) Emission is not simply the reversal of absorption. In fact, the two phenomena are thermodynamically irreversible and therefore, emission spectroscopy furnishes data not available in absorption.
2) Emission is easier to measure than absorption, since its intensity depends on the intensity of excitation.
3) The applications of emission from solids, such as those of fluorescent lights and television, far outnumber the application of absorption.

In Section 12.7.5, we discuss the photon emission processes in solids.

12.7.4 Doped Semiconductors

Two types of impurities are particularly important when considering the optical behavior of semiconductors.

Donors: As we have seen in Section 12.6.4, when a material made of group IV atoms, like Si or Ge, is doped with a small amount of group V atoms like As, the extra electrons of these atoms continue to reside in the parent atoms, loosely bound to them. The binding energy, called E_D, is typically around 0.01 eV. It is in fact 0.014 eV for As, 0.0098 eV for Sb and 0.0128 eV for P. E_D is also called the *ionization energy of the donor atom*. The electrons that a donor puts in the conduction band, because of thermal excitation, cannot produce luminescence, because this process needs, besides an excited electron, a hole where the electron can go, and the valence band, being filled with electrons, has no holes.

Acceptors: If a material made of group IV atoms, such as Si or Ge, is doped with group III atoms, such as Ga or Al, a hole for each of these atoms forms and remains loosely bound to the parent atom. The amount of energy necessary to move an electron from the top of the valence band to one of these holes is labeled E_A and is typically around 0.03 eV. E_A may also be called the *ionization energy of the acceptor*.

Both types of impurities can be doped into the same crystal, deliberately, or they may be due to the fact that it is practically impossible to fabricate semiconductor crystals of perfect purity.

12.7.5 Radiative Transitions Across the Band Gap

We shall now examine, following Elliott and Gibson [47], the radiative processes that can take place across the band gap of a semiconductor (see Figure 12.26).

Processes A and B: An electron excited to a level in the conduction band will thermalize quickly with the lattice and reside in a region ~1 kT wide at the bottom of the conduction band. Thermalization is generally achieved by phonon emission, but also, less frequently, by phonon-assisted radiative transitions. If such photons have energy exceeding E_g, they can be reabsorbed and promote another electron to the conduction band

Process C: The recombination of electrons and holes with photon emission, the reverse process to absorption, is possible, but not very likely, because of competing processes. It may be present only in high purity single crystals. The widths of the related emission bands are expected to be ~1 kT because the thermalized electrons and holes reside at the band edges in this range of energy.

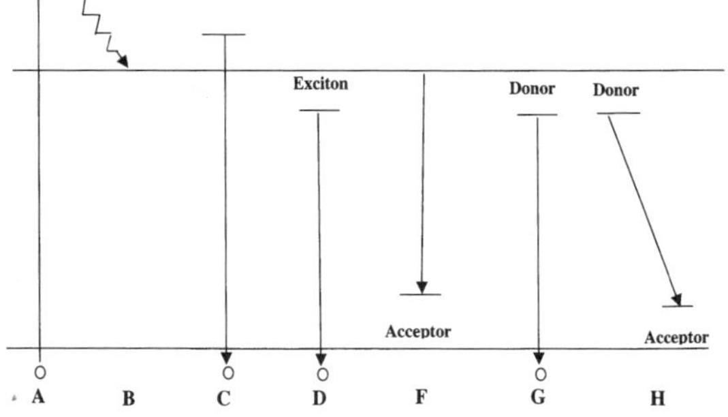

Figure 12.26 Transitions producing emission of photons in solids.

Process D: The radiative decay of the exciton can be observed at low temperature in very pure crystals. There are two types of decay: 1) the decay of the free exciton, and 2) the decay of an exciton bound to an impurity.

Transitions of the first type are observed at low temperatures. Since the exciton levels are well-defined, a sharply structured emission can be expected.

As for the transitions of the second type, they may be observed in materials of high purity into which impurities are purposely doped. An exciton may bind itself to one such impurity; the energy of the *bound exciton* is lower than the $n = 1$ energy by an amount equal to the binding energy of the exciton to the impurity. It may be noted that emission from bound excitons in indirect gap materials can take place without the assistance of phonons because the localization of a bound exciton negates the requirement of wavevector conservation. Bound electron emission is observed at low temperature and is generally much sharper than free electron emission.

Processes F and G: The transitions related to these processes are between the band edges and donors and acceptors and are commonly observed in solids. In particular, we may have conduction band to neutral acceptor (F) and neutral donor to valence band (G) transitions. They may be phonon-assisted.

Process H: In the transitions related to these processes, an electron leaves a neutral donor and moves to a neutral acceptor. After such a transition both donor and acceptor are ionized and have a binding energy equal to

$$E_{\text{b}} = -\frac{e^2}{4\pi\varepsilon_o kr} \tag{12.130}$$

where r is the donor-acceptor distance. The energy of the transition is then

$$\hbar\omega(r) = E_{\text{g}} - E_{\text{A}} - E_{\text{D}} + \frac{e^2}{4\pi\varepsilon_o kr} \tag{12.131}$$

An example of such a transition is given by GaP-containing sulfur donors and silicon acceptors, both set in phosphor sites.

12.7.6 Non-Radiative Processes

In the great majority of cases, recombination of electrons and holes takes place by emission of phonons. Since the probability of such processes decreases with the number of phonons emitted, these processes are favored by the presence of intermediate levels between the valence and the conduction bands produced by impurities or defects.

An additional mechanism, known as the *Auger process* could be responsible for the non-radiative recombination of electrons and holes. In an Auger process, an electron undergoes an interband transition and gives the corresponding energy to another conduction band electron, which is then brought to a higher level in the same band. The latter electron decays then to the bottom of the band with the phonon emission facilitated by the near continuum of states. In most cases, however, the mechanism of non-radiative decay has not been identified with certainty.

12.7.7 *p-n* Junctions

12.7.7.1 Basic Properties

A *p-n* junction consists of a semiconductor crystal doped in one region with donors and in an adjacent region with acceptors. Assume, for simplicity's sake, that the junction has been formed mechanically by pushing towards each other a bar of *n*-type semiconductor and a bar of *p*-type semiconductor, so that a *junction plane* divides the two regions (see Figure 12.27).

Let us now examine the motion of the electrons (majority carriers of the *n*-type bar) and of holes (majority carriers of the *p*-type bar). Electrons on the *n*-side of the junction plane tend to diffuse (from right to left in the figure) across this plane and go to the *p*-side, where there are only very few electrons. On the other hand, holes on the *p*-side tend to diffuse (from left to right in the figure) and go to the *n*-side, where there are only very few holes.

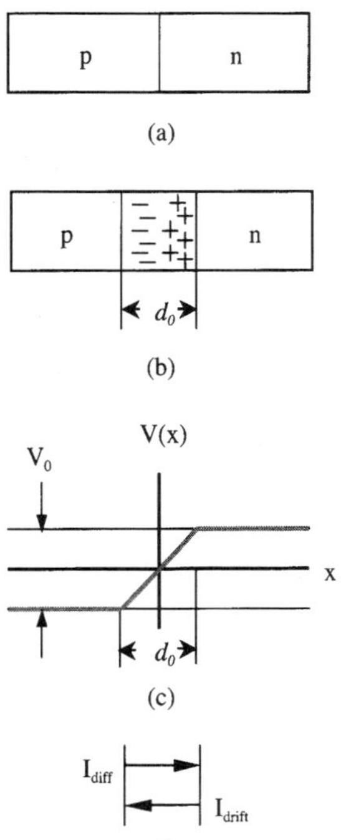

Figure 12.27 a) An *n*-type material and a *p*-type material joined to form a *p-n* junction. b) Space charge associated with uncompensated donor ions at the right of the junction plane and acceptor ions at the left of the plane. c) Contact potential difference associated with the space charge. d) Diffusion current I_{diff} made up by majority carriers, both electrons and holes, compensated in an isolated *p-n* junction by a current I_{drift} made up by minority carriers.

The n-side region is full with positively charged donor ions. If this region is isolated, the positive charge of each donor ion is compensated by the negative charge of an electron in the conduction band. But, when an n-side electron diffuses towards the p-side, a donor ion, having lost its compensating electron, remains positively charged, thus introducing a fixed positive charge near the junction plane. An electron arriving to the p-side quickly combines with an acceptor ion and introduces a fixed negative charge near the junction plane on the p-side. Holes also diffuse, moving from the p-side to the n-side, and have the same effect as the electrons.

Both electrons and holes, with their motion, contribute to a *diffusion current* I_{diff}, that is conventionally directed from the p-side to the n-side. An effect of the motion of electrons and holes across the junction plane is the formation of two space charge regions, one negative and one positive. These two regions together form a *depletion zone* of width d_o in Figure 12.27, so called because it is relatively free of mobile charge carriers. The space charge has associated with it a *contact potential difference* V_o across the depletion zone, which limits the further diffusion of electrons and holes.

Let us now examine the motion of the minority carriers: electrons on the p-side and holes in the n-side. The potential V_o set by the space charges represents a barrier for the majority carriers, but favors the diffusion of minority carriers across the junction plane. Together both types of minority carriers produce with their motion a drift current I_{drift} across the junction plane in the sense contrary to that of I_{diff}. An isolated p-n junction in equilibrium presents a contact potential difference V_o between its two ends. The average diffusion current I_{diff} that moves from the p-side to the n-side is balanced by the average drift current I_{drift} that moves in the opposite direction.

Note the following:

a) The net current due to holes, both majority and minority carriers, is zero.
b) The net current due to electrons, both majority and minority carriers, is zero.
c) The net current due to both holes and electrons, both majority and minority carriers included, is zero.

12.7.7.2 The Junction Rectifier

When a potential difference is applied across a p-n junction with such a polarity that the higher potential is on the p-side and the lower potential is on the n-side—an arrangement called *forward-bias connection* (Figure 12.28a)—a current flows through the junction. The reason for this phenomenon is that the p-side becomes more positive than it was before and the n-side more negative, with the result that the potential barrier V_o decreases, making it easier for the majority carriers to move through the junction plane and increasing considerably the diffusion current I_{diff}. The minority carriers sense no barrier and are not affected, and the current I_{drift} does not change.

Another effect that accompanies the setting of a forward bias connection is the narrowing of the depletion zone, due to the fact that the lowering of the potential barrier must be associated with a smaller space charge. The space charge is due to ions fixed in their lattice sites, and a reduction of their number produces a reduction of the width of the depletion zone. If the polarity is reversed in a *backward-bias* connection (Figure 12.28b) with the lower potential on the *p*-side and the higher potential on the *n*-side of the *p-n* junction, the applied voltage increases the contact potential difference and, consequently, I_{diff} decreases while I_{drift} remains unchanged. The result is a very small back current I_B.

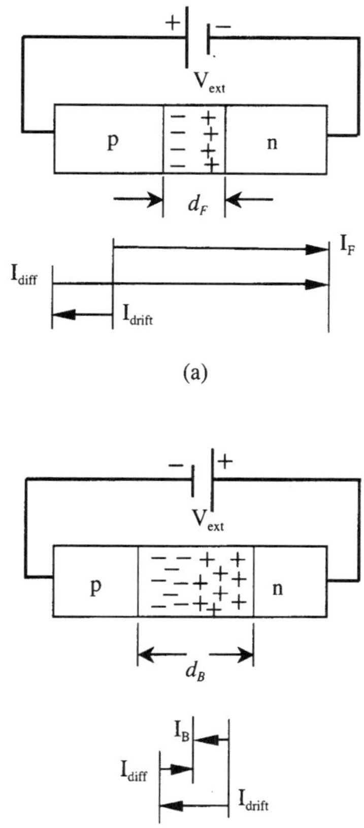

Figure 12.28 a) Forward-bias connection of a *p-n* junction, showing the narrowed depletion zone and the large forward current I_F. b) Backward-bias connection of a *p-n* junction, showing the widened depletion zone and the small back current.

12.7.7.3 Radiative Processes in *p-n* Junctions and Applications

In a simple semiconductor, one electron-hole pair may combine with the effect of releasing an energy E_g corresponding to the band gap. This energy in silicon, germanium, and other simple semiconductors is transformed into thermal energy, i.e., vibrational energy of the lattice. In certain semi-conductors, such as GaAs, the energy of a recombined electron-hole pair can be released as a photon of energy E_g. However, due to the limited number of electron-hole possible recombinations at room temperature, pure semi-conductors are not apt to be good emitters.

Doped semiconductors also do not provide an adequate large number of electron-hole pairs, with the *n*-type not having enough holes and the *p*-type not enough electrons.

A semiconductor system with a large number of electrons in the conduction band and a large number of holes in the valence band can be provided by a heavily doped *p-n* junction. In such systems, a current can be used in a forward-bias connection to inject electrons in the *n*-type part of the junction and holes in the *p*-part. With large dopings and intense currents the depletion zone becomes very narrow, perhaps a few microns wide, and a great number of electrons are in the *n*-type material and a large number of holes in the *p*-type material. The radiative recombination of electrons and holes produces a light emission called *electroluminescence*, or, more aptly, *injection electroluminescence*.

The materials used for *light emitting diodes* (LEDs) comprise such alloys as $GaAs_{1-x}P_x$, in which the band gap can be varied by changing the concentration x of the P atoms. For $x \approx 0.4$ the material is a direct-gap semiconductor and emits red light. Almost pure GaP produces green light, but, since it is an indirect-gap semiconductor, it has a low transition probability.

The passage of current through a properly arranged *p-n* junction can generate light. The reverse process is also possible, where a beam of light impinging on a suitable *p-n* junction can generate a current. This principle is at the basis of the *photo-diode*.

A remote TV control consists of an LED that sends a coded sequence of infrared light pulses. These pulses are detected by a photo-diode that produces the electrical signals that perform such various tasks as change of volume or channel.

In a forward biased *p-n* junction, a situation may be created in which there are more electrons in the conduction band of the *n*-type material than holes in the valence band of the *p*-type material. Such a situation of *population inversion* is essential for the production of *laser action*. Of course, in addition to this condition the appropriate geometry for the *p-n* junction is necessary in order to allow the light to be reflected back and forth and produce the chain reaction of stimulated emission. In this way, a *p-n* junction can act as a *p-n junction laser*, with a coherent and monochromatic light emission.

12.8 DIRECTION OF FUTURE EFFORTS

12.8.1 Why Luminescence?

Before projecting into the future, it seems worthwhile to ask Why has luminescence spectroscopy been such a fruitful field? Some possible responses to this question are:

1. The basic study of luminescent systems has been a precursor to the development of the laser (laser diodes and doped insulator-based lasers) and several other devices/materials that enjoy widespread application.
2. The study is good in itself, in that it acts as a probe to the understanding of very complex systems, lending insight into their static and dynamic properties.

Static: Luminescence allows us to (1) study the energy levels that cannot be identified solely by absorption measurements, (2) help probe the symmetry of the host material, and (3) reveal the existence of excitonic levels and to provide the means to study them.

Dynamic: Through luminescence, we (1) probe collective excitations of solids, such as phonons, magnons, etc. and (2) track the energy as the system decays from an excited level to the ground level, revealing the existence of processes such as energy transfer, upconversion, and non-radiative decay.

From the essentially inexhaustible source of luminescent systems, there emerges a continuous stream of new data and with it a deeper understanding of the processes relevant to those systems.

Luminescence spectroscopy remains a vibrant field of research, even 400 years since the investigation into the Bologna stone, over 150 years since Stokes's observations on luminescence from fluorspar, and after more than a half century of modern, scientific inquiry. The body of past work has revealed only some of the dynamics, complexity, and possibilities of luminescent systems. As testament to the validity of past investigations, we have been educated in the workings of these systems, and presented with more questions and challenges as we move forward.

12.8.2 Challenges and Future Work

Though luminescence spectroscopy has much to offer, the complexity of these systems renders the task of establishing general criteria for luminescent systems very difficult, posing continuoal challenges to workers in the field. The ultimate goal, of course, is to improve the predictive value of our knowledge so that we can accurately predict the response of a system to an impulse of excitation.

The realization of this goal is an extremely difficult task. The luminescence properties depend not just on the general features of the solid, but also on its more nuanced features—breaks in symmetry, impurities, clustering of ions, local charge distributions, phonon properties of the solid, etc. Indeed, an accurate prediction of even the most fundamental process of luminescence, the radiative transition probability, requires precise knowledge of the wavefunctions of the system, something that has heretofore proven elusive.

This complexity, which provides the challenges of the field, is also the source of the usefulness of luminescent materials. These systems are so varied and complex that they can be designed for a wide variety of applications. Even after several decades of successes, developments of optical materials and their applications continues to proceed rapidly.

Advances in the field of luminescence spectroscopy will continue to be aided by the following:

1. The emergence of new materials: These include the historically important single crystals, glasses, and thin films, as well as novel materials such as ceramics, nanoparticles, nano-layered systems, quantum wires, and quantum dots.

2. Advances in technology: This includes femtosecond lasers, high intensity lasers, and more sophisticated electronics. Advances in computing have allowed for easier manipulation of data and made theoretical calculations accessible to those with even modest computing facilities by modern standards.

3. The demands of the market: For numerous applications, there is a desire for light sources—lasers, in particular—that are more efficient, of higher intensity, more compact, cheaper, faster, and that operate over a wide bandwidth. Recently, for example, the areas of reading, storing, and transmitting data have been particularly instrumental in providing impetus for the development of new materials and devices.

The following list presents a few areas of research that should be particularly fruitful to the field of luminescence, both in the development of luminescence-related technologies and in the understanding of basic physical phenomena.

1. The understanding and control of non-radiative losses: Following excitation, a luminescent system often releases at least part of its energy as heat. By controlling these losses one can, in principle increase or otherwise tailor the radiative emission.

One system in which the control over non-radiative processes can be exercised is nanomaterials. Because of the small particle size involved, the phonon spectrum is no longer a continuum, but is discrete. The lower frequency modes, in particular, are not present [48]. Thus, phonon-related processes may be suppressed in these systems, making radiative processes more probable.

Another possibility for controlling non-radiative decay is through the use of femtosecond laser pulse-shaping techniques. For example, a method called *coherent wavepacket excitation* can create a superposition state of vibrational modes in an upper electronic state. The ensuing motion directs the excitation energy to the opposite side of the parabola (in a configurational coordinate model), instead of simply decaying via phonon emission to the base of the parabola, offering the possibility of altering the normal decay process [49].

Also, Auzel et al. [10] have shown that under condition of high excitation density, the multiphonon decay rate in rare earth-doped glasses is inhibited. Such an effect could have important consequences in the development of certain lasers where high excitation densities can be achieved.

2. Extension to higher energy regions: To date, work on rare earth ion-doped systems has concentrated mainly on energy levels roughly below 40,000 cm^{-1}. Though these represent only a small fraction of the 4f-levels, they have been utilized for many applications. A continuation into the higher 4f-and 5d energy levels seems promising for both scientific and technological reasons.

3. Shorter time scales: The use of femtosecond lasers has yet to be exploited in luminescent materials, particularly in the investigation of the decay of an excited system by phonon emission. With the period of lattice vibrations on the order of 50 femtoseconds, and phonon emission occurring over several vibrational periods, femtosecond laser pulses could be used to study phonon emission processes while they are in progress. The pump-probe technique seems especially suitable for such investigations.

4. Nonlinear processes: The importance of nonlinear processes, such as upconversion and two-photon excitation, will continue to expand, particularly as high power lasers become more accessible. UV lasers pumped by visible lasers and relying on upconversion could provide an application for nonlinear effects. Using upconversion or two-photon absorption, one can even access the vacuum ultraviolet region without the need for special optics or an evacuated excitation system.

From the current vantage point, we see two major societal factors that will drive research in the field. First, the scarcity of nonrenewable energy resources, the growing demand for these resources, and concerns about the environment will provide an impetus for the development of more energy efficient lighting. Second, the rapid a growth in communications will require improved optical systems, one important component of which will be solid state lasers.

The response to these and other forces will be the continued research into the fundamental processes of luminescence, resulting in more answers and, equally likely, more challenges. Using the past as an example, our scientific curiosity and imagination will lead us to reconfirm, time and again, the observation attributed to van Vleck: "Every crystal, a surprise."

ACKNOWLEDGEMENT

The authors would like to thank Professor Claus Klingshirn for very helpful discussions and clarifications.

REFERENCES

[1] Goethe, J.W. *The Sorrows of Young Werther*, trans Mayer, E. & Brogan, L. New York: The Modern Library.

[2] Reif, F. (1965) *Statistical Thermal Physics*. New York McGraw.

[3] DiBartolo, B. & Peccei, R. (1965) *Phys. Rev. A* **137**, 1770.

[4] Dieke, G.H. (1968) In (eds.) *Spectra and Energy Levels of Rare Earth Ions in Crystals*, Crosswhite, H.M. & Crosswhite, H., (New York: Interscience.)

[5] Carnall, W.T., Goodman, G.L., Rajnak, K. & Rana, R.S. (1988) *A systematic analysis of the spectra of the lanthanides doped into single crystal LaF3,* Argonne National Laboratory Report AN-88-8.

[6] Morrison, C.A., & Leavitt, R.P. (1982) *Handbook on the Physics and Chemistry of Rare Earths*. Gschneidner, K.A., & Eyring, L. (eds.) (Amsterdam: North-Holland).

[7] Wegh, R.T., Meijerink, A., Lamminmaki, R.-J., Holsa, J. (2000) *J. Luminescence* 87-89 1002-4

[8] Wegh, R.T., Van Loef, E.V., Burdick, D., G.W. & Meijerink, A. (2003) *Molecular Physics* **101**(7), 1047-56.

[9] Wang, X., Ganem, J., Dennis, W.M. & Yen, W.M. (1991) *Phys. Rev. B,* **44**(2), 900-2.

[10] Auzel, F. & Pelle, F. (1996) *J. Luminescence* **69**, 249-55.

[11] Imbusch, G.F. (1978) In *Luminescence Spectroscopy* Lumb, M.D., ed. (London, New York: Academic) p 31,.

[12] Moulton, P.F. (1986) *J. Opt. Soc. Amer. B* **3**, 125-32.

[13] Tanabe, Y. & Sugano, S. (1954) *J. Phys. Soc. Japan* **9**, 753-766.

[14] Rebane, K. (1970) *Impurity Spectra of Solids* (New York: Plenum).

[15] Struck, C.W. & Fonger, W.H. (1991) *Understanding Luminescence Spectra and Efficiency Using Wp and Related Functions* Berlin Heidelberg: Springer-Verlag.

[16] Di Bartolo, B. (1978) In *Luminescence of Inorganic Solids* Di Bartolo D. ed (New York: Plenum).

[17] Toyazawa, Y. & Kamimura, H. (1967) *Dynamical Processes in Solid State Lasers* Kubo R. ed (New York: W.A. Benjamin) p 90.

[18] Struck, C.W. & Fonger, W.H. (1991) *Understanding Luminescence Spectra and Efficiency Using Wp and Related Functions* Berlin Heidelberg: Springer-Verlag.

[19] Fowler, W.B. (1968) in Physics of Color Centers, Fowler, W.B., ed (New York: Academic) p 55.

[20] Gebhardt, W. & Kuhnert, H. (1964) *Phys. Letters* **11**, 15.

[21] Baldacchini, G. (2001) *Advances in Energy Transfer Processes* Di Bartolo, B., ed (New Jersey London Singapore Hong Kong: World Scientific).

[22] Henderson, B. (1987) *Spectroscopy of Solid State Laser Materials* Di Bartolo, B., ed (New York: Plenum).

[23] Judd, B.R. (1962) *Phys. Rev.* **127,** 750-761.

[24] Ofelt, G.S. (1962) *J. Chem. Phys.* **37,** 511-20.

[25] Carnall, W.T., Fields, P.R. & Rajnak, K. (1968) *J. Chem. Phys.* **49,** 4424.

[26] Walsh, B., Barnes, N. & Di Bartolo, B. (1998) *J. Appl. Phys.* **83** 8 2772-87.

[27] Imbusch, G.F. (1978) *Luminescence Spectroscopy* Lumb, M.D., ed (London, New York: Academic).

[28] Weber, M.J. (1973) *Phys. Rev. B* **8,** 54-64.

[29] Auzel, F. (1978) *Luminescence of Inorganic Solids*, Di Bartolo B. ed (New York: Plenum).

[30] Moos, H.W. (1970) *J. Luminescence* 1,2 106.

[31] Struck, C.W. & Fonger, W.H. (1991) *Understanding Luminescence Spectra and Efficiency Using Wp and Related Functions* (Berlin Heidelberg: Springer-Verlag).

[32] Auzel, F. (1980) *Radiationless Processes* Di Bartolo B. ed (New York: Plenum).

[33] de Mello Donega, C., Meijerink, A. & Blasse, G. (1992) *J. Phys. Condens. Matter* **4**. 8889-902.

[34] Förster, T. (1948) *Ann. Phys.* **21,** 55.
[35] Dexter, D.L. (1953) *J. Chem. Phys.* **21,** 836-850.
[36] Inokuti, M. & Hirayama, F. (1965) *J. Chem. Phys.* **43,** 1978.
[37] Förster, T. (1949) *Z. Naturforch.* **4a,** 321.
[38] Yamada, N., Shionoya, S. & Kushida, T. (1972) *J. Phys. Soc. Japan* **32,** 1577.
[39] Collins, J.M. & Di Bartolo, B. (1996) *J. Luminescence* **69,** 335-341.
[40] Auzel, F. (2004) *Chem. Rev.* **104,** 139-173.
[41] Di Bartolo, B. (1968) *Optical Interactions in Solids* (New York: Wiley.
[42] Chen, X., Di Bartolo, B., Barnes, N.P. & Walsh, B.M. (2004) *Physica Status Solidi* (b) **241** (8) 1957-76.
[43] Dekker, A.J. (1957) *Solid State Physics* (Englewood Cliffs, NJ: Prentice Hall) p 312.
[44] Elliott, R.J. & Gibson, A.F. (1974) *An Introduction to Solid State Physics and Its Applications* (London and Basingstoke: MacMillan.
[45] Flaherty, J.M. & Di Bartolo, B. (1973) *Phys. Rev.* B **8,** 5232.
[46] Baumesteir, P.W. (1961) *Phys. Rev.* **121,** 359.
[47] Elliott, R.J. & Gibson, A.F. (1974) *An Introduction to Solid State Physics and Its Applications* (London and Basingstoke: MacMillan).
[48] Liu, G.K., Zhuang, H.Z. & Chen, X.Y. (2002) *Nano Letters* **2,** (5) 535-9.
[49] Collins, J.M. (1998) *Ultrafast Dynamics of Quantum Systems* Di Bartolo B. editor (New York: Plenum).

BIBLIOGRAPHY

Dekker, A.J. (1957) *Solid State Physics* Englewood Cliffs, NJ: Prentice Hall.
Di Bartolo, B. (1968) *Optical Interactions in Solids* New York, London, Sydney, Toronto: John Wiley & Sons.
Elliott, R.J., & Gibbon, A.F. (1974) *An Introduction to Solid State Physics and Its Applica- tions*, London and Basingstoke: Macmillan.
Lumb, M.D. ed (1978) *Luminescence Spectroscopy* New York: Academic.
Kaplianskii, A.A. & Macfarlane, R.M. (1987) *Spectroscopy of Solids Containing Rare Earth Ions* Amsterdam Oxford New York: North-Holland.
Henderson, B. & Imbusch, G.F. (1989) *Optical Spectroscopy of Inorganic Solids* Oxford: Clarendon Press.
Rebane, K. (1970) *Impurity Spectra of Solids* New York: Plenum.
Liu, G.K. & Jacquier, B. eds (2005) *Spectroscopic Properties of Rare Earths in Optical Materials* Berlin Heidelberg New York: Springer.
Klingshirn, C.F. (1997) *Semiconductor Optics* Berlin, Heidelberg, New York: Springer.
Yu, P.Y. & *Cardonna M., Fundamentals of Semiconductors* Berlin, Heidelberg, New York: Springer

CHAPTER 13

LASER-INDUCED FLUORESCENCE SPECTROSCOPY

G. Geipel
Institute of Radiochemistry, Forschungszentrum Rossendorf, P.O. Box 510119, D-01314 Dresden

13.1 INTRODUCTION

The emission of so-called *cold light* is named "luminescence". According to the origin of the luminescence, four categories can be distinguished: photoluminescence, electroluminescence, chemiluminescence, and bioluminescence. Photoluminescence is generated by absorption of radiation. This category is divided into two phenomena, fluorescence and phosphorescence. Formerly, lifetime was used to discriminate between fluorescence and phosphorescence. Fluorescence is generated by a spin-allowed transition, where no change in multiplicity occurs. In contrast to this, the radiative decay to the ground state responsible for phosphorescence properties is a spin-forbidden intercombination transition.

As one can see on the periodic table of elements, many of the lanthanides and actinides show photoluminescence properties in solution and in solid state. In addition, photoluminescence can be observed in complex systems containing point defects, like color centers in alkali halides, single crystals, and many others.

Laser-induced spectroscopic methods were also used in actinide chemistry as a tool to study the speciation of the actinide elements in aquatic environments. The main goal of the application of time-resolved laser-induced fluorescence spectroscopy (TRLFS) and laser-induced photoacoustic spectroscopy (LIPAS) is to achieve detection limits of the actinide species as low as possible or in concentration ranges expected under environmental conditions.

With laser sources it is possible to provide any wavelength from the near UV to the NIR (Near Infrared) wavelength range. The development of tunable solid-state lasers overcame some disadvantages of the dye laser systems with respect to short tunable wavelength range and the use of hazardous chemicals.

The application of low temperatures for samples measured with fluorescence spectroscopic methods brought success, especially in the detection of carbonate species.

It is not possible to include all literature in this field, therefore this contribution includes only a selection of papers. The focus here will be mainly on the fluorescence spectroscopy of the lower actinides like uranium (VI), americium (III) and curium (III).

13.2 EXPERIMENTAL SETUP

A typical setup for fluorescence measurements is shown in Figure 13.1. Fluorescence measurements of solid samples are mainly performed by use of intensified camera systems and diode arrays. In the case of time-resolved measurements this results in a time-domain setup. An overview of used excitation wavelength for fluorescence spectroscopy of actinides is given in Table 13.1. According to the monochromatic radiation of laser sources no excitation monochromator for the excitation of samples is necessary.

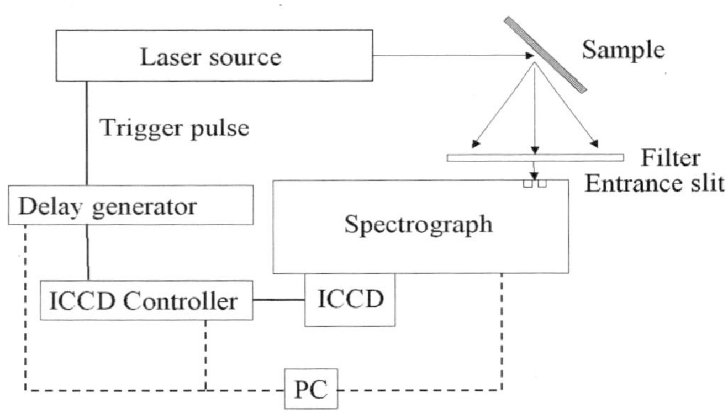

Figure 13.1 Typical scheme for time-resolved laser-induced fluorescence spectroscopy with solid samples.

Solid samples generate a lot of stray light off the exciting laser beam. To minimize this stray light two different methods are applicable. First, the emitted fluorescence photons may be focused in other than a rectangular setup into the spectrograph. However, for samples that do not have planar surfaces, stray light may be generated at each surface plane, so this leads only to a reduction of the stray light. Second, a filter between the excited sample and the entrance of the spectrograph can be used. The edge of this filter should block all wavelengths below the emission spectrum. Some additional remarks on stray light can be found in the literature [1].

The spectrograph should be a conventional imaging monochromator system with no exit slit. The width of the entrance slit influences the spectral resolution. It should be selected as small as possible. The ICCD normally consists of a multichannel plate (an intensifier with a factor of about 1:10^6), an optical connector, and a CCD camera system. The optical connector usually is a fiberoptical taper or an imaging telescope. Telescopes do not generate interferences in the image, therefore their use should be preferred. The delay generator shifts the time of the measurements against the laser pulse. The experiment can be triggered either by a TTL pulse provided by the laser system or by the delay generator if the laser can be fired by an external trigger pulse. The PC controls delay, spectrograph, and ICCD, and it stores the received spectra.

In the case of uranium (VI) as a detected fluorescent species, often a laser wavelength of 266 nm is used as excitation source. It is necessary to point out that in such an arrangement the selection of the filter has to be made very carefully. Normal glass filters absorb the UV laser light completely and therefore they are a nearly ideal filter material. Nevertheless these glass filters generate so-called "white light." This increases the background of the spectrum, and the calculation of fluorescence decay times becomes more difficult. Therefore, for each application of laser-induced fluorescence measurements, the laser stray light has to be looked for and its second order diffraction in the spectrograph.

Table 13.1 Excitation wavelength applicable to fluorescent oxidation states of the actinide series from protactinium to curium.

Ion	Excitation Wavelength (nm)	Reference
Pa^{4+}	278 (308)	[2]
U^{4+}	248	[3]
UO_2^{2+}	266 337 355 380–440	[4, 5] [6] [7] [8]
Am^{3+}	504 (aqueous solution) 337/355 (solid sample)	[9] [10]
Cm^{3+}	395	[11]

13.3 FLUORESCENCE SPECTROSCOPY OF MINERALS

Uranium as well as many other heavy metals, form a wide variety of minerals. It is known that minerals containing uranium (VI) show more or less intense fluorescence emission after excitation with UV light. Therefore, fluorescence measurements on these minerals are a possible tool to determine their chemical composition. However, fluorescence spectroscopy is not often used in mineralogy. The description of minerals remains only in remarks on fluorescent or nonfluorescent minerals and sometimes information on the fluorescence color can also be found [12]. A systematic study of the fluorescence of uranium minerals and its comparison to that of synthetic compounds can contribute to better precision in the determination of natural samples. Therefore, from the point of view of actinide chemistry the uranium minerals are of the greatest of interest. About 180 uranium minerals are known. However, not all of them show fluorescence properties.

A publication on the selective detection of uranium in geologic samples was contributed in the 1980s by deNeufville [13]. Unfortunately no spectral information about the emission maxima of the uranyl species was supplied. Some fluorescence decay times are summarized. For most of the studied minerals decay times in the μs range were detected.

Among uranium (VI) minerals the phosphates show a very intense fluorescence emission. Therefore, it is not surprising that the fluorescence spectrum of autunite is one of the first published fluorescence spectra of a uranium mineral [14]. Later, a series of fluorescence spectra of several uranium phosphates and arsenates were measured [15]. From data on the mineral troegerite, $H_2(UO_2)_2(AsO_4)_2 \times 8H_2O$ researchers were later able to identify a solution species. In addition to this, the synthesis of solid arsenate phases $UO_2(H_2AsO_4)_2 \times 1H_2O$ and $UO_2HAsO_4 \times 4H_2O$ brought success in the determination of the aqueous species. From these fluorescence spectra of minerals and solid phases, the fluorescence spectra in solution could be deconvoluted and the concentration of the species determined. This led last but not least to the determination of complex stability constants [16] already reviewed in the thermodynamic database of uranium [17].

The problem of false determination of species from a *natural* mineral sample may be shown by the example of the mineral chernikovite, $(H_3O)_2(UO_2)_2(PO_4)_2 \times 6H_2O$. The mineral used from the minerals collection of the Technical University Mining Academy, Freiberg (collections number 72753) did not show any fluorescence spectrum. However, it could be expected from extrapolations of other phosphate minerals and by comparison with real arsenate mineral troegerite that the mineral should show intense fluorescence properties. The *synthetic* sample of this mineral did show a very intense fluorescence spectrum as expected [18]. The fluorescence spectrum of this sample is shown in Figure 13.2.

The centers of the fluorescence maxima are located at 502.3, 523.6, 547.4, 573.1, and 603.2 nm, respectively. The fluorescence decay time was found to be multiexponential. However, no spectral shift with increasing delay time was observed. Therefore this behavior can only be explained by the existence of species with a different number of water molecules in the crystal lattice or by a loss of water during the measurement. The longest detected fluorescence decay time was determined to be 27.6 ± 0.05 μs.

From solution chemistry it is known that uranyl di- and tri-carbonate do not show fluorescence properties at room temperature. However, in seepage waters from a mining area in Saxony/Germany, fluorescence spectra could be obtained [19]. This unexpected result led to several studies in the alkaline earth uranium (VI) carbonates. At first in could be shown that the fluorescence spectrum results from a dissolved $Ca_2UO_2(CO_3)_3$ (aq.) species [20–22]. The fluorescence spectrum of this solution species agrees very well with the fluorescence spectrum of the mineral liebigite [23]. They differ only in the determined fluorescence decay times. For the aqueous species the fluorescence decay time was determined to be about 50 ns and for the mineral species 320 μs. In Table 13.2 spectroscopic parameters of alkaline earth uranyl carbonates are summarized [23–25].

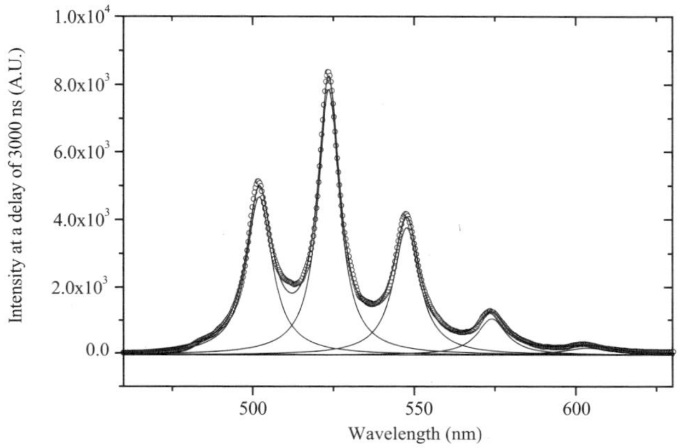

Figure 13.2 Fluorescence spectrum of synthetic chernikovite.

Many other uranium minerals show also fluorescence properties. About 150 have been tested for their fluorescence spectra. For example, the spectrum of johannite, a copper-containing uranium mineral, is shown in Figure 13.3. The data obtained from TRLFS measurements at room temperature for this and some selected other uranium bearing minerals are summarized in Table 13.3 [26].

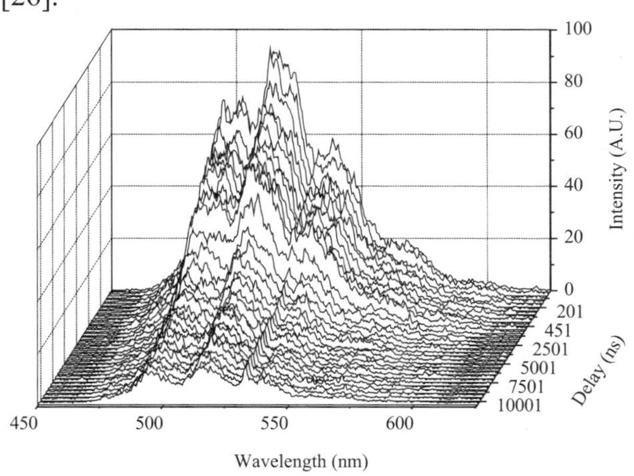

Figure 13.3 Time-resolved fluorescence spectrum of the mineral johannite $Cu(UO_2)_2(SO_4)_2(OH)_2 \times 8\ H_2O$.

Table 13.2 Fluorescence spectroscopic data of alkaline earth uranyl carbonates.

Name	Formula	Emission Maxima	Fluorescence Decay Time	Remark	Reference
Liebigite	$Ca_2UO_2(CO_3)_3$ $\times 10\ H_2O$	466.9 483.1 502.7 524.1 545.9 571.9	313 ± 10 μs	Natural	[23]
Andersonite	$Na_2CaUO_2(CO_3)_3$ $\times 6\ H_2O$	468.4 485.2 504.8 526.2 549.6 575.4	33.1 ± 3.9 μs	Natural	[24]
Bayleyite	$Mg_2UO_2(CO_3)_3$ $\times 18\ H_2O$	471.3 487.7 506.9 527.6 551.4 578.3	17.9 ± 0.5 μs	Synthetic	[23]
Liebigite	$Ca_2UO_2(CO_3)_3$ $\times 10\ H_2O$	465.4 482.9 502.7 524.5 545.4 571.5	145 ± 5 μs	Synthetic	[23]
	$Sr_2UO_2(CO_3)_3$ $\times 8\ H_2O$	482.5 488.8 502.8 522.9 545.4 569.2	77 ± 1 μs	Synthetic	[23]
	$Ba_2UO_2(CO_3)_3$ $\times 6\ H_2O$	469.9 487.7 507.3 528.9 552.2 570.5	16.3 ± 0.4 μs	Synthetic	[23]
Andersonite	$Na_2CaUO_2(CO_3)_3$ $\times 6\ H_2O$	470.6. 486.1 505.4 526.7 549.6 573.9	65.2 ± 0.6 μs	Synthetic	[23, 24]
Swartzite	$MgCaUO_2(CO_3)_3$ $\times 12\ H_2O$	472.3 488.9 509.0 531.1 554.7 578.9	59.4 ± 0.1 μs	Synthetic	[25]

Table 13.3 Fluorescence properties of some uranium minerals at room temperature.

Mineral	Formula	Centers of Fluorescence Emission, nm	Fluorescence Decay Time at Room Temperature	Remark (Number in Minerals Collection in Freiberg)
Asselbornite	$(Pb,Ba)(UO_2)_6(BiO)_4(AsO_4)_2(OH)_{12} \times 3\ H_2O$	488.6 502.2 525.0 549.1 570.6 595.6	3.54 ± 0.15 µs	(76217)
Billietite	$6[UO_2/(OH)_2] \times Ba(OH)_2 \times 4\ H_2O$	478.8 495.7 514.3 535.0 559.0 574.3	163 ± 7 ns	(67259)
Johannite	$Cu(UO_2)_2(SO_4)_2(OH)_2 \times 8\ H_2O$	503.0 524.5 547.2	5.2 ± 0.6 µs	(78557)
Kahlerite	$Fe[UO_2/AsO_4]_2 \times 10\ H_2O$	417.5 429.3 443.0 447.4	0.75 ± 0.02 µs	(37925)
Metalodevite	$Zn(UO_2)_2(AsO_4)_2 \times 10\ H_2O$	487.6 502.1 524.2 548.3 572.0 600.1	6.3 ± 0.5 µs 0.7 ± 0.1 µs	(69929)
Ranunculite	$HAl(UO_2)(PO_4)(OH)_3 \times 4\ H_2O$	486.5 501.2 521.7 544.1 569.3 594.7	2.8 ± 1.1 µs	(78816)
Triangulite	$Al_3(UO_2)_4(PO_4)_4(OH)_5 \times 5\ H_2O$	478.8 495.7 514.3 535.0 559.0 574.3	1.3 ± 0.3 µs	(75824)

The fluorescence lifetimes are given in Table 13.2 for some natural and synthetic samples. It can be seen that these data differ from each other. However, there is no trend between the natural and synthetic samples. It can only be deduced that impurities and lattice effects may influence the fluorescence lifetime.

Comparing the data from Table 13.3 two conclusions can be drawn:

1. The centers of the fluorescence emissions vary significantly. It can be expected that the fluorescent minerals should be distinguished by their fluorescence properties.
2. The fluorescence decay times at room temperature also vary in a wide range from nanoseconds to microseconds. At the present time there is no common relationship between the fluorescence decay constant (reciprocal fluorescence decay time) and the environment of the uranyl ion as known, for instance, for fluorescent trivalent actinides.

13.4 FLUORESCENCE SPECTROSCOPY OF SURFACE SPECIES AND IN SOLID PHASES

Gabriel [27] studied the interaction of uranium (VI) on the surface of silica particles. The studies were performed in the presence of atmospheric CO_2. The initial concentration of uranium was 1×10^{-6} and 1×10^{-7} M UO_2^{2+}. By TRLFS two surface complexes were identified. These species are characterized by fluorescence decay times of 170 ± 25 μs and 360 ± 50 μs. These complexes should be a mononuclear species. During the formation of these species occurs a release of 2 and 3 protons, respectively. Nevertheless, by comparison of the fluorescence spectra with sorption data, a third uranyl-silicate-carbonate surface species could be postulated, existing in a lower amount in the pH region 8.0 to 9.0. The three species were assigned to be $\equiv SiO_2UO_2°$, $\equiv SiO_2UO_2OH,^-$ and $\equiv SiO_2UO_2OHCO_3^{3-}$.

Spectroscopic techniques were used by Kowal-Fouchard [28] to study the interaction of uranium with montmorillonite and to model this behavior. The authors suggest at least three uranium-bearing species. At the aluminol site the species $\equiv Al(OH)_2UO_2^{2+}$ is postulated. At silanol groups two species were assigned to be $\equiv SiO_2UO_2°$ and $\equiv SiO_2(UO_2)_3(OH)_5^-$. The formation constants for the sorption equlibria for these species were derived to be log $K°_{Al} = 14.9 \pm 0.2$, $\log K°_{Si1} = -3.8 \pm 0.2$ and $\log K°_{Si2} = -20.0 \pm 0.1$, respectively. Unfortunately no information about the number of the released protons is available. Second the samples were dried before the spectroscopic measurements were performed. Therefore, it is possible that the species composition was changed during this process.

The sorption of uranium (VI) onto gibbsite was studied by Baumann and coworkers [29] by TRLFS. The sorption experiments were carried out under ambient conditions in a pH range from 5.0 to 8.5 and a uranium concentration of 1.0×10^{-5} M. At the surface of gibbsite two uranyl species were detected. These species are characterized by fluorescence decay times of 330 ± 115 ns and 5600 ± 1640 ns, respectively. The first species dominates the lower pH range, whereas the influence of the second species increases at higher pH values. The maxima of the center of the fluorescence emission of both species are located at 479.5, 497.4, 518.7, 541.6, 563.9, and 585.8 nm. The authors conclude that the species with the shorter fluorescence decay time is attributed to a bidentate mononuclear inner sphere surface complex. The uranium (VI) is bound in this complex to two reactive OH^- groups at the broken edge linked to on Al. The other species is assigned to be a small sorbed cluster of a polynuclear surface species.

To study the surface of a depleted uranium (DU) disc immersed in a Ca-phosphate solution for 182 days, time-resolved laser-induced fluorescence spectroscopy was used [30]. The weathering solution contained among other components 2.49×10^{-3} M calcium and 1.05×10^{-3} M phosphate, which represent pore water concentrations of agricultural soils. Six fluorescence emission bands at 486, 501, 522, 546, 573, and 601 nm, and two fluorescence decay times of 50 ± 5 ns and 700 ± 25 ns characterize the TRLFS spectrum. By comparison of the obtained TRLFS spectra it could be shown clearly that meta-autunite, a U(VI) phosphate, had formed during the low temperature alteration on the DU disc. This secondary uranium(VI) mineral phase was identified at least in a fingerprinting procedure by comparing it with TRLFS spectra from an in-house U(VI) TRLFS database, including U(VI) oxides, U(VI) hydroxides, U(VI) sulphates, and also U(VI) phosphates. The observed spectrum agreed completely in the centers of emission bands, and also the determined fluorescence decay times agreed very well.

Theuvenot and coworkers [10] reported on the fluorescence of americium in a ThO_2 matrix. They doped ThO_2 with 5×10^{-2}% Am^{3+} and excited this sample with laser light of 337 nm. The resulting fluorescence spectrum consisted in two main groups of lines. One group was observed in the wavelength range 670–750 nm; the other one in the range 800–850 nm. Additional experiments were carried out in order to get information on the quantum yield. These more diluted samples (10^{-6}–10^{-3} atom % Am^{3+}) were excited with 355-nm laser pulses. The resulting fluorescence spectrum was less resolved and did show only two broad emissions centered at 712 nm and 820 nm. The fluorescence decay was found to be monoexponential with a decay time of 420 μs at 300 K. With decreasing temperature the fluorescence decay time increased, resulting in a value of 870 μs at 10 K. These values were much larger than the fluorescence decay times observed in aqueous solution, where values of 24.6 ns were observed [9].

Stumpf [31] studied the sorption of curium onto smectite and kaolinite. An outer sphere complex was detected at low pH by fluorescence spectroscopy. Increasing pH leads to an inner-sphere adsorption that occurs via aluminol edge sites. The species detected on the surface of clay minerals were assigned to be $\equiv$Al-O-Cm^{2+}-$(H_2O)_5$ and at higher pH $\equiv$Al-O-Cm^{+}(OH)-$(H_2O)_4$, or probably a bidentate species $\equiv$(Al-O)$_2$-Cm^{+}-$(H_2O)_5$. The sorption behavior of Cm (III) onto γ-alumina was studied also by [32]. Here the same types of species were detected. However, the peak maxima for the first species differed slightly (598.8 nm and 601.2 nm, respectively). In the case of the interaction with calcite, it could be shown that curium was incorporated into the crystal lattice of calcite [33]. The authors came to this conclusion by studying the spectral shift of the fluorescence signal and the increase of the fluorescence decay time of the formed curium species. As clearly multi-exponential fluorescence decays were observed, changes of the equilibria between the several species in the excited state were not expected. Some structural suggestions on the environment of the curium ion on the surface and included in the calcite lattice were given, explained by the increase of the fluorescence decay time to 314 μs and 1302 μs. These fluorescence lifetimes correlated to Cm species having one and no water molecules in the solvation shell, respectively.

13.5 FLUORESCENCE SPECTROSCOPY OF FROZEN SAMPLES

In the series of the several solid sample forms, the frozen samples have drawn interest, since it is known that *quench effects* decrease at lower temperatures. It is known that the $UO_2(CO_3)_2^{2-}$ and the $UO_2(CO_3)_3^{4-}$ species do not show any fluorescence properties at room temperature. This is caused by the quench effect of the carbonate ions and the water molecules in the solvation shell. With decreasing temperature this dynamic quench effect is decreased as the energy transfer from the excited uranyl ion decreases.

The fluorescence of the $UO_2(CO_3)_3^{4-}$ species can be observed at temperatures lower than about 235 K. The temperature dependence of the fluorescence of this species is shown in Figure 13.4 for the temperature range 180 K to 273 K. All other experimental conditions for the measurements were the same in order to compare the intensity of the observed spectra directly. Plotting the integrated fluorescence intensity as function of the reciprocal temperature a straight line can be observed. This is shown for the data obtained from the above spectra in Figure 13.5 [34].

This behavior has already been exploited in order to study the uranium speciation of samples especially from the Hanford Site in the state of Washington, USA. Wang and coworkers [35] measured the fluorescence of aqueous pore water samples from the vadose zone at the site. By cooling the samples to the temperature of liquid helium it was possible to observe for the first time the fluorescence of the species $UO_2(CO_3)_2^{2-}$, $UO_2(CO_3)_3^{4-}$ and $(UO_2)_2(OH)_3CO_3^{-}$. The spectral data for these species are summarized in Table 13.4.

By comparison of the fluorescence properties of these synthetic samples with the data obtained from the pore water samples it was shown that the $UO_2(CO_3)_3^{4-}$ species was the dominating species in the studied samples.

In 2005 Wang [8] published studies on the interaction of uranium (VI) with calcite. It was found that two uranium (VI) species were formed during the interaction. One of these species was clearly detected to be a

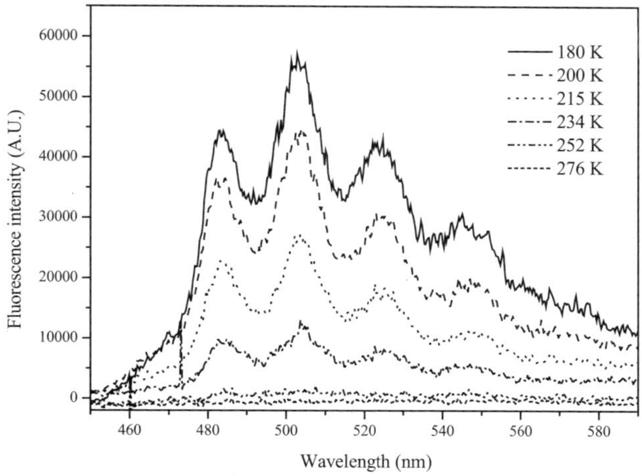

Figure 13.4 Fluorescence spectra of the $UO_2(CO_3)_3^{4-}$ species as a function of the temperature (5.2×10^{-7} M UO_2^{2+}).

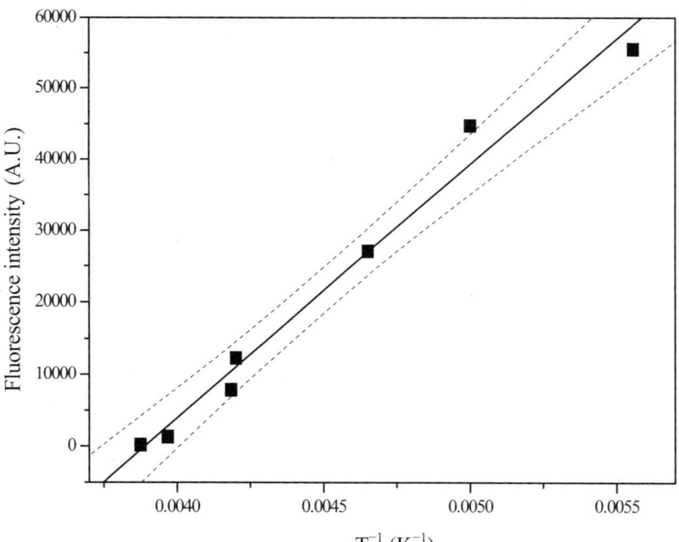

Figure 13.5 Fluorescence intensity of the $UO_2(CO_3)_3^{4-}$ species as a function of the reciprocal temperature.

$Ca_2UO_2(CO_3)_3$ species. On the other component, a less resolved fluorescence spectrum, no assumptions were made.

The fluorescence of a series of uranium minerals was studied by Wang and coworkers [36] at temperatures of 6 K. Some of the results are summarized in Table 13.5.

Table 13.4 Fluorescence spectroscopic data uranyl carbonates at temperature of liquid helium (synthetic samples).

Formula	Emission Maxima	Fluorescence Decay Time	Remark	Reference
$UO_2(CO_3)_2^{2-}$	477.4 496.4 517.2 539.8 563.5	962 μs		[35]
$UO_2(CO_3)_3^{4-}$	479.6 499.2 519.9 542.4 565.6	883 μs		[35]
$(UO_2)_2(OH)_3CO_3^-$	523.0 542.3 561.3	144 μs	Not all emissions determined	[35]

Table 13.5 Fluorescence properties of synthetic uranyl silicate minerals.

Mineral	Formula	Centers of Fluorescence Emission	Fluorescence Decay Time at 4.2 K	Reference
Boltwoodite	$HK(UO_2)SiO_4$ $\times 1.5\ H_2O$	505.4 535.2 558.0 579.4	266 μs	[36]
Cuprosklodowskite	$(H3O)_2Cu(UO_2)_2(SiO_2)_2$ $\times 2\ H_2O$	502.9 524.7 547.5 573.5	196 μs	[36]
Kasolite	$Pb(UO_2)(SiO_4)$ $\times H_2O$	531.2 552.2 574.2	25 μs	[36]
Sklodowskite	$(H_3O)_2Mg(UO_2)_2(SiO_4)_2$ $\times 4\ H_2O$	496.6 516.7 538.7 561.5	207 μs	[36]
Soddyite	$(UO_2)_{15}[(OH)_{20}/Si_6O_{17}$ $\times 8$ H2O	504.5 527.4 551.0 576.3	300 μs	[36]
Uranophane	$Ca(UO_2)_2[SiO_3(OH)]_2$ $\times 5\ H_2O$	503.1 525.5 547.0 570.5	192 μs	[36]
Haiweeite	$Ca(UO_2)_2Si_6O_{15}$ $\times 5\ H_2O$	494.7 501.0 519.2 540.1	246 μs	[36]
Weeksite	$K_2(UO_2)_2Si_6O_{15}$ $\times 4\ H_2O$	513.8 535.0 558.6 582.6	134 μs	[36]
	$UO_2OSi(OH)_3^+$	-	19.0 ± 4.0 μs	[37]

13.6 FLUORESCENCE SPECTROSCOPY OF NON-ACTINIDE SOLID MATRICES

It is not the main aim of this chapter to deal with fluorescence spectroscopy of solid samples containing nonactinide elements. However, some short remarks should be made. Besides the uranium minerals many other fluorescent minerals exist. It is known that already some disorder in the crystal lattice can cause intense fluorescence of the sample, and so many fluorites show intense fluorescence.

As an example of fluorescence properties not caused by the central ion, the spectrum of the mineral ferghanite $LiH[(UO_2)_4/(OH)_4/(VO_4)_2] \times 2H_2O$ (minerals collection Freiberg No. 20959) is shown in Figure 13.6 [26]. The sample was excited with laser pulses of 266 nm. The resulting fluorescence spectrum is strongly shifted to the nẹar UV region if compared with common known uranyl fluorescence spectra. However, it is clear that this fluorescence light cannot be emitted by de-excitation of the uranyl ion. Taking into account the extremely short fluorescence decay time of 25 ± 2 ns, this fluorescence emission should be caused by defects in the crystal lattice.

The luminescence of the rare earth elements was used by Gaft [38] to determine several minerals containing these elements and to distinguish between them. The method of time-resolved measurements was applied to these minerals, which contain rare earth elements with emissions in the same spectral range. The elements Tm^{3+}, Pr^{3+}, Er^{3+}, and Ho^{3+} were studied in the minerals apatite, scheelite, zircon, calcite, and fluorite. Additionally, the luminescence properties of the element Eu^{3+} were studied in the minerals apatite, zircon, calcite, fluorite, danburite, and datolite, respectively. Also results on measurements of Dy^{3+}, Sm^{3+}, Ce^{3+}, Yb^{3+}, Gd^{3+}, Mn^{2+}, and Eu^{2+} were reported. The decay times vary depending on the lanthanide from the nanosecond range (Ce^{3+} in apatite) to the millisecond range (Yb^{3+} in apatite and scheelite). Also the gate (exposure time of the CCD camera controlled by the multichannel plate system after one laser shot) was varied for several measurements.

On the luminescence obtained from zircon specimens, Gaft and coworkers reported in details [39]. Usually, lanthanides can easily be included in the $ZrSiO_4$ lattice where the Zr^{4+} is replaced by the lanthanides. Besides, the fluorescence emitted from the included lanthanides, several other fluorescence emissions were observed in the zircon minerals. The authors were able clarify the nature of some of these additional luminescence bands in minerals of different origin. The yellow emission at around 575 nm and a decay time of ~32μs should be caused by radiation effects with neutrons and α-particles. The applied dose rates for neutrons and α-particles were not reported. The emission at 505 nm was only observed in synthetic samples and was linked to the presence of uranyl ions and phosphorous. At 750 nm an emission with decay time of more than 3 ms was observed. This emission should be connected with Fe^{3+}. Other emissions (480 nm / ~300 μs; 515 nm / ~500 μs; 605 nm / ~10 μs) were not detected in synthetic samples and could not yet be interpreted.

Of interest is also another contribution of Gaft in [40]. The authors reported on the luminescence of Pb^{2+}. As the spectrum of this luminescence is very similar to that of Ce^{3+} and Eu^{2+}, it is not easy to assign the spectrum. The use of time-resolved methods allows the differentiation between such luminescence centers. An intense emission at 312 nm with a decay time of 120 ns was observed in calcite. The authors assigned this spectrum to the presence of Pb^{2+} impurities in the host lattice. The excitation maximum was found at 240 nm.

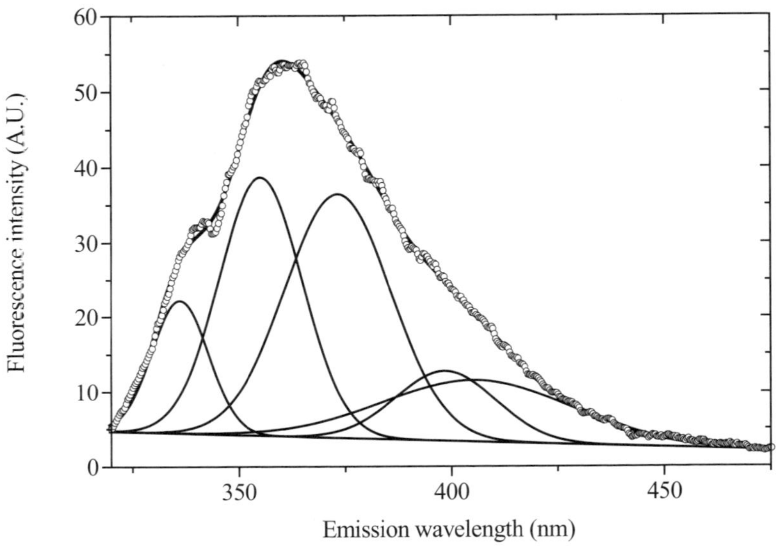

Figure 13.6 Fluorescence spectrum of the mineral ferghanite.

Piriou [41] has investigated the sorption behavior of europium on calcite by TRLFS. The experiments were carried out at an elevated temperature of 323 K at the site and specific fluorescence measurements were made at 15 K. Three different sorption sites were assigned. One type was found in all samples and was assigned to be a species containing H_2O or OH^- groups. The spectrum of these species consisted in broad bands and a fluorescence decay time between 0.55 and 0.85 ms was observed. The second type of species also had a hydroxylated or hydrated environment. However, the fluorescence lifetime is shorter and resulted in a value of 0.45 ms. The third species was characterized by a fluorescence decay time of 9.0 ms and a narrow emission line at 590.1 nm. Due to its long fluorescence lifetime this species should have lost the water molecules from the salvation shell and should be included in the calcite lattice. Such a type of species was also assigned for Cm^{3+} by Stumpf [33].

13.7 OUTLOOK

The improvement of the detection, especially the introduction of measurements at low temperatures, shown by Wang [3] as an example, increases the number of species detectable by fluorescence methods. Complexes with mixed ligands, which were not detected before, like $(UO_2)_2(OH)_3CO_3^-$, could be determined by their spectroscopic properties. The increase in quality was demonstrated by the detection of uranium (VI) carbonate species in pore waters. It is expected that in the future an enhancement of such measurements will occur.

The description of new fluorescent species like uranium (IV) and protactinium (IV) motivates a strong development in the study of the behavior of tetravalent actinides towards solids. The introduction of laser systems providing ultrashort pulses (fs-scale), in combination with appropriate detection systems, extends the possibilities for studying the systems with very short fluorescence lifetimes

Laser-induced time-resolved fluorescence spectroscopy has become a very important tool in the study of solid samples of different origin during the last years. From the spectra a lot of information about the sample regarding its composition and structure can be gathered.

ACKNOWLEDGEMENTS

The author thanks Prof. G. Bernhard for much helpful discussion during the preparation of the manuscript. The Technical University Mining Academy, Freiberg, is acknowledged for providing the minerals used in fluorescence measurements.

Thanks are also due to Prof. emeritus R. Vochten from University of Leuven, Belgium, and Dr. A. Scheinost, Institute of Radiochemistry, Forschungs-zentrum, Rossendorf, for providing with some synthetic uranium phosphate samples.

REFERENCES

[1] Lakowics, J.R. (1999) *Principles of Fluorescence Spectroscopy* Kluwer Academic/ Plenum Publishers New York: 2nd Edition.

[2] Marquardt, C.M., Panak, P.J., Apostolidis C., Morgenstern, A., Walther, C., Klenze, R. & Fanghänel, Th. (2004) *Radiochim. Acta* **92,** 445-446.

[3] Kirishima, A., Kimura, T., Tochiyama, O. & Yoshida, Z. (2003) *Chem. Comm.* 910-911.

[4] Billard, I., Ansoborlo, E., Apperson, K., Arpigny, S., Azenha, M.E., Birch, D., Bros, P., Burrows, H.D., Choppin, G., Coustin, L., Dubois, V., Fanghänel, Th., Geipel, G.,

Hubert, S., Kim, J.I., Kimura, T., Klenze, R., Kronenberg, A., Kumke, M., Lagarde, G., Lamarque, G., Lis, S., Madic, Ch., Meinrath, G., Moulin, Chr., Nagaishi, R., Parker, D., Planque, G., Scherbaum, F., Simoni, E., Sinkov, S. & Viallesoubranne, C. (2003) *Applied Spectroscopy* **57**, 1027-1031.

[5] Geipel, G., Brachmann, A., Brendler, V., Bernhard, G. & Nitsche, H. (1996) *Radiochim. Acta* **75,** 199-204.

[6] Deniau, H., Decambox, P., Mauchien, P. & Moulin, C. (1993) *Radiochim. Acta* **61,** 23-28.

[7] Brendler, V., Geipel, G., Bernhard, G. & Nitsche, H. (1996) *Radiochim. Acta* **74,** 75-79.

[8] Wang, Z., Zachara, J.M., McKinley, J.P. (2005) *Environ. Sci. Technol.* **39,** 2651-2659.

[9] Kimura, T. & Kato, Y. (1998) *J. Alloys Comp.* 271-273; 867-871.

[10] Thouvenot, P., Hubert, S., Moulin, C., Decambox, P. & Mauchien, P. (1993) *Radiochim. Acta* **61,** 15-21.

[11] Wimmer, H., Kim, J.I., Klenze, R. (1992) *Radiochim. Acta* **58/59,** 165-171.

[12] Frondel, C. (1958) *Systematic Mineralogy of Uranium and Thorium Geological Survey Bulletin 1064*, Washington, DC.

[13] deNeufille, J.P., Kasdan, A. & Chimenti, R.J.L. (1981) *Appl Optics* **20,** 1279-1296.

[14] Dunham, J. (1991) http://cat.middlebury.edu/~PHManual/laser.html.

[15] Geipel, G., Bernhard, G., Rutsch, M., Brendler, V. & Nitsche, H. (2000) *Radiochim. Acta* **88,** 757-762.

[16] Rutsch, M., Geipel, G., Brendler, V., Bernhard, G. & Nitsche, H. (1999) *Radiochim. Acta* **86,** 135-141.

[17] Guillaumont, R., Fanghänel, Th., Fuger, J., Grenthe, I., Neck, V., Palmer, D.A. & Rand, M.H. (2003) *Chemical Thermodynamics* **5**, Update of the chemical thermodynamics of uranium, neptunium, plutonium, americium, and technetium. Amsterdam: Elsevier.

[18] Geipel, G. (2005) Internal report FZR.

[19] Bernhard, G., Geipel, G., Brendler, V. & Nitsche, H. (1996) *Radiochim. Acta* **74**, 87-90.

[20] Geipel, G., Bernhard, G., Rutsch, M., Brendler, V. & Nitsche, H. (2000) in Baca, T.E. & Florkowski, T. (eds.), *The Environmental Challenges of Nuclear Disarmament*, City Kluwer Academic Publishers.

[21] Bernhard, G., Geipel, G., Reich, T., Brendler, V., Amayri, S. & Nitsche, H. (2001) *Radiochim. Acta* **89,** 511-518.

[22] Kalmykov, S.N. & Choppin, G.R. (2000) *Radiochim. Acta* **88,** 603-606.

[23] Amayri, S., Reich, T., Arnold, Th., Geipel, G. & Bernhard, G. (2005) *Journal of Solid State Chemistry* **178,** 567-577.

[24] Amayri, S., Arnold, T., Reich, T., Foerstendorf, H., Geipel, G., Bernhard, G. & Massanek, A. (2004) *Environ. Sci. Technol* **38,** 6032-6036.

[25] Amayri, S., Arnold, T., Foerstendorf, H., Geipel, G. & Bernhard, G. (2004) *The Canadian Mineralogist* **42,** 953-962.

[26] Geipel, G. (2000) Internal report FZR.

[27] Gabriel, U., Charlet, L., Schlaepfer, C.W., Vial, J.C., Brachmann, A. & Geipel, G. (2001) *J. Colloid Interface Sci.* **239,** 358-368.

[28] Kowal-Fouchard, A., Drot, R., Simoni, E., Ehrhardt, J.J. (2004) *Environmental Science and Technology* **38,** 1399-1407.

[29] Baumann, N., Brendler, V., Arnold, T., Geipel, G. & Bernhard, G. (2005) *J Colloid Interface Sci* **290,** 318-324.

[30] Baumann, N., Arnold, Th., Geipel, G., Trueman, E., Black, S. & Read, D. (2006), Technical note, *Journal of the Total Environment,* Publication in progress.

[31] Stumpf, Th., Bauer, S., Coppin, F. & Kim, J.I. (2001) *Environmental Science and Technology* **35,** 3691-3694.

[32] Stumpf, Th., Rabung, Th., Klenze, R., Geckeis, H. & Kim, J.I. (2001)*J Colloid Interface Sci.* **238,** 219-224.

[33] Stumpf, Th. & Fanghänel, Th. (2002) *J Colloid Interface Sci.* **249,** 119-122.

[34] Geipel, G. (2005) Internal report FZR.

[35] Wang, Z., Zachara, J.M., Yantasee, W., Gassmann, P.L., Chongxuan, L. & Joly, A. (2004) *Environ. Sci. Technol.* **38,** 5591-5597.

[36] Wang, Z., Zachara, J.M., Gassman, P.L., Chongxuan, L., Qafoku, O., Yantasee, W. & Catalano, J.G. (2005) *Geochim. Cosmochim. Acta* **69**, 1391-1403.

[37] Moll, H., Geipel, G., Brendler, V., Bernhard, G. & Nitsche, H. (1998) *J. Alloys and Compounds* 271-273, 765-768.

[38] Gaft, M., Panczer, G., Reisfeld, R. & Uspensky, E. (2001) *Physicsand Chemistry of Minerals* **28,** 347-363.

[39] Gaft, M., Shinno, I., Panczer, G. & Reisfeld, R. (2002) *Mineralogyand Petrology***76,** 235-246.

[40] Gaft, M., Seigel, H., Panczer, G. & Reisfeld, R. (2001) *Bull. Liaison S.F.M.C.* **13,** 76-77.

[41] Piriou, B., Fedoroff, M., Jean, J. & Berics, L. (1997) *J. Colloid Interface Sci.* **194**, 1-8.

CHAPTER 14

SOFT X-RAY EMISSION AND RESONANT INELASTIC X-RAY SCATTERING SPECTROSCOPY

[1]E.J. Nordgren, S.M. Butorin, L.-C. Duda, and [2]J.-H. Guo
[1]Uppsala University, Department of Physics, Uppsala, Sweden
[2]Lawrence Berkeley National Laboratory, Berkeley, CA 94720, USA

14.1 INTRODUCTION

X-ray spectroscopy was a key technique to elucidate the atomic structure of matter, starting in the early 20th century through the discovery of X-ray diffraction and the observation that X-rays have characteristic energies for different elements of the periodic system. Later, soft X-ray emission spectroscopy offered an experimental basis for the exploration of the electronic structure of solids. Over the years X-ray spectroscopy has been widely used both for elemental analysis in various applications, and as a method for basic research in physics and chemistry.

The introduction of synchrotron radiation, and in particular the subsequent continued development of this kind of X-ray sources, has paved the way for new applications of X-ray spectroscopy. Early on the availability of tunable X-rays from synchrotron radiation storage rings promoted the field of X-ray absorption spectroscopy to more general applicability. The notion that total yield methods replicate X-ray absorption quite well meant that one had a new and simple method at hand for many different kinds of samples. The use of near edge fine structure in X-ray absorption (NEXAFS or XANES) to reveal chemical states is widespread [1], and the use of extended features (EXAFS) provides a standard technique today for determining bond lengths and local order [2].

X-ray emission spectroscopy (XES) (often called *fluorescence for photon-in photon-out processes*), with its low yield being more dependent on intense sources, had to await the third generation synchrotron source development before seeing a substantially increased interest and use. This was especially true for soft X-ray emission where yields typically are in the 1% range [3].

X-ray emission possesses an inherent local character due to the involvement of core electrons. For valence-core transitions (K_β, L_β, etc.) this means that direct probing of the valence band at a particular atomic species is conducted. Furthermore, the transitions are governed by selection rules, usually electric dipole rules that impose restrictions on the states involved. This approach provides valuable selectivity with respect to orbital symmetry. For molecules, it can be used to describe molecular orbitals in terms of atomic composition (LCAO, *linear combinations of molecular orbitals*), and for solids, partial densities of states can be used to describe the electronic structure of the bands. In either picture, soft X-ray emission offers a means to determine the weight of the respective components. For an overview of soft X-ray emission for molecular applications, see, e.g., [4].

In addition to the local character and symmetry-probing capability of X-ray emission the use of tunable synchrotron radiation provides a further means of obtaining detailed information about the electronic structure and dynamics of various systems. The tunability of synchrotron radiation facilitates excitation at specific chemical sites of a compound, since electron-binding energies differ between atoms in different chemical environments. The well-defined polarization of synchrotron radiation can also be used to obtain information on local order and orientation as well as magnetic state contrast. Detuning from resonant excitation of X-ray emission provides a means to obtain timing information in the sub-femtosecond range in an indirect way, since the detuning leads to an effective shortening of the X-ray scattering timescale.

Resonant X-ray emission (RXE) is a common term for describing emission that is excited by monochromatized photons at a specific resonance. The process is often seen as a two-step process where the first step is the X-ray absorption and the second step is the emission. The absorption step offers selectivity with respect to, for example, site, and the emission probes the local electronic structure at the particular site selected. Usually emission spectra are recorded at a number of different excitation energies and plotted together with an X-ray absorption spectrum in which the energy positions of the various excitations have been indicated.

It is important to note that in many cases resonant X-ray emission cannot be treated in terms of two consecutive, disconnected steps: excitation and de-excitation. Instead, one needs to regard the entire process as one scattering process where the excitation and emission steps are interlinked. This is because the cross-section for inelastic scattering (which is very small for soft X-rays in nonresonant cases) can become very large at resonances, and this has a number of important consequences. Resonant X-ray emission is therefore often termed *resonant inelastic X-ray scattering* (RIXS) when the inelastic scattering process becomes significant.

Applications of soft X-ray emission spectroscopy are found in many areas of solid state physics and materials science, and the number of applications is constantly growing. Many synchrotron radiation laboratories in the world provide soft X-ray fluorescence beam lines, and there are a number of laboratories that pursue soft X-ray emission using other means of excitation, notably electron beams. The use of RIXS to study low energy excitations in strongly correlated materials, compounds of transition metals and rare earths, has become a widespread technique. The feasibility of investigating buried structures owing to the finite penetration of soft X-rays compared to electrons makes soft X-ray emission a powerful tool, also, for various *in situ* studies of processes where the sample needs to be separated from the vacuum system by a thin window. For instance, molecular properties of liquids and *in situ* studies of chemical reactions are examples where the direct probing of the valence band as offered by soft X-ray emission allows interesting studies to be made.

14.2 PROPERTIES OF X-RAY SPECTRA

Spectroscopy is one of the most powerful methods for studying matter at the atomic level as it directly probes the atomic energy levels and the probability for transitions between these levels. Valence band X-ray spectroscopy, and soft X-ray spectroscopy in particular, involves both inner electron shells and the outermost shell that hosts the weakly bound electrons, i.e., the electrons that take direct part in chemical bonding between atoms and in transport phenomena. Therefore, X-ray spectra can directly probe the chemically active part of the electronic structure and provide partial density of states distributions or molecular orbital compositions. The use of tunable excitation as provided by synchrotron radiation yields a trivial but important feature, namely the ability to separate the parts of spectra that originate in multiple excitations (so called "satellites"). By tuning the excitation energy to just above the threshold of the core-state in question, one quenches multiple excitations by energy discrimination. This ability is important since spectral features due to satellite transitions do not reflect density of states in the simple manner that the main (diagram) spectra do.

Owing to the finite attenuation length of soft X-rays [5] probe depths of up to several hundred nanometers can be attained. The bulk-probing capability of X-ray spectroscopy is often a useful asset in materials research, since the decisive structure for a certain property often lies buried at an interface or below a capping layer. An example demonstrating the depth-probing capability is shown in Figure 14.1. An ultrathin layer of silicon was buried at 10 nm depth in a gallium arsenide matrix, and it was investigated by means of Si L-emission excited by synchrotron radiation. Three different samples were studied, 1 monolayer (respectively 3 monolayers) of Si buried in GaAs, and, as a reference, bulk Si. One observes clear differences between the spectra. For the 1-monolayer (ML) case all Si atoms neighbor Ga or As atoms and their states are found to mix strongly into the Si states, while for the 3-ML Si layer, where 1/3 of the Si atoms have only Si atoms as nearest neighbor, this effect is fairly small. This example demonstrates the feasibility to study electronic structure and chemical bonding in buried structures, even for very dilute samples [6].

Whereas X-ray spectroscopy can be used to probe bulk properties and buried structures, photoemission spectroscopy is extremely surface-sensitive. X-ray spectroscopy has a complementary relation to photoemission and Auger electron emission spectroscopy, both in terms of probing capabilities and the underlying atomic processes. The states involved in the different spectroscopic methods are related, as illustrated in Figure 14.2.

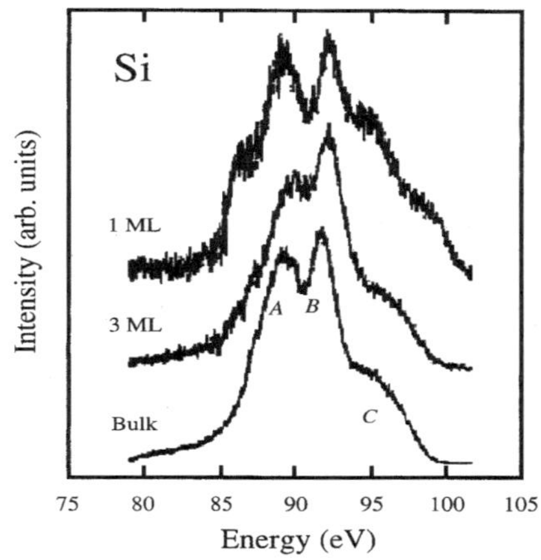

Figure 14.1 Si L-emission spectra of 1- and 3-ML Si buried at 10 nm in GaAs, compared with bulk Si L-emission. 120 eV photons were used for the excitation.

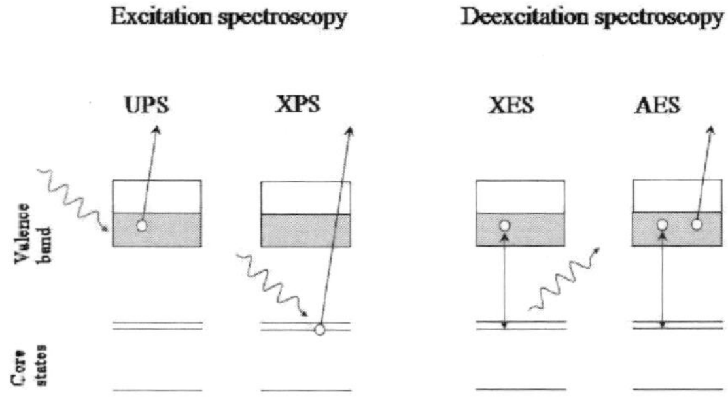

Figure 14.2 Schematic representation of UV and X-ray photoemission, and X-ray *and* Auger emission spectroscopy.

X-ray photoemission, UV photoemission, X-ray emission, and Auger emission are represented as diagrams in the figure. X-ray emission spectroscopy (XPS) measures the binding energy of inner electrons, identifying the emitting atomic element and yielding information about its chemical environment. UV photoemission spectroscopy (UPS) measures the valence band electronic structure (molecular orbitals or density of states distribution), and, in particular, when pursued in *angular resolved mode* (ARUPS or ARPES), it measures the band dispersion in periodic solids. Both XPS and UPS end up in final states that are ionized, and the XPS final state decays by either Auger electron emission or X-ray emission. As mentioned earlier, for soft X-rays the Auger yield dominates by typically a factor of 100 over X-ray emission. Auger electron spectroscopy (AES) leads to doubly ionized states, whereas in the XES case the transition takes place in the singly ionized system. In XES one measures transitions between a core hole state and a valence hole state, i.e., the two different types of states that are probed by XPS (respectively UPS). One must not take this comparison too far, though, since the XPS and UPS processes are prompt, whereas XES is a de-excitation process following possible relaxation of the core hole. Therefore, spectral shapes may deviate from those that could be inferred from results of XPS and UPS measurements, respectively, depending on the lifetime of the core state.

A different situation is encountered when the initial photon is absorbed into a discrete state instead of leading to ionization. Figure 14.3 illustrates *X-ray absorption* (XAS), *UV absorption* (UAS), *resonant X-ray emission* (RXES), and *resonant Auger emission spectroscopy* (RAES). Again, the possibility for relaxation of the core excited state may affect the emission spectrum, although, as we will see, one may control this time lag by detuning the excitation energy in certain cases. Additionally, the UV photon absorption process is associated with a parity change (one-photon transition), whereas the X-ray emission decay of the XAS state implies a second parity change. Thus, UV absorption (one-photon process) probes a different set of states than resonant X-ray emission (two-photon process). Important examples of this are the dd and ff excitations that are of importance for strongly correlated systems, excitations that RXES (or RIXS) is able to probe but the UV absorption technique is not able to examine. It is important to note that, in the case of resonant X-ray emission, the transitions take place in the neutral system, as opposed to traditional X-ray emission and photoemission or Auger emission.

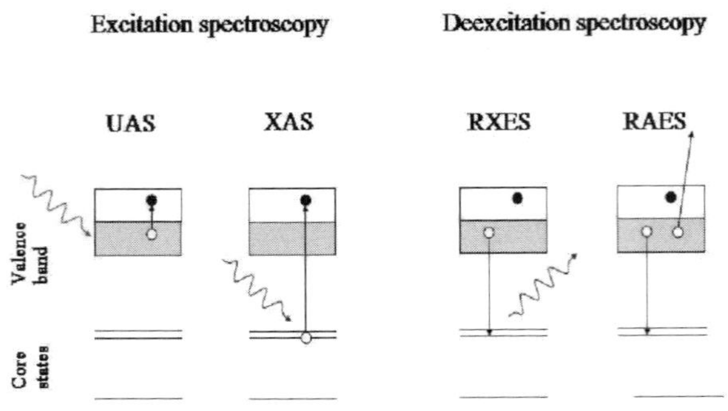

Figure 14.3 Schematic representation of UV and X-ray absorption spectroscopy and resonant X-ray emission and resonant Auger emission spectroscopy. The final result of the RXES process is an excitation of a valence electron to an unoccupied state.

The inner electron binding energies are varying slightly depending on the chemical environment of the atom, a property that is the basis for the ESCA photoelectron spectroscopy method. In traditional (i.e., nonselective excitation) soft X-ray emission atoms in nonequivalent sites (chemically shifted) give rise to largely overlapping spectra, since generally valence band-widths are larger than chemical shifts of the core states. Using monochromatic synchrotron radiation for the excitation solves this problem. Tunable synchrotron radiation allows selective excitation and one can single out a specific atomic species at a particular site of a multi-element system for study in X-ray emission. Therefore, the use of tunable synchrotron radiation for excitation of X-ray emission takes the inherent element selectivity of normal X-ray emission one step further, to chemical site selectivity.

One can achieve further selectivity by using the well-defined polarization of synchrotron radiation in a way that creates contrast with respect to alignment and local order of the sample. Note that the sample need not possess global order since the quasi-alignment imposed by the propensity of excitation of certain suitably aligned states is very long-lived in relation to core-state lifetimes, and this applies even for gaseous samples (rotational periods are in the picosecond range and core-state lifetimes are of femtosecond order). It means, e.g., that a polycrystalline sample can be preferentially excited to states that happen to be aligned in a suitable way. By detecting emission at various directions with respect to the incident polarization, the anisotropy of the emission can be used to reveal information about the local order and state symmetries. In order to carry out experiments like the ones described above one needs to have means to rotate the spectrometer in the vertical plane, in addition to being able to independently rotate the sample to achieve the desired excitation. This is a challenging task for traditional designs of grating spectrometers, but with compact design it is quite feasible. Recent developments in design of insertion devices for synchrotron sources provide polarization state control of the emitted radiation. Beamlines of this kind dispense with the need to rotate the spectrometer in order to measure the anisotropy of the emission.

The site and polarization selectivity of resonant soft X-ray emission spectroscopy is clearly demonstrated by an investigation of the oxygen K-emission of superconducting cuprates [7]. Figure 14.4 displays the oxygen K-emission spectra of polycrystalline $La_{2-x}Sr_xCuO_4$ (x = 0.07) recorded at different excitation energies across the oxygen 1s absorption edge. Significant variations in the spectral shape are observed. When the excitation energy is set to a doping-induced pre-edge peak at 528.5 eV in the X-ray absorption spectrum, the emission spectrum is wide and exhibits a double-structure profile with the main maximum at ~526.4 eV and an intense shoulder on the low-energy side. For 532.8 eV-photon excitation, the spectrum becomes a narrow and nearly symmetric line with a maximum at ~525.5 eV.

The origin of these changes is a large difference in the occupied oxygen 2p density of states (DOS) between in-plane oxygen O(1) and apical oxygen O(2) atoms together with variations in X-ray absorption cross-section for the inequivalent oxygen sites across the K edge. This demonstrates that the changes in the local electronic structure at the inequivalent O sites can be probed separately by tuning the excitation energy to the respective resonance. The present resonant X-ray emission data support predominant creation of electronic holes by Sr doping on O(1) atoms. Furthermore, by making use of polarized excitation and detection of emission anisotropy, further information can be obtained, as it allows distinguishing between the O 2p-orbital components.

Traditional X-ray emission is governed by Fermi's golden rule, i.e., the transition probability is determined by the square of the matrix element over the dipole operator between the core state and the final valence state. The core ionized state has no memory of the excitation and only the population of the core state matters with regard to the excitation. Since the matrix element is nonzero only if parity is changed by unity in the transition, owing to the dipole selection rules, the X-ray intensity distribution will provide a partial density of state distribution pertaining to the symmetry of the intermediate hole state, i.e., p partial DOS for s holes, s and d DOS for p holes, and so forth. Similarly, for a molecular system the X-ray intensity will measure the weight of local, symmetry-allowed atomic orbitals in a MO-LCAO expansion.

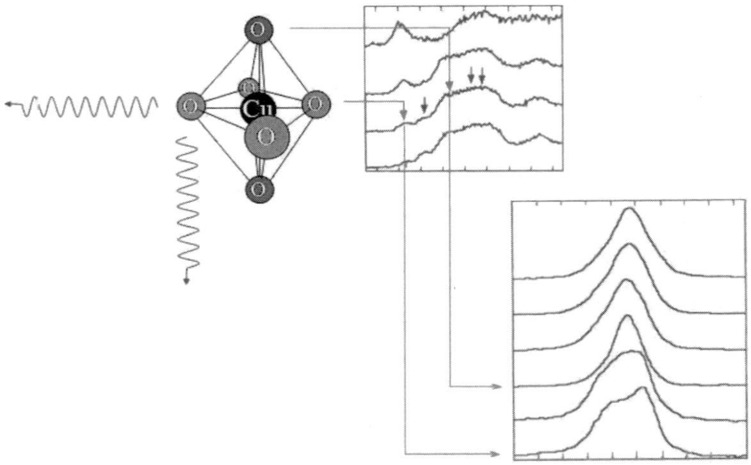

Figure 14.4 By tuning the excitation to certain features in the XAS spectrum one can separate the contributions from different oxygen sites in the X-ray emission. Also, by detecting anisotropy one can determine the orbital components of the doping-induced oxygen 2p states.

14.3 RESONANT INELASTIC X-RAY SCATTERING

When excitation of X-ray emission is resonant we can no longer assume that the intermediate core excited state does not carry any memory of the excitation. On the contrary, the memory of the excitation can be very strong, and the entire process has to be treated in the more general framework of inelastic scattering. In this view the incident photon is inelastically scattered leaving behind a low energy valence excitation. The normally very small cross-section for such processes at soft X-ray energies can be very much enhanced near core-state edges, as seen in the Kramers-Heisenberg formula governing this kind of process [8]:

$$I(E_m) \sim \sum_f \left| \frac{\sum_n \left(\langle f | \mathbf{D} | n \rangle \langle n | \mathbf{D} | g \rangle \right)}{E_{inc} - \left(E_n - E_g \right) + \frac{i\Gamma}{2}} \right|^2$$

$I(E_m)$ is the resonantly scattered intensity distribution. The ground state is represented by g, the intermediate core state by n, and the final state by f. **D** is the dipole operator. E_{inc} is the incident photon energy, and E_n and E_g are the energies of the intermediate and ground state, respectively. Γ stands for the lifetime width of the intermediate core excited state.

Inspecting the denominator of this expression one realizes that the intensity can increase very much near a core excitation edge. Also, one can notice that cross terms appear by the squaring, since the summation over the intermediate state is done for amplitudes rather than intensities. These cross terms represent interference between different paths of the scattering when degenerate intermediate states are present.

Dipole transition amplitudes for exciting and de-exciting intermediate core states govern the intensity. In the time-independent, two-step model the probability for scattering to a final state is the product of the probability for exciting the intermediate states and the probability for de-excitation of those intermediate states to the specific final state. This is an approximation, which is not valid when the final states can be reached via several excitation paths. Then, one cannot assume that one particular intermediate state is excited, and instead the various channels interfere.

The two-step approximation is good, however, when one single intermediate state dominates. It may be intuitively obvious that not much can be said about the dynamics when only one path is involved: Only one thing may (or may not) happen and there are no amplitudes to compare.

There is one conceptually simple but important aspect of the dynamics in which the scattering process could be considered a three-step, rather than a two-step process. In this language the intermediate state may undergo a transition in the time interval between the excitation and the emission of the soft X-ray photon, e.g., an electron or a phonon could be emitted or absorbed. The rate for this transition can then be directly compared to the core hole decay rate, and information about the time-development in the range of the core hole lifetime is achieved, i.e., in the femtosecond and sub-femtosecond range.

The cross terms in the time-independent framework describe the interference between the various excitation-emission paths. This interference contains information about the dynamics in the scattering process. The time-independent theory is successful in predicting experimental results, and the mathematical description is relatively simple (at least in principle). There seems to be little need for alternative descriptions. On the other hand, it has been shown that a time-dependent framework equally well reproduces the data. Mathematically such a treatment becomes more involved than time-independent theory, but the advantage is that one can relate the observations to the time development of the wave functions in a direct way. Thus, the time development is not probed in the same way as in pump-probe experiments. Instead, the duration time of the scattering process is used to get information on the time development.

An interesting consequence of the notion that resonant X-ray emission is essentially a photon-scattering process is that the lifetime broadening of the intermediate core excited state does not lead to broadening in the scattering spectra. From traditional X-ray spectrometry one is used to observing a resolution limitation corresponding to the short lifetime, for soft X-rays typically 0.1–1 eV. In RIXS spectra, however, only the sharpness of the excitation and the resolution of the recording instrument determine the detail to which the final states can be studied. This has significant consequences for the application of resonant X-ray emission spectroscopy.

The independence of the intermediate core state lifetime broadening on the resolution of the RIXS spectrum is clearly demonstrated in a study of low-energy excitations in $Sr_2CuO_2Cl_2$, where the role of dd excitations for the mechanism of superconductivity was investigated [9]. The energies of the lowest excitations in the Cu-based superconductors are basic quantities of interest to determine the underlying electronic structure and elementary excitations, and among other things the local on-site dd excitations have been a topic of debate. The study was carried out at the Cu 3p edge, i.e., M edge RIXS, and although the natural lifetime width of the Cu 3p hole state is 1.5 eV, it was possible to obtain 0.2 eV resolution in the resonant SXES spectrum, as shown in Figure 14.5. Incidentally, the study in question ruled out the alleged importance of on-site dd excitations for the superconductivity mechanism.

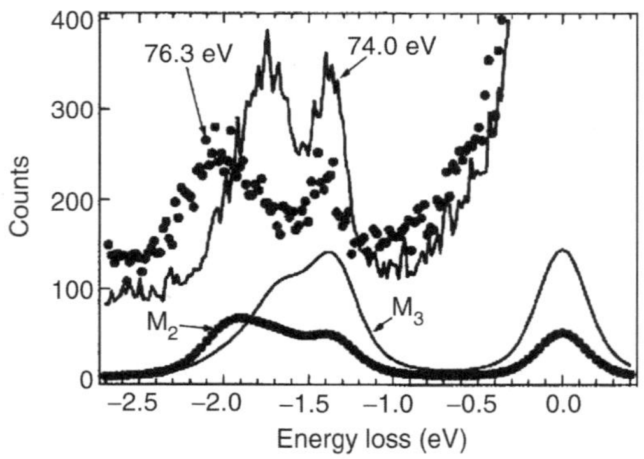

Figure 14.5 RIXS spectra of $Sr_2CuO_2Cl_2$ at the M_2 and M_3 edges. The observed features are about 0.2 eV despite the 1.5 eV lifetime width of the $M_{2,3}$ hole state. Lower part shows calculated spectra.

Strong dependence on excitation energy for valence band soft X-ray emission in broad band materials was observed for crystalline silicon [10], and an explanation was later found in the analysis of resonant X-ray emission spectra of diamond [11]. The observations made for the resonant X-ray emission of diamond that there is strong excitation energy dependence could be explained in terms of RIXS with momentum conservation. It was recognized that momentum conservation leads to restrictions for the absorption-emission scattering process. The analysis demonstrated that RIXS can be used as a photon-based, band-mapping technique, as a complement to angular-resolved photoemission. This suggests a wide range of applications and several attempts to apply the method to semi-conducting broad band materials have been made [12]. For high symmetry points of the Brillouin zone, band structure information has been extracted to a certain extent. The feasibility of quantitative photon-based band structure mapping for TiC has been demonstrated [13].

14.4 EXPERIMENTAL TECHNIQUES

In order to study details of valence electronic structure in soft X-ray emission spectra one needs resolutions substantially better than 1 eV, and for the lower energy range there is presently no alternative to using gratings as dispersive elements. Bragg diffraction requires crystal periods of at least half the wavelength and natural crystals are not available for practical use for softer X-rays than oxygen K emission at about 500 eV. Artificial Bragg crystals are available in the form of synthetic multilayer structures, but to date sufficient resolution has not been obtained due to limited interface quality and number of periods effectively used in the structures. Energy dispersive detectors offer an efficient means for elemental analysis, but *not* for chemical analysis. Detectors based on superconducting tunnel junctions offer the highest energy resolution among energy dispersive detectors and they are constantly improving. However, they are still not a feasible alternative for high resolution spectroscopic work.

For harder X-rays than about 500 eV, natural Bragg crystals with sufficiently large period exist. Bragg diffraction occurs at one particular wavelength at a time for a specific angle of incidence, as opposed to grating diffraction. For small sources, usually encountered for synchrotron radiation, this gives grating diffraction an advantage when using spatially sensitive detectors. For large sources (or virtual sources at off-Rowland geometry), on the other hand, Bragg diffraction works with area detectors generating a multichannel spectrum, each part of the source only contributing to one specific wavelength.

14.4.1 Grating Spectrometers for Soft X-Ray Emission

The advancement in soft X-ray fluorescence spectroscopy is strongly linked to the development of synchrotron radiation sources, although new grating instrument concepts have also contributed. The basic optical and electronic design of the first soft X-ray spectrometer [14] to be used with monochromatized synchrotron radiation, briefly outlined below, was subsequently adopted by a number of researchers in the field. This instrument is based on the use *of* spherical gratings at Rowland mounting. Recently, other schemes have been developed, e.g., designs that utilize plane gratings with variable line spacing [15, 16, 17].

A grating for soft X-rays has to work at grazing incidence in order to achieve total reflection, a necessity due to the low reflectivity at vacuum wavelengths. The resolving power is determined by the number of grating grooves that are coherently illuminated, and therefore one has to have a small source or choose a substantial distance between source and grating, alternatively to work at every grazing angle in order to attain high resolution. This leads to a small angle of acceptance and, hence low efficiency of the instrument. Increasing the groove frequency can obviously also offer higher resolution, but only to a limit, as grating efficiency becomes strongly grating period-dependent for high groove frequencies.

The energy resolution of a Rowland instrument, like the present one, is wavelength-dependent, unlike its wavelength resolution, which is constant. An instrument configured for high energy resolution at high energies may therefore become unnecessarily long if low energies also need to be covered. Therefore, in order to cover a large energy range, 50 – 1000 eV, the present spherical grating spectrometer uses several individually blazed gratings of different radii and ruling density in order to optimize the overall performance of the instrument. The instrument is compact in order to allow it to be mounted on a UHV chamber that can rotate around the synchrotron radiation beam. This facilitates the use of the defined polarization properties of synchrotron radiation. Since it is difficult to measure the polarization state of the emitted X-rays one instead measures the anisotropy of the emission by rotation of the spectrometer. Recently, one has designed beamlines that allow full control of the polarization of the synchrotron radiation. This makes it possible to perform polarized experiments without rotating the experiment chamber.

Figure 14.6 shows the optical outline of the instrument. It consists of an entrance slit, a movable and adjustable shutter for selecting grating and illumination area, three gratings mounted fixed on a precision slab using gauge pieces, and an area detector that can be moved in a three-axis coordinate system (two translations and one rotation). The optical arrangement follows Rowland geometry, and the concept can be seen as three different Rowland spectrometers merged into each other such that they have a common entrance slit. The different Rowland circles run relatively close to each other so that a detector moving in a three-axis coordinate system can be positioned and aligned to the focal curve of the grating in use. Gratings with different ruling density as well as radii are mixed in order to accomplish optimal performance. Illumination of a defined area of a particular grating is done by the shutter. In this way no moving parts of crucial nature for the

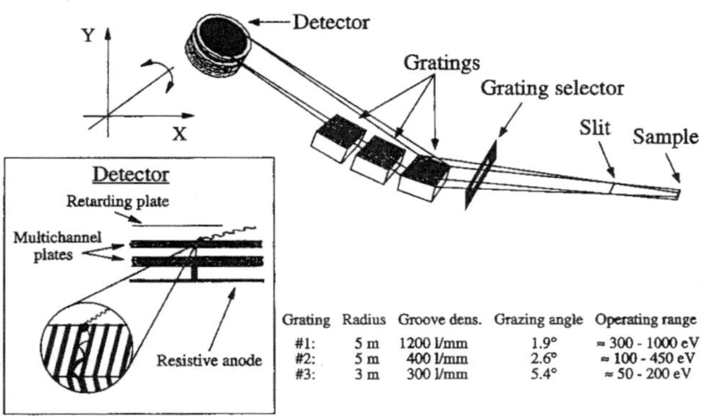

Figure 14.6 Outline of the compact Rowland multigrating spectrometer. Each grating defines a Rowland circle that can be reached by the two-dimensional detector. The resolution is about 1000–3000, depending on energy range.

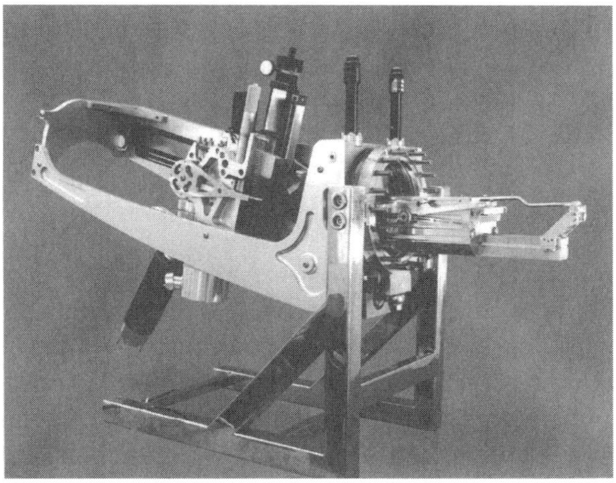

Figure 14.7 Photograph of flange mounted compact soft X-ray emission spectrometer [14].

alignment of the instrument are necessary. Figure 14.7 shows the spectrometer in a version that is presently marketed by Gammadata/Scienta [18].

The detector is based on multichannel plates, coated with CsI for enhanced efficiency and resistive anode readout. The spatial resolution is 1/400, i.e., 50 micron for a 20 mm long detector. A negative-potential electrode at the input of the detector reduces the intensity loss due to escaping secondary electrons. This is particularly important when multichannel plates are used for detection at the grazing angle of incidence since the interstitial surfaces rather than the channels will provide the effective photoemissive surface. It is estimated that the overall efficiency of the detector is up to about 20% depending on photon energy.

The requirements on synchrotron radiation beamlines suitable for high resolution soft X-ray emission spectroscopy are high due to the low yield (low fluorescence yield and limited instrument sensitivity). It is especially important to be able to focus the excitation beam onto the sample. This is nowadays quite feasible down to about 10 microns, and improved focusing schemes suggest that it will be feasible to work with even smaller beam spots on a routine basis in the near future at the fluxes required for soft X-ray emission work. Since fluorescence yields are typically of the order of 1% or less and the sensitivity of spectrometers of the order of 10^{-6} we require an available incident flux on the sample of more than 10^{10} photons per second to get 100 cps count rate. "Available flux" means the flux on the sample that is seen by the instrument. These requirements are fulfilled or nearly fulfilled today at third generation synchrotron radiation sources, and one is witnessing a continuous improvement. The upcoming linear accelerator-based photon sources, notably the SASE-type free electron lasers are likely to provide new and interesting scientific opportunities for soft X-ray emission spectroscopy, not only because of the very much increased intensities but also because of the ultrafast time structure and coherence properties.

14.4.2 Samples at Ambient Conditions

As already mentioned, soft X-ray fluorescence spectroscopy is essentially bulk-sensitive owing to the finite photon attenuation length. The penetration depth offers some special experimental opportunities not present in electron-based spectroscopy since electron escape depths are typically two orders of magnitude less than photon attenuation lengths. Apart from the obvious advantage of relaxed requirements for ultrahigh vacuum and cleanliness, there is the feasibility for studying buried structures since windows withstanding atmospheric pressure can be used to separate the sample from the vacuum system. This offers the possibility of studying many systems *in situ*, such as a solid exposed to an ambient gas or a liquid, or the gas or liquid itself.

New systems for soft X-ray emission studies of various systems of *in situ* character, such as liquids, solids exposed to liquids or gases, and ultrathin liquid layers on solids are under development in various laboratories. One system built at Uppsala [18] consists of a dedicated manipulator that can host one of two different sample heads. The manipulator provides movements and rotation of the sample head, as well as electrical connections and transfer lines for gases and liquids from the outside. The heads are of two different types.

One is provided with a thin window that can be removed and transferred to a preparation chamber under vacuum. This allows freshly deposited thin films to be exposed to a gas or a liquid in the sample head. The other head has a fixed window mounted at 0.1 mm distance from a removable substrate. The substrate, which constitutes a quartz microbalance crystal (QCM), can be removed under vacuum and transferred to a preparation chamber for sample deposition. Mounted in position with a particular sample film, humidified air (or any other gas or gas mixture) is flowed past the sample, and measurements are performed *in situ*. In this way atmospheric corrosion can be studied under realistic conditions.

The sample system lends itself to other kinds of studies, like liquids and liquid mixtures and solutions, properties of materials in the presence of gases, like hydrogen uptake, and valence electronic structure modification of battery electrodes upon cycling of Li ion batteries. In an *in situ* study of a thin film of yttrium exposed to hydrogen at different pressures, soft X-ray absorption and emission spectra revealed upon increasing H_2 pressure that a transition from metallic state to insulator took place. This could be seen as a gap opening up to about 2.5 eV in the compound soft X-ray emission/absorption spectrum. The observation was in line with the fact that yttrium becomes transparent at hydrogen exposure. The results indicate that H and Y states hybridize strongly, causing breakdown of the rigid band model and redistribution of electron states [20].

14.5 APPLICATIONS

14.5.1 Surfaces, Interfaces, and Thin Films

Soft X-ray fluorescence spectroscopy applied in different ways can be used to obtain information about bulk properties as well as to study more surface-related phenomena, depending on which type of excitation and geometry is used. X-rays are absorbed in the sample and produce an exponential depth distribution where the attenuation length depends solely on the material and the X-ray energy. The attenuation length can be calculated from tabulated values [5]. Depending on the energies of the exciting photon beam and the fluorescence, the probe depth can vary from several nanometers at low X-ray energies to hundreds of nanometers at higher energies. The angular variations are strictly geometrical for near normal angles but since the refractive index of materials for X-ray wavelengths is smaller than unity total external reflection conditions can be reached where the incident radiation penetrates only several nanometers into the sample. This is the basis for many analytical techniques where for example trace element analysis on surfaces is performed with X-ray fluorescence.

In the case of studying an adsorbed overlayer of material different than the substrate's, like an absorbed layer of atoms or molecules, SXES can be used even for submonolayer thicknesses. This is due to the fact that different atomic species have well-separated X-ray energies. The introduction of SXES as a tool for the study of surface adsorbates about a decade ago has already contributed to considerable progress in terms of understanding the chemical bonding of atoms and molecules on surfaces. The element selectivity of SXES allows one to separate the spectral features pertaining to the adsorbate from those of the substrate. In photoemission studies of atoms and molecules adsorbed on transition metal surfaces, the valence band spectrum is often quite dominated by the 3d density of states (DOS) of the metal. Figure 14.8 demonstrates this by showing photoemission spectra (UPS) of a Cu(100) surface with and without a nitrogen overlayer, compared with an SXES spectrum. In the UPS spectrum one observes some modification on each side of the Cu 3d peak going to the adsorbate system. In the SXES spectrum, on the other hand, the corresponding parts of the valence band appear as two intense bands reflecting the N 2p distribution and the hybridization with the Cu 3d electrons, respectively [21]. For a comprehensive review on the subject of SXES studies of surface adsorbates we refer to a recent publication [22].

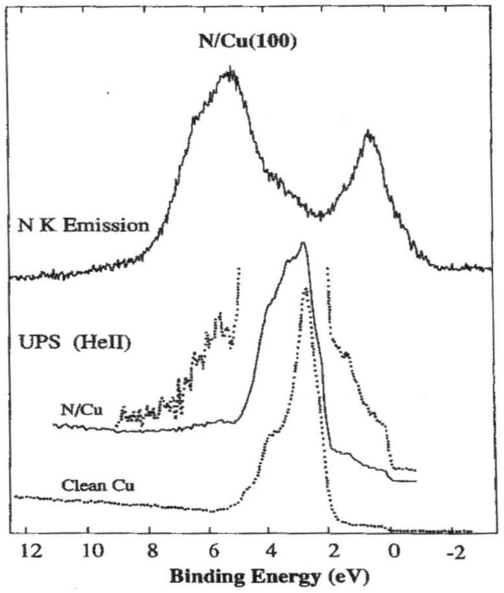

Figure 14.8 UV photoemission spectra of clean Cu respectively a c(2 × 2)N/Cu overlayer system, together with a nitrogen K emission spectrum of the overlayer system.

In thin film applications, and particularly for *in situ* characterization, electron beam-excited SXES has shown interesting potential [23, 24, 25]. An approach to enhance the surface sensitivity of SXES has been to employ grazing exit geometry to achieve a higher ratio of surface-to-bulk signal in thin film growth applications. This can also increase the intensity due to obvious geometrical effects of probing a larger surface area, thus the changes in the film composition due to changes in the deposition process can be detected more quickly with soft X-ray fluorescence spectroscopy.

The angular dependence of the emitted radiation is very large at small angles as illustrated by Figure 14.9. The angular dependence of the emission from three different layers in an iron (50 Å), copper (100 Å), and vanadium (100 Å) multilayer structure has been measured. In the left panel two spectra are shown, one recorded at grazing exit (2.5°) and one at a larger angle (20°). There is a clear difference between them, in particular regarding the intensities of the emission lines originating from different elements. The identification of different emission lines is made, and some of the transitions are recorded simultaneously in different order of diffraction. At grazing exit almost all the intensity originates from the top iron layer, whereas at larger exit angle, intensity from all three layers can be observed. The X-ray emission peak of the silicon substrate falls outside the present window of the spectrometer and is therefore not seen in the spectrum. The Cu L emission that is dominant in the top spectrum has decreased more than an order of magnitude by the change in exit angle.

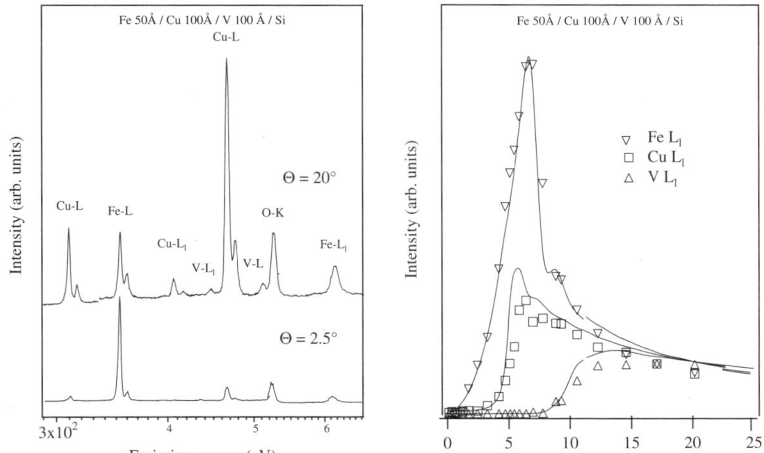

Figure 14.9 Illustration of depth changes with the exit angle. Left: SXES spectrum of a Fe 50 Å/Cu 100 Å/V 100 Å/Si sample at emission angles 20° and 2.5°. The labels represent identified X-ray transitions. Right: Angular variation of the Fe, Cu, and V L_l core-core transitions. The solid lines represent simulated intensities. The intensities are normalized to the intensity at exit angle of 20°.

Intensity plots of the L_l inner transitions of the metals as a function of the emission angle are presented in Figure 14.9 (right panel). There are benefits of using the core-core transitions as compared to using the valence transitions in a study comparing experiments and calculations, since the scattering factors are better determined for the former. The solid lines are simulated profiles using tabulated optical constants and known layer thicknesses. The effect observed originates from the difference in refractive index in the material and in vacuum. At sufficiently low emission angles, below the total reflection angle, the emission originates from a depth determined by the range of the evanescent field, in this case about 5 (UPS) nm [26, 27, 28].

The determination of the chemistry and microscopic structure of carbon nitride films is of considerable interest because of the scientific and technological importance of such materials [29, 30, 31, 32, 33]. In resonant soft X-ray fluorescence spectroscopy, core holes can be created in selected atoms by tuning the excitation energy to a specific 1s to π^* or 1s to σ^* transition [34]. Nitrogen K emission spectra are shown in Figure 14.10. The excitation levels correspond to those absorption resonances. The fact that N

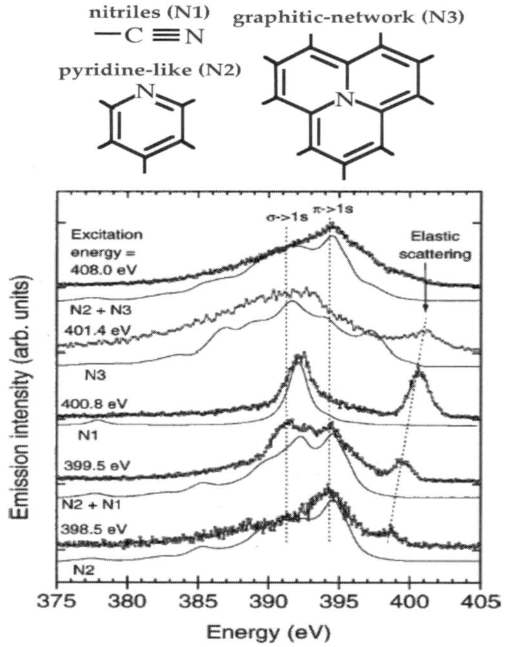

Figure 14.10 Experimental N K emission spectra from the same sample, achieved using different excitation energies and calculated spectra (solid lines) from nitrile structures (N1), pyridine-like N (N2), and graphite-like N (N3).

atoms in different bonding environments are excited depending on the photon energy is clearly reflected by the differences in the SXES spectra. For the lowest excitation energy (398.5 eV), presumably corresponding to the 1s to π^* transition of pyridine-like N, the main emission line is centered at 394.5 eV. This peak can be attributed to a π to 1s transitions while the 392-eV shoulder is mainly due to σ to 1s. This emission pattern is similar to the calculated spectra for N2, which supports the assignment to pyridine-like N.

When the excitation energy is increased to 399.5 eV, mainly nitrogen in the N1 structure should be excited. However, the emission spectra indicate that also a large fraction of the pyridine-like N atoms are excited at this energy. This can be explained by a relatively broad signal corresponding to pyridine-like N due to shifts depending on the second nearest neighbors. Thus, the emission spectrum can be modeled by a superposition of the N1 and N2 spectra. The excitation energy of 400.8 eV, however, is just below the excitation threshold of the N3 structure and the second peak of N2, and therefore mainly the N1 structure is excited, which is also evident from the sharp emission spectrum corresponding to the nitriles.

The calculated spectrum of graphite-like N (N3) resembles well the broad experimental spectrum for the excitation energy of 401.4 eV, as we would expect based on the SXES results. The 408.0-eV spectrum corresponds to electrons excited to the σ^* levels, and since, according to the simulated X-ray absorption spectra, mainly pyridine-like and graphite-like N have intensity in this region, the spectrum can be modeled well by a sum of the N2 and N3 spectra.

The finite attenuation length makes soft X-ray fluorescence spectroscopy a tempting choice for studying buried interfaces and modifications of solid surfaces and thin films in a gas atmosphere, where conventional surface techniques, such as X-ray photoemission spectroscopy, ultraviolet photoemission spectroscopy, and Auger electron spectroscopy are not usable. The bulk-probing property of X-ray spectroscopy is often a very useful asset in materials research, as quite often the decisive structure for a certain property lies buried in an interface or below a capping layer. In the study of Si(100) buried in GaAs mentioned earlier, it was demonstrated that it is possible to extract detailed information about DOS distributions [35]. *Ab initio* calculations [36, 37] showed very good agreement with the experimental spectra and allowed for a quantitative analysis. The layer thickness and the distribution over atomic sites were concluded to be very important for the total spectra.

A fundamental research area of current interest concerns the lattice-mismatched Si/Ge system. Depending on growth conditions, Ge islands or wetting layers are formed on the Si(100) substrate [38, 39, 40]. Through soft X-ray emission spectroscopy combined with *ab initio density functional theory* (DFT) calculations, fine details in the partial local density of states of buried layers can be extracted [41]. The bulk DOS of Si and Ge display three features, each labeled S_1, S_2, S_3, and G_1, G_2, and G_3, respectively. The 1 multilayer (ML) spectrum ("0% mix") shows six features. It appears that the G_1 state is split into two states G_1^{xy} and G_1^{z}, and similarly for G_2 and G_3. The reason for the splitting is the changed symmetry in the Ge layer. Bulk Ge has a larger lattice constant (5.58 Å) than bulk Si (5.39 Å) and the epitaxial Ge monolayer is therefore compressed within its plane. As there is a tendency to keep the unit volume unchanged the Ge-Si spacing perpendicular to the plane (z direction) increases by 3.5%, which is somewhat lower than that required for a constant unit volume. The monolayer Ge atoms are thus surrounded by a tetragonal structure, instead of the bulk cubic structure. The s-states are, however, not directly influenced by this effect. On the other hand, the three p-levels, degenerate in cubic symmetry, are split, creating two degenerate levels directed within the monolayer plane (x,y) and one level directed perpendicular (z) to the monolayer plane. As we have a mixing of s and p-states in the solid, the crystal field splitting of the p-states will be reflected indirectly also in the s-states.

To test this geometrical argument the (s + d) DOS of the buried Ge layer was recalculated, but with the Si atoms substituted by Ge atoms, keeping all the previous distances unchanged. Indeed it was found, as summarized in Figure 14.11, that the low-lying peak G_1 in bulk Ge splits into two peaks $G_1'^{xy}$ and $G_1'^{z}$ in the Ge monolayer. The peak G_2 is similarly split. This shows that the splitting of the levels in the Ge layer in bulk Si is a geometrical effect. A further shift of the levels occurs due to hybridization between the Ge and Si states, as illustrated in Figure 14.11.

In the real system the Ge atoms will not form a perfect smooth monolayer, but due to intermixing the Ge atoms will, to some extent, occur on neighboring Si sites, and this situation has been simulated by "25% mix." An entirely new, complicated structure appears, however, as compared to "0% mix." We note, though, that two new strong peaks appear in the 11–12 eV region (I_1 and I_2), i.e., below any structure in the "0% mix" spectrum.

The agreement between experiment and theory is very good, in particular for the case of 1 ML Ge. There is a strong indication that the layers are of high perfection, but that some intermixing with the neighboring Si atoms occurs. The electronic states in the Ge layer can be viewed as the Si bulk states tunneling through the Ge layer and resonating on the Ge atoms. Thus the situation resembles the "virtual crystal approximation" for alloys, where an averaged electronic structure is observed. Due to the reduced symmetry in the Ge layer two Ge resonances are observed for each Si bulk state.

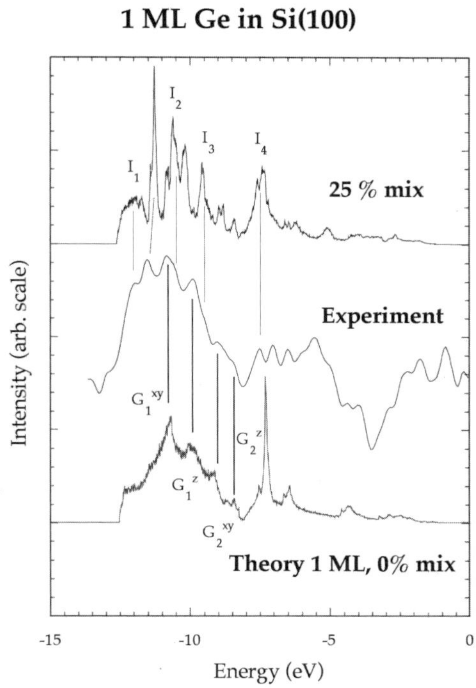

Figure 14.11 Comparison between experimental SXE spectra and (s + d) DOS theory for 1 ML Ge in Si (100). The experimental spectrum has been convoluted by 0.3 eV Gaussian.

In short, soft X-ray emission spectroscopy has been used as a chemical probe in the characterization of thin film materials. Some examples of the studies are presented to show the chemical sensitivity, site selectivity, and depth-dependence of this method. The finite penetration of X-rays offers true bulk-probing and facilitates studies of interfaces and buried structures.

14.5.2 Nano Structures

The properties of matter at nanoscale dimensions can be dramatically different from the bulk or the constituent molecules. The differences arise through quantum confinement, altered thermodynamics, or changed chemical reactivity. In general, electronic structure ultimately determines the properties of matter, and it is therefore natural to anticipate that a complete understanding of the electronic structure of nanoscale systems will lead to progress in nanoscience and bioscience, not inferior to the progress we have seen in other fields in recent years.

In the nano-regime, two effects dominate the chemical and physical properties: (i) Increasing contribution of surface atoms (surface effect): Surface atom-to-bulk atom ratio increases dramatically as the particle size decreases. For a 30-nm particle, about 5% of all the atoms are on the surface, while for 3-nm particle, up to 50% are surface atoms. (ii) Quantum confinement as the nanoclusters approach a few nanometers (size effect), when electrons will begin to feel the effects of quantum confinement as the size of the structure becomes comparable to the electron wavelengths.

The energy position of the band edge can be controlled by the electronegativity of the dopants, solution pH (flatband potential variation of 60 mV per pH unit), as well as by quantum confinement effects. Accordingly, band edges and bandgap can be tailored to achieve specific electronic, optical, or photocatalytic properties. A very important application is found in the generation of H_2 from direct photo-oxidation of water without external bias [42]. Quantum confinement effects on bandgap profiling in similar arrays has been studied by resonant inelastic X-ray scattering of synchrotron radiation for potential application of nanomaterials for direct photo-oxidation of water by solar irradiation.

The measurements were done on synthetic α-Fe_2O_3 nanorods grown by *controlled aqueous chemical growth*. The samples investigated are thin films that consist of crystalline arrays of hematite nanorod bundles of 50-nm diameter and 500-nm length perpendicularly oriented onto the substrate. Each bundle was found to consist of self-assembled nanorods 3–5 nm in diameter [43]. The samples were prepared by heteronucleation growth and thermodynamic stabilization of akaganeite (α-FeOOH) in solution at 90°C onto the substrate and subsequently heated in air to 550°C to allow the crystal phase transition to hematite (α-Fe_2O_3) as confirmed by RIXS.

The Fe L-edge absorption spectrum of α-Fe_2O_3 nanorods is shown in Fig. 14.12 (upper). The spectral shape is very similar to the previous XAS measurements of powder [44] and single-crystal [45] α-Fe_2O_3. The spectrum shows the spin-orbit interaction of the 2p core level that splits the L_2 ($2p_{1/2}$) and L_3 ($2p_{3/2}$) edges, and the p-d and d-d Coulomb and exchange interactions that cause multiplets within the edges. The ligand field splitting of 3d transition metal, being of the same order of magnitude as p-d and d-d interactions (1–2 eV), gives a 1.4-eV energy splitting between t_{2g} (xy, yz, xz) and e_g (x^2-y^2, $3z^2-r^2$) orbitals. The charge transfer has two main effects on spectral shape [46]. First, it splits d orbitals further than the atomic ones by the formation of molecular orbitals. Second, charge transfer gives rise to the tail shape that can be seen on the higher energy side of the edges, especially at L_3 (in the region of 711–718 eV).

The resonant RIXS spectra at the Fe L-edge of α-Fe_2O_3 nanorods were measured and are shown in the bottom of Figure 14.12. Several energy loss features are clearly resolved. The low energy excitations, such as the strong d-d d and charge-transfer excitations, are identified in the region from 1 to 5 eV. The 1-eV energy loss features originate from multiple excitation transitions. The 4.1 and 6.4 eV excitations originate from charge-transfer between oxygen 2p and iron 3d orbitals. The 2.5-eV excitation, which corresponds to the bandgap transition of hematite, appears significantly blue-shifted as compared to the reported 1.9–2.2 eV bandgap of single-crystal and polycrystalline samples. This finding indicates that such designed nanomaterials would meet the bandgap [47] requirement for the photocatalytic oxidation of water without an applied bias.

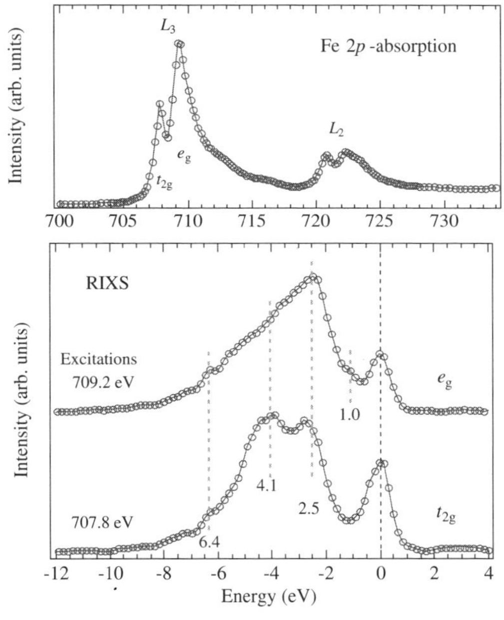

Figure 14.12 Fe L-edge absorption spectrum of α-Fe_2O_3 nanorods (upper); energy-dependent RIXS at Fe L-edge of α-Fe_2O_3 nanorods (lower).

ZnO, a wide bandgap semiconductor, has attracted considerable attention during the last years due to its potential technological applications [48, 49, 50, 51, 52] such as, for instance, high efficient vacuum fluorescent displays (VFD) and field-emission displays. ZnO has also been used for short wavelength laser devices.

The X-ray emission spectra of bulk and nanostructured ZnO are displayed together with the corresponding X-ray absorption spectrum in Figure 14.13 [53]. The oxygen K emission spectrum reflects the oxygen 2p occupied states (valence band), and the oxygen K absorption spectrum reflects the O 2p unoccupied states (conduction band). In the photon energy region of 530–539 eV, the X-ray absorption can be mainly assigned to O 2p hybridized with Zn 4s states. In the region of 539–550 eV the spectrum is mainly attributed to O 2p hybridized with Zn 4p states. Above 550 eV, the contribution is mainly coming from O 2p-Zn 4d mixed states. Stronger s-p-d hybridization was revealed in nanostructured ZnO since the contributions of features at 520 eV and 523 eV are enhanced. A well-defined bandgap can be observed between the valence band maximum and conduction band minimum. The absorption-emission spectrum yields the fundamental bandgap energy of 3.3 eV.

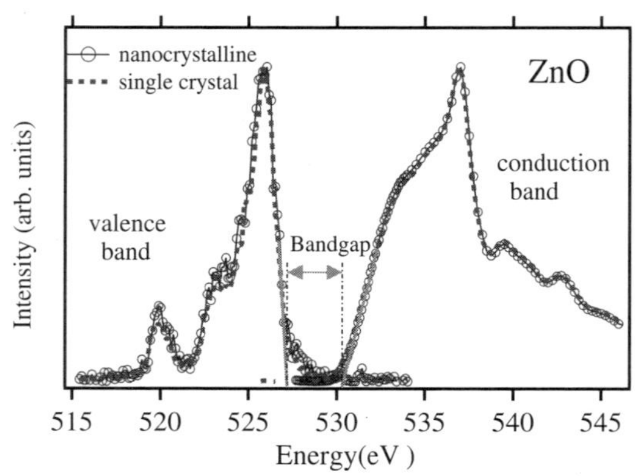

Figure 14.13 Oxygen X-ray absorption and emission spectra reflecting, respectively, the conduction band and valence band near the Fermi-level of ZnO nanoparticles in comparison with bulk ZnO.

In order to emphasize the differences of spectra, the spectrum of bulk ZnO has been subtracted from the spectrum of nanostructured ZnO by using a weighting factor, as shown in Fig. 14.14. The weighting factor was chosen to avoid negative intensity, obtained when subtracting the maximum bulk contribution. The difference spectrum was then multiplied by 3. The feature C has pronounced intensity compared to features A and B, which can be correlated to enhanced p-d hybridization in nanostructured ZnO.

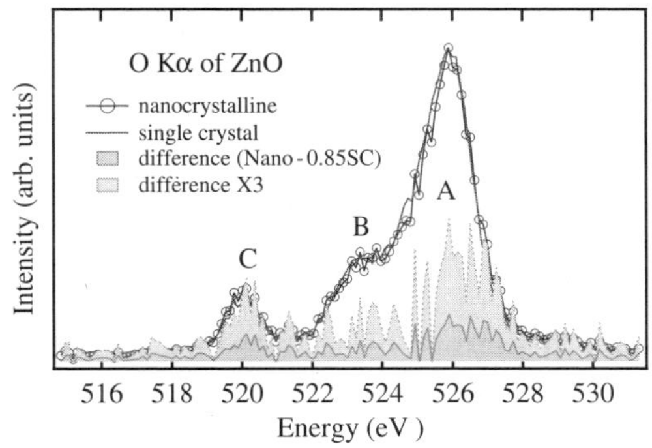

Figure 14.14 The difference spectrum of bulk and nanoparticles ZnO, from the X-ray emission spectra in previous figure.

14.5.3 Transition Metal Systems

14.5.3.1 Cu 2p- and 3p-RIXS of Cuprates

For correlated 3d-transition metal oxides, resonantly excited X-ray emission spectra differ significantly from high energy excited (fluorescence-like) spectra. As was pointed out previously the inelastic scattering cross-section becomes very large at core level absorption resonances. Also, since the scattering process conserves energy locally the spectra are not subject to intermediate state lifetime broadening, which otherwise can be a limiting factor for the spectral resolution. The excited electron that is expelled from the core level remains close to the atom in a bound state and influences the spectrum. Local energy conservation entails that the outgoing photon energy either corresponds to a specific atomic-like excitation (excitonic spectrum) within the same sub-shell, i.e., a dd-excitation, or more molecular-like excitations that involve electron-transfer from the excited atom to a ligand, e.g., an oxygen atom. In particular, for cuprates that have a $|3d^9\rangle$ ground state configuration, possible final states for excitation at a p-resonance include excited $|3d^9\rangle$-states and $|3d^{10}\underline{L}\rangle$-states (where $\underline{L}$ denotes a missing electron in a ligand atom). These states are separated in energy (typically in the 1-eV range) by the crystal field that is due to electrostatic fields and hybridization effects, therefore, these excitations are also called *crystal field excitations*.

For cuprates dd-excitations involve the rearrangement of merely one empty orbital (in octahedral symmetry grouped into a triplet of e_g-orbitals: xy, xz, yz, and a doublet of t_{2g}-orbitals: x^2-y^2, $3r^2-z^2$) and thus only 9 possible final states exist. In practice, the narrow spacing between the orbitals only allows one to resolve two or three RIXS peaks separated by some tenths of an electron volt. This brings us to the question of instrumental resolution: present spectrometers offer an ultimate resolution of about one part in several thousand. Therefore it is difficult to resolve many of the features related to these low energy excitations at the transition metal L-resonances (metal 2p-excitation), which are located in the approximate range 400–1000 eV, where the energy resolution is about 0.5 eV at best. Instead, instrumental resolution better than 0.1 eV is readily obtained with excitation at the M-resonances (metal 3p-excitation), which are located below 75 eV. A caveat is the high cross-section for elastic scattering of low energy X-rays that leads to an enhancement of the tails of the monochromator function so that weak inelastic scattering features less than about 0.5 eV below the excitation energy can be obscured.

The scattering nature of the RIXS process entails that binding anisotropies lead to angular radiation anisotropy. This in turn can be recorded by varying the angle between the incident X-ray; polarization vector and the direction of the detected X-rays; see Figure 14.15. The polarization-dependence of the dd-excitation multiplet was investigated in a pioneering work on the Cu 2p-RIXS of La_2CuO_4 [54, 55]. Although the spectral resolution was not better than 2 eV, the polarization analysis enabled resolving spectral shifts due to the dependence of transition probabilities to dd-excitations on orbital symmetry. Following this example, other RIXS experiments [56] have taken advantage of the extra information that can be extracted by comparing spectra taken in the so-called "depolarized" geometry (corresponding to the "vertical position" in Figure 14.15 versus in the "polarized" geometry (corresponding to the "horizontal position").

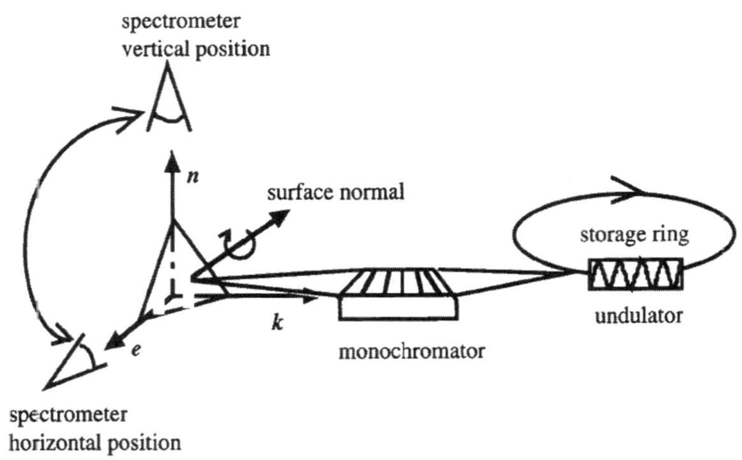

Figure 14.15 Experimental geometries for RIXS measurements. Linearly polarized monochromatic synchrotron radiation with the electric field vector in the horizontal plan is directed towards the sample. The detection direction can be varied continuously between the horizontal (so-called "depolarized") and the vertical (so-called "polarized") spectrometer positions. A manipulator can orient the sample axes in any desired direction.

As pointed out above, energy resolution is much higher at lower energies. Cu 2p-RIXS requires excitation energies above 900 eV compared to 75 eV for Cu 3p-RIXS. Figure 14.16 shows a comparison of Cu 2p-RIXS and Cu 3p-RIXS for the cuprate $CuGeO_3$ using the best instrumental resolution available [57]. The peaks at about 1.6 eV are crystal field excitations that have splittings of a few tenths of an electron volt. Crystal field excitations have been studied for a long time at very high energy resolution in infrared spectroscopy or Raman scattering spectroscopy. However, interpretation and calculations of optical spectra are much more intricate than for RIXS (which in analogy has also been called resonant Raman X-ray scattering). RIXS is a dipole allowed two-photon (i.e., second order) scattering process described by the Kramers-Heisenberg formula connecting, in the first step, the ground state with a core hole-excited, charge-neutral excitation (intermediate state) and, in the second step, connecting the intermediate state with a charge-neutral,

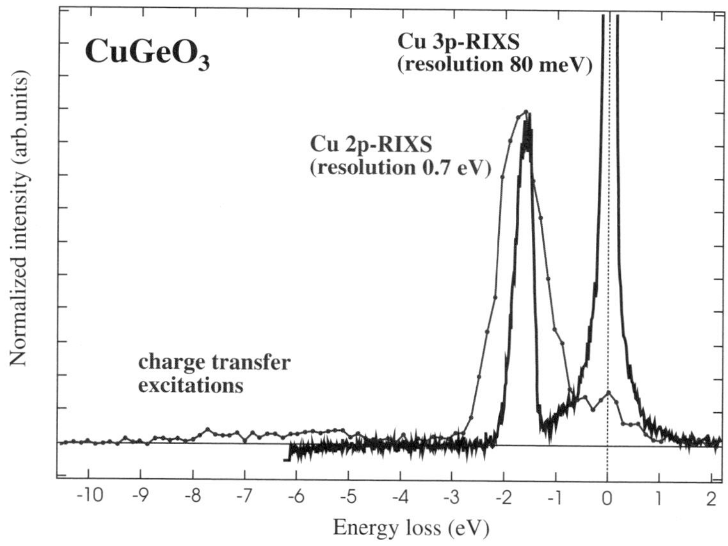

Figure 14.16 Cu 2p-RIXS (solid line) and Cu 3p-RIXS (dashed line) of $CuGeO_3$ shown on an energy loss scale.

valence-excited state obeying the selection rule $\Delta l = 0, \pm 2$ (which includes, e.g., the ground state and dd-excitations). More on theoretical calculations can be found in the review by Kotani and Shin [58].

The first demonstration of 3p-RIXS of a cuprate was performed on $Sr_2CuO_2Cl_2$ by Kuiper et al. [59]. The lifetime broadening of the 3p core hole state of approximately 1.5 eV notwithstanding, features separated by a few tenths of an eV could be resolved, as noted in a previous section. In particular, this study addressed the controversy [60] whether one has found dd-excitations at sufficiently low energies to have relevance in the mechanism for high-T_c superconductivity. This would require that excitations as low as 0.5 eV or below exist, something for which some groups claim evidence exists from optical Raman spectroscopy. However, no excitation below 1.5 eV is found in Cu 3p-RIXS and therefore it is likely that dd-excitations at such low energies do not exist.

Also, other types of excitations appear in Cu p-RIXS, e.g., *charge transfer* (CT) excitations that leave the atom in a $|\underline{L}d^{10}\rangle$ final state. RIXS measurements of CT-excitation energies in connection with model calculations are highly interesting with respect to magnetic interactions because the CT energy is the dominating term of the superexchange mechanism parameter *J*. CT states require higher excitation energies and are spread out reflecting the oxygen 2p bandwidth: CT states are seen in Figure 14.16 as a weak band between about 4 eV and 8 eV energy loss. Since CT states have a valence hole in the oxygen ligand it is expected that these excitations are much stronger in O 1s-RIXS (see next Section, 14.5.3.2).

14.5.3.2 Oxygen 1s RIXS of Cuprates

The electronic structure of cuprates is determined by the interplay of the narrow Cu 3d band localized on the Cu sites and the predominantly itinerant O 2p band. The bonding of the Cu-O octahedra themselves and their coupling, e.g., edge- or vertex-sharing plates or chains carry key information on the type of macroscopic low temperature behavior of these compounds, be it superconductivity or, as in the case of $CuGeO_3$, a *spin-Peierls state* (T_{SP} = 14 K). Although a vast amount of literature has been dedicated to the understanding of high-T_c superconductivity, the underlying mechanism is still not fully understood. Spin-Peierls states are unusual for inorganic materials and it is interesting to note that the one-dimensional edge-sharing CuO_4

chains of $CuGeO_3$ are coupled along its crystallographic *c*-axis, in contrast to high-T_c cuprates that have vertex-sharing chains. Therefore it is important to gain detailed experimental knowledge of the electronic structure and, in particular, of the valence band of these Cu-O units. $CuGeO_3$ is an orthorhombic antiferromagnetic insulator at room temperature. Early band structure calculations [61] predicted a metallic ground state, whereas more sophisticated techniques [62] yielded gap values comparable to the value found in optical absorption (3.7 eV). Although the O 2p electrons are much less correlated than the Cu 3d electrons, the theoretical *partial density of states* (PDOS) appears to depend strongly on the computational methods used. The O 2p electrons are expected to have an itinerant character forming a wide valence band but also some localized character due to their bonding to the central Cu atom and thus their hybridization with 3d electrons.

For $CuGeO_3$ it was shown for the first time that SXES at the O 1s-resonance from a cuprate can, apart from showing PDOS features, reflect oxygen-specific valence states with *localized* character [63]. The localized states are observed as RIXS features (Figure 14.17) that resonate (and disperse linearly with excitation energy) *only* close to the O 1s pre-edge absorption-peak, which is associated with the *upper Hubbard band* (UHB). Moreover, the analysis of excitation-energy and polarization dependence of these O 1s *soft X-ray fluorescence* (SXF) spectra allowed researchers to separate features derived from the symmetry-projected 2p DOS at different oxygen sites. A high energy shoulder of in-plane excited spectra emerges at higher excitation energies and is evident of the occurrence of a theoretically expected two-hole state (Zhang-Rice singlet), but it has not been possible to identify this state in photoemission spectra of $CuGeO_3$.

Oxygen 1s-RIXS can be described as an excitation into the upper Hubbard band, i.e., a Cu 3d-O 2p hybridized band with large Cu 3d character: $|\underline{L}d^{10}\rangle$ - $\rangle$ $|cd^{10}\rangle$ where the ground state is assumed to be a hybridization of $|d^9\rangle$ and $|\underline{L}d^{10}\rangle$ wavefunctions. The intermediate state can decay into several different final states whereby the localization of the core hole on the oxygen site leads to the strongest overlap with states such as $|\underline{L}d^{10}\rangle$, i.e., charge-transfer states. However, also dd-excitations are observed due to the strong hybridization of the UHB. This has been observed in several insulating cuprates such as CuO, $SrCuO_2$, and La_2CuO_4 [64]. Some of the dd-excitations are shown to have unusually large values (more than 2 eV) in some compounds, e.g., $SrCuO_2$.

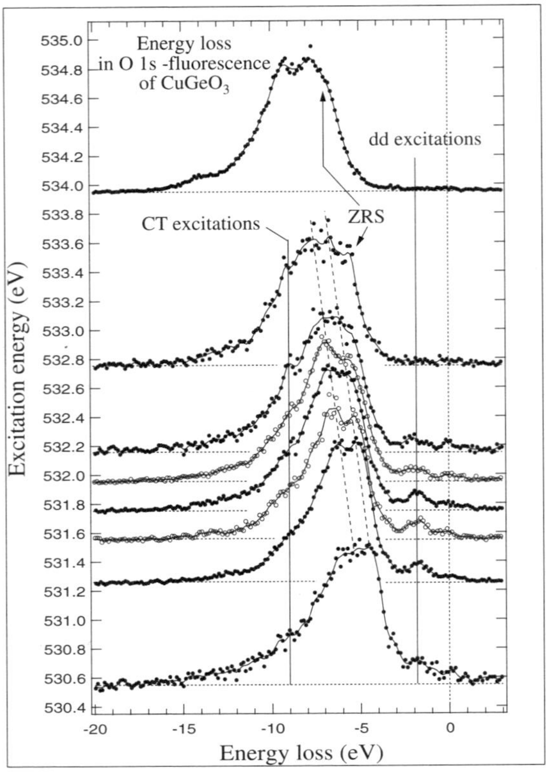

Figure 14.17 Energy-dependent O 1s-RIXS of $CuGeO_3$ taken in the depolarized geometry shown on an energy loss scale. The excitation energy for the spectra shown (with vertical offset) increases from bottom to top and corresponds to the point where the baselines intersect the left hand scale. Several different excitation features are marked.

Other possible excitations in O 1s-RIXS can be attributed, for instance, to a delocalized hole in the ligands around a specific Cu site. In that case, one needs to distinguish between final states with a ligand hole around the original Cu d^{10} site ($|d^{10}\underline{L}; d^9\rangle$), on one hand, and a ligand hole around a neighboring site with Cu $3d^9$ wavefunction ($|d^{10}; d^9\underline{L}\rangle$), on the other hand. The latter state forms a *spin singlet* on a copper-oxygen plaquette due to spin-conserving dipole transitions and the fact that Cu sites are antiferromagnetically ordered in the ground state. This is a nonlocal CT excitation and can be considered the charge-neutral RIXS analogy to the *Zhang-Rice singlet* (ZRS) that forms as the first excitation state upon electron removal (photoemission). A ZRS is a two-hole state of mainly $|d^9\underline{L}\rangle$ character with antiferromagnetic spin alignment of the holes, i.e., a spin-singlet state. Okada and Kotani [65] proposed that a ZRS can appear in O 1s-RIXS of edge-sharing cuprates. Its energy is sensitive to the charge-transfer energy and lies in the same energy region as the crystal field excitations. Harada et al. [66] claimed that they found evidence for an RIXS-ZRS in $Sr_2CuO_2Cl_2$ by observing changes in RIXS spectra between the polarized and the depolarized geometry.

ZRS states can appear as "by-products" in other core-level spectroscopies as well. For instance, van Veenendal et al. [67] investigated cuprate X-ray photoemission spectra (XPS) of the Cu 2p-level theoretically. Supported by cluster calculations using the Anderson impurity model [68], they interpreted the main line of the Cu 2p-XPS as being due to the excitation of $|d^{10}; d^9\underline{L}\rangle$, i.e., the creation of a nonlocal ZRS on the neighbor site of the central Cu site, instead of the previously accepted $|d^{10}\underline{L}^2\rangle$ assignment, which is now attributed to the high energy shoulder. This emphasizes the importance of nonlocal screening effects in the 2p core-hole creation of correlated solids. O 1s-RIXS offers the possibility of measuring the excitation energy of nonlocal ZRS states without the presence of the core hole.

14.5.3.3 Vanadium Oxide Systems

Vanadium oxides display a broad variety of electronic, magnetic, and structural material properties, many of which show *metal-to-insulator transitions* (MITs) [69] and other interesting phase transitions. NaV_2O_5 is an insulator that has a charge ordering or spin-Peierls-like phase transition at 34 K [70]; also V_2O_3 [71], VO_2 [72], and V_6O_{13} [73] exhibit an MIT at very low temperatures. The metallic monoclinic MoO_2 is isostructural to the low temperature semiconducting phase of VO_2. By doping VO_2 with Mo the MIT temperature for the $Mo_xV_{1-x}O_2$ compound system can be shifted towards lower temperatures [74]. V_6O_{13} doped with Li [75], VO_2 doped with Mo [76], and vanadium oxide nanotubes doped with Li [77] have been studied using RIXS at the V 2p- and the O 1s-resonances. V_2O_3, VO_2, V_6O_{13} undergo MITs at 160 K, 340 K, and 150 K, respectively, whereas doping with Mo leads to a concentration-dependent reduction of the MIT in VO_2. The detailed origin of the MITs is still under debate and therefore studies of the electronic structure are very valuable. RIXS studies show that both localized and itinerant states are present in all compounds and therefore a theoretical approach using only one of these pictures is insufficient to fully describe the electronic structure.

In order to understand the underlying mechanisms for the various phase transitions displayed by these vanadium oxides it is essential to obtain better understanding of their electronic structure also at room temperature. NaV_2O_5 has been subject to intense experimental and theoretical studies of its electronic structure. NaV_2O_5 is an insulator at room temperature with a quasi-one-dimensional (chain-like) crystal structure. The phase transition at 34 K is due to a dimerization along the chains. NaV_2O_5 possesses an orthorhombic crystal structure at room temperature and is centro-symmetric (Pmmn) with only one distinct V ion in a mixed valence state (average valence $V^{+4.5}$) [78]. The spins are carried by V-O-V molecular orbitals (rungs), the basic building blocks of the electronic structure of this unconventional spin-Peierls material. The electronic structure of insulating NaV_2O_5 has been the subject of many experimental and theoretical investigations [79, 80]. By comparing model calculations of the electronic structure with experimental spectra from conventional optical techniques the key parameters of the electronic structure can be determined.

The molecular like V-O-V-rungs of NaV_2O_5 that run along the chains build up a ladder-like electronic system. Each of these rungs takes up one of the electrons that the sodium atoms, separating the chains, donate. Formally this leads to a $V^{4.5+}$-valency and both the possibility of distinct V^{4+}/V^{5+}sites (charge-ordering) and a delocalization of this electron (mixed valence) has been discussed. X-ray diffraction studies [79] support the model of a mixed valence structure. Density functional theory band structure calculations using the local density approximation of this quarter-filled ladder system cannot reproduce the insulating state but show that the lowest d-states are of *xy*-character with considerable dispersion across the Fermi level. Moreover, the bands belonging to the antibonding d_{xy}-states are separated by about 1.2 eV from the occupied d_{xy}-orbital. LDA + U calculations show that the on-site Coulomb interaction (correlation effect) actually splits the lowest d_{xy}-bands into a spin-up and spin-down component with a gap of less than 1 eV. The remaining empty d-orbital states have energies of some 1.6 eV and more above the occupied d_{xy}-orbital. In a localized picture this can be described by the crystal field splitting.

V 2p-RIXS of NaV_2O_5 experiments show a rather strong energy loss peak around 1.6 eV [80]. There exists a controversy regarding which orbital symmetry the corresponding final states have and whether the energy loss reflects the on-site correlation energy (Mott gap) or the crystal field splitting 10 Dq. Duda et al. [81] have shown that in O 1s-RIXS of NaV_2O_5 there exists an RIXS peak at 1.2 eV energy loss (see Figure 14.18), when excited into the upper Hubbard band state (i.e., the hybridized O 2p-V 3d-state). This is a non-local excitation that transfers a d-electron from the excited V site to the other V site on the same rung (*on-rung excitation*) via an O 1s-excitation, i.e., an excitation to the antibonding d_{xy}-states. It is therefore likely that the spin-split empty d_{xy}-states are found at even lower excitation energies, i.e., less than 1 eV, in accord with theoretical predictions and optical measurements. Note also that an excitation requiring the transfer of an electron from another rung to the core-excited one (interrung hopping) entails a small overlap between the involved wavefunctions and in turn a low excitation probability. It is therefore unlikely for such a spin-flip excitation across the correlation gap to have a substantially larger spectral weight than the crystal field excitations.

Moreover, van Veenendal et al. [82] pointed out that the excitation into this state would be forbidden in the applied depolarized geometry (outgoing X-rays parallel with the incident X-ray polarization). Thus, more experiments using higher instrumental resolution and high detection efficiency are required to eventually detect this nonlocal, low energy excitation.

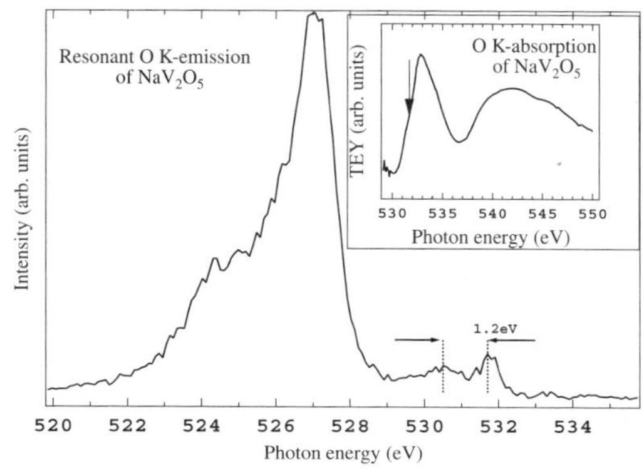

Figure 14.18 O 1s-RIXS of α'-NaV_2O_5 taken in the depolarized geometry. The inset shows the O 1s-absorption spectrum of α'-NaV_2O_5 and the arrow marks the excitation energy that corresponds to the transition of an O 1s-electron into the upper Hubbard band.

Vanadates are also being studied because of their useful properties as Li-battery anodes. These vanadates have hollow and stable crystal structures, which readily can take up many Li atoms per unit cell, a property is desirable for high capacity batteries. The electronic structure of cycled battery anodes made of Li-doped V_6O_{13} and Li-doped vanadium oxide nanotubes have been investigated with X-ray absorption and RIXS. The studies [83] show how the valence and conduction band states change with discharging, i.e., at different Li stages. This is clearly observed in Figure 14.19, which shows V L-emission spectra for lithiated vanadium oxide nanotubes used as a Li-battery electrode and discharged to various cell potentials. Several features may be distinguished. Starting from the high energy side, we see a sharp feature that arises from elastically scattered X-rays; the shoulder on the low energy side (marked dd-exc) reflects that a small amount of crystal field excitations are created, then a peak marked V 3d is found and attributed to the top of the vanadium valence band and of predominant V 3d-character, and finally a large broad band (O 2p-h) is seen, which is attributed to the bottom of the vanadium valence band that is strongly hybridized with O 2p-states. The spectral evolution that mainly takes place in the V 3d-peak is explained due to electron donation by the Li atoms into the top of the valence band. Moreover, a comparison of the V L-absorption spectrum with those of reference vanadium oxide compounds reveals the oxidation states at the different stages and the most likely crystal structure that is assumed.

Electron-doping in both $Li_xV_6O_{13}$ (Figure 14.20), and Mo_xVO_2 (Figure 14.21) leads to apparent shifts in the positions of features attributed to V 3d-derived and O 2p-derived features in the V L-emission spectra. This can be interpreted as a band-filling effect due to the extra charge that enters the vanadium 3d-band. This is also manifested in the relative weights of the spectral features, where the top of the valence band increases in intensity relative to the filled bottom of the valence band, i.e., a rather rigid band behavior of the V 3d-band structure (Figure 14.20). The donation of electrons by Li and Mo is consistent with the results of *ab initio* band structure calculations [84]. In contrast, it is interesting to note that Li-doped TiO_2-based compounds behave in a way that is quite correlated [85]. There the Ti 3d-states appear as localized band gap states and strong correlation effects limit the stability of the crystal structure.

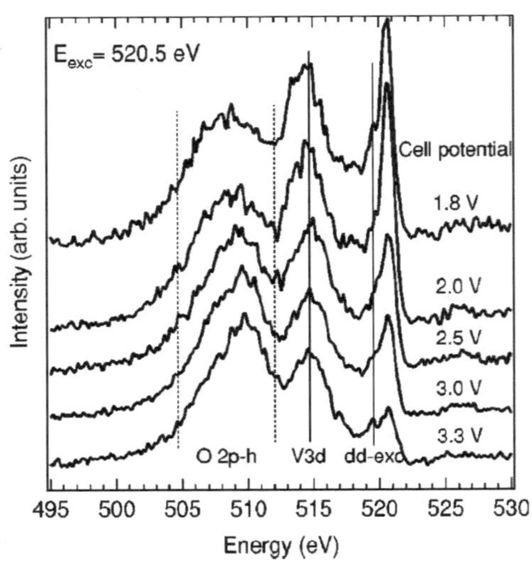

Figure 14.19 V L_α-emission of lithiated vanadium oxide nanotubes excited at the L_3-resonance (520.5 eV). Spectra for 5 different cell potentials are compared (with vertical offset) and several spectral features are marked.

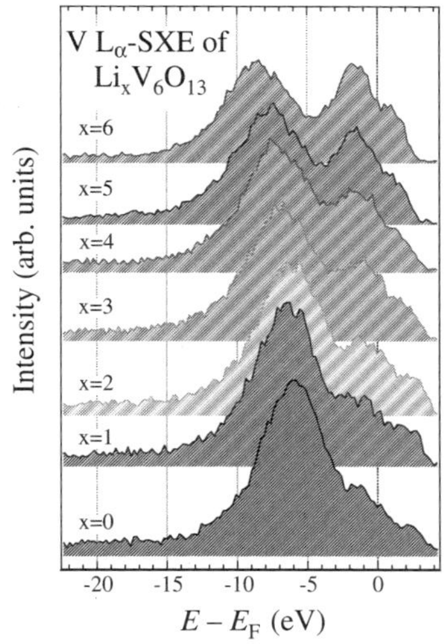

Figure 14.20 V L_{α}-emission of $Li_xV_6O_{13}$ excited at the L_3-resonance shown on a scale relative to the top of the valence band (Fermi level). Spectra for 7 different Li concentrations ($x = 0 - 6$) are compared (with vertical offset).

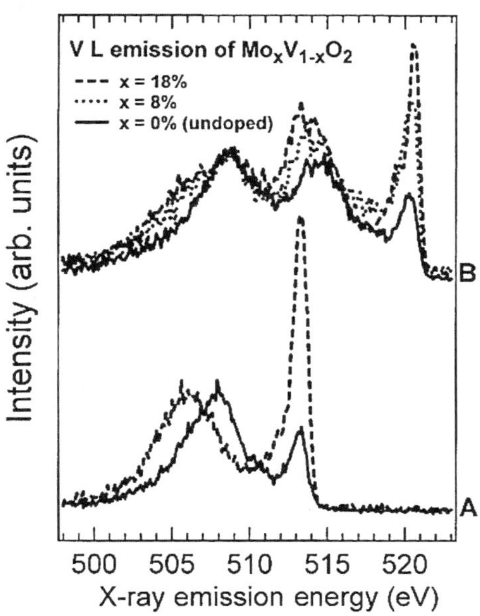

Figure 14.21 (A) V L_{α}-emission excited at the L_3-resonance and (B) V L_{β}-emission of $Mo_xV_{1-x}O_2$ excited at the L_2-resonance. Spectra for several different Mo-concentrations are compared.

14.5.3.4 Magnetic Dichroism

Magnetic circular dichroism in soft X-ray absorption (SXA-MCD) is nowadays a standard technique for determining the atomic specific magnetic moments in ferromagnets. Magnetic circular dichroism in soft X-ray emission (SXE-MCD) for iron was predicted by Strange et al. [86]. With SXE-MCD it had been hoped that a method to determine the spin-specific density of states would develop. Experimentally, Hague et al. [87] showed that the spectral shape of Fe $L_{\alpha,\beta}$ spectra excited with circularly polarized bending magnet synchrotron radiation indeed depended on the relative alignment of sample magnetization and the photon helicity. Moreover, the difference spectrum resembled the difference of the spin-specific density of states. On the other hand, the asymmetry was only on the order of 1%. Energy-dependent excitation of thin film samples of Fe, Co, and Ni with circularly polarized X-rays then showed sensitivity of the spectral shape on the chosen excitation energy and the sample thickness, due to the coupling of SXA-MCD and SXE-MCD in the scattering process. Subsequent SXE-MCD investigations studied the dichroic differential self-absorption effect in more detail [88] and the conclusion was very similar: In order to reduce unwanted geometrical effects, it is essential to use as grazing incidence as possible and an exit angle perpendicular to the magnetization. Only excitation at energies where the absorption cross-section is virtually helicity-independent can eliminate effects from SXA-MCD.

This knowledge notwithstanding, it has been difficult to understand the discrepancy in the theoretically predicted magnitude of the dichroic asymmetry and the experimental results. This problem has been addressed in several publications: Duda [89] shows a comparison between the measured dichroic asymmetry for different atomic species (Mn, Fe, Co, Ni in different compounds/metals) and a simple model assuming that screening reduces the asymmetry. Nevertheless, Mn with the originally half-filled 3d-shell would still show a 15% (30%) asymmetry in the L_α-(L_β-)spectra, whereas experimentally only about 8% asymmetry is observed (Figure 14.22). Kuiper [90] pointed out the importance of taking into account the spin orbit coupling that renders spin no longer as a good quantum number, but that one should instead consider the projection of the j-states onto the magnetization axis (m_j). This reduces the discrepancy considerably but not completely. Therefore, it is likely that the original polarization of the core-hole is subject to rapid depolarization and/or there exist fast spin-flip processes in the valence state that reduce the asymmetry. The latter process can only be significant in metallic systems where spin-flip transition energies across the Fermi level are negligible und therefore virtually undetectable in the spectra. On the other hand, in an insulator with significant excitation gap this channel would be quenched and such an experiment would help decide which effect is of greatest significance for the reduction of SXE-MCD asymmetry. A recent unpublished work on Mn-ferrite by Magnuson et al. shows that in an insulator there is also a large loss in core-hole polarization that cannot be explained by valence electron rearrangements.

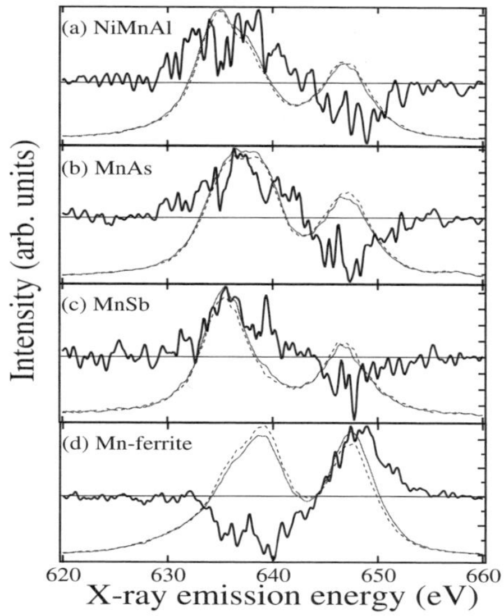

Figure 14.22 Four panels showing dichroic Mn $L_{\alpha,\beta}$-emission spectra (full and dashed lines, respectively) and the corresponding difference spectrum (MCD, heavy lines). (a) NiMnAl, (b) MnAs, (c) MnSb, (d) $(MnZn)Fe_2O_4$ (Mn-ferrite).

Heusler alloys are interesting magnetic materials in which the majority spin 3d-band is almost completely occupied and the minority spin 3d-band is unoccupied with an energy gap at the Fermi level. This leads to an unusual half-metallic ferromagnetic state, which is of considerable interest for spintronics applications. SXE-MCD experiments on Mn-based Heusler alloys indeed show unusually large asymmetries for certain energies [91]. This is found to be a consequence of the half-metallic electronic structure. Other SXE-MCD studies concern, for instance, magnetic impurities [92], ferromagnetic compounds [93], and multilayers [94] as well as the use of other experimental geometries (perpendicular geometry) that promise the extraction of information about higher order magnetic multipole moments [93].

It has been shown that specular reflectivity close to the L-absorption resonances of ferromagnets, e.g., Co [95] or Fe $L_{2,3}$-resonances, is strongly helicity-dependent. Based on this type of magnetic circular dichroism Duda et al. [96] proposed and built a prototype of a helicity-resolving X-ray spectrometer. This was made by exchanging the ordinary entrance slit by a knife edge close to a magnetized iron mirror (magnetic Codling slit) which served both the purpose of an entrance slit and a helicity analyzer (Figure 14.23). Figure 14.24 shows the response of such a spectrometer when off-plane elliptically polarized bending magnet synchrotron radiation is shone onto the entrance slit. Reversal of the magnetization of the mirror leads to a difference in reflectivity (see the bottom curve of Figure 14.24).

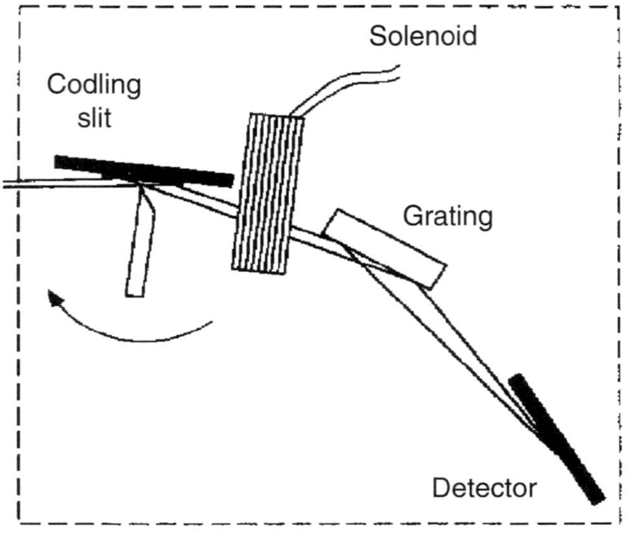

Figure 14.23 Schematic of a helicity-resolving X-ray spectrometer. The Codling slit is a knife edge close to a magnetized iron mirror. The solenoid behind the mirror can be pulsed to switch the magnetization to the desired direction.

There is a considerable difference between 705 eV and 715 eV, which can be used to analyze the Fe L-emission line of a suitably magnetized iron sample. The trade-off that one has to do between magnetic contrast and specular reflectivity, and hence the count rate, is significant and therefore it is a flux-demanding experiment. However, future availability of very much increased photon fluxes from free electron lasers could compensate for this drawback and enable a wider use of this technique.

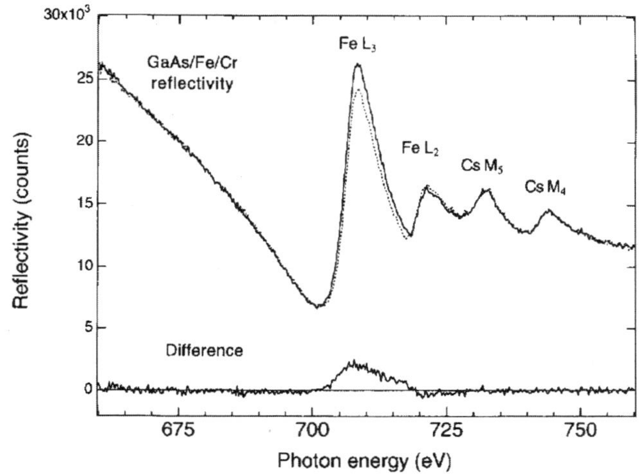

Figure 14.24 Circular dichroism of the specular reflection at the Fe L-resonances from a thin iron film. The top spectra are the raw data and the bottom spectrum is the corresponding difference spectrum (MCD).

14.5.3.5 Nitrides

Group III nitrides are wide band-gap semiconductor materials which are of considerable technological interest. The range of the band gaps in InN, GaN, AlN is 1.9–6.2 eV, which corresponds to photon energies covering the entire visible to ultraviolet parts of the electromagnetic spectrum. Therefore, knowledge of the electronic structure of valence and conduction bands of these materials is of great importance. Compared to photoemission spectroscopy (PES), which is a common method used to study the valence states of materials, SXES has the advantage of directly probing the bulk material. SXES is less prone to suffer from surface-related ambiguities in the interpretation than the highly surface-sensitive PES, especially since cleaning procedures can be time-consuming and intricate.

Combined SXAS and SXES studies of pure GaN [97] and the alloy system $Al_xGa_{1-x}N$ [98] showed early on how valuable information about the electronic structure can be extracted. Due to the atomic and orbital sensitivity of either methods the spectra can be interpreted as partial occupied (SXES, Figure 14.25) or unoccupied (SXAS, Figure 14.26) DOS. By plotting the onset of the high energy edge of the N K-emission spectrum and the low energy onset of the N K-absorption spectrum, one can extract information on the band gap behavior as a function of the relative concentration of Al and Ga in the $Al_xGa_{1-x}N$ system, and it was found that the band gap essentially has a linear dependence with no detectable band bowing. Hybridization effects allow the determination of binding energies of important shallow core-level states such as Ga 3d-states, which are found some 19 eV below the top of the valence band maximum of $Al_xGa_{1-x}N$, independent of the Al concentration (Figure 14.27). Similar hybridization features were investigated recently in $In_xGa_{1-x}N$ [99] as well as post-transition metal oxides [100]. Polarization-dependent SXAS is expected to show a strongly anisotropic behavior of the bonding in single crystalline nitride films.

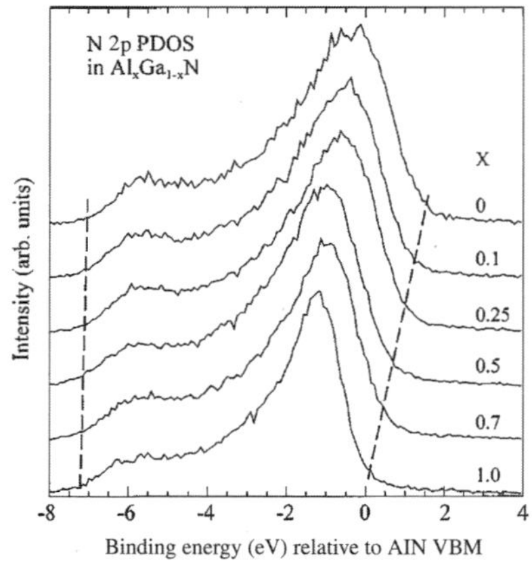

Figure 14.25 N K-emission of $Al_xGa_{1-x}N$ of different Al concentrations x shown on a scale relative to the valence band maximum. Six different Al concentrations (x = 0 –1) are compared. Dashed lines indicate the approximate extent of the N 2p-band and emphasize the movement of the top of the valence band as a function of Al concentration.

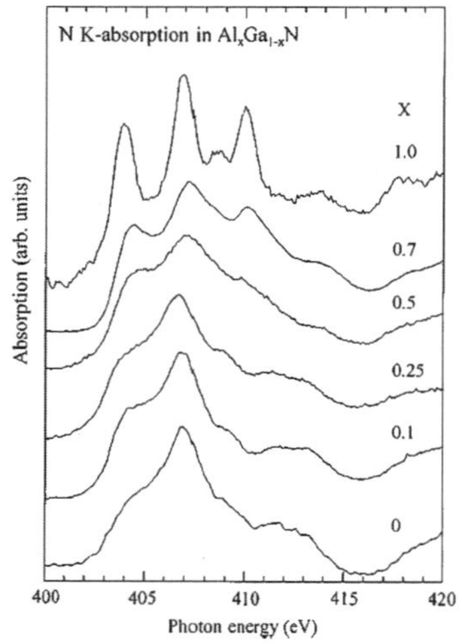

Figure 14.26 N K-absorption of $Al_xGa_{1-x}N$ of different Al concentrations x. Spectra for 6 different Al concentrations (x = 0−7) are compared (with vertical offset).

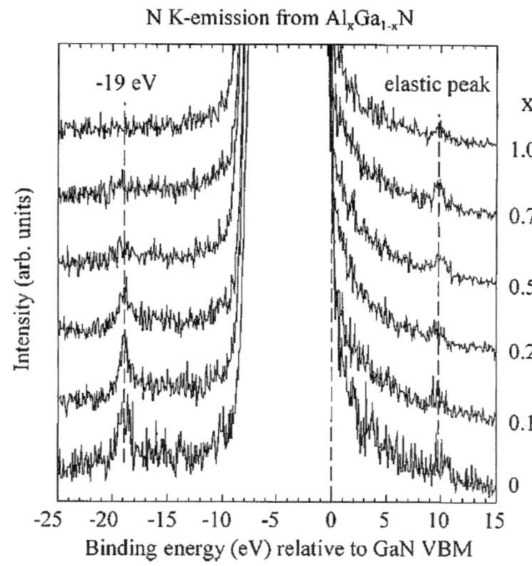

Figure 14.27 Blown-up view of the N K-emission of $Al_xGa_{1-x}N$ of different Al concentrations x shown on a scale relative to the valence band maximum. The elastic peak contribution and the Ga 3d-hybridization peak at 19 eV below the VBM are marked.

14.5.3.6 Lanthanides

Since the nature of the ground state defines various physical properties of a system, knowing the ground state character is a key issue, especially in materials science. It turns out that for rare earth compounds the RIXS technique can provide information about J-mixing in the ground state through studies of intra-atomic f→f excitations.

When the symmetry in a solid is not spherical, the angular momentum of f electrons is not conserved, and therefore J is not a good quantum number. However, the crystal field is an order of magnitude smaller than the spin-orbit splitting and can be treated as a perturbation. In the case of weak hybridization effects, the interlevel coupling and consequently J-mixing in the ground state of the system are often disregarded in the interpretation of experimental data by the application of a pure atomic approximation (mainly for high-energy spectroscopies) or by using a first-order crystal field theory where the crystal field interaction is assumed to act only within the separate J manifolds. This is partly due to complications in extracting information about the ground state J-mixing directly from the data. For example, the estimation of the J-mixing degree in high order crystal field theory by adjusting the crystal field parameters from the fit of optical absorption or low energy electron-energy-loss spectra [101, 102] may result in a large uncertainty originating from difficulties in calculating the intensities of dipole-forbidden transitions. In turn, the possible influence of the weak metal-ligand hybridization is difficult to analyze quantitatively in the absence of so-called charge transfer satellites in high energy spectroscopic data.

In this situation, the use of alternative spectroscopic means to obtain ground state J-mixing information is essential. Recently, Finazzi et al. [103] showed that this mixing could be studied by taking advantage of dichroic properties of rare earth 3d X-ray absorption. However, the method is limited to magnetically ordered systems. Here, we discuss the potential of resonant X-ray scattering spectroscopy in studying of the ground-state J-mixing when applied to compounds without distinct, long-range, magnetic order and significant metal-ligand hybridization.

Similar to optical absorption and electron-energy-loss spectroscopy with respect to probing the optical scale excitations, RIXS at the same time provides an additional level of the transition selectivity due to the element specificity and dipole selection rules. In contrast to systems with strong metal-ligand hybridization where the charge-transfer process leads to appearance of intense lines in RIXS spectra as a result of interionic excitations, J-mixing in systems with weak hybridization effects is expected to manifest itself in an intensity gain of some intra-ionic (f→f) transitions which are disallowed for the pure Hund rule ground state. In other words, transitions with ΔJ other than 0, ± 1, and ± 2 are probed in the resonant excitation-deexcitation process. Although J is not a good quantum number in the J-mixing case, we use this terminology for simplicity.

A discussion about the RIXS potential to probe the ground state J-mixing is based on the analysis of data obtained at the Dy 4d threshold of DyF_3 at room temperature. Measurements at the 4d threshold of rare earths provide naturally higher resolution than those at the 3d threshold, thus allowing one to study elementary excitations in greater detail (see, e.g., [104]). Experimental spectra of DyF_3 are displayed in Figures 14.28 and 14.29 on, respectively, photon-energy and energy-loss scales. Two distinct groups of pronounced inelastic-scattering peaks are observed in these spectra. The first group is distinguished by small energy losses on the tail of the elastic line, whereas the second is characterized by energy losses of more than 2.5 eV. When the excitation approaches the energy of the main broad maximum of the Dy 4d absorption edge, the first group still possesses significant intensity while the structures of the second group become relatively faint. Regarding the energy scale on which the spectral variations occur, the observed transitions can be attributed to intra-ionic f→f excitations. The energy gap between two groups of inelastic X-ray scattering structures reflects the separation between sextets and quartets of trivalent Dy [106, 107], which can be reached due to the excitation-deexcitation process.

Indeed, the overall spectral shapes and their behavior with varying excitation energies are reasonably well-reproduced by atomic multiplet calculations for the Dy^{3+} ion [108]. The results of calculations show that the dominant elastic peak in all of the RIXS spectra of DyF_3 is to a large extent a consequence of strong interference effects in the intermediate state of the coherent second-order optical process. The states constituting the main 4d absorption edge have a lifetime broadening of about 2 eV largely because of the 4d-4f4f Coster-Koenig decay.

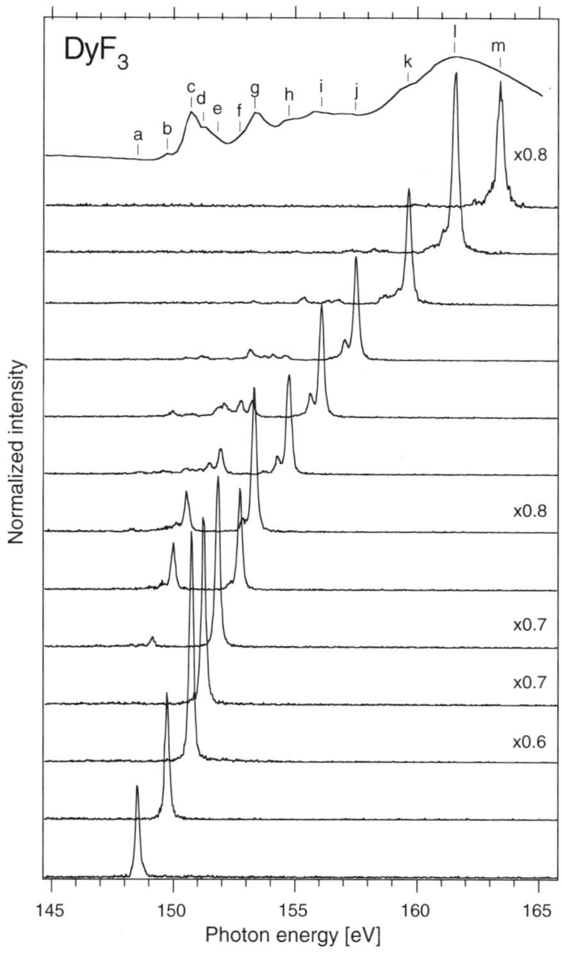

Figure 14.28 The total electron yield spectrum at the Dy 4d edge and resonant X-ray scattering spectra of DyF_3 normalized to the incident photon flux [105]. The letters correspond to the excitation energies indicated in the absorption spectrum.

However, a close inspection of experimental RIXS spectra shows that there are some spectral structures that are not revealed in calculations within the pure atomic approximation. Thus, the feature with the energy loss of about 1.17 eV is observed in spectra *h*, *k*, and *l*, presented in detail in Figure 14.29. While atomic multiplet theory predicts the nonzero intensities for resonant inelastic X-ray scattering transitions to the $^6H_{13/2}$ and $^6H_{11/2}$, and $^6F_{11/2}$ sextets of the $4f^9$ configuration (the Hund rule's ground state is $^6H_{15/2}$), the energy of the extra-feature in experimental spectra *h*, *k*, and *l* is close to those of $^6F_{9/2}$ and $^6H_{7/2}$ manifolds of Dy^{3+} in LaF_3 [109]. This is an indication of J-mixing and the presence of J = 13/2 and J = 11/2 components in the ground state of DyF_3. Indications of other extra-structures missing from atomic multiplet calculations can be seen in the energy loss range between 1.2 and 2.0 eV, as in spectra *l* and *m*.

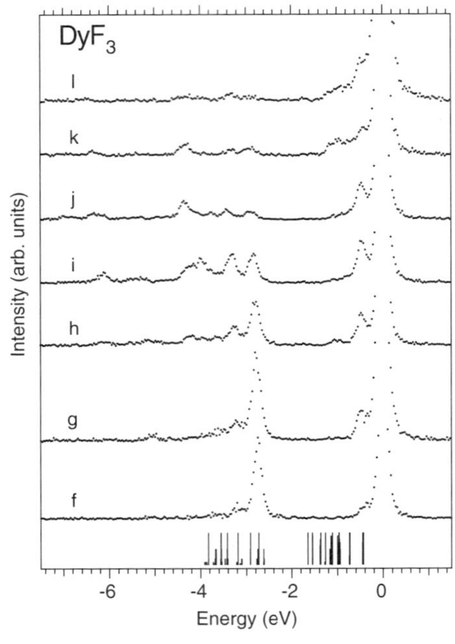

Figure 14.29 RIXS spectrum of DyF_3 excited at various energies across the 4d edge, and plotted on an energy loss scale.

To simulate the effect of J-mixing, we calculated transition intensities within the pure atomic approximation by choosing $^6H_{13/2}$ and $^6H_{11/2}$ as initial states for the scattering process. The excitation energy was taken to be the same as that for experimental spectrum *l* (since the intensity of extra-structures is higher for excitation energies at the main absorption edge) and the core hole lifetime broadening was set to 2.0 eV. According to these calculations, the spectral weight at an energy loss of about 1.17 eV (transitions to $^6F_{9/2}$ and $^6H_{7/2}$) should constitute 56% and 217% of that at 1.0 eV (transitions to $^6F_{11/2}$) for $^6H_{13/2}$ and $^6H_{11/2}$ as initial states, respectively. Although accounting for weak but finite Dy 4f-F 2p hybridization may give rise to some changes in calculated relative intensities in both cases, it is clear that an admixture of the J = 11/2 component in the ground state results in a stronger effect than for the J = 13/2 component. Since the degree of *J*-mixing in DyF_3 is relatively low, the J = 11/2 contribution to the ground states is expected to be comparable with the J = 13/2 contribution in order to explain the noticeable weight of "forbidden" transitions in the inelastic scattering spectra in Figure 14.29. This is not unusual. For example, the J = 11/2 component has been found to be comparable to the J = 13/2 component in the ground state of Dy^{3+}-doped yttrium scandium gallium garnet [102] as a result of the crystal field interaction.

To estimate the effect of this interaction on the shape of RIXS spectra, model crystal field multiplet calculations for the Dy^{3+} ion in the crystal field of O_h symmetry with a strength of 35 meV were made. Figure 14.30 shows the 1.0–2.2 eV energy-loss region of the spectra calculated using a pure atomic approximation and crystal field multiplet theory. It is clear that switching on the crystal field gives rise to additional transitions. The calculated intensities are too low to fully account for the observed spectral weight in experimental data at the corresponding energy loss. This suggests that interatomic coupling is also important for the description of the inelastic-scattering profile in the energy loss range between 1.0 and 2.0 eV and that the appearance of additional structures is rather a combined effect of the crystal field interaction and Dy 4f-F 2p hybridization. However, calculations taking into account both the crystal field and F 2p→Dy 4f charge transfer excitations are complicated by a huge number of multiplets and require very large computational resources.

The existence of J-mixing in the intermediate state raises a question about how strongly the inelastic scattering intensity at energy losses between 1.0 eV and 2.0 eV is related to J-mixing in the ground state of DyF_3. To estimate this, crystal field multiplet calculations with the crystal field interaction switched off in the intermediate state have been performed. Thus, any J-mixing in the core excited state was disallowed. A comparison of the results of calculations with, respectively without, crystal field interaction in the intermediate state (Figure 14.30) does not reveal any significant changes in the inelastic scattering intensities of "forbidden" structures on switching off J-mixing in the core excited state. One of the main reasons for that is a large core hole lifetime broadening. As a whole, the calculations indicate that the spectral weight in the energy loss region between 1.0 and 2.0 eV is largely determined by J-mixing in the ground state of DyF_3.

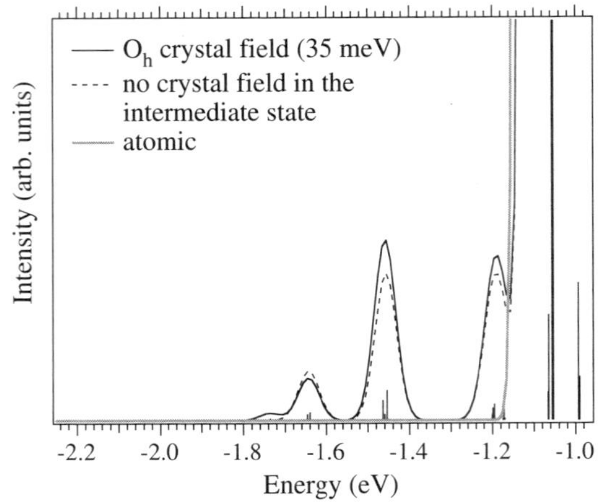

Figure 14.30 A 1.0–2.2 eV energy-loss region of inelastic X-ray scattering spectra of the Dy^{3+} system calculated using a pure atomic approach (intermediate coupling) and crystal field multiplet theory (O_h symmetry). In the calculations Slater integrals were scaled down to 80%, 75%, and 66%, respectively, from the Hartree-Fock values. Excitation energy was set to the main absorption maximum (spectrum *l* in Fig. 14.28) and the core hole lifetime broadening was taken to be 2.0 eV.

14.5.3.6 Actinides

The case of uranium is quite illustrative for what the RIXS technique can do. RIXS measurements at the U 5d threshold provide an opportunity to study in detail elementary excitations in U compounds due to the higher resolution of such experiments in comparison with those at the U 3d and 4d thresholds. It turns out that the technique, in contrast to X-ray absorption spectroscopy, is very sensitive to the valency and the chemical state of uranium. Figure 14.31 shows X-ray absorption spectra of a number of U compounds recorded across the U 5d edge in the total electron yield mode. The 5d core hole lifetime broadening is quite large, thus reducing the usability of XAS.

As a result, the U 5d absorption spectra in the figure do not exhibit sharp features. The substantial smearing of spectral structures hampers the analysis of the chemical state and the chemical environment of uranium in various compounds. In particular, it is difficult to distinguish between uranium species with different oxidation states, especially in the case when one of the species has a much lower concentration than other.

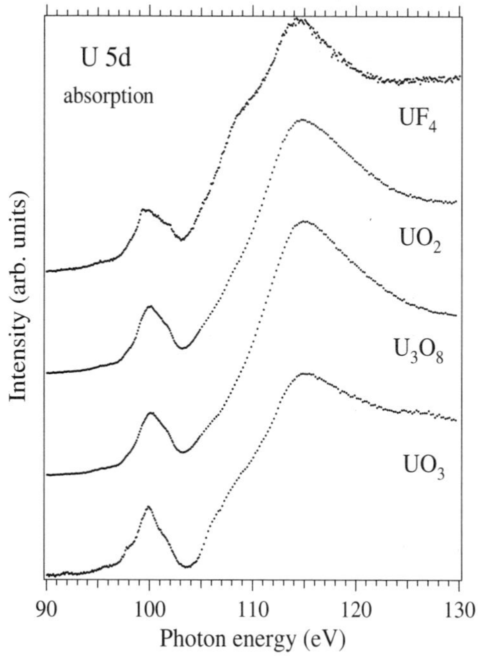

Figure 14.31 Total electron-yield spectra of UF_4, UO_2, U_3O_8, and UO_3 near the U 5d absorption edges.

In this situation, the principally unlimited resolution (defined by the excitation sharpness and the response function of the instrument) of the RIXS technique and its ability to enhance transitions to specific excited states are especially useful. The RIXS profile strongly depends on the choice of the excitation energy. The spectral weight corresponding to electronic transitions within the 5f shell is enhanced at excitation energies close to 100 eV, while at higher energies of the incident photon beam set to the main 5d absorption edge, interionic excitations of charge-transfer character, such as ligand 2p→U 5f charge-transfer, dominate the RIXS spectra. This has been established by measurements on a set of model U compounds and by model calculations. The spectral pattern of intra-ionic f→f excitations is mainly determined by the formal valency of U; in turn, the charge-transfer transitions strongly depend on the chemical environment of U ions.

An example of probing the f→f excitations in U systems is illustrated in Figure 14.32 where the RIXS spectra of solid UF_4, recorded for various incident photon energies in the pre-5d threshold region, are displayed. The assignment of sharp inelastic scattering structures to the f→f transitions is supported by atomic multiplet calculations for the U^{4+} ion. The spectra were calculated using a model where the varying lifetime of core excited states due to the autoionization via the 5d-5f5f super Coster-Koenig decay was taken into account. The autoionization into the continuum of *g* symmetry only was considered since it is the most dominant path. Matrix elements were obtained from Cowan's programs [110] so that Slater integrals were scaled down to 75%, 75%, 66%, and 80%, respectively, from the Hartree-Fock values. The reduction factors for Slater integrals represent the effect of the intra-atomic configuration interaction; the same factors were used in a number of publications to successfully reproduce 4d X-ray absorption and 4d photoemission spectra of various rare earth systems [111, 112, 113]. The density of states of the continuum was assumed to be constant and the kinetic energy of the continuum electron was set to the value that made the average energies of $5d^9f^3$ and $5d^{10}5f^1\,\varepsilon g$ equal.

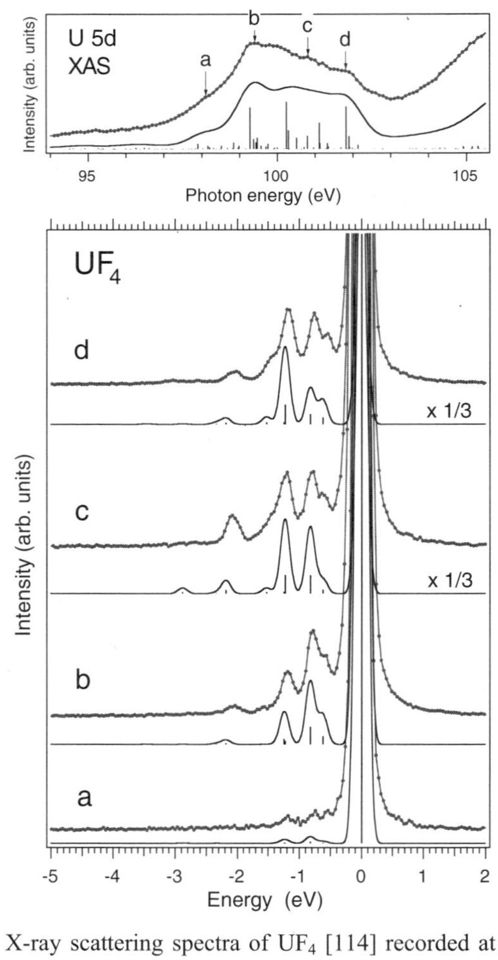

Figure 14.32 Resonant X-ray scattering spectra of UF_4 [114] recorded at different excitation energies close to the U 5d threshold (lines with markers) together with the results of atomic multiplet calculations (sticks with thin lines) for the U(IV) ion. Excitation energies are indicated by arrows on the total electron yield spectrum at the U 5d absorption edge shown in the top panel.

The calculations reproduce all of the spectral structures very well, especially the enhancement of the peak at about 1.2 eV with increasing excitation energies. The growth of the peak is due to enhanced transitions into the $^1G_4$ state. Changes in absolute intensities of inelastic scattering structures corresponding to the f→f transitions are reproduced on going from spectrum *a* to spectrum *b*. For spectra *c* and *d*, such changes in calculated intensities are about three times higher as compared to those in the experiment. The discrepancy may originate from the normalization procedure for the experimental spectra to account for variations in the incident photon flux. The intensity of the elastic peak was used as a reference in this procedure. However, the elastic peak contains some contribution of diffuse scattering, which may vary with varying excitation energies.

RIXS profiles, corresponding to the f→f excitations, are found to be very sensitive to the chemical state of U in different systems [114]. For example, it is a matter of the presence or absence of these excitations when going from U^{4+} to U^{6+} compounds. Therefore, RIXS measurements near the U 5d threshold provide good fingerprints for the chemical state of U in different systems in contrast to X-ray absorption spectra, which show only small differences at the U 5d edge.

One interesting example concerns the chemical state of U in U_3O_8 and reduced UO_3. In particular, for U_3O_8, it was discussed in the literature that the chemical state of U is described as either $U^{VI}U^{V,2}O_8$ or $U^{VI,2}U^{IV}O_8$. To our knowledge there is no clear and convincing answer to this question and it is still under debate. Thus, core level photoemission [115, 116, 117] and electron spin resonance [118, 119] data have been interpreted in favor of either situation by different groups, and therefore these techniques cannot really provide an unambiguous answer. It turns out that the RIXS technique can. RIXS spectra of f→f excitations are a good indicator of whether uranium is in the U(IV), U(V) or U(VI) state. The data are easy to interpret and not very difficult to calculate. Establishing the real chemical state of U in oxides is important for applied, environmental and fundamental science (e.g., a development of the theory of nonstoichiometry is a fundamental problem).

Figure 14.33 displays RIXS spectra of f→f transitions for a number of U oxides. Close similarity of the f→f transitions profile in U_3O_8 to those in UO_2 and UF_4, as well as to that calculated for the U(IV) ion (see Figure 14.32), unambiguously indicates the presence of the U(IV) fraction in U_3O_8, thus favoring the $U^{VI,2}U^{IV}O_8$ description of the uranium chemical state in this oxide. Furthermore, a similar pattern is observed for reduced UO_3, thus indicating that oxygen deficiency leads to the creation of U(IV) species in the compound.

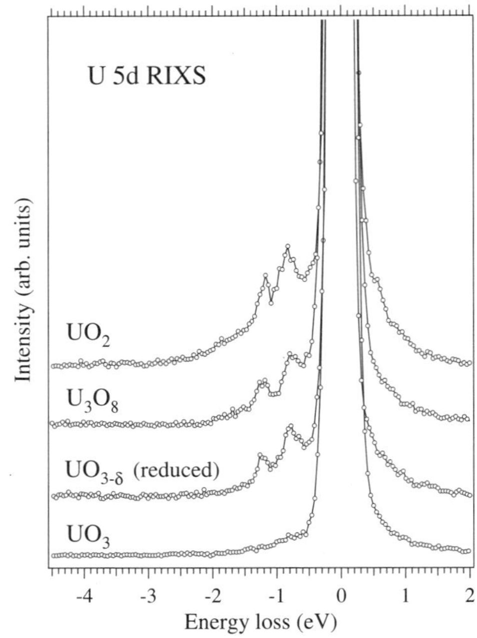

Figure 14.33 Resonant inelastic X-ray scattering spectra of U oxides recorded at the incident photon energy of 99.9 eV.

Charge-transfer effects are expected to be significant in actinide compounds as a result of metal 5f-ligand 2p hybridization. The analysis of our data [114] obtained at the U $3d_{5/2}$ threshold shows that the ligand 2p → U 5f charge-transfer plays an important role in uranium compounds, such as UO_2, $UO_2(NO_3)_2 \times 6H_2O$, and even in UF_4.

This is also supported by theoretical studies. Molecular orbital calculations by several research groups [120, 121, 122, 123] gave values for the 5f occupancy, which range from 2.3 to 2.9 electrons, while this occupancy was estimated at about 2.3 electrons from the analysis of X-ray absorption and photoemission data within an Anderson impurity model [124, 125]. These results indicate a significant degree of covalency for U–O chemical bonds in UO_2. For UF_4, a 5f contribution of ~0.3 electron to the bonding orbitals was also predicted from relativistic Dirac-Slater local density calculations [126]. For compounds containing U^{6+}, the degree of covalency for metal-ligand bonds is expected to be even higher than that for U^{4+} systems. For example, molecular orbital calculations yielded the 5f occupancy of ~2.6 electrons for the uranyl ion $UO^{2+,2}$ [127, 128]. Although the values for the 5f occupancy obtained from molecular orbital calculations seem to be overestimated [129], one cannot rule out the importance of the U 5f-ligand 2p hybridization even in a compound with "ionic" bonds such as UF_4.

The consequence of high covalency and hybridization in the ground state is an appearance of charge-transfer satellites in high energy spectroscopic data. Figure 14.34 illustrates the dependence of the RIXS spectra on the chemical environment of U atoms in various compounds.

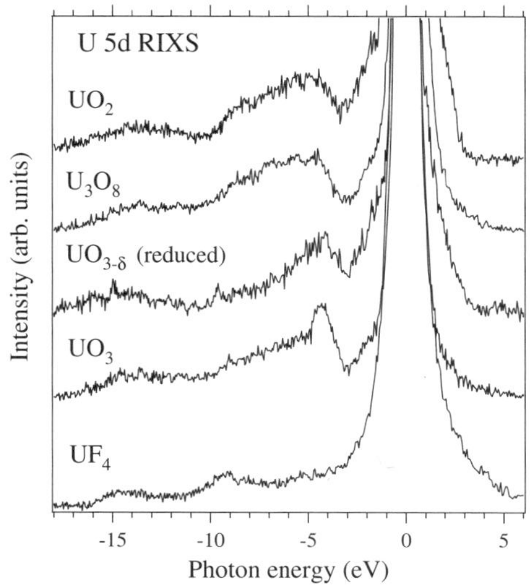

Figure 14.34 Resonant inelastic X-ray scattering spectra of a number of U compounds recorded at the incident photon energy of 115.0 eV.

The spectra were measured under conditions when the inelastic RIXS cross-section is enhanced for the ligand 2p → U 5f charge-transfer transitions, in particular, when the energy of the incident photon beam is tuned to the main edge of the U 5d X-ray absorption spectrum (see Figure 14.31). At these energies, the charge-transfer excitations dominate RIXS spectra, thus defining the RIXS profile. The profiles can be clearly divided into three groups: (i) UF_4, (ii) UO_3 and UO_{3-d}, (iii) U_3O_8 and UO_2. The differences between profiles are determined by the character of the bonding and the local geometrical arrangement of ligands, i.e., local crystal structure.

14.5.3.7 Molecular Systems and Liquids

Free Molecules: During the last decade, it became possible to measure sub-keV X-ray emission of molecules in gas phase using excitation by monochromatized synchrotron radiation. This was demonstrated by [130], which presented resonant X-ray spectra of free molecular oxygen [130], where the results gave conclusive proof that parity is conserved in the two-photon, absorption-emission X-ray process. Cases where vibronic coupling can lower the symmetry upon core excitation have been observed, notably CO_2. It has been demonstrated that by detuning the excitation energy down from resonance one effectively restores the symmetry by virtue of an effective shortening of the X-ray scattering time [131]. Further, the symmetry-selective character of this spectroscopy, spectator shifts, angular anisotropy, vibronic coupling leading to symmetry breaking, and different types of interference effects in the resonant and nonresonant spectra of molecular N_2, O_2, CO, CO_2 have been studied. Some of the molecular studies serve as illustrative examples of the properties of resonant soft X-ray emission spectroscopy.

Figure 14.35 shows the nitrogen K emission spectra recorded at various excitation energies, which resonantly promote a core electron to various unoccupied levels of different symmetry. The top spectrum represents the nonresonant case where continuum states above the ionization threshold are excited. These are infinitely degenerate with respect to all symmetries. The spectrum shows three bands, representing emission from the three outermost valence orbitals of both *gerade* (g) and *ungerade* (u) symmetry. When, on the other hand, the excitation is to a level with a defined symmetry the emission spectrum appears different, not showing emission from all three levels. Instead, two, or even a single band, are observed in this case. When exciting to a Rydberg state of *ungerade* symmetry, the $3p\pi_u$, only emission from the two *ungerade* levels are observed. The $3s\sigma_g$ excitation leads to a single emission band, which corresponds to the $3\sigma_g$ state. This also holds for the $1\pi_g$ excitation where the $3\sigma_g^{-1}1\pi_g^{1}\pi_g$ final state is seen. In that spectrum, the high energy band at ~400.5 eV is the so-called *participator transition*, where the excited electron fills the core hole. One can note energy shifts of the $3\sigma_g$-related emission in these spectra that are caused by the presence of the excited electron in the intermediate and final state. The solid lines in Figure 14.35 represent simulations based on potential energy curves taken from the literature and lifetime vibrational interference theory describing the vibrational bands [132].

For molecules having inversion symmetry experimental spectra obey a $g \rightarrow g$ selection rule with respect to unoccupied (respectively, occupied) orbitals [130, 132, 133, 134, 135, 136]. In a molecular orbital picture describing also the core orbitals this comes out naturally because the dipole nature of the transitions implies that the parity has to change at the absorption or the emission of a photon, e.g., $g \rightarrow u$ or $u \rightarrow g$, as described earlier. In the full absorption-emission transition we have a $g \rightarrow u \rightarrow g$ or $u \rightarrow g \rightarrow u$ rule, which explains the observations for N_2. For polyatomic molecules, where vibrations of non-total symmetry are possible, selection rules are observed to relax because of dynamical (vibronic) symmetry breaking in the intermediate core state. This effect can be observed in one of the simplest molecules of this kind, CO_2, where only two non-totally symmetric vibrations are present [131].

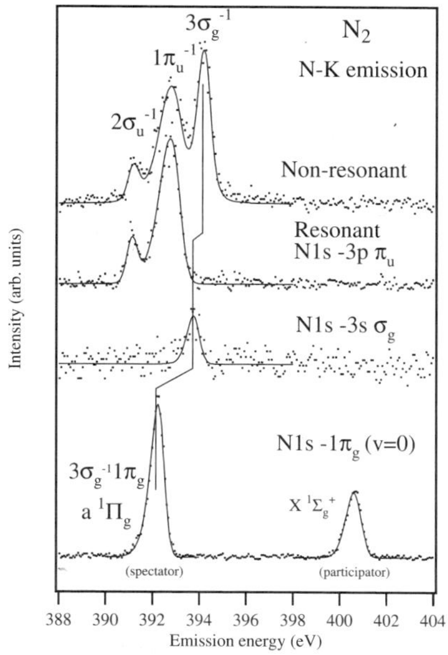

Figure 14.35 X-ray emission spectra of molecular N_2. From the top, nonresonant spectrum, resonantly excited to the Rydberg $3p\pi_u$ level, $3s\sigma_g$ and to the $1\pi_g$ valence level, respectively. Note the *g-g* selection rule, which holds for the resonant spectra.

The strict symmetry selection rules appropriate for a particular molecule as described above can in some cases be broken [137]. For larger molecules symmetry-forbidden transitions have been observed [131, 138]. As these molecules possess a vast number of vibrations, the results are difficult to describe in detail. Just as N_2 or O_2 are good cases for demonstrating that the parity selection rule is valid, CO_2 is as good a candidate to study the effect of symmetry breaking. The central carbon atom between the two oxygen atoms adds an antisymmetric stretch mode that can couple the *gerade* and *ungerade* oxygen core excited states and therefore break the parity selection rule [137].

Figure 14.36 shows resonant and nonresonant oxygen K emission spectra of CO_2 and O_2. The resonant excitations are to the first empty molecular orbitals, $2\pi_u$ and $1\pi_g$ respectively. In the O_2 case only symmetry-allowed transitions are observed [130], in a similar way as was described for N_2 [132], i.e., only transitions that give *gerade* final states. The resonant O_2 spectrum consists of only two bands, a spectator and a single participator. Especially the absence of the $^3\Sigma_u^-$ final state proves the validity of the selection rules.

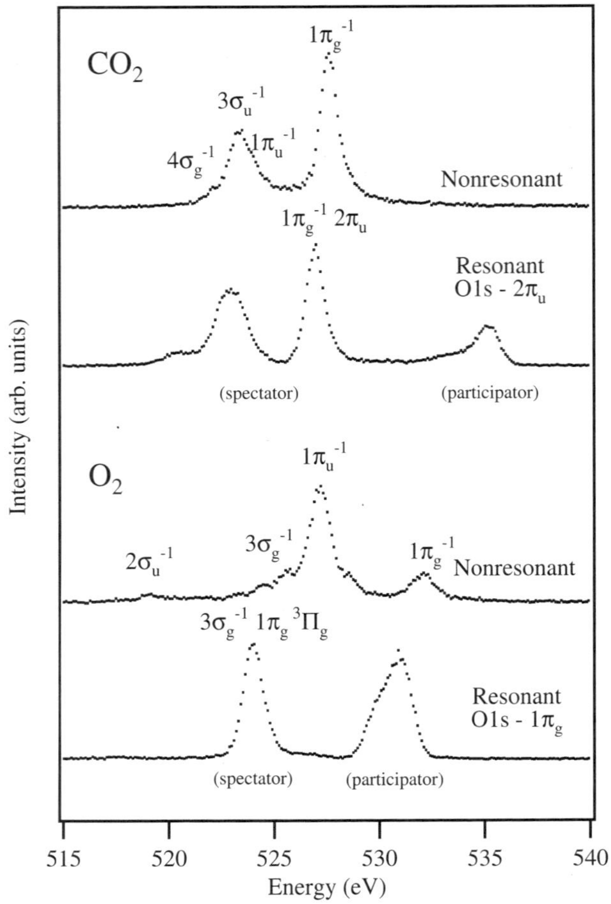

Figure 14.36 π*-excited and nonresonant oxygen K emission spectra of CO_2 and O_2 illustrating the effect of symmetry-breaking in the intermediate core excited states of CO_2. Symmetry-forbidden transitions in ground state symmetry are given intensity by non-totally symmetric vibrational excitations lowering the symmetry of the intermediate state. The resonant O_2 spectrum only shows intensity from symmetry-allowed transitions similar to the case of N_2.

In contrast to how different the nonresonant and resonant spectra of O_2 are the two CO_2 spectra are rather similar. Here, intensity from all valence orbitals is observed in both spectra, despite the symmetry-selective excitation to an *ungerade* state, providing proof that the symmetry of the intermediate, core-excited state is broken. In particular, the appearance of the high energy spectator transition involving the outer valence orbital, $1\pi_g$, would be strictly forbidden if the inversion symmetry would be retained in the core-excited state. The relative intensities between the high and low energy bands are not the same in the nonresonant and resonant spectra. The deviation can give information about the strength of the vibronic coupling and hence how strong the symmetry breaking is [131]. The fact that some effect of the selection rule pertains can be used in the determination of the parity of specific empty states probed by resonant excitation [138]. The effect of symmetry-breaking is very clear for CO_2 but observations of "forbidden" lines in other molecular systems have been made, like in C_6H_6 [136], and C_{60} [133], which can also be explained by the vibronic coupling.

It has been observed that the symmetry selection rule in polyatomic molecules is more strict when the narrow bandpass excitation is detuned below the first resonance [135, 136]. This is discussed in more detail for the CO_2 molecule [131] where the above discussed forbidden high energy band decreases monotonically with the detuning energy. Figure 14.37 shows the detuning spectra referring to resonant excitation to the first unoccupied level, $2\pi_u$, of CO_2 giving the forbidden $|g\rangle = 1\sigma_u^{-1}2\pi_u^{1}\Pi_g$ and allowed $|g\rangle = 1\sigma_g^{-1}2\pi_u^{1}\Pi_u$ core-excited states.

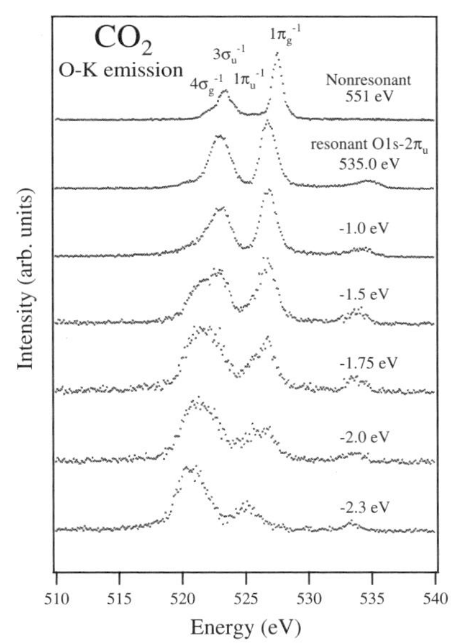

Figure 14.37 Oxygen K $2\pi_u$ resonant X-ray emission spectra of CO_2 with different detuning energies below the $2\pi_u$ resonance.

The spectra consist of two peaks; the high energy peak derives from transitions to the forbidden $|f(u)\rangle = |1\pi_g^{-1}2\pi_u\rangle^{1}\Pi_u$ final state, and the low energy peak to the allowed $|f(g)\rangle = |1\pi_u^{-1}2\pi_u\rangle^{1}\Pi_g$ final state (configuration state splitting and contribution from the weak $4\sigma_g$ emission can be neglected). As seen in the figure, the high energy peak receives significant intensity in violation of the selection rule and appears quite like the nonresonant X-ray emission spectrum of CO_2. Thus, the core-excited state cannot be described as a pure $|1\pi_g^{-1}2\pi_u\rangle^1\Pi_u$ state, but must be mixed with the $|1\pi_u^{-1}2\pi_u\rangle^1\Pi_g$ state. By detuning the energy it is clear that the spectrum becomes symmetry-purified. In a time-dependent picture this can be seen as an effective shortening of the scattering time, allowing sub-femtosecond study of dynamics.

Liquids: Our microscopic understanding of a liquid is very much based on the study of spatial and spatio-temporal correlation functions. The study of correlations allows us to appreciate the local organization of one molecule surrounded by others, and to unravel the microscopic dynamics. Neutron diffraction and scattering have played and continue to play a major role in these studies. X-ray diffraction is important as well, and recently X-ray inelastic scattering has become available. Using X-ray absorption and selectively excited X-ray emission spectroscopy to probe unoccupied and occupied electronic states, one can establish a firm interpretation for the unusual thermodynamic properties of molecular liquids. Furthermore, one can elucidate finer details of the structural properties of molecular liquids. XAS and XES spectra reflect the local electronic structure of various conformations; in this case, the oxygen X-ray absorption and emission line-shapes are sensitive to the hydrogen bonding configurations.

Water is a very abundant substance on our planet, and it is the principal constituent of all living organisms. Chemical reactions taking place in liquid water are essential for many important processes in electrochemistry, environmental science, pharmaceutical science, and biology in general. Many models have been proposed to view the details of how liquid water is geometrically organized by hydrogen bond networks. The hydrogen bond is an attractive interaction forming a link of a hydrogen atom with a highly electronegative and nonmetallic element that contains a lone pair of electrons. Although H-bonds are much weaker than conventional chemical bonds, they have important consequences on the properties of water. Diffraction of X-rays [139] and neutrons [140] provide strong evidence that tetrahedral hydrogen-bond order persists beyond the melting transition, but with substantial disorder present [141]. Important questions remain about the precise nature of the disorder and how it is spatially manifested.

The local structure of liquid water is still under debate. It has been demonstrated that soft X-ray emission spectra, emanating from the radiative decay, subsequent to core excitation, can be useful in assigning structures in X-ray absorption spectra [138, 142, 143]. Especially, it has been shown that the resonantly excited SXE measurements on liquid water are compatible with the traditional view that three and four hydrogen bonds dominate the structure [142].

An X-ray absorption study of liquid water [144] suggests that the four hydrogen-bonding networks mainly contribute to a single broad feature, while a shoulder located at 534.7 eV suggests the presence of broken hydrogen bonds. Indeed, recent theoretical simulations assign this pre-edge structure to a particular three-hydrogen bond structure with one missing hydrogen bond at the hydrogen site.

This assignment is fully confirmed by X-ray emission spectra, excited on this particular pre-edge shoulder (Figure 14.38). Different local structures provide different nonresonant spectra because of their different valence band structures. The nonresonant spectrum should be a sum of the ones from all different structures. However, since the SYM structure is the dominating structure in the liquid water, its nonresonant spectrum is close to the experimental one. Compared to the nonresonant spectrum we observe a substantial narrowing of the $1b_1$ peak at 526 eV, and a further attenuation of the $3a_1$-associated structure at 524.5 eV. This is in excellent agreement with the predictions for the DASYM configuration.

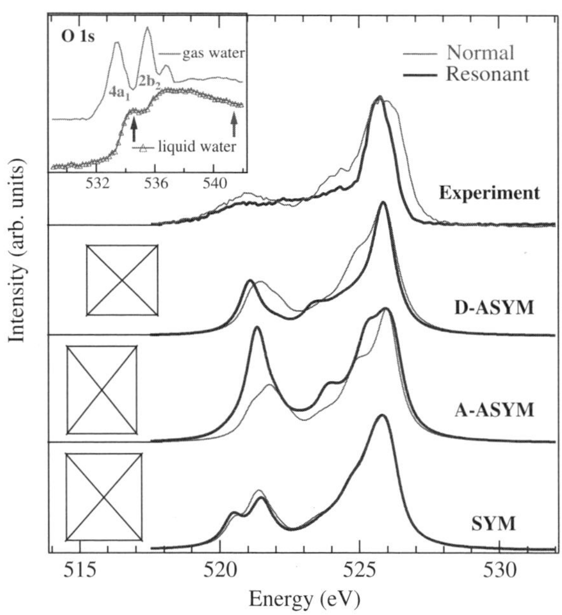

Figure 14.38 The sharp structures in the X-ray absorption spectrum of the water molecule (inset) are smeared out in the liquid water due to the diffuse nature of the corresponding molecular orbitals. A pre-peak is resolved at 534.7 eV in the spectrum of liquid water. The arrows indicate the photon energies used for resonantly and normally excited X-ray emission spectra. The resonantly excited spectrum is in good agreement with the prediction for the three-bonded D-ASYM configuration.

The early hypothesis of cyclic structures by Pauling [145] has been both supported [146] and contested [147] by neutron diffraction analysis, the competing interpretation being that the majority of liquid molecules are ordered in chains with up to 10 members [147, 148] or linear trimer-tetramer chains [149]. This uncertainty is accentuated when one considers alcohol-water solutions. The observed entropy increase upon mixing alcohol and water is much smaller than expected for an ideal solution. Early neutron diffraction data provide structure information of water cages around hydrophobic headgroups in solution [150, 151, 152, 153]. Recently, a new neutron diffraction analysis demonstrated that incomplete mixing at the molecular level is essential to understanding the smaller than expected entropy increase [154]. This contrasts with the prevailing view that a rearrangement of the water network around hydrophobic methyl groups of the methanol clusters is the main feature behind the unusual thermodynamic properties [155].

A comparison of the theoretical and experimental X-ray emission spectra (in Figure 14.39) associated with possible molecular arrangements makes it obvious that both rings and chains are present in liquid methanol. The maxima at 526 eV and 527 eV can be assigned to ring and chain structures, respectively. In addition, we find that the measured spectra are typical for rings and chains consisting of 6 and 8 molecules. Configurations with fewer than 6 molecules do not give representative predictions, and odd-numbered configurations yield an unobserved high energy feature at 528 eV. Chains with more than 10 molecules are unlikely in light of neutron diffraction data [147, 148]. Therefore, the 6- and/or 8-membered rings and chains provide the predominant structures in liquid methanol. From the excitation energy-dependence we can state that the pre-shoulder in the XA spectrum is due to ring structures (*A*); that the onset of absorption in the chains is located at the main absorption band (*B*); from the emission spectrum corresponding to high energy excitations (*C*) we estimate that the relative abundance of chains and rings in liquid methanol is equal.

Details in the water-methanol interaction are revealed in the X-ray emission spectra generated at excitation energies near threshold. The X-ray emission spectra that are excited in this region (Figure 14.40) are significantly different from the emission spectra of the pure liquids. The 0.8-eV linewidth of the main peak is narrower than any X-ray emission feature of either liquid phase: for water (1.0 eV), for methanol (1.9 eV). The main peak E_1 shifts towards higher energy as the excitation energy is lowered from 532.5 eV to 531.5 eV and 530.6 eV. We also note that the spectrum excited at 530.6 eV shows depletion of intensity in the region around 525 eV. One can conclude that specific bonded structures involving water as well as methanol are responsible for this spectral behavior.

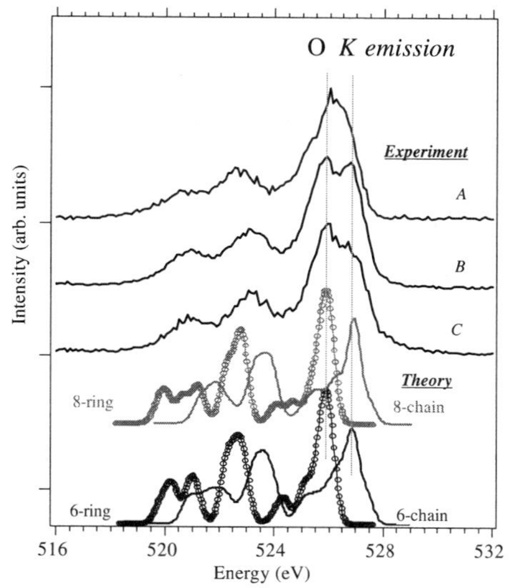

Figure 14.39 Experimental oxygen XE spectra of methanol recorded at 534.5 (*A*), 537 (*B*), and 560 eV (*C*) excitation energies compared with theoretical XE spectra of different structures: 6 – 8-chains and rings.

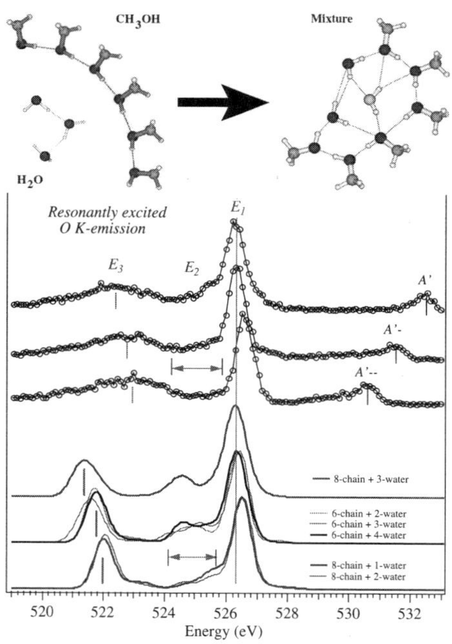

Figure 14.40 The experimental resonant XE spectra are in good agreement with the prediction for the water clusters associated with methanol 6- and 8-chain configurations. The top diagram illustrate how methanol chain interacts with water molecules to form the new mixing structure.

The narrow feature observed in the resonant emission spectra holds the key for identifying the structures. Based on the simulations of resonant emission spectra of water and methanol in both gas and liquid phases, one can conclude that this narrow spectral feature can only be generated from the emission of a water molecule that is "isolated" from the rest of the water network. The emission originates from water molecules that bridge methanol chains to form rings as illustrated in Figure 14.40.

14.6 SUMMARY

During the last decade soft X-ray emission with resonant excitation using synchrotron radiation has emerged as a new and powerful tool for studies of the electronic structure of materials. The selective probing of this spectroscopy together with its ability to probe bulk properties and buried structures, which also offers *in situ*-probing capabilities, makes it a unique method for many problems in materials science and chemical physics. In particular, the finding and recognition of the consequences—that resonant X-ray emission is essentially an inelastic scattering process (RIXS)—has led to new and important opportunities for this spectroscopic method. For example, direct probing of dd, ff and charge transfer low energy excitations in strongly correlated materials, using virtual core holes for element and site specificity without suffering from core-hole lifetime broadening, is elegantly offered by this method.

The present overview of soft X-ray emission spectroscopy applications in solid state physics and materials science included a brief description of the experimental technique, and a number of study cases from different fields. Investigations of thin films and buried structures, nano materials, transition metal and rare earth systems, as well as lanthanides and actinides were discussed. Furthermore, molecular systems were discussed in terms of basic probing properties, and, finally, applications to the study of liquids were presented and discussed.

REFERENCES

[1] Stöhr, J. (1992) *NEXAFS Spectroscopy*, Springer-Verlag Series in Surface Science.

[2] Koningsberger, D.C. & Prins, R. (Eds.) (1996) *X-Ray Absorption. Principles, Applications, Techniques of EXAFS, SEXAFS and XANES*, World Scientific Publishing Company.

[3] Krause, M.O. (1979) *J. Phys. Chem.* Ref. Data (USA), **8**, 307-27.

[4] Meisel, A., Leonhardt, G. & Szargan, R. (1989) *X-Ray Spectra and Chemical Binding*, Springer Series in Chemical Physics, *Vol. 37*.

[5] Henke, B.L., Gullikson, E.M. & Davis, J.C. (1993) *At. Data Nucl. Data Tables*, **54**, 181.

[6] Nilsson, P.O., Kanski, J., Thordson, J.V., Andersson, T., Nordgren, J., Guo, J. & Magnuson, M. (1995) *Phys. Rev. B*, **52**, 8643.

[7] Guo, J.-H., Butorin, S.M., Wassdahl, N., Skytt, P., Nordgren, J. & Ma, Y. (2000) *Phys. Rev. B*, **49**, 1376 (1994); Butorin, S. et al., *J. Electr. Spectr.* 110-111.

[8] Sakurai, J.J. (1967) "*Advanced Quantum Mechanics*, Addison Wesley: Menlo Park, Ca; T. Åberg & B, Crasemann, *Resonant Anomalous Scattering*: Theory and Applications, edited by Materlik, G., Sparks, C.J. & Fisher, K. (North-Holland, Amsterdam 1994), p. 431.

[9] Kuiper, P., Guo, J.H., Såthe, C. & Nordgren, J., et al., (1998) *Phys. Rev. Lett.* **80**, 5204.

[10] Rubensson, J.-E., Mueller, D., Shuker, R., Ederer, D. L., Zhang, C. H. & Callcott, T. A. (1988) *Phys. Rev. Lett.* **60**, 1759.

[11] Ma, Y., Wassdahl, N., Skytt, P., Guo, J.-H., Nordgren, J., Johnson, P.D., Rubensson, J.-E. Böske, T., Eberhardt, W. & Kevan, S.D. (1992) *Phys. Rev. Lett.* **69**, 2598.

[12] Carlisle, J.A., Shirley, E.L., Hudson, E.A., Terminello, L.J., Calcott, T.A., Jia, J.J., Ederer, D.L., Perera, R.C.C. & Himpsel, F.J. (1995) *Phys. Rev. Lett.* **74**, 1234; Agui, A., Shin, S., Fujisawa, M., Tezuka, Y., Ishii, T., Mishima, O., Era, K., Shigemasa, E. & Yagishita, A. (1996) *J. Electron Spectrosc. Relat. Phenom.*, **79**, 191; Miyano, K.E., Ederer, D.L., Callcott, T.A., O'Brien, W.L., Jia, J.J., Zhou, L., Dong, Q.-Y., Ma, Y., Woicik, J.C. & Mueller, D.R., (1993) *Phys. Rev. B,* **48**, 1918; Bruhwiler, P.A., Kuiper, P., Eriksson, O., Ahuja, R. & Svensson, S. (1996) *Phys. Rev. Lett.* **76**, 1761; Carlisle, J.A., Shirley, E.L., Hudson, E.A., Terminello, L.J., Callcott, T.A., Jia, J.J., Ederer, D., Perera, R. & Himpsel, F. (1996) *Phys. Rev*. Lett. **76**, 1762.

[13] Luning, J., Rubensson, J.-E., Ellmers, C., Eisebitt, S. & Eberhardt, W. (1997) *Phys. Rev. B*, **56**, 13147.

[14] Nordgren, J. & Nyholm, R. (1986) *Nucl. Instr. Meth. A* **242**, 246; Nordgren, J., Bray, G., Cramm, S., Nyholm, R., Rubensson, J.-E. & Wassdahl, N. (1989) *Rev. Sci. Instrum.* **60**, 1690.

[15] Hague, C.F., Underwood J.H. & Avila, A., et al. (2005) *Rev. Sci. Instr.* **76** (2): Art. No. 023110.

[16] Tokoshima, T., Harada, Y., Watanabe, M., Takata, Y., Ishiguro, E., Hiraya, A. & Shin, S. (2002) *Surface Review and Letters* **9**, 503.

[17] Cocco, D., Matteucci, M. & Prince, K. (2001) *SPIE Proc.* Vol. 4506.

[18] http://www.gammadata.se.

[19] http://usxs.fysik.uu.se.

[20] Hjörvarsson, B., Guo, J., Ahuja, R. & Ward, R.C.C., et al. (1999) *J. Phys. – Cond. Matter* **11**, 119.

[21] Wiell, T., Tillborg, H., Nilsson, A., Wassdahl, N., Mårtensson, N. & Nordgren, J. (1994) *Surf. Sci.* **304**, L451.

[22] Nilsson, A. & Pettersson, L.G.M. (2004) *Surface Science Reports*, **55** (2-5): 49-167.

[23] Hecq M. & Leleux, J. (1987) *Anal. Chem.* **59**, 440.

[24] Georgsson, M., Bray, G., Claesson, Y., Nordgren, J., Ribbing, C.-G. & Wassdahl, N. (1991) *J. Vac. Sci. Technol.* **9**, 638.

[25] Legrand, P.B., Dauchot, J.P., Hecq, M., Charbonnier, M. & Romand, M. (1994) *J. Vac. Sci. Technol.* **12**, 1551.

[26] Skytt, P., Gålnander, B., Nyberg, T., Nordgren, J. & Isberg, P. (1997) *Nucl. Instr. and Meth. in Phys. Res.* ***A*** **384**, 558.

[27] Gålnander, B., Käämbre, T., Blomquist, P., Nilsson, E., Guo, J.-H., Rubensson, J.-E. & Nordgren, J. (1999) *Thin Solid Films* **344**, 35-38.

[28] Becker, R.S., Golovchenko, J.A. & Patel, J.R. (1983) *Phys. Rev. Lett.* **50**, 153.

[29] Liu, A.M. & Cohen, M.L. (1989) *Science* **245**, 841.

[30] Niu, C., Lu, Y.Z. & Lieber, C.M. (1993) *Science* **261**, 334.
[31] Li, D., Chung, Y.-W., Wong, M.-S. & Sproul, W.D. (1993) *J. Appl. Phys.* **74**, 219.
[32] Sjostrom, H., Stafstrom, S. & Boman, M., et al. (1995) *Phys. Rev. Lett.* **75**, 1336.
[33] Sjostrom, H., Stafstrom, S., Boman, M. & Sundgren, J.-E. (1995) *Phys. Rev. Lett.* **75**, 1336.
[34] Hellgren, N., Guo, J.-H., Såthe, C., Agui, A., Nordgren, J., Luo, Y., Ågren, H. & Sundgren, J.-E. (2001) *Appl. Phys. Lett.* **79**, 4348.
[35] Nilsson, P.O., Kanski, J., Thordson, J.T., Andersson, T.G., Nordgren, J., Guo, J.-H. & Magnuson, M. (1995) *Phys. Rev. B* **52**, R8643.
[36] Mankefors, S., Nilsson, P.O., Kanski, J. & Karlsson, K. (1998) *Vacuum* **49**, 181.
[37] Mankefors, S., Nilsson, P.O., Kauski, J. & Karlsson, K. (1998) *Phys. Rev.* ***B*** **58**, 10551.
[38] Capellini, G., Di Gaspare, L., Evangelisti, F., Palange, E., Notargiacomo, A., Spinella, C. & Lombardo, S. (1999) Supercond. *Sci. Technol.* **14**, L21.
[39] Darhuber, A.A., Schittenhelm, P., Holy, V., Stangl, J., Bauer, G. & Abstreiter, G. (1997) *Phys. Rev. B* **55**, 15652.
[40] Sunamura, H., Usami, N., Shiraki, Y. & Fukatsu, S. (1995) *Appl. Phys. Lett.* **66**, 3024.
[41] Nilsson, P.O., Mankefors, S., Guo, J.-H., Nordgren, J., Debowska-Nilsson, D., Ni, W.-X. & Hansson, G.V. (2001) *Phys. Rev. B* **64**, 1153061.
[42] Zhou, Z., Ye, J., Sayama, K. & Arakawa, H. (2001) *Nature*, **414**, 625.
[43] Vayssieres, L., Beermann, H., Lindquist, S.-E. & Hagfeldt, A. (2001) *Chem. Mater* **13**, 233.
[44] Kuiper, P., Searle, B.G., Rudolf, P., Tjeng, L.H. & Chen, C.T. (1993) *Phys. Rev. Lett.* **70**, 1549.
[45] Duda, L.-C., Nordgren, J., Dräger, G., Bocharov, S. & Kirchner, Th. (2000) *J. Electr. Spectros. and Related Phenom.*, **275**, 110-111.
[46] Crocombette, J.P., Pollak, M., Jollet, F., Thromat, N. & Gautier-Soyer, M. (1995) *Phys. Rev. B*, **52**, 3143.
[47] Matsumoto, Y. (1996) J. *Solid-State Chem.* **126**, 227.
[48] Shionoya, S. & Yen, W.M (Eds.) (1999) *Phosphor Handbook*, Boca Raton, FL: CRC Press.
[49] Shin M.W. & Trew, R.J. (1995) *Electron. Lett.* **31**, 489.
[50] Hamdani, F., Botchkarev, A.E., Tang, H., Kim, W. & Morkoc, H. (1997) *Appl. Phys. Lett.* **71**, 3111.
[51] Ohtomo A., Kawasaki, M., Koida, T., Koinuma, H., Sakurai, T., Yoshida, Y., Sumiya, M. Fuke, S., Yasuda, T. & Segawa, Y. (1998) *Mater. Sci. Forum.* **264**, 1463.
[52] Sharma, P., Gupta, A., Rao, K.V., Owens, F.J., Sharma, R., Ahuja, R., Osorio Guillen, J.M., Johansson, B. & Gehring, G.A. (2003) *Nature Mater.* **2**, 673.
[53] Dong, C.L., Persson, C., Vayssieres, L., Augustsson, A., Schmitt, T., Mattesini, M., Ahuja, R., Chang, C.L. & Guo, J.-H. (2004) *Phys. Rev. B* **70**, 195325.
[54] Duda, L.-C. et al. (1998) *J. Phys. Soc. Japan.* **67**, 416.
[55] Duda, L.-C. et al. (2000) *J. El. Spectrosc. Rel. Phenom.* **110**, 275.
[56] Matsubara, M. et al. (2002) *J. Phys. Soc. Japan.* **71**, 347.
[57] Duda, L.-C. et al. (2004) *J. Alloys Comp.* **362**, 116.
[58] Kotani, A. & Shin, S. (2001) *Rev. Mod. Phys.* **73**, 201.
[59] Kuiper, P. et al. (1998) *Phys. Rev. Lett.* **80**, 5204.
[60] Perkins, J.D. et al. (1993) *Phys. Rev. Lett.* **71**, 1621; Lorenzana, J. et al. (1995) *Phys. Rev. Lett.* **74**, 1867; Liu R. et al. (1993) *Phys. Rev. Lett.* **71**, 3709; Salamon D. et al. (1995) *Phys. Rev. B* **51**, 6617.
[61] Mattheiss, L.F. et al. (1994) *Phys. Rev. B*, **49**, 14 050; Z. S. Popovic et al. (1995) *J. Phys. Condens. Matter* **7**, 4549.
[62] Zagoulaev S. & Tupitsyn, I.I. (1997) *Phys. Rev. B* **55**, 13 528; Z. V. Sljivancanin et al. (2000) *Phys. Rev. B* **56**, 4186.

[63] Duda, L.-C. et al. (1997) *Phys. Rev. B* **61**, 4432.
[64] Duda, L.-C. et al. (2002) *Surf. Rev. Lett.* **9**, 1103.
[65] Okada K. & Kotani, A. (2001) *Phys. Rev. B* **63**, 04103.
[66] Harada, Y. et al. (2002) *Phys. Rev. B* **66**, 165104.
[67] van Veenendal M. et al. (1993) *Phys. Rev. B* **47**, 11467.
[68] Gunnarsson, O. & Schönhammer, K. (1983) *Phys. Rev. B* **23**, 4315; Zaanen, J., Westra, C., Sawatzky, G.A. (1986) *Phys. Rev. B* **33**, 8060.
[69] Imada M., Fujimori, A. & Tokura, Y. (1998) *Rev. Mod. Phys.* **70,** 1039; Mott, N.F. (1990) *Metal–Insulator Transitions, 2nd Ed.,* Taylor & Francis.
[70] Isobe, M. & Ueda, Y. (1996) *J. Phys. Soc. Japan.* **65**, 1178; Lohmann M. et al. (2000) *Phys. Rev. Lett.* **85**, 1742.
[71] Föex, M. & Hebd, C.R. (1946) *Sean. Acad. Sci. B* **223**, 1126.
[72] M. Marezio et al. (1972) *Phys. Rev. B* **5**, 2541; Morin, F.J. (1959) *Phys. Rev. Lett.* **8** 34; Allen, P.B. et al. (1993) *Phys.Rev. B* **48**, 4359.
[73] Wilhelmi, K.-A., Waltersson, K. & Kihlborg, L. (1971) *Acta Chem. Scand.* **25**, 2675; Dernier, P.D. (1974) *Mater. Res. Bull.* **9**, 955; Kimizuka, N., Nakano-Onoda, M., Kato, K. (1978) *Acta Crystallogr. Sect. B* **34**, 1037.
[74] Hörlin, T., Niklewski, T. & Nygren, M. (1973) *Mater. Res. Bull.* **8**, 179; Nygren, M. (1973) *Chem. Commun.* **11**, 1.
[75] Schmitt T. et al. (2004) *J. Appl. Phys.* **95**, 6444; Duda, L.-C. et al. (2005) *Appl. Phys. Lett.* **86**, 064101.
[76] Schmitt, T. et al. (2004) *J. Alloys Comp.* **362**, 143.
[77] Nordlinder, S. et al. (2003) *Chem. Mater.* **15**, 3227.
[78] Smolinski, H. et al. (1998) *Phys. Rev. Lett.* **80**, 5164.
[79] Golubchik, S.A. et al. (1997) *J. Phys. Soc. Japan* **66**, 4042; Long, V.C. et al. (1999) *Phys. Rev. B* **60**, 15721; Damascelli, A. et al. (2000) *Phys. Rev. B* **61**, 2535; Presuraet, C. et al. (2000) *Phys. Rev. B* **62**, 16522; Konstantinovic, M.J. et al. (2001) *Phys. Rev. B* **63**, R121102; Popovic, Z.S., Vukajlovic, F.R. (1999) *Phys. Rev. B* **59**, 5333; Wu, H., Zheng, Q. (1999) *Phys. Rev. B* **59**, 15027; Yaresko, A.N. et al. (2000) *Phys. Rev. B* **62**, 15538; Atzkern, S. et al. (2001) *Phys. Rev. B* **63**, 165113.
[80] Zhang, et al. (2002) *Phys. Rev. Lett.* **88**, 077401; Schmitt, T. et al. (2002), *Surf. Rev. Lett.* **9**, 1369.
[81] Duda, L.-C. et al. (2004) *Phys. Rev. Lett.* **93**, 169701.
[82] van Veenendaal, M. & Fedro, A.J. (2004) *Phys. Rev. Lett.* **92**, 219701.
[83] Augustsson, A. et al. (2003) *J. Appl. Phys.* **94**, 5083.
[84] de Groot, F.M.F., Fuggle, J.C., Thole, B.T., Sawatzky, G.A. (1990) *Phys. Rev. B* **42**, 5459; Strange, et al. (1991) *Phys. Rev. Lett.* **67**, 3590.
[85] Augustsson, A. et al. (2003) *J. Chem. Phys.* **119**, 3983.
[86] Hague, C.F. et al. (1993) *Phys. Rev. B* **48**, 3560.
[87] Braicovich, et al. (1997) *Phys. Rev. B* **55**, R14729.
[88] Duda, L.-C. (2000) *J. El. Spectrosc. Rel. Phenom.* **110**, 287.
[89] Kuiper, P. (2000) *J. Phys. Soc. Japan* **69**, 874.
[90] Yablonskikh, M.V. et al. (2000) *Solid State Comm.* **117**, 79; Yablonskikh, M.V. et al. (2001) *Phys. Rev. B* **63**, 235117.
[91] Borgatti, F. et al. (2002) *Phys. Rev. B* **65**, 094406.
[92] Braicovich, L. et al. (2002) *Phys. Rev. B* **66**, 174435.
[93] Sacchi, M. et al. (1999) *Phys. Rev. B* **60**, R12569.
[94] Kao, C.C. et al. (1994) *Phys. Rev. B* **50**, 9599–9602.
[95] Duda, L.-C. et al. (1996) *Nucl. Instrum. Meth. Phys. Res.* ***A*** **376**, 291.
[96] Stagarescu, C.B. et al. (1996) *Phys. Rev. B* **54**, R17335.
[97] Stagarescu, C.B. et al. (1999) *J. Solid State Chem.* **143**, 1.
[98] Ryan, P. et al. (2002) *Phys. Rev. B* **65**, 205201.

[99] McGuinness, C. et al. (2003) *Phys. Rev. B* **68**, 165104.
[100] Axe, J.D. & Dieke, G.H. (1962) *J. Chem. Phys.* **37**, 2364.
[101] Seltzer, M.D., Wright, A.O., Morrison, C.A., Wortman, D.E., Gruber, J.B. & Filer, E.D. (1996) *J. Phys. Chem. Solids* **57** 1175.
[102] Finazzi, M., de Groot, F.M.F., Dias, A.-M., Kierren, B., Bertran, F., Sainctavit, Ph., Kappler, J.-P., Schulte, O., Felsch, W. & Krill, G. (1995) *Phys. Rev. Lett.* **75**, 4654.
[103] Moewes, A., Eskildsen, T., Ederer, D.L., Wang, J., McGuire, J. & Callcott, T.A. (1998) *Phys. Rev. B*, **57**, R8059.
[104] Butorin, S.M., Guo, J.-H., Shuh, D. & Nordgren, J. (1997) *ALS Compendium of User Abstracts and Technical Reports* (LBNL, University of California, Berkeley, 1998) p. 143.
[105] van der Marel, D. & Sawatzky, G.A. (1988) *Phys. Rev. B* **37**, 10674.
[106] Carnall, W.T., Fields, P.R. & Rajnak, K. (1968) *J. Chem. Phys.*, **49**, 4424.
[107] Kotani, A. (1999) *Technical Report of ISSP, Ser. A*, No. 3456, Proceedings of the Second International Conference on Synchrotron Radiation in Materials Science (Kobe, 1998).
[108] Fry, J.L., Caspers, H.H., Rast, H.E. & Miller, S.A. (1968) *J. Chem. Phys.* **48**, 2342.
[109] Cowan, R.D. (1981) *The Theory of Atomic Structure and Spectra* University of California Press, Berkeley.
[110] Sugar, J. (1972) *Phys. Rev. B* **5** 1785.
[111] Kotani, A. & Ogasawara, H. (1992) *J. Electr. Spectrosc. & Rlt. Phenom.*, **60**, 257.
[112] Ogasawara, H. & Kotani, A. (1995) *J. Phys. Soc. Japan* **64**, 1394.
[113] Butorin, S.M. (2000) *J. Electr. Spectrosc.* **110-111,** 233.
[114] Pireaux, J.J., Riga, J. Thibaut, E., Tenret-Noël, C., Caudino, R. & Verbist, J.J. (1977) *Chem. Phys.* **22**, 113.
[115] Teterin, Yu.A., Kulakov, V.M., Baev, A.S., Nevzorov, N.B., Melnikov, I.V., Streltsov, V.A., Mashirov, L.G., Suglobov, D.N. & Zelenkov, A.G. (1981) *Phys. Chem. Minerals,* **7**, 151.
[116] Madhavaram, H., Buchanan, P. & Idriss, H. (1997) *J. Vac. Sci. Technol. A* **15**, 1685.
[117] Verbist, J.J., Riga, J., Tenret-Noël, C., Pireaux, J.J. Dúrsel, G., Caudino, R. & Derouane, E.G. (1976) in *Plutonium and OtherAactinides,* Eds. H. Blank & R. Lindner (North-Holland, Amsterdam), p. 409.
[118] Allen, G.C., Griffiths, A.J. & Suckling, C.W. (1978) *Chem. Phys. Lett.* **53**, 309.
[119] Gubanov, V. A., Rosén, A. & Ellis, D.E. (1977) *Solid State Commun.* **22**, 219.
[120] Heera, V., Seifert, G. & Ziesche, P. (1983) *Phys. Stat. Sol. (b)*, **118**, K107.
[121] Ellis, D.E. & Goodman, G.L. (1984) *Int. J. Quant. Chem.* **25**, 185.
[122] Goodman, G.L. (1992) *J. Alloys and Compounds,* **181**, 33.
[123] Gunnarsson, O., Sarma, D.D., Hillebrecht, F.U. & Schönhammer, K. (1988) *J. Appl. Phys.* **63**, 3676; Gunnarsson, O.& Li, T.C. (1987) *Phys. Rev. B* **36**, 9488.
[124] Kotani, A. & Ogasawara, H. (1993) *Physica C,* **186-188**, 16.
[125] Pierloot, K., Reinders, A., Goodman, G.L., Devoghel, D., Görller-Walrand, C. & Vanquickenborne, L.G. (1991) *J. Chem. Phys.* **94**, 2928.
[126] Walsh, P.F., Ellis, D.E. (1976) *J. Chem. Phys.* **65**, 2387.
[127] Wood, J.H., Boring, M. & Woodruff, S.B. (1981) *J. Chem. Phys.* **74**, 5225 and references therein.
[128] Cox, L.E. (1982) *J. Electr. Spectrosc.* **26**, 167.
[129] Glans, P., Gunnelin, K., Skytt, P., Guo, J.-H., Wassdahl, N., Nordgren, J., Ågren, H., Gel'mukhanov, F. Kh. & Warwick, T. (1996) *Phys. Rev. Lett.* **76**, 2448.
[130] Skytt, P., Glans, P., Guo, J.-H., Gunnelin, K., Såthe, C., Nordgren, J., Gel'mukhanov, F.Kh., Cesar, A. & Ågren, H. (1996) *Phys. Rev. Lett.* **77**, 5035-5038.
[131] Glans, P., Skytt, P., Gunnelin, K., Guo, J.-H., Nordgren, J., Gelmukhanov, F., Cesar, A. & Ågren, H. (1996) *J. Electr. Spectr. Rel. Phenom.* **82**, 3.

[132] Guo, J.-H., Glans, P., Skytt, P., Wassdahl, N., Nordgren, J., Luo, Y., Ågren, H., Ma, Y., Warwick, T., Heimenn, P., Rotenberger, E., & Denlinger, J. (1995) *Phys. Rev. B,* **52**, 10681.

[133] Guo, J.-H., Skytt, P., Wassdahl, N., Nordgren, J., Luo, Y., Vahtras, O. & Ågren, H. (1995) *Chem. Phys. Lett.* **235**, 152.

[134] Luo, Y., Ågren, H., Gel'mukhanov, F.Kh., Guo, J.-H., Skytt, P., Wassdahl, N., & Nordgren, J. (1995) *Phys. Rev. B,* **52**, 14479.

[135] Skytt, P., Guo, J.-H., Wassdahl, N., Nordgren, J., Luo, Y. & Ågren, H. (1995) *Phys. Rev. A*, **52**, 3572.

[136] Cederbaum, L.S. (1995) *J. Chem. Phys.* **103**, 562.

[137] Gunnelin, K., Glans, P., Skytt, P., Guo, J.-H., Nordgren, J. & Ågren, H. (1998) *Phys. Rev.* A **57**, 864.

[138] Narten, A.H. & Levy, H.A. (1971) *J. Chem. Phys.* **55**, 2263.

[139] Narten, A.H. (1972) *J. Chem. Phys.* **56**, 5681.

[140] Stillinger, F.H. (1980) *Science* **209**, 451.

[141] Guo, J.-H., Luo, Y., Augustsson, A., Rubensson, J.-E., Såthe, C., Ågren, H., Siegbahn, H. & Nordgren, J. (2002) *Phys. Rev. Lett.* **89**, 137402.

[142] Guo, J.-H., Luo, Y., Augustsson, A., Kashtanov, S., Rubensson, J.-E., Shuh, D.K., Agren, H. & Nordgren, J. (2003) *Phys. Rev. Lett.* **91**, 157401.

[143] Myneni, S., Luo, Y., Naslund, L.Å., Cavalleri, M., Ojamae, L., Ogasawara, H., Pelmenschikov, A., Wernet, Ph., Vaterlein, P., Heske, C., Hussain, Z., Pettersson, L.G.M. & Nilsson, A. (2002) *J. Phys. Condensed Matter* **14**, L213.

[144] Pauling, L. (1960) *The Nature of the Chemical Bond*, Cornell Univ. Press.

[145] Sarkar, S. et al. (1993) *J. Chem. Phys.* **99**, 2032.

[146] Yamaguchi, T. et al. (1999) *Mol. Phys.* **96**, 1159, and Erratum (1999) *ibid*, **97**, 603.

[147] Haughney, M. et al. (1987) *J. Phys. Chem.* **91**, 4934; Svishchev, I.M. et al. (1994) *J. Chem. Phys.* **100**, 5165.

[148] Tanaka, Y. et al. (1985) *Bull. Chem. Soc. Japan* **58**, 270.

[149] Soper, A.K. et al. (1993) *Phys. Rev. Lett.* **71**, 4346.

[150] Tsai, J. et al. (1996) *J. Chem. Phys.* **104**, 9417.

[151] Dixit, S. et al. (1999) *J. Phys. Condens. Matter* **12**, L323.

[152] Wakisaka, A. et al. (2001) *J. Mol. Liq.* **90**, 175–184.

[153] Dixit, S. et al. (2002) *Nature* **416**, 829.

[154] Frank, H.S. et al. (1945) *J. Chem. Phys.* **13**, 507.

CHAPTER 15

LASER RAMAN SPECTROSCOPY

Alfons Schulte and Yu Guo
Department of Physics, and College of Optics and Photonics - CREOL
University of Central Florida, Orlando, FL 32816-2385, USA

15.1 INTRODUCTION

Inelastic scattering of light—or Raman scattering—from elementary excitations in a material yields structural and dynamic information on a molecular level. The Raman spectrum can be analyzed in terms of the molecular components or functional groups thus providing a "fingerprint" of the molecule. The Raman effect was first observed in 1928 by C.V. Raman and K.S. Krishnan [1], but wide application was delayed until the development of the laser. The nondestructive nature of the probe, flexibility in sampling arrangements, and a technical revolution [2, 3, 4, 5] in multi-channel detection and Rayleigh filters opened up many new areas where Raman measurements have proven to be very informative [6, 7]. Applications in the electronics and chemical industries are increasing and range from process control in semiconductor and polymer production to microanalysis of integrated circuits [8].

The interplay between microstructure and desirable properties of devices may be illustrated in materials suitable for use in high speed optical communication applications. These require all-optical processing and switching capabilities that must be compatible with current system configurations, possess ultrafast broadband response time, as well as low linear and nonlinear loss. Chalcogenide glasses (ChGs) have shown promise in that they exhibit properties compatible with the above requirements at 1.3 and 1.55 mm wavelengths [9]. Efforts to optimize film properties and device performance have focused on identifying the chemical and structural origin of the linear and nonlinear response in terms of the material processing conditions used in creating the optical element [10]. ChGs are photosensitive when exposed to band gap energy (~2.35 eV for As_2S_3) [11]. Taking advantage of these photosensitive effects (photo-darkening and photo-expansion) in ChGs allows the creation of bulk waveguide structures [12], or the patterning of photo-induced relief gratings and guided wave structures in

ChG films [13, 14]. Waveguide structures fabricated in ChG can be combined with proteins such as bacteriorhodopsin with potential applications in spectroscopy and switching of biomolecules on surfaces [15].

Near-infrared Raman spectroscopy affords new opportunities in the non-destructive analysis of materials, which are strongly absorbing in the visible range. A distinct advantage over the more conventional approach using the visible range of the spectrum is the ability to obtain the Raman spectrum of photosensitive compounds without interference from photoreactions caused by the probe beam. In chalcogenide glasses shifting the excitation wavelength to 840 nm (below the band gap) allows one to obtain high quality Raman spectra and to correlate the underlying structure with nonlinear optical properties. The higher spatial resolution necessary to characterize planar films can be achieved with a microscope attachment. In the microstructural analysis of single and multilayer waveguide devices, Raman spectroscopy employing integrated optical techniques can be extremely powerful. The material of interest is cast into a slab waveguide, thereby significantly increasing both the scattering volume and the electrical field intensity within the film. Waveguide Raman spectroscopy (WRS) using guided mode excitation [16] has recently been applied to the structural characterization of chalcogenide glasses [17]. We discuss such experiments where Raman spectra were measured in the chalcogenide waveguide itself and in composite structures with protein layers excited by the evanescent field.

Through the incorporation of a microscope in Raman systems, spatially resolved compositional information is obtained. This opens up a wide range applications where only a minuscule amount of sample ($< 10^{-12}$ cm^3) is available or rely on imaging such as the characterization of internal stresses in semiconductor circuits. Forensic applications and trace detection fall into this category as well. Following recent technological innovations micro-spectroscopic techniques are now gaining widespread acceptance in research and industry, and applications are growing rapidly [18].

There are a number of excellent books and review articles on Raman spectroscopy, see [2, 3, 7, 19, 20] to reference just a few. In this chapter we put some emphasis on recent applications, which were developed in our own laboratory. This chapter is organized as follows. First a brief background on Raman scattering is presented with examples of current experimental techniques. Experimental aspects of dispersive near-infrared Raman and Fourier-transform infrared Raman spectroscopy are presented. We continue with the use of these techniques for novel Raman gain materials, and on investigations of As_2S_3 bulk glasses and films, depending on composition and processing conditions. Another application is the use of site-specific Raman bands to probe conformational changes in proteins under high pressure. Finally, micro-Raman spectroscopy applications are illustrated.

15.2 SPONTANEOUS RAMAN SCATTERING

In a typical Raman experiment, the excitation source is a laser and the scattered light is analyzed by a spectrometer and a detector with sensitivity near the single photon level. The inelastically scattered light contains information on vibrational states of the sample, which manifests itself by a frequency shift from the incident light. The underlying physics is that vibrations (or other excitations) modulate the polarizability tensor and cause the induced dipole moment to radiate at frequencies different from that of the electric field vector of the incoming light wave. For most applications the spontaneous Raman scattering originating from the linear response to the electric field is measured. The experimental challenge is to detect the weak Raman signal while rejecting the intense Rayleigh background from elastic scattering.

A classical treatment of Raman scattering illustrates the basic physics. Assuming a simple diatomic molecule the binding effect of the electronic charge distribution is approximated by a spring between its pointlike nuclei [21, 22]. For linear molecules the number of possible vibrations is $(3N - 5)$ (for nonlinear molecules $(3N - 6)$) where N is the number of atoms. Accordingly, the nuclei are space- and rotation-fixed about their equilibrium positions but free to vibrate in simple harmonic motion along one normal coordinate q.

If the excitation is a vibration, then the dynamical variable is the vibrational displacement q. The polarizability of the molecule can be expanded as a Taylor series:

$$\alpha = \alpha_0 + \left(\frac{\partial \alpha}{\partial q_k}\right)_0 q_k + \left(\frac{\partial^2 \alpha}{\partial {q_k}^2}\right) \frac{{q_k}^2}{2} + \cdots \tag{15.1}$$

Here q_k is the displacement of the kth normal coordinate, which can be represented as: $q_k = q_k^0 \cdot \cos(\omega_m t)$ for a molecule oscillating at frequency ω_m.

The induced dipole moment can be written as $P = \alpha E$, where $E = E_0 \cos(\omega t)$ is the incoming monochromatic electromagnetic wave. Neglecting the high power terms, we obtain

$$P = \alpha_0 E_0 \cos(\omega t) + \left(\frac{\partial \alpha}{\partial q_k}\right)_0 E_0 {q^0}_k \cos(\omega t)\cos(\omega_m t)$$

or

$$P = \alpha_0 E_0 \cos(\omega t) + \left(\frac{\partial \alpha}{\partial q_k}\right)_0 \mathrm{E}_0 {q^0}_k \frac{1}{2}[\cos(\omega + \omega_m)t + \cos(\omega - \omega_m)t]. \tag{15.2}$$

A Hertzian dipole emits electromagnetic radiation. Its intensity S is proportional to the square of the absolute value of the second time derivative of the induced dipole moment: $S \propto |\ddot{P}|^2$. Thus, the first of the terms in equation (15.2) describes Rayleigh scattering. The second term and third terms concern frequency-shifted (i.e., inelastically scattered) light. These are also known as *anti-Stokes* and the *Stokes-Raman scattering*. The above equations can be generalized to the case where the molecular polarizabilities are not isotropic and the induced dipole moment vector points in a direction different from the electric field vector's [21]. A quantum mechanical treatment relates the polarizability tensor to the wavefunctions and the scattering levels of the scattering system [21]. Similar to Rayleigh scattering, in the absence of nonlinear effects, the intensity of the scattered Raman light increases with the fourth power of the frequency of the exciting radiation. The intensity ratio of the anti-Stokes to Stokes line is given by a Boltzmann factor, $\exp(-\hbar\omega / kT)$, where $\hbar$ is Planck's number and k the Boltzmann constant. This is illustrated by the Raman spectrum of carbon tetrachloride in Figure 15.1. Note the strong polarization-dependence of the totally symmetric mode at 13.8 THz (or 459 cm^{-1}). Both the polarized scattering (vv) and depolarized (vh) scattering intensities are shown. Here, the first v means that the incoming polarization is perpendicular to the scattering plane and that the polarization of the scattered light is selected either in the scattering plane (vh) or perpendicular to it (vv). Due to its higher intensity in spontaneous Raman scattering, mostly the Stokes side of the spectrum is recorded.

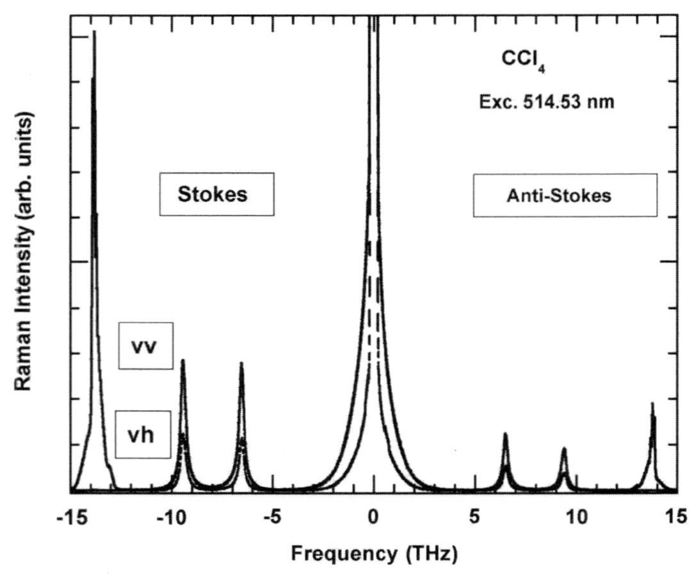

Figure 15.1 Raman spectrum of CCl_4 measured with a double monochromator on the Stokes and anti-Stokes side for two polarizations. Vibrational frequencies are given in Terahertz (0.03 THz $\approx$ 1 cm^{-1}). Spectral resolution is 25 GHz.

In general, excitations modulate the electric susceptibility and consequently the induced polarization through fluctuations in their dynamical variables ξ. Such dynamical variables can include vibrational displacement for phonons, magnetization for spin waves and spin fluctuations, and electron (quasiparticle) density for electronic (superconducting) excitations.

The modulation of the susceptibility by ξ adds an additional term to the polarization:

$$P = \varepsilon_o(\chi E + \chi'\xi E) \tag{15.3}$$

where $\chi' = d\chi / d\xi$ is the susceptibility derivative with respect to the dynamical variable ξ. The first term in equation (15.3) drives the polarization at the incident field frequency, and therefore contributes to simple elastic scattering. However, since ξ is itself time-dependent, reflecting the characteristic fluctuations of the excitation, the second term modulates the induced polarization at frequencies different from the incident field. This term therefore contributes inelastic features to the spectral response.

Within this framework the differential Raman-scattering cross-section associated with an elementary excitation can be written as [23]

$$\frac{d^2\sigma}{d\Omega\ d\omega_S} = \frac{V\omega_I\ \omega_S^3}{(4\pi\varepsilon_o)^2 c^4 |E_I|^2} \left\langle \hat{\varepsilon}_S \cdot P_S^* \hat{\varepsilon}_S \cdot P_S \right\rangle_{\omega_S} \tag{15.4}$$

where $\left\langle \hat{\varepsilon}_S \cdot P_S^* \hat{\varepsilon}_S \cdot P_S \right\rangle_{\omega_S}$ is the spectral density of polarization fluctuations with frequency ω_S, V is the light-scattering volume, c is the speed of light, and ε_S is the polarization of the scattered light.

The power spectrum due to polarization fluctuations is given by

$$\frac{d^2\sigma}{d\Omega\ d\omega_S} = \frac{V\ \omega_I\ \omega_S^3}{(4\pi\ \varepsilon_o)^2\ c^4} |(\varepsilon_S\chi\ \varepsilon_I)|^2 \left\langle \xi\xi^* \right\rangle_\omega . \tag{15.5}$$

Here ε_S and ε_I are the scattered and incident polarization directions, and $\left\langle \xi\xi^* \right\rangle_\omega$ is the thermally averaged correlation function for the dynamical variable ξ.

The light-scattering cross-section is related both to the correlation fuction of the relevant dynamical variable ξ and to the light-scattering volume V. Futhermore, the light scattering cross-section depends on the symmetry of the susceptibility derivative tensor $\boldsymbol{\chi}$, and indeed one can obtain excitation symmetry information by varying the scattering geometry defined by the polarization directions ε_I and ε_S. This is an extremely powerful feature of light-scattering techniques, since it allows identification of excitation symmetries [23].

When light-scattering is used to probe relaxations in a liquid the spectrum is mostly presented in the form of the susceptibility $\chi(\omega)$, which is obtained from the scattered intensity (up to a constant) by division with the thermal population factor $n(\nu) + 1$. Here, $n(\nu,T) = [\exp(h\nu/kT) - 1]^{-1}$, ν is the frequency, T the temperature, and h and k are Planck and Boltzmann constants [24].

As an example, the light-scattering spectrum of toluene over a wide frequency range is shown in Figure 15.2 on a double-logarithmic scale. The molecular Raman active vibrations are visible as sharp bands over the frequency range from 5×10^3 to 10^5 GHz. Below 5×10^3 GHz a broad band due to structural relaxation in the liquid is discernible followed by the Brillouin line near 10 GHz. Other spectral representations, which are in use for low-frequency Raman spectroscopy but are related to the susceptibility, are discussed in [25]. For a more extensive treatment of the theoretical background on light-scattering the reader is referred to the literature [23, 24].

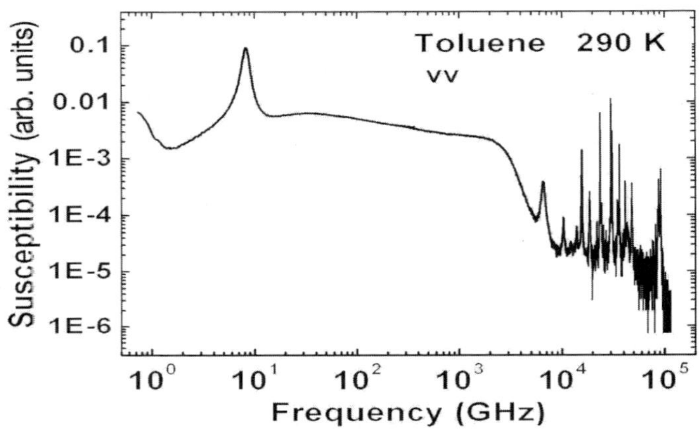

Figure 15.2 Susceptibility of toluene showing signatures of molecular vibrations, structural relaxation, and the Brillouin line. The horizontal axis displays the frequency of the excitation relative to the Rayleigh on the Stokes side. The excitation wavelength is 514.53 nm. The spectrum shown combines light-scattering data measured with a double monochromator and a six-pass tandem Fabry-Perot interferometer.

15.3 EXPERIMENTAL APPROACHES

Over the past 15 years the development of efficient filters for Rayleigh rejection and the availability of multichannel detectors have considerably simplified the experimental setup with the additional benefits of increased optical throughput and shorter acquisition time. The spectrometer can be reduced to a single spectrograph stage with a notch [26, 27] or sharp cut-off filter [28] selected for high extinction at the Rayleigh line [18]. Thus, Raman spectroscopy has become even more accessible as a scientific tool.

From an analytical point of view it is often desirable to characterize samples in nano- or picogram quantities. These requirements can be met by combining an optical microscope with a Raman system. Then the excitation spot has a dimension in the micron range, and Raman spectroscopy provides molecular compositional information with high spatial resolution. A schematic of a Raman setup is shown in Figure 15.3. Raman scattering is excited by either an Ar ion or a Ti:sapphire laser. The scattered light is collected with a low f-number lens and focused with a second lens on the entrance slit of a single-grating spectrograph (typical dispersion: 1.2 nm/mm in the focal plane). Multichannel detection at the single photon level is achieved with a backthinned *charge-coupled device* (CCD) detector.

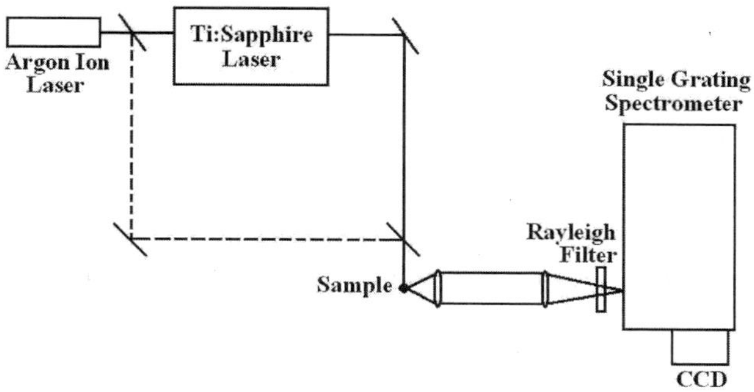

Figure 15.3 Schematic of an experimental setup for Raman spectroscopy using a single-grating spectrograph and Rayleigh rejection filter.

The CCD itself is a Si-based array detector. Absorbed photons are converted to electron-hole pairs. The electrons are collected in potential wells created by a depletion layer. The wells can be addressed individually and digitized with the readout electronics. CCDs are characterized by a high dynamic range (~10^5), high quantum efficiency (~ 90%), wide spectral range (400–1050 nm) and low readout noise (~2–5 e^-/pixel as seen in Figure 15.4).

Traditionally one works in the visible region of the spectrum, sometimes also to exploit resonance enhancement to extract information on the

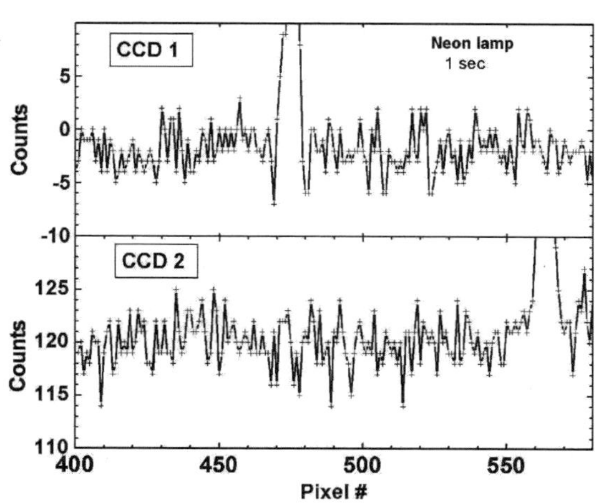

Figure 15.4 Readout noise characteristics of two commercial CCD detectors.

chromophore or the prosthetic group of a protein [29]. At the same time this approach has drawbacks since strongly absorbing samples such as many polymers and biological molecules can degrade or undergo undesired photochemistry during exposure to the laser beam. More important, a fluorescence background frequently obscures the Raman signal. One way to overcome these problems is to shift the laser excitation into the near-infrared.

Fourier transform (FT) Raman spectroscopy [30, 31, 20, 3] has provided a means of measuring the Raman spectrum of strong visible absorbers in the absence of fluorescence and resonance enhancement. When using the 1064-nm line of a Nd:YAG laser as an excitation source the Stokes-shifted Raman spectrum occurs in the near infrared, typically between 6000 and 10000 cm^{-1}. The scattered radiation is focused on the entrance port of a conventional FT-IR spectrometer and effectively replaces the internal light source for absorption spectroscopy. The analysis of the Raman spectrum via Fourier transform benefits not only from the multiplex and throughput advantages, but also from the inherently higher wavenumber accuracy of the interferometric method.

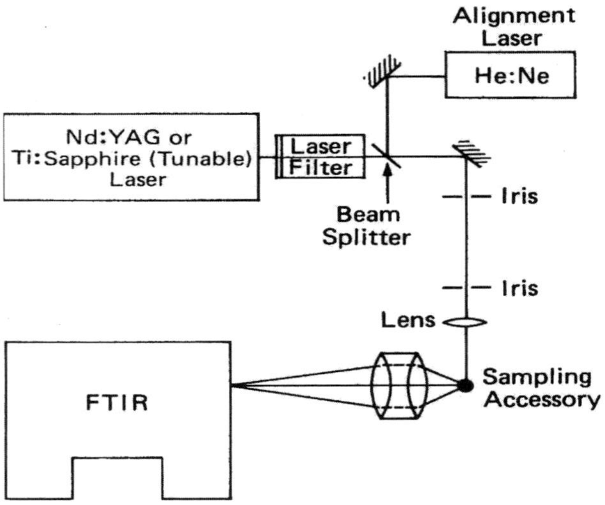

Figure 15.5 Schematic diagram of a FT-Raman experiment.

In polymer science FT-Raman spectroscopy has been used to probe conformation and side-chain packing in polysilanes and nonlinear optical materials [32]. Applications to macromolecules of biological interest concentrated on a synthetic polypeptide [33], polyene antibiotics, and lipid bilayers [34]. Of particular interest in biology is the vibrational structure and photochemistry of light-harvesting proteins such as bacteriorhodopsin and visual pigments. Due to the photolabile nature of these compounds great care has to be taken to avoid sample deterioration. FT-Raman spectroscopy using excitation beyond a wavelength of 1 micron often can overcome these problems. For instance, high quality spectra of the retinal isomers and the protein bacteriorhodopsin can be obtained within minutes.

Figure 15.6 displays the FT-Raman spectrum of bacteriorhodopsin between 400 and 3500 cm^{-1} [35]. FT-Raman spectra were measured in a 90° scattering with excitation by the 1064-nm line from a cw Nd:YAG laser [35]. The scattered light was analyzed with a Bomem model DA3.02 Fourier transform interferometer, which was equipped with a cooled (–35°C) indium-gallium-arsenide photodiode detector. The sample was in solution and no smoothing or baseline correction was performed. To obtain the resonance enhanced Raman spectrum with excitation in the visible, rather tedious sampling handling methods such as molecular flow or spinning cells were required [36]. The CH stretch mode was near 2900 cm^{-1}. The most intense band at 1531 cm^{-1} could be assigned to the C=C due to the polyene part of the retinal. The chromophore vibrations dominated the spectrum. This may have been caused by the high polarizability of the conjugated bonds in the retinal. The FT-Raman technique referenced the measured frequencies to the frequency of an internal He:Ne laser. Therefore, the absolute frequency could be determined to better than 0.01 cm^{-1} and the band positions reported were limited by the collection parameters. Figure 15.7 displays the spectra in the ethylenic mode region of bacteriorhodopsin, 13-*cis*, and all-*trans* retinal. Some bands of the isomers differ by frequency shifts of a few wavenumbers only, but are clearly resolved. The interaction with the protein causes a frequency shift of the chromophore bands in bacteriorhodopsin.

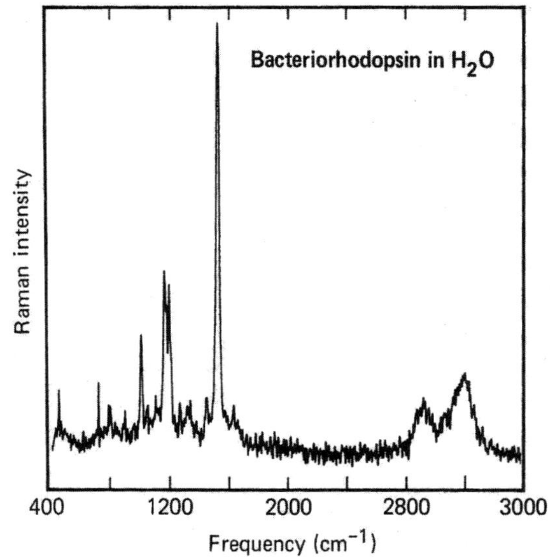

Figure 15.6 FT-Raman spectrum of a bacteriorhodopsin solution obtained at 1064 nm with a laser power of 700 mW.

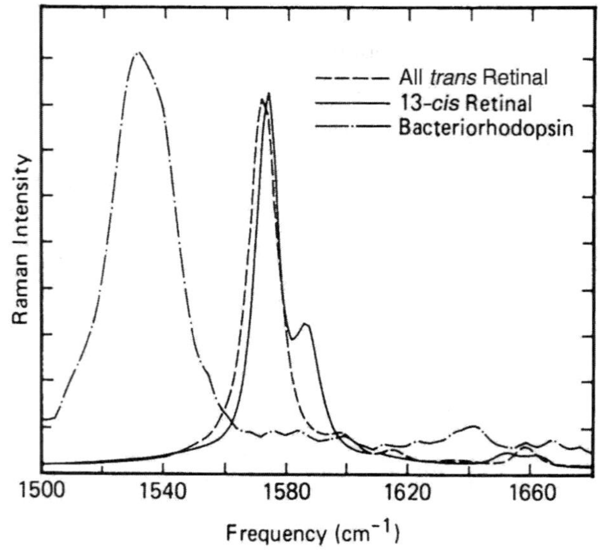

Figure 15.7 The ethylenic mode in bacteriorhodopsin, all-*trans*, and 13-*cis* retinal.

The examples presented above demonstrate the potential of FT-Raman spectroscopy as a nondestructive technique for molecular characteri-zation of photolabile chromophores and biopolymers. Among the limitations of FT-Raman spectroscopy are the loss in signal intensity due to the dependence of the scattering cross-section on the fourth power of the wavenumber and the need for Rayleigh line rejection filters. By employing a tunable Ti:sapphire laser in the range between 700 and 1000 nm the excitation can be shifted to shorter wavelengths, yet the excitation wavelength can be chosen long enough to avoid fluorescence or undesired photochemistry. An excellent long pass filter can be realized by a semiconductor single crystal with a band gap in this energy range [37]. The alignment can be performed in the visible before the Ti:sapphire laser is tuned to the desired excitation wavelength in the near infrared.

Another approach is to use a near-infrared laser source (most commonly a diode laser at 785 nm or a tunable Ti:sapphire laser) and a backthinned CCD detector in combination with a dispersive instrument [38, 28, 2]. Some applications using this technique are included in the next section.

15.4 APPLICATIONS

In the following we illustrate recent applications of Raman spectroscopy to materials of current technological interest. To begin with we discuss recent examples of advanced materials for Raman gain applications. The elucidation of structural properties in chalcogenides glass and the use of waveguide Raman spectroscopy follows. Then, we describe experiments which use

Raman spectroscopy to probe conformational changes in heme proteins. Finally, we touch on a few applications of micro-Raman spectroscopy.

15.4.1 Glasses for Raman Gain

One the most crucial components in optical communication systems is the optical amplifier. Recent progress in the fabrication of glass fibers have significantly increased the available transmission window for optical communication [39]. The increase in bandwidth has caused researchers great interest in the employment of Raman amplifiers. This is due to their potentially much larger bandwidth as compared with erbium-doped fiber amplifiers (EDFA). It is important, therefore, to find a material with wide bandwidth for Raman gain. On theoretical grounds, the coefficient for Raman gain depends linearly on the spontaneous Raman scattering cross-section [40]. Experimentally, a direct comparison between spontaneous and Raman gain spectra in two TeO_2-based glasses has recently shown a peak gain 30 times that of fused silica and twice its spectral bandwidth. It was also demonstrated that the Raman gain profile and intensity mimic that of the spontaneous Raman spectrum. [41].

The Raman spectrum of a tellurite-based glass is shown over an extended frequency range from 6–1500 cm^{-1} in Figure 15.8, together with the spectrum of fused silica (SiO_2). The spontaneous Raman spectra were measured using 514.53 nm excitation and a double monochromator. The top spectrum is that of the glass with composition 85% TeO_2-15% WO_3. Note that the fused slica spectrum has been multiplied by a factor of 17. Fused silica is employed as a standard material to quantify Raman gain. The high intensity and large bandwidth of the tellurite glass compared to fused silica predicts favorable properties for Raman gain application.

Raman bands in the high frequency region originate from the vibrations of molecular bonds. In the tellurium oxide system, increasing TeO_2 concentrations are determinants for the intensity of the main peaks. The intensity of the bands between 610 and 670 cm^{-1} associated with trigonal bipyramides increases with TeO_2 content. Significant features in the Raman spectra are bands at 770 and 950 cm^{-1}, attributable to vibrations of distorted WO_3 units. This band is highly polarized indicating that WO_3 units are preserved with small intermolecular coupling.

The depolarization ratio is indicative of the symmetry of the vibrations involved in the scattering process. A highly symmetric vibration will have a depolarization ratio close to zero. In the low frequency region, large Raman scattering is observed. The intense band near 40 cm^{-1} (Figure 15.8) is attributed to the Boson peak. The larger depolarization ratio in the frequency

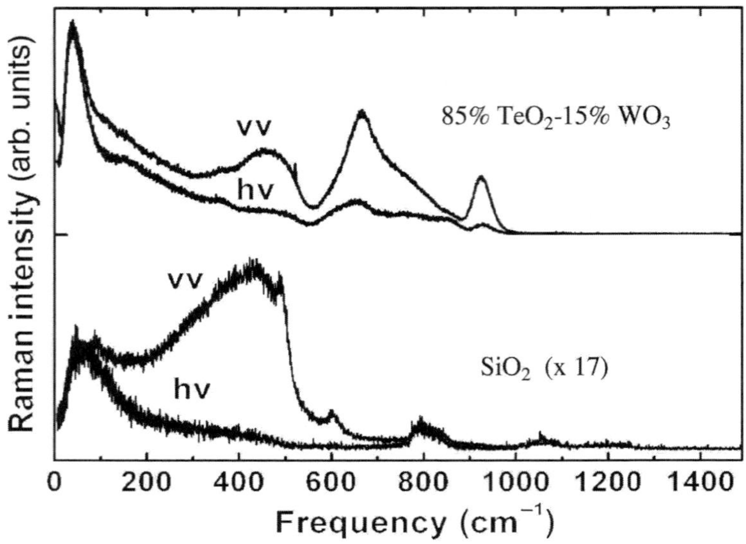

Figure 15.8 Polarized (vv) and depolarized (*h*v) Raman spectra of a tellurite glass, and fused silica measured with 1.5 cm^{-1} spectral resolution. The excitation wavelength is 514.53 nm. Depolarized spectra are multiplied by factors between 2 and 5 due to lower intensity to allow comparison. Note that the scale for the fused silica spectra has been expanded by a factor of 17.

region below 400 cm^{-1} suggests that in this range Raman amplification is much less polarization-dependent.

In the low frequency range the scattered intensity needs to be corrected for the thermal population factor, $n(\nu) + 1$ where $n(\nu, T) = [\exp(h\nu/kT) - 1]^{-1}$, ν is the frequency, *T* the temperature, and *h* and *k* are Planck and Boltzmann constants [2]. The vv polarized spectra divided by $n(\nu) + 1$ are shown in Figure 15.9 [42]. These correspond to the predicted Raman gain curves and they show a very broad band in the low frequency region (50–400cm^{-1}). This indicates that in the tellurite glasses a flat Raman gain profile down to very low wavenumber can be obtained while the gain is more than 10 times that of fused silica. the bandwidths of both glasses are almost twice as wide as that of fused silica, and the Raman intensity (which is proportional to the Raman scattering cross-section) of the telluride glass is several times higher than that of the fused silica [42].

A universal feature of the Raman spectrum of glasses is the so-called *Boson peak*. We can see from Figure 15.9 that after correction for the thermal occupation factor there is broad and intense Raman scattering in the low frequency region due to excess vibrational excitations.

We know that, for crystals, the density of states follows the Debye's law (e.g., $g(\omega) \propto \omega^2$) in the low frequency region, but there are deviations for amorphous materials. For amorphous materials, there is an additional contribution to the density of states as compared with the Debye's law. This

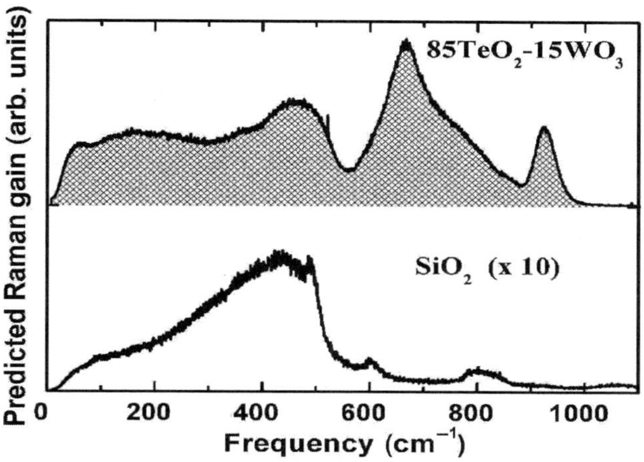

Figure 15.9 Predicted Raman gain curves after correcting for the thermal population factor of tellurite glass (85 TeO_2-15 WO_3) and fused silica. The fused silica spectrum has been multiplied by a factor 10 to allow better comparison with the spectrum of 85 TeO_2-15 WO_3.

excess density of states is characteristic for the Boson peak. Sometimes the representation $g(\omega)/\omega^2$ is chosen which peaks at a frequency ω_{BP}.

In amorphous materials the density of states $g(\omega)$ and the Raman spectrum can be connected using a relation by Shuker and Gammon [43]:

$$I(\omega) = \frac{C(\omega)g(\omega)[n(\omega)+1]}{\omega} \tag{15.6}$$

where $I(\omega)$ denotes the Raman intensity for the Stokes side of the spectrum, $[n(\omega) + 1]$ the Bose-Einstein factor, and $C(\omega)$ is the light-vibration coupling coefficient. The density of states can be independently measured by neutron-scattering. Based on such combined data it has been suggested that in many glasses the coupling coefficient varies nearly linearly with frequency [44].

The origin of the Boson peak or excess of density of states in amorphous materials is still under debate, and several models including localized vibrational states and medium range disorder effects have been proposed [45]. Excess density of states due to disorder can be considered to arise from the atomic positions or a distribution of force constants [46]. Computer simulations have shown that force constant disorder alone can give rise to the Boson peak [47].

15.4.2 Chalcogenide Glasses

Two characteristics of As-S-(Se) compounds—a large glass-forming region and a wide optical transmission band, with potentially low loss for the 1.3–1.55 mm telecommunications window—make them excellent candidates for infrared guiding configurations. The availability of these glasses in substantial quantities and the capability we have to fabricate good optical quality thin films by thermal evaporation and other deposition techniques enables us to realize relatively low cost As-S-(Se) integrated optical devices. Another attractive feature is our capability to create integrated components with one- and two-photon laser writing [48, 49].

Figure 15.10 illustrates the near-infrared Raman spectra (incident and scattered polarization resolved along the *z*-axis) for a series of binary and ternary compounds. The spectra were obtained at a spectral resolution of 1.5 cm^{-1}. The bulk spectra clearly show that each of the dominant bands consists of several overlapping components [50, 51]. The dominant feature in the binary sulfide and selenide compounds are bands at 345 cm^{-1} ($As_{40}S_{60}$) and 230 cm^{-1} ($As_{40}Se_{60}$), respectively. These spectra are in good agreement with other studies [52, 53], and the strong, broad band is attributed to an antisymmetric As-(S,Se)-As stretching vibration in the $As(S,Se)_3$ pyramids. According to the analysis of Lucovsky and Martin [54], the normal modes of the bulk glasses (e.g., clusters of $As(S,Se)_3$ molecules with weak inter-molecular coupling) are obtained by treating the molecular pyramid modes ($As(S,Se)_3$) and bridging chain modes (As-(S,Se)-As) independently.

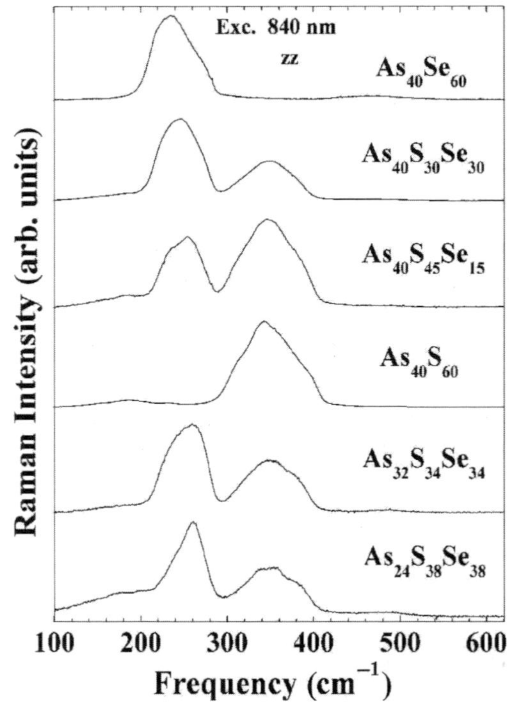

Figure 15.10 Raman spectra of a series of bulk chalcogenide glasses obtained with near-infrared excitation.

In the ternary compounds with S/Se = 1 molar ratio and decreasing As content, a progressive decrease of these broad bands is observed, indicative of a decrease in the number of As-containing pyramidal sites. New bands appearing around 255 cm^{-1}. and 440–480 cm^{-1} form in the now chalcogen-rich glasses and are attributed to Se-Se and S-S homo-polar bonds. These may be correlated with the enhancement of nonlinear optical properties (n_2) in ternary compounds with S/Se = 1 molar ratio and decreasing as content [10]. These units serve as chalcogen chains connecting the remaining pyramidal units. The small number of S-S bonds indicated by a weak band near 495 cm^{-1} for equal concentrations of S and Se suggests that the S stays with the remaining pyramids, and that it is the Se that dominates the connecting chain units.

15.4.3 Chalcogenide Thin Films –Waveguide Raman

Waveguide Raman spectroscopy (WRS) using guided mode excitation has been applied to thin organic and polymeric films to probe spontaneous [55, 16] and coherent [56] scattering, and very recently, to sol-gel-derived planar germano-silicate [57] waveguides and lead titanate [58] films. The relative low refractive index n of these organic and oxide materials allows the use of glass prisms such as LaSF5 (n ~ 1.8) for coupling a range of propagation vectors into the waveguide structure. In spite of its sensitivity WRS has not been applied to the structural characterization of chalcogenide glasses until recently [17], most likely due to their high index (n ~ 2.45), the lack of suitable prism couplers, and difficulties associated with working in the near-infrared.

Cleaved silicon substrates were employed for high efficiency endface coupling of the near-infrared laser beam into a single-layer channel waveguide structure [17]. The waveguides were nominally 1.75–2 μm thick. Figure 15.11 displays the Raman spectrum of As_2S_3 and Figure 15.12 the spectrum obtained from bacteriorhodopsin layered on the waveguide substrate [15]. As shown in Figure 15.13, As_2S_3 has Raman active vibrational bands below 500 cm^{-1}. The vibrational frequencies below 300 cm^{-1} are attributed to S-S interactions and the vibrational frequencies between 300–400 cm^{-1} are attributed to AsS_3 pyramidal units and their interactions [52].

For the integration of waveguide structures with photosensitive proteins, bacteriorhodopsin (bR) is of particular interest, since it has applications in molecular electronic devices and optical switching [59, 60]. As with all proteins and organic assemblies, the characteristic vibrational frequencies of bR are in the 1000–4000 cm^{-1} range. This allows for no interference in the protein's signal from that of the waveguide. The bR spectrum in Figure 15.12 is indicative of the light-adapted state [61]. The light-adapted form of bR is the initial state for the proton pumping cycle. At 785 nm the Raman spectrum is still associated with the vibrations of the atoms that make up the chromophore. Resonance Raman, nuclear magnetic resonance (NMR), and chemical extraction studies have established that the chromophore in bR_{558} is a C_{13}=C_{14} *trans*, C_{15}=NHR *trans* protonated Schiff base of retinal. As shown in Figure 15.13, bR has several vibrational frequencies that are identifiable. The band at 1012 cm^{-1} is assigned to the rocking vibrations of a C-CH_3 group of the bR molecule. The 1100–1300 cm^{-1} region of the Raman spectrum is the fingerprint region for C-C bonds and is very sensitive to isomerization. The intense band at 1528 cm^{-1} is attributed to the ethylenic stretching of C=C bonds [61]. The data demonstrate that evanescently excited near-infrared Raman spectra can be measured with high signal-to-noise ratio providing an *in situ* probe of the native state of the protein.

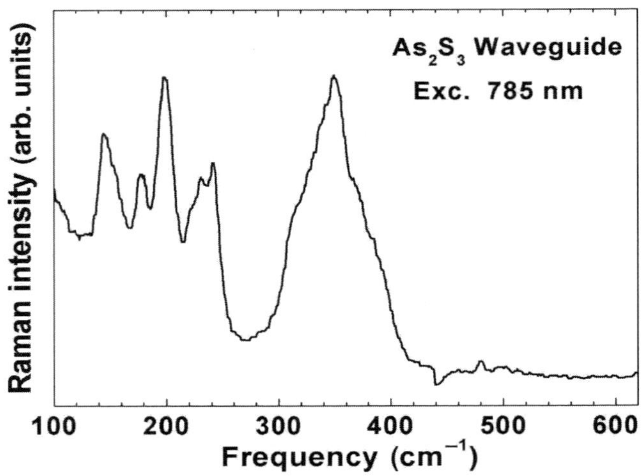

Figure 15.11 Raman spectrum of a chalcogenide thin film obtained with waveguide excitation and a power of 20 mW.

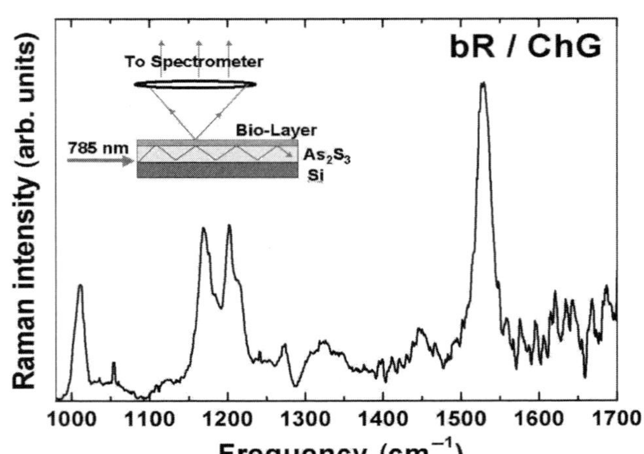

Figure 15.12 Evanescent wave-excited Raman spectrum of a bacteriorhodopsin layer on a As_2S_3 waveguide.

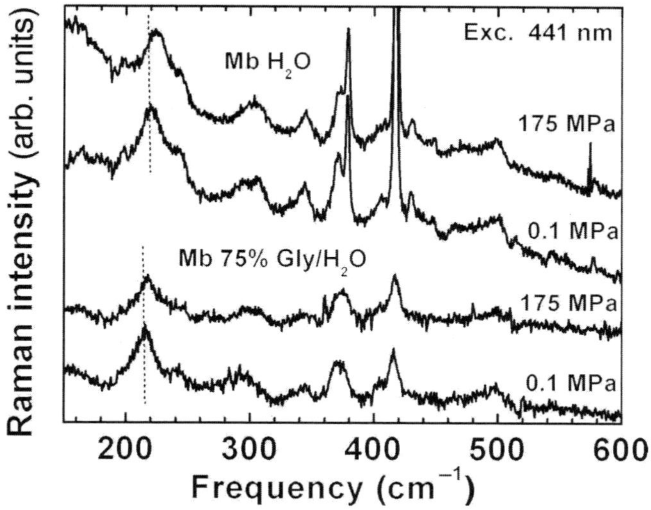

Figure 15.13 Low frequency region of the Raman spectra of deoxygenated myoglobin at ambient and high pressure. Samples are in aqueous solution or glycerol-water mixtures at pH of 7. Note the shift of the band near 220 cm^{-1} to higher frequency with increasing pressure.

ChG waveguides are optimized for near-infrared excitation allowing one to obtain the Raman spectra of biological compounds at minimal background fluorescence. This feature of the substrates can be a useful tool in the study of cells and microorganisms. Obtaining the Raman spectra at wavelengths less than 1 μm allows for the resolution of small spatial features compared to mid-infrared absorption wavelengths of 5–10 μm. Integrated optical components fabricated with ChG can be combined with proteins such as bR. Evanescent wave excitation may be employed for optical switching and spectroscopy of bio-assemblies on patterned ChG semiconductors. There are also potential applications involving biomolecular sensors.

15.4.3 High Pressure Raman Spectroscopy of Proteins

Resonance Raman spectroscopy is one of the few techniques that can probe the local environment of the active site inside a large biological system. The laser excitation wavelength is chosen near an electronic transition of the chromophore, and the Raman scattering cross-sections for vibrational modes that couple to this transition are selectively enhanced [29]. Raman spectra of heme proteins reveal a number of bands that have been well characterized and yield information on the spin state, coordination, and environment of the heme [29]. The small, globular heme protein myoglobin has served as a model system for extensive experimental and theoretical studies of protein dynamics. During the process of reversibly binding small ligands such as O_2, CO, or NO to the heme iron, both the chromophore and the protein undergo conformational changes. The bond between the iron and the proximal histidine imidazole nitrogen is the only covalent linkage between the heme group and protein. Resonance Raman studies at ambient pressure support the view that modulation of the iron by the protein through the proximal histidine exerts control at the level of reactivity [62, 63, 64].

In vivo, functional properties of proteins are affected by environmental parameters such as viscosity, pH, temperature, and pressure. For instance, sea animals survive over a wide range of pressure, from sea level to extreme depths. On the other hand, pressure can deactivate enzymes and kill bacteria [65]. For a description on a molecular level the effect of high pressure on prototype reactions of isolated proteins must be understood [66, 67]. The approach to combine high pressure and vibrational spectroscopy is motivated by the following observations: Spectral band parameters (frequencies, intensities, lineshapes and linewidths) are sensitive to dynamic and structural changes of biomolecules [68] at the sub-Angstrom level, a length scale where small, yet significant conformational changes for enzyme activity occur. From changes in the Raman spectra, pressure effects on protein function can be correlated with structural changes, for instance at the chromophore-protein interface [69] and compared with theoretical models [70]. Deoxymyoglobin is used as a reference structure since the reaction process is absent, and pressure-induced changes of the conformation can be separated from those along the reaction coordinate.

Resonance Raman scattering was excited with the frequency doubled output of a Ti:sapphire laser tunable from 441 to 425 nm or by the 457.9-nm line of an Ar ion laser. Detection of the backscattered Raman radiation is accomplished using a thin back-illuminated, charge-coupled in conjunction with a single-grating spectrograph and a Rayleigh line rejection filter. The pressure cell is constructed of beryllium-copper that combines the ability to resist high pressure (up to 400 MPa) with good thermal conductivity. Sapphire windows allow measurements from the near UV to the near infrared region. The high pressure Raman setup has been described in more detail in [71, 72].

The resonance Raman spectra of horse deoxy myoglobin (Mb) in the frequency range from 150 to 600 cm^{-1} are shown for ambient and high pressure in Figure 15.13. The band at 220 cm^{-1} has been assigned to the iron-histidine (Fe-His) stretching mode [73, 74, 75]. The other lines have been classified as follows [75]: The band near 241 cm^{-1} is a pyrole ring tilting mode. The modes from 250 to 420 cm^{-1} all involve peripheral substituents, and those from 420 to 520 cm^{-1} are attributed to out-of-plane distortions of the pyrole rings [76, 75].

The most significant spectral change with pressure is a shift of the peak frequency $\nu_{Fe\text{-}His}$ of the iron histidine mode. $\nu_{Fe\text{-}His}$ shifts to higher wavenumber by ~3 cm^{-1} between 0.1 and 175 MPa. The shift of the Fe-His mode has been observed in different solvents (75% gly/H_2O) and in Mb from sperm whale. The peak position of $\nu_{Fe\text{-}His}$ as a function of pressure is plotted in Figure 15.14. The error bars correspond to a precision of 0.6 cm^{-1}.

A smaller shift is apparent in the band near 343 cm^{-1}. Choi and Spiro [76] assigned this band to out-of-plane modes of propionate porphyrin macrocycle substituent groups. Since the hydrogen bonding partner of the carboxyl group of the propionic acid attached to the D pyrole [77] is Arg 45 (CD3), changes in the 343 cm^{-1} band may indicate motion of the protein helices.

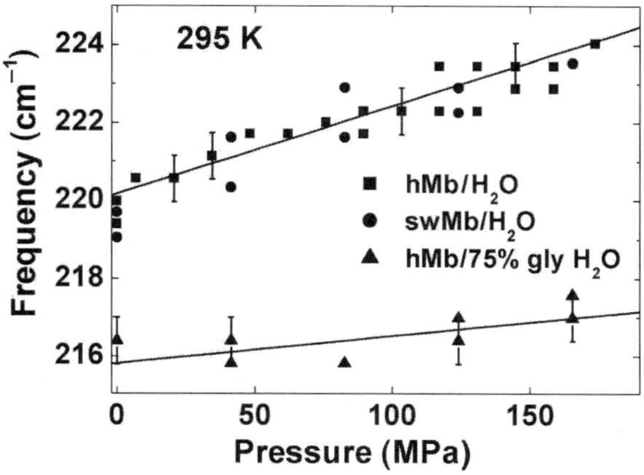

Figure 15.14 Pressure dependence of the peak frequency of the iron-histidine mode in myoglobin in aqueous solution and 75% gly/H_2O.

We attribute the observed frequency shift to a conformational change, which alters the tilt angle between the heme plane and the proximal histidine and the out-of-plane iron position. The geometry of the proximal histidine influences both the frequency and the intensity of the Fe-His stretch band [74]. An important factor affecting the global protein conformation, the heme pocket structure, and the iron-histidine mode is water activity [78]. The altered water activity in a glycerol/H_2O mixture causes a 2.6 cm^{-1} downshift of $n_{Fe\text{-}His}$ as compared to aqueous solution. Apparently glycerol influences the protein-water interaction with a possible release of bound water molecules from the surface, though this perturbation of the Fe-His frequency is opposite to the perturbation from pressure.

The main conclusion from the shift of the iron-histidine mode is that pressure causes global conformational changes in the protein as well as rearrangements of the active site environment. Indeed, a very recent high-pressure crystallographic study [79] of myoglobin confirms that the change in protein structure due to pressure is not purely compressive but involves conformational changes. Large collective displacements are observed in six regions including sliding of the F-helix towards the E-helix [79].

In the case of ligand bound myoglobin (MbCO, MbO_2), photolysis by the laser beam during the acquisition of a Raman spectrum creates a stationary mixture of bound and photolyzed molecules. Photostationary experiments demonstrate a significant pressure-dependence of the ligand rebinding rate in myoglobin [80]. The photolysis of ligated myoglobin by the laser beam during the acquisition of a Raman spectrum (100 seconds) creates a stationary mixture of bound and photolyzed molecules (Figure 15.15).

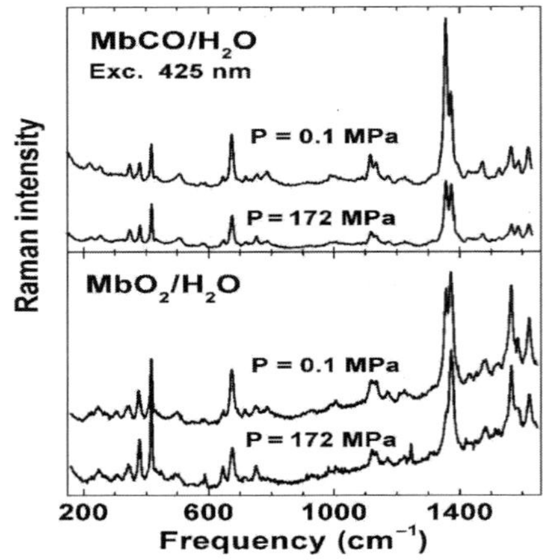

Figure 15.15 Resonance Raman spectra of MbCO (top) and MbO_2 (bottom) in aqueous solution as a function of pressure at 295 K. The sample is in photostationary equilibrium. As the pressure is increased the intensity of the ν_2 band of MbCO (1374 cm^{-1}) increases relative to that of Mb (1356 cm^{-1}).

This is evident from the oxidation state marker band (n_4), which appears at 1354 cm^{-1} for deligated Mb and 1372 cm^{-1} for MbCO [62]. The spectrum at high pressure (keeping all other parameters fixed) shows a significant increase in intensity of the peak at 1372 cm^{-1} relative to that at 1354 cm,$^{-1}$ reflecting an increase in population of the bound state [71]. This population increase is due to a speedup of the overall rebinding. Consistent changes between bound and unbound states are also seen in the core size marker band n_2 (1560 and 1582 cm^{-1}) and the vinyl modes (1618 and 1632 cm^{-1}). The lower amount of photolysis at high pressure is also indicated by the reduced intensity of the iron-histidine mode at 220 cm^{-1} [71].

15.4.4 Micro-Raman Spectroscopy

For the measurement of a minuscule amount of sample or when spatial resolution is required (in addition to spectral resolution) various combinations of a Raman spectrometer with an optical microscope have been developed [81, 18]. This is known as *Raman microscopy* or *micro-Raman spectroscopy*.

A typical setup is shown in Figure 15.16. The excitation laser is reflected by a beam splitter, goes through a microscope objective, and is focused on the sample. The backscattered Raman is then collected with the same objective. The major part of the collimated beam is reflected from the beam splitter and is focused on the entrance slit of the monochromator. The small sample area, which is at the focusing point, is thus imaged through the entrance slit. It is important to understand that a narrow slit width will determine a well-defined sharp image, and thus result in a high resolution.

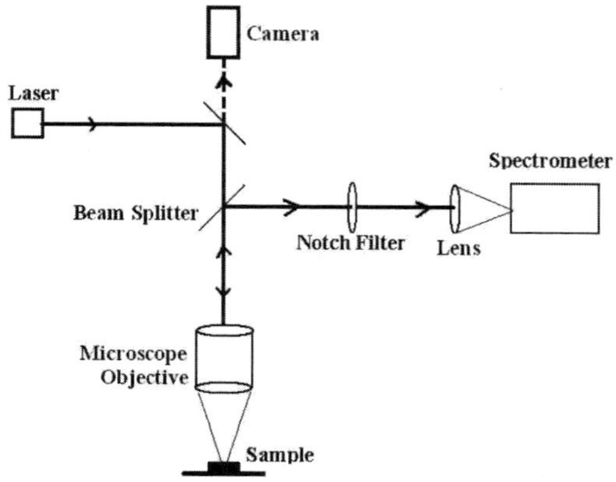

Figure 15.16 Schematics of micro-Raman spectrometer. The sample is located on an 3-dimensional positioning stage.

Applications in an industrial environment or for routine testing of samples require an experimental technique that requires no specialized sample preparation. Raman spectroscopy is a nondestructive probe and spectra can be measured on materials in solid or liquid form. Using a Raman microscope the sample can simply be placed on a slide and the area of interest is selected optically by using a viewing system. Figure 15.17 shows the Raman spectrum of a piece of silicon and an aspirin sample obtained with a commercial micro-Raman instrument. With micro-Raman, spectroscopy vibrational spectra can be measured from micron-sized particles, which makes the technique well suited as an analytical tool in chemistry and biotechnology.

Raman spectroscopy has proven to be an informative and nondestructive technique in III-V material characterization including local structure determination and stress analysis [82]. One of the key issues for the performance of wide band gap semiconductor-based device structures is the control of growth-induced defects and their impact on optoelectronic and transport properties. Despite the impressive progress in device applications, a deeper understanding of defect and impurity issues is necessary for continued rapid development in the areas of LEDs, laser diodes, UV detectors, and high voltage unipolar and bipolar [83].

We present an example where micro-Raman spectroscopy was used to probe optical phonons in gallium nitride close to the GaN/sapphire substrate interface [84]. Frequency shifts in vibrational modes correlate with independently obtained data on the dislocation density and are connected to the strain due to lattice mismatch at the interface. Micro-Raman spectra were measured in a backscattering geometry using the 514.53-nm line of an Ar^{-} ion laser. An infinity-corrected microscope objective focused the laser beam to a spot size of about 1 μm in diameter. A very narrow slit width (5 μm) in combination with binning in the vertical direction of the CCD chip acts as a confocal aperture which leads to a 1–2 μm depth resolution. A translation stage with submicron sensitivity was used for laser beam positioning at the predetermined distance from the GaN/sapphire interface. The excitation beam was parallel to the surface of the interface and perpendicular to grows direction of the GaN layers.

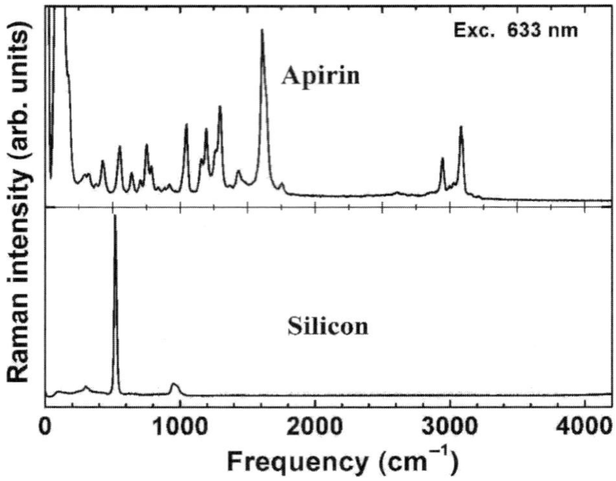

Figure 15.17 Raman spectra of aspirin and silicon obtained with a commercial micro-Raman instrument (LabRam HRUV). Integration time is 1 sec, HeNe laser power 10 mW, and slit width 100 μm.

Figure 15.18 depicts the Raman spectra at various distances from the GaN/sapphire interface over a range of 60 μm. The data were measured with a backscattering geometry corresponding to x(..)x configuration in Porto notation. The propagation direction of the laser beam was perpendicular to the *c*-axis but not along one of the principal axes so the A_1(TO), E_1(TO), E_1(LO), and E_2(high) modes [85] are observed at 535 cm^{-1}, 562 cm^{-1}, 745 cm^{-1} and 569 cm^{-1}, respectively. Due to limitations of the Rayleigh filter the E_2(low) mode (144 cm^{-1}) was not investigated. The spectral resolution of 1 cm^{-1} was achieved by recording the Raman spectrum in second order. The raw data presented in Figure 15.18 show a shift to lower frequency with increasing distance from the sapphire substrate. The precision in the determination of the frequency shifts could be further increased by fitting Lorentzian lines to the Raman peaks. Since the lineshape remains constant we estimate that spectral shifts of 0.1 cm^{-1} can be reliably detected [86].

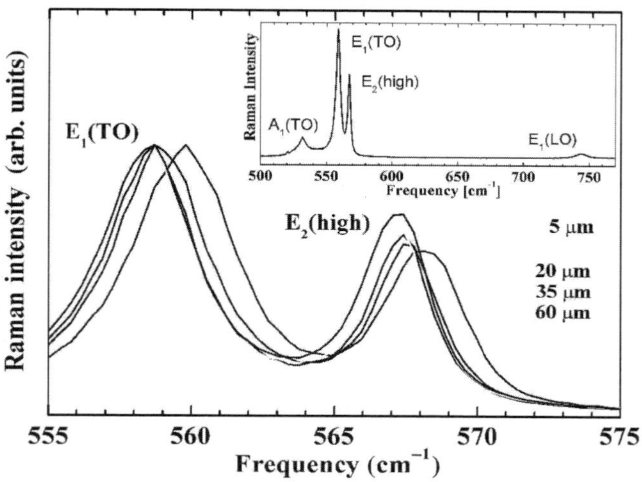

Figure 15.18 Raman spectra of a 64-mm thick GaN film measured with 514.5-nm excitation. Insert shows E_1(TO) and E_2(high) modes for various distances from the GaN interface.

The peak frequency of the Raman modes are displayed as a function of the distance from the GaN/sapphire interface in Figure 15.19. To verify, the reproducibility measurements were made in both directions: from the interface to the surface and then backwards. For the A_1 (TO) mode no shift could be observed. Also there were no significant changes in *full width at half maximum* (FWHM) or intensity of the spectral bands. The E_2 Raman mode is known to be shifted by stress. In GaN films large compressive stresses causing frequency shifts of +4.5 cm^{-1} compared to single crystals have been observed [87, 88]. The small shifts in the data presented here indicate the good quality of our films. The absence of a shift in the A_1 phonon can be explained by the direction of the optical phonon eigenvectors with respect to the *c*-axis. Among the Raman active modes A_1 is the only one with displacements along the *c*-axis as opposed to the E_2(high) and E_1(TO) vibrations [82, 85]. Thus it appears that the stresses are perpendicular to the *c*-axis.

Raman microscopy enables measurement of spectra from a single "point" defined by the optical resolution limit of the microscope. It thus allows point-to-point mapping whereby the spectroscopic information from different points of a relatively large sample can be detected and compared. Raman imaging allows one to noninvasively visualize chemical heterogeneity through the integration of microscopy and spectroscopy. It provides information that is useful in the fabrication of new materials, evaluation of the performance of existing materials, and control of product quality.

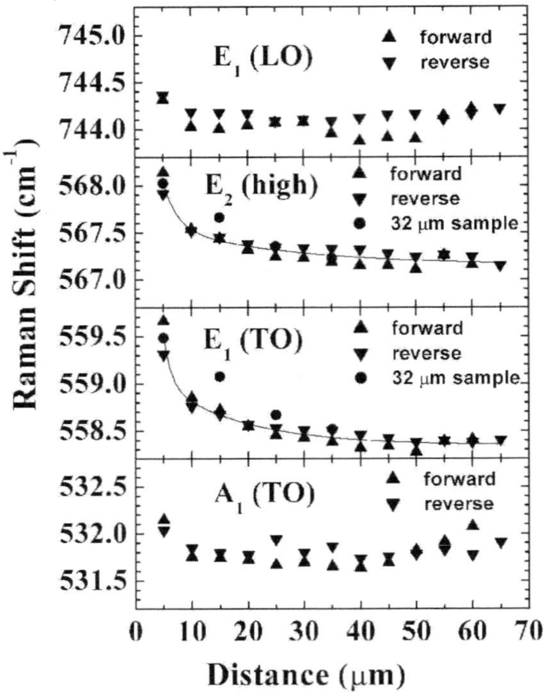

Figure 15.19 Peak frequencies of the Raman E_2(high) and E_1(TO) modes as a function of distance to the substrate interface. The solid lines represent exponential fits to the data points with asymptotic values of 567.2 and 558.3 cm^{-1}, respectively. The open symbols show results from a second experiment on a sample grown under the same conditions.

Figure 15.20 shows an optical image of an integrated circuit and a Raman intensity image of the region of the circuit indicated by the square. The stepsize was 1.75 μm at an integration time of 10 sec per point spectrum. From the spectra the Raman intensity map on the right of Figure 15.20 was constructed. Different gray levels are used to label different intensity of the 520 cm^{-1} silicon mode.

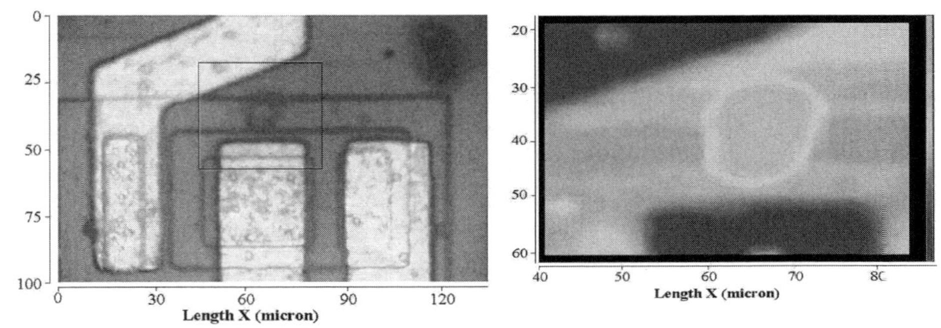

Figure 15.20 Optical image (left) and Raman intensity map of the region indicatd by the square (right) of an integrated circuit.

The frequency of the Si phonon is sensitive to local stress and thus Raman microscopy is employed in strain analysis of silicon materials [89]. Raman imaging is a rapidly developing field that extends into cell imaging and biomedical diagnostics. We refer to some excellent reviews in the literature [6, 90, 91] which describe the development and detailed applications of this technique.

15.5 CONCLUSIONS AND OUTLOOK

Raman spectroscopy combined with flexible sampling arrangement such as that provided by a microscope attachment has developed into a powerful analytical technique providing molecular information on materials with high spatial resolution. Changes in chemical bonding as well as structural variations depending on processing and illumination conditions can be characterized. The micro-Raman configuration provides depth-resolved data suitable for probing interfacial structural differences within multilayer structures, whereas the waveguide Raman technique enables longitudinal sampling.

Small, all-solid state excitations sources are likely to continue to have an impact on extremely compact, portable microspectroscopy systems for material analysis. This applies particularly to near-infrared diode or miniaturized Ti:sapphire lasers. Another recent development combines two of the principal probes of molecular structure—Raman scattering and infrared absorption—in a united instrument [92]. Because their selection rules complement one another, the two vibrational spectroscopies provide a powerful tool in combination.

ACKNOWLEDGMENTS

It is a pleasure to acknowledge a host of colleagues, students, and collaborators, in particular T. Cardinal, L. Chernyak, O. Galkin, N. Ho, C. Lopez, S. Nonnenmann, G. Nootz, W. Petry, A. Pope, C. Rivero, K. Richardson, J. Rabolt, S. Seal, W. Schirmacher, R. Vallee, and T. Winningham. We appreciate financial support from NSF grants ECS-0123484 and DMR-0421253, and the University of Central Florida.

REFERENCES

[1] Raman, C.V. & Krishnan, K.S. (1928) *Nature* **121**, 501.
[2] McCreery, R.L. (2000) *Raman Specctroscopy for Chemical Analysis* New York:Wiley.
[3] Chase, B. (1994) *Appl. Spectrosc.* **48**, 14A.
[4] Pelletier, M. (1999) *Analytical applications of Raman spectroscopy* Blackwell Science.
[5] Laserna, J.J. (Ed.) (1996) *Modern Techniques in Raman Spectroscopy* New York: Wiley.

[6] Colomban, Ph. & Corset, J. (Eds.) (1999) *J. Raman Spectrosc.* **30** Special Issue: Raman (Micro) Spectrometry and Materials Science.
[7] Weber, W.H. & Merlin, R. (Eds.) (2000) *Raman Scattering in Materials Science* New York: Springer.
[8] Zhang, S.-L. & Zhu, B.-F. (Eds.) (2000) *Proc. XXVIIth Intl. Conf. on Raman Spectroscopy* New York: Wiley.
[9] Schulte, A. & Richardson, K. (2002) In Eds. M. Kawasaki, N. Ashgriz, & R. Anthony, *Recent Research Developments in Non-Crystalline Solids 2*, p 143-158 Vol 1 Trivandrum: Transworld Research Network.
[10] Cardinal, T., Richardson, K.A., Shim, G., Stegeman, G., Beathy, R., Schulte, A., Meghini, C., Viens, J.-F., Le Foulgoc, K. & Villeneuve, A. (1999) *J. Non-Cryst. Solids* **256 & 257**, 353.
[11] Ramachandran, S., Bishop, S., Guo, J.P. & Brady, D.J. (1996) *IEEE Phot. Tech. Lett.* **8,** 1041.
[12] Efimov, O.M., Glebov, L.B., Richardson, K.A., Van Stryland, E., Cardinal, T., Park, S.H., Couzi, M. & Bruneel, J.L. (2001) *J. Opt. Mater.* **17**, 379.
[13] Andriesh, A.M., Bykovskii, Yu.A., Kolomeiko, E.P., Makovkin, A.V., Smirnov, V.L. & Shmal'ko, A.V. (1977) *Sov. J. Quantum Electron.* **7**, 347.
[14] Asobe, M., Kanamori, T. & Kubodera, K. (1993) *IEEE J. Quant. Electr.* **29,** 2325.
[15] Pope, A., Schulte, A., Guo, Y., Ono, L.K., Roldan-Cuenya, B., Lopez, C., Richardson, K., Kitanowski, C. & Winningham, T. (2005) *Vibrationaal Spectroscopy* in press.
[16] Rabolt, J.F. (1994) In *Fourier Transform Raman Spectroscopy* p 133 Eds. Chase, D.B. & Rabolt, J. F. San Diego: Academic Press.
[17] Schulte, A., Rivero, C., Richardson, K., Turcotte, K., Laniel, J., Hamel, V., Villeneuve, A., Saliminia, A. & Galstian, T. (2001) *Optics Comm.* **198**, 125-128.
[18] Adar, F. (2001) *Evolution and Revolution of Raman Instrumentation.* In Eds. Lewis, I.R. & Edwards, H.G.M. *Handbook of Raman Spectroscopy* p 11-40 New York: Dekker.
[19] Lewis, I.R. & Edwards, H.G.M. (Eds.) (2001) *Handbook of Raman Spectroscopy.* New York: Dekker.
[20] Chase, D.B. & Rabolt, J.F. (Eds.) (1994) *Fourier Transform Raman Spectroscopy* San Diego: Academic Press.
[21] Long, D.A. (1977) *Raman Spectroscopy* New York: McGraw Hill.
[22] Schrader, B. (2001) *Infrared and Raman Spectroscopy: Methods and Applications* New York: VCH Publishers.
[23] Hayes, W. & Loudon, R. (1978) *Scattering of Light by Crystals* New York: John Wiley & Sons
[24] Berne, B.J. & Pecora, (1976) *Dynamic Light Scattering* New York: Wiley.
[25] Faurskov Nielsen, O. (2001) *Low-Frequency Raman Spectroscopy and Biomolecular Dynamics,* In Eds. Lewis, I.R. & Edwards, H.G.M. *Handbook of Raman Spectroscopy* p 593-615 (New York: Dekker)
[26] Carraba, M.M., Spencer, K.M., Rich, C. & Rauh, D. (1990) *Appl. Spectrosc.* **44**, 1558-1561.
[27] Yang, B., Morris, M. & Owen, H. (1991) *Appl. Spectrosc.* **45**, 1533-1536.
[28] Schulte, A. (1992) *Appl. Spectrosc.* **46**, 891-893.
[29] Spiro, T.G. (Ed.) (1998) *Biological Applications of Raman Spectroscopy* New York: Wiley.
[30] Hirschfeld, T. & Chase, B. (1986) *Appl. Spectrosc.* **40**, 1079.
[31] Zimba, C.G., Hallmark, V.M., Swalen, J.D., Rabolt, J.F. (1987) *Appl. Spectrosc.* **41**, 721.
[32] Hallmark, V.M., Sooriyakumaran, R., Miller, R.D. & Rabolt, J.F. (1989) *J. Chem. Phys.* **90** , 2486.
[33] Hallmark, V.M. & Rabolt, J.F. (1989) *Macromolecules* **22**, 500-502.
[34] Lewis, E.N., Kalasinsky, V.F. & Levin, I.W. (1988) *Appl. Spectrosc.* **42**, 1188.
[35] Schulte, A., Hallmark, V.M. & Rabolt, J.F., unpublished.
[36] Myers, A.B., Harris, R.A. & Mathies, R.A. (1983) *J. Chem. Phys.* **79**, 603.

[37] Schulte, A., Lenk, T., Hallmark, V. & Rabolt, J.F. (1991) *Appl. Spectrosc.* **45,** 325-327.
[38] Williamson, J.M., Bowling, R.J. & McCreery, R.L. (1989) *Appl. Spectrosc.* **43**, 372.
[39] Refi, J.J. (1999) *Bell Labs Techn. J.* Jan-March 246-260.
[40] Hopf, F.A. & Stegeman, G.I. (1986) In Applied Classical Electrodynamics, Chapter 20 New York: Wiley Interscience.
[41] Stegeman, R., Jankovic, L., Kim, H., Rivero, C., Stegeman, G., Richardson, K., Delfyett, P., Guo, Y., Schulte, A. & Cardinal, T. (2003) *Optics Lett.* **28**, 1126-1128.
[42] Guo, Y., Nonnenmann, S., Schulte, A., Rivero, C., Richardson, K., Stegeman, R., Stegeman, G. & Cardinal, T. (2004) *Conference on Lasers and Electro-Optics (CLEO) Technical Digest (OSA)* CThP2.
[43] Shuker, R.V. & Gammon, R.W. (1970) *Phys. Rev. Lett.* **25**, 222.
[44] Surovtsev, N.V. & Sokolov, A.P. (2002) *Phys. Rev. B* **66**, 054205.
[45] Elliott, S.R. (2001) *Amorphous Materials*, In Encyclopedia of Materials Science and Technology p 171-174, Elsevier.
[46] Schirmacher, W., Diezemann, G. & Ganter, C. (1998) *Phys. Rev. Lett.* **81**, 136-139.
[47] Schirmacher, W., Pöhlmann, M. & Maurer, E. (2002) *Phys. Stat. Sol. (b)* **230**, 31.
[48] Asobe, M., Kanamori, T. & Kubodera, K. (1993) *IEEE J. Quant. Electr.* **29**, 2325.
[49] Zoubir, A., Richardson, M., Rivero, C., Schulte, A., Richardson, K., Ho, N. & Vallee, R. (2004) *Opt. Lett.* **29**, 748-750.
[50] Pope, A. (2005) Honor's thesis, University of Central Florida.
[51] Li, W., Rivero, C., Pope, A., Myneni, S., Lopez, C., Schulte, A., Richardson, K., Seal, S., Jain, H., Antoine, K. & Miller, A. (2005) *J. Appl. Phys.* in press.
[52] Frumar, M., Vlcek, M., Cernosek, Z., Polak, Z. & Wagner, T. (1997) *J Non-Cryst. Solids* **213 & 214**, 215.
[53] Krecmer, P., Vlcek, M. & Elliot, S.R. (1998) *J. Non-Cryst. Solids* **227-228**, 682.
[54] Lucovsky, G. & Martin, R. (1972) *J. Non-Cryst. Solids* **8-10**, 185.
[55] Levy, Y., Imbert, C., Cipriani, J., Racine, S. & Dupeyrat, R. (1974) *Opt. Comm.* **11**, 66.
[56] Hetherington III, W.M., Van Wyck, N.E., Koenig, E.W., Stegeman, G.I. & Fortenberry, R.M. (1984) *Optics Lett.* **8**, 88.
[57] Duverger, C., Nedelec, J.-M., Benatsou, M., Bouazoui, M., Capoen, B., Ferrari, M. & Turrell, S. (1999) *J. Mol. Struct.* **480-481**, 169.
[58] Urlacher, C., Marty, O., Plenet, J.C., Serughetti, J. & Mugnier, J. (1999) *Thin Solid Films* **349**, 63.
[59] Wise, K.J., Gillespie, N.B., Stuart, J.A., Krebs, M.P. & Birge, R.R. (2002) *Trends in Biotech.* **20**, 387.
[60] Hampp, N. (2000) *Chem. Rev.* **100**, 1755-1776.
[61] Mathies, R.A., Smith, S.O. & Palings, I. (1987) In Spiro T G (Ed.) *Biological Applications of Raman Spectroscopy Vol 2* p. 59-108 (New York: Wiley).
[62] Friedman, J., Rousseau, D.L. & Ondrias, M.R. (1982) *Annu. Rev. Phys. Chem.* **33**, 471-491.
[63] Bangcharoenpaurpong, O.K., Schomaker, T. & Champion, P.M. (1984) *J. Am. Chem. Soc.* **108**, 1163-1167.
[64] Ahmed, A.M., Campbell, B.F., Caruso, D., Chance, M.R., Chavez, M.D., Courtney, S.H., Friedman, J.M., Iben, I.E.T., Ondrias, M.R. & Yang, M. (1991) *Chem. Phys.* **158**, 329-352.
[65] Weber, G. & Drickamer, H.G. (1983) *Quart. Rev. Biophys.* **16**, 89-112.
[66] Kauzmann, W. (1987) *Nature* **325**, 763.
[67] Winter, R. (2003) (Ed.), *Advances in High Pressure Bioscience and Biotechnology II*, New York: Springer.
[68] Wong, P.T. (1984) *Ann. Rev. Biophys. Bioeng.* **13**, 1 (1984).
[69] Schulte, A., Buchter, S., Galkin, O. & Williams, W. (1995) *J. Am. Chem. Soc.* **117**, 10149-10150.
[70] Morild, E. (1981) Theory of the pressure effects on enzymes in *Adv. Protein Chem.* **34**, 93.
[71] Galkin, O., Buchter, S., Tabirian, A. & Schulte (1997) *Biophys. J.* **73**, 2752 -2763.

[72] Schulte, A. & Bradley II, L. (1995). *Biophys. J.* **69**, 1554-1562.
[73] Kitagawa, T. (1988) Heme protein structure and the iron histidine stretching mode. In Spiro, T.G. (Ed.) *Biological Applications of Raman Spectroscopy*. 97-132, vol 3 (New York: Wiley).
[74] Rousseau, D.L. & Friedman, J.M. (1988) Transient and cryogenic studies of photodissociated hemoglobin and myoglobin. In Spiro, T.G. (Ed.) *Biological Applications of Raman Spectroscopy*. vol. 3 133-216 (New York: Wiley).
[75] Sassaroli, M.S., Dasgupta & Rousseau, D.L. (1986) *J. Biol. Chem.* **261**, 13704-13703.
[76] Choi, S. & Spiro, T.G. (1983) *J. Am. Chem. Soc.* **105**, 3683-3692.
[77] Schlichting, I., Berendzen, J., Phillips, Jr. G.N. & Sweet, R.N. (1994) *Nature* **371**, 808-812.
[78] Sage, J.T., Schomacker, K.T. & Champion, P.M. (1995) *J. Phys. Chem.* **99**, 3394-3405.
[79] Urayama, P., Phillips, G.N. & Gruner, S. (2002) Probing substates in spermwhale myoglobin using high-pressure x-ray crystallography. *Structure* **10**, 51-60.
[80] Schulte, A. & Galkin, O. (2003) *In Ed. Winter R. Advances In High Pressure Bioscience and Biotechnology II*, (New York: Springer) p 121-124.
[81] Delhaye, M. & Dhamelincourt, P. (1975) *J. Raman Spectrosc.* **3,** 33-43.
[82] Bergman, L., Dutta, M. & Nemanich, R.J. (2000) In Eds. Weber, W.H. & Merling, R. *Raman Scattering in Materials Science* (Berlin: Springer).
[83] Pearton, S.J., Zoiper, J.C., Shul, R.J. & Ren, F. (1999) *J. Appl. Phys.* **86**, 1.
[84] Nootz, G., Schulte, A., Chernyak, L., Osinsky, A., Jasinski, J., Benamara, M. & Liliental-Weber (2002) *Appl. Phys. Lett.* **80**, 1355-1357.
[85] Zhang, J.M., Ruf, T., Cardona, M., Ambacher, O., Stutzmann, M., Wagner, J.-M. & Bechstedt, F. (1997) *Phys. Rev. B* **56**, 14399.
[86] Rousseau, D.L. (1981) *J. Raman Spectrosc.* **10**, 94.
[87] Perlin, P., Jauberbie-Carillon, C., Itie, A.S., Miguel, I., Grzegory, I. & Polian, A. (1992) *Phys. Rev. B* **45**, 83.
[88] Kiesielowski, C., Krueger, J., Ravimov, S., Suski, T., Ager III, J.W., Jones, E., Liliental-Weber, Z., Rubin, M., Weber, E.R., Bremser, M.D. & Davis, R.F. (1996) *Phys. Rev. B* **54**, 17745.
[89] De Wolf, I. (1999) *J. Raman Spectrosc.* **30**, 877-883.
[90] Puppels, G. et al. (2001) In-vivo Raman spectroscopy, In Eds. Lewis I.R. & Edwards, H.G.M. *Handbook of Raman Spectroscopy* p 549-574 (New York: Dekker).
[91] Diem, M., Romeo, M., Boydston-White, S., Miljovic, M. & Matthäus, C. (2004) *Analyst* **129**, 880-885.
[92] Adar, F., LeBurdon, G., Reffner, J. & Whitley, A. (2003) *Spectroscopy* **16**, 34-40.

CHAPTER 16

POLARIZATION SPECTROSCOPY OF ORDERED SAMPLES

[1] Peter W. Thulstrup and [2] Erik W. Thulstrup
[1] Dept. of Natural Sciences, The Royal Veterinary and Agricultural University, Frederiksberg, Denmark
[2] Dept. of Life Sciences and Chemistry, Roskilde University, Roskilde, Denmark

16.1 INTRODUCTION

16.1.1 Linearly Polarized Light

Light around us is often, at least partially, linearly polarized; this is, for example, true for scattered light. Since light from the blue sky is also scattered light, it, too, is partially linearly polarized. The polarization direction of this light is perpendicular to the plane defined by the point on the blue sky that is being observed, the eye of the observer, and the sun. Clouds may reduce or eliminate completely the light polarization.

This striking property of light from the blue sky is used for navigation by insects, such as bees and desert ants [1], and has also been used by humans for that purpose. When the Vikings traveled from Scandinavia to North America more than 1000 years ago (and 500 years before the travels of Columbus) they used a navigational instrument called a *sunstone*, according to the Icelandic sagas. A sunstone is believed to be a naturally occurring linear polarizer (several candidates for this may be found on Scandinavian beaches) that makes it possible to determine the polarization of the light from the sky and thus also the position of the sun, even when it is below the horizon or covered by clouds. Since this technology was not research-based, the sunstone was later forgotten, but it was reinvented in the early 1900s and called a *twilight compass* [2]. This compass was used for flights in the Polar region where magnetic compasses do not work.

Just like light polarization may serve navigational purposes in the macroscopic world, it may help determine directional properties in the molecular world. The first use of linearly polarized light in optical spectroscopy goes back to Ambronn [3], who studied dyed (and aligned) membranes this way. Ambronn's samples were dichroic, i.e., light absorption changed when the polarization direction of the light beam was rotated, specifically 90°.

The *linear dichroism* (LD) of a sample is a function of the wavelength and is defined as the change in absorption (absorbance) that is observed when the electric vector of the light beam is rotated 90 degrees. LD spectroscopy is only of interest when the samples studied are not isotropic, as liquid solutions are. Many other samples (from biological samples to industrial products) have a molecular alignment that makes them dichroic. For such samples the use of linearly polarized light in spectroscopic experiments may provide two (in the general case, but rarely in practice, three) linearly independent spectra instead of just one, and this is likely to lead to additional, and sometimes extremely useful, information [2, 4–6].

16.1.2 Transition Moment Directions

In the following we shall concentrate on optical spectroscopy, primarily absorption spectroscopy, in the infrared (IR), visible, and ultraviolet (UV) regions. The molecular transitions in question are thus vibrational transitions, electronic transitions, and a combination of these, vibronic transitions, as illustrated in Figure 16.1.

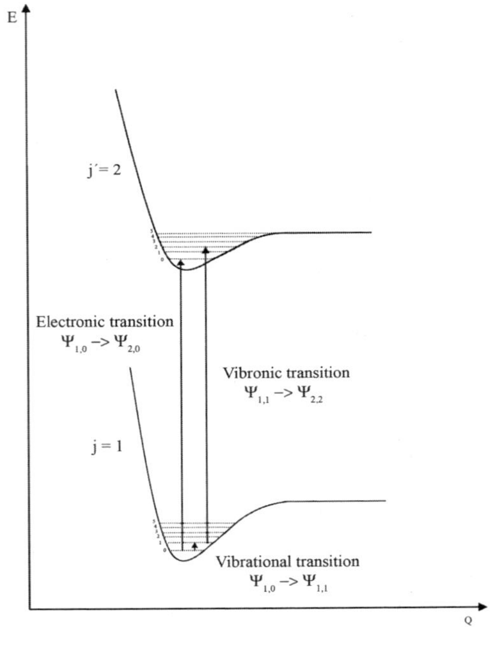

Figure 16.1 Simplified potential energy curves depicting energy (*E*) as a function of a coordinate (*Q*), with indication of vibrational, electronic, and vibronic transitions.

A spectroscopic transition (in absorption) is characterized by energy, intensity, and direction. The observed transition energy is the energy difference between final state $\langle i |$, and the initial (ground) state $| 0 \rangle$; the observed intensity is roughly proportional to the square of the *transition moment* between the two states:

$$\mathbf{M_i} = \langle i | \mathbf{M_{op}} | 0 \rangle.$$

Here, $\mathbf{M_{op}}$ is the electric dipole operator, a sum over all charged particles in the molecule (nuclei, electrons) of their position vector times their charge. $\mathbf{M_{op}}$ is a vector operator; therefore, $\mathbf{M_i}$ is also a vector, and this gives rise to the third property of light absorption, in addition to transition energy and intensity: the direction of $\mathbf{M_i}$ in the molecule.

Unfortunately, this directional property is usually overlooked, although it may often be very useful. The specific direction of $\mathbf{M_i}$ in the molecular framework is called the transition moment direction for the transition between states 0 and i. $\mathbf{M_i}$ may be thought of as an antenna, able to catch energy from electromagnetic radiation (light). The probability for absorption of linear polarized light, polarized along direction $\boldsymbol{\varepsilon}$, and thus the observed intensity, is proportional to [2]:

$$(\mathbf{M_i} \cdot \boldsymbol{\varepsilon})^2$$

In other words, when the transition moment in a molecule is perpendicular to $\boldsymbol{\varepsilon}$, the molecule will not be able to absorb light, but when they are parallel, light absorption by the molecule will reach a maximum. In more precise terms, the transition probability is proportional to $\cos^2(\mathbf{M_i}, \boldsymbol{\varepsilon})$, where $(\mathbf{M_i}, \boldsymbol{\varepsilon})$ is the angle between the two vectors. For a general sample with differently aligned molecules the observed absorption will be proportional to $\langle \cos^2(\mathbf{M_i}, \boldsymbol{\varepsilon}) \rangle$, where $\langle\ \rangle$ indicates an average over all sample molecules.

We are now able to understand the science behind a piece of simple consumer technology. As mentioned above, light from the blue sky is scattered light and therefore partially polarized. In our environment there are many other kinds of scattered, and thus linearly polarized, light than that from the blue sky. Much of this comes from horizontal surfaces, and is therefore polarized horizontally. Such light is responsible for most unpleasant glare in nature. This is why Polaroid sunglasses became so popular. They use linear polarizers as lenses and, when worn in the traditional way, these polarizers are able to remove the horizontally polarized glare. Polaroid sunglasses consist of extruded poly(vinyl) alcohol) containing iodine, which forms long chains in the direction of the extrusion (here the horizontal direction). The iodine chains absorb all visible wavelengths (colors), but since their relevant transition moments are all aligned along the chain direction, and thus horizontal, they are unable to absorb vertically polarized components of the light, since for vertically polarized light, $\cos^2(\mathbf{M_i}, \boldsymbol{\varepsilon}) = 0$. Thus, the sunglasses are still transparent, but are nevertheless able to remove the unwanted, horizontally polarized light, as illustrated in Figure 16.2 (for the exciting story about Land and the start of Polaroid Corporation, see [6, 7]).

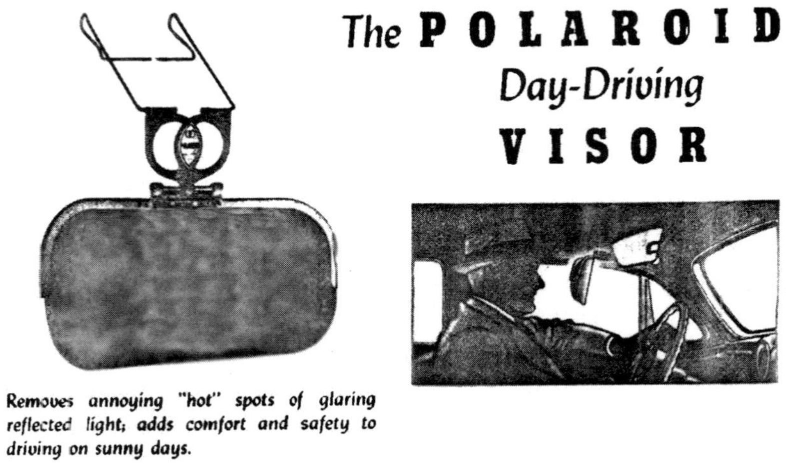

Figure 16.2 An old advertisement for a Polaroid product, which took advantage of the ability of a polarizer to eliminate glare from scattered sunlight.

The transition moment directions within the molecular framework are in many ways dependent on molecular symmetry. For example, for a molecule of C_{2v} symmetry, such as water, *cis*-butadiene, or phenanthrene (Figure 16.3), only three different transition moments are possible: along the symmetry axis (z) or perpendicular to it, either in the molecular plane (y) or perpendicular to the molecular plane (x). This is because the dipole moment operator (with components along x, y, and z) forms a basis for only those irreducible representations that contain the symmetry-determined axes, in this case, A_1 (z), B_1 (x), and B_2 (y). In order for the transition moment to be different from zero, the integral $\langle i | \mathbf{M}_{op} | 0 \rangle$ must have a totally symmetric integrand—in the case of non-totally symmetric integrands, contributions to the integral from different parts of space will cancel [2]. Thus, if we assume that the ground state is totally symmetric, as is usually the case, only excited states of A_1, B_1, and B_2 symmetry may be reached through absorption of a single photon. Transitions to states of A_2 symmetry cannot be caused by electromagnetic radiation and such transitions are said to be symmetry-forbidden.

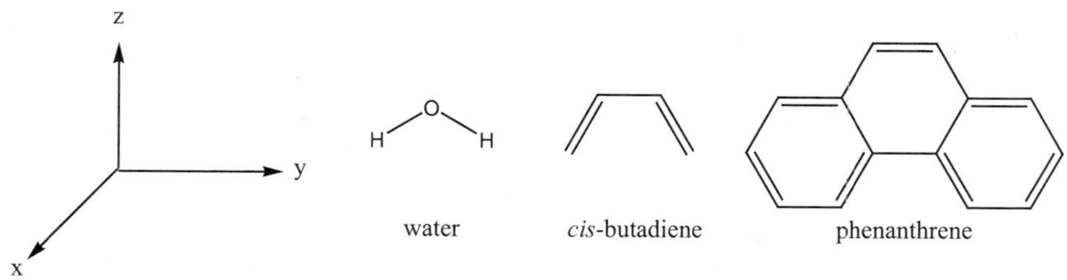

Figure 16.3 Examples of molecules belonging to the C_{2v} point group; *z* is the symmetry axis, *z* and *y* compose the molecular plane, while *x* is perpendicular to the molecular plane.

In other words, light polarized along the molecular *z*-axis may be able to excite (electronically and/or vibrationally) the molecule from a totally symmetric ground state (A_1) to another totally symmetric state. Light polarized along *x* may excite the molecule from the ground state to states of B_1 symmetry, and light polarized along *y* may make it possible to reach states of B_2 symmetry from the ground state, while excited states of A_2 symmetry cannot be reached by single photon light absorption. Figure 16.4 shows the similar conditions for a molecule of D_{2h} symmetry [2], where also three symmetry classes can be reached by light absorption.

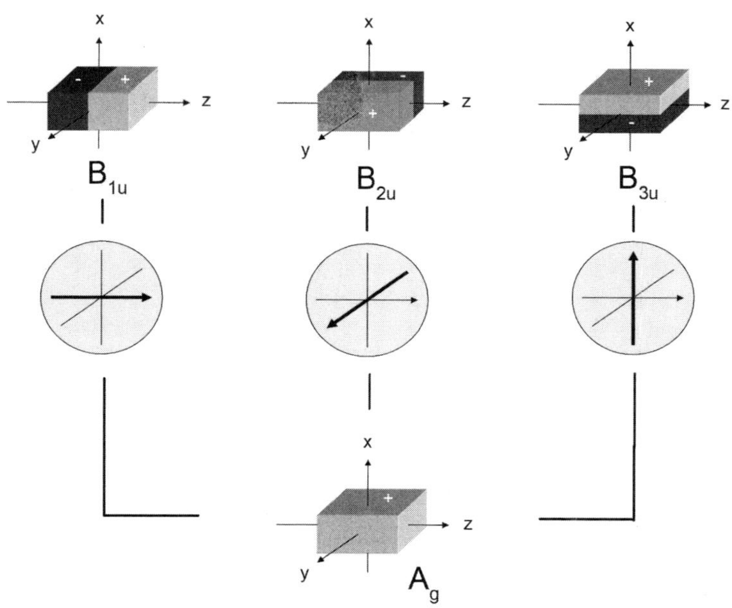

Figure 16.4 A molecule belonging to the D_{2h} point group having three symmetry planes; *xy*, *xz*, and *yz*. From the totally symmetric ground state (A_g), three different excited states can be reached through absorption of a linearly polarized photon, oriented along each of the three molecular axes. Illustration redrawn from [2].

16.1.3 Spectroscopy with Linearly Polarized Light

One of the important applications of LD spectroscopy is now clear. It provides a directional (or symmetry) criterion for spectral assignments, which often may be crucial, as illustrated later. In the case of molecules with symmetry elements the transition moment direction often acts like an *absolute criterion*, i.e., it specifies the correct symmetry without any deviations, and in that sense it is better than, e.g., the energy or intensity criteria.

Another important application of the directional information provided by LD spectroscopy is the determination of sample structures, both intramolecular—for example determination of the pyramidilization angle around N in aniline [8]—as well as intermolecular—for example the bonding of carcinogenic molecules to DNA [9]. The latter case is illustrated in Figure 16.5, where the difference in DNA binding of syn- and anti- forms of benzo[*a*]pyrene-diol-epoxides is reflected in the LD spectra. There exists a corresponding difference in carcinogenicity of these two stereoisomers, the anti-form being the stronger carcinogen [9].

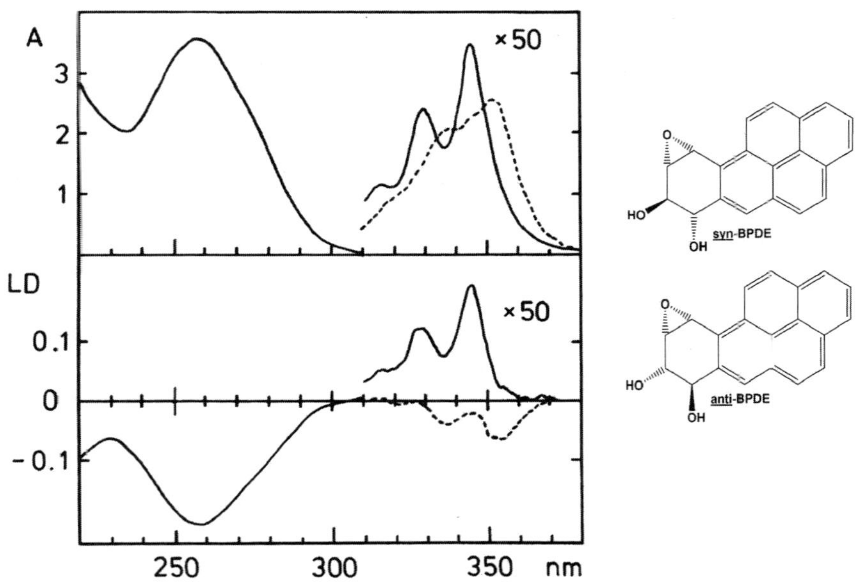

Figure 16.5 Absorbance (A) and linear dichroism (LD) spectra of anti(------)- and syn(------)-BPDE-DNA in aqueous solution (10 mM cacodylate buffer, pH = 7.0).The strong negative LD band at 258 nm can be attributed to DNA, while the longer wavelength signals correspond to π–π* transitions in the two benzo[*a*]pyrene-diol-epoxides. The data is reproduced from [9].

Clearly, sample structure will influence the result of an LD experiment and this gives an additional possibility for extracting information from an LD experiment. When, for example, the directions of one or more transition moments in the framework of a given sample molecule are known, they may be used to identify the position (orientation) of the molecule in the sample. It is, however, important to stress from the start that an LD experiment cannot provide anywhere near complete information on the sample structure, such as a complete orientation distribution function, except in the simple case where the alignment is perfect.

The information that is available from LD spectroscopy on the molecular orientation distribution in the sample is restricted to information about average values over all sample molecules of cosine squares of the angles between transition moments in molecules and the sample (laboratory) axes, as we shall see later. In the general case, a given set of such averages will correspond to an infinite number of possible orientation distribution functions for the sample molecules.

It is interesting to note that often neither the transition moment directions, nor the sample structure are known before the LD spectrum is recorded, but may be uncovered through initial assumptions on both kinds of information, which later may be verified.

In addition to absorption (LD) spectroscopy, other kinds of spectroscopy with linearly polarized light may also be performed on aligned samples. The most important of these involve two photons (see Table 16.1). In Raman spectroscopy and luminescence spectroscopy (fluorescence and phosphorescence), one photon is absorbed and another one, with different energy, is created. Other two-photon experiments, in which both photons are absorbed, are two-photon absorption and photoinduced dichroism. In Raman scattering and two-photon absorption spectroscopy the two photons act simultaneously, and new optical properties may be revealed: the Raman polarizibility tensor and the two-photon absorption tensor. In luminescence spectroscopy and photoinduced dichroism experiments the two photon events are successive and the spectra may be described in terms of transition moments.

Table 16.1 Overview of two-photon events and the corresponding physical processes [2].

Photon 1	annihilated	annihilated	annihilated	annihilated
Photon 2	annihilated	annihilated	created	created
Timing	successive	simultaneous	simultaneous	successive
Physical Process	photoinduced dichroism	2-photon absorption	Raman scattering	photolumi-pescence

In all two-photon experiments, new information may be obtained, both on the sample structure and on the optical properties of individual molecules. While the only information on the sample structure that may be provided by absorption spectroscopy are average values over the sample molecules of cosine squares of angles between transition moments and the sample axis, two-photon spectroscopy may in addition provide averages of the fourth powers of such angles. This is important information; one may say that while LD absorption spectroscopy provides information on the average molecular alignment, the addition knowledge from two-photon spectroscopy may provide information about the spread of the molecular orientation distribution function around these average alignments.

16.2 OCCURRENCE, PRODUCTION, AND OPTICAL PROPERTIES OF ALIGNED, SOLID SAMPLES

16.2.1 Perfectly and Partially Aligned Samples

While the molecular orientation distribution in liquid solutions is isotropic, many naturally occurring samples have some kind of order (are aligned). This is not only the case for crystals; many biological samples, for example, from plants or animals, have alignment, although this alignment tends to be only partial, not perfect. Also, many industrial products are aligned samples: the strength of common plastic bags, for example, is due to the fact that they consist of two crossed layers of stretched or extruded, in other words, aligned polyethylene. Such polyethylene material cannot be stretched further in the direction of stretching; it is therefore very strong in this direction, and by crossing two layers a very sturdy material is produced. The alignment in the polyethylene bags and in many naturally occurring or industrial samples is not perfect, but only partial. However, it turns out that such samples, especially those produced by means of aligning (solid) solvents, are often the most useful. Therefore, after a brief discussion of other alignment techniques, we shall concentrate on solutes that are partially aligned in solid solvents.

Intuitively, one might expect that perfectly aligned crystals would provide ideal samples for LD spectroscopy. It is also true that both single crystals and mixed crystals have been used for polarization spectroscopic purposes [2]. However, spectroscopic experiments with polarized light on crystalline samples have often been difficult, for example, because very thin slices of single crystals are required, especially for studies of electronic transitions. An alternative possibility, to measure reflection spectra, leads to new complications. In addition, interpretation of the spectra in terms of single molecular transitions requires knowledge of the crystal structure, as well as of the often very strong intermolecular interactions (Davydov splitting).

Instead, mixed crystal methods have sometimes been used. This requires that a host crystal be found that does not absorb in the region of interest and in which the molecule to be studied fits perfectly, without changing the crystal structure (or possibly changes it in a known fashion). The intermolecular interactions are often quite limited in such mixed crystals, but it is frequently very difficult to find suitable host crystals [2].

A very elegant way of orienting molecules, or rather of selecting an oriented subsample is photo-orientation. For example, if a sample absorbs light, linearly polarized along axis *U*, the subset that absorbs will have a symmetrical orientation distribution of the absorbing transition moment $\mathbf{M}_i$ around axis *U*, with a probability of excitation that is proportional to $\cos^2(\mathbf{M}_i, U)$—assuming that saturation effects are not important. The set of remaining molecules will also be aligned; if the absorption experiment runs to near completion the alignment of the remaining molecules may become extremely good (near perfect)—then only molecules for which $\mathbf{M}_i$ and *U* are nearly perpendicular will be left.

If the absorbing molecules just returned to their ground state, little would be achieved. They must undergo a transformation (chemical reaction) in order to produce a useful sample. Similarly, they must also be fixed in their positions; if they were allowed to rotate, the alignment of the subset would be lost. Therefore the samples are typically fixed in a rare gas matrix at low temperatures, as we shall see in an example given below.

Also, electric or magnetic (static) fields may be used to align samples in which the molecules are not completely fixed. Molecules with a permanent electric dipole or an anisotropic electric polarizability may be aligned in strong electric fields. Similarly, molecules with a permanent magnetic dipole or an anisotropic electric susceptibility may be aligned in magnetic fields, although this latter effect is usually very weak and the methods are primarily relevant for large molecules. The importance of these techniques for solid samples is, however, limited. See, for example the book by Fredericq and Houssier [10] and the work by Liptay [11], the latter deals primarily with smaller molecules. In recent years, alignment by strong laser pulses have become of interest [12], but again the methods are of very limited interest in the present context.

16.2.2 Solutes in Partially Aligning Solvents

In the following, we shall concentrate on the production of aligned samples by means of solid solvents with an ability to align solutes. In particular we shall look at stretched polymers, mostly polyethylene, but nematic liquid crystals are also able to align solutes. However, they are hardly solids and are thus a borderline case for this discussion. However, they may in many ways be used in the same way as stretched polymers, as we shall see in a later example.

The main properties of aligning solvents to be considered for use in spectroscopy are related to their baseline absorption (or scattering), their ability to dissolve the solutes to be studied, and their ability to align them. Also, other properties are of interest, for example, the ability to withstand low temperatures and thus to be used in low temperature work. Partial alignment of solutes in stretched polymers, especially polyethylene and to some extent also poly(vinylalcohol), has been studied in great detail [2] and the discussion below will most specifically refer to these two solvents.

It is extremely important that the aligning solvent (e.g., the stretched polymer) not absorb or scatter light excessively in the spectral region to be studied. Both polyethylene and poly(vinylalcohol) have good transmission properties in the visible region and down to around 200 nm in the ultraviolet (UV) region. In addition, polyethylene has the added advantage of being quite transparent in most of the infrared (IR) region. The main exceptions are small regions around 720 cm^{-1}, 1450 cm^{-1}, and 2900 cm^{-1} [2]. These regions may even be included in the investigation, through the recording of spectra with per-deuterated polyethylene as the aligning solvent (see, for example, [13]). For the deuterated solvent, the corresponding absorption bands are located at lower wavenumbers; thus, the regions blocked in spectra of normal polyethylene are visible. However, commercially available per-deuterated polyethylene is expensive and for some qualities, the production of high quality stretched sheets is not always simple.

Poly(vinylalcohol) has a much more complicated IR spectrum than polyethylene, the baseline problems are considerable in this region, and it is usually avoided in IR LD spectroscopy. Nevertheless, a few examples exist of successful use of poly(vinylalcohol) in IR spectroscopy (see Holmén [14]).

The reason for the wish to use poly(vinylalcohol) is that it is able to dissolve and align quite large amounts of many important polar molecules, in contrast to polyethylene. It is also possible to obtain high concentrations in a poly(vinylalcohol) sheet by mixing poly(vinylalcohol) powder with the solute in a water-alcohol solution, casting the thick solution on a glass plate and letting the solvent evaporate [2, 14]. In general, spectral intensity in absorption is roughly proportional to the transition energy (wavenumber) of the transition. This means that in the infrared region, where transition energies are about one tenth of those in the UV-visible regions, a much larger amount of solute is required. This makes polyethylene less well suited for studies of strongly polar molecules, especially in the IR region.

Both polyethylene and poly(vinylalcohol) are able to align solutes sufficiently for LD spectroscopy, unless these are near-spherical in shape. In particular, the alignment of molecules in polyethylene has been studied in some detail [15–19]. Polyethylene consists of crystalline and amorphous regions; it is known that a very high crystallinity results in poor solubility for most solutes and that a very low crystallinity reduces the aligning power of the stretched polymer. This is understandable from LD IR measurements of the pure polyethylene sheet [18], where molecules associated with the amorphous regions were shown to have essentially no alignment (the transitions had K-values around 0.33, corresponding to an isotropic sample, see below). In contrast, molecules associated with crystalline regions turned out to be well aligned, although not perfectly aligned (K-values between 0.60 and 0.90, see later) [18]. Based on such measurements it has been concluded [15–19] that the most likely location of solute molecules in stretched polyethylene is the surface of lateral crystalline regions.

It is important that the sample alignment be reasonably simple to handle in terms of the mathematical treatment of the spectra. Thus, it is tempting to assume that samples aligned in stretched polymers are uniaxial, i.e., that one sample axis (*U*, the stretching direction for the polymer) is unique and all directions perpendicular to *U* (including the other laboratory axes *V* and *W*) are equivalent. This greatly simplifies the interpretation of the spectra. Uniaxiality has been assumed without proof for many important samples. This has even been done for solutes in thin stretched polyethylene sheets (with *U* as the stretching direction, *V* as the other axis in the sheet plane, and *W* perpendicular to the sheet), for which it may seem surprising that *V* and *W* are equivalent. However, the assumption of uniaxiality has been studied in detail, both in the case of undoped, thin stretched polyethylene sheets (by X-ray diffraction [15, 16]) and for solutes in stretched polyethylene (by the IR optical spectroscopic methods mentioned above [18]), and has been shown to hold within experimental accuracy. This is a major advantage in connection with the mathematical treatment of the spectra.

16.3 ONE-PHOTON SPECTROSCOPY: LINEAR DICHROISM

16.3.1 Optical Spectroscopy with Linearly Polarized Light: Experimental Needs

Linear dichroism absorption spectroscopy may be performed in the ultraviolet, visible, near-infrared, infrared, and far-infrared spectral regions on the proper ordinary spectroscopic instruments that have been equipped with an adjustable linear polarizer or other polarizing element (see later) for the spectral region in question. The only nontrivial instrumental demands exist in the infrared region where the use of Fourier transform instruments in reality is necessary, but today FTIR instruments have become both inexpensive and widely available. The light polarization may be ensured by either a static polarizer or by means of a polarization modulator. It may be added that measurements may even be performed without polarizers by taking advantage of the fact the electric vectors of natural light are not isotropic, but always perpendicular to the direction of the light beam [20] (see later).

For the UV-visible region the most commonly used quality linear polarizers have been calcite Glan prisms. They are efficient and transparent throughout the visible region and down to about 215 nm. Much less expensive sheet polarizers are available, in particular for the visible region, where even Polaroid sunglasses or polarizers taken from discarded liquid crystal displays (each contain two of such polarizers) may be used. For measurements around

200 nm and below more expensive and less convenient polarizers are required. They may be of Rochon or Wollaston type and consist of magnesium fluoride or quartz.

In the infrared region aluminum or gold grid polarizers have been widely used, but today also higher quality sheet polarizers are available. Usually several polarizers are required to cover the whole IR region.

In order to produce the two LD spectra either the sample or the polarizer must be rotated. It is often most convenient to rotate the polarizer; however, if the instrument in itself produces partially polarized light, as is often the case, the polarizer must be rotated between positions that are equivalent with respect the this polarization (e.g., be rotated from –45 to +45 degrees relative to the vertical) in order to ensure identical light intensity for the two spectra. In the light path, the polarizer may be placed either before or after the sample, but it should not be separated from the sample by any other optical elements, if at all possible, in order to avoid depolarization. However, both depolarization and scattering within the sample is quite common in the UV region (but is much less so in the IR). For example, aligning solvents such as stretched polymers or liquid crystals scatter UV light much more than most ordinary liquid solvents. What is usually referred to as "baseline absorption" due to a stretched polymer or liquid crystal should often really be called "baseline scattering."

Many of the uniaxial samples used are birefringent; this means that the light polarization will change when the beam passes the sample, unless it is polarized either along or perpendicular to the sample axis U (or is progressing along U). For this reason it is recommended that all spectra be recorded with light polarized along U, V, or W. For a survey of actual experimental setups, with both static polarization and modulated polarization, see [2].

16.3.2 Mathematical Descriptions of Aligned, Uniaxial Samples

Over the years, it has often been assumed that a full interpretation of LD spectra of a partially aligned molecular sample requires a complete knowledge of the orientation distribution function of the molecules in the sample. Fortunately this is not so. Only those properties of the orientation distribution function that influences the absorption probability are required.

Let us look at a single molecule in a uniaxial sample with sample axis U and a transition in this molecule between states 0 and i caused by absorption of light polarized along $\boldsymbol{\varepsilon}$ the transition moment is $\mathbf{M}_i$. We have already seen that the transition probability P_i in a single molecule is proportional to $(\mathbf{M}_i \cdot \boldsymbol{\varepsilon})^2$ or to $\cos^2(\mathbf{M}_i, \boldsymbol{\varepsilon})$, where $(\mathbf{M}_i, \boldsymbol{\varepsilon})$ is the angle between $\mathbf{M}_i$ and $\boldsymbol{\varepsilon}$. For a study of the same transition in the whole sample we have to average over all sample molecules and we obtain:

$$P_i = \langle \cos^2(\mathbf{M}_i, \boldsymbol{\varepsilon}) \rangle$$

This quantity contains all the orientational information required to describe the orientational dependence of the light absorption by the transition between states 0 and i.

The interpretation of the LD spectra will be greatly facilitated if the light polarization (**ε**) is chosen either along the sample axis U or perpendicular to it, along laboratory axes V and W. Therefore, we will introduce the following short notation for the quantities needed in former case:

$$\langle \cos^2 (\mathbf{M}_i, U) \rangle = K_{i.}$$

Basic geometry tells us that for each individual molecule we have

$$\cos^2 (\mathbf{M}_i, U) + \cos^2 (\mathbf{M}_i, V) + \cos^2 (\mathbf{M}_i, W) = 1.$$

This is thus also true for the averages:

$$\langle \cos^2 (\mathbf{M}_i, U) \rangle + \langle \cos^2 (\mathbf{M}_i, \mathrm{V}) \rangle + \langle \cos^2 (\mathbf{M}_i, \mathrm{W}) \rangle = 1.$$

Since the sample is uniaxial the absorption probability for light polarized along V and W will be the same, and we obtain for the averages:

$$\langle \cos^2 (\mathbf{M}_i, V) \rangle = \langle \cos^2 (\mathbf{M}_i, W) \rangle = (1 - K_i)/2.$$

In other words, all orientational information required for a description of the relevant experiments that may be performed on the transition from the ground state to state i is contained in the quantity K_i. Note that if the sample has no alignment (is isotropic) all K_i will be equal to $\frac{1}{3}$. This description contains no assumptions at all, except that the sample is uniaxial. It does not define the orientation distribution function (except in the special case of perfect alignment—K_i being equal to 1), only aspects of it, but it contains exactly that (minimum) information which is required to describe the orientational effects on spectroscopic observations of transition i.

The literature on mathematical descriptions of partially aligned samples used for LD spectroscopy has over the years demonstrated some unfortunate approximations, or even mistakes; some of these led to incorrect conclusions [2]. Most of these mathematical models contain (unfortunately usually implicit) assumptions about the orientation distribution of molecules in the sample. This is, for example, the case for one of the early models, the Tanizaki model [21].

Tanizaki's model is mathematically quite complicated. It produces a complete orientation distribution function for stretched polymer sheets and solutes in it. It was originally primarily applied for UV transitions in samples of stretched poly(vinylalcohol) sheets and solutes in these; it was assumed that, at an infinite stretch ratio, all molecules would be aligned similarly, with a specific molecular axis (not necessarily the long axis) perfectly aligned with the stretching direction, and that the orientation for actual stretch ratios would follow a specific (and quite complicated) formula. A detailed analysis of the assumptions contained in the Tanizaki model, when they are valid, and when they fail, may be found in [2].

A simpler model that also attempts a determination of the complete orientation distribution function is the Fraser-Beer model [22, 23]. It was originally designed to describe the alignment of stretched polymers. However, it has frequently been applied also to solutes in stretched polymers. In the Fraser-Beer model it is simply assumed that that a fraction f of the sample molecules are perfectly aligned and that orientation of the remaining fraction, $1 - f$, is perfectly isotropic. Although this orientation distribution is hardly ever realistic, the Fraser-Beer model may still yield correct results for the description of absorption experiments, but only as long as the molecules studied are aligned like rods (see later). For other alignments than this very special case the model will fail; see [2] for further details.

16.3.3 LD Spectra of Aligned, Uniaxial Samples

In the following, all the spectra are assumed to be corrected for baseline effects, especially absorption and light scattering by the solvent (see later). We can now write the observed solute absorbance in a spectroscopic absorption experiment with light polarized along the sample axis U:

$$E_U(\tilde{\nu}) = \Sigma\langle\cos^2(\mathbf{M}_i, U)\rangle A_i(\tilde{\nu}) = \Sigma K_i A_i(\tilde{\nu}) \qquad (16.1)$$

where the sum goes over all relevant transitions to states i and $A_i(\tilde{\nu})$ is equal to 3 times the contribution from transition i to the absorbance that would be observed if the sample were isotropic, in other words, the absorbance that would be observed if all transition moments $\mathbf{M}_i$ were perfectly aligned with U.

Similarly we have for the observed absorbance when the light is polarized perpendicularly to U (along V or W):

$$E_V(\tilde{\nu}) = E_W(\tilde{\nu}) = \Sigma\, \tfrac{1}{2}(1 - K_i)A_i(\tilde{\nu}). \qquad (16.2)$$

Now, K_j for a nonoverlapping transition j at wavenumber $\tilde{\nu}_j$ may easily be determined. We first define the dichroic ratio $d(\tilde{\nu})$

$$d(\tilde{\nu}) = \frac{E_U(\tilde{\nu})}{E_V(\tilde{\nu})}$$

and we obtain for the non-overlapping region of transition j:

$$K_j = \frac{d(\tilde{\nu}_j)}{2 + d(\tilde{\nu}_j)}.$$

Spectral regions with only a single transition (nonoverlapping regions) are common in IR spectroscopy, but much rarer in UV spectroscopy. However, usually K_j values may be determined quite accurately, also in the case of overlapping transitions. The condition is that transition j may be recognized in the LD spectra as a spectral feature (peak or shoulder). If so, the *trial and error method* (the *TEM method* [2, 24, 25]) may be used. In this highly efficient method, linear combinations of the observed spectra $E_U(\tilde{\nu})$ and $E_V(\tilde{\nu})$ are formed:

$$E_U(\tilde{\nu}) - xE_V(\tilde{\nu}).$$

For a given value of x, x_j, the spectral feature that is due to transition j, will disappear. We now have, according to equations (16.1) and (16.2):

$$K_j = x_j \frac{1 - K_j}{2}$$

and thus

$$K_j = \frac{x_j}{2 + x_j}.$$

In some cases it is not $E_U(\tilde{\nu})$ and $E_V(\tilde{\nu})$ that are measured, but related quantities, such as $E_U(\tilde{\nu})$ or $E_V(\tilde{\nu})$ and $E_{iso}(\tilde{\nu})$ measured on an isotropic sample [2], or $E_U(\tilde{\nu})$ or $E_V(\tilde{\nu})$ and $E_{U,V}(\tilde{\nu})$, where $E_{U,V}(\tilde{\nu})$ is measured with natural light on the aligned sample (electric vectors along U and V).

In instruments where the polarization may be modulated, e.g., by a photoelastic modulator one may obtain the linear dichroism *LD* directly:

$$LD(\tilde{\nu}) = E_U(\tilde{\nu}) - E_V(\tilde{\nu}).$$

The observed $LD(\tilde{\nu})$ may then be combined with $E_{U,V}(\tilde{\nu})$ (the absorbance of the aligned sample measured with natural light) in order to obtain the available orientational information. In particular, direct measurement of $LD(\tilde{\nu})$ may provide a significantly better accuracy when the difference between $E_U(\tilde{\nu})$ and $E_V(\tilde{\nu})$—the linear dichroism—is very small and baseline inaccuracy becomes significant.

Finally, as a low cost solution, since especially good IR polarizers are expensive, it may be mentioned that similar information may be obtained without polarizers. The fact that the electric vectors of a light beam are not isotropic, but located in only two dimensions, perpendicular to the direction of the beam, may be used for "polarizer-free polarization spectroscopy" [20]. Spectra measured with natural light, first on an aligned sample, then on an isotropic sample of the same molecule, reveal a similar difference, although less pronounced, as ordinary LD spectra. The two samples may be a squeezed polyethylene tablet and an unstretched polyethylene sheet. This technique also makes a distinction between different transition moment directions possible, at least for somewhat elongated molecules.

16.3.4 Transition Moment Directions and Reduced Spectra: Symmetrical Molecules

The situation is particularly simple for molecules of C_{2v}, D_{2h}, and D_2 symmetry (here labeled "symmetrical'' cases, fortunately this group includes numerous important molecules) for which only three different transition moment directions are possible (along the symmetry-determined molecular axes x, y, and z). Thus also only three different K_i may be observed. The three K are not independent, but (by definition) add up to 1

$$K_x + K_y + K_z = 1.$$

While it is possible to determine any number of different K-values from the two observed spectra, a separation of the observed absorbance into contributions from different transition moment directions (a determination of the $A_i(\tilde{\nu})$ curves) is much more complicated. This separation is particularly useful for UV spectra where it, among other, may help reveal otherwise "hidden" transitions (usually hidden due to overlap with other, stronger transitions). If only y- and z-polarized intensity is present in a spectral region ($A_x(\tilde{\nu}) = 0$), we have for that region:

$$A_z(\tilde{\nu}) = \frac{(1-K_Y)E_U(\tilde{\nu}) - 2K_Y E_V(\tilde{\nu})}{K_Z - K_Y} \tag{16.3}$$

$$A_y(\tilde{\nu}) = \frac{2K_Z E_V(\tilde{\nu}) - (1-K_Z)E_U(\tilde{\nu})}{K_Z - K_Y}. \tag{16.4}$$

We see that in a spectral region with intensity exclusively from transitions polarized along the y- and z-axes, the two observed spectra $E_U(\tilde{\nu})$ and $E_V(\tilde{\nu})$ provide sufficient information for a mathematical construction of $A_y(\tilde{\nu})$ and $A_z(\tilde{\nu})$, the sums of transitions polarized along molecular axes y and z, respectively. Construction of spectral functions like $A_y(\tilde{\nu})$ and $A_z(\tilde{\nu})$ is only possible in some special cases. Even in symmetrical molecules, three different transition moment directions (along x, y, z) will be present in the general case and the wanted three spectral functions $A_x(\tilde{\nu})$, $A_y(\tilde{\nu})$, and $A_z(\tilde{\nu})$ cannot be constructed since only two observed spectra are available.

In vibrational spectroscopy of symmetrical molecules, where all three transition moment directions usually are present, a separation into $A_x(\tilde{\nu})$, $A_y(\tilde{\nu})$, and $A_z(\tilde{\nu})$ is usually not considered necessary due to the very limited extent of overlap between different IR transitions. However, as shown below in an example, it is possible to produce spectral curves in which contributions from any of the three transition moment directions have been removed, that is, contributions from either $A_x(\tilde{\nu})$, $A_y(\tilde{\nu})$, or $A_z(\tilde{\nu})$. This is often a useful option, although the construction of $A_x(\tilde{\nu})$, $A_y(\tilde{\nu})$, or $A_z(\tilde{\nu})$, whenever possible, is the best option, especially in the case of the strongly overlapping UV transitions, as we shall see later.

Spectral regions with only two different transition moment directions are common for "symmetrical" planar, polycyclic hydrocarbons, for which the low-energy transitions are dominated by the in-plane polarized π-π* transitions (see below). For these molecules, out-of-plane polarized intensity usually only occurs in or near the vacuum UV region. The benefits of a separation of spectral intensity according to the two in-plane transition moments have been demonstrated numerous times (see, for example, [2, 15, 23, 24]), especially for UV spectra with strongly overlapping bands.

In a few other cases the molecular alignment is such that only two different K exist. This is true for molecules that have a C_3 or higher axis. For such molecules the absorption may be separated in two parts, one with transition moments along the high symmetry molecular axis, and another with transition moments perpendicular to the axis. This is all the information that is available and all that is required for this kind of experiment.

An interesting case is that of molecules which for reasons other than symmetry, are either aligned like "rods" (the two smaller K are identical) or like "disks" (the two larger K are identical) [2]. For these molecules the absorption may also be separated into two parts, one with transition moments along the molecular axis with the unique K value and another with transition moments perpendicular to this axis. For a rod-like aligned symmetrical molecule this would mean that, for example, $A_z(\tilde{\nu})$ and $A_x(\tilde{\nu}) + A_y(\tilde{\nu})$ may be determined. However, a separation of $A_x(\tilde{\nu})$ and $A_y(\tilde{\nu})$ is not possible.

16.3.4.1 The Orientation Triangle

If we define the K so that $K_x \leq K_y \leq K_z$, keeping in mind that $K_x + K_y + K_z = 1$, a plot of, e.g., (K_y, K_z), in this case for planar, symmetrical molecules (see below), will all be located within a triangle with corners $(K_y, K_z) = (0, 1)$, $(\frac{1}{3}, \frac{1}{3})$, and $(\frac{1}{2}, \frac{1}{2})$; see Figure 16.6. The sides of the triangle are defined by:

$$K_z + K_y = 1$$

This side corresponds to $K_x = 0$; in other words, the (y, z)-plane is parallel to the sample U-axis:

$$K_z = K_y$$

Points on this side have a rod-like alignment with an equal (but not necessarily identical) alignment of the two least aligned molecular dimensions (x and y):

$$K_x = K_y \quad \text{or} \quad K_z = 1 - 2K_y.$$

Points located on the last side of the triangle correspond to a disk-like alignment with an equal (but not necessarily identical) alignment of the two most well-aligned molecular dimensions (y and z).

This orientation triangle provides a convenient overview of molecular alignment in stretched polymer sheets. By inspection of the figure it is clear that alignment is closely related to molecular shape, and it is actually possible to

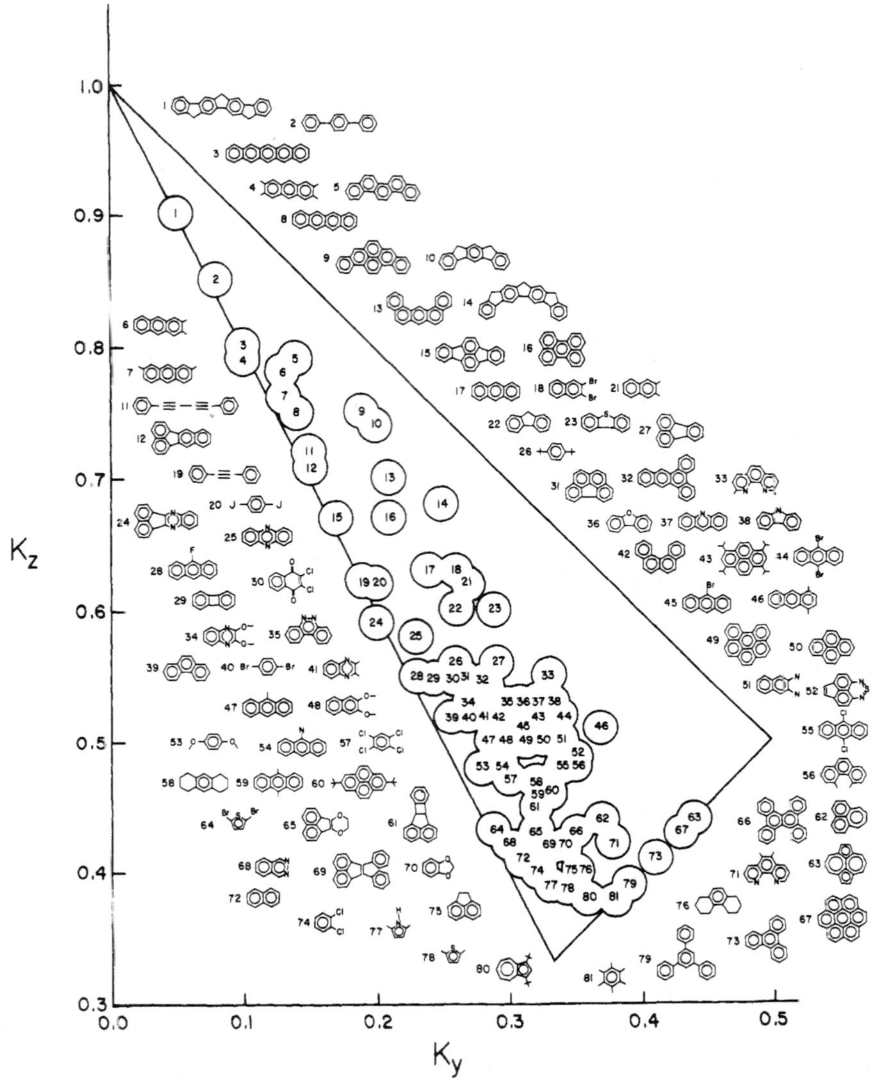

Figure 16.6 The orientation aromatic molecules in stretched polyethylene at room temperature. Reproduced from [25].

make a reasonable guess on the *K* values for a new compound based on inspection of the known points in the triangle.

Example 1: The IR Spectrum of 9,10-Anthraquinone

In connection with a study of colored anthraquinone derivatives of interest in liquid crystal displays, an IR spectrum of the parent compound in stretched polyethylene sheets were recorded, mainly to check the out-of-plane axis (*x*) alignment independently [26]. It turned out that this simple spectrum was able to solve a few long-time problems within the assignment of the vibrational transitions of 9,10-anthraquinone. 9,10-anthraquinone is a molecule of D_{2h}

symmetry (see Fig. 16.4). Therefore, all transition moments for both electronic and vibrational transitions can only be directed along the three symmetry-determined axes, x, y, and z. The two baseline corrected spectra, $E_U(\tilde{\nu})$ and $E_V(\tilde{\nu})$ covering the mid-IR region and measured at room temperature, are shown in Figure 16.7.

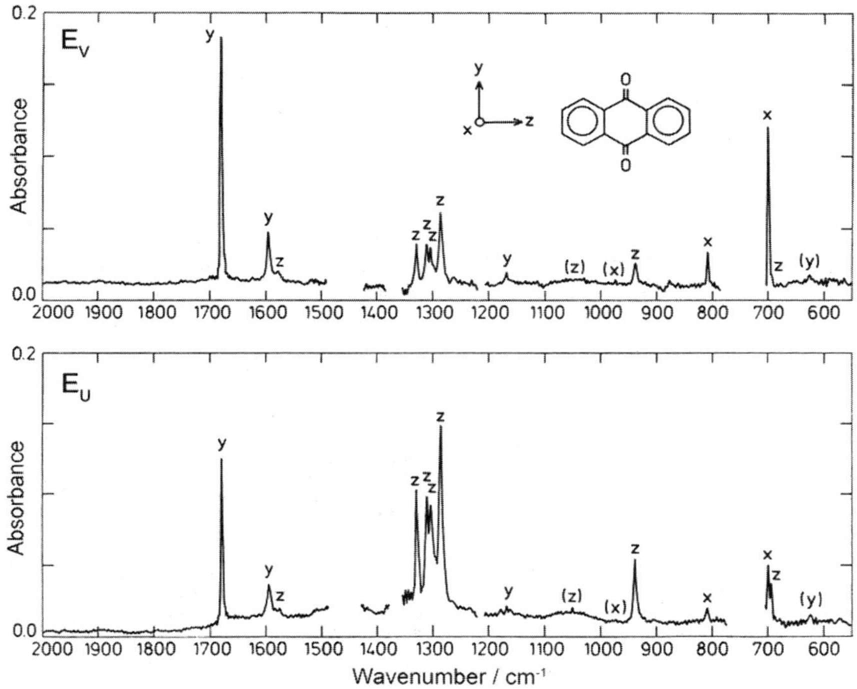

Figure 16.7 FT-IR LD spectra of 9,10-anthraquinone in stretched polyethylene with an indication of the assignment of transition moment directions of individual bands. Reproduced from [26].

Using the TEM method, the K values for a little more than a dozen peaks and a single shoulder were easily determined. As expected, only three different K-values were registered: $K_x = 0.15$, $K_y = 0.25$, and $K_z = 0.59$, with a sum of 0.99, in other words equal to 1 within the experimental accuracy, as they should be. These K-values immediately provide symmetry assignments for the transitions, those with $K = 0.15$ (out-of-plane, x) are b_{3u}, those with $K = 0.25$ (short, in-plane direction, y) are b_{1u}, and those with $K = 0.59$ (long axis, z) are of b_{2u} symmetry. The transitions that were observed between 600 and 2000 cm^{-1} are only about half of the total number of allowed transitions; of the 66 (from $24 \times 3 - 6$) fundamental vibrations, 28 are allowed (11 b_{1u}, 11 b_{2u}, and 6 b_{3u}). The remaining 38 transitions of other symmetries cannot be observed in a simple absorption experiment; they are symmetry forbidden.

One unexpected bonus was the information about the identical b_{2u} symmetry (and long-axis, z-, polarization) of the members of the quartet of peaks around 1300 cm^{-1}. This quartet had never been properly assigned to z-polarized transitions. Now it became clear that only the outer two transitions in the quartet are b_{2u} fundamentals (ν_{44} at 1284 cm^{-1} and ν_{43} at 1329 cm^{-1}). The two center peaks are combination bands of $\nu_{10} + \nu_{48}$ and $\nu_{12} + \nu_{47}$ (both $a_g + b_{2u}$ combinations of b_{2u} symmetry) that gain unexpected strong intensity through Fermi resonance caused by the two intense surrounding fundamentals.

In addition to the highly useful information about the z-polarized quartet, a second new and valuable piece of information appeared. From the E_U spectrum in Figure 16.7 it is clear that the peak around 700 cm^{-1} is really a double peak. This is not obvious from the E_V spectrum or from an ordinary solution spectrum. The reason for the double peak is the occurrence at almost the same wavenumber of both a strong b_{3u} transition (ν_{63} at 699 cm^{-1}) and a weak b_{2u} fundamental (ν_{48} at 693 cm^{-1}). The latter had never been observed before because of the overlap with the stronger ν_{63}, and various incorrect assignments had been attempted. The reason for the observation of the double peak in E_U, and not in the other spectra, is that long axis polarized transitions are amplified in the E_U spectrum; similarly short axis polarized transitions are amplified in E_V.

Example 2: The UV Spectrum of the Pyrene Molecule

As an example, let us look at the UV spectrum of the important, symmetrical pyrene molecule [2, 27]. The low energy part of the electronic spectrum of this conjugated, planar molecule is quite well described in a π-π^* model, where excitation only leads to changes in the orbitals of π-type; these are the molecular orbitals that are antisymmetric with respect to the molecular plane. This symmetry property restricts the transition moments for singlet-singlet π-π^* transitions to the molecular plane. Furthermore, the D_{2h} symmetry of pyrene only allows two transition moment directions in the molecular plane, along the symmetry axes z (the long axis) and y (the short in-plane axis). Other transitions, for example σ-π^*, will have transition moments perpendicular to the molecular plane. But for the low energy transitions it is safe to assume that only in-plane transition moments are of interest.

In a simple MO picture, the first four singlet-singlet transitions of pyrene may be described as transitions to excited electron configurations. These four configurations correspond to a single electron moving from one of the two highest occupied molecular orbitals (1 and 2) into one of the two lowest unoccupied orbitals (–1 and –2), in other words:

$$|1 \rightarrow -1\rangle, \quad |1 \rightarrow -2\rangle, \quad |2 \rightarrow -1\rangle, \quad |2 \rightarrow -2\rangle$$

In the simple MO model the configurations $|1\rightarrow-2\rangle$ and $|2\rightarrow-1\rangle$ have the same energy (are degenerate) and will be poor approximations to the excited states. The + and – combinations of the two are much closer to reality. Thus we end up with the following simple description of the four excited states (for a discussion of the labeling that originally is due to Platt; see [2]):

$$\begin{aligned}
L_a &= |1 \to -1\rangle, \\
L_b &= |1 \to -2\rangle - |2 \to -1\rangle, \\
B_b &= |1 \to -2\rangle + |2 \to -1\rangle, \\
B_a &= |2 \to -2\rangle
\end{aligned}$$

In the case of pyrene, the singlet L_b state turns out to have lower energy than the singlet L_a state. It also corresponds to an extremely weak transition, while the other three are strong. The transition to L_b is only seen because it appears at slightly lower energy than the transition to L_a. The transition moments to a-states are all along the molecular z-axis (they go through **a**toms) while transition moments to b-states are along the molecular y-axis (they go through **b**onds). It may be added that the b-states correspond to molecular symmetry B_{2u}, while the a-states correspond to B_{1u} (Fig. 16.4).

This picture is confirmed by the observed K-values of 0-0 transitions corresponding to the three strong transitions: $K(B_b) = K_y = 0.33$, $K(L_a) = K(B_a) = K_z = 0.58$; see Figure 16.8. Furthermore, because of the sum rule for the K-values we may conclude that $K_x = 0.09$. However, in the case of pyrene there is a rare exception that has to do with the very weak L_b transition, and the fact that it is located very close to L_a.

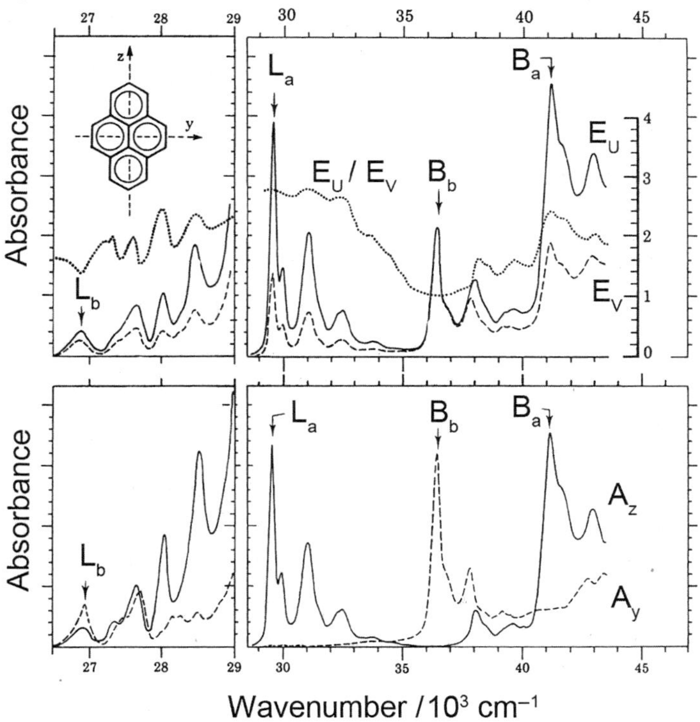

Figure 16.8 Polarized UV absorption of pyrene in stretched polyethylene at 77 K. Bottom: Reduced absorption spectra $A_z = E_U - 1.0E_V$ and $A_y = 0.625(2.8E_V - E_U)$. Top: Baseline-corrected polarized absorption spectra and the dichroic ratio E_U/E_V. The absorbance scale is in arbitrary units, different for the different sections of the spectrum. Note the change in the horizontal scale in the left section. Reproduced from [2].

A slight perturbation of the molecular structure and symmetry, e.g., from a solvent, and even more from a substituent that destroys the symmetry, may result in an excited state that is a mixture of two close-lying states. This happens regularly for the L_b and L_a states of conjugated molecules. If one of the two unperturbed transitions, say the L_a transition, is much stronger than the L_b transition, it may dominate the transition moment for a resulting transition, and the transition moment will turn from the *y*-axis to a position closer to the *z*-axis. Because of the large difference in the original intensities, even a small dose of L_a mixed into L_b will rotate the transition moment significantly.

A remarkable example of this solvent effect has been observed for pyrene, aligned in stretched polyethylene [2, 28]. The low symmetry interaction with the polymer solvent is able to mix a small amount of L_a into L_b. The result is that the transition moment for the 0-0 transition of L_b is located almost in the middle between the *y*- and *z*-axes, and not along *y* as expected according to the molecular symmetry. This is demonstrated by the observed *K*-value for the 0-0 transition that is neither 0.33, nor 0.58, but between the two values, actually corresponding to a transition moment angle of around 37° with the *y*-axis.

It may be added that the complicated picture of vibronic peaks between the L_b 0-0 transition and L_a are quite dominated by *z*-polarized intensity, gained through vibronic mixing of the two states through non-totally symmetric vibrations of B_{3g} symmetry. It is said that the L_b transition "borrows" or "steals" intensity from L_a. This is a common phenomenon in such spectra, and LD spectroscopy is particularly well suited to study these processes. However, the twisting of a transition moment for the (L_b) 0-0 transition due to solvent effects is very rare. The best known other case of this kind is the appearance of the 0-0 band of the symmetry forbidden first singlet-singlet transition (to B_{2u}) in benzene in various solvents (it is not seen in the gas phase).

For the overall analysis of the pyrene UV spectrum, the TEM procedure is required because of the extensive spectral overlap between two or more transitions. The TEM treatment results in the expected *K*-values for the other 0-0 peaks, as indicated above. The case of symmetrical π-electron systems in spectral regions with only in-plane polarized π-π* transitions makes a very detailed analysis possible, as discussed above.

With only intensity from transitions polarized along the *y*- and *z*-axes, the two observed spectra $E_U(\tilde{\nu})$ and $E_V(\tilde{\nu})$ provide sufficient information for a mathematical construction of $A_y(\tilde{\nu})$ and $A_z(\tilde{\nu})$, the sums of transitions polarized along molecular axes *y* and *z*, respectively. Equations (16.3) and (16.4) provide the results that are shown in the bottom of Figure 16.8.

We immediately notice that there is a medium-strong z-polarized transition around 38,000 cm^{-1} and one or more y-polarized transitions under the strong B_a band. These correspond to excited states with large contributions from other excited configurations than those above, involving only the two highest occupied and the two lowest unoccupied molecular orbitals. For a test of theoretical models such information is of considerable interest and LD spectroscopy has been widely used in this connection; see [27, 29, 30].

In the general case of symmetrical molecules, the LD spectra will reveal three different K: K_x, K_y, and K_z. In regions with little overlap (as in IR spectra) this usually allows a simple interpretation, in particular assignment of transitions to specific symmetry classes. In the case of spectral overlap by two differently polarized transitions, the situation may often be handled as shown above. But in the case of spectral overlap by three differently polarized transitions, the interpretation is usually not simple.

Example 3: A Comparison of LD Spectra with Stretched Polymers and Liquid Crystals as the Aligning Solvent

The infrared spectrum of fluorene has been studied by LD spectroscopy, using both a stretched polymer (polyethylene) and a nematic liquid crystal (4-cyano-4'-alkylbicyclohexyl mixture) as the aligning solvent [31]. Fluorene is a symmetrical molecule, point group C_{2v}; it has 63 normal modes: $22a_1 + 10a_2 + 11b_1 + 20b_2$. Of these, the 10 a_2 vibrations are symmetry-forbidden and cannot be observed in absorption spectra.

The LD spectra, measured at room temperature in stretched polyethylene, are shown in Figure 16.9. The Figure shows the assignments that are revealed almost instantly, by inspection: A peak with $E_U >> E_V$ corresponds to a long axis (y) polarized transition, one with $E_U \approx E_V$ to a short in-plane axis (z) polarized transition, and a peak with $E_U << E_V$ corresponds to an out-of-plane (x) polarized transition.

The K-values obtained are $(K_x, K_y, K_z) = (0.17, 0.49, 0.31)$ with polyethylene as the aligning solvent and (0.11, 0.64, 0.25) with the nematic liquid crystal as the aligning force. It may be added that in an earlier experiment [2], performed in the UV region and with a lower solute concentration in a thin polyethylene sheet, the K were measured to be (0.14, 0.60, 0.26). Still, the fluorene molecules in the nematic liquid crystal demonstrate the best alignment of all. However, the spectra obtained with polyethylene as solvent are sharper and show more details.

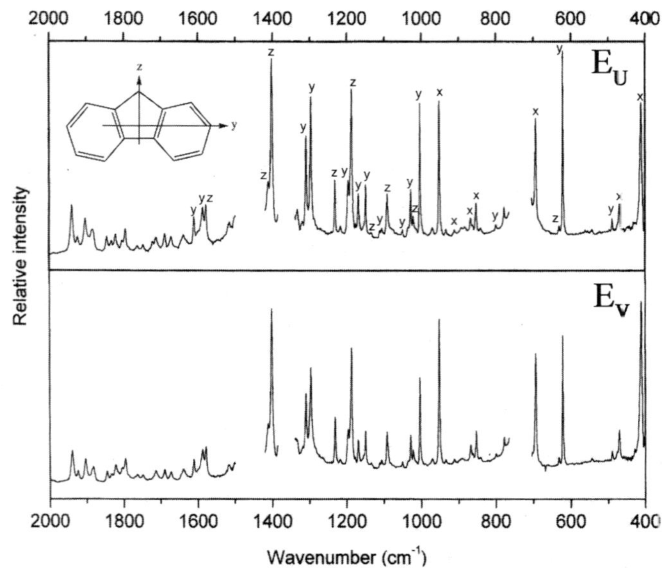

Figure 16.9 LD absorbance spectra of fluorene in stretched polyethylene recorded at room temperature. Reproduced with permission from [31].

In Figure 16.10 the spectra with the nematic liquid crystal as the aligning solvent are shown in a format that even more directly reveals the transition moment directions.

Although individual curves A_x, A_y, and A_z cannot be constructed from only two observed spectra, it is possible to construct curves in which either A_x, A_y, or A_z is removed! If we want to remove all contributions from y-polarized transitions, $A_y(\tilde{\nu})$, we have to form the linear combination:

$$A_{-y}(\tilde{\nu}) = (1-K_y)\,E_U(\tilde{\nu}) - 2K_yE_V(\tilde{\nu}) = (1-K_y)\,\Sigma K_iA_i(\tilde{\nu}) - 2K_y(\tfrac{1}{2})\Sigma(1-K_i)A(\tilde{\nu})$$
$$= (K_x-K_y)A_x(\tilde{\nu}) + (K_z-K_y)A_z(\tilde{\nu})\,.$$

The curve shown in the top part of Figure 16.10 is equal to $-A_{-y}(\tilde{\nu})$ (minus is used because both coefficients to the two $A_i(\tilde{\nu})$; are negative) and is thus a linear combination of only $A_x(\tilde{\nu})$; and $A_z(\tilde{\nu})$;. Note that the two contributions occur with different coefficients (unless $K_z = K_x$—a less interesting case of isotropic-like alignment). In other words, the contributions from transitions

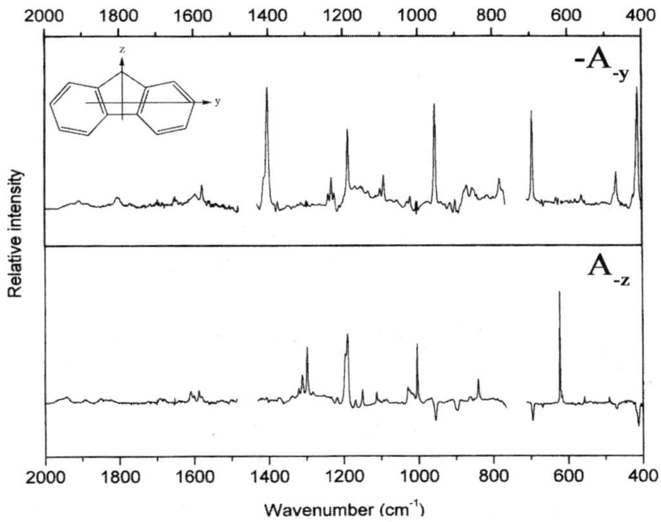

Figure 16.10 Linear combinations of absorbance spectra of fluorene at room temperature, aligned in a nematic liquid crystal. Top: $-A_{-y}(\tilde{\nu})$; bottom $A_{-z}(\tilde{\nu})$. See text. Figure adapted by permission from [31].

polarized along different axes are not necessarily on the same scale! This is even clearer from the bottom of Figure 16.10, in which all z-polarized intensity has been removed:

$$A_{-z}(\tilde{\nu}) = (1-K_z)\,E_U(\tilde{\nu}) - 2K_zE_V(\tilde{\nu}) = (K_x-K_z)A_x(\tilde{\nu}) + (K_y-K_z)A_y(\tilde{\nu})$$

Here, y-polarized intensity occurs with a positive sign (the coefficient on $A_y(\tilde{\nu})$ is positive), while x-polarized intensity appears with a negative sign. This may easily be seen from a comparison with Figure 16.9. The removal of spectral intensity with a given transition moment direction may be applied to any set of LD spectra, and is commonly used also for spectra obtained from stretched polymer samples.

From the information obtained on transition moment directions and excited state symmetries it was possible to correct several mistakes in earlier assignments of the vibrational transitions of fluorene [31].

Example 4: Photoinduced Dichroism in β-Thioxoketones in a Frozen Matrix

Photoinduced dichroism can be observed when a photochemical reaction is made to occur through a polarized light source on an immobilized group of sample molecules. By using a polarized light source, only a subset of the sample molecules are phototransformed, namely those with the proper orientation for interaction with the light source. Simple β-thioxoketones, such as thio-(*p*-methylbenzoyl)acetone, undergo a photochemical reaction upon irradiation with UV-VIS light [32]. A transformation occurs both via n-π^* excitation (at 514 nm) and π-π^* excitation (at 410 nm). Figure 16.11 illustrates the effect of 410-nm irradiation on the UV-VIS spectrum of thio-(*p*-methylbenzoyl)acetone in 3-methylpentane at 110 K. The phototransformation leads to hyperchromism around 240 nm and 400 nm, and hypochromism at 320 nm.

The product of this type of photoreaction has been discussed on several occasions. Recently, it has been demonstrated through polarized IR spectroscopy and density functional calculations, that the transformation proceeds from the (Z)-enol form, where the H is chelated between O and S, to the Z-(enethiol) form where the –SH bond is oriented away from O [32]. In the case of thio-(*p*-methylbenzoyl)acetone, the initial form has fundamental vibrations at 1464, 1423, and 888.6 cm^{-1}, while the photoproduct has fundamentals at 1679, 1509, 1357, and 1323 cm^{-1}, as detailed in Table 16.2.

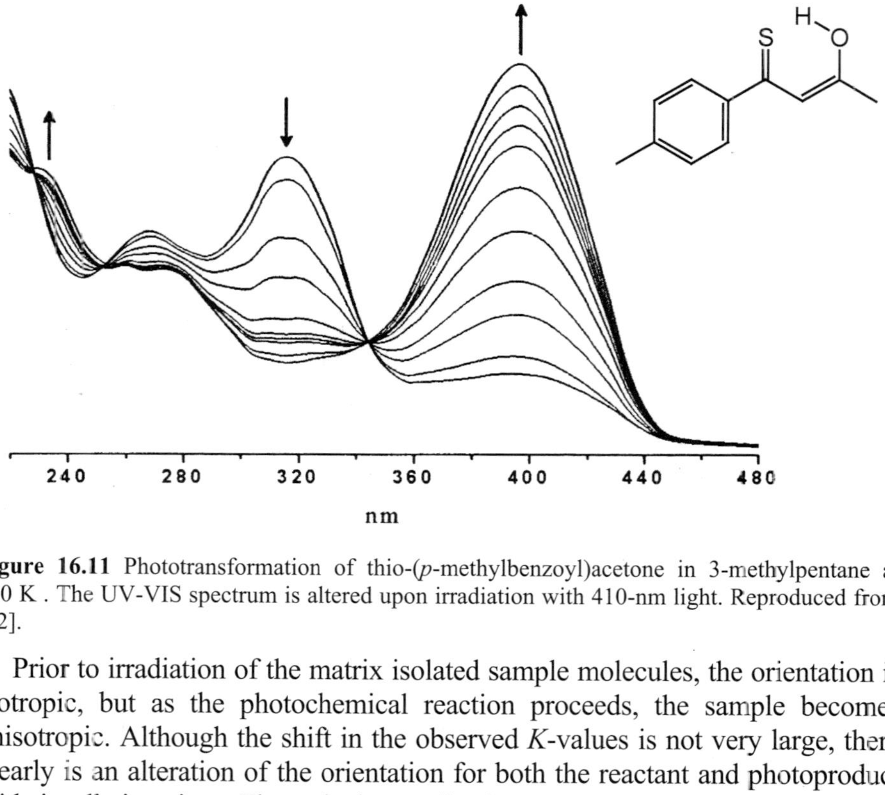

Figure 16.11 Phototransformation of thio-(*p*-methylbenzoyl)acetone in 3-methylpentane at 110 K . The UV-VIS spectrum is altered upon irradiation with 410-nm light. Reproduced from [32].

Prior to irradiation of the matrix isolated sample molecules, the orientation is isotropic, but as the photochemical reaction proceeds, the sample becomes anisotropic. Although the shift in the observed *K*-values is not very large, there clearly is an alteration of the orientation for both the reactant and photoproduct with irradiation time. Through the application of polarization spectroscopy of and the use of quantum mechanical calculations it was possible to determine the identity of the photoproduct for several β-thioxoketones [32].

Table 16.2 IR Linear dichroism IR measurements of thio-(*p*-methylbenzoyl)acetone in an Ar matrix at 20 K upon irradiation at 410-nm. The K_i values change as a function of irradiation time. Reproduced from [32].

Initial form (enol)						
IR band, Cm^{-1}	Vibrational mode	Sign of LD	Tendency of K	K_i	K_i'	K_i''
1464	C–O str., C–H bend	–	↓	0.3	0.27	0.25
1423	CH_3 bend	+	↑	0.35	0.37	0.39
888.6	O–H *l* oop bend		↑	0.33	0.36	0.39
Photoproduct (enethiol)						
IR band Cm^{-1}	Vibrational mode	Sign of LD	Tendency of K	K_i	K_i'	K_i''
1679	C=C, C=O str.	–	↑	0.30	0.31	0.33
1509	C–C str., C–H bend	+	↓	0.35	0.36	0.343
1357	CH_3 bend, C–CH_3 str.	+	↓	0.4	0.40	0.36
1323	CH_3 bend, C–CH_3 str.	–	↑	0.29	0.29	0.30

16.3.5 Transition Moment Directions: Molecules of Lower Symmetry

In molecules of lower symmetry there are fewer restrictions on the possible transition moment directions. A very important case is that of a molecule with a single plane of symmetry; for such molecules the transition moments are restricted to be in the symmetry plane (*x*, *y*) or along the direction *z* directed perpendicularly to this plane. Usually, the out-of-plane (*z*) polarized transitions are easily recognizable—they are characterized by the same $K = K_z$. In the very common case when alignment depends on molecular shape, this is the lowest *K*-value observed among all transitions, since the out-of-plane direction is least well aligned with the sample axis and thus has the smallest $K = \langle \cos^2 (\mathbf{M}_i, U)\rangle = K_z = \langle \cos^2 (z, U)\rangle$.

The in-plane polarized transitions may have their transition moments anywhere in the molecular (*x*, *y*)-plane. Assume that the axes are chosen so that *x* corresponds to the smallest possible in-plane value for *K*, $K_x = \langle \cos^2 (\mathrm{x}, U)\rangle$ and *y* so that it corresponds to the largest, $K_y = \langle \cos^2 (x, U)\rangle$. It may then be shown [2] that the angle that $\mathbf{M}_i$ forms with *y*, φ_i, satisfies the equation:

$$\tan^2 \varphi_i = (K_y - K_i)/(K_i - K_x). \qquad (16.5)$$

It must be kept in mind that the values of (K_x, K_y), the position in the molecular plane of (x, y), and the "sign" of the angle φ_i are initially unknown (see below). The direction in the molecule corresponding to the largest K-value is called the molecular orientation axis. In the present case the orientation axis will be located in the molecular plane and equal to axis y, $K_{\text{max}} = K_y$.

The above expression makes it possible to determine the angle between transition moment $\mathbf{M}_i$ and the orientation axis y. There are, however, some conditions and some restrictions to keep in mind:

First, the K-values for these axes (K_x, K_y) must be determined. Often, the out-of-plane K-value, K_z, is easily recognized and determined; this provides important information on K_x and K_y:

$$K_x + K_y = 1 - K_z.$$

This limits the possible values for (K_x, K_y). Assume that the in-plane polarized transitions have produced observed K-values from K_{min} to K_{max} within a range above K_z. It is known that

$$K_{\text{max}} \leq K_y$$
$$K_{\text{min}} \geq K_x.$$

These restrictions may be sufficient for a determination of the K-values, especially when it is possible to measure a whole range of K-values. For this reason, the determination of K-values in IR LD spectra of larger molecules is often relatively simple. Some of these correspond to (out-of-plane) z-polarized transitions, they have the same, low K-value, $K = K_z$, and are thus easy to identify. The knowledge of $K_z = 1 - K_x - K_y$ and of K_{max} and K_{min} often makes it possible to estimate the in-plane K values (K_x, K_y) quite accurately. In other cases specific transition moment directions are known; this may provide an additional clue. Figure 16.12 shows an example of the possible distribution of K-values for a planar molecule.

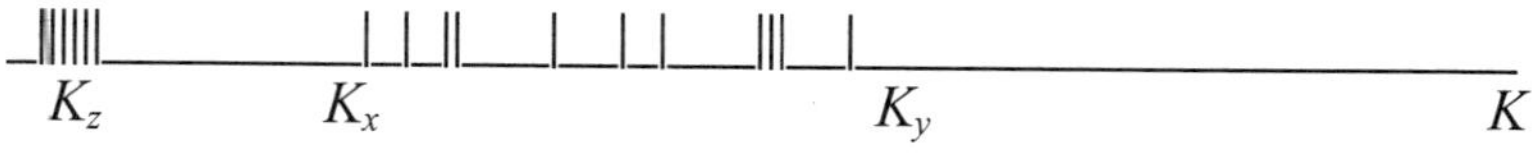

Figure 16.12 A typical set of observed K values in an IR LD spectrum of a planar molecule. Within experimental accuracy, there is a single out-of-plane K value, while the in-plane K values are spread over a larger region.

Now only two tasks are left, the determination of the position of the orientation axis and the determination of the "sign" of φ_i, (see below). The determination of the position of the orientation axis (and thus axis x) in the molecular plane may be done in several ways; by inspection, for example, in the common case when alignment is related to molecular shape, or with the help of transitions with known transition moment directions.

Finally, the determination of $\tan^2\varphi_i$ can only provide information on the numerical value of the angle with the orientation axis; $\mathbf{M}_i$ may be located on either side of y, corresponding to a sign change of φ_i. Thus, there will be two possible directions, unless φ_i is zero or 90°. In many cases it has been possible to determine which of the two directions is the correct one by introduction of additional information from fluorescence polarization spectroscopy; see, for example, [2, 33]. This technique may provide information about relative angles between transition moments for the absorbing transition and the transition moment directions for the emitting transitions in the molecular framework. See below for a more complete discussion.

Also in a few other cases than π-π* transitions it may be possible to determine transition moment directions accurately, namely for rod-like ($K_x = K_z$) and disk-like ($K_x = K_y$) alignments.

If $K_x = K_z$ we obtain [2]

$$\tan^2 \varphi_i = K_y - K_i / K_i - K_x \quad (16.6)$$

which may be used for a determination of the angle between $\mathbf{M}_\mathrm{i}$ and y.

If $K_x = K_y$ the result is

$$\tan^2 \varphi_i = K_i - K_z / K_y - K_i \quad (16.7)$$

which may be used for a determination of the angle between $\mathbf{M}_\mathrm{i}$ and z.

Example 1: The UV Spectrum of Chrysene

The UV spectrum of the polycyclic aromatic hydrocarbon chrysene has been studied extensively over the years. These studies also include polarization fluorescence work [34] on an isotropic sample, which may be used to estimate relative angles between the emitting transition and different absorbing transitions, as discussed later.

Chrysene is a planar molecule of C_{2h} symmetry with the following symmetry elements: a C_2 axis, a symmetry plane and an inversion center. The symmetry classes are A_g, B_g, A_u, and B_u. All transitions between a ground state of "g" symmetry and other g-states are symmetry-forbidden, so only excited states of A_u and B_u symmetry may be observed. These are polarized along the symmetry axis (out-of-plane) and in the molecular plane, respectively. In the π-electron approximation only in-plane transitions are expected to be observed in the UV region down to 200 nm, so we expect all observed transitions to be in-plane polarized. The K-values for these are expected to lie between K_y for the best aligned in-plane axis and K_x for the worst aligned in-plane axis. All out-of-plane K will be equal to the (low) value that is observed for transitions to A_u states.

The LD spectra in the UV region of chrysene, obtained in stretched polyethylene at room temperature, are shown in Figure 16.13 [35]. Figure 16.13 also shows an application of the TEM procedure which results in K-values ranging from 0.25 to 0.70. We thus have: $K_x \leq 0.25$ and $K_y \geq 0.70$. An IR LD spectrum of chrysene in a similar stretched polyethylene sheet show several nice out-of-plane polarized transitions, all with $K_z = 0.09$. This tells us that: $K_x + K_y =$ 0.91, and $K_x \geq 0.09$, and thus: $0.70 \leq K_y \leq 0.82$.

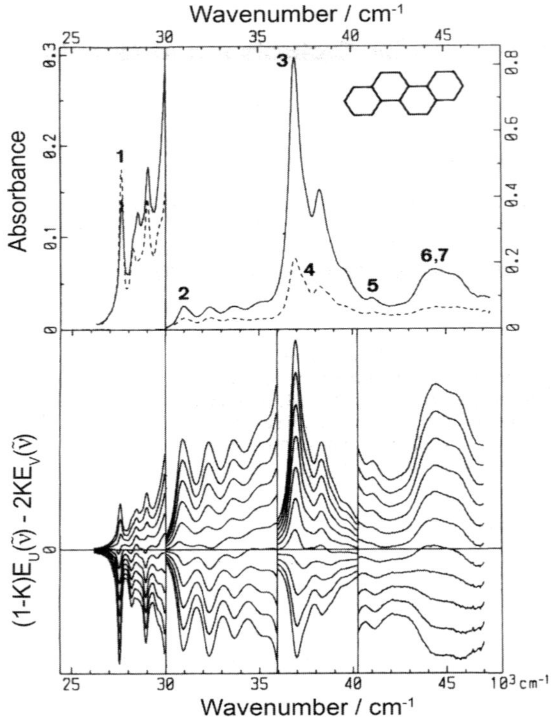

Figure 16.13 UV-VIS linear dichroic absorption curves for chrysene aligned in stretched polyethylene at room temperature. The figure shows $E_U(\tilde{\nu})$ (full line) and $E_V(\tilde{\nu})$ (broken line) obtained with the electric vector of polarized light parallel and perpendicular, respectively, to the stretching direction of the sample. The bottom graph shows linear combinations of the dichroic curves, obtained by varying K from 0 to 1 in steps of 0.1 in the relation $(1-K)\,E_U(\tilde{\nu})$ $-2K\,E_V(\tilde{\nu})$. Reproduced from [35].

The lacking equation is obtained from fluorescence polarization studies [34] that strongly indicate that the angle between the first two UV transitions is 69°. By inserting the observed K-values in an expression for this relative angle it is concluded that the K-values for the symmetry-determined axes are [35]:

$$(K_y, K_x, K_z) = (0.70, 0.21, 0.09).$$

Now equation (16.5) may be used and the angles between the in-plane transition moments and the molecular long axis y may be determined. However, the results we obtain are only relative angles; the transition moment $\mathbf{M_i}$ may be twisted at an angle φ_i to one or the other side of y. A complete description of the applied procedure is given in reference [35].

Example 2: The IR Spectrum of Ethylene-*D*

While the planar molecule ethylene C_2H_4 and perdeuterated ethylene C_2D_4 both have D_{2h} symmetry and thus only three different transition moment directions, isotopomers like C_2H_3D have only the molecular plane as symmetry element (point group C_s) and all transition moments in this plane are possible (symmetry class a'). The out-of-plane polarized transitions, however, are limited to the direction *z*, perpendicular to the molecular plane (symmetry class a"). The possible *K*-values for the molecule correspond to the situation illustrated in Figure 16.14 [8].

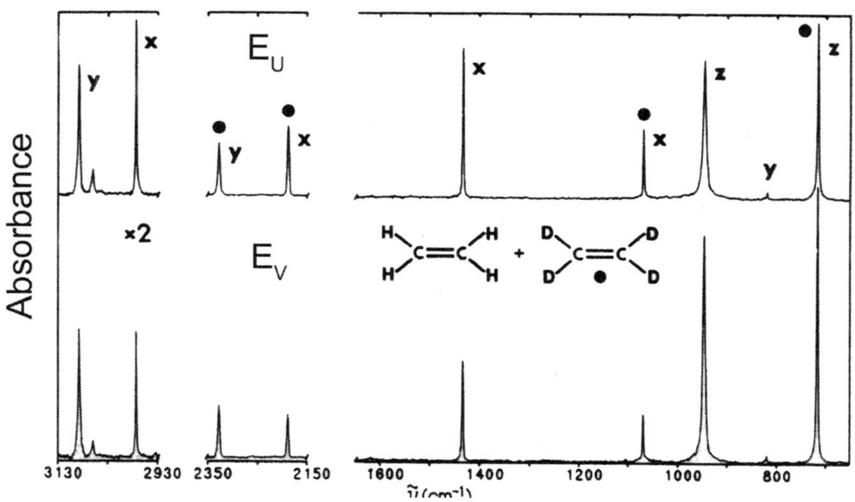

Figure 16.14 Polarized IR spectra of C_2H_4 and C_2D_4 (the latter marked by dots in the E_U spectrum) simultaneously dissolved in a polyethylene matrix at 10 K. E_U (E_V) is obtained with the electric vector parallel (perpendicular) to the stretching direction. Parts of the spectrum (3130 – 2930 cm^{-1}, 1480 – 1420 cm^{-1}, and 780–650 cm^{-1} were obtained using polyethylene-d_4. Figure reproduced from [8].

The three different ethylenes were trapped at low temperature in a stretched polyethylene sheet kept in a closed cycle helium cryostat. It is remarkable that polyethylene sheets are stable and sturdy, also under these conditions. In order to cover the whole spectral region without and gaps due to baseline absorption, both ordinary and perdeuterated polyethylene were used.

Figure 16.14 shows the resulting E_U and E_V curves obtained in normal polyethylene for the two symmetrical molecules. There are no surprises; as expected, perdeuteration shifts essentially all transitions significantly to lower energies, and assignments to symmetry allowed b_{1u}, b_{2u} and b_{3u} transitions can easily be made [8]. Many vibrational transitions are symmetry-forbidden and only a few fundamental vibrations are observed for each molecule. The *K*-values are essentially the same for the two molecules, $(K_x, K_y, K_z) = (0.43, 0.34, 0.23)$ in ordinary polyethylene, and (0.42, 0.33, 0.25) in perdeuterated polyethylene. In the following, we will assume that the *K*-values for the molecular axes in C_2H_3D, (K_x, K_y, K_z), are the same as those for the two symmetrical molecules.

Figure 16.15 shows the results for C_2H_3D. Now, in the C_s symmetry, the number of symmetry-allowed transitions is much higher. Three of them, with *K*-values around 0.25 are immediately recognized as *z*- or out-of-plane polarized transitions. The remaining transitions are located in the molecular plane. The angles their transition moments form with the C–C axis (*x*) may be easily calculated from equation (16.5) above and the results are given in the Figure 16.15. Notice, however, that the direction of the deviation from the C–C axis cannot be obtained.

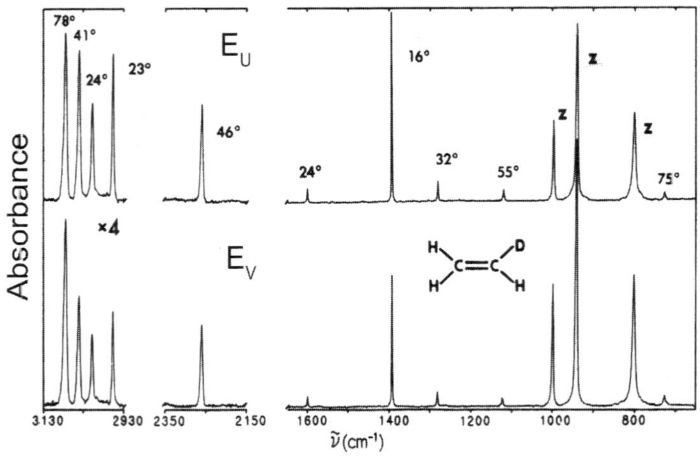

Figure 16.15 Polarized infrared spectra of ethylene-d in polyethylene matrix at 10 K. Angles shown by each line indicate experimental transition moment directions. Parts of the spectrum (3130– 2930 cm^{-1} and 780 – 650 cm^{-1} were obtained using polyethylene-d_4. Figure reproduced from [8].

The results obtained for transition moment angles not only provide a demanding and useful test of theoretical models for vibrational spectra. The changes of the transition moment directions as a result of the symmetry-lowering also give insight into the nature of vibrational molecular movement. Some directions hardly change, while others change drastically. For a further discussion of this, see [8].

16.3.6 Non-Uniaxial Samples

Finally a few words on samples that are not uniaxial. When uniaxiality cannot be assumed one must introduce orientational averages (*K*) referring not only to a single sample axis U, but also to the other sample axes, V and W, and a mixture of these [2], e.g.,

$$K_{UVxy} = \langle \cos(x,U)\cos(y,V)\rangle.$$

Although both molecular and sample alignment symmetry, as well as restriction of the experiments to polarization along only two sample axes, will simplify the expressions somewhat in individual cases, the advantages connected with uniaxial samples remain high.

16.4 TWO-PHOTON SPECTROSCOPY

An overview of two-photon spectroscopy is given in Table 16.1. In photoinduced dichroism and two-photon absorption two photons are absorbed; in the former these events are separated in time, while they are simultaneous in the latter. So far, the most useful two-photon experiments for studies of aligned samples have been luminescence and Raman spectroscopy. In both cases an initial photon is annihilated and a second, with a different energy, is created. In luminescence spectroscopy the two events are separated in time, usually moreso in the case of phosphorescence than in fluorescence; in Raman scattering the two events are simultaneous. This means that in Raman studies there is no movement of the sample molecules between the two events, whereas such mobility must be taken into account in luminescence spectroscopy unless the molecules are fixed in a solvent of high viscosity (e.g., at very low temperatures).

Luminescence polarization measurements require two polarizers, one in the exciting beam, and another in the emitted one. Fluorescence polarization measurements have been performed for years and have provided numerous useful results. The early work by Perrin [36] and others goes back to the 1920s; later work by Jablonski [37] gave name to the simple, yet famous diagram of luminescence processes, the "Jablonski diagram". The literature is extensive, for summaries see the work by Lakowicz [38]; for literature specializing in luminescence polarization applied to aligned samples, see [2].

Ordinary luminescence polarization spectroscopy is performed on an isotropic sample in which molecular movement between the initial absorption and later emission processes is prevented by the use of a proper solvent and cooling to liquid nitrogen or lower temperatures. One may say that the first linearly polarized photon selects an aligned subset of the sample (this time with a known orientation distribution), while the second corresponds to emission from this aligned subset. In luminescence polarization spectroscopy two linearly independent measurements may be performed; they correspond to the following polarizer settings:

$$U/U = V/V = W/W$$

and

$$U/V = U/W = V/U = W/U = V/W = W/V$$

Different angles between the directions of the exciting and emitted beams may be used. Often an angle of 90 degrees is preferred, but smaller angles may also be useful. A commonly used expression for the correction required in luminescence experiments with different angles between the two beams was derived by Gallivan, Brinen, and Koren [39].

Just like in LD spectroscopy two linearly independent spectra may be recorded, one in which the two polarizers are parallel, $E_{\parallel}(i,j)$, and one in which they are perpendicular, $E_{\perp}(i,j)$. These two spectra might be, for example, $E_{UU}(i, j)$ and $E_{UV}(i,j)$. We have for the ratio between these observed intensities [2]:

$$\frac{E_{UU}(i,j)}{E_{UV}(i,j)} = \frac{E_{\parallel}(i,j)}{E_{\perp}(i,j)} = \frac{(1+2\cos^2\varphi)}{(2-\cos^2\varphi)}$$

where φ is the angle between the absorbing and emitting transition moments $\mathbf{M}_i$ and $\mathbf{M}_j$. Thus, ordinary photoselection experiments may provide information about the angle between the transition moments for the two separate photon processes.

The observed results are usually condensed in a single expression, the degree of polarization, $P(i,j)$, which is defined as

$$P(i,j) = \frac{E_{\parallel}(i,j) - E_{\perp}(i,j)}{E_{\parallel}(i,j) + E_{\perp}(i,j)}$$

or the degree of anisotropy, $R(i, j)$, which is defined as

$$R(i,j) = \frac{E_{\parallel}(i,j) - E_{\perp}(i,j)}{E_{\parallel}(i,j) + 2E_{\perp}(i,j)}$$

$P(i, j)$ and $R(i, j)$ may be expressed in terms of the angle between the absorbing and transmitting transition moments, φ. We have

$$P(i,j) = \frac{3\cos^2\varphi - 1}{\cos^2\varphi + 3}$$

$$R(i,j) = \frac{3\cos^2\varphi - 1}{5}.$$

Values of P and R will be between those obtained for two parallel transition moments ($\varphi = 0°$) and those obtained for two perpendicular transition moments ($\varphi = 90°$):

$$\tfrac{1}{2} > P(i,j) > -\tfrac{1}{3}$$
$$\tfrac{2}{5} > R(i,j) > -\tfrac{1}{5}.$$

Ordinary fluorescence polarization spectroscopy is often a very valuable supplement to LD spectroscopy. As we have seen, LD spectroscopy may provide the numerical value of angles between a given molecular axis (the orientation axis) and transition moments for individual transitions. Information about φ may help determine the signs of these angles (see above).

The description of ordinary luminescence polarization spectroscopy is simple since it in reality consists of the (photo)selection of an aligned subset, followed by an emission spectrum from this subset. In order to describe two-photon experiments performed on aligned samples a considerably more extensive treatment is required. In addition to the cosine square averages, the *K*, used in LD spectroscopy, also averaged over the fourth power of the directional cosines, are required; e.g.,

$$L_{\mathrm{stuv}} = \langle \cos s \cos t \cos u \cos v \rangle$$

where the labels *s*, *t*, *u*, and *v* correspond to any selection of the molecular axes, *x*, *y*, and *z*.

The number of different experiments that may be performed with linearly polarized light is also larger in the case of two-photon spectroscopy applied to aligned samples. For a uniaxial, aligned sample (sample axis *U*) the following five experiments are (at least in the general case) different:

$$U/U\,,\ U/V = U/W,\ V/U = W/U,\ V/V = W/W,\ V/W = W/V$$

Here the laboratory axes *U*, *V*, *W* indicate the polarization of (photon 1/photon 2). However, for practical reasons it is not always possible to record all five independent spectra. Another variable is again the angle between the directions of photon 1 and photon 2; the choice depends on a number of experimental conditions. It is important that the observed intensities are corrected for the actual angle used, according to Gallivan, Brinen, and Koren [39].

The observed intensity $E_{PQ}(i,j)$ in a two-photon experiment, involving the process (i, j), with the polarization direction of the two photons along laboratory axes *P* and *Q* (= *U*, *V*, *W*) may now be written

$$E_{PQ}(i,j) = \Sigma\, [F_{PQ}]_{stuv}[G(i,j)]_{stuv}$$

where the sum goes over *s*, *t*, *u*, and $v = x, y, z$. The matrix element $[\mathrm{F}_{\mathrm{PQ}}]_{stuv}$ is expressed in terms of the *K* and *L* and describe the orientational effects, including the polarizer positions *P* and *Q*, while $[G(i,j)]_{stuv}$ describes the molecular properties and the specific process involved. For example, in a Raman experiment, with simultaneous photon processes, *G* is the Raman scattering tensor, while in a luminescence experiment, with successive photon processes, G is simply the product of the transition moments for the two transitions involved, $\mathbf{M}_\mathrm{i}$ and $\mathbf{M}_\mathrm{j}$. For more details, including the general, complete expressions for the tensors F_{PQ}, see [2].

As an example, we shall list some of the expressions for luminescence spectroscopy applied to "symmetrical" molecules (i.e., with only three perpendicular transition moments). With a proper choice of molecular axes ($s, t, u, v = x, y, z$,; see the discussion in connection with the theory for linear dichroism above) only elements $[F_{PQ}]_{stuv}$ with $u = s$ and $v = t$ are nonzero, and we obtain [2]:

$$[F_{UU}]_{stst} = L_{stst}$$

$$[F_{VV}]_{stst} = \frac{[(1+2\delta_{st})(1-K_s-K_t)+3L_{stst}]}{8}$$

$$[F_{VU}]_{stst} = \tfrac{1}{2}(K_t - L_{stst})$$

$$[F_{UV}]_{stst} = \tfrac{1}{2}(K_s - L_{stst})$$

$$[F_{VW}]_{stst} = \frac{[(3-2\delta_{st})(1-K_s-K_t)+L_{stst}]}{8}$$

It is worth noticing that the K and L are related. Similarly to

$$K_x + K_y + K_z = 1$$

we have

$$L_{xxss} + L_{yyss} + L_{zzss} = K_s$$

$$L_{stst} = \frac{K_s - L_{ssss} + Kt - L_{tttt} + L_{uuuu} - K_u}{2}$$

where s, t, and u are three different of the molecular axes x, y, and z.

We have already seen that for an isotropic sample all K are equal to ⅓; similarly we have that all $L_{ssss} = 1/5$ and all $L_{stst} = 1/15$ for $s \neq t$, or, in general,

$$L_{stst} = \frac{1 + 2\delta_{st}}{15}$$

in the case of no alignment. These values are of interest in ordinary luminescence polarization spectroscopy (measured on an unaligned sample; see above).

Example 1: The Fluorescence Polarization Spectra of Aligned Molecules

As an example let us look at a fluorescence experiment performed on an aligned sample of symmetrical molecules. Figure 16.16 shows four out of the five independent fluorescence polarization spectra of pyrene, that in principle it is possible to measure (for experimental details, such as the actual angle between the beams, etc., see [40]).

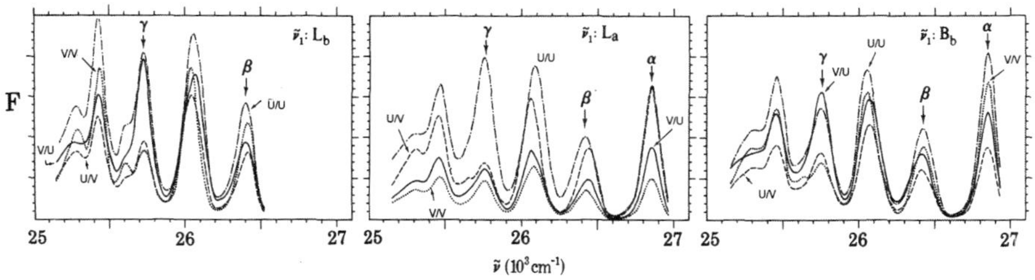

Figure 16.16 Polarized fluorescence (*U/U, V/V, U/V*, and *V/U*) of pyrene in stretched polyethylene at 77 K [40]. Reproduced from [2].

The three parts of the figure correspond to initial excitation at the three first singlet-singlet absorption bands of the molecule: The weak L_b band and the stronger L_a and B_b bands. The four spectra recorded had the following polarizer settings:

$$U/V,\ V/V,\ U/V, \text{ and } V/U.$$

Thus, the "missing spectrum" out of the five that might have been recorded, was *V/W*. For practical reasons it could not be recorded [40]. In a completely corresponding experiment, the D_{2h} symmetrical 2-fluoropyrene was aligned in a stretched polyethylene film and kept at 77 K in order to prevent molecular rotation between the absorption and emission processes. All of the above shown expressions for the fluorescence intensities may now be used for a determination of the *L*, for example the relatively simple ones:

$$[F_{UU}]_{stst} = L_{stst}$$
$$[F_{VU}]_{stst} = ½(K_t - L_{stst})$$
$$[F_{UV}]_{stst} = ½(K_s - L_{stst})$$

The *K*-values are known from absorption, $(K_x, K_y, K_z) = (0.08, 0.29, 0.63)$ [40]. Since the absolute value of the "common part" for each choice of absorbing and emitting transitions $[G(i, j)]_{stuv}$ is not known, it must be removed by division, for example by $[F_{UV}]_{stst}$. Even then, $[F_{UU}]_{stst}/[F_{UV}]_{stst}$ and $[F_{VU}]_{stst}/[F_{UV}]_{stst}$ in addition to the two above expressions relating the *K* and *L* provides more than enough information for a determination of the *L* for 2-fluoropyrene.

If we consider excitation at the *z*-polarized L_b and B_b ṡ in the above expressions becomes *z*, and since emission is along either *y* or *z* we obtain information related to the *L* values L_{zzzz} and L_{zyzy}. Excitation at the *y*-polarized L_a band similarly provides us with the *L*-values L_{yyyy} and L_{yzyz} and thus the set of *L* is actually overdetermined. We get:

$$L_{xxxx} = 0.00$$
$$L_{yyyy} = 0.12$$
$$L_{zzzz} = 0.42$$
$$L_{xyxy} = 0.15$$

$$L_{yzyz} = 0.06$$
$$L_{xzxz} = 0.02.$$

It is remarkable that the squares of the K values $(K_x^2, K_y^2, K_z^2) = (0.00, 0.09, 0.40)$ are very close to the values of the L: $(L_{xxxx}, L_{yyyy}, L_{zzzz}) = (0.00, 0.12, 0.42)$. This means that the angles the three molecular axes form with the sample U-axis are almost the same for all molecules (a very small spread). If these angles were the same for all molecules, we would have

$$(K_x^2, K_y^2, K_z^2) = (L_{xxxx}, L_{yyyy}, L_{zzzz}).$$

This spread of the alignment of individual molecules is very low for example, compared with anthracene, which shows a much larger spread. The maximum spread, where all angles are either 0° or 90°, would correspond to

$$(K_x, K_y, K_z) = (L_{xxxx}, L_{yyyy}, L_{zzzz})$$

This illustrates how two-photon spectroscopy may add information on the molecular alignment to that obtained in absorption (LD spectroscopy), which only provides average values for the angles formed by the molecular axes with the sample axis. The K values for 2-fluoropyrene (0.08, 0.29, 0.63) corresponds to average angles (calculated as $\cos^{-1} K_u^{1/2}$) for $u = (x, y, z)$ which are (42°, 55°, 72°). The fluorescence polarization spectra tell us that most molecules are close to these positions.

16.5 CONCLUSIONS

Optical spectroscopy with linearly polarized light is a convenient and useful method that provides essential information usually overlooked. The technical requirements for application of polarization spectroscopy need not be advanced, and for uniaxial samples the interpretation of polarized spectra can be greatly simplified. The main requirement for an aligned sample is very often inherent in biological systems and nanotechnological devices. Polarization spectroscopy is also a profoundly pedagogical method that provides deeper insight into the basic aspects of light-molecule interactions.

REFERENCES

[1] Wehner, R. (1976) *Sci. Am.* July 106-115.

[2] Michl, J. & Thulstrup, E.W. (1995) Spectroscopy with polarized light – Solute alignment by photoselection, In *Liquid Crystals, Polymers, and Membranes*, 2nd *ed.* Wiley, New York.

[3] Ambronn, H. (1888) *Ber. Deutsch. Botan. Ges.* **6**, 85.

[4] Kliger, D.S., Lewis, J.W. & Randall, C.E. (1990) *Polarized Light in Optics and Spectroscopy* Boston Academic Press.
[5] Rodger, A. & Nordén, B. (1997) *Circular Dichroism and Linear Dichroism* Oxford University Press, UK.
[6] Thulstrup, E.W. & Michl, J. (1989) *Elementary Polarization Spectroscopy* New York Wiley:
[7] Land, E.H. (1940) *J. Opt. Soc. Am.* **30**, 230-238.
[8] Radziszewski, J.G. & Michl, J. (1986) *J. Am. Chem. Soc.* **108**, 3289-3297.
[9] Tjerneld, F., (1982) DNA – Structure and interactions studied with linear and circular dichroism, Ph.D. thesis. Chalmers University of Technology, Göteborg, Sweden. See also Undeman, P.O., Lycksell, A., Gräslund, A., Astlind, T., Ehrenberg, A., Jernstrom, B., Tjerneld, F., & Norden, B. (1983) *Cancer Res.* **43**, 1851-1860.
[10] Fredericq, E. & Houssier, C. (1973) *Electric Dichroism and Electric Birefringence* Oxford, UK Clarendon Press.
[11] Liptay, W. (1974) In Lim, E.C . Ed. *Excited States*, New York Academic Press.
[12] Stapelfeldt, H. & Seideman, T. (2003) *Rev. Mod. Phys.* **75**, 543-557.
[13] Radziszewski, J.G., Balaji, V., Čársky, P. & Thulstrup, E.W. (1991) *J. Phys. Chem.* **95**, 50645068.
[14] Holmén, A., Broo, A. & Albinsson, B. (1994) *J. Phys. Chem.* **98**, 4998-5009.
[15] Thulstrup, E.W., Michl, J. & Eggers, J.H. (1970) *J. Phys. Chem.* **74**, 3868-3877.
[16] Michl, J., Eggers, J.H. & Thulstrup, E.W. (1970) *J. Phys. Chem.* **74**, 3878-3884.
[17] Yang, Y.T., Phillips, P.J. & Thulstrup, E.W. (1982) *Chem. Phys. Lett.* **93**, 66-73.
[18] Steenstrup, F.R., Christensen, K., Svane, C. & Thulstrup, E.W. (1997) *J. Mol. Struct.* **408**, 139-149.
[19] Phillips, P.J. (1990) *Chem. Rev.* 90 425-436.
[20] Armony, J. & Thulstrup, E.W. (1989) *Chem. Phys. Letters* 158 107-110.
[21] Tanizaki, Y. & Kubodera, S.-I. (1967) *J. Mol. Spectr.* **24**, 1-18.
[22] Fraser, R.B.D. (1953) *J. Chem. Phys.* **21**, 1511-1515; and Fraser, R.B.D. 1956 *J. Chem. Phys.* **24**, 89-95.
[23] Beer, M. (1956) *Proc. Roy. Soc. London Ser. A* 236 136-140.
[24] Thulstrup, E.W. & Eggers, J.H. (1968) *Chem Phys. Lett.* **1**, 690-692.
[25] Thulstrup, E.W. & Michl, J. (1982) *J. Am. Chem. Soc.* **104**, 5594-5604.
[26] Spanget-Larsen, J., Christensen, D.H. & Thulstrup, E.W. (1987) *Spectrochim. Acta* **43A**, 331-335.
[27] Thulstrup, E.W., Downing, J.W., & Michl, J. (1977) *J Chem. Phys.* **23**, 307-319.
[28] Langkilde, F.W., Thulstrup, E.W. & Michl, J. (1983) *J. Chem. Phys.* 78 3372-3381.
[29] Michl, J., Thulstrup, E.W. & Eggers, J.H. (1974) *Ber. Bunsenges. Physik.Chem.* **78**, 575-586.
[30] Thulstrup, E.W., Spanget-Larsen, J. & Waluk, J. (1994) *Theoret. Chim. Acta* **89**, 301-309.
[31] Thormann, T., Rogojerov, M., Jordanov, B. & Thulstrup, E.W. (1999) *J. Mol. Struct.* **509**, 93-104.
[32] Gorski, A. (2005) *Applications of polarized spectroscopy and matrix isolation techniques*. Ph. D. thesis Polish' Academy of Sciences, Warsaw, Poland.
[33] Waluk, J., Mordzinski, A., Spanget-Larsen, J. & Thulstrup, E.W. (1987) *Chem. Phys.* **116**, 411-420.
[34] Zimmermann, H. & Joop, N. (1961) *Z. Elektrochem.* **65**, 66-70.
[35] Spanget-Larsen, J., Waluk, J. & Thulstrup, E.W. (1990) *J. Phys. Chem.* **94**, 1800-1806.
[36] Perrin, F. (1926) *J. Phys. Radium* **7**, 390-401.
[37] Jablonski, A. (1935) *Z. Physik* **96**, 236-246.
[38] Lakowicz, J.R. (1999) *Principles of Fluorescence Spectroscopy 2nd ed.* Kluwer Academic/ Plenum Publishers, New York.
[39] Gallivan, J.B., Brinen, J.S. & Koren, J.G. (1968) *J. Mol. Spectros.* **26**, 24-35.
[40] Langkilde, F.W., Gisin, M., Thulstrup, E.W. & Michl, J. (1983) *J. Phys. Chem.* **87**, 2901-2911.

Index

C

D

E

F

G

O

P

Q

R

S

T

Y

Z